Generation of Multivariate Hermite Interpolating Polynomials

PURE AND APPLIED MATHEMATICS

A Program of Monographs, Textbooks, and Lecture Notes

Generation of Multivariate Hermite Interpolating Polynomials

Santiago Alves Tavares

University of Florida

Gainesville, FL, U.S.A.

CRC Press
Taylor & Francis Group
Boca Raton London New York

CRC Press is an imprint of the
Taylor & Francis Group, an **informa** business
A CHAPMAN & HALL BOOK

CRC Press
Taylor & Francis Group
6000 Broken Sound Parkway NW, Suite 300
Boca Raton, FL 33487-2742

First issued in paperback 2019

© 2006 by Taylor & Francis Group, LLC
CRC Press is an imprint of Taylor & Francis Group, an Informa business

No claim to original U.S. Government works

ISBN-13: 978-1-58488-572-6 (hbk)
ISBN-13: 978-0-367-39226-0 (pbk)
Library of Congress Card Number 2005049366

Library of Congress Cataloging-in-Publication Data

Alves Tavares, Santiago.
  Generation of multivariate hermite interplating polynomials / Santiago Alves Tavares.
    p. cm.
  Includes bibliographical references and index.
  ISBN 1-58488-572-6 (alk. paper)
  1. Hermite polynomials. 2. Multivariate analysis. I. Title.

QA404.5.A565 2005
515'.55--dc22   2005049366

**Visit the Taylor & Francis Web site at
http://www.taylorandfrancis.com**

**and the CRC Press Web site at
http://www.crcpress.com**

To my wife Dea,
our sons Michel and Mauricio,
and
our daughter Samia

Preface

This book contains a contribution to the approximate solution of differential equations assuming that the solution can be expanded on a basis of polynomials, which have the properties that all derivatives up to a predefined order can be specified at the boundaries. If the domain is divided into finite elements they guarantee the inter element continuity of the function and their derivatives, including all the cross derivatives. It also provides algorithms to expand the derivative of the product of functions of several variables. If these functions are polynomial the algorithms yield the derivatives.

The material developed in this book can be classified in the fields of Approximation Theory, Interpolation, Numerical Solution of Differential Equations, and Finite Element Methods, among others. This book may be of interest to graduate students, researchers, and software developers in mathematics, physics, or engineering.

How the book evolved

The beginning of this book can be traced back to 1995 when I was trying to solve heat-transfer differential equations using Chebyshev polynomials. I was looking for one challenging problem to apply the methodology when I had the idea of asking NASA LARC if they had some heat-transfer problem that I could use to test my approach. They sent me the problem of the heating of the lower surface of the Space Shuttle during reentry into the atmosphere.

It is a two-dimensional problem and I was using the product of two Chebyshev polynomials to generate a polynomial in two variables. But the solution was not presenting a good behavior – something was missing. I came to the conclusion that the product of Chebyshev polynomials was not covering the space of polynomials in the same way that a cartesian product of two one-dimensional real spaces generates the two-dimensional real space, that is, the roots of the polynomial product are the roots of each multiplier polynomial. The two-dimensional polynomial should have roots that do not belong to either of the multiplier polynomials and these roots could be real or complex. A second problem that I faced was the continuity inter-elements. Then I started thinking about the Hermite Interpolating Polynomials.

A computer program that does algebraic manipulation could be used to generate the Hermite Interpolating Polynomials, but how could I integrate it with the numerical computation? To be able to apply the boundary conditions, polynomials that assume the value one at the point intersection of boundaries and zero at all other boundaries would be useful. If the polynomials would have these properties then some roots were already known and the polynomial could be written in the form $P(x)Q(x)f(x)$ where $P(x)$ contains the roots related to the other boundary conditions and $Q(x)$ contains the roots related to the point intersection of the boundaries. The next step was to obtain $f(x)$ as a complete polynomial. If the value of the polynomial is defined as one at the intersection of the boundaries then it gives one relation to be used, that is, $P(x)Q(x)f(x) = 1$ at that point. After several attempts to find out how to evaluate the coefficients of $f(x)$, I came to the idea of solving the polynomial for $f(x)$, that is, $f(x) = 1/(P(x)Q(x))$ where $P(x)$ and $Q(x)$ are known since their roots are known, and then perform the derivative of order equal to the degree of $f(x)$. The result was an expression for the computation of the coefficient of the highest term. This technique is no more than the back substitution solution of an upper triangular system of equations. Therefore, I came up with an algorithm to construct the polynomial without using any algebraic manipulation software. It was found that $f(x)$, in general, has complex roots.

The next step was how to generate the Hermite Interpolating Polynomials in several variables. It was observed that the powers of the terms belonging to each degree are permutations of the same set of numbers and that the sum of these powers equals to the degree. That is, the powers of each terms could be considered as a compound number whose sum of the components or coordinate numbers equals to the degree. In other words, they formed a set of numbers whose sum was constrained to a fixed value. This is why they have been called *Constrained Numbers*. In analogy to the level curves in topography those fixed values were called the *the value of the level* or, in short, *level* of the set.

The identification of the constrained numbers led to the problem of ordering relation, which would allow me to write the polynomials in a sequence appropriate for the construction of algorithms to be used in computers. To obtain ordering functions it was necessary to find out how to count the number of elements in each set with fixed levels. The solution of this problem led to a more formal definition of the constrained numbers. In doing so I observed that the definition constructed for the set of nonnegative integers could be extended to the set of integers. Later on I saw that the coordinate numbers could be real or complex or matrices. To be able to perform the operations of derivative of polynomials it was necessary to define some operations with the constrained numbers. As a consequence the definition of the neutral element of addition and multiplication and the inverse element of the addition came naturally, and then, the definition of the space of constrained numbers. The next step was to define a geometric representation for the constrained numbers, which, later on, permitted me to relate the constrained numbers to the natural numbers that are well known in the Finite Element Method. The last problem to be solved with the constrained numbers was to obtain algorithms to generate them.

With the support of the constrained numbers it was not difficult to obtain algorithms to generate the Multivariate Hermite Interpolating Polynomials if I had at hand an algorithm to compute the derivative of the product of functions, which in this case were the functions $P(x)$, $Q(x)$, and $f(x)$. Again these algorithms were developed with the aid of the constrained numbers. Two new operations emerged that were called *Constrained Multiplication* and *Cartesian Product of Constrained Numbers*.

With all difficulties to generate the Multivariate Hermite Interpolating Polynomials solved many new questions appeared: What are the properties of these polynomials? What are their behavior? Are they appropriate for interpolations? What are the polynomials for some known finite elements? Can they be used to solve linear and non-linear differential equations? Has the optimization of the nonlinear system of the equations to calculate the coordinate numbers or the expansion of the solution global minimum or has it several local minima? Some of these questions are answered in this book.

I observed that the one-dimensional Multivariate Hermite Interpolating Polynomials have complex roots and that they gave excellent accuracy to the linear and nonlinear one-dimensional differential equations used as test. When applied to a two-dimensional problem they give good approximation when close to the boundaries but the error is not uniform in the domain. The rational fraction of polynomials permitted to increase the accuracy of the solution.

To extend the Multivariate Hermite Interpolating Polynomials to higher dimensions the one-dimensional boundary condition was analyzed to identify the properties that a two-dimensional boundary condition must satisfy. A one-dimensional problem with one boundary condition, usually, it is a propagation problem, that is, the other boundary is in the infinity. The same do not happen to a two-dimensional problem, that is, a closed two-dimensional region has only one boundary condition. Therefore, the property that the Multivariate Hermite Interpolating Polynomials must vanish on all other boundaries no longer exists. Therefore, the defining polynomial expression for the two-dimensional case is different from the one for the one-dimensional case, but both have the property of satisfying the boundary conditions.

The next challenge for the numerical solution of differential equations in higher dimension is the choice of the structure of the polynomials. Several authors suggest the use of complete polynomials to obtain better results. Many authors generate a two-dimensional polynomial performing the product of two one-dimensional polynomials, but the result is not a complete polynomial. It was observed that gathering the terms that contains one variable multiplied by the same power of the other variable shows a rule for the construction of complete higher-dimensional polynomials from one-dimensional polynomials. A generalization of this rule permits the construction of polynomials with respect to any function basis based on polynomials such as Chebyshev, Gegenbauer, Jacob, etc. If the basis is infinity, the resultant polynomial, using this construction, coincides with the expression of the product of one-dimensional polynomials.

It is known that the expansion of the one-dimensional solution in power series, that is, the usual polynomials, gives good approximation in the neighborhood of the reference point, which, in general, is one of the boundaries. If the solution is written in the form of rational fraction of polynomials expanded in power series it gives good approximation even when taking the limit to infinity. This result is not surprising as the set of rational polynomials is a superset of the set of polynomials and, therefore, one should expect a better approximation.

Recalling that the power series basis functions are almost linearly dependent it is preferable to use polynomials that are orthogonal or almost orthogonal such Chebyshev, Gegenbauer, Jacobi, etc. Therefore, a natural choice for the interpolating polynomials to be used in the solution of higher-dimensional problems is the one with the following structure: rational fraction whose numerator and denominator are complete polynomials constructed using a finite basis of one-variable elements.

Some of the problems that have not yet been analyzed are: (1) the convergence when dividing the domain into finite sub domains, that is, finite elements; (2) the trade between number of elements and number of basis elements; (3) the relation between the Hermite Interpolating Polynomials and the Hermite Polynomials; (4) the generation of the Multivariate Hermite Interpolating Polynomials where the powers of the variables are real or complex numbers; and (5) how to increase the rate of convergence of an optimization problem that is slow because it is limited by the size of the word in a computer. These problems are intended to be treated in a next publication.

What the book contains

The book is divided into three parts:

1. Constrained Numbers, chapters 1 to 6

2. Hermite Interpolating Polynomials, chapters 7 to 15

3. Selected Applications, chapter 17

Chapter 1 contains the definition, operations, and geometric visualization of the constrained numbers. Chapter 2 presents the rules to write algorithms to generate the constrained numbers based on the integers and nonnegative integers. It also contains the algorithms in pseudo code. Chapter 3 defines Natural Coordinates as used in the Finite Element Method in terms of the constrained numbers. Chapter 4 develops expressions for the computation of the number of elements in each set of constrained numbers. Chapter 5 defines an Ordering Relation for the constrained numbers and its properties. Chapter 6 applies the concept of constrained numbers to develop a notation for a complete polynomial in several variables, as well as for its derivative and integration. Then follows the definition of order of derivative of a function in terms of constrained numbers. With these tools, algorithms are developed to obtain, symbolically, the derivative of the product and sum of functions of several variables.

Chapter 7 starts defining the properties that characterize the Hermite Interpolating Polynomials. Based on these properties an expression is proposed for the generation of those polynomials. Then an algorithm is shown to generate the Hermite Interpolating Polynomials. During the proof of the algorithm it is also shown that there are other polynomials, which I called Modified Hermite Interpolating Polynomials, that can be generated with the same properties. The last part of this chapter contains several properties that are useful to reduce the amount of computation during the generation of the polynomials. Chapter 8 shows several techniques for the generation of the Hermite Interpolating Polynomials. Chapter 9 compares the proposed approach to the classical approach in the case of one variable. It also shows the advantages of the present technique.

Chapter 10 contains a detailed computation of the polynomials, which intends not only to make clear the algorithm but also to permit one to troubleshoot the development of computer codes. Chapter 11 makes another application to a generic case to show the flexibility of the algorithm. Chapter 12 shows how to apply the algorithm to any region. Chapter 13 compares the proposed approach with the method of Divided Differences. It also shows that the constrained numbers can use the field of the Real Numbers. Chapter 14 shows that the constrained numbers can be based on the complex numbers or on matrices. Chapter 15 shows the graphs of a select set of Hermite Interpolating Polynomials in one variable along a one-dimensional domain.

Chapter 16 shows how to generate the approximate solution for problems with one or more variables. It also shows how to analyze the residual of the approximate solution. It also suggests a procedure to generate complete polynomials in any basis such as Chebyshev, Gegenbauer, etc. including the case of rational fractions of polynomials. Chapter 17 solves some linear and nonlinear differential equations with one variable considering the solution expanded in a basis of Hermite Interpolation Polynomials. These same differential equations have been solved in the literature using a different approach for each, but the method presented here used the same technique for all of them with the same or better accuracy. Chapter 18 solves several differential equations, linear and nonlinear, with more than one variable. Chapter 19 analyzes the distribution of the heat along a line in the lower surface of the space shuttle during the reentry to the earth atmosphere. It shows several approximate solutions and their accuracy in solving the problem. Chapter 19 also comments on what may happen if there is a overheating of the surface.

Acknowledgments

It is impossible to forget a few friends at work: Seymour Block who constantly pushed me; James Bosworth for his incentive and for allowing me to hide in his office in order to complete the book; Spyros Svoronos for his incentive; I thank Kim S. Bey from the Thermal-structural Branch, NASA Langley Research Center for providing the data and useful comments. The reviewers, and Maria Alegra, Marcel Dekker editor, that pressured to me add more examples, which led to another form of treating the case with several variables and the addition of the example about the space shuttle aerodynamic heating during the reentry. The one who helped me most by putting the ideas into my head and, on one occasion, saying in my ear "The error is in the computer program. Do not go elsewhere because it will be a waste of time." I double checked and it really was in the computer program. Finally my wife, sons, and daughter that have been without my presence for many weekends to give me time to work on the book. Without their support this book would have been impossible.

Notation

σ	number of hypersurfaces in a domain
δ	dimension of a constrained number where $\delta = d - 1$
λ_0	value of the level of a constrained number
λ_i	value of the sublevel i of a constrained number
$(\lambda)_n$	Pochhammer symbol
η	maximum order of derivative of the interpolating polynomial
μ	generic order of derivative of the interpolating polynomial
ζ	the order of derivative for which the polynomial equals one at the reference node
Ω	a bounded domain
$\partial\Omega$	boundary of the bounded domain
$\varphi_j(x) = 0$	hyper surface j
χ_e	a subset, of set of hypersurfaces, containing the reference node e
ρ	number of hypersurfaces intersecting at the reference node e
α	coordinate of the reference node e where $\alpha = (\alpha_0, \alpha_1, \ldots, \alpha_\delta)$
τ	number of elements in visibility $\mathcal{V}_e$ of the reference node e
$\Phi_{e,\zeta}^{(\mu)}(x)$	multidimensional Hermite's interpolating polynomial such that if $\mu = \zeta$ if the polynomial equals one at the reference node e
$\mathcal{V}_e$	set of hypersurfaces that do not contain the reference node e, called the visibility of the reference node e
I	set of indices
d	number of coordinate numbers
i, j, k	dummy indices
e	a generic reference node
n	order of derivative of the interpolating polynomial
q	degree of the polynomial $f_{e,n}(a, x)$
t	number of functions in a product $A_0 A_1 \cdots A_{t-1}$
$u!$	factorial of a constrained number if $u = (u_0, u_1, \ldots, u_\delta)$ then $u! = u_0! \, u_1! \ldots u_\delta!$
$\mathcal{S}$	set of hypersurfaces
$\mathbb{F}$	a field
$\mathbb{Z}$	the set of integer numbers
$\mathbb{Z}^+$	the set of nonnegative integer numbers
$\mathbb{R}$	the set of real numbers
$\mathbb{R}^+$	the set of the nonnegative real numbers

Contents

III Selected applications   535

16 Construction of the approximate solution   537

Part I

Constrained Numbers

Chapter 1

Constrained coordinate system

> "The more we find,
> the more we find,
> the more there is to find"

1.1 Definition

Notation 1.1.1

$\mathbb{Z} = \{\ldots, -2, -1, 0, 1, 2, \ldots\}$, as the set of integer numbers.

$\mathbb{Z}^+ = \{0, 1, 2, \ldots\}$, as the set of nonnegative integer numbers.

$\mathbb{R}$ as the set of real numbers.

$\mathbb{R}^+$ as the set of the nonnegative real numbers.

end of notation 1.1.1

Definition 1.1.2 *Constrained number system*

Let $w = (w_0, w_1, \ldots, w_\delta) \in \mathbb{F}^{\delta+1}$ be an ordered $\delta + 1$-tuple whose elements $w_k, k = 0, 1, \ldots, \delta$ belong to the set $\mathbb{F}$. The union

$$L = \bigcup_{i=0}^{\delta} \bigcup_{\ell=0}^{\lambda} \left\{ w \in \mathbb{F}^{i+1} \,\middle|\, \sum_{j=0}^{i} |w_j| = \ell \in \mathbb{G} \right\} \tag{1.1}$$

generates a set of constrained numbers called here *constrained number system*.

end of definition 1.1.2

Definition 1.1.3 *Underling set*

The set $\mathbb{F}$ is called the underling set of the constrained number system.

end of definition 1.1.3

The pair $(\mathbb{F}, \mathbb{G})$ can assume, among others, one of the following values $(\mathbb{Z}, \mathbb{Z}^+)$, $(\mathbb{Z}^+, \mathbb{Z}^+)$, or $(\mathbb{R}, \mathbb{R}^+)$. Therefore $\mathbb{Z}$, $\mathbb{Z}^+$, or $\mathbb{R}$ can be chosen as underlying set. The set $\mathbb{G}$ contains only nonnegative values to be able to satisfy the definition 1.1.2.

Definition 1.1.4 *Constrained number*

The δ-tuple $(w_0, w_1, \ldots, w_\delta)$ such that $\sum_{i=0}^{\delta} |w_i| = \ell$, $0 \leq \ell \leq \lambda$ is called a *constrained number*.

end of definition 1.1.4

Definition 1.1.5 *Constrained coordinate number*

Given an $\ell \in \mathbb{G}$, the elements w_k of the constrained number $(w_0, w_1, \ldots, w_\delta)$ are called *constrained coordinate numbers*.

end of definition 1.1.5

Definition 1.1.6 *Dimension*

The number δ of independent coordinates is called the *dimension* of the constrained coordinate system.

end of definition 1.1.6

Note that the number of coordinate numbers is given by $\delta + 1$.

Notation 1.1.7 *Set of the constrained numbers*

The set of the constrained numbers can be denoted, according to the definition 1.1.2 in page 3, as the union of subsets as follows

$$L = \bigcup_{i=0}^{\delta} L^i = \{L^0 \cup L^1 \cup L^2 \cup \cdots \cup L^\delta \cup \cdots\} \tag{1.2}$$

where

$$L^\delta = \bigcup_{\ell=0}^{\lambda} \left\{ w \in \mathbb{F}^{\delta+1} \ \middle| \ \sum_{j=0}^{\delta} |w_j| = \ell \in \mathbb{G} \right\} \tag{1.3}$$

Each subset in (1.2) is the union of subsets

$$L^\delta = \bigcup_{\ell=0}^{\lambda} L_\ell^\delta = \{L_0^\delta \cup L_1^\delta \cup L_2^\delta \cup \cdots \cup L_\ell^\delta \cup \cdots\} \tag{1.4}$$

where

$$L_\ell^\delta = \left\{ w \in \mathbb{F}^{\delta+1} \,\middle|\, \sum_{j=0}^{\delta} |w_j| = \ell \in \mathbb{G} \right\} \tag{1.5}$$

The notation L_ℓ^δ represents a set that belongs to the δ-dimensional constrained number system and whose sum of absolute value of coordinate numbers of each element is constrained to ℓ.

Since the set $\mathbb{G}$ contains only nonnegative numbers then ℓ can assume any value such that $0 \le \ell < \infty$.

The notation

$$L^\delta(\ell) = \{L_0^\delta \cup L_1^\delta \cup L_2^\delta \cup \cdots \cup L_j^\delta \cup \cdots \cup L_\ell^\delta\} \tag{1.6}$$

is used when the number ℓ of subsets $L_j^\delta, j = 0, 1, \ldots, \ell$ is finite.

The notation

$$L^\delta(\ell_1, \ell_2) = \{L_{\ell_1}^\delta \cup L_{\ell_1+1}^\delta \cup L_{\ell_1+2}^\delta \cup \cdots \cup L_{\ell_1+j}^\delta \cup \cdots \cup L_{\ell_2}^\delta\} \tag{1.7}$$

is used to denote a set that contains the subsets from the level ℓ_1 until the level ℓ_2 where $\ell_1 < \ell_2 \le \ell$.

end of notation 1.1.7

The figurate numbers can be obtained as a particular case of if $(\mathbb{F}, \mathbb{G}) = (\mathbb{Z}^+, \mathbb{Z}^+)$. The set L_2^2, having $\mathbb{Z}^+$ as its underlying set, can be represented geometrically in a triangular form known as Pascal Triangle.

Example 1.1.8 *Expand the equation* (1.1) *in definition 1.1.2 page 3 for $\delta = 2$ using the notation 1.1.7.*

Using the notation in equation (1.2) the equation (1.1) in page 3 writes

$$L = \bigcup_{i=0}^{2} \bigcup_{\ell=0}^{\lambda} \left\{ (w_0, w_1, \ldots, w_i) \in (\mathbb{F})^{i+1} \,\middle|\, \sum_{j=0}^{i} |w_j| = \ell \in \mathbb{G} \right\} = L^0 \cup L^1 \cup L^2 \tag{1.8}$$

where

$$L^0 = \bigcup_{\ell=0}^{\lambda} \left\{ (w_0) \in (\mathbb{F})^1 \,\middle|\, |w_0| = \ell \in \mathbb{G} \right\}$$

$$L^1 = \bigcup_{\ell=0}^{\lambda} \left\{ (w_0, w_1) \in (\mathbb{F})^2 \,\middle|\, |w_0| + |w_1| = \ell \in \mathbb{G} \right\}$$

$$L^2 = \bigcup_{\ell=0}^{\lambda} \left\{ (w_0, w_1, w_2) \in (\mathbb{F})^3 \,\middle|\, |w_0| + |w_1| + |w_2| = \ell \in \mathbb{G} \right\} \tag{1.9}$$

end of example 1.1.8

Example 1.1.9 *In addition to $\delta = 2$ in example 1.1.8, assume that $\lambda = 1$, then expand further the subsets L^0, L^1, and L^2.*

Using the notation in equation (1.2) the equation (1.1) writes

1. For the subset L^0 it follows

$$L^0 = \bigcup_{\ell=0}^{1} \left\{ (w_0) \in \mathbb{F}^1 \,\middle|\, |w_0| = \ell \in \mathbb{G} \right\} = L_0^0 \cup L_1^0 \qquad (1.10)$$

where

$$L_0^0 = \left\{ (w_0) \in \mathbb{F}^1 \,\middle|\, |w_0| = 0 \right\}$$
$$L_1^0 = \left\{ (w_0) \in \mathbb{F}^1 \,\middle|\, |w_0| = 1 \right\} \qquad (1.11)$$

2. The elements of the subset L^1 are ·

$$L^1 = \bigcup_{\ell=0}^{1} \left\{ (w_0) \in \mathbb{F}^2 \,\middle|\, |w_0| = \ell \in \mathbb{G} \right\} = L_0^1 \cup L_1^1 \qquad (1.12)$$

where

$$L_0^1 = \left\{ (w_0, w_1) \in \mathbb{F}^2 \,\middle|\, |w_0| + |w_1| = 0 \right\}$$
$$L_1^1 = \left\{ (w_0, w_1) \in \mathbb{F}^2 \,\middle|\, |w_0| + |w_1| = 1 \right\} \qquad (1.13)$$

3. The elements of the subset L^2 are

$$L^2 = \bigcup_{\ell=0}^{1} \left\{ (w_0) \in \mathbb{F}^3 \,\middle|\, |w_0| = \ell \in \mathbb{G} \right\} = L_0^2 \cup L_1^2 \qquad (1.14)$$

where

$$L_0^2 = \left\{ (w_0, w_1, w_2) \in \mathbb{F}^3 \,\middle|\, |w_0| + |w_1| + |w_3| = 0 \right\}$$
$$L_1^2 = \left\{ (w_0, w_1, w_2) \in \mathbb{F}^3 \,\middle|\, |w_0| + |w_1| + |w_3| = 1 \right\} \qquad (1.15)$$

end of example 1.1.9

Example 1.1.10 *Obtain the sets in the example 1.1.9 for the underlying set* $\mathbb{Z}$.

1. Computation of $L^0 = L_0^0 \cup L_1^0$

   (a) $L_0^0 = \{(0)\}$, since there exists only the solution $w_0 = 0$ for the equation $|w_0| = 0$.

   (b) $L_1^0 = \{(1), (-1)\}$. Since the equation $|w_0| = 1$ has two solutions in the set of the integers, namely ± 1.

   Then, from equation (1.10)

   $$L^0 = L_0^0 \cup L_1^0 = \{\{(0)\} \cup \{(1), (-1)\}\} = \{(0), (1), (-1)\} \qquad (1.16)$$

2. Computation of $L^1 = L_0^1 \cup L_1^1$

   $$L_0^1 = \{(0,0)\} \qquad (1.17)$$
   $$L_1^1 = \{(1,0), (0,1), (0,-1), (-1,0)\} \qquad (1.18)$$

   Which are the solutions for the equations $|w_0| + |w_1| = 0$ and $|w_0| + |w_1| = 1$, respectively. Then, from equation (1.12)

   $$\begin{aligned} L^1 &= L_0^1 \cup L_1^1 \\ &= \{\{(0,0)\} \cup \{(1,0), (0,1), (0,-1), (-1,0)\}\} \\ &= \{(0,0), (1,0), (0,1), (0,-1), (-1,0)\} \qquad (1.19) \end{aligned}$$

3. Computation of $L^2 = L_0^2 \cup L_1^2$

   $$L_0^2 = \{(0,0,0)\} \qquad (1.20)$$
   $$L_1^2 = \{(-1,0,0), (0,-1,0), (0,0,-1), (0,0,1), (0,1,0), (1,0,0)\} \qquad (1.21)$$

   Which are the solutions for the equation $|w_0| + |w_1| + |w_2| = 0$ and $|w_0| + |w_1| + |w_2| = 1$, respectively. Then, from equation (1.14)

   $$\begin{aligned} L^2 &= L_0^2 \cup L_1^2 \\ &= \{\ \{(0,0,0)\} \cup \{(-1,0,0), (0,-1,0), (0,0,-1), (0,0,1), \\ &\qquad (0,1,0), (1,0,0)\}\ \} \\ &= \{\ (0,0,0)), (-1,0,0), (0,-1,0), (0,0,-1), (0,0,1), (0,1,0), (1,0,0)\ \} \\ &\qquad\qquad (1.22) \end{aligned}$$

   end of example 1.1.10

Example 1.1.11 *Obtain the sets in the example 1.1.9 for the underlying set* $\mathbb{Z}^+$.

1. Computation of $L^0 = L_0^0 \cup L_1^0$

   The equations $|w_0| = 0$ and $|w_0| = 1$ have only one solution in the set if the nonnegative integers, which are 0 and 1, respectively. Therefore

$$L_0^0 = \{(0)\} \tag{1.23}$$
$$L_1^0 = \{(1)\} \tag{1.24}$$

   And from equation (1.10) it follows

$$L^0 = L_0^0 \cup L_1^0 \{\{(0)\} \cup \{(1)\}\} = \{(0), (1)\} \tag{1.25}$$

2. Computation of $L^1 = L_0^1 \cup L_1^1$

$$L_0^1 = \{(0,0)\} \tag{1.26}$$
$$L_1^1 = \{(1,0),(0,1)\} \tag{1.27}$$

   Corresponding to the solutions for the equation $|w_0| + |w_1| = 0$ and $|w_0| + |w_1| = 1$, respectively, in the set if the nonnegative integers. Then, from equation (1.12), it is obtained

$$
\begin{aligned}
L^1 &= L_0^1 \cup L_1^1 \\
&= \{\{(0,0)\} \cup \{(1,0),(0,1)\}\} \\
&= \{(0,0),(1,0),(0,1)\}
\end{aligned} \tag{1.28}
$$

3. Computation of $L^2 = L_0^2 \cup L_1^2$

$$L_0^2 = \{(0,0,0)\} \tag{1.29}$$
$$L_1^2 = \{(1,0,0),(0,1,0),(0,0,1)\} \tag{1.30}$$

   which are the solutions for the equations $|w_0| + |w_1| + |w_2| = 0$ and $|w_0| + |w_1| + |w_2| = 1$. Then, from (1.14)

$$
\begin{aligned}
L^2 &= L_0^2 \cup L_1^2 \\
&= \{\{(0,0,0)\} \cup \{(1,0,0),(0,1,0),(0,0,1)\}\} \\
&= \{(0,0,0)),(1,0,0),(0,1,0),(0,0,1)\}
\end{aligned} \tag{1.31}
$$

end of example 1.1.11

Definition 1.1.12 *Level*

Given a constrained coordinate system of dimension δ and an $\ell \in \mathbb{G}$, the constrained numbers w in the set

$$L_\ell^\delta = \left\{ w \in \mathbb{F}^{\delta+1} \,\Big|\, \sum_{j=0}^{\delta} |w_j| = \ell \right\} \qquad (1.32)$$

are said to be in the level ℓ. It will be used $\lambda_0(w) = \ell$ to denote that the number w belongs to the level ℓ. The value of the level equals to the sum of the absolute value of the $\delta + 1$ coordinate numbers of w.

end of definition 1.1.12

Definition 1.1.13 *Constraint equation*

The expression

$$\sum_{j=0}^{\delta} |w_j| = \ell \in \mathbb{G} \qquad (1.33)$$

for the computation of the level is called the constraint equation. Here $\mathbb{G}$ can be $\mathbb{Z}^+$ or $\mathbb{R}^+$.

This definition can also be understood as the norm of the absolute value for the set L^δ.

end of definition 1.1.13

The level ℓ for the constrained coordinate numbers plays the same role as the geodesic curves on a surface. Recall that geodesic are curves whose level is constant.

Definition 1.1.14 *Sublevel*

Given a constrained coordinate system of dimension δ and an $\ell \in \mathbb{G}$, the constrained numbers ω in the set

$$L_{\lambda_i}^{\delta-i} = \left\{ \omega \in \mathbb{F}^{\delta+1-i} \,\Big|\, \sum_{j=i}^{\delta} |w_j| = \lambda_i \leq \ell \right\} \qquad (1.34)$$

are said to be in the sublevel $\lambda_i(w)$ of the level ℓ. The value of the sublevel $\lambda_i(w)$ equals to the sum of the absolute values of the coordinate numbers of w from the i-th until the δ-th, that is $\lambda_i(w) = |w_i| + |w_{i+1}| + \cdots + |w_\delta|$.

The notation λ_i may be used to represent the value of the level $\lambda_i(w)$ of the number $\omega = (w_i + w_{i+1} + \cdots + w_\delta)$, that is $\lambda_i = \lambda_i(w)$. If $i = 0$ then $\lambda_0(w) = \ell$. Recall that $w = (w_0 + w_1 + \cdots + w_\delta)$ denotes the constrained number while $\omega = (w_i + w_{i+1} + \cdots + w_\delta)$ is a number in the sublevel i of of the constrained number w.

The constraint equation is

$$\lambda_i = \sum_{j=i}^{\delta} |w_j| \tag{1.35}$$

If $j = 0$ it gives the constraint equation (1.33) for the level.

end of definition 1.1.14

Definition 1.1.15 *Dual level and sublevel*

Given a constrained coordinate system of dimension δ and an $\ell \in \mathbb{G}$, the constrained numbers ω in the set

$$L_{\lambda^i}^{\delta-i} = \left\{ \omega \in \mathbb{F}^{\delta+1-i} \mid \sum_{j=0}^{\delta-i} |w_j| = \lambda^i \le \ell \right\} \tag{1.36}$$

are said to be in the dual sublevel $\lambda^i(w)$ of the level ℓ which corresponds the the numbers of the form $\omega = (w_0, w_1, \ldots, w_{\delta-i})$. The value of the dual sublevel $\lambda^i(w)$ equals to the sum of the absolute values of the coordinate numbers of w from the 0-th until the $(\delta - i)$-th, that is, $\lambda^i(w) = |w_0| + |w_1| + \cdots + |w_{\delta-i}|$. If $i = 0$ then $\lambda^0(w) = \lambda_0(w) = \ell$.

The constraint equation is

$$\lambda^i = \sum_{j=0}^{\delta-i} |w_j| \tag{1.37}$$

If $j = 0$ it gives the constraint equation (1.33) for the level and the constraint equation (1.35) for the sublevel. Therefore, it can be written that

$$\ell = \lambda_0 = \lambda^0 \qquad \text{if} \quad j = 0 \tag{1.38}$$

end of definition 1.1.15

Note that the sublevel is obtained by the sum of the absolute values from the coordinate number w_i until the last coordinate number which is w_δ. The dual sublevel is obtained performing the sum of the absolute values of the coordinate numbers from the first, which is w_0, until the coordinate number $w_{\delta-i}$.

Example 1.1.16 *Level and sublevel*

Let $w = (w_0, w_1, w_2, w_3) = (1, 2, 0, 1)$ be a constrained number whose underlying set is $\mathbb{Z}^+$. It belongs to the set of the 3-dimensional constrained numbers, whose dimension is denoted by $\delta = 3$. The constrained number w has four coordinate numbers, which makes it an element of a subset of the set of numbers represented using the four-dimensional ($d = 4$) cartesian coordinates system.

The level of $w = (1, 2, 0, 1)$ is evaluated as $\ell = |w_0| + |w_1| + |w_2| + |w_3| = 1 + 2 + 0 + 1 = 4$.

1. Sublevel λ_0.

   For the sublevel λ_0 it follows that $i = 0$ then $\delta - i = 3 - 0 = 3$ and the value of the sublevel is obtained from $\lambda_0(w) = |w_0| + |w_1| + |w_2| + |w_3| = 4$, which is equivalent to the definition of level. As a consequence of the equivalence, the value of the sublevel λ_0 equals the value of the level, which permits to write $\lambda_0 = \ell = 4$.

2. Sublevel λ_1.

   The value of first sublevel is computed as $\lambda_1(w) = |w_1| + |w_2| + |w_3| = 2 + 0 + 1 = 3$, its dimension is given by $\delta - i = 3 - 1 = 2$, and the number is given by $(w_1, w_2, w_3) = (2, 0, 1)$.

3. Sublevel λ_2.

   The value of second sublevel is obtained from $\lambda_2(w) = |w_2| + |w_3| = 0 + 1 = 1$ giving the number $(w_2, w_3) = (0, 1)$ whose dimension is $\delta - i = 3 - 2 = 1$.

4. Sublevel λ_3.

   The dimension of the third sublevel is given by $\delta - i = 3 - 3 = 0$ and its value is obtained from $\lambda_3(w) = |w_3| = 1$.

For the number $w = (-1, -2, 0, 1)$ whose underlying set is $\mathbb{Z}$ its the sublevels are

1. $\lambda_0(w) = |-1| + |-2| + |0| + |1| = 4$.

2. $\lambda_1(w) = |-2| + |0| + |1| = 3$.

3. $\lambda_2(w) = |0| + |1| = 1$.

4. $\lambda_3(w) = |1| = 1$.

end of example 1.1.16

Example 1.1.17 *Dual level and dual sublevel*

For the same number used in example 1.1.16 if follows

1. Dual sublevel $\lambda^0(w)$.

   The value of the dual sublevel of order $i = 0$ of the constrained number $w = (w_0, w_1, w_2, w_3) = (1, 2, 0, 1)$ is given by $\lambda^0(w) = |w_0| + |w_1| + |w_2| + |w_3| = 1 + 2 + 0 + 1 = 4$ and its dimension is evaluated from $\delta - i = 3 - 0 = 3$. Recalling that the level of the number $(1, 2, 0, 1)$ is obtained from $\lambda_0(w) = |w_0| + |w_1| + |w_2| + |w_3| = 4$, it can be seen that the value of the sublevel $\lambda^0(w)$ equals to the value of the level sublevel $\lambda_0(w)$ that is $\lambda^0(w) = \lambda_0(w) = \ell$.

2. Dual sublevel $\lambda^1(w)$.

   The value of first dual sub-level is obtained as $\lambda^1(w) = |w_0| + |w_1| + |w_2| = 1 + 2 + 0 = 3$ where $\delta - i = 3 - 1 = 2$ and the corresponding number is $(w_0, w_1, w_2) = (1, 2, 0)$.

3. Dual sublevel $\lambda^2(w)$.

   The dimension of the second sublevel is given by $\delta - i = 3 - 2 = 1$ and its value is obtained from $\lambda^2(w) = w_0 + w_1 = 1 + 2 = 2$ giving the number $(w_0, w_1) = (1, 2)$.

4. Dual sublevel $\lambda^3(w)$.

   The value of third dual sublevel is evaluated as $\lambda^3(w) = w_0 = 1$ where $\delta - i = 3 - 3 = 0$ the number is $(w_0) = (1)$.

   end of example 1.1.17

1.1.1   Operations with constrained numbers

Definition 1.1.18 *Addition of constrained numbers*

Given two constrained numbers $w \in L^\delta$ and $v \in L^\delta$ the operation of addition is defined as

$$w + v := (w_0 + v_0, w_1 + v_1, \ldots, w_d + v_\delta) \tag{1.39}$$

end of definition 1.1.18

As consequence of the definition 1.1.18 if $w \in L^\delta_{\ell_1}$ and $v \in L^\delta_{\ell_2}$, then $w + v \in L^\delta_{\ell_1 + \ell_2}$, since from the definition of level $\ell_1 = \sum_{i=0}^{\delta} w_i$, and $\ell_2 = \sum_{i=0}^{\delta} v_i$, therefore

$$\sum_{i=0}^{\delta} w_i + \sum_{i=0}^{\delta} v_i = \ell_1 + \ell_2 \tag{1.40}$$

Definition 1.1.19 *Identity element of the addition*

The identity element of the addition of constrained numbers is given by

$$\theta = (0, 0, \ldots, 0) \in L^{\delta} \tag{1.41}$$

which satisfies the property $w + \theta = w$, $\forall w \in L^{\delta}$ according to the definition of addition of constrained numbers.

end of definition 1.1.19

Definition 1.1.20 *Inverse element of the addition*

The inverse element $v \in L^{\delta}$ of $w \in L^{\delta}$ for the addition is defined as

$$v := (-w_0, -w_1, \ldots, -w_d) \tag{1.42}$$

and denoted by $v = -w = (-w_0, -w_1, \ldots, -w_d)$.

end of definition 1.1.20

Definition 1.1.21 *Inverse of the addition*

Given two numbers $w \in L^{\delta}$ and $v \in L^{\delta}$ the operation inverse of addition of constrained numbers is given by

$$w - v := (w_0 - v_0, w_1 - v_1, \ldots, w_{\delta} - v_{\delta}) \tag{1.43}$$

Observe that $w - v := w + (-v)$.

end of definition 1.1.21

Definition 1.1.22 *Multiplication of a constrained number by a scalar*

Given a constrained number $w \in L^{\delta}$ and a scalar α the operation of multiplication of a constrained number by a scalar is defined as

$$\alpha w := (\alpha w_0, \alpha w_1 + \ldots, \alpha w_d) \tag{1.44}$$

end of definition 1.1.22

Definition 1.1.23 *Constrained space*

The set of constrained numbers with the above operations is called a *constrained space*.

end of definition 1.1.23

The constrained space is a linear space, Madox [26], since it satisfy the following properties,

1. $v + w = w + v$

2. $(u + v) + w = v + (u + w)$

3. There exists a $\theta \in L^\delta$ such that $w + \theta = w$.

4. There exists a $-w \in L^\delta$ such that $w + (-w) = \theta$.

5. $1 \cdot w = w$.

6. $\alpha(v + w) = \alpha v + \alpha w$

7. $(\alpha + \beta)w = \alpha w + \beta w$.

8. $\alpha(\beta w) = (\alpha\beta)w$.

for all scalars α, β and constrained numbers u, v, w in L^δ. Considering the equation (1.33) the space of constrained numbers is a normed space where the definition of level is equivalent to a norm function, the absolute value norm.

1.1.2   Operations with sets of constrained numbers

Definition 1.1.24 *Constrained multiplication*

Given two sets of constrained numbers $L_a^\delta = \{w_0, w_1, \ldots, w_\delta\}$ and $L_b^\delta = \{u_0, u_1, \ldots, u_\delta\}$ the constrained multiplication is defined as

$$L_a^\delta \star L_b^\delta := \{\, w_0 + u_0, w_0 + u_1, \ldots, w_0 + u_\delta +$$
$$w_1 + u_0, w_1 + u_1, \ldots, w_1 + u_\delta +$$
$$\vdots$$
$$w_\delta + u_0, w_\delta + u_1, \ldots, w_\delta + u_\delta\} \tag{1.45}$$

or

$$L_a^\delta \star L_b^\delta := \{w_i + u_j \mid i = 0, 1, \ldots, \delta, \quad j = 0, 1, \ldots, \delta\} \tag{1.46}$$

end of definition 1.1.22

As an example of the application of the above definition consider to expand $(x + y)^2$. It can be written as $(x^1 y^0 + x^0 y^1)^2$ where the exponents belong to the set $L_1^1 = \{(1, 0), (0, 1)\}$. The exponents of the square of $(x + y)$ can be obtained as

$$\{(1, 0), (0, 1)\} \star \{(1, 0), (0, 1)\} = \{(2, 0), (1, 1), (1, 1), (0, 2)\}$$
$$= \{(2, 0), (1, 1), (0, 2)\} \tag{1.47}$$

which gives as result for the squaring operation

$$(x+y)^2 = (x^1y^0 + x^0y^1)^2 = x^2y^0 + 2x^1y^1 + x^0y^2 = x^2 + 2xy + y^2 \quad (1.48)$$

An algorithm for the computation of the coefficients in the expansion is shown in chapter 6 section 6.9 on page 263.

Definition 1.1.25 *Cartesian product of constrained numbers*

1. Constrained cartesian product of two constrained numbers

   Let $L_{a_0}^\delta$ and $L_{a_1}^\delta$ be two sets of constrained numbers whose elements have the dimension δ and are in the levels a_0 and a_1. Consider two elements $w_i = (w_{i0}, w_{i1}, \ldots, w_{i\delta}) \in L_{a_0}^\delta$ and $u_j = (u_{j0}, u_{j1}, \ldots, u_{j\delta}) \in L_{a_1}^\delta$. The cartesian product of w_i and u_j is given by the set

   $$\{w_i \times u_j\} = \Big\{ (w_{i0}, u_{j0}), (w_{i1}, u_{j1}), \ldots, (w_{i\delta}, u_{j\delta}) \Big\} \quad (1.49)$$

   The first constrained number of the product is obtained taking the first coordinate number of the multiplicand w_{i0} and the first coordinate number of the multiplier u_{j0}. The second contains the second coordinate number of the multiplicand w_{i1} and the second coordinate number of the multiplier u_{j1}, and so on.

2. Constrained cartesian product of n constrained numbers

   The above definition can be extended to the product of n constrained numbers. Let $L_{a_0}^\delta$, $L_{a_1}^\delta$, $\cdots$, $L_{a_{n-1}}^\delta$ be n sets of constrained numbers whose elements have the dimension δ and are in the levels $a_0, a_1, \ldots, a_{n-1}$. Consider the constrained numbers

   $$w_i = (w_{i0}, w_{i1}, \ldots, w_{i\delta}) \in L_{a_0}^\delta$$
   $$u_j = (u_{j0}, u_{j1}, \ldots, u_{j\delta}) \in L_{a_1}^\delta$$
   $$\vdots$$
   $$v_k = (v_{k0}, v_{k1}, \ldots, v_{k\delta}) \in L_{a_{n-1}}^\delta \quad (1.50)$$

   The cartesian product of $w_1 \in L_{a_0}^\delta$, $u_j \in L_{a_1}^\delta$, $\ldots$, $v_k \in L_{a_{n-1}}^\delta$ is given by the set

   $$\{w_i \times u_j \times \cdots \times v_k\} =$$
   $$\Big\{ (w_{i0}, u_{j0}, \cdots, v_{k0}), (w_{i0}, u_{j0}, \cdots, v_{k1}), \ldots,$$
   $$(w_{i0}, u_{j1}, \cdots, v_{k0}), (w_{i0}, u_{j1}, \cdots, v_{k1}), \ldots,$$
   $$(w_{i1}, u_{j0}, \cdots, v_{k0}), (w_{i1}, u_{j0}, \cdots, v_{k1}), \ldots,$$
   $$(w_{i\delta}, u_{j\delta}, \ldots, v_{k\delta}) \Big\} \quad (1.51)$$

end of definition 1.1.25

Definition 1.1.26 *Constrained cartesian product of sets of constrained numbers*

1. Cartesian product of two sets

   Let

   $$L^\delta_{a_0} = \{w_0, w_1, \ldots, w_p\}$$
   $$L^\delta_{a_1} = \{u_0, u_1, \ldots, u_q\} \tag{1.52}$$

   be two sets of constrained numbers whose elements have the dimension δ and are in the levels a_0 and a_1. The cartesian product of two sets $L^\delta_{a_0}$ and $L^\delta_{a_1}$, denoted as $L^\delta_{a_0} \times L^\delta_{a_1}$ is defined by

   $$L^\delta_{a_0} \times L^\delta_{a_1} = \{w_0 \times u_0\} \cup \{w_0 \times u_1\} \cup \cdots \cup \{w_p \times u_q\} \tag{1.53}$$

   where $\{w_i \times u_j\}$ is given by (1.49).

2. Cartesian product of m sets

   Let

   $$L^\delta_{a_0} = \{w_0, w_1, \ldots, w_p\}$$
   $$L^\delta_{a_1} = \{u_0, u_1, \ldots, u_q\}$$
   $$\vdots$$
   $$L^\delta_{a_{m-1}} = \{v_0, v_1, \ldots, v_r\} \tag{1.54}$$

   be sets of constrained numbers whose elements have the dimension δ and are in the levels $a_0, a_1, \ldots, a_{m-1}$. The cartesian product of the sets $L^\delta_{a_0}, L^\delta_{a_1}, \ldots, L^\delta_{a_{m-1}}$ denoted as

   $$L^\delta_{a_0} \times L^\delta_{a_1} \times \cdots \times L^\delta_{a_{m-1}} \tag{1.55}$$

   is defined by

   $$L^\delta_{a_0} \times L^\delta_{a_1} \times \cdots \times L^\delta_{a_{m-1}} =$$
   $$\{w_0 \times u_0 \times \cdots \times v_0\} \cup \{w_0 \times u_0 \times \cdots \times v_1\} \cup \cdots \cup$$
   $$\{w_0 \times u_1 \times \cdots \times v_0\} \cup \{w_0 \times u_1 \times \cdots \times v_1\} \cup \cdots \cup$$
   $$\{w_p \times u_q \times \cdots \times v_r\} \tag{1.56}$$

   where $\{w_i \times u_j \times \cdots \times v_k\}$ is given by (1.51).

end of definition 1.1.26

Example 1.1.27 *Constrained cartesian product*

1. Constrained cartesian product of two constrained numbers

$$\{w_0 \times u_0\} = \{(1,0) \times (3,1)\} = \{(1,3),(0,1)\} \qquad (1.57)$$

2. Constrained cartesian product of four constrained numbers

$$\{w_0 \times u_0 \times v_0 \times z_0\}$$
$$= \{(2,0,0) \times (1,1,0) \times (1,0,1) \times (0,2,0)\}$$
$$= \{(2,1,1,0),(0,1,0,2),(0,0,1,0)\} \qquad (1.58)$$

3. Constrained cartesian product of two sets of constrained numbers

$$L_2^1 \times L_3^1 = \{(2,0),(1,1),(0,2)\} \times \{(3,0),(2,1),(1,2),(0,3)\}$$
$$= \Big\{ (2,3),(0,0),(2,2),(0,1),(2,1),(0,2),(1,0),(1,3),$$
$$(1,3),(1,0),(1,2),(1,1),(1,1),(1,2),(1,0),(1,3),$$
$$(0,3),(2,0),(0,2),(2,1),(0,1),(2,2),(0,0),(2,3) \Big\} \qquad (1.59)$$

4. Constrained cartesian product of four sets of constrained numbers

   Let

$$L_1^2 = \{(1,0,0),(0,1,0),(0,0,1)\} \qquad (1.60)$$
$$L_3^2 = \{(3,0,0),(2,1,0),(2,0,1),(1,2,0),(1,1,1),(1,0,2),$$
$$(0,3,0),(0,2,1),(0,1,2),(0,0,3)\} \qquad (1.61)$$
$$L_0^2 = \{(0,0,0)\} \qquad (1.62)$$
$$L_2^2 = \{(2,0,0),(1,1,0),(1,0,1),(0,2,0),(0,1,1),(0,0,2)\} \qquad (1.63)$$

   be four sets of constrained numbers. The constrained cartesian product
   of these sets is given by

$$L_1^2 \times L_3^2 \times L_0^2 \times L_2^2 = \{(1,0,0) \times (3,0,0) \times (0,0,0) \times (2,0,0)\}$$
$$\cup \{(1,0,0) \times (3,0,0) \times (0,0,0) \times (1,1,0)\}$$
$$\cup \{(1,0,0) \times (3,0,0) \times (0,0,0) \times (1,0,1)\}$$
$$\cup \{(1,0,0) \times (3,0,0) \times (0,0,0) \times (0,2,0)\}$$
$$\cup \{(1,0,0) \times (3,0,0) \times (0,0,0) \times (0,1,1)\}$$
$$\cup \{(1,0,0) \times (3,0,0) \times (0,0,0) \times (0,0,2)\}$$
$$\cup \cdots \qquad (1.64)$$

Expanding each partial constrained cartesian product using (1.51) it follows

$$L_1^2 \times L_3^2 \times L_0^2 \times L_2^3 = [(1,3,0,2),(0,0,0,0),(0,0,0,0)],$$
$$[(1,3,0,1),(0,0,0,1),(0,0,0,0)],$$
$$[(1,3,0,1),(0,0,0,0),(0,0,0,1)],$$
$$[(1,3,0,0),(0,0,0,2),(0,0,0,0)],$$
$$[(1,3,0,0),(0,0,0,1),(0,0,0,1)],$$
$$[(1,3,0,0),(0,0,0,0),(0,0,0,2)], \cdots \quad (1.65)$$

end of example 1.1.27

Definition 1.1.28 *Addition of a set with a constrained number*

The operation of addition a set $L_{\ell_1}^\delta$ with a constrained number $w \in L^\delta$ denoted as $L_{\ell_2}^\delta = L_{\ell_1}^\delta + \{(w)\}$ gives the following new set

$$L_{\ell_2}^\delta = \{(u) \mid \forall v \in L_{\ell_1}^\delta, u = v + w\} \quad (1.66)$$

If $v, w \in \mathbb{Z}^+$ then any $u_i < 0$ implies that u_i does not belong to the set $\mathbb{Z}^+$ and so the number u is removed from the resultant set.

The same definition holds for the inverse of the addition.

end of definition 1.1.28

These operations will be used in the computation of the derivative and integration of polynomials.

Example 1.1.29

1. Example of addition $L_1^2 + \{(w)\}$

   Let

   $$L_1^2 = \{(1,0,0),(0,1,0),(0,0,1)\} \quad (1.67)$$

   and $w = (0,1,1)$ then

   $$L_1^2 + \{(0,1,1)\} = \{(1,1,1),(0,2,1),(0,1,2)\} \quad (1.68)$$

2. Example of subtraction $L_4^2 - \{(w)\}$

   Let

   $$L_4^2 = \{(4,0,0),(3,1,0),(3,0,1),(2,2,0),(2,1,1),(2,0,2),$$
   $$(1,3,0),(1,2,1),(1,2,1),(1,1,2),(1,0,3),(0,4,0),$$
   $$(0,3,1),(0,2,2),(0,1,3),(0,0,4)\} \quad (1.69)$$

and $w = (1, 1, 0)$ then

$$L_4^2 - \{(1, 1, 0)\} = \big\{\, (2, 0, 0), (1, 1, 0), (1, 0, 1), (0, 2, 0), (0, 1, 1), (0, 0, 2) \,\big\} \tag{1.70}$$

end of example 1.1.27

1.2 Geometric representation

Theorem 1.2.1 *Geometric representation of level*

Let $\mathbb{R}$ be the underlying set. The constrained numbers such that

$$\sum_{j=0}^{\delta} |x_j| = \ell \in \mathbb{R}^+ \tag{1.71}$$

span a hyperplane for each $\ell \in \mathbb{R}^+$.

Proof: The constrain condition $\sum_{j=0}^{\delta} |x_j| = \ell$ represents the equation of a hyperplane intercepting each coordinate axis X_j at a distance ℓ along the positive direction, that is, in the first quadrant.

end of theorem 1.2.1

Note that the negative numbers are mapped to positive numbers by the constraining equation (1.71).

If $\mathbb{Z}^+$ is the underlying set then the numbers satisfying $\sum_{j=0}^{\delta} |x_j| = \ell \in \mathbb{Z}^+$ are on a hyperplane.

1.2.1 Contravariant coordinate representation

Let X, Y be a two-dimensional contravariant coordinate system, Krogdahl [24], as shown in figure 1.2.2. Let P be a point whose coordinate numbers are (x, y). Construct lines parallel to the coordinate axes and passing through the point P. The distance x from the origin to the X-axis intercept is called the x coordinate number. The y coordinate number is defined analogously.

The location of a point P, whose coordinate numbers are (x, y), in a contravariant cartesian coordinate system, and whose axes are (X, Y), is obtained by the intersection of two lines. One is parallel to the axis X at a distance y from the origin along the Y axis, and the other is parallel to the axis Y at a distance x from the origin along the X axis, as shown in figure 1.2.2.

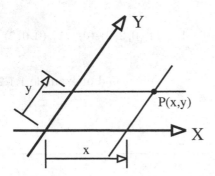

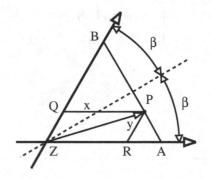

Figure 1.2.2 *Contravariant cartesian coordinate system representation of a point P which is defined by the intersection of lines parallel to each coordinate axis.*

Figure 1.2.3 *The line AB represents the constraint equation in a contravariant system of coordinates. The sum of the coordinate numbers equals to a constant value, that is,*
$$x + y = ZR + RA = \ell.$$

To show the geometric representation of the constraint condition observe that the coordinate numbers of the point A are $(w_0, 0)$. Since $w_0 + 0 = \ell$, then the $(w_0, 0) = (\ell, 0)$ which implies that $ZA = \ell$. Analogously, the coordinate numbers of the point B are $(0, w_1) = (0, \ell)$. Then $ZB = \ell$. Therefore, $ZA = ZB = \ell$ and the triangle AZB is isosceles having AB as its basis, as shown in in figure 1.2.3. This implies that the angles $\widehat{ZAB}$ and $\widehat{ZBA}$ are equal. Since PR is parallel to ZB then the angles $\widehat{ZBA}$ and $\widehat{RPA}$ are equal. This implies the equality of the angles $\widehat{ARP}$ and $\widehat{APR}$. Consequently $RA = RP = y$ and then $ZA = ZR + RA = x + y = \ell$.

Conclusion: the segment AB represents geometrically the constraint equation (1.71).

1.2.2   Covariant coordinate representation (dual system)

The coordinate numbers in a covariant coordinate system, Krogdahl [24], representation are obtained by the construction of lines passing through the point P and perpendicular to each coordinate axes as shown in figure 1.2.4. Similarly to the contravariant system, the distance x from the origin to the X-axis intercept is called the x coordinate number and the distance y from the origin to the Y-axis intercept is called the y coordinate number.

The location of a point P, whose coordinate numbers are (x, y), in a covariant cartesian coordinate system is obtained by the intersection of two lines, one perpendicular to the axis X and the other perpendicular to the axis Y, as shown in figure 1.2.4.

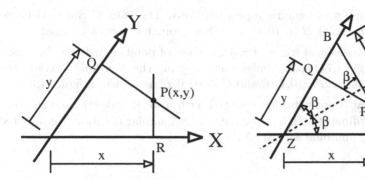

Figure 1.2.4 *Covariant coordinate system representation (dual system) of a point P which is defined by the intersection of lines perpendicular to each coordinate axis.*

Figure 1.2.5 *The line AB represents the constraint equation in a covariant system of coordinates. The sum of the coordinate numbers equals to a constant value, that is,*
$$x + y = 2\ell - k \sin \beta.$$

If the angle between the coordinate axes is $\pi/2$ then the contravariant and covariant systems have the same coordinate numbers and the same graphical representation.

To show geometrically the constraint condition (1.71) represents a line segment for the covariant coordinate system observe in figure 1.2.5 that the condition (1.71) implies $ZA = ZB = \ell$ and so $\widehat{ZAB} = \widehat{ZBA}$.

The triangles $\triangle QPB$ and $\triangle RPA$ are similar since they are square and $\widehat{QBP} = \widehat{RAP}$ since the triangle $\triangle ABZ$ is isosceles. Noting that $\widehat{RAP} \equiv \widehat{ZAB} = \widehat{ZBA} \equiv \widehat{QBP}$ then $\widehat{RPA} = \widehat{QPB} = \beta$.

Observing that

$$RA = m \sin \beta \qquad \text{and} \qquad QB = n \sin \beta \qquad (1.72)$$

the sum of the coordinate numbers writes

$$x + y = (ZB - QB) + (ZA - RA) = (\ell - RA) + (\ell - QB) = 2\ell - (RA + QB) \tag{1.73}$$

Introducing RA and QB given by equation (1.72) into equation (1.73) it follows

$$x + y = 2\ell - (m \sin \beta + n \sin \beta) = 2\ell - (m + n) \sin \beta = 2\ell - k \sin \beta \quad (1.74)$$

Since ℓ, k, and β are constants then $x + y$ is constant and the line AB is the locus of the points such that $x + y = 2\ell - k \sin \beta$.

The condition $ZA = ZA = \ell$ is the same either for the contravariant representation or covariant showing that the equation (1.71) represents the same line segment on both geometric representations. The point A can be denoted by $(\ell, 0)$ and the point B by $(0, \ell)$ on either geometric representation.

In a three-dimensional space $(d = 3)$, the set of points satisfying the condition in equation (1.71) spans a plane intercepting the coordinate axes at the points $A = (\ell, 0, 0)$, $B = (0, \ell, 0)$ and $C = (0, 0, \ell)$ as shown in figure 1.3.2.

The covariant coordinates are denoted with a lower index x_i and the contravariant coordinates with a upper index x^i. A similar notation was adopted for the level λ_i and dual sublevel λ^i.

1.3   Visualization of the coordinate axes

1.3.1   Orthogonal representation of the coordinate axes

One of the most common visualizations of the coordinate axes show them as orthogonal axes. An one-dimensional constraint line is shown in figure 1.3.1 in a two-cartesian axes representation and a two-dimensional constraint plane is shown in figure 1.3.2 in a three-cartesian axes representation.

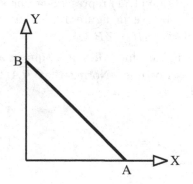

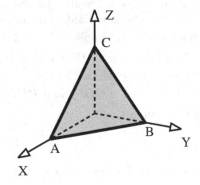

Figure 1.3.1 *The visualization of an one-dimensional set of constrained numbers is given by the line segment AB.*

Figure 1.3.2 *The visualization of a two-dimensional set of constrained numbers is given by the triangle ABC.*

1.3.2   Oblique representation of the coordinate axes

In several cases it is more convenient to rotate the coordinate system and use oblique coordinate axes. The coordinate axes in figure 1.3.1 are shown rotated in 1.3.3 for the one-dimensional case, which is the traditional representation of the Pascal Triangle. The coordinate axes in figure 1.3.3 are shown rotated in figure 1.3.4 for the two-dimensional case.

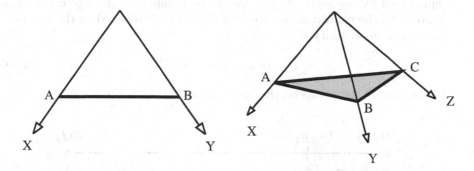

Figure 1.3.3 *Coordinate axes representation of a set of one-dimensional constrained numbers, which set is given by the line AB, in a system with oblique downward axes.*

Figure 1.3.4 *Coordinate axes representation of a set of two-dimensional constrained numbers, which set is given by the triangle ABC, in a system with oblique downward axes.*

If $ZA = ZB = \ell$ in figures 1.3.1 and 1.3.3, that is, if the segment AB belongs to the level ℓ then the coordinates of the points A and B are $A = (\ell, 0)$ and $B = (0, \ell)$. If $ZA = ZB = ZC = \ell$ in figure 1.3.2 and figure 1.3.4, that is, if the plane ABC belongs to the level ℓ then the coordinates of the points A, B and C are $A = (\ell, 0, 0)$, $B = (0, \ell, 0)$, and $C = (0, 0, \ell)$, which shows that $AB = BC = AC$.

1.4 Geometric location of the constrained numbers

1.4.1 One-dimensional constrained numbers

Consider the line segment AB shown if figure 1.2.3 and figure 1.2.5 for two different coordinate definitions. For both cases the coordinate numbers of A are $(\ell, 0)$ and of B are $(0, \ell)$.

Let the interval $[0, \ell] \in \mathbb{R}$ be represented by the line segment AB as shown in figure 1.4.1 where $A = (\ell, 0)$ and $B = (0, \ell)$.

From figure 1.2.3 or figure 1.2.5 it can be seen that the first coordinate number of the point P goes from $x = \ell$ on the axis X to $x = 0$ on the axis Y. That is, on the line segment AB, the first coordinate number of the point

P has origin at the point B and end at the point A. As a consequence the first coordinate number of the point P can be represented by the segment BP. Since the second coordinate number has origin at the point A it can be represented by the segment AP. Using the notation in the figure 1.2.3 and figure 1.2.5 the coordinate numbers of the point P referred to the points A and B can be written as

$$P = (w_0, w_1) = (BP, AP) \tag{1.75}$$

Recall that all the points on the line segment AB have the same level ℓ.

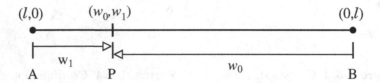

Figure 1.4.1 *Geometric presentation of the coordinate numbers of the point P as $w_0 = BC$ and $w_1 = AC$.*

1.4.2   Two-dimensional constrained numbers

Each of sides AB, BC, and AC of the triangle $\triangle ABC$ in figure 1.4.2 is equivalent to the constrained line segment in figure 1.4.1.

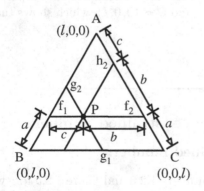

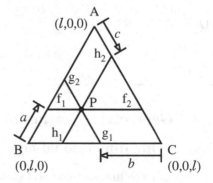

Figure 1.4.2 *Constraint condition for a two-dimensional constrained number system in the level ℓ.*

Figure 1.4.3 *Location of a point P given the coordinate numbers $w_0 = a$, $w_1 = b$, and $w_2 = c$.*

The constraint condition is represented geometrically by the property that the sum of the segment lines representing each coordinate number equals to the length of the side of the triangle. To prove this property first recall, from subsection 1.3, figure 1.3.2 and figure 1.3.4 that the triangle $\triangle ABC$ is equilateral. This same triangle is shown in figure 1.4.2 in a top view. Then note that the segment line $Bf_1 = a$, which represents the first coordinate number w_0, equals Cf_2 along the side AC. The line segment $Pf_2 = b$, which represents the second coordinate number w_1, equals h_2f_2 since the triangle $\triangle Pf_2h_2$ is equilateral. The triangle $\triangle Pf_1g_2$ is equilateral, then $Pf_1 = Pg_2 = c$. The line segments g_2A and Ph_2 are parallel as well as Pg_2 and Ah_2, therefore, $Pg_2 = Ah_2 = c$ which represents the third coordinate number w_2. Then the geometric constraint condition can be written as $AC = a+b+c = \ell$, which is a constant for any point inside the triangle $\triangle ABC$. This implies that all points in the triangle $\triangle ABC$ belong to the level ℓ.

The point P can be located, as shown if figure 1.4.3, by construction lines parallel to the sides of the triangle as follows:

1. from the coordinate number $w_0 = a$ construct the line f_1f_2 parallel to the side BC, which is the line whose values of w_0 are constant,

2. from the coordinate number $w_1 = b$ construct the line g_1g_2 parallel to the side AC, which is the line whose values of w_1 are constant,

The intersection of these two lines determines the point P. Observe that if from the coordinate number $w_2 = c$ a line h_1h_2 is constructed parallel to the side AB this line has the values of w_2 constant and, consequently, it passes by the point P.

1.5   Two-dimensional representation of n-dimensional co-ordinate axes

In the two-dimensional representation of an n-dimensional coordinate axes, called *floating coordinate axes*, the coordinate axes are drawn horizontally or vertically such that if one axis is horizontal the following is vertical and the next is horizontal. The first axis can be horizontal of vertical according to the convenience of the representation.

Figure 1.5.1 shows the floating coordinate axes representation of an one-dimensional system ($\delta = 1, d = 2$). The origin of the second axis is on the first axis. The position of the last coordinate y defines the position of the point $P(x, y)$.

Figure 1.5.2 shows the floating coordinate axes representation of a two-dimensional system ($\delta = 2, d = 3$). The origin of the second axis is on the first axis and the origin of the third axis is on the second one. The position of the last coordinate z defines the position of the point $P(x, y, z)$.

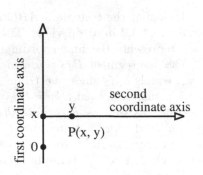

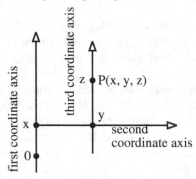

Figure 1.5.1 *Floating coordinate axes representation for the one-dimensional* $(\delta = 1, d = 2)$.

Figure 1.5.2 *Floating coordinate axes representation for the two-dimensional* $(\delta = 2, d = 3)$.

This type of representation can be used for any dimension. The higher the dimension the higher will be the complexity of the figure.

1.6 Zero-dimensional constrained space

1.6.1 Definitions

If $\delta = 0$ in equation 1.3 page 4, it follows that

$$L^0 = \left\{ w_0 \in \mathbb{F} \,\middle|\, |w_0| = \ell \in \mathbb{G} \right\} \tag{1.76}$$

and the constraint equation is given by

$$\sum_{j=0}^{0} |w_j| = |w_0| = \ell \in \mathbb{G} \tag{1.77}$$

The solution of the constraint equation (1.77) is $w_0 = \pm \ell$, $\ell \in \mathbb{G}$ giving two constrained numbers

$$L_\ell^0 = \{(-\ell), (\ell)\} \tag{1.78}$$

for each level, which permits to write the set of zero-dimensional constrained number system as follows

$$L^0 = \left\{ \bigcup_{\ell \in \mathbb{G}} \{(-\ell), (\ell)\} \right\} \tag{1.79}$$

For the level $\ell = 0$ the set L_0^0 contains only one element, namely the zero number, which is the origin of the zero-dimensional constrained number system. Actually, it should be written $L_0^0 = \{(-0), (0)\}$ to be consistent with equation (1.78).

The zero-dimensional constrained numbers in equation (1.78) writes

$$
L_\ell^0 = \begin{cases}
\{(0)\} & \text{if } \ell = 0 \text{ for } \mathbb{R}, \mathbb{Z}, \mathbb{R}^+, \text{ or } \mathbb{Z}^+ \\
\{(-\ell), (\ell)\} & \text{if } \ell > 0 \text{ for } \mathbb{R}, \text{ or } \mathbb{Z} \\
\{(\ell)\} & \text{if } \ell > 0 \text{ for } \mathbb{R}^+, \text{ or } \mathbb{Z}^+
\end{cases}
\tag{1.80}
$$

The expansion of (1.79) using equation (1.78) with $\ell \in \mathbb{Z}$, that is, $\mathbb{G} = \mathbb{Z}$, gives the sequence of sets

$$
\begin{aligned}
L^0 &= \{L_0^0 \cup L_1^0 \cup L_2^0 \cup \cdots\} \\
&= \{\{(0)\} \cup \{(-1), (1)\} \cup \{(-2), (2)\} \cup \ldots \cup \{(-\ell), (\ell)\} \cup \ldots\} \\
&= \{(0), (-1), (1), (-2), (2), \ldots, (-\ell), (\ell), \ldots\}
\end{aligned}
\tag{1.81}
$$

For the particular case of the zero-dimensional constrained whose underlying set is $\mathbb{Z}^+$ the equation (1.79) becomes

$$
L^0 = \left\{ \bigcup_{\ell=0}^{\infty} \{(\ell)\} \right\}
\tag{1.82}
$$

and equation (1.81) simplifies to

$$
\begin{aligned}
L^0 &= \{L_0^0 \cup L_1^0 \cup L_2^0 \cup \cdots\} \\
&= \{\{(0)\} \cup \{(1)\} \cup \{(2)\} \cup \ldots \cup \{(\ell)\} \cup \ldots\} \\
&= \{(0), (1), (2), \ldots, (\ell), \ldots\}
\end{aligned}
\tag{1.83}
$$

1.6.2 Geometric representation of the zero-dimensional constrained numbers

Several geometric representations can be created. For example, in figure 1.6.1(a) each level is represented by a segment line. The line $X(-)$ contains the first element, which is the negative number while the line $X(+)$ contains the second element, which is the positive number. Each thick line represents the set of constrained numbers in the level defined by the absolute value of the number. Thus the points -2 and 2 define the set of elements $[-2, 2]$ constrained to the level $\ell = |-2| = 2$. Each set contains only two elements, for example, if $\ell = 1.33$ the set contains the elements -1.33 and $+1.33$, that is, $[-1.33, 1.33]$ since all other points in the set have absolute value different from 1.33.

A second representation can be obtained rotating the lines $X(-)$ and $X(+)$ until they are coincident as shown in figure 1.6.1(b).

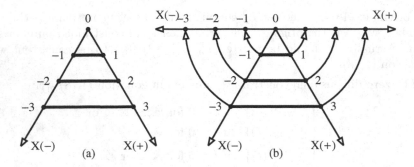

Figure 1.6.1 *Possible geometrical representations for the zero-dimensional constrained space. (a) in a triangular form, (b) in an axial form.*

The representation in figure 1.6.1(b) is equivalent to the one-dimensional cartesian space and can be pictured on a line as shown in figure 1.6.2.

Figure 1.6.2 *Zero-dimensional constrained number system represented using the one-dimensional cartesian axis.*

Another possible geometric representations consists in showing the sets in (1.81) on an axis, ordered by the value of their respective levels as it is shown in figure 1.6.3. Note the use of the symbol $\prec$ to indicate that the elements of the level L_i^0 precedes the ones of the level L_{i+1}^0.

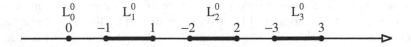

Figure 1.6.3 *Representation of ordered levels $L_0^0 \prec L_1^0 \prec L_2^0 \prec \cdots$ in a zero-dimensional constrained number system.*

For the case of the zero-dimensional constrained whose underlying set is $\mathbb{Z}^+$ the sets represented in figure 1.6.3 simplify to the ones shown in figure 1.6.4.

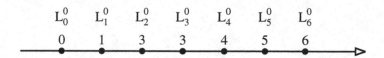

Figure 1.6.4 *The zero-dimensional constrained space can be represented by points on a line.*

where

$$L_0^0 = \{(0)\}, \quad L_1^0 = \{(1)\}, \quad L_2^0 = \{(2)\}, \quad \ldots, \quad L_k^0 = \{(\ell)\}, \quad \ldots \quad (1.84)$$

Note 1.6.5

1. The above representations suggest a natural way to define an ordering relation for the constrained number system.
2. The zero-dimensional space of constrained numbers is the fundamental space for the construction of all the other.

end of note 1.6.5

Chapter 2

Generation of the coordinate system

2.1 Generation as a cartesian product of two sets

2.1.1 One-dimensional constrained numbers

The set of one-dimensional constrained numbers can be obtained by performing the constrained cartesian product, defined 1.1.26, of two sets of zero-dimensional constrained numbers such as L_i^0 and L_j^0. The element $w_0 \in L_i^0$ is called the first coordinate number and, analogously, $w_1 \in L_j^0$ is called the second coordinate number.

According to the constraint equation (1.33) in definition 1.1.13 of level a one-dimensional number needs to satisfy the condition $|w_0| + |w_1| = \ell$ to belong to the level ℓ. Since the coordinate numbers w_0 and w_1 are constrained by the equation (1.33), one of the sets L_i^0 or L_j^0 in the product $L_i^0 \times L_j^0$ is dependent. Choosing the set L_j^0 as the dependent one, then the second coordinate is obtained from the first with the aid of the constraint equation as $w_1 = \ell - w_0$.

The level of the cartesian product of two zero-dimensional sets L_i^0 and L_j^0 is $\ell = i + j$. There are two possible cases, $i = \ell - j$ and $j = \ell - i$, which give the following two expressions

$$L_i^0 \times L_{\ell-i}^0 \qquad i = 0, 1, \ldots, \ell \qquad (2.1)$$

$$L_{\ell-j}^0 \times L_j^0 \qquad j = 0, 1, \ldots, \ell \qquad (2.2)$$

for the generation of the elements of the partial sets. The set product is obtained by the union of the partial sets as

$$L_\ell^1 = \bigcup_{i=0}^{\ell} \left\{ L_i^0 \times L_{\ell-i}^0 \right\} \qquad (2.3)$$

if the equation (2.1) is used.

Example 2.1.1 *Obtain the elements of the constrained one-dimensional set* L_2^1 *in the level* $\ell = 2$ *for the underlying set* $\mathbb{Z}$.

Substituting $\ell = 2$ into equation (2.3) one dimension L_2^1 is obtained as

$$L_2^1 = \bigcup_{i=0}^{\ell=2} \left\{ L_i^0 \times L_{\ell-i}^0 \right\} = \left\{ L_0^0 \times L_1^0 \right\} \cup \left\{ L_1^0 \times L_1^0 \right\} \cup \left\{ L_2^0 \times L_0^0 \right\} \quad (2.4)$$

Computation of each partial product.

1. For $\ell = 2$, and $i = 0$ the first partial cartesian product is

$$\left\{ L_0^0 \times L_2^0 \right\} = \left\{ (0) \right\} \times \left\{ (-2), (2) \right\} = \left\{ (0, -2), (0, 2) \right\} \quad (2.5)$$

   where the numbers $(0, -2), (0, 2)$ are constrained to the level $\ell = 2$, which can be verified from $|-2| + |0| = |2| + |0| = 2$.

2. Similarly for the second partial product is obtained with $i = 1$

$$\left\{ L_1^0 \times L_1^0 \right\} = \left\{ (-1), (1) \right\} \times \left\{ (-1), (1) \right\}$$
$$= \left\{ (-1, -1), (-1, 1), (1, -1), (1, 1) \right\} \quad (2.6)$$

   where the numbers $(-1, -1), (-1, 1), (1, -1), (1, 1)$ are constrained to the level $\ell = 2$, which can be verified from $|-1| + |-1| = |-1| + |1| = |1| + |-1| = |1| + |1| = 2$.

3. And the third partial product is obtained with $i = 2$

$$\left\{ L_2^0 \times L_0^0 \right\} = \left\{ (-2), (2) \right\} \times \left\{ (0) \right\} = \left\{ (-2, 0), (2, 0) \right\} \quad (2.7)$$

   where the numbers $(-2, 0), (2, 0)$ are constrained to the level $\ell = 2$ which can be verified from $|0| + |-2| = |0| + |2| = 2$.

Substituting the partial products obtained in equations (2.5)–(2.7) into the expression (2.4) it gives the elements of the set L_2^1 as follows

$$L_2^1 = \left\{ (-2, 0), (2, 0)(-1, -1), (-1, 1), (1, -1), (1, 1)(0, -2), (0, 2) \right\} \quad (2.8)$$

end of example 2.1.1

2.1.2   Two-dimensional constrained numbers

Similarly a two-dimensional constrained space can be constructed by the union of the cartesian products $L_\ell^2 = L_i^0 \times L_j^1$ as

$$L_\ell^2 = \bigcup_{i=0}^{\ell} \left\{ L_i^0 \times L_{\ell-i}^1 \right\} \quad (2.9)$$

The level of the product can be obtained as $\ell = i + j$ if one of the factors of the cartesian product is a zero-dimensional set of constrained numbers L_i^0.

2.1.3   δ-dimensional constrained numbers

The set of elements of a δ-dimensional space can be obtained using the expression

$$L_\ell^\delta = \bigcup_{i=0}^{\ell} \{ L_i^0 \times L_{\ell-i}^{\delta-1} \} \qquad \delta > 0 \tag{2.10}$$

2.2   Generation of the elements of the set L_ℓ^δ for the underlying set $\mathbb{Z}$

The definition of sublevel 1.1.14 is based in the constraint equation (1.35) on page 10 and the definition of dual sublevel 1.1.15 is related to the equation (1.37). For $i = 0$ both definitions coincide leading to the following expression

$$\lambda_i = \sum_{j=i}^{\delta} |w_j| = \lambda^i = \sum_{j=0}^{\delta-i} |w_j| \tag{2.11}$$

which implies the equality $\lambda_0 = \lambda^0 = \ell$. This permits us to use the notations ℓ, λ_0 for level, and λ^0 for dual level indistinguishable. In general ℓ will be used to refer to the highest level or dual level in a set and λ_i as well as λ^i when it may refer to any level or dual level in a set.

2.2.1   Rules for the generation

Given ℓ and the constraint equation (1.35) the following rules may be used to compute the constrained numbers in the level ℓ and dimension δ.

Rule 2.2.1 *Generation by recurrence if the underlying set is* $\mathbb{Z}$.

1. Given $\ell > 0$ and $\lambda_0 = \ell$, the coordinate number w_0 takes the sequence of values

$$w_0 = \lambda_0, (\lambda_0 - 1), \ldots, 1, 0, -1, \ldots, -(\lambda_0 - 1), -\lambda_0 \tag{2.12}$$

   where λ_0 is the value of the level.

2. For each w_0 coordinate number there is associated a set $L_{\lambda_1}^{\delta-1}$ given by

$$L_{\lambda_1}^{\delta-1} = \left\{ \omega \in \mathbb{F}^{\delta-1} \,\middle|\, \sum_{j=1}^{\delta} |w_j| = \lambda_1 \le \lambda_0 \right\} \tag{2.13}$$

From the definition of sublevel 1.1.14 on page 9 the constraint equation in (2.13) can be rewritten as

$$\lambda_1 = \sum_{j=1}^{\delta} |w_j| = \sum_{j=0}^{\delta} |w_j| - |w_0| = \lambda_0 - |w_0| \qquad (2.14)$$

Recalling equation (2.10) the elements in the set L_ℓ^δ can be obtained by the union of the cartesian product

$$L_{\lambda_0}^\delta = \bigcup_{w_0=\lambda_0}^{-\lambda_0} \{(w_0)\} \times L_{\lambda_1}^{\delta-1} \qquad (2.15)$$

3. For each w_1 coordinate number there is associated a set $L_{\lambda_2}^{\delta-2}$ given by

$$L_{\lambda_2}^{\delta-2} = \left\{ \omega \in \mathbb{F}^{\delta-2} \,\Big|\, \sum_{j=2}^{\delta} |w_j| = \lambda_2 \leq \lambda_1 \right\} \qquad (2.16)$$

From the definition of sublevel 1.1.14 the constraint equation in (2.16) can be rewritten, recalling equation (2.13), as

$$\lambda_2 = \sum_{j=2}^{\delta} |w_j| = \left(\sum_{j=0}^{\delta} |w_j| - \sum_{j=0}^{0} |w_j| \right) - |w_1| = \lambda_1 - |w_1| \qquad (2.17)$$

Then the elements in the set $L_{\lambda_1}^{\delta-1}$ can be obtained by the union

$$L_{\lambda_1}^{\delta-1} = \bigcup_{w_1=\lambda_1}^{-\lambda_1} \{(w_1)\} \times L_{\lambda_2}^{\delta-2} \qquad (2.18)$$

Introducing $L_{\lambda_1}^{\delta-1}$ into equation (2.15) it follows

$$L_{\lambda_0}^\delta = \bigcup_{w_0=\lambda_0}^{-\lambda_0} \{(w_0)\} \times \left[\bigcup_{w_1=\lambda_1}^{-\lambda_1} \{(w_1)\} \times L_{\lambda_2}^{\delta-2} \right] \qquad (2.19)$$

which are nested unions. The value of λ_1 must be evaluated just after the computation of w_0 to be used in the next union.

4. For each w_{k-1} coordinate number there is associated a set $L_{\lambda_k}^{\delta-k}$ given by

$$L_{\lambda_k}^{\delta-k} = \left\{ \omega \in \mathbb{F}^{\delta-k} \,\Big|\, \sum_{j=k}^{\delta} |w_j| = \lambda_k \leq \lambda_{k-1} \right\} \qquad (2.20)$$

where

$$\lambda_k = \sum_{j=k}^{\delta} |w_j|$$

$$= \left(\sum_{j=0}^{\delta} |w_j| - \sum_{j=0}^{k-2} |w_j| \right) - |w_{k-1}|$$

$$= \lambda_{k-1} - |w_{k-1}| \tag{2.21}$$

Then the elements in the set $L^1_{\lambda_{\delta-1}}$ can be obtained by the union

$$L^{\delta-(k-1)}_{\lambda_{k-1}} = \bigcup_{w_{k-1}\lambda_{k-1}}^{-\lambda_{k-1}} \{(w_{k-1})\} \times L^{\delta-k}_{\lambda_k} \tag{2.22}$$

5. For each $w_{\delta-1}$ coordinate number there is associated a set $L^0_{\lambda_\delta}$ given by

$$L^0_{\lambda_\delta} = \left\{ w \in \mathbb{F}^0 \;\middle|\; \sum_{j=\delta}^{\delta} |w_j| = \lambda_\delta \le \lambda_{\delta-1} \right\} \tag{2.23}$$

where

$$\lambda_\delta = \sum_{j=\delta}^{\delta} |w_j|$$

$$= \left(\sum_{j=0}^{\delta} |w_j| - \sum_{j=0}^{\delta-2} |w_j| \right) - |w_{\delta-1}|$$

$$= \lambda_{\delta-1} - |w_{\delta-1}| \tag{2.24}$$

Then the elements in the set $L^1_{\lambda_{\delta-1}}$ can be obtained by the union

$$L^1_{\lambda_{\delta-1}} = \bigcup_{A=a}^{b} \{(w_{\delta-1})\} \times L^0_{\lambda_\delta} \tag{2.25}$$

where $A = w_{\delta-1}$, $a = \lambda_{\delta-1}$, and $b = -\lambda_{\delta-1}$.

6. Since

$$L^0_{\lambda_\delta} = \{(-\lambda_\delta), (\lambda_\delta)\} \quad \text{and} \quad \lambda_\delta = \lambda_{\delta-1} - |w_{\delta-1}| = |w_\delta| \tag{2.26}$$

then

$$w_\delta = \pm\lambda_\delta \tag{2.27}$$

Performing all substitutions analogously to what was done in equation (2.19) the algorithm can be expressed in one equation as below

$$
L_{\lambda_0}^{\delta} = \bigcup_{w_0=\lambda_0}^{-\lambda_0} \{(w_0)\} \times
$$

$$
\left[\bigcup_{w_1=\lambda_1}^{-\lambda_1} \{(w_1)\} \times \cdots \left[\bigcup_{w_{k-1}=\lambda_{k-1}}^{-\lambda_{k-1}} \{(w_{k-1})\} \times \cdots \right.\right.
$$

$$
\left.\left.\left[\bigcup_{w_{\delta-1}=\lambda_{\delta-1}}^{-\lambda_{\delta-1}} \{(w_{\delta-1})\} \times w_\delta \right]\right]\right] \tag{2.28}
$$

which are nested unions. The value of λ_k must be evaluated just after the computation of w_k to be used in the next union. The value of the sub-level λ_δ in the last union must be obtained to allow the computation of $w_{\delta-1}$ and then to perform the cartesian product.

end of rule 2.2.1

2.2.2 Rules for the particular cases

Rule 2.2.2 *Particular cases of the generation by recurrence if the underlying set is $\mathbb{Z}$.*

1. The constrained number zero

   If $\delta > 0$ and $\ell = \lambda_0 = 0$ then from $\ell = |w_0| + |w_1| + \cdots + |w_\delta|$ it implies that $|w_0| = |w_1| = \cdots = |w_\delta| = 0$, and the equation (2.15) reduces to

$$
L_\ell^\delta = \{(0)\} \times L_0^{\delta-1}
$$
$$
= \{0, 0, \ldots, 0\} \qquad (\delta+1) \text{ zeros} \tag{2.29}
$$

   If in addition to $\ell = 0$ also $\delta = 0$ it follows

$$
L_0^0 = \{0\} \tag{2.30}
$$

2. If $\delta = 0$ and $\ell = \lambda_0 > 0$

$$
L_\ell^0 = \{(\ell), (-\ell)\} \tag{2.31}
$$

end of rule 2.2.2

The following algorithm can be constructed based on rules 2.2.1 and 2.2.2

2.2.3   Algorithm for the generation of the elements

Algorithm 2.2.3 *Generation of the elements w_j of the set $L_{\lambda_0}^{\delta}$ where w_j belongs to the underlying set $\mathbb{Z}$.*

The equation (2.28) written as an algorithm takes the form below

for $w_0 = \lambda_0$ **to** $w_0 = -\lambda_0$ **do**
$\{\,\lambda_1 = \lambda_0 - |w_0|$
    for $w_1 = \lambda_1$ **to** $w_1 = -\lambda_1$ **do**
    $\{\,\lambda_2 = \lambda_1 - |w_1|$
       for $w_2 = \lambda_2$ **to** $w_2 = -\lambda_2$ **do**
       $\{\,\lambda_3 = \lambda_2 - |w_2|$
          $\vdots$

          for $w_{k-1} = \lambda_{k-1}$ **to** $w_{k-1} = -\lambda_{k-1}$ **do**
          $\{\,\lambda_k = \lambda_{k-1} - |w_{k-1}|$
             for $w_k = \lambda_k$ **to** $w_k = -\lambda_k$ **do**
             $\{\,\lambda_{k+1} = \lambda_k - |w_k|$
                for $w_{k+1} = \lambda_{k+1}$ **to** $w_{k+1} = -\lambda_{k+1}$ **do**
                $\{\,\lambda_{k+2} = \lambda_{k+1} - |w_{k+1}|$

                   $\vdots$

                   for $w_{\delta-2} = \lambda_{\delta-2}$ **to** $w_{\delta-2} = -\lambda_{\delta-2}$ **do**
                   $\{\,\lambda_{\delta-1} = \lambda_{\delta-2} - |w_{\delta-2}|$
                      for $w_{\delta-1} = \lambda_{\delta-1}$ **to** $w_{\delta-1} = -\lambda_{\delta-1}$ **do**
                      $\{\,\lambda_{\delta} = \lambda_{\delta-1} - |w_{\delta-1}|$
                         if $\lambda_{\delta} = 0$ **then** $w_{\delta} = 0$ **else** $w_{\delta} = \pm\lambda_{\delta}$
                      $\}$ end of the loop for $w_{\delta-1}$
                   $\}$ end of the loop for $w_{\delta-2}$

                   $\vdots$

                $\}$ end of the loop for w_{k+1}
             $\}$ end of the loop for w_k
          $\}$ end of the loop for w_{k-1}

          $\vdots$

$\}$ end of the loop for w_2

} end of the loop for w_1

} end of the loop for w_0

end of algorithm 2.2.3

2.3   Generation of the elements of the set $L^\delta(\ell)$ for the underlying set $\mathbb{Z}$

The set $L^\delta(\ell)$ was defined in equation (1.6) on page 5 as the set

$$L^\delta(\ell) = \bigcup_{\lambda_0=0}^{\ell} L_{\lambda_0}^\delta \qquad (2.32)$$

which denotes the union of the sets from level $\lambda_0 = 0$ until the level $\lambda_0 = \ell$. To generate the elements of this set it is sufficient to use the algorithm 2.2.3 and add $\lambda_0 = 0, \ldots, \ell$ as the first loop. Substituting $L_{\lambda_0}^\delta$ given by equation (2.28) into equation (2.32) the expression for the generation of the elements of the set $L^\delta(\ell)$ becomes

$$L^\delta(\ell) = \bigcup_{\lambda_0=0}^{\ell} \left[\bigcup_{w_0=\lambda_0}^{-\lambda_0} \{(w_0)\} \times \left[\bigcup_{w_1=\lambda_1}^{-\lambda_1} \{(w_1)\} \times \cdots \left[\bigcup_{w_{k-1}=\lambda_{k-1}}^{-\lambda_{k-1}} \{(w_{k-1})\} \times \cdots \right. \right. \right.$$
$$\left. \left. \left. \left[\bigcup_{w_{\delta-1}=\lambda_{\delta-1}}^{-\lambda_{\delta-1}} \{(w_{\delta-1})\} \times w_\delta \right] \right] \right] \right] \qquad (2.33)$$

Below is shown the equation (2.33) in algorithm.

2.3.1   Algorithm for the generation of the elements

Algorithm 2.3.1 *Generation of the elements w_j of the set $L^\delta(\ell)$ where $w_j, j = 0, \ldots, \delta$ belongs to the underlying set $\mathbb{Z}$.*

for $\lambda_0 = 0$ **to** $\lambda_0 = \ell$ **do**

    for $w_0 = \lambda_0$ **to** $w_0 = -\lambda_0$ **do**

    $\{ \lambda_1 = \lambda_0 - |w_0|$

        for $w_1 = \lambda_1$ **to** $w_1 = -\lambda_1$ **do**

        $\{ \lambda_2 = \lambda_1 - |w_1|$

        for $w_2 = \lambda_2$ **to** $w_2 = -\lambda_2$ **do**

$\left\{ \lambda_3 = \lambda_2 - |w_2| \right.$

$\vdots$

for $w_{k-1} = \lambda_{k-1}$ **to** $w_{k-1} = -\lambda_{k-1}$ **do**

$\left\{ \lambda_k = \lambda_{k-1} - |w_{k-1}| \right.$

  for $w_k = \lambda_k$ **to** $w_k = -\lambda_k$ **do**

  $\left\{ \lambda_{k+1} = \lambda_k - |w_k| \right.$

    for $w_{k+1} = \lambda_{k+1}$ **to** $w_{k+1} = -\lambda_{k+1}$ **do**

    $\left\{ \lambda_{k+2} = \lambda_{k+1} - |w_{k+1}| \right.$

$\vdots$

      for $w_{\delta-2} = \lambda_{\delta-2}$ **to** $w_{\delta-2} = -\lambda_{\delta-2}$ **do**

      $\left\{ \lambda_{\delta-1} = \lambda_{\delta-2} - |w_{\delta-2}| \right.$

        for $w_{\delta-1} = \lambda_{\delta-1}$ **to** $w_{\delta-1} = -\lambda_{\delta-1}$ **do**

        $\left\{ \lambda_{\delta} = \lambda_{\delta-1} - |w_{\delta-1}| \right.$

          if $\lambda_{\delta} = 0$ **then** $w_{\delta} = 0$ **else** $w_{\delta} = \pm\lambda_{\delta}$

        $\}$ end of the loop for $w_{\delta-1}$

      $\}$ end of the loop for $w_{\delta-2}$

$\vdots$

    $\}$ end of the loop for w_{k+1}

  $\}$ end of the loop for w_k

$\}$ end of the loop for w_{k-1}

$\vdots$

$\}$ end of the loop for w_2

$\}$ end of the loop for w_1

$\}$ end of the loop for w_0

$\}$ end of the loop for λ_0

                                         end of algorithm 2.3.1

2.4   Generation of the elements of the set L_ℓ^δ for the underlying set $\mathbb{Z}^+$

The generation of the elements of the set L_ℓ^δ is the problem of generating combinations as described in Rosen [31].

2.4.1   Rules for the generation

The rules for the underlying set $\mathbb{Z}^+$ are similar to the rules for the underlying set $\mathbb{Z}$ when the negative numbers are removed. Since $w_j \in \mathbb{Z}^+$ then $w_j = |w_j|$. With these modifications the rules in subsection 2.2.1 become

Rule 2.4.1 *Generation by recurrence if the underlying set is* $\mathbb{Z}^+$.

1. Given $\ell > 0$ and $\lambda_0 = \ell$, the coordinate number w_0 takes the sequence of values

$$w_0 = \lambda_0, (\lambda_0 - 1), \ldots, 1, 0 \qquad (2.34)$$

where λ_0 is the value of the level.

2. For each w_0 coordinate number there is associated a set $L_{\lambda_1}^{\delta-1}$ given by

$$L_{\lambda_1}^{\delta-1} = \left\{ \omega \in \mathbb{F}^{\delta-1} \,\middle|\, \sum_{j=1}^{\delta} w_j = \lambda_1 \leq \lambda_0 \right\} \qquad (2.35)$$

From the definition of sublevel 1.1.14 on page 9 the constraint equation in (2.35) can be rewritten as

$$\lambda_1 = \sum_{j=1}^{\delta} w_j = \sum_{j=0}^{\delta} w_j - w_0 = \lambda_0 - w_0 \qquad (2.36)$$

Recalling equation (2.10) the elements in the set L_ℓ^δ can be obtained by the union of the cartesian product

$$L_{\lambda_0}^{\delta} = \bigcup_{w_0=\lambda_0}^{0} \{(w_0)\} \times L_{\lambda_1}^{\delta-1} \qquad (2.37)$$

Solving equation (2.36) for w_0 it follows $w_0 = \lambda_0 - \lambda_1$. For $w_0 = \lambda_0$ it implies that $\lambda_1 = 0$ and for $w_0 = 0$ it gives $\lambda_1 = \lambda_0$. Substituting this index transformation into equation (2.37) it becomes

$$L_{\lambda_0}^{\delta} = \bigcup_{\lambda_1=0}^{\lambda_0} \{(\lambda_0 - \lambda_1)\} \times L_{\lambda_1}^{\delta-1} \qquad \delta > 0, \quad \ell > 0 \qquad (2.38)$$

3. For each w_1 coordinate number there is associated a set $L_{\lambda_2}^{\delta-2}$ given by

$$L_{\lambda_2}^{\delta-2} = \left\{ \omega \in \mathbb{F}^{\delta-2} \,\middle|\, \sum_{j=2}^{\delta} w_j = \lambda_2 \leq \lambda_1 \right\} \qquad (2.39)$$

From the definition of sublevel 1.1.14 the constraint equation in (2.39) can be rewritten, recalling equation (2.35), as

$$\lambda_2 = \sum_{j=2}^{\delta} w_j = \left(\sum_{j=0}^{\delta} w_j - \sum_{j=0}^{0} w_j \right) - w_1 = \lambda_1 - w_1 \qquad (2.40)$$

Then the elements in the set $L_{\lambda_1}^{\delta-1}$ can be obtained by the union

$$L_{\lambda_1}^{\delta-1} = \bigcup_{w_1=\lambda_1}^{0} \{(w_1)\} \times L_{\lambda_2}^{\delta-2} \qquad (2.41)$$

Solving equation (2.41) for w_1 it follows $w_1 = \lambda_1 - \lambda_2$. For $w_1 = \lambda_1$ it implies that $\lambda_2 = 0$ and for $w_1 = 0$ it gives $\lambda_2 = \lambda_1$. Substituting this index transformation into equation (2.41) it becomes

$$L_{\lambda_1}^{\delta-1} = \bigcup_{\lambda_2=0}^{\lambda_1} \{(\lambda_1 - \lambda_2)\} \times L_{\lambda_2}^{\delta-2} \qquad \delta > 0, \quad \ell > 0 \qquad (2.42)$$

4. For each w_{k-1} coordinate number there is associated a set $L_{\lambda_k}^{\delta-k}$ given by

$$L_{\lambda_k}^{\delta-k} = \left\{ \omega \in \mathbb{F}^{\delta-k} \,\middle|\, \sum_{j=k}^{\delta} w_j = \lambda_k \leq \lambda_{k-1} \right\} \qquad (2.43)$$

where

$$\lambda_k = \sum_{j=k}^{\delta} w_j = \left(\sum_{j=0}^{\delta} w_j - \sum_{j=0}^{k-2} w_j \right) - w_{k-1} = \lambda_{k-1} - w_{k-1} \qquad (2.44)$$

Then the elements in the set $L_{\lambda_{\delta-1}}^{1}$ can be obtained by the union

$$L_{\lambda_{k-1}}^{\delta-(k-1)} = \bigcup_{w_{k-1}=\lambda_{k-1}}^{0} \{(w_{k-1})\} \times L_{\lambda_k}^{\delta-k} \qquad (2.45)$$

Solving equation (2.45) for w_{k-1} it follows $w_{k-1} = \lambda_{k-1} - \lambda_k$. For $w_{k-1} = \lambda_{k-1}$ it implies that $\lambda_k = 0$ and for $w_{k-1} = 0$ it gives $\lambda_k = \lambda_{k-1}$. Substituting this index transformation into equation (2.45) it becomes

$$L_{\lambda_{k-1}}^{\delta-(k-1)} = \bigcup_{\lambda_k=0}^{\lambda_{k-1}} \{(\lambda_{k-1} - \lambda_k)\} \times L_{\lambda_k}^{\delta-k} \qquad \delta > 0, \quad \ell > 0 \qquad (2.46)$$

5. For each $w_{\delta-1}$ coordinate number there is associated a set $L^0_{\lambda_\delta}$ given by

$$L^0_{\lambda_\delta} = \left\{ \omega \in \mathbb{F}^0 \mid \sum_{j=\delta}^{\delta} w_j = \lambda_\delta \leq \lambda_{\delta-1} \right\} \qquad (2.47)$$

where

$$\lambda_\delta = \sum_{j=\delta}^{\delta} w_j = \left(\sum_{j=0}^{\delta} w_j - \sum_{j=0}^{\delta-2} w_j \right) - w_{\delta-1} = \lambda_{\delta-1} - w_{\delta-1} \qquad (2.48)$$

Then the elements in the set $L^1_{\lambda_{\delta-1}}$ can be obtained by the union

$$L^1_{\lambda_{\delta-1}} = \bigcup_{w_{\delta-1}=\lambda_{\delta-1}}^{0} \{(w_{\delta-1})\} \times L^0_{\lambda_\delta} \qquad (2.49)$$

Solving equation (2.49) for $w_{\delta-1}$ it follows $w_{\delta-1} = \lambda_{\delta-1} - \lambda_\delta$. For $w_{\delta-1} = \lambda_{\delta-1}$ it implies that $\lambda_\delta = 0$ and for $w_{\delta-1} = 0$ it gives $\lambda_\delta = \lambda_{\delta-1}$. Substituting this index transformation into equation (2.49) it becomes

$$L^1_{\lambda_{\delta-1}} = \bigcup_{\lambda_\delta=0}^{\lambda_{\delta-1}} \{(\lambda_{\delta-1} - \lambda_\delta)\} \times L^{\delta-\delta}_{\lambda_\delta} \qquad \delta > 0, \quad \ell > 0 \qquad (2.50)$$

6. Since

$$L^0_{\lambda_\delta} = \{(\lambda_\delta)\} \qquad \text{and} \qquad \lambda_\delta = \lambda_{\delta-1} - |w_{\delta-1}| = w_\delta \qquad (2.51)$$

then

$$w_\delta = \lambda_\delta \qquad (2.52)$$

end of rule 2.4.1

Performing all substitutions analogously to what was done in equation (2.28) the algorithm can be expressed in one equation as below

$$L^\delta_{\lambda_0} = \bigcup_{w_0=\lambda_0}^{0} \{(w_0)\} \times \left[\bigcup_{w_1=\lambda_1}^{0} \{(w_1)\} \times \cdots \left[\bigcup_{w_{k-1}=\lambda_{k-1}}^{0} \{(w_{k-1})\} \times \cdots \right. \right.$$
$$\left. \left. \left[\bigcup_{w_{\delta-1}=\lambda_{\delta-1}}^{0} \{(w_{\delta-1})\} \times w_\delta \right] \right] \right] \qquad (2.53)$$

which are nested unions. The value of λ_k must be evaluated just after the computation of w_k to be used in the next union. The value of the sub-level

λ_δ in the last union must be obtained to allow the computation of $w_{\delta-1}$ and then to perform the cartesian product.

In terms of λ_k the equation (2.53) writes

$$
L_{\lambda_0}^\delta = \bigcup_{\lambda_1=0}^{\lambda_0} \{(\lambda_0 - \lambda_1)\} \times \left[\bigcup_{\lambda_2=0}^{\lambda_1} \{(\lambda_1 - \lambda_2)\} \times \cdots \left[\bigcup_{\lambda_k=0}^{\lambda_{k-1}} \{(\lambda_{k-1} - \lambda_k)\} \times \cdots \right. \right.
$$
$$
\left. \left. \left[\bigcup_{\lambda_\delta=0}^{\lambda_{\delta-1}} \{(\lambda_{\delta-1} - \lambda_\delta)\} \times \lambda_\delta) \right] \right] \right] \tag{2.54}
$$

2.4.2 Rules for the particular cases

Rule 2.4.2 *Particular cases of the generation by recurrence if the underlying set is* $\mathbb{Z}^+$.

1. The constrained number zero

   If $\delta > 0$ and $\ell = \lambda_0 = 0$ then from $\ell = |w_0| + |w_1| + \cdots + |w_\delta|$ it implies that $|w_0| = |w_1| = \cdots = |w_\delta| = 0$, and the equation (2.37) reduces to

   $$
   L_\ell^\delta = \{(0)\} \times L_0^{\delta-1}
   $$
   $$
   = \{0, 0, \ldots, 0\} \quad (\delta + 1) \text{ zeros} \tag{2.55}
   $$

   If in addition to $\ell = 0$ also $\delta = 0$ it follows

   $$
   L_0^0 = \{0\} \tag{2.56}
   $$

2. If $\delta = 0$ and $\ell = \lambda_0 > 0$

   $$
   L_\ell^0 = \{(\ell)\} \tag{2.57}
   $$

   end of rule 2.4.2

2.4.3 Algorithm for the generation of the elements of the set L_ℓ^δ using the coordinate numbers

The following algorithm can be constructed based on rules 2.4.1 and 2.4.2

Algorithm 2.4.3 *Generation of the elements* w_j *of the set* $L_{\lambda_0}^\delta = L_\ell^\delta$ *where* w_j *belongs to the underlying set* $\mathbb{Z}^+$.

The equation (2.53) when written in algorithm form becomes

    for $w_0 = \lambda_0$ **to** $w_0 = 0$ **do**

$\{\, \lambda_1 = \lambda_0 - w_0$

   for $w_1 = \lambda_1$ **to** $w_1 = 0$ **do**

   $\{\, \lambda_2 = \lambda_1 - w_1$

      for $w_2 = \lambda_2$ **to** $w_2 = 0$ **do**

      $\{\, \lambda_3 = \lambda_2 - w_2$

$$\vdots$$

        for $w_{k-1} = \lambda_{k-1}$ **to** $w_{k-1} = 0$ **do**

        $\{\, \lambda_k = \lambda_{k-1} - w_{k-1}$

           for $w_k = \lambda_k$ **to** $w_k = 0$ **do**

           $\{\, \lambda_{k+1} = \lambda_k - w_k$

              for $w_{k+1} = \lambda_{k+1}$ **to** $w_{k+1} = 0$ **do**

              $\{\, \lambda_{k+2} = \lambda_{k+1} - w_{k+1}$

$$\vdots$$

              for $w_{\delta-2} = \lambda_{\delta-2}$ **to** $w_{\delta-2} = 0$ **do**

              $\{\, \lambda_{\delta-1} = \lambda_{\delta-2} - w_{\delta-2}$

                  for $w_{\delta-1} = \lambda_{\delta-1}$ **to** $w_{\delta-1} = 0$ **do**

                  $\{\, \lambda_{\delta} = \lambda_{\delta-1} - w_{\delta-1}$

                  $w_{\delta} = \lambda_{\delta}$

                  $\}$ end of the loop for $w_{\delta-1}$

              $\}$ end of the loop for $w_{\delta-2}$

$$\vdots$$

              $\}$ end of the loop for w_{k+1}

           $\}$ end of the loop for w_k

        $\}$ end of the loop for w_{k-1}

$$\vdots$$

      $\}$ end of the loop for w_2

   $\}$ end of the loop for w_1

$\}$ end of the loop for w_0

                              end of algorithm 2.4.3

2.4.4 Algorithm for the generation of the elements of the set L_ℓ^δ using sublevels

The algorithm 2.4.3 can be rewritten with the variable λ_i in the loop, and w_j as the dependent variable as follows

Algorithm 2.4.4 *Generation of the elements w_j of the set $L_{\lambda_0}^\delta$ where $\lambda_0 = \ell$ and w_j belongs to the underlying set $\mathbb{Z}^+$.*

The equation (2.54) when written in algorithm form becomes

for $\lambda_1 = 0$ **to** $\lambda_1 = \lambda_0$ **do**
$\{\, w_0 = \lambda_0 - \lambda_1$
   for $\lambda_2 = 0$ **to** $\lambda_2 = \lambda_1$ **do**
   $\{\, w_1 = \lambda_1 - \lambda_2$
      for $\lambda_3 = 0$ **to** $\lambda_3 = \lambda_2$ **do**
      $\{\, w_2 = \lambda_2 - \lambda_3$

$\vdots$

      for $\lambda_k = 0$ **to** $\lambda_k = \lambda_{k-1}$ **do**
      $\{\, w_{k-1} = \lambda_{k-1} - \lambda_k$
         for $\lambda_{k+1} = 0$ **to** $\lambda_{k+1} = \lambda_k$ **do**
         $\{\, w_k = \lambda_k - \lambda_{k+1}$
            for $\lambda_{k+2} = 0$ **to** $\lambda_{k+2} = \lambda_{k+1}$ **do**
            $\{\, w_{k+1} = \lambda_{k+1} - \lambda_{k+2}$

$\vdots$

            for $\lambda_{\delta-1} = 0$ **to** $\lambda_{\delta-1} = \lambda_{\delta-2}$ **do**
            $\{\, w_{\delta-2} = \lambda_{\delta-2} - \lambda_{\delta-1}$
               for $\lambda_\delta = 0$ **to** $\lambda_\delta = \lambda_{\delta-1}$ **do**
               $\{\, w_{\delta-1} = \lambda_{\delta-1} - \lambda_\delta$
                  $w_\delta = \lambda_\delta$
               $\}$ end of the loop for λ_δ
            $\}$ end of the loop for $\lambda_{\delta-1}$

$\vdots$

            $\}$ end of the loop for λ_{k+2}
         $\}$ end of the loop for λ_{k+1}
      $\}$ end of the loop for λ_k

$$\vdots$$

    } end of the loop for λ_3

   } end of the loop for λ_2

  } end of the loop for λ_1

end of algorithm 2.4.4

2.5   Generation of the elements of the set $L^\delta(\ell)$ for the underlying set $\mathbb{Z}^+$

The set $L^\delta(\ell)$ was defined in equation (1.6) as the set

$$L^\delta(\ell) = \bigcup_{i=0}^{\ell} L_i^\delta \tag{2.58}$$

that is, it is the union of the sets from level $i = 0$ until the level $i = \ell$. To generate the elements of this set it is sufficient to use the algorithm 2.4.3 or 2.4.4 for each i; that is, adding the loop $\lambda_0 = 0, \ldots, \ell$ as the first loop for the algorithm, which is shown below for the algorithm 2.4.4, which gives

$$L_{\lambda_0}^\delta = \bigcup_{\lambda_0=0}^{\ell} \left[\bigcup_{w_0=\lambda_0}^{0} \{(w_0)\} \times \left[\bigcup_{w_1=\lambda_1}^{0} \{(w_1)\} \times \cdots \left[\bigcup_{w_{k-1}=\lambda_{k-1}}^{0} \{(w_{k-1})\} \times \cdots \right.\right.\right.$$
$$\left.\left.\left. \left[\bigcup_{w_{\delta-1}=\lambda_{\delta-1}}^{0} \{(w_{\delta-1})\} \times w_\delta \right] \right] \right] \right] \tag{2.59}$$

which are nested unions. The value of λ_k must be evaluated just after the computation of w_k to be used in the next union. The value of the sublevel λ_δ in the last union must be obtained to allow the computation of $w_{\delta-1}$ and then to perform the cartesian product.

To generate the elements of the set $L^\delta(\ell_1, \ell_2)$ the first union must be from ℓ_1 to ℓ_2.

In terms of λ_k the equation (2.59) writes

$$L_{\lambda_0}^\delta = \bigcup_{\lambda_0=0}^{\ell} \left[\bigcup_{\lambda_1=0}^{\lambda_0} \{(\lambda_0 - \lambda_1)\} \times \left[\bigcup_{\lambda_2=0}^{\lambda_1} \{(\lambda_1 - \lambda_2)\} \times \cdots \right. \right.$$
$$\left[\bigcup_{\lambda_k=0}^{\lambda_{k-1}} \{(\lambda_{k-1} - \lambda_k)\} \times \cdots \right.$$
$$\left. \left. \left[\bigcup_{\lambda_\delta=0}^{\lambda_{\delta-1}} \{(\lambda_{\delta-1} - \lambda_\delta)\} \times \lambda_\delta) \right] \right] \right] \right] \quad (2.60)$$

2.5.1 Algorithm for the generation of the elements of the set $L^\delta(\ell)$ using the coordinate numbers

Algorithm 2.5.1 *Generation of the elements w_j of the set $L^\delta(\ell)$ where w_j belongs to the underlying set $\mathbb{Z}^+$.*

The equation (2.59) written in algorithm becomes

for $\lambda_0 = 0$ **to** $\lambda_0 = \ell$ **do**
  for $w_0 = \lambda_0$ **to** $w_0 = 0$ **do**
  $\{\lambda_1 = \lambda_0 - w_0$
    for $w_1 = \lambda_1$ **to** $w_1 = 0$ **do**
    $\{\lambda_2 = \lambda_1 - w_1$
      for $w_2 = \lambda_2$ **to** $w_2 = 0$ **do**
      $\{\lambda_3 = \lambda_2 - w_2$

          $\vdots$

      for $w_{k-1} = \lambda_{k-1}$ **to** $w_{k-1} = 0$ **do**
      $\{\lambda_k = \lambda_{k-1} - w_{k-1}$
        for $w_k = \lambda_k$ **to** $w_k = 0$ **do**
        $\{\lambda_{k+1} = \lambda_k - w_k$
          for $w_{k+1} = \lambda_{k+1}$ **to** $w_{k+1} = 0$ **do**
          $\{\lambda_{k+2} = \lambda_{k+1} - w_{k+1}$

            $\vdots$

        for $w_{\delta-2} = \lambda_{\delta-2}$ **to** $w_{\delta-2} = 0$ **do**
        $\{\lambda_{\delta-1} = \lambda_{\delta-2} - w_{\delta-2}$
          for $w_{\delta-1} = \lambda_{\delta-1}$ **to** $w_{\delta-1} = 0$ **do**

$$\{ \lambda_\delta = \lambda_{\delta-1} - w_{\delta-1}$$
$$w_\delta = \lambda_\delta$$
$\}$ end of the loop for $w_{\delta-1}$
$\}$ end of the loop for $w_{\delta-2}$

$$\vdots$$

$\}$ end of the loop for w_{k+1}
$\}$ end of the loop for w_k
$\}$ end of the loop for w_{k-1}

$$\vdots$$

$\}$ end of the loop for w_2
$\}$ end of the loop for w_1
$\}$ end of the loop for w_0
$\}$ end of the loop for $\lambda_0 = 0$

end of algorithm 2.5.1

To generate the set elements of the set $L^\delta(\ell_1, \ell_2)$ the first loop must be
for $\lambda_0 = \ell_1$ **to** $\lambda_0 = \ell_2$ **do**

2.5.2 Algorithm for the generation of the elements of the set $L^\delta(\ell)$ using sublevels

Algorithm 2.5.2 *Generation of the elements w_j of the set $L^\delta(\ell)$ where w_j belongs to the underlying set $\mathbb{Z}^+$.*

The equation (2.60) written in algorithm becomes

for $\lambda_0 = 0$ **to** $\lambda_0 = \ell$ **do**
  for $\lambda_1 = 0$ **to** $\lambda_1 = \lambda_0$ **do**
  $\{ w_0 = \lambda_0 - \lambda_1$
    for $\lambda_2 = 0$ **to** $\lambda_2 = \lambda_1$ **do**
    $\{ w_1 = \lambda_1 - \lambda_2$
      for $\lambda_3 = 0$ **to** $\lambda_3 = \lambda_2$ **do**
      $\{ w_2 = \lambda_2 - \lambda_3$

$$\vdots$$

        for $\lambda_k = 0$ **to** $\lambda_k = \lambda_{k-1}$ **do**

$$\{ w_{k-1} = \lambda_{k-1} - \lambda_k$$

$\quad$ **for** $\lambda_{k+1} = 0$ **to** $\lambda_{k+1} = \lambda_k$ **do**

$$\{ w_k = \lambda_k - \lambda_{k+1}$$

$\quad\quad$ **for** $\lambda_{k+2} = 0$ **to** $\lambda_{k+2} = \lambda_{k+1}$ **do**

$$\{ w_{k+1} = \lambda_{k+1} - \lambda_{k+2}$$

$$\vdots$$

$\quad\quad\quad$ **for** $\lambda_{\delta-1} = 0$ **to** $\lambda_{\delta-1} = \lambda_{\delta-2}$ **do**

$$\{ w_{\delta-2} = \lambda_{\delta-2} - \lambda_{\delta-1}$$

$\quad\quad\quad\quad$ **for** $\lambda_\delta = 0$ **to** $\lambda_\delta = \lambda_{\delta-1}$ **do**

$$\{ w_{\delta-1} = \lambda_{\delta-1} - \lambda_\delta$$

$$w_\delta = \lambda_\delta$$

$\quad\quad\quad\quad\quad$ } end of the loop for λ_δ

$\quad\quad\quad\quad$ } end of the loop for $\lambda_{\delta-1}$

$$\vdots$$

$\quad\quad$ } end of the loop for λ_{k+2}

$\quad\quad$ } end of the loop for λ_{k+1}

$\quad\quad$ } end of the loop for λ_k

$$\vdots$$

$\quad$ } end of the loop for λ_3

$\quad$ } end of the loop for λ_2

$\quad$ } end of the loop for λ_1

} end of the loop for ℓ

$\hfill$ end of algorithm 2.5.2

2.6 Examples of zero-dimensional constrained space

2.6.1 Generation of the elements

The zero-dimensional constrained space was defined in the section 1.6 on page 26. For the constraint equation $|w_0| = \ell$ and the underlying set $\mathbb{Z}$ there are two elements per level. If it is $\mathbb{Z}^+$ then there is only one element on each level. For the zero-dimensional constrained system the constraint equation is $|w_0| = \ell$, whose solutions are given in equation (1.80), and they are

$$L_\ell^0 = \begin{cases} \{(0)\} & \text{if } \ell = 0 \text{ for } \mathbb{R}, \mathbb{Z}, \mathbb{R}^+, \text{ or } \mathbb{Z}^+ \\ \{(\ell), (-\ell)\} & \text{if } \ell > 0 \text{ for } \mathbb{R}, \text{ or } \mathbb{Z} \\ \{(\ell)\} & \text{if } \ell > 0 \text{ for } \mathbb{R}^+, \text{ or } \mathbb{Z}^+ \end{cases} \qquad (2.61)$$

2.6.2   Elements of the set L_0^0 and $L^0(0)$

The set L_0^0, defined in equation (1.80) on page 27 with $\ell = 0$, contains the zero element of the 0-dimensional constrained number system, therefore

$$\{(0)\} \quad \text{for } \mathbb{R}, \mathbb{Z}, \mathbb{R}^+, \text{ or } \mathbb{Z}^+ \qquad (2.62)$$

as the underlying set. For the zero element the sets L_0^0 and $L^0(0)$ are identical.

Any of the algorithm shown before can be used to evaluate zero element of the 0-dimensional constrained number system, for example, the algorithm 2.2.3 gives for $\ell = \lambda_0 = 0$

$$\textbf{if } \lambda_0 = 0 \textbf{ then } w_0 = 0 \textbf{ else } w_0 = \pm 0 \qquad (2.63)$$

Below there are some examples of the element zero of the 0-dimensional constrained number system in function of the underlying set

1. for the underlying set $\mathbb{Z}$ or $\mathbb{R}$

$$L_0^0 = L^0(0) = \{(0), (-0)\} = \{(0)\} \qquad (2.64)$$

2. for the underlying set $\mathbb{Z}^+$ or or $\mathbb{R}^+$

$$L_0^0 = L^0(0) = \{(0)\} \qquad (2.65)$$

2.6.3   Elements of the set L_ℓ^0

From equation (1.80) on page 27, with $\ell > 0$, it follows

1. for the underlying set $\mathbb{Z}$ or $\mathbb{R}$

$$L_\ell^0 := \{(\ell), (-\ell)\} \qquad (2.66)$$

If $\ell = 3$, for example, then the set L_3^0 is given by $L_3^0 = \{(3), (-3)\}$. This same result can be obtained with the algorithm 2.2.3 applied to $\delta = 0$ and $\lambda_0 = 3$ as shown below

$$\textbf{if } \lambda_0 = 0 \textbf{ then } w_0 = 0 \textbf{ else } w_0 = \pm \lambda_0 = \pm 3 \qquad (2.67)$$

2. for the underlying set $\mathbb{Z}^+$ or $\mathbb{R}^+$

$$L_\ell^0 := \{(\ell)\} \tag{2.68}$$

If $\ell = 3$ for example then the set L_3^0 is given by $L_3^0 = \{(3)\}$. The algorithm 2.4.4 applied to $\delta = 0$ and $\lambda_0 = 3$ gives

$$w_0 = \lambda_0 = 3 \tag{2.69}$$

the same result.

2.6.4 Elements of the set $L^0(\ell)$

From the notation 1.1.7, equation (1.6) on page 5, the set $L^0(\ell)$ is given by

$$L^0(\ell) = \bigcup_{i=0}^{\ell} L^0(i) = \{L_0^0 \cup L_1^0 \cup L_2^0 \cup \cdots \cup L_\ell^0\} \tag{2.70}$$

If $\ell = 3$, for example, then the elements of the set $L^0(3)$ are obtained as

1. for the underlying set $\mathbb{Z}$

$$L^0(3) = \bigcup_{i=0}^{3} L^0(i) = L^0(0) \cup L^0(1) \cup L^0(2) \cup L^0(3)$$
$$= \{(0), (1), (-1), (2), (-2), (3), (-3)\} \tag{2.71}$$

The algorithm 2.3.1 reduces to

> **for** $\lambda_0 = 0$ **to** $\lambda_0 = \ell$ **do**
> > **if** $\lambda_0 = 0$ **then** $w_\delta = 0$ **else** $w_\delta = \pm\lambda_0$ $\qquad$ (2.72)

and with $\ell = 3$ it gives the same results.

Figure 2.6.3 shows the geometric representation of the set in equation (2.71).

2. for the underlying set $\mathbb{Z}^+$

$$L^0(3) = \bigcup_{i=0}^{3} L^0(i) = L^0(0) \cup L^0(1) \cup L^0(2) \cup L^0(3)$$
$$= \{(0), (1), (2), (3)\} \tag{2.73}$$

The algorithm 2.5.1 for this case reduces to

> **for** $\lambda_0 = 0$ **to** $\lambda_0 = \ell$ **do**
> > $w_\delta = \lambda_0$ $\qquad$ (2.74)

and with $\ell = 3$ it gives the same results.

Figure 2.6.4 shows the geometric representation of the set in equation (2.73).

2.6.5 Geometric representation of the set L_ℓ^0

<table>
<tr><td>

2
●————————
1
●————————
0
●————————

$L_2^0 = \{(2)\}$

$L_1^0 = \{(1)\}$

$L_0^0 = \{(0)\}$

</td><td>

2   −2
●————————●
1   −1
●————————●
0
●————————

$L_2^0 = \{(2),(-2)\}$

$L_1^0 = \{(1),(-1)\}$

$L_0^0 = \{(0)\}$

</td></tr>
</table>

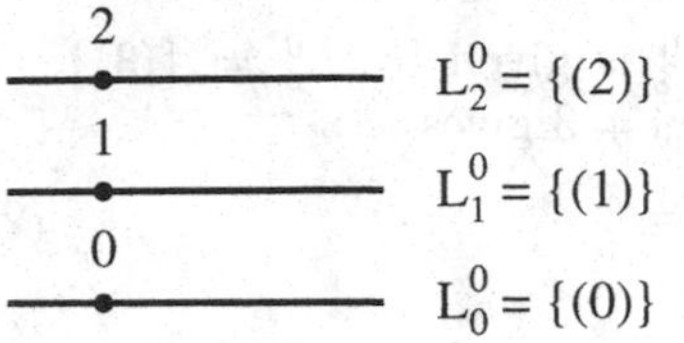

Figure 2.6.1 *Geometric representation of the elements of the sets $L_0^0, L_1^0,$ and L_2^0 for $\mathbb{Z}^+$ as the underlying set.*

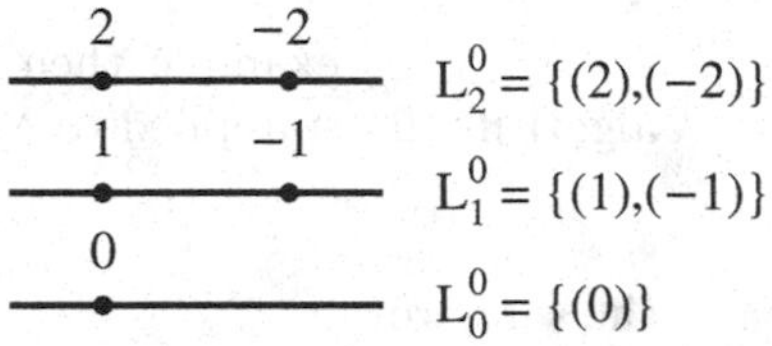

Figure 2.6.2 *Geometric representation of the elements of the sets $L_0^0, L_1^0,$ and L_2^0 for $\mathbb{Z}$ as the underlying set.*

The sets in the zero-dimensional space for the underlying set $\mathbb{Z}^+$ contain one element each and they are represented on a line. Figure 2.6.1 shows examples for the levels $\ell = 0, 1, 2$ and their respective sets $L_0^0 = \{(0)\}$, $L_1^0 = \{(1)\}$ and $L_1^0 = \{(2)\}$.

If $\mathbb{Z}$ is the underlying set then the sets in the zero-dimensional space contain two elements and they are represented by two points on a line. For example, the sets $L_0^0 = \{(0)\}$, $L_1^0 = \{(1),(-1)\}$, and $L_1^0 = \{(2),(-2)\}$ are shown in figure 2.6.2.

2.6.6 Geometric representation of the set $L^0(\ell)$

Figure 2.6.3 shows the geometric representation of the set in equation (2.71).

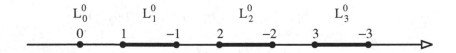

Figure 2.6.3 *Representation of set $L^0(3) = L_0^0 \cup L_1^0 \cup L_2^0 \cup L_3^0$ in a zero-dimensional constrained number system for the underlying set $\mathbb{Z}$. The arrow defines the order of the levels, which is*
$$L_0^0 \prec L_1^0 \prec L_2^0 \prec L_3^0.$$

Figure 2.6.4 shows the geometric representation of the set in equation (2.73).

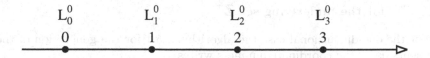

Figure 2.6.4 *Representation of set* $L^0(3) = L_0^0 \cup L_1^0 \cup L_2^0 \cup L_3^0$ *in a zero-dimensional constrained number system for the underlying set* $\mathbb{Z}^+$. *The arrow defines the order of the levels, which is*
$$L_0^0 \prec L_1^0 \prec L_2^0 \prec L_3^0.$$

2.7 Examples of one-dimensional constrained space

The constraint equation for the one-dimensional system is $|w_0| + |w_1| = \ell$. Each element in the set is denoted by two coordinate numbers. Considering w_0 as the dependent variable, the independent one must satisfy the inequality $|w_1| \leq \ell$.

For the one-dimensional case the algorithm 2.2.3 writes

2.7.1 Generation of the elements

2.7.1.1 Algorithm for the generation of the elements of the set L_ℓ^1 for the underlying set $\mathbb{Z}$

For the one-dimensional case the algorithm 2.2.3 writes

$$
\begin{aligned}
&\textbf{for } w_0 = \lambda_0 \textbf{ to } w_0 = -\lambda_0 \textbf{ do} \\
&\quad \{\, \lambda_1 = \lambda_0 - |w_0| \\
&\qquad \textbf{if } \lambda_1 = 0 \textbf{ then } w_1 = 0 \textbf{ else } w_1 = \pm\lambda_1 \\
&\quad \} \text{ end of the loop for } w_0
\end{aligned}
\tag{2.75}
$$

2.7.1.2   Algorithm for the generation of the elements of the set L_ℓ^1 for the underlying set $\mathbb{Z}^+$

For the one-dimensional case the algorithm 2.5.1 for the generation of the elements using the coordinate numbers writes

$$\textbf{for } w_0 = \lambda_0 \textbf{ to } w_0 = 0 \textbf{ do}$$
$$\{\, \lambda_1 = \lambda_0 - w_0$$
$$w_1 = \lambda_1$$
$$\} \text{ end of the loop for } w_0 \qquad\qquad (2.76)$$

For the one-dimensional case the algorithm 2.4.4 for the generation of the elements using sublevels writes

$$\textbf{for } \lambda_1 = 0 \textbf{ to } \lambda_1 = 0 = \lambda_0 \textbf{ do}$$
$$\{\, w_0 = \lambda_0 - \lambda_1$$
$$w_1 = \lambda_1$$
$$\} \text{ end of the loop for } \lambda_1 \qquad\qquad (2.77)$$

2.7.1.3   Algorithm for the generation of the elements of the set $L^1(\ell)$ for the underlying set $\mathbb{Z}$

For the one-dimensional case the algorithm 2.3.1 writes

$$\textbf{for } \lambda_0 = 0 \textbf{ to } \lambda_0 = \ell \textbf{ do}$$
$$\textbf{for } w_0 = \lambda_0 \textbf{ to } w_0 = -\lambda_0 \textbf{ do}$$
$$\{\, \lambda_1 = \lambda_0 - |w_0|$$
$$\textbf{if } \lambda_1 = 0 \textbf{ then } w_1 = 0 \textbf{ else } w_1 = \pm\lambda_1$$
$$\} \text{ end of the loop for } w_0 \qquad\qquad (2.78)$$

2.7.1.4   Algorithm for the generation of the elements of the set $L^1(\ell)$ for the underlying set $\mathbb{Z}^+$

For the one-dimensional case the algorithm 2.5.1 for the generation of the elements using the coordinate numbers writes

$$\textbf{for } \lambda_0 = 0 \textbf{ to } \lambda_0 = \ell \textbf{ do}$$
$$\textbf{for } w_0 = \lambda_0 \textbf{ to } w_0 = 0 \textbf{ do}$$
$$\{\, \lambda_1 = \lambda_0 - w_0$$
$$w_1 = \lambda_1$$
$$\} \text{ end of the loop for } w_0 \qquad\qquad (2.79)$$

For the one-dimensional case the algorithm 2.4.4 for the generation of the elements using sublevels writes

$$
\begin{aligned}
&\textbf{for } \lambda_0 = 0 \textbf{ to } \lambda_0 = \ell \textbf{ do} \\
&\quad \textbf{for } \lambda_1 = 0 \textbf{ to } \lambda_1 = \lambda_0 \textbf{ do} \\
&\quad \{ w_0 = \lambda_0 - \lambda_1 \\
&\quad\ \ w_1 = \lambda_1 \\
&\quad \} \text{ end of the loop for } \lambda_1
\end{aligned}
\tag{2.80}
$$

2.7.2 Elements of the set L_0^1 and $L^1(0)$

2.7.2.1 Underlying set $\mathbb{Z}$ or $\mathbb{Z}^+$

Generation

The generation of the elements of the set L_0^1, is obtained using the algorithm in equation (2.75) and for the set $L^1(0)$ it is obtained using the algorithm in equation (2.78). The result is the same using either algorithm and it is shown in the table 2.7.1 below

Table 2.7.1 *Elements of the sets L_0^1 or $L^1(0)$ for $\mathbb{Z}^+$ or $\mathbb{Z}$ as the underlying set.*

w_0	w_1	(w_0, w_1)
0	0	$(0,0)$

Set notation

The result is denoted as

$$
L_0^1 = L^1(0) = \{ (0,0) \}
\tag{2.81}
$$

Geometric representation of elements of the set $L_0^1 = L^1(0)$

Figure 2.7.2 shows the geometric representation of the set $L_0^1 = L^1(0) = \{(0,0)\}$ using an oblique two-dimensional cartesian axes. Figure 2.7.3 represents the same sets with a two-dimensional orthogonal cartesian axes w_0 and w_1.

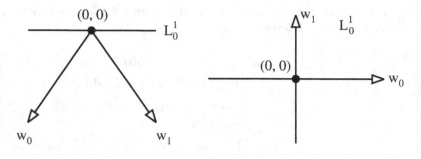

Figure 2.7.2 *Elements of the set L_0^1 in the level $\ell = 0$ represented using two oblique cartesian axes.*

Figure 2.7.3 *Elements of the set L_0^1 in the level $\ell = 0$ represented using two orthogonal cartesian axes.*

2.7.3   Elements of the sets L_1^1 and $L^1(1)$

2.7.3.1   Elements of the sets L_1^1 for the underlying set $\mathbb{Z}$

Generation

The generation of the elements of the set L_1^1 is obtained using the algorithm in equation (2.75) and the result is shown in the table 2.7.4 below

Table 2.7.4 *Generation of the elements of the set L_1^1 for $\mathbb{Z}$ as the underlying set.*

w_0	w_1	(w_0, w_1)
1	0	$(1, 0)$
0	1	$(0, 1)$
0	-1	$(0, -1)$
-1	0	$(-1, 0)$

Set notation

The result is denoted as

$$L_1^1 = \left\{ (1, 0), (0, 1), (0, -1), (-1, 0) \right\} \tag{2.82}$$

Geometric representation of elements of the set L_1^1

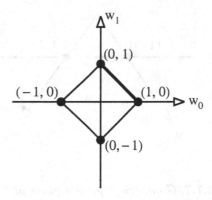

Figure 2.7.5 *Geometric representation of the elements of the set L_1^1 for the underlying set $\mathbb{Z}$.*

The thicker line in figure 2.7.5 represents the elements of the set whose underlying set is $\mathbb{Z}^+$. These elements are also shown in figure 2.7.7.

2.7.3.2 Elements of the sets L_1^1 for the underlying set $\mathbb{Z}^+$

Generation

The generation of the elements of the set L_1^1, is obtained using the algorithm in equation (2.76) and the result is shown in the table 2.7.6 below

Table 2.7.6 *Generation of the elements of the set L_1^1 for $\mathbb{Z}^+$ as the underlying set.*

w_0	w_1	(w_0, w_1)
1	0	$(1, 0)$
0	1	$(0, 1)$

Set notation

The result is denoted as

$$L_1^1 = \big\{\, (1,0), (0,1) \,\big\} \tag{2.83}$$

Geometric representation of elements of the set L_1^1

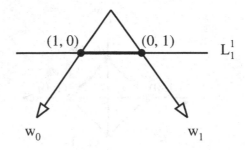

Figure 2.7.7 *Geometric representation of the elements of the set L_1^1 for the underlying set $\mathbb{Z}^+$.*

The set L_1^1 for the underlying set is $\mathbb{Z}^+$ is a subset of L_1^1 for the underlying set $\mathbb{Z}$ as the thicker line in figure 2.7.7 shows.

2.7.3.3 Elements of the sets $L^1(1)$ for the underlying set $\mathbb{Z}$

Generation

The generation of the elements of the set L_1^1, is obtained using the algorithm in equation (2.78) and the result is shown in the table 2.7.8 below

Table 2.7.8 *Generation of the elements of the set $L^1(1)$ for $\mathbb{Z}$ as the underlying set.*

w_0	w_1	(w_0, w_1)
0	0	$(0,0)$
1	0	$(1,0)$
0	1	$(0,1)$
0	-1	$(0,-1)$
-1	0	$(-1,0)$

Set notation

The result is denoted as

$$L^1(1) = L_0^1 \cup L_1^1 = \{(0,0),(1,0),(0,1),(0,-1),(-1,0)\} \qquad (2.84)$$

Geometric representation of elements of the set $L^1(1)$

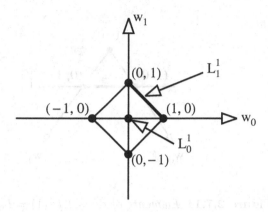

Figure 2.7.9 *Elements of the set $L^1(1) = L_0^1 \cup L_1^1$ for the underlying set $\mathbb{Z}$.*

2.7.3.4 Elements of the sets $L^1(1)$ for the underlying set $\mathbb{Z}^+$

Generation

The generation of the elements of the set L_1^1, is obtained using the algorithm in equation (2.80) and the result is shown in the table 2.7.10 and geometrically represented in figure 2.7.11.

Table 2.7.10 *Generation of the elements of the set $L^1(1)$ for $\mathbb{Z}^+$ as the underlying set.*

w_0	w_1	(w_0, w_1)
0	0	$(0,0)$
1	0	$(1,0)$
0	1	$(0,1)$

Set notation

The result is denoted as

$$L^1(1) = L_0^1 \cup L_1^1 = \{(0,0),(1,0),(0,1)\} \tag{2.85}$$

Geometric representation of elements of the set $L^1(1)$

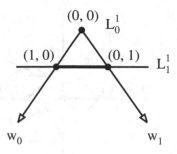

Figure 2.7.11 *Elements of the set $L^1(1) = L_0^1 \cup L_1^1$ for the underlying set $\mathbb{Z}^+$.*

The set $L^1(1)$ for the underlying set is $\mathbb{Z}^+$ is a sub set of $L^1(1)$ for the underlying set $\mathbb{Z}$ as the thicker line in figure 2.7.11 shows.

2.7.4   Elements of the set L_2^1 and $L^1(2)$

2.7.4.1   Elements of the sets L_2^1 for the underlying set $\mathbb{Z}$

Generation

The generation of the elements of the set L_2^1, is obtained using the algorithm in equation (2.75) and the result is shown in the table 2.7.12 below

Table 2.7.12 *Generation of the elements of the set L_2^1 for $\mathbb{Z}$ as the underlying set.*

w_0	w_1	(w_0, w_1)
2	0	$(2,0)$
1	1	$(1,1)$
1	-1	$(1,-1)$
0	2	$(0,2)$
0	-2	$(0,-2)$
-1	1	$(-1,1)$
-1	-1	$(-1,-1)$
-2	0	$(-2,0)$

Set notation

The result is denoted as

$$L_2^1 = \{\, (2,0), (1,1), (1,-1), (0,2), (0,-2),$$
$$(-1,1), (-1,-1), (-2,0) \,\} \tag{2.86}$$

Geometric representation of elements of the set L_2^1

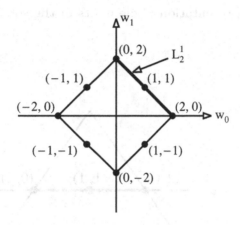

Figure 2.7.13 *Elements of the set L_2^1 for the underlying set $\mathbb{Z}$.*

The thicker line in figure 2.7.13 represents the elements of the set whose underlying set is $\mathbb{Z}^+$. These elements are also shown in figure 2.7.15.

2.7.4.2 Elements of the sets L_2^1 underlying set $\mathbb{Z}^+$

Generation

The generation of the elements of the set L_2^1, is obtained using the algorithm in equation (2.76) and the result is shown in the table 2.7.14 below

Table 2.7.14 *Generation of the elements of the set L_2^1 for $\mathbb{Z}^+$ as the underlying set.*

w_0	w_1	(w_0, w_1)
2	0	$(2,0)$
1	0	$(1,1)$
0	2	$(0,2)$

Set notation

The result is denoted as

$$L_2^1 = \left\{\, (2,0), (1,1), (0,2) \,\right\} \tag{2.87}$$

Geometric representation of elements of the set L_2^1

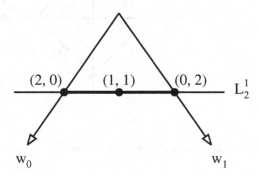

Figure 2.7.15 *Elements of the set L_2^1 for the underlying set $\mathbb{Z}^+$.*

The set L_2^1 for the underlying set is $\mathbb{Z}^+$ is a subset of L_2^1 for the underlying set $\mathbb{Z}$ as the thicker line in figure 2.7.13 shows.

2.7.4.3   Elements of the sets $L^1(2)$ for the underlying set $\mathbb{Z}$

Generation

The generation of the elements of the set $L^1(2)$, is obtained using the algorithm in equation (2.78) and the result is shown in the table 2.7.16 below

Table 2.7.16 *Generation of the elements of the set* $L^1(2) = L_0^1 \cup L_1^1 \cup L_2^1$ *for* $\mathbb{Z}$ *as the underlying set.*

set	w_0	w_1	(w_0, w_1)
L_0^1	0	0	$(0,0)$
	1	0	$(1,0)$
L_1^1	0	1	$(0,1)$
	0	-1	$(0,-1)$
	-1	0	$(-1,0)$
	2	0	$(2,0)$
	1	1	$(1,1)$
	1	-1	$(1,-1)$
	0	2	$(0,2)$
L_2^1	0	-2	$(0,-2)$
	-1	1	$(-1,1)$
	-1	-1	$(-1,-1)$
	-2	0	$(-2,0)$

Set notation

The result is denoted as

$$L^1(2) = L_0^1 \cup L_1^1 \cup L_2^1 = \{\, (0,0), (1,0), (0,1), (0,-1), (-1,0)$$
$$= (2,0), (1,1), (1,-1), (0,2), (0,-2),$$
$$(-1,1), (-1,-1), (-2,0)\,\} \qquad (2.88)$$

Geometric representation of elements of the set $L^1(2)$

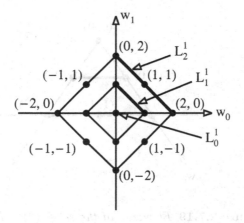

Figure 2.7.17 *Elements of the set* $L^1(2) = L_0^1 \cup L_1^1 \cup L_2^1$ *for the underlying set* $\mathbb{Z}$.

The thicker line in figure 2.7.17 represents the elements of the set whose underlying set is $\mathbb{Z}^+$. These elements are also shown in figure 2.7.19.

2.7.4.4   Elements of the sets $L^1(2)$ underlying set $\mathbb{Z}^+$

Generation

The generation of the elements of the set $L^1(2)$, is obtained using the algorithm in equation (2.80) and the result is shown in the table 2.7.18 below

Table 2.7.18 *Generation of the elements of the set $L^1(2) = L_0^1 \cup L_1^1 \cup L_2^1$ for $\mathbb{Z}^+$ as the underlying set.*

set	w_0	w_1	(w_0, w_1)
L_0^1	0	0	$(0,0)$
	1	0	$(1,0)$
L_1^1	0	1	$(0,1)$
	2	0	$(2,0)$
L_2^1	1	1	$(1,1)$
	0	2	$(0,2)$

Set notation

The result is denoted as

$$L^1(2) = L_0^1 \cup L_1^1 \cup L_2^1 = \big\{ (0,0), (1,0), (0,1), (2,0), (1,1), (0,2) \big\} \quad (2.89)$$

Geometric representation of elements of the set $L^1(2)$

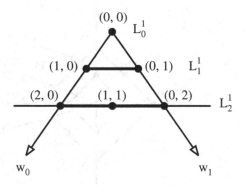

Figure 2.7.19 *Elements of the set $L^1(2) = L_0^1 \cup L_1^1 \cup L_2^1$ for the underlying set $\mathbb{Z}^+$.*

The set L_2^1 for the underlying set is $\mathbb{Z}^+$, shown in figure 2.7.19, is a subset of L_2^1 for the underlying set $\mathbb{Z}$ as the thicker line in figure 2.7.17 shows.

2.7.5 Elements of the set L_3^1 and $L^1(3)$

2.7.5.1 Elements of the sets L_3^1 for the underlying set $\mathbb{Z}$

Generation

The generation of the elements of the set L_3^1, is obtained using the algorithm in equation (2.75) and the result is shown in the table 2.7.20 below

Table 2.7.20 *Generation of the elements of the set L_3^1 for $\mathbb{Z}$ as the underlying set.*

w_0	w_1	(w_0, w_1)
3	0	$(3,0)$
2	1	$(2,1)$
2	−1	$(2,−1)$
1	2	$(1,2)$
1	−2	$(1,−2)$
0	3	$(0,3)$
0	−3	$(0,−3)$
−1	2	$(−1,2)$
−1	−2	$(−1,−2)$
−2	1	$(−2,1)$
−2	−1	$(−2,−1)$
−3	0	$(−3,0)$

Geometric representation of elements of the set L_3^1

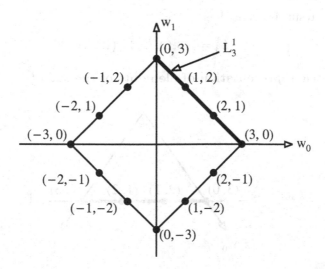

Figure 2.7.21 *Elements of the set L_3^1 for the underlying set $\mathbb{Z}$.*

The thicker line in figure 2.7.21 represents the elements of the set whose underlying set is $\mathbb{Z}^+$. These elements are also shown in figure 2.7.23.

Set notation

The result is denoted as

$$L_3^1 = \big\{\, (3,0), (2,1), (2,-1), (1,2), (1,-2), (0,3),$$
$$(0,-3), (-1,2), (-1,-2), (-2,1), (-2,-1), (-3,0) \,\big\} \qquad (2.90)$$

2.7.5.2   Elements of the sets L_3^1 underlying set $\mathbb{Z}^+$

Generation

The generation of the elements of the set L_3^1, is obtained using the algorithm in equation (2.76) and the result is shown in the table 2.7.22 below

Table 2.7.22 *Generation of the elements of the set L_3^1 for $\mathbb{Z}^+$ as the underlying set.*

w_0	w_1	(w_0, w_1)
3	0	$(3,0)$
2	1	$(2,1)$
1	2	$(1,2)$
0	3	$(0,3)$

Set notation

The result is denoted as

$$L_3^1 = \big\{\, (3,0), (2,1), (1,2), (0,3) \,\big\} \qquad (2.91)$$

Geometric representation of elements of the set L_3^1

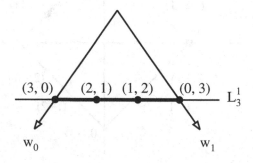

Figure 2.7.23 *Elements of the set L_3^1 in the underlying set $\mathbb{Z}^+$.*

The set L_3^1 for the underlying set is $\mathbb{Z}^+$, shown in figure 2.7.23, is a subset of L_3^1 for the underlying set $\mathbb{Z}$ as the thicker line in figure 2.7.21 shows.

2.7.5.3 Elements of the sets $L^1(3)$ for the underlying set $\mathbb{Z}$

Generation

The generation of the elements of the set $L^1(3)$, is obtained using the algorithm in equation (2.78) and the result is shown in the table 2.7.24 below

Table 2.7.24 *Generation of the elements of the set $L^1(3) = L_0^1 \cup L_1^1 \cup L_2^1 \cup L_3^1$ for $\mathbb{Z}$ as the underlying set.*

set	w_0	w_1	(w_0, w_1)
L_0^1	0	0	$(0,0)$
	1	0	$(1,0)$
L_1^1	0	1	$(0,1)$
	0	-1	$(0,-1)$
	-1	0	$(-1,0)$
	2	0	$(2,0)$
	1	1	$(1,1)$
	1	-1	$(1,-1)$
	0	2	$(0,2)$
L_2^1	0	-2	$(0,-2)$
	-1	1	$(-1,1)$
	-1	-1	$(-1,-1)$
	-2	0	$(-2,0)$
	3	0	$(3,0)$
	2	1	$(2,1)$
	2	-1	$(2,-1)$
	1	2	$(1,2)$
	1	-2	$(1,-2)$
	0	3	$(0,3)$
L_3^1	0	-3	$(0,-3)$
	-1	2	$(-1,2)$
	-1	-2	$(-1,-2)$
	-2	1	$(-2,1)$
	-2	-1	$(-2,-1)$
	-3	0	$(-3,0)$

Geometric representation of elements of the set $L^1(3)$

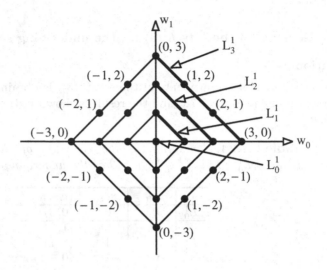

Figure 2.7.25 *Elements of the set $L^1(3) = L_0^1 \cup L_1^1 \cup L_2^1 \cup L_3^1$ for the underlying set $\mathbb{Z}$.*

The thicker line in figure 2.7.25 represents the elements of the set whose underlying set is $\mathbb{Z}^+$. These elements are also shown in figure 2.7.27.

Set notation

The result is denoted as

$$
\begin{aligned}
L^1(2) = {}& L_0^1 \cup L_1^1 \cup L_2^1 \\
= {}& \big\{ (0,0), (1,0), (0,1), (0,-1), (-1,0), \\
& (2,0), (1,1), (1,-1), (0,2), (0,-2), (-1,1), (-1,-1), (-2,0) \\
& (3,0), (2,1), (2,-1), (1,2), (1,-2), (0,3), (0,-3), \\
& (-1,2), (-1,-2), (-2,1), (-2,-1), (-3,0) \big\} \qquad (2.92)
\end{aligned}
$$

2.7.5.4   Elements of the sets $L^1(3)$ underlying set $\mathbb{Z}^+$

Generation

The generation of the elements of the set $L^1(3)$, is obtained using the algorithm in equation (2.80) and the result is shown in the table 2.7.26 below

Table 2.7.26 *Generation of the elements of the set* $L^1(3) = L_0^1 \cup L_1^1 \cup L_2^1 \cup L_3^1$ *for* $\mathbb{Z}^+$ *as the underlying set.*

set	w_0	w_1	(w_0, w_1)
L_0^1	0	0	$(0,0)$
	1	0	$(1,0)$
L_1^1	0	1	$(0,1)$
	2	0	$(2,0)$
L_2^1	1	1	$(1,1)$
	0	2	$(0,2)$
	3	0	$(3,0)$
	2	1	$(2,1)$
L_3^1	1	2	$(1,2)$
	0	2	$(0,3)$

Set notation

The result is denoted as

$$L^1(2) = L_0^1 \cup L_1^1 \cup L_2^1 \cup L_3^1$$
$$= \big\{ (0,0), (1,0), (0,1), (2,0), (1,1), (0,2), (3,0), (2,1), (1,2), (0,3) \big\} \tag{2.93}$$

Geometric representation of elements of the set $L^1(3)$

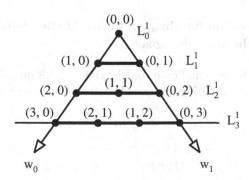

Figure 2.7.27 *Elements of the set* $L^1(3) = L_0^1 \cup L_1^1 \cup L_2^1 \cup L_3^1$ *for the underlying set* $\mathbb{Z}^+$.

The set L_3^1 for the underlying set is $\mathbb{Z}^+$ is a subset of L_3^1 for the underlying set $\mathbb{Z}$ as the thicker line in figure 2.7.25 shows.

2.8   Examples of two-dimensional constrained space

The constraint equation for the two-dimensional system is $|w_0| + |w_1| + |w_2| = \ell$. Each element in the set is denoted by two coordinate numbers. Considering w_0 as the dependent variable, the independent one must satisfy the inequality $|w_1| + |w_2| \leq \ell$.

2.8.1   Generation of the elements

2.8.1.1   Algorithm for the generation of the elements of the set L_ℓ^2 for the underlying set $\mathbb{Z}$

For the two-dimensional case the algorithm 2.2.3 on page 37 writes

$$
\begin{aligned}
&\textbf{for } w_0 = \lambda_0 \textbf{ to } w_0 = -\lambda_0 \textbf{ do} \\
&\quad \left\{ \lambda_1 = \lambda_0 - |w_0| \right. \\
&\quad \textbf{for } w_1 = \lambda_1 \textbf{ to } w_1 = -\lambda_1 \textbf{ do} \\
&\quad\quad \left\{ \lambda_2 = \lambda_1 - |w_1| \right. \\
&\quad\quad \textbf{if } \lambda_2 = 0 \textbf{ then } w_2 = 0 \textbf{ else } w_2 = \pm\lambda_2 \\
&\quad \left. \right\} \text{ end of the loop for } w_1 \\
&\left. \right\} \text{ end of the loop for } w_0
\end{aligned}
$$

(2.94)

2.8.1.2   Algorithm for the generation of the elements of the set L_ℓ^2 for the underlying set $\mathbb{Z}^+$

For the two-dimensional case the algorithm 2.4.3 on page 43 for the generation of the elements using the coordinate numbers writes

$$
\begin{aligned}
&\textbf{for } w_0 = \lambda_0 \textbf{ to } w_0 = 0 \textbf{ do} \\
&\quad \left\{ \lambda_1 = \lambda_0 - w_0 \right. \\
&\quad \textbf{for } w_1 = \lambda_1 \textbf{ to } w_1 = 0 \textbf{ do} \\
&\quad\quad \left\{ \lambda_2 = \lambda_1 - w_1 \right. \\
&\quad\quad w_2 = \lambda_2 \\
&\quad \left. \right\} \text{ end of the loop for } w_1 \\
&\left. \right\} \text{ end of the loop for } w_0
\end{aligned}
$$

(2.95)

For the two-dimensional case the algorithm 2.4.4 for the generation of the elements using sublevels writes

$$\text{for } \lambda_1 = 0 \text{ to } \lambda_1 = \lambda_0 \text{ do}$$
$$\{ w_0 = \lambda_0 - \lambda_1$$
$$\text{for } \lambda_2 = 0 \text{ to } \lambda_2 = \lambda_1 \text{ do}$$
$$\{ w_1 = \lambda_1 - \lambda_2$$
$$w_2 = \lambda_2$$
$$\} \text{ end of the loop for } \lambda_2$$
$$\} \text{ end of the loop for } \lambda_1 \qquad (2.96)$$

2.8.1.3 Algorithm for the generation of the elements of the set $L^2(\ell)$ for the underlying set $\mathbb{Z}$

For the two-dimensional case the algorithm 2.3.1 writes

$$\text{for } \lambda_0 = 0 \text{ to } \lambda_0 = \ell \text{ do}$$
$$\text{for } w_0 = \lambda_0 \text{ to } w_0 = -\lambda_0 \text{ do}$$
$$\{ \lambda_1 = \lambda_0 - |w_0|$$
$$\text{for } w_1 = \lambda_1 \text{ to } w_1 = -\lambda_1 \text{ do}$$
$$\{ \lambda_2 = \lambda_1 - |w_1|$$
$$\text{if } \lambda_2 = 0 \text{ then } w_2 = 0 \text{ else } w_2 = \pm\lambda_2$$
$$\} \text{ end of the loop for } w_1$$
$$\} \text{ end of the loop for } w_0 \qquad (2.97)$$

2.8.1.4 Algorithm for the generation of the elements of the set $L^2(\ell)$ for the underlying set $\mathbb{Z}^+$

For the two-dimensional case the algorithm 2.5.1 for the generation of the elements using the coordinate numbers writes

$$\text{for } \lambda_0 = 0 \text{ to } \lambda_0 = \ell \text{ do}$$
$$\text{for } w_0 = \lambda_0 \text{ to } w_0 = 0 \text{ do}$$
$$\{ \lambda_1 = \lambda_0 - w_0$$
$$\text{for } w_1 = \lambda_1 \text{ to } w_1 = 0 \text{ do}$$
$$\{ \lambda_2 = \lambda_1 - w_1$$
$$w_2 = \lambda_2$$
$$\} \text{ end of the loop for } w_1$$
$$\} \text{ end of the loop for } w_0 \qquad (2.98)$$

For the two-dimensional case the algorithm 2.4.4 for the generation of the elements using sublevels writes

$$\text{for } \lambda_0 = 0 \text{ to } \lambda_0 = \ell \text{ do}$$
$$\text{for } \lambda_1 = 0 \text{ to } \lambda_1 = \lambda_0 \text{ do}$$
$$\{ w_0 = \lambda_0 - \lambda_1$$
$$\text{for } \lambda_2 = 0 \text{ to } \lambda_2 = \lambda_1 \text{ do}$$
$$\{ w_1 = \lambda_1 - \lambda_2$$
$$w_2 = \lambda_2$$
$$\} \text{ end of the loop for } \lambda_2$$
$$\} \text{ end of the loop for } \lambda_1 \qquad (2.99)$$

2.8.2   Elements of the set L_0^2 and $L^2(0)$

2.8.2.1   Underlying set $\mathbb{Z}$ or $\mathbb{Z}^+$

Generation

The generation of the elements of the set L_0^2, is obtained using the algorithm in equation (2.94) and for the set $L^2(0)$ it is obtained using the algorithm (2.97). The result is the same using either algorithm and it is shown in the table 2.8.1 below

Table 2.8.1 *Elements of the sets L_0^2 or $L^2(0)$ for $\mathbb{Z}^+$ or $\mathbb{Z}$ as the underlying set.*

w_0	w_1	w_2	(w_0, w_1, w_2)
0	0	0	$(0,0,0)$

Set notation

The result is denoted as

$$L_0^2 = L^2(0) = \{ (0,0,0) \} \qquad (2.100)$$

Geometric representation of elements of the set $L_0^2 = L^2(0)$

Figure 2.8.2 shows the geometric representation of the set $L_0^2 = L^2(0) = \{(0,0,0)\}$ using an oblique two-dimensional cartesian axes. Figure 2.8.3 represents the same sets with a two-dimensional orthogonal cartesian axes w_0, w_1, and w_2.

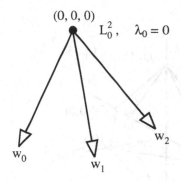

Figure 2.8.2 *Elements of the set L_0^2 in the level $\ell = 0$ represented using two oblique cartesian axes.*

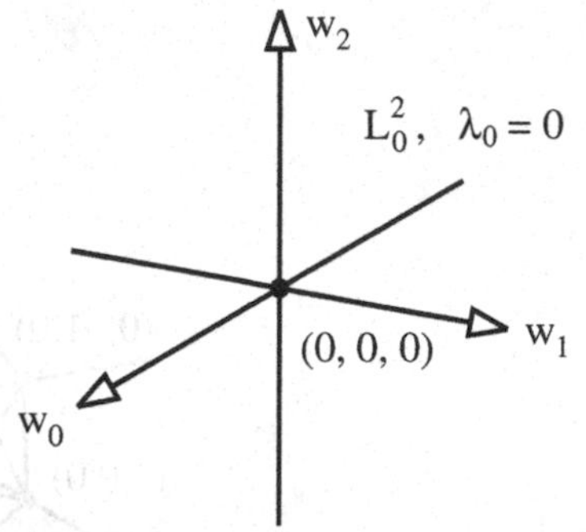

Figure 2.8.3 *Elements of the set L_0^2 in the level $\ell = 0$ represented using two orthogonal cartesian axes.*

2.8.3   Elements of the sets L_1^2 and $L^2(1)$

2.8.3.1   Elements of the sets L_1^2 for the underlying set $\mathbb{Z}$

Computation

The generation of the elements of the set L_1^2, is obtained using the algorithm in equation (2.94) and the result is shown in the table 2.8.4 below

Table 2.8.4 *Generation of the elements of the set L_1^2 for $\mathbb{Z}$ as the underlying set.*

w_0	w_1	(w_0, w_1, w_2)	
1	0	0	$(1,0,0)$
0	1	0	$(0,1,0)$
0	0	1	$(0,0,1)$
0	0	-1	$(0,0,-1)$
0	-1	0	$(0,-1,0)$
-1	0	0	$(-1,0,0)$

Set notation

The result is denoted as

$$L_1^2 = \big\{ (1,0,0), (0,1,0), (0,0,1), (0,0,-1), (0,-1,0), (-1,0,0) \big\} \tag{2.101}$$

Geometric representation of elements of the set L_1^2

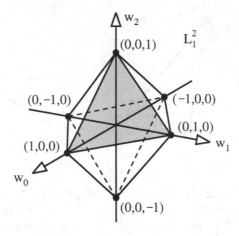

Figure 2.8.5 *Geometric representation of the elements of the set L_1^2 for the underlying set $\mathbb{Z}$.*

The thicker line in figure 2.8.5 represents the elements of the set whose underlying set is $\mathbb{Z}^+$. These elements are also shown in figure 2.8.7.

2.8.3.2   Elements of the sets L_1^2 for the underlying set $\mathbb{Z}^+$

Generation

The generation of the elements of the set L_1^2, is obtained using the algorithm in equation (2.95) and the result is shown in the table 2.8.6 below

Table 2.8.6 *Generation of the elements of the set L_1^2 for $\mathbb{Z}^+$ as the underlying set.*

w_0	w_1	w_2	(w_0, w_1, w_2)
1	0	0	$(1, 0, 0)$
0	1	0	$(0, 1, 0)$
0	0	1	$(0, 0, 1)$

Set notation

The result is denoted as

$$L_1^2 = \big\{ (1, 0, 0), (0, 1, 0), (0, 0, 1) \big\} \tag{2.102}$$

Geometric representation of elements of the set L_1^2

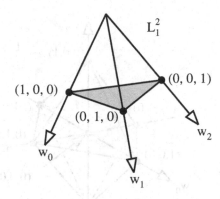

Figure 2.8.7 *Geometric representation of the elements of the set L_1^2 for the underlying set $\mathbb{Z}^+$.*

The set L_1^2 for the underlying set is $\mathbb{Z}^+$ is a subset of L_1^2 for the underlying set $\mathbb{Z}$ as the thicker line in figure 2.8.5 shows.

2.8.3.3 Elements of the sets $L^2(1)$ for the underlying set $\mathbb{Z}$

Generation

The generation of the elements of the set L_1^1, is obtained using the algorithm in equation (2.97) and the result is shown in the table 2.8.8 below

Table 2.8.8 *Generation of the elements of the set $L^2(1)$ for $\mathbb{Z}$ as the underlying set.*

set	w_0	w_1	w_2	(w_0, w_1, w_2)
L_0^2	0	0	0	$(0,0,0)$
	1	0	0	$(1,0,0)$
	0	1	0	$(0,1,0)$
	0	0	1	$(0,0,1)$
L_1^2	0	0	-1	$(0,0,-1)$
	0	-1	0	$(0,-1,0)$
	-1	0	0	$(-1,0,0)$

Set notation

The result is denoted as

$$L^2(1) = L_0^2 \cup L_1^2$$
$$= \left\{ (0,0,0), (1,0,0), (0,1,0), (0,0,1), (0,0,-1), (0,-1,0), (-1,0,0) \right\} \tag{2.103}$$

Geometric representation of elements of the set $L^2(1)$

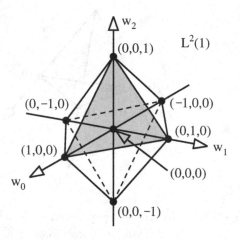

Figure 2.8.9 *Elements of the set $L^2(1) = L_0^2 \cup L_1^2$ for the underlying set $\mathbb{Z}$.*

2.8.3.4   Elements of the sets $L^2(1)$ for the underlying set $\mathbb{Z}^+$

Generation

The generation of the elements of the set $L^2(1)$, is obtained using the algorithm in equation (2.99) and the result is shown in the table 2.8.10 below

Table 2.8.10 *Generation of the elements of the set $L^2(1)$ for $\mathbb{Z}^+$ as the underlying set.*

set	w_0	w_1	w_2	(w_0, w_1, w_2)
L_0^2	0	0	0	$(0,0,0)$
	1	0	0	$(1,0,0)$
L_1^2	0	1	0	$(0,1,0)$
	0	0	1	$(0,0,1)$

Set notation

The result is denoted as

$$L^2(1) = L_0^2 \cup L_1^2 = \big\{\, (0,0,0), (1,0,0), (0,1,0), (0,0,1) \,\big\} \qquad (2.104)$$

Geometric representation of elements of the set $L^2(1)$

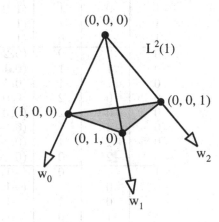

Figure 2.8.11 *Elements of the set $L^2(1) = L_0^2 \cup L_1^2$ for the underlying set $\mathbb{Z}^+$.*

The set $L^2(1)$ for the underlying set is $\mathbb{Z}^+$, shown in figure 2.8.11, is a subset of $L^2(1)$ for the underlying set $\mathbb{Z}$ as the thicker line in figure 2.8.9 shows.

2.8.4 Elements of the set L_2^2 and $L^2(2)$

2.8.4.1 Elements of the sets L_2^2 for the underlying set $\mathbb{Z}$

Generation

The generation of the elements of the set L_2^2, is obtained using the algorithm in equation (2.94) and the result is shown in the table 2.8.12 below

Table 2.8.12 *Generation of the elements of the set L_2^2 for $\mathbb{Z}$ as the underlying set.*

w_0	w_1	w_2	(w_0, w_1, w_2)
2	0	0	$(2, 0, 0)$
1	1	0	$(1, 1, 0)$
1	0	1	$(1, 0, 1)$
1	0	−1	$(1, 0, -1)$
1	−1	0	$(1, -1, 0)$
0	2	0	$(0, 2, 0)$
0	1	1	$(0, 1, 1)$
0	1	−1	$(0, 1, -1)$
0	0	2	$(0, 0, 2)$
0	0	−2	$(0, 0, -2)$
0	−1	1	$(0, -1, 1)$
0	−1	−1	$(0, -1, -1)$
0	−2	0	$(0, -2, 0)$
−1	1	0	$(-1, 1, 0)$
−1	0	1	$(-1, 0, 1)$
−1	0	−1	$(-1, 0, -1)$
−1	−1	0	$(-1, -1, 0)$
−2	0	0	$(-2, 0, 0)$

Set notation

The result is denoted as

$$
\begin{aligned}
L_2^2 = \big\{\, &(2,0,0), (1,1,0), (1,0,1), (1,0,-1), (1,-1,0), \\
&(0,2,0), (0,1,1), (0,1,-1), (0,0,2), \\
&(0,0,-2), (0,-1,1), (0,-1,-1), \\
&(0,-2,0), (-1,1,0), (-1,0,1), \\
&(-1,0,-1), (-1,-1,0), (-2,0,0) \,\big\}
\end{aligned}
\tag{2.105}
$$

Geometric representation of elements of the set L_2^2 for the underlying set $\mathbb{Z}$

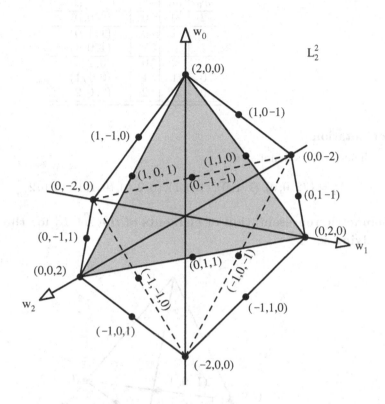

Figure 2.8.13 *Elements in the set L_2^2 for $\mathbb{Z}$ as the underlying set.*

The shaded triangle in figure 2.8.13 represents the elements of the set whose underlying set is $\mathbb{Z}^+$. These elements are also shown in figure 2.8.15.

2.8.4.2 Elements of the sets L_2^2 for the underlying set $\mathbb{Z}^+$

Generation

The generation of the elements of the set L_2^2, is obtained using the algorithm in equation (2.95) and the result is shown in the table 2.8.14 below

Table 2.8.14 *Generation of the elements of the set L_2^2 for $\mathbb{Z}^+$ as the underlying set.*

w_0	w_1	w_2	(w_0, w_1, w_2)
2	0	0	$(2,0,0)$
1	1	0	$(1,1,0)$
1	0	1	$(1,0,1)$
0	2	0	$(0,2,0)$
0	1	1	$(0,1,1)$
0	0	2	$(0,0,2)$

Set notation

The result is denoted as

$$L_2^2 = \big\{ (2,0,0), (1,1,0), (1,0,1), (0,2,0), (0,1,1), (0,0,2) \big\} \quad (2.106)$$

Geometric representation of elements of the set L_2^2 for the underlying set $\mathbb{Z}^+$

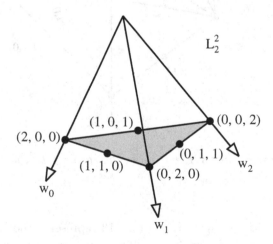

Figure 2.8.15 *Elements of the set L_2^2 for the underlying set $\mathbb{Z}^+$.*

The set L_2^2 for the underlying set is $\mathbb{Z}^+$, shown in figure 2.8.15, is a subset of L_2^2 for the underlying set $\mathbb{Z}$ shown by the shaded triangle in figure 2.8.13 shows.

2.8.4.3 Elements of the sets $L^2(2)$ for the underlying set $\mathbb{Z}$

Generation

The generation of the elements of the set $L^2(2)$ is obtained using the algorithm in equation (2.97) and the result is shown in the table 2.8.16 below

Table 2.8.16 *Generation of the elements of the set* $L^2(2) = L_0^2 \cup L_1^2 \cup L_2^2$ *for* $\mathbb{Z}$ *as the underlying set.*

set	w_0	w_1	w_2	(w_0, w_1, w_2)
L_0^2	0	0	0	$(0,0,0)$
	1	0	0	$(1,0,0)$
	0	1	0	$(0,1,0)$
	0	0	1	$(0,0,1)$
L_1^2	0	0	-1	$(0,0,-1)$
	0	-1	0	$(0,-1,0)$
	-1	0	0	$(-1,0,0)$
	2	0	0	$(2,0,0)$
	1	1	0	$(1,1,0)$
	1	0	1	$(1,0,1)$
	1	0	-1	$(1,0,-1)$
	1	-1	0	$(1,-1,0)$
	0	2	0	$(0,2,0)$
	0	1	1	$(0,1,1)$
	0	1	-1	$(0,1,-1)$
	0	0	2	$(0,0,2)$
L_2^2	0	0	-2	$(0,0,-2)$
	0	-1	1	$(0,-1,1)$
	0	-1	-1	$(0,-1,-1)$
	0	-2	0	$(0,-2,0)$
	-1	1	0	$(-1,1,0)$
	-1	0	1	$(-1,0,1)$
	-1	0	-1	$(-1,0,-1)$
	-1	-1	0	$(-1,-1,0)$
	-2	0	0	$(-2,0,0)$

Set notation

The result is denoted as

$$L^2(2) = L_0^2 \cup L_1^2 \cup L_2^2$$
$$= \big\{ (0,0,0), (1,0,0), (0,1,0), (0,0,1),$$
$$(0,0,-1), (0,-1,0), (-1,0,0)$$
$$(2,0,0), (1,1,0), (1,0,1), (1,0,-1), (1,-1,0),$$
$$(0,2,0), (0,1,1), (0,1,-1), (0,0,2),$$
$$(0,0,-2), (0,-1,1), (0,-1,-1), (0,-2,0),$$
$$(-1,1,0), (-1,0,1), (-1,0,-1), (-1,-1,0), (-2,0,0) \big\}$$
$$(2.107)$$

Geometric representation of elements of the set $L^2(2)$ for the underlying set $\mathbb{Z}$

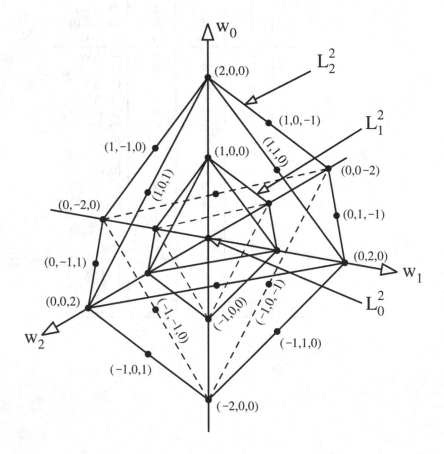

Figure 2.8.17 *Elements in the set $L^2(2) = L_0^2 \cup L_1^2 \cup L_2^2$ for $\mathbb{Z}$ as the underlying set.*

Figure 2.8.17 shows the elements of the sets L_0^2, L_1^2, and L_2^2 referenced to the coordinate axes w_0, w_1, and w_2. Figure 2.8.18 shows the same sets independent of the coordinate axes. The value of the level indicates how far from the origin each level is. Each set L_0^2, L_1^2, L_2^2 in the levels $0, 1$, and 2, respectively, is a two-dimensional constrained system.

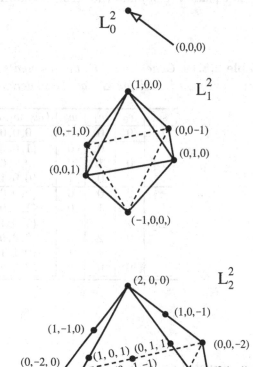

Figure 2.8.18 *The sets L_0^2, L_1^2, and L_2^2 in the levels $\ell = 0, 1, 2$ are subsets of the set $L^2(2)$ for $\mathbb{Z}$ as the underlying set.*

2.8.4.4   Elements of the sets $L^2(2)$ for the underlying set $\mathbb{Z}^+$

Generation

The generation of the elements of the set $L^2(2)$, is obtained using the algorithm in equation (2.99) and the result is shown in the table 2.8.19 below

Table 2.8.19 *Generation of the elements of the set $L^2(2) = L_0^2 \cup L_1^2 \cup L_2^2$ for $\mathbb{Z}^+$ as the underlying set.*

set	w_0	w_1	w_2	(w_0, w_1, w_2)
L_0^2	0	0	0	$(0,0,0)$
	1	0	0	$(1,0,0)$
L_1^2	0	1	0	$(0,1,0)$
	0	0	1	$(0,0,1)$
	2	0	0	$(2,0,0)$
	1	1	0	$(1,1,0)$
L_2^2	1	0	1	$(1,0,1)$
	0	2	0	$(0,2,0)$
	0	1	1	$(0,1,1)$
	0	0	2	$(0,0,2)$

Set notation

The result is denoted as

$$L^1(2) = L_0^2 \cup L_1^2 \cup L_2^2$$
$$= \{\,(0,0,0), (1,0,0), (0,1,0), (0,0,1),$$
$$(2,0,0), (1,1,0), (1,0,1), (0,2,0), (0,1,1), (0,0,2)\,\} \quad (2.108)$$

Geometric representation of elements of the set $L^2(2)$ for the underlying set $\mathbb{Z}^+$

Figure 2.8.20(a) shows the levels L_0^2, L_1^2, and L_2^2 referenced to the coordinate axes w_0, w_1, and w_2. Figure 2.8.20(b) shows the same levels independent of the coordinate axes. The value of the level indicates how far from the origin that level is. Each set L_0^2, L_1^2, L_2^2 in the levels $0, 1$, and 2, respectively, is a two-dimensional constrained system.

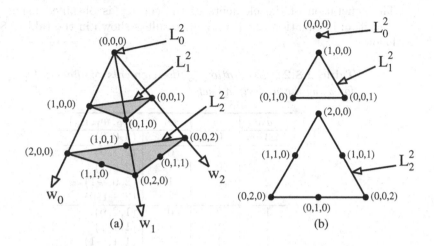

Figure 2.8.20 *Elements of the set $L^2(2) = L_0^2 \cup L_1^2 \cup L_2^2$ for the underlying set $\mathbb{Z}^+$.*

2.8.5 Elements of the set L_3^2 and $L^2(3)$

2.8.5.1 Elements of the sets L_3^2 for the underlying set $\mathbb{Z}$

Set notation

The result is denoted as

$$
\begin{aligned}
L_3^1 = \big\{ & (3,0,0), (2,1,0), (2,0,1), (2,0,-1), (2,-1,0), \\
& (1,2,0), (1,1,1), (1,1,-1), (1,0,2), (1,0,-2), \\
& (1,-1,1), (1,-1,-1), (1,-2,0), \\
& (0,3,0), (0,2,1), (0,2,-1), (0,1,2), (0,1,-2), (0,0,3), \\
& (0,0,-3), (0,-1,2), (0,-1,-2), (0,-2,1), (0,-2,-1), (0,-3,0), \\
& (-1,2,0), (-1,1,1), (-1,1,-1), (-1,0,2), \\
& (-1,0,-2), (-1,-1,1), (-1,-1,-1), (-1,-2,0), \\
& (-2,1,0), (-2,0,1), (-2,0,-1), (-2,-1,0), (-3,0,0) \big\} \quad (2.109)
\end{aligned}
$$

Generation

The generation of the elements of the set L_3^2, is obtained using the algorithm in equation (2.94) and the result is shown in the table 2.8.21 below

Table 2.8.21 *Generation of the elements of the set L_3^2 for $\mathbb{Z}$ as the underlying set.*

w_0	w_1	w_1	(w_0, w_1, w_3)
3	0	0	$(3, 0, 0)$
2	1	0	$(2, 1, 0)$
2	0	1	$(2, 0, 1)$
2	0	−1	$(2, 0, −1)$
2	−1	0	$(2, −1, 0)$
1	2	0	$(1, 2, 0)$
1	1	1	$(1, 1, 1)$
1	1	−1	$(1, 1, −1)$
1	0	2	$(1, 0, 2)$
1	0	−2	$(1, 0, −2)$
1	−1	1	$(1, −1, 1)$
1	−1	−1	$(1, −1, −1)$
1	−2	0	$(1, −2, 0)$
0	3	0	$(0, 3, 0)$
0	2	1	$(0, 2, 1)$
0	2	−1	$(0, 2, −1)$
0	1	2	$(0, 1, 2)$
0	1	−2	$(0, 1, −2)$
0	0	3	$(0, 0, 3)$
0	0	−3	$(0, 0, −3)$
0	−1	2	$(0, −1, 2)$
0	−1	−2	$(0, −1, −2)$
0	−2	1	$(0, −2, 1)$
0	−2	−1	$(0, −2, −1)$
0	−3	0	$(0, −3, 0)$
−1	2	0	$(−1, 2, 0)$
−1	1	1	$(−1, 1, 1)$
−1	1	−1	$(−1, 1, −1)$
−1	0	2	$(−1, 0, 2)$
−1	0	−2	$(−1, 0, −2)$
−1	−1	1	$(−1, −1, 1)$
−1	−1	−1	$(−1, −1, −1)$
−1	−2	0	$(−1, −2, 0)$
−2	1	0	$(−2, 1, 0)$
−2	0	1	$(−2, 0, 1)$
−2	0	−1	$(−2, 0, −1)$
−2	−1	0	$(−2, −1, 0)$
−3	0	0	$(−3, 0, 0)$

Geometric representation of elements of the set L_3^2 for the under-lying set $\mathbb{Z}$

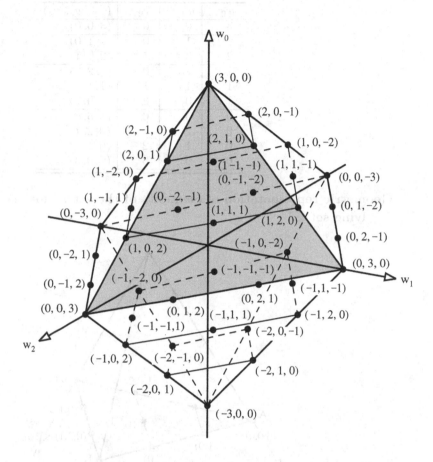

Figure 2.8.22 *Elements of the set L_3^2 for the underlying set $\mathbb{Z}$.*

The shaded triangle in figure 2.8.22 represents the elements in the set L_3^2 for $\mathbb{Z}^+$ as the underlying set. These elements are also shown in figure 2.8.24.

2.8.5.2 Elements of the sets L_3^2 underlying set $\mathbb{Z}^+$

Generation

The generation of the elements of the set L_3^2, is obtained using the algorithm in equation (2.95) and the result is shown in the table 2.8.23 below

Table 2.8.23 *Generation of the elements of the set L_3^2 for $\mathbb{Z}^+$ as the underlying set.*

w_0	w_1	w_2	(w_0, w_1, w_2)
3	0	0	$(3,0,0)$
2	1	0	$(2,1,0)$
2	0	1	$(2,0,1)$
1	2	0	$(1,2,0)$
1	1	1	$(1,1,1)$
1	0	2	$(1,0,2)$
0	3	0	$(0,3,0)$
0	2	1	$(0,2,1)$
0	1	2	$(0,1,2)$
0	0	3	$(0,0,3)$

Geometric representation of elements of the set L_3^2 for the underlying set $\mathbb{Z}^+$

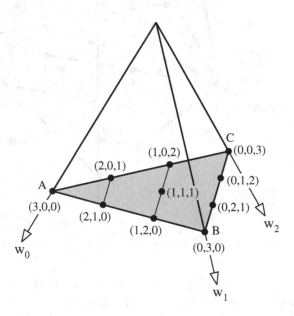

Figure 2.8.24 *Elements in the set L_3^2 in the level $\ell = 3$ for $\mathbb{Z}^+$ as the underlying set.*

Set notation

The result is denoted as

$$L_3^2 = \big\{ (3,0,0), (2,1,0), (2,0,1), (1,2,0), (1,1,1), (1,0,2),$$
$$(0,3,0), (0,2,1), (0,1,2), (0,0,3) \big\} \tag{2.110}$$

The set in equation (2.110) is represented geometrically in figure 2.8.24.

2.8.5.3 Elements of the sets $L^2(3)$ for the underlying set $\mathbb{Z}$

Generation

The generation of the elements of the set $L^2(3)$, is obtained using the algorithm in equation (2.97). The table is long and it follows the same procedures as the previous cases, therefore it is not exhibited.

Set notation

The result is denoted as

$$
\begin{aligned}
L^2(3) = L_0^2 \cup L_1^2 \cup L_2^2 \\
= \big\{ (0,0,0), (1,0,0), (0,1,0), (0,0,1), (0,0,-1), (0,-1,0), \\
(-1,0,0) \\
(2,0,0), (1,1,0), (1,0,1), (1,0,-1), (1,-1,0), \\
(0,2,0), (0,1,1), (0,1,-1), (0,0,2), (0,0,-2), (0,-1,1), \\
(0,-1,-1), (0,-2,0), \\
(-1,1,0), (-1,0,1), (-1,0,-1), (-1,-1,0), (-2,0,0) \\
(3,0,0), (2,1,0), (2,0,1), (2,0,-1), (2,-1,0), \\
(1,2,0), (1,1,1), (1,1,-1), (1,0,2), (1,0,-2), (1,-1,1), \\
(1,-1,-1), (1,-2,0), \\
(0,3,0), (0,2,1), (0,2,-1), (0,1,2), (0,1,-2), (0,0,3), \\
(0,0,-3), (0,-1,2), (0,-1,-2), (0,-2,1), (0,-2,-1), \\
(0,-3,0), \\
(-1,2,0), (-1,1,1), (-1,1,-1), (-1,0,2), \\
(-1,0,-2), (-1,-1,1), (-1,-1,-1), (-1,-2,0), \\
(-2,1,0), (-2,0,1), (-2,0,-1), (-2,-1,0), (-3,0,0) \big\}
\end{aligned}
$$
$$(2.111)$$

2.8.5.4 Elements of the sets $L^2(3)$ for the underlying set $\mathbb{Z}^+$

Generation

The generation of the elements of the set $L^2(3)$, is obtained using the algorithm in equation (2.99) and the result is shown in the table 2.8.25 below

Table 2.8.25 *Generation of the elements of the set* $L^2(3) = L_0^2 \cup L_1^2 \cup L_2^2 \cup L_3^2$ *for* $\mathbb{Z}^+$ *as the underlying set.*

set	w_0	w_1	w_1	(w_0, w_1, w_2)
L_0^2	0	0	0	$(0,0,0)$
	1	0	0	$(1,0,0)$
L_1^2	0	1	0	$(0,1,0)$
	0	0	1	$(0,0,1)$
	2	0	0	$(2,0,0)$
L_2^2	1	1	0	$(1,1,0)$
	1	0	1	$(1,0,1)$
	0	2	0	$(0,2,0)$
	0	1	1	$(0,1,1)$
	0	0	2	$(0,0,2)$
	3	0	0	$(3,0,0)$
	2	1	0	$(2,1,0)$
	2	0	1	$(2,0,1)$
	1	2	0	$(1,2,0)$
	1	1	1	$(1,1,1)$
L_3^2	1	0	2	$(1,0,2)$
	0	3	0	$(0,3,0)$
	0	2	1	$(0,2,1)$
	0	1	2	$(0,1,2)$
	0	0	3	$(0,0,3)$

Set notation

The result is denoted as

$$L^2(3) = L_0^2 \cup L_1^2 \cup L_2^2 \cup L_3^2$$
$$= \big\{\, (0,0,0), (1,0,0), (0,1,0), (0,0,1),$$
$$(2,0,0), (1,1,0), (1,0,1), (0,1,1), (0,0,2),$$
$$(3,0,0), (2,1,0), (2,0,1), (1,2,0), (1,1,1), (1,0,2), (0,3,0),$$
$$(0,2,1), (0,1,2), (0,0,3) \,\big\} \qquad\qquad (2.112)$$

The set in equation (2.112) is represented geometrically in figure 2.8.26.

Geometric representation of elements of the set $L^2(3)$ for the underlying set $\mathbb{Z}^+$

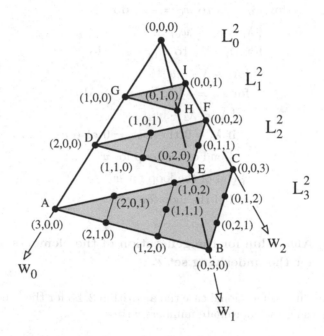

Figure 2.8.26 *Elements of the set $L^2(3) = L_0^2 \cup L_1^2 \cup L_2^2 \cup L_3^1$ for the underlying set $\mathbb{Z}^+$.*

2.9   Examples of three-dimensional constrained space

The constraint equation for the two-dimensional system is $|w_0| + |w_1| + |w_2| + |w_3| = \ell$. Each element in the set is denoted by two coordinate numbers. Considering w_0 as the dependent variable, the independent one must satisfy the inequality $|w_1| + |w_2| + |w_3| \le \ell$.

2.9.1   Generation of the elements

2.9.1.1   Algorithm for the generation of the elements of the set L_ℓ^3 for the underlying set $\mathbb{Z}$

For the three-dimensional case the algorithm 2.2.3 writes

$$
\begin{aligned}
&\textbf{for } w_0 = \lambda_0 \textbf{ to } w_0 = -\lambda_0 \textbf{ do} \\
&\quad \{ \lambda_1 = \lambda_0 - |w_0| \\
&\quad \textbf{for } w_1 = \lambda_1 \textbf{ to } w_1 = -\lambda_1 \textbf{ do} \\
&\quad\quad \{ \lambda_2 = \lambda_1 - |w_1| \\
&\quad\quad \textbf{for } w_2 = \lambda_2 \textbf{ to } w_2 = -\lambda_2 \textbf{ do} \\
&\quad\quad\quad \{ \lambda_3 = \lambda_2 - |w_2| \\
&\quad\quad\quad \textbf{if } \lambda_3 = 0 \textbf{ then } w_3 = 0 \textbf{ else } w_3 = \pm\lambda_3 \\
&\quad\quad \} \text{ end of the loop for } w_2 \\
&\quad \} \text{ end of the loop for } w_1 \\
&\} \text{ end of the loop for } w_0
\end{aligned}
\tag{2.113}
$$

2.9.1.2   Algorithm for the generation of the elements of the set L_ℓ^3 for the underlying set $\mathbb{Z}^+$

For the three-dimensional case the algorithm 2.4.3 for the generation of the elements using the coordinate numbers writes

$$
\begin{aligned}
&\textbf{for } w_0 = \lambda_0 \textbf{ to } w_0 = 0 \textbf{ do} \\
&\quad \{ \lambda_1 = \lambda_0 - w_0 \\
&\quad \textbf{for } w_1 = \lambda_1 \textbf{ to } w_1 = 0 \textbf{ do} \\
&\quad\quad \{ \lambda_2 = \lambda_1 - w_1 \\
&\quad\quad \textbf{for } w_2 = \lambda_2 \textbf{ to } w_2 = 0 \textbf{ do} \\
&\quad\quad\quad \{ \lambda_3 = \lambda_2 - w_2 \\
&\quad\quad\quad\quad w_3 = \lambda_3 \\
&\quad\quad\quad \} \text{ end of the loop for } w_3 \\
&\quad\quad \} \text{ end of the loop for } w_1 \\
&\quad \} \text{ end of the loop for } w_0
\end{aligned}
\tag{2.114}
$$

For the three-dimensional case the algorithm 2.4.4 for the generation of the elements using sublevels writes

$$
\begin{aligned}
&\textbf{for } \lambda_1 = 0 \textbf{ to } \lambda_1 = \lambda_0 \textbf{ do} \\
&\quad \{\, w_0 = \lambda_0 - \lambda_1 \\
&\qquad \textbf{for } \lambda_2 = 0 \textbf{ to } \lambda_2 = \lambda_1 \textbf{ do} \\
&\qquad\quad \{\, w_1 = \lambda_1 - \lambda_2 \\
&\qquad\qquad \textbf{for } \lambda_3 = 0 \textbf{ to } \lambda_3 = \lambda_2 \textbf{ do} \\
&\qquad\qquad\quad \{\, w_2 = \lambda_2 - \lambda_3 \\
&\qquad\qquad\qquad w_3 = \lambda_3 \\
&\qquad\qquad \} \text{ end of the loop for } \lambda_3 \\
&\qquad \} \text{ end of the loop for } \lambda_2 \\
&\quad \} \text{ end of the loop for } \lambda_1
\end{aligned}
\tag{2.115}
$$

2.9.1.3 Algorithm for the generation of the elements of the set $L^3(\ell)$ for the underlying set $\mathbb{Z}$

For the three-dimensional case the algorithm 2.3.1 writes

$$
\begin{aligned}
&\textbf{for } \lambda_0 = 0 \textbf{ to } \lambda_0 = \ell \textbf{ do} \\
&\quad \textbf{for } w_0 = \lambda_0 \textbf{ to } w_0 = -\lambda_0 \textbf{ do} \\
&\quad \{\, \lambda_1 = \lambda_0 - |w_0| \\
&\quad\; \textbf{for } w_1 = \lambda_1 \textbf{ to } w_1 = -\lambda_1 \textbf{ do} \\
&\quad\;\; \{\, \lambda_2 = \lambda_1 - |w_1| \\
&\qquad\; \textbf{for } w_2 = \lambda_2 \textbf{ to } w_2 = -\lambda_2 \textbf{ do} \\
&\qquad\;\; \{\, \lambda_3 = \lambda_2 - |w_2| \\
&\qquad\;\;\;\, \textbf{if } \lambda_3 = 0 \textbf{ then } w_3 = 0 \textbf{ else } w_3 = \pm\lambda_3 \\
&\qquad\;\; \} \text{ end of the loop for } w_2 \\
&\quad\;\; \} \text{ end of the loop for } w_1 \\
&\quad \} \text{ end of the loop for } w_0
\end{aligned}
\tag{2.116}
$$

2.9.1.4 Algorithm for the generation of the elements of the set $L^3(\ell)$ for the underlying set $\mathbb{Z}^+$

For the three-dimensional case the algorithm 2.5.1 for the generation of the elements using the coordinate numbers writes

$$
\begin{aligned}
&\textbf{for } \lambda_0 = 0 \textbf{ to } \lambda_0 = \ell \textbf{ do} \\
&\quad \textbf{for } w_0 = \lambda_0 \textbf{ to } w_0 = 0 \textbf{ do} \\
&\quad \big\{\, \lambda_1 = \lambda_0 - w_0 \\
&\quad \textbf{for } w_1 = \lambda_1 \textbf{ to } w_1 = 0 \textbf{ do} \\
&\qquad \big\{\, \lambda_2 = \lambda_1 - w_1 \\
&\qquad \textbf{for } w_2 = \lambda_2 \textbf{ to } w_2 = 0 \textbf{ do} \\
&\qquad\quad \big\{\, \lambda_3 = \lambda_2 - w_2 \\
&\qquad\qquad w_3 = \lambda_3 \\
&\qquad \big\} \text{ end of the loop for } w_2 \\
&\qquad \big\} \text{ end of the loop for } w_1 \\
&\quad \big\} \text{ end of the loop for } w_0
\end{aligned}
\tag{2.117}
$$

For the three-dimensional case the algorithm 2.4.4 for the generation of the elements using sublevels writes

$$
\begin{aligned}
&\textbf{for } \lambda_0 = 0 \textbf{ to } \lambda_0 = \ell \textbf{ do} \\
&\quad \textbf{for } \lambda_1 = 0 \textbf{ to } \lambda_1 = \lambda_0 \textbf{ do} \\
&\quad \big\{\, w_0 = \lambda_0 - \lambda_1 \\
&\quad \textbf{for } \lambda_2 = 0 \textbf{ to } \lambda_2 = \lambda_1 \textbf{ do} \\
&\qquad \big\{\, w_1 = \lambda_1 - \lambda_2 \\
&\qquad \textbf{for } \lambda_3 = 0 \textbf{ to } \lambda_3 = \lambda_2 \textbf{ do} \\
&\qquad\quad \big\{\, w_2 = \lambda_2 - \lambda_3 \\
&\qquad\qquad w_3 = \lambda_3 \\
&\qquad \big\} \text{ end of the loop for } \lambda_2 \\
&\qquad \big\} \text{ end of the loop for } \lambda_2 \\
&\quad \big\} \text{ end of the loop for } \lambda_1
\end{aligned}
\tag{2.118}
$$

2.9.2 Elements of the set L_0^3 and $L^3(0)$

2.9.2.1 Underlying set $\mathbb{Z}$ or $\mathbb{Z}^+$

Generation

The generation of the elements of the set L_0^3, is obtained using the algorithm in equation (2.113) and for the set $L^3(0)$ it is obtained using the algorithm (2.116). The result is the same using either algorithm and it is shown in the table 2.9.1 below

Table 2.9.1 *Elements of the sets L_0^3 or $L^3(0)$ for $\mathbb{Z}^+$ or $\mathbb{Z}$ as the underlying set.*

w_0	w_1	w_2	w_3	(w_0, w_1, w_2, w_3)
0	0	0	0	$(0,0,0,0)$

Set notation

The result is denoted as

$$L_0^3 = L^3(0) = \{\,(0,0,0,0)\,\} \qquad (2.119)$$

2.9.3 Elements of the sets L_1^3 and $L^3(1)$

2.9.3.1 Elements of the sets L_1^3 for the underlying set $\mathbb{Z}$

Generation

The generation of the elements of the set L_1^3, is obtained using the algorithm in equation (2.113) and the result is shown in the table 2.9.2 below

Table 2.9.2 *Generation of the elements of the set L_1^3 for $\mathbb{Z}$ as the underlying set.*

w_0	w_1	w_2	w_3	(w_0, w_1, w_2, w_3)
1	0	0	0	$(1,0,0,0)$
0	1	0	0	$(0,1,0,0)$
0	0	1	0	$(0,0,1,0)$
0	0	0	1	$(0,0,0,1)$
0	0	0	-1	$(0,0,0,-1)$
0	0	-1	0	$(0,0,-1,0)$
0	-1	0	0	$(0,-1,0,0)$
-1	0	0	0	$(-1,0,0,0)$

Set notation

The result is denoted as

$$L_1^2 = \big\{ (1,0,0,0), (0,1,0,0), (0,0,1,0), (0,0,0,1),$$
$$(0,0,0,-1), (0,0,-1,0), (0,-1,0,0), (-1,0,0,0) \big\} \qquad (2.120)$$

2.9.3.2 Elements of the sets L_1^3 for the underlying set $\mathbb{Z}^+$

Generation

The generation of the elements of the set L_1^3, is obtained using the algorithm 2.114 and the result is shown in the table 2.9.3 below

Table 2.9.3 *Generation of the elements of the set L_1^3 for $\mathbb{Z}^+$ as the underlying set.*

w_0	w_1	w_2	(w_0, w_1, w_2)
1	0	0	$(1,0,0,0)$
0	1	0	$(0,1,0,0)$
0	0	1	$(0,0,1,0)$
0	0	1	$(0,0,0,1)$

Set notation

The result is denoted as

$$L_1^3 = \big\{ (1,0,0,0), (0,1,0,0), (0,0,1,0), (0,0,0,1) \big\} \qquad (2.121)$$

Geometric representation of elements of the set L_1^3 for the underlying set $\mathbb{Z}^+$

The set L_1^3 is in the sublevel $\lambda_1 = 1$ of the level $\ell = 3$ therefore it requires three components to be represented, which is shown in figure 2.9.4 below.

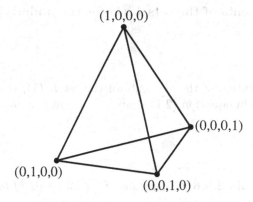

Figure 2.9.4 *Geometric representation of the elements of the set L^3_1 for the underlying set $\mathbb{Z}^+$.*

2.9.3.3 Elements of the sets $L^3(1)$ for the underlying set $\mathbb{Z}$

Generation

The generation of the elements of the set L^1_1, is obtained using the algorithm in equation (2.116) and the result is shown in the table 2.9.5 below

Table 2.9.5 *Generation of the elements of the set $L^3(1)$ for $\mathbb{Z}$ as the underlying set.*

set	w_0	w_1	w_2	w_3	(w_0, w_1, w_2, w_3)
L^3_0	0	0	0	0	$(0,0,0,0)$
	1	0	0	0	$(1,0,0,0)$
	0	1	0	0	$(0,1,0,0)$
	0	0	1	0	$(0,0,1,0)$
	0	0	1	0	$(0,0,0,1)$
L^3_1	0	0	0	−1	$(0,0,0,-1)$
	0	−1	−1	0	$(0,0,-1,0)$
	0	−1	0	0	$(0,-1,0,0)$
	−1	0	0	0	$(-1,0,0,0)$

Set notation

The result is denoted as

$$L^3(1) = L^3_0 \cup L^3_1$$
$$= \{\, (0,0,0,0), (1,0,0,0), (0,1,0,0), (0,0,1,0), (0,0,0,1),$$
$$(0,0,0,-1), (0,0,-1,0), (0,-1,0,0), (-1,0,0,0) \,\} \quad (2.122)$$

2.9.3.4   Elements of the sets $L^3(1)$ for the underlying set $\mathbb{Z}^+$

Generation

The generation of the elements of the set $L^3(1)$, is obtained using the algorithm in equation (2.118) and the result is shown in the table 2.9.6 below

Table 2.9.6 *Generation of the elements of the set $L^3(1)$ for $\mathbb{Z}^+$ as the underlying set.*

set	w_0	w_1	w_2	w_3	(w_0, w_1, w_2, w_3)
L^3_0	0	0	0	0	$(0,0,0,0)$
	1	0	0	0	$(1,0,0,0)$
	0	1	0	0	$(0,1,0,0)$
L^3_1	0	0	1	0	$(0,0,1,0)$
	0	0	1	0	$(0,0,0,1)$

Set notation

The result is denoted as

$$L^3(1) = L^3_0 \cup L^3_1$$
$$= \big\{ (0,0,0,0), (1,0,0,0), (0,1,0,0), (0,0,1,0), (0,0,0,1) \big\}$$

$$(2.123)$$

2.9.4   Elements of the set L^3_2 and $L^3(2)$

2.9.4.1   Elements of the sets L^3_2 for the underlying set $\mathbb{Z}$

Generation

The generation of the elements of the set L^3_2, is obtained using the algorithm in equation (2.113) and the result is shown in the table 2.9.7 below

Table 2.9.7 *Generation of the elements of the set L_2^3 for $\mathbb{Z}$ as the underlying set.*

w_0	w_1	w_1	w_1	(w_0, w_1, w_2, w_3)
2	0	0	0	$(2,0,0,0)$
1	1	0	0	$(1,1,0,0)$
1	0	1	0	$(1,0,1,0)$
1	0	0	1	$(1,0,0,1)$
1	0	0	−1	$(1,0,0,-1)$
1	0	−1	0	$(1,0,-1,0)$
1	−1	0	0	$(1,-1,0,0)$
0	2	0	0	$(0,2,0,0)$
0	1	1	0	$(0,1,1,0)$
0	1	0	1	$(0,1,0,1)$
0	1	0	−1	$(0,1,0,-1)$
0	1	−1	0	$(0,1,-1,0)$
0	0	2	0	$(0,0,2,0)$
0	0	1	1	$(0,0,1,1)$
0	0	1	−1	$(0,0,1,-1)$
0	0	0	2	$(0,0,0,2)$
0	0	0	−2	$(0,0,0,-2)$
0	0	−1	1	$(0,0,-1,1)$
0	0	−1	−1	$(0,0,-1,-1)$
0	0	−2	0	$(0,0,-2,0)$
0	−1	1	0	$(0,-1,1,0)$
0	−1	−1	0	$(0,-1,-1,0)$
0	−1	0	1	$(0,-1,0,1)$
0	−1	0	−1	$(0,-1,0,-1)$
0	−2	0	0	$(0,-2,0,0)$
−1	1	0	0	$(-1,1,0,0)$
−1	−1	0	0	$(-1,-1,0,0)$
−1	0	1	0	$(-1,0,1,0)$
−1	0	−1	0	$(-1,0,-1,0)$
−1	0	0	1	$(-1,0,0,1)$
−1	0	0	−1	$(-1,0,0,-1)$
−2	0	0	0	$(-2,0,0,0)$

Set notation

The result is denoted as

$$
\begin{aligned}
L_2^3 = \big\{\, &(2,0,0,0), (1,1,0,0), (1,0,1,0), (1,0,0,1), \\
&(1,0,0,-1), (1,0,-1,0), (1,-1,0,0), \\
&(0,2,0,0), (0,1,1,0), (0,1,0,1), (0,1,0,-1), (0,1,-1,0), \\
&(0,0,2,0), (0,0,1,1), (0,0,1,-1), (0,0,0,2), \\
&(0,0,0,-2), (0,0,-1,1), (0,0,-1,-1), \\
&(0,0,-2,0), (0,-1,1,0), (0,-1,-1,0), (0,-1,0,1), (0,-1,0,-1), \\
&(0,-2,0,0), (-1,1,0,0), (-1,-1,0,0), (-1,0,1,0), (-1,0,-1,0), \\
&(-1,0,0,1), (-1,0,0,-1), \\
&(-2,0,0,0)\,\big\}
\end{aligned}
\tag{2.124}
$$

2.9.4.2 Elements of the sets L_2^3 for the underlying set $\mathbb{Z}^+$

Generation

The generation of the elements of the set L_2^3, is obtained using the algorithm in equation (2.114) and the result is shown in the table 2.9.8 below

Table 2.9.8 *Generation of the elements of the set L_2^3 for $\mathbb{Z}^+$ as the underlying set.*

w_0	w_1	w_2	w_3	(w_0, w_1, w_2, w_3)
2	0	0	0	$(2,0,0,0)$
1	1	0	0	$(1,1,0,0)$
1	0	1	0	$(1,0,1,0)$
1	0	0	1	$(1,0,0,1)$
0	2	0	0	$(0,2,0,0)$
0	1	1	0	$(0,1,1,0)$
0	1	0	1	$(0,1,0,1)$
0	1	2	0	$(0,0,2,0)$
0	0	1	1	$(0,0,1,1)$
0	0	0	2	$(0,0,0,2)$

Set notation

The result is denoted as

$$
\begin{aligned}
L_2^3 = \big\{\, &(2,0,0,0), (1,1,0,0), (1,0,1,0), (1,0,0,1), (0,2,0,0), \\
&(0,1,1,0), (0,1,0,1), (0,0,2,0), (0,0,1,1), (0,0,0,2)\,\big\}
\end{aligned}
\tag{2.125}
$$

Geometric representation of elements of the set L_2^3 for the underlying set $\mathbb{Z}^+$

Figure 2.9.9 shows the geometric representation of the set in equation (2.125).

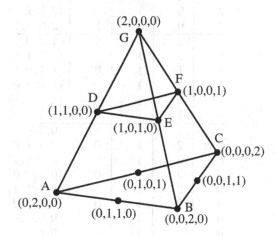

Figure 2.9.9 *Elements of the set L_2^3 for the underlying set $\mathbb{Z}^+$.*

Observe that the points on the basis ABC of the pyramid have the coordinate number $w_3 = 0$ that is they are elements in the set L_2^2 in the first dual sublevel $\lambda^1 = 2$. That is, the triangle ABC contain the elements $w = (w_0, w_1, w_2)$ for $w_3 = 0$ as

$$L_2^2 = \big\{\, (2,0,0), (1,1,0), (0,2,0), (1,0,1), (0,1,1), (0,0,2) \,\big\} \qquad (2.126)$$

The same for the triangle DEF whose elements are in set L_1^2 the first dual sub-level $\lambda^1 = 1$ given by $w = (w_0, w_1, w_2)$ for $w_3 = 1$ as

$$L_1^2 = \big\{\, (1,0,0), (0,1,0), (0,0,1) \,\big\} \qquad (2.127)$$

and the vertex G contains the element in the set L_0^2 in the first dual sub-level $\lambda^1 = 0$, which is $w = (w_0, w_1, w_2) = (0,0,0)$ for for $w_3 = 2$.

2.9.5   Elements of the set L_3^3 and $L^3(3)$

2.9.5.1   Elements of the sets L_3^3 underlying set $\mathbb{Z}^+$

Generation

The generation of the elements of the set L_3^3, is obtained using the algorithm in equation (2.114) and the result is shown in the table 2.9.10 below

Table 2.9.10 *Generation of the elements of the set L_3^3 for $\mathbb{Z}^+$ as the underlying set.*

w_0	w_1	w_2	w_3	(w_0, w_1, w_2, w_3)
3	0	0	0	$(3,0,0,0)$
2	1	0	0	$(2,1,0,0)$
2	0	1	0	$(2,0,1,0)$
2	0	0	1	$(2,0,0,1)$
1	2	0	0	$(1,2,0,0)$
1	1	1	0	$(1,1,1,0)$
1	1	1	0	$(1,1,0,1)$
1	1	1	0	$(1,0,2,0)$
1	1	1	0	$(1,0,1,1)$
1	1	1	0	$(1,0,0,2)$
0	3	0	0	$(0,3,0,0)$
0	2	1	0	$(0,2,1,0)$
0	2	1	0	$(0,2,0,1)$
0	2	1	0	$(0,1,2,0)$
0	2	1	0	$(0,1,1,1)$
0	2	1	0	$(0,1,0,2)$
0	2	1	0	$(0,0,3,0)$
0	2	1	0	$(0,0,2,1)$
0	2	1	0	$(0,0,1,2)$
0	0	3	0	$(0,0,0,3)$

Set notation

The result is denoted as

$$
\begin{aligned}
L_3^3 = \{\, & (3,0,0,0), (2,1,0,0), (2,0,1,0), (2,0,0,1), \\
& (1,2,0,0), (1,1,1,0), (1,1,0,1), (1,0,2,0), (1,0,1,1), (1,0,0,2), \\
& (0,3,0,0), (0,2,1,0), (0,2,0,1), (0,1,2,0), (0,1,1,1), (0,1,0,2), \\
& (0,0,3,0), (0,0,2,1), (0,0,1,2), (0,0,0,3), \}
\end{aligned}
\tag{2.128}
$$

Geometric representation of elements of the set L_3^3 for the underlying set $\mathbb{Z}^+$

Figure 2.9.11 shows the geometric representation of the set in equation (2.128).

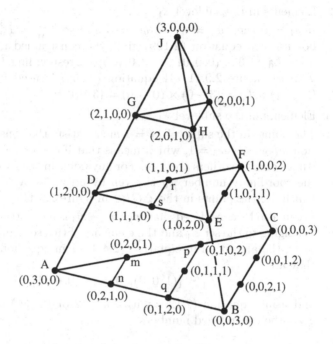

Figure 2.9.11 *Elements in the set L_3^3 in the level $\ell = 3$ for $\mathbb{Z}^+$ as the underlying set.*

Comparing the rules to the generation of the elements of the set L_3^3 with the points in the figure 2.9.11 it can be seen that

1. Values of the coordinate number w_0.

   The coordinate number w_0 assume the values $\lambda_0 = 3, 2, 1, 0$. Associated to the coordinate number w_0 there are the sublevels $\lambda_1 = \ell - |w_0|$, which are $\lambda_1 = 0, 1, 2, 3$.

   Equation (2.37) when applied to generation of the elements of the set L_3^3 gives

$$L_3^3 = \bigcup_{w_0=3}^{0} \{(w_0)\} \times L_{\lambda_1}^2 \qquad (2.129)$$

in terms of the coordinate number w_0. The equation (2.38) can be written in terms of λ_1 as follows

$$L_3^3 = \bigcup_{\lambda_1=0}^{3} \{(3-\lambda_1)\} \times L_{\lambda_1}^2 \qquad (2.130)$$

2. Elements in the sublevel $\lambda_1 = 0$.

If $\lambda_1 = 0$ then $w_1 = w_2 = w_3 = 0$ and so $L_{\lambda_1}^2 = L_0^2 = (0,0,0)$. For $w_0 = 3$ equation (2.129) gives the constrained number $w = w_0 \times L_0^2 = 3 \times (0,0,0) = (3,0,0,0)$ corresponding to the point J in the figure 2.9.11. If equation (2.130) is used it gives $w = (3-\lambda_1) \times L_0^2 = (3-0) \times (0,0,0) = (3,0,0,0)$.

3. Elements in the sublevel $\lambda_1 = 1$.

The points in the sublevel $\lambda_1 = 1$ must satisfy the constraint equation $w_1 + w_2 + w_3 = 1$, which means that if one of the coordinates equals one, the others are zero. For any point in the sublevel $\lambda_1 = 1$ the coordinate number w_0 is given by $w_0 = \ell - \lambda_1 = 3 - 1 = 2$, which are the points in the $\triangle GHI$ in figure 2.9.11.

It can also be considered that $w_1 + w_2 + w_3$ is a constrained number belonging to the set L_1^2 and the elements of this set can be obtained using the algorithm in equation (2.4.3) or in equation (2.4.4) and they are

$$L_1^2 = \{(1,0,0),(0,1,0),(0,0,1)\} \qquad (2.131)$$

and using equation (2.129) with $w_0 = 2$ or (2.130) with $\lambda_1 = 1$ gives the constrained numbers

$$(2,1,0,0),(2,0,1,0),(2,0,0,1) \qquad (2.132)$$

The generation of L_1^2 is performed using equation (2.42), which gives

$$L_1^2 = \bigcup_{\lambda_2=0}^{\lambda_1} \{(\lambda_1-\lambda_2)\} \times L_{\lambda_2}^1 = \bigcup_{\lambda_2=0}^{1} \{(1-\lambda_2)\} \times L_{\lambda_2}^1 \qquad (2.133)$$

where for $\lambda_2 = 0$ it follows $L_0^1 = \{(0,0)\}$ and for $\lambda_2 = 1$ it gives the set $L_0^1 = \{(1,0),(0,1)\}$. Expanding equation (2.133)

$$\begin{aligned}
L_1^2 &= \bigcup_{\lambda_2=0}^{1} \{(1-\lambda_2)\} \times L_{\lambda_2}^1 \\
&= (1-0) \times L_0^1 \cup (1-1) \times L_1^1 \\
&= [(1) \times (0,0)] \cup [(0) \times \{(1,0),(0,1)\}] \\
&= \{(1,0,0),(0,1,0),(0,0,1)\} \qquad (2.134)
\end{aligned}$$

4. Elements in the sublevel $\lambda_1 = 2$.

The points in the sublevel $\lambda_1 = 2$ must satisfy the constraint equation $w_1 + w_2 + w_3 = 2$. For any point in the sublevel $\lambda_1 = 2$ the coordinate number w_0 is given by $w_0 = \ell - \lambda_1 = 3 - 2 = 1$ which are the points in the $\triangle DEF$ in figure 2.9.11.

It can also be considered that $w_1 + w_2 + w_3$ is a constrained number belonging to the set L_2^2 and the elements of this set can be obtained using the algorithm in equation (2.4.3) or 2.4.4 and they are

$$L_2^2 = \{(2,0,0),(1,1,0),(1,0,1),(0,2,0),(0,1,1),(0,0,2)\} \quad (2.135)$$

The generation of L_2^2 is performed using equation (2.42) which gives

$$L_2^2 = \bigcup_{\lambda_2=0}^{\lambda_1} \{(\lambda_1 - \lambda_2)\} \times L_{\lambda_2}^1 = \bigcup_{\lambda_2=0}^{2} \{(2 - \lambda_2)\} \times L_{\lambda_2}^1 \quad (2.136)$$

where for $\lambda_2 = 0$ it follows $L_0^1 = \{(0,0)\}$, for $\lambda_2 = 1$ it gives the set $L_0^1 = \{(1,0),(0,1)\}$, and for $\lambda_2 = 2$ the set is $L_2^1 = \{(2,0),(1,1),(0,2)\}$, . Expanding equation (2.136)

$$L_1^2 = \bigcup_{\lambda_2=0}^{1} \{(1 - \lambda_2)\} \times L_{\lambda_2}^1$$
$$= (2 - 0) \times L_0^1 \cup (2 - 1) \times L_1^1 \cup (2 - 2) \times L_2^1$$
$$= [(2) \times (0,0)] \cup [(1) \times \{(1,0),(0,1)\}] \cup$$
$$[(0) \times \{(2,0),(1,1),(0,2)\}]$$
$$= \{(2,0,0),(1,1,0),(1,0,1),(0,2,0),$$
$$(0,1,1),(0,0,2)\} \quad (2.137)$$

5. Elements in the sub-level $\lambda_1 = 3$.

The points in the sub-level $\lambda_1 = 3$ must satisfy the constraint equation $w_1 + w_2 + w_3 = 3$. For any point in the sub-level $\lambda_1 = 3$ the coordinate number w_0 is given by $w_0 = \ell - \lambda_1 = 3 - 3 = 0$ which are the points in the $\triangle ABC$ in figure 2.9.11.

It can also be considered that $w_1 + w_2 + w_3$ is a constrained number belonging to the set L_3^2 and the elements of this set can be obtained using the algorithm in equation (2.4.3) or in equation (2.4.4) and they are

$$L_3^2 = \{(3,0,0),(2,1,0),(2,0,1),(1,2,0),(0,1,1),(1,0,2),(0,3,0),$$
$$(0,2,1),(0,1,2),(0,0,3)\} \quad (2.138)$$

2.10   Properties of sets of constrained numbers

Definition 2.10.1 *Complete set*

A set L_ℓ^{d-1} of constrained numbers $w = (w_0, \ldots, w_{d-1})$ in the level ℓ and with d coordinate numbers is said to be complete if it contains all combinations of the d numbers that satisfies the equation $\sum_0^{d-1} w_i = \ell$ otherwise it is said to be incomplete.

end of definition 2.10.1

As consequence of the definition 2.10.1 the set

$$L^{d-1}(q) = \bigcup_{i=0}^{q} L_i^{d-1} \tag{2.139}$$

is complete since each subset L_i^{d-1} is complete.

Example 2.10.2

The sets

$$L_3^2 = \big\{ (3,0,0), (2,1,0), (2,0,1), (1,2,0), (1,1,1), (1,0,2),$$
$$(0,3,0), (0,2,1), (0,1,2), (0,0,3) \big\}$$
$$L^1(3) = \big\{ (0,0), (1,0), (0,1), (2,0), (1,1), (0,2), (3,0), (2,1), (1,2), (0,3) \big\}$$
$$L_0^2 = \big\{ (0,0,0) \big\}$$
$$L_2^0 = \big\{ (2) \big\}$$

are complete, and

$$L_3^2 = \big\{ (2,1,0), (2,0,1), (1,2,0), (1,0,2), (0,2,1), (0,1,2) \big\}$$

is an incomplete set of numbers constrained to $\ell = 3$.

end of example 2.10.2

Theorem 2.10.3 *The set of constrained numbers L_ℓ^δ or $L^\delta(\ell)$ do not change if the same permutation is applied to any two coordinate numbers of all of its elements*

If in the rule 2.2.1 or in the rule 2.4 the sequence $w_0, w_1, \ldots, w_\delta$ of computation of the coordinate numbers is changed to $w_k, w_0, \ldots, w_\delta$ then it generates the same set where the values of the coordinate number w_0 are in the position of the coordinate number w_k and vice versa.

Proof:

The numbers generated by the first loop in the algorithm are always the same independent of the variable under consideration. If those numbers are assigned to the k-th position, that is to w_k, then it is equivalent to a permutation between the coordinate numbers w_0 and w_k for each element of the set.

end of theorem 2.10.3

Example 2.10.4

Let

$$L_3^2 = \big\{ (3,0,0), (2,1,0), (2,0,1), (1,2,0), (1,1,1), (1,0,2),$$
$$(0,3,0), (0,2,1), (0,1,2), (0,0,3) \big\} \tag{2.140}$$

and consider the permutation $T : (1,2,3) \longrightarrow (2,1,3)$, then the set L_3^2 transforms into

$$\mathcal{L}_3^2 = \big\{ (0,3,0), (1,2,0), (0,2,1), (2,1,0), (1,1,1), (0,1,2),$$
$$(3,0,0), (2,0,1), (1,0,2), (0,0,3) \big\} \tag{2.141}$$

which has the same elements in different order therefore $\mathcal{L}_3^2 = L_3^2$.

end of example 2.10.4

Corollary 2.10.5 *The set of constrained numbers L_ℓ^δ or $L^\delta(\ell)$ do not change if the same permutation is applied to all of its elements*

Given the permutation

$$(w_0, w_1, \ldots, w_\delta) \longrightarrow (w_i, w_j, \ldots, w_k) \tag{2.142}$$

then coordinate numbers of all elements of the set generated $w_i, w_j, \ldots, w_k$ and the coordinate numbers of elements of the set generated by $w_0, w_1, \ldots, w_\delta$ satisfy the same permutation.

Proof:

Apply theorem 2.10.3 to each pair w_i, w_k of coordinate numbers for all elements of the set.

end of corollary 2.10.5

Example 2.10.6

Let

$$L_3^2 = \big\{ (3,0,0), (2,1,0), (2,0,1), (1,2,0), (1,1,1), (1,0,2),$$
$$(0,3,0), (0,2,1), (0,1,2), (0,0,3) \big\} \qquad (2.143)$$

and consider the permutation $T : (1,2,3) \longrightarrow (3,2,1)$, then the set L_3^2 transforms into

$$\mathcal{L}_3^2 = \big\{ (0,0,3), (0,1,2), (1,0,2), (0,2,1), (1,1,1), (2,0,1),$$
$$(0,3,0), (1,2,0), (2,1,0), (3,0,0) \big\} \qquad (2.144)$$

which has the same elements in different order therefore $\mathcal{L}_3^2 = L_3^2$.

end of example 2.10.6

Theorem 2.10.7

Let L_q^δ be a set of constrained numbers in the level q whose coordinate numbers belong to the set $\mathbb{Z}^+$ and let $v \in L_q^\delta$ then $L_{(q-\lambda_0(v))}^\delta = L_q^\delta - \{(v)\}$ is complete set of constrained numbers in the level $q - \lambda_0(v)$.

Proof:

Let $w \in L_q^\delta$ and $u \in L_{(q-\lambda_0(v))}^\delta$ where $w = (w_0, w_1, \ldots, w_\delta)$ and $u = (u_0, u_1, \ldots, u_\delta)$.

The elements of the set L_q^δ can be generated by the rule 2.4.1 for the underlying set $\mathbb{Z}^+$. From equation (2.38) it follows

$$L_q^\delta = \bigcup_{\lambda_1=0}^{q} \{(\lambda_0 - \lambda_1)\} \times L_{\lambda_1}^{\delta-1} \qquad \delta > 0, \quad q > 0 \qquad (2.145)$$

where $\lambda_0 - \lambda_1 = w_0$. After the subtraction $u_0 = w_0 - v_0$ if the coordinate number u_0 is such that $u_0 = w_0 - v_0 < 0$, that is, such that $u_0 = \lambda_0 - \lambda_1 < 0$ then u is not an element of the resulting set since $u_0 \notin \mathbb{Z}^+$. As consequence, all constrained numbers $\{(\lambda_0 - \lambda_1)\} \times L_{\lambda_1}^{\delta-1}$ will not be elements of the resultant set. The values of $\lambda_0 = 0, \ldots, \ell$ are transformed into $\lambda_0 = 0 - v_0, \ldots, v_0 - v_0, \ldots, q - v_0$, that is, into $\lambda_0 = 0, \ldots, m_0$ where $m_0 = q - v_0$ and the new set is generated by the algorithm

$$L_{m_0}^\delta = \bigcup_{\lambda_1=0}^{m_0} \{(\lambda_0 - \lambda_1)\} \times L_{\lambda_1}^{\delta-1} \qquad \delta > 0, \quad m_0 > 0 \qquad (2.146)$$

By corollary 2.10.5 the permutation of the coordinate numbers do not change the elements of the set. Then the first coordinate number of the elements of the set $L^\delta(m_0)$ can be permuted with the second and then the operation $u_1 = w_1 - v_1$ will give the set $L^\delta(m_1)$ where $m_1 = q - (v_0 + v_1)$. Repeating the same procedure until $u_\delta = w_\delta - v_\delta$ it gives $L^\delta(m_\delta)$ where $m_\delta = q - (v_0 + v_1 + \cdots + v_\delta) = q - \lambda_0(v)$ generated by the algorithm

$$L^\delta_{m_\delta} = \bigcup_{\lambda_1=0}^{m_\delta} \{(\lambda_0 - \lambda_1)\} \times L^{\delta-1}_{\lambda_1} \qquad \delta > 0, \quad m_\delta = q - \lambda_0(v) > 0 \qquad (2.147)$$

where $L^\delta_{m_\delta}$ can be rewritten as $L^\delta_{q-\lambda_0(v)}$. Since equation (2.147) generates all elements of the set $L^\delta_{q-\lambda_0(v)}$ then it is complete, which completes the proof.

end of theorem 2.10.7

Corollary 2.10.8 *Corollary to the theorem 2.10.7*

Let L^δ_q be a set of constrained numbers in the level q whose coordinate numbers belong to $\mathbb{Z}^+$ and let $v \in L^\delta(q)$ then $L^\delta_{(q-\lambda_0(v))} = L^\delta_q - \{(v)\}$ is complete set of constrained numbers in the level $q - \lambda_0(v)$.

Proof:
From equation (1.5) $L^\delta(q)$ it can be written that

$$L^\delta(q) = \{L^\delta_0 \cup L^\delta_1 \cup L^\delta_2 \cup \cdots \cup L^\delta_q\} \qquad (2.148)$$

Performing the operation $L^\delta(q) - \{(v)\}$ it follows

$$L^\delta(q) - \{(v)\} = \{L^\delta_0 \cup L^\delta_1 \cup L^\delta_2 \cup \cdots \cup L^\delta_q\} - \{(v)\} \qquad (2.149)$$

Recalling the definition 1.1.28 of the addition of a set with a constrained number the equation (2.149) can be expanded as

$$L^\delta(q) - \{(v)\} = \{L^\delta_0 - \{(v)\}\} \cup \{L^\delta_1 - \{(v)\}\} \cup \cdots$$
$$\cdots \cup \{L^\delta_{\lambda_0(v)} - \{(v)\}\} \cup \cdots$$
$$\cdots \cup \{L^\delta_q - \{(v)\}\} \qquad (2.150)$$

where the sets from $L^\delta_0 - \{(v)\}$ until $\{L^\delta_{\lambda_0(v)-1} - \{(v)\}\}$ are void sets then

$$L^\delta(q) - \{(v)\} = L^\delta_0 \cup L^\delta_1 \cup \cdots \cup L^\delta_{q-\lambda_0(v)} \qquad (2.151)$$

and from theorem 2.10.7 each $L^\delta_i, i = 0, \ldots, i = q - \lambda_0(v)$ is a complete set, therefore $L^\delta(q - \lambda_0(v))$ is a complete set, which completes the proof.

end of corollary 2.10.8

Example 2.10.9

Let

$$L^2(3) = \{ (0,0,0), (1,0,0), (0,1,0), (0,0,1),$$
$$(2,0,0), (1,1,0), (1,0,1), (0,2,0), (0,1,1), (0,0,2),$$
$$(3,0,0), (2,1,0), (2,0,1), (1,2,0), (1,1,1), (1,0,2),$$
$$(0,3,0), (0,2,1), (0,1,2), (0,0,3) \}$$

and $v = (0,1,1)$ then

$$L^\delta(q - \lambda_0(v)) = L^2(3 - \lambda_0(0,1,1)) = L^2(1)$$

which gives

$$L^\delta(q) - \{(v)\} = L^2(3) - ((0,1,1))$$

Performing the operation it follows that

$$L^2(1) = \{ (0,0,0) - (0,1,1), \quad (1,0,0) - (0,1,1),$$
$$(0,1,0) - (0,1,1), \quad (0,0,1) - (0,1,1),$$
$$(2,0,0) - (0,1,1), \quad (1,1,0) - (0,1,1),$$
$$(1,0,1) - (0,1,1), \quad (0,2,0) - (0,1,1),$$
$$(0,1,1) - (0,1,1), \quad (0,0,2) - (0,1,1),$$
$$(3,0,0) - (0,1,1), \quad (2,1,0) - (0,1,1),$$
$$(2,0,1) - (0,1,1), \quad (1,2,0) - (0,1,1),$$
$$(1,1,1) - (0,1,1), \quad (1,0,2) - (0,1,1),$$
$$(0,3,0) - (0,1,1), \quad (0,2,1) - (0,1,1),$$
$$(0,1,2) - (0,1,1), \quad (0,0,3) - (0,1,1) \}$$

gives

$$L^2(1) = \{ (0,0,0), (1,0,0), (0,1,0), (0,0,1) \}$$

which is complete.

end of example 2.10.9

Theorem 2.10.10

Let L_q^δ be a set of constrained numbers in the level q whose coordinate numbers belong to $\mathbb{Z}^+$ and let $v, r \in L^\delta(q)$ where r is a permutation of v then

$$\forall \mu \in \{ L_{q-\lambda_0(v)}^\delta + \{(v)\} \} \quad \exists \rho \in \{ L_{q-\lambda_0(r)}^\delta + \{(r)\} \} \mid T : \rho \longrightarrow \mu \quad (2.152)$$

Proof:

Let $\mu = (\mu_0, \ldots, \mu_\delta)$ and $\rho = (\rho_0, \ldots, \rho_\delta)$.

If r is a permutation of v then $\lambda_0(v) = \lambda_0(r)$, which implies the equality of the two complete sets $L^\delta_{q-\lambda_0(v)} = L^\delta_{q-\lambda_0(r)}$.

A permutation function satisfies the property if $T : u \longrightarrow w$ and $T : v \longrightarrow r$ then $T : (u + v) \longrightarrow (w + r)$.

Given a u in a complete set there exists a w in the same set such that $T : u \longrightarrow w$. By hypothesis $T : v \longrightarrow r$ which, by the above permutation property, implies that $T : (u + v) \longrightarrow (w + r)$

Note that the sets $L^\delta_{q-\lambda_0(v)} + \{(v)\}$ and $L^\delta_{q-\lambda_0(r)} + \{(r)\}$ are not complete.

end of theorem 2.10.10

Example 2.10.11

Let

$$L^2_4 = \big\{ (4,0,0), (3,1,0), (3,0,1), (2,2,0),$$
$$(2,1,1), (2,0,2), (1,3,0), (1,2,1), (1,1,2), (1,0,3),$$
$$(0,4,0), (0,3,1), (0,2,2), (0,1,3), (0,0,4) \big\}$$

Let $v = (1,1,0)$ and $r = (0,1,1)$ where $T : v \longrightarrow r$ whose permutation is defined as

$$T = \begin{pmatrix} 1 & 2 & 3 \\ 3 & 1 & 2 \end{pmatrix}$$

Performing the set-constrained number subtraction it gives

$$L^\delta_{q-\lambda_0(v)} = L^2_{4-\lambda_0(1,1,0)} = L^2_2$$

where

$$L^2_2 = \big\{ (2,0,0), (1,1,0), (1,0,1), (0,2,0), (0,1,1), (0,0,2) \big\}$$

Performing the set-constrained number addition it gives

$$L^2_2 + \{(1,1,0)\} = \big\{ (3,1,0), (2,1,0), (2,1,1), (1,3,0),$$
$$(1,2,1), (1,1,2) \big\}$$

and

$$L^2_2 + \{(0,1,1)\} = \big\{ (2,1,1), (1,2,1), (1,1,2), (0,3,1),$$
$$(0,2,2), (0,1,3) \big\}$$

Where can be seen that for each element a in the set $L_2^2 + \{(1,1,0)\}$ there is an element b in the set $L_2^2 + \{(0,1,1)\}$ such that $T : a \longrightarrow b$ where

$$T = \begin{pmatrix} 1 & 1 & 0 \\ 0 & 1 & 1 \end{pmatrix}$$

end of example 2.10.11

Example 2.10.12

Let

$$L_2^2 = \big\{ (2,0,0), (1,1,0), (1,0,1), (0,2,0), (0,1,1), (0,0,2) \big\}$$

Let $v = (1,1,0)$ and $r = (0,1,1)$ where $T : v \longrightarrow r$ whose permutation is defined as

$$T = \begin{pmatrix} 1 & 2 & 3 \\ 3 & 1 & 2 \end{pmatrix}$$

Performing the set-constrained number subtraction it gives

$$L_{q-\lambda_0(v)}^\delta = L_{4-\lambda_0(1,1,0)}^2 = L_1^2$$

where

$$L_1^2 = \big\{ (1,0,0), (0,1,0), (0,0,1) \big\}$$

Performing the set-constrained number addition it gives

$$L_1^2 + \{(1,1,0)\} = \big\{ (2,1,0), (1,2,0), (1,1,1) \big\}$$

and

$$L_1^2 + \{(0,1,1)\} = \big\{ (1,1,1), (0,2,1), (0,1,2) \big\}$$

end of example 2.10.12

Chapter 3

Natural coordinates

Definition 3.0.13 *Natural coordinate numbers*

The natural coordinate numbers are defined here as the constrained coordinate numbers in the level $\ell = 1$, which implies that they are elements in the set L_1^δ.

end of definition 3.0.13

3.1 One-dimensional natural coordinate numbers

3.1.1 Location of a point

The one-dimensional natural coordinate numbers can be obtained either from subsection 1.4.1, figure 1.4.1, by taking $\ell = 1$, or from example 2.7, geometric representation of elements of the set L_1^1 for the underlying set $\mathbb{Z}^+$ shown in figure 2.7.7.

From subsection 1.4.1 figure 1.4.1 it is known that

$$w_0 = BC \quad \text{and} \quad w_1 = AC \tag{3.1}$$

therefore

$$w_0 = BP = \ell\xi \quad \text{and} \quad w_1 = AP = \ell(1 - \xi) \tag{3.2}$$

where ℓ is the value of the level of the points in the segment $\overline{AB}$ and, geometrically, it is equal to the length of the segment, that is, $\ell = \overline{AB}$.

Denote by ξ_0, ξ_1 the constrained coordinate numbers for $\ell = 1$ then it can be written that

$$\xi_0 = \xi \quad \text{and} \quad \xi_1 = 1 - \xi \tag{3.3}$$

which are exhibited in figure 3.1.2.

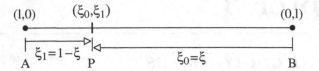

Figure 3.1.1 *Interpolation of a constrained number.*

The natural coordinates can also be seen as the coordinate numbers in a normalized line segment, that is, a segment whose length equals $\overline{AB} = \ell = 1$ and evaluated as

$$\xi_1 := \frac{\overline{AP}}{\overline{AB}} = \overline{AP} \quad \text{and} \quad \xi_0 := \frac{\overline{BP}}{\overline{AB}} = \overline{BP} \tag{3.4}$$

3.1.2   Relation between the one-dimensional natural coordinate numbers and the one-dimensional cartesian coordinate numbers

The bijection between the points of interval $[A, B] = [(1, 0), (0, 1)]$ in figure 3.1.1 and the points of the interval $[M, N] = [x_m, x_n]$ in figure 3.1.2 allows to locate a point in $[M, N]$ in terms of the a point in $[A, B]$ and vice versa. The function relating the points can be written as

$$\xi_1 = \frac{\overline{AP}}{\overline{AB}} = \frac{\overline{MR}}{\overline{MN}} = \frac{x_r - x_m}{x_n - x_m} \quad \text{and} \quad \xi_0 = \frac{\overline{BP}}{\overline{AB}} = \frac{\overline{NR}}{\overline{MN}} = \frac{x_n - x_r}{x_n - x_m} \tag{3.5}$$

and then the constrained coordinate numbers (w_0, w_1) locating the point R with respect to the end points of the interval $[M, N]$, which frequently are called local coordinates, are evaluated as

$$w_1 = \overline{MR} = \xi_1 \overline{MN} = \ell \xi_1 \quad \text{and} \quad w_0 = \overline{NR} = \xi_0 \overline{MN} = \ell \xi_0 \tag{3.6}$$

where ℓ is the level of the points in the segment line $\overline{MN}$.

The cartesian coordinate numbers of the point R are given by

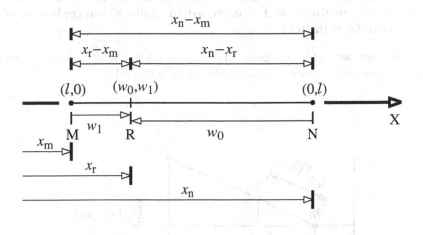

Figure 3.1.2 *Relation between the one-dimensional constrained coordinate numbers and the one-dimensional cartesian coordinate numbers.*

The cartesian coordinate number x_r of the point R can be obtained as $x_r = x_m + \overline{AP}$ then

$$
\begin{aligned}
x_r = x_m + \overline{AP} &= x_m + (x_n - x_m)\xi_1 \\
&= x_m + x_n\xi_1 - x_m\xi_1 \\
&= (1 - \xi_1)x_n + x_n\xi_1 \\
&= \xi_0 x_m + \xi_1 x_n
\end{aligned}
\tag{3.7}
$$

The above result shows that the cartesian coordinate number x_r of a point R in figure 3.1.2 is a linear combination of the cartesian coordinates x_m and x_n of the points M and N, respectively, where the scalars ξ_0 and ξ_1 are the one-dimensional constrained numbers locating the point P in the level $\ell = 1$ as shown in figure 3.1.1.

Figure 3.1.2 shows that the one-dimensional constrained coordinate system needs two coordinate numbers to define a point P while the one-dimensional cartesian coordinate system needs only one coordinate number to define a point R, analogous of P, as shown in figure 3.1.2. In other words, the one-dimensional constrained coordinate system is represented geometrically in a two-dimensional cartesian coordinate system where the origin of each coordinate is at the end points of the interval.

3.1.3 Relation between the one-dimensional natural coordinate numbers and the two-dimensional cartesian coordinate numbers

The line segment MN in figure 3.1.2 can be represented in a two dimensional cartesian coordinate system as shown in figure 3.1.3.

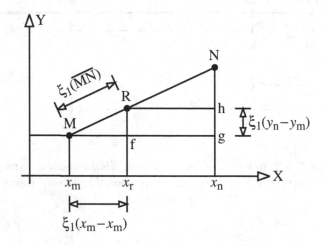

Figure 3.1.3 *Relation between the one-dimensional constrained coordinate numbers and the two-dimensional cartesian coordinate numbers.*

To show the relation among both coordinate systems observe that the triangles $\triangle MRf$ and $\triangle MNg$ are similar, thus line segments MR and Mf as well as MN and Mg in figure 3.1.3 are between two parallel lines, thus

$$\xi_1 = \frac{MR}{MN} = \frac{Mf}{Mg} = \frac{x_r - x_m}{x_n - x_m} \quad \text{and} \quad \xi_1 = \frac{MR}{MN} = \frac{hg}{Ng} = \frac{y_r - y_n}{y_n - y_m}$$

$$(3.8)$$

From the first equation in (3.8) it is obtained

$$x_r = x_m + \xi_1 (x_n - x_m)$$
$$x_r = (1 - \xi_1) x_m + \xi_1 x_n$$
$$x_r = \xi_0 x_m + \xi_1 x_n \qquad (3.9)$$

and similarly

$$y_r = \xi_0 y_m + \xi_1 y_n \qquad (3.10)$$

Therefore the cartesian coordinates x_r and y_r of a point R on a line segment MN represented in a two-dimensional cartesian coordinate system is given by a linear combination of the coordinates along each direction, that is $x_r = \xi_0 x_m + \xi_1 x_n$ and $y_r = \xi_0 y_m + \xi_1 y_n$. This result can be extended to a d-dimensional cartesian coordinate system.

Example 3.1.4 *Assume that $\lambda = 4$ in figure 3.1.2. Find the coordinates of a point $R = (w_0, w_1)$ if $\xi_0 = 3/4$ and $\xi_0 = 9/10$.*

Solution for the case $\xi = 3/4$

From the constraint equation $\xi_0 + \xi_1 = 1$ then and $\xi_1 = 1 - \xi_0 = 1/4$. If $\ell = 4$ then $MN = \ell = 4$ and the constrained coordinate numbers of the point R are

$$w_0 = \xi_0 \ell = \frac{3}{4} \times 4 = 3 \quad \text{and} \quad w_1 = \xi_1 \ell = \frac{1}{4} \times 4 = 1 \quad (3.11)$$

therefore $(w_0, w_1) = (3, 1)$, which is shown in figure 3.1.5.

Solution for the case $\xi = 9/10$

From the constraint equation $\xi_0 + \xi_1 = 1$, it follows $\xi_1 = 1 - \xi_0 = 1/10$. The constrained coordinate numbers of the point R are

$$w_0 = \xi_0 \ell = \frac{9}{10} \times 4 = 3.6 \quad \text{and} \quad w_1 = \xi_1 \ell = \frac{1}{10} \times 4 = 0.4 \quad (3.12)$$

and $(w_0, w_1) = (3.6, 0.4)$

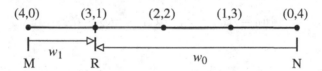

Figure 3.1.5 *Geometrical representation of the constrained number $(w_0, w_1) = (3, 1)$.*

end of example 3.1.4

3.2 Two-dimensional natural coordinate numbers

3.2.1 Location of a point

The two-dimensional natural coordinates can be obtained from example 2.8 on page 70, geometric representation of elements of the set L_1^2 for the underlying set $\mathbb{Z}^+$ shown in figure 2.8.7 on page 75.

Each of the sides $AB = 1$, $BC = 1$, and $AC = 1$ of the triangle $\triangle ABC$ if figure 3.2.1 is equivalent to the constrained line segment in figure 3.1.1. Let

1. ξ_0 be the coordinate number along the side AB defining the point f_1,

2. ξ_1 be the coordinate number along the side BC defining the point g_1,

3. ξ_2 be the coordinate number along the side AC defining the point h_1.

In subsection 1.4.2, it was shown that the sum of the constrained coordinate numbers equals, geometrically, to the side of a equilateral triangle, that is, $w_0 + w_1 + w_2 = \ell$. The coordinate numbers of a point P in the level ℓ of a two-dimensional constrained number system is given by $P = (w_0, w_1, w_2)$.

Since the coordinate numbers for a point P in the level $\ell = 1$ are defined as (ξ_0, ξ_1, ξ_2) then the coordinate numbers for any level ℓ are related to the coordinate numbers in the level $\ell = 1$, with the aid of equation (3.6), by

$$w_0 = \ell\xi_0 \qquad w_1 = \ell\xi_1 \qquad w_2 = \ell\xi_2 \tag{3.13}$$

As consequence, any point P in the triangle $\triangle ABC$ has the coordinates $P = (\xi_0, \xi_1, \xi_2)$ and is in the level $\ell = 1$ where

$$\xi_0 = \frac{Bf_1}{AB} \qquad \xi_1 = \frac{Cg_1}{BC} \qquad \xi_2 = \frac{Ah_1}{AC} \tag{3.14}$$

and they are shown in figure 3.2.1.

If the constrained numbers belong to the set L_ℓ^2 their coordinate numbers w_0, w_1, w_2 in terms of ξ_0, ξ_1, ξ_2 can be obtained from equation (3.14) as

$$w_0 = Bf_1 = AB \times \xi_0 = \ell\xi_0$$
$$w_1 = Cg_1 = BC \times \xi_1 = \ell\xi_1$$
$$w_2 = Ah_1 = AC \times \xi_2 = \ell\xi_2 \tag{3.15}$$

and they are shown in figure 3.2.2.

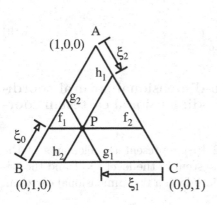

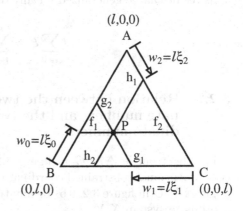

Figure 3.2.1 *Location of a two-dimensional constrained number in the level $\ell = 1$.*

Figure 3.2.2 *Location of a two-dimensional constrained number in the level ℓ.*

Example 3.2.3 *Assume that $\ell = 4$ in figure 3.2.2. Obtain the coordinates (w_0, w_1, w_2) of the point P if $\xi_0 = 1/2$ and $\xi_1 = 3/10$.*

Solution From the constraint equation $\xi_0 + \xi_1 + \xi_2 = 1$ it follows that $\xi_2 = 1 - 1/2 - 3/10 = 1/5$.

From equation (3.15) it follows that the coordinate numbers of the point P are

$$w_0 = \ell\xi_0 = 4 \times \frac{1}{2} = 2$$

$$w_1 = \ell\xi_1 = 4 \times \frac{3}{10} = \frac{6}{5} \tag{3.16}$$

$$w_2 = \ell\xi_2 = 4 \times \frac{1}{5} = \frac{4}{5}$$

thus the constrained coordinate numbers of the point P are $(w_0, w_1, w_2) = (2, 6/5, 4/5)$, which satisfy the constraint equation $2 + 6/5 + 4/5 = 4$.

end of example 3.2.3

The constraint equation for the level L_1^2 writes

$$\xi_0 + \xi_1 + \xi_2 = 1 \tag{3.17}$$

Multiplying (3.17) by ℓ and considering equation (3.15) it follows

$$w_0 + w_1 + w_2 = \ell \tag{3.18}$$

This result can be generalized for any dimension δ as

$$\ell \sum_{i=0}^{\delta} \xi_i = \sum_{i=0}^{\delta} w_i = \ell \qquad (3.19)$$

3.2.2   Relation between the two-dimensional natural coordinate numbers and the two-dimensional cartesian coordinate numbers

Let the triangle $\triangle ABC$ in figure 3.2.4(a) represent the elements of the two-dimensional constrained coordinate system in the level $\ell = 1$ and the triangle $\triangle ABC$ in figure 3.2.4(b) be any triangle in a two-dimensional cartesian coordinate system X, Y.

Let $\xi_0, \xi_1, \xi_2 \in L_1^2$ be the constrained coordinate numbers of the point P in figure 3.2.4(a) where $\xi_0 + \xi_1 + \xi_2 = 1$.

The coordinate number x_p of the point P in figure 3.2.4(b) is given by

$$x_p = x_b + (\overline{Bf_1})_x + (\overline{f_1P})_x = x_a + (\overline{Bf_1})_x + (\overline{Bh_2})_x \qquad (3.20)$$

where $(\overline{Bf_1})_x$ and $(\overline{Bh_2})_x$ denotes the projection of the line segment $\overline{Bf_1}$ and $\overline{Bh_2}$ on the X axis.

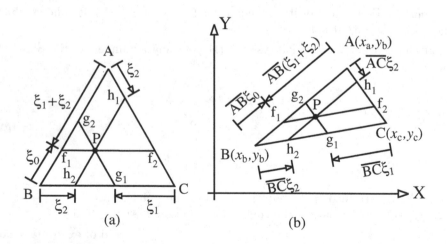

Figure 3.2.4 *Relation between the two-dimensional constrained coordinate numbers and the two-dimensional cartesian coordinate numbers.*

The sides of the triangles in figure 3.2.4(a) and 3.2.4(b) are homologous with respect to ξ_0, ξ_1, ξ_2 as was shown in figure 3.1.3. Therefore, the projection of the segment lines $\overline{Bf_1}$ and $\overline{Bh_2}$ can be computed as

$$(\overline{Bf_1})_x = \xi_0 \overline{AB} = \xi_0 (x_a - x_b)$$
$$(\overline{Bh_2})_x = \xi_2 \overline{BC} = \xi_2 (x_c - x_b) \tag{3.21}$$

Introducing (3.21) into equation (3.20) it follows

$$x_p = x_b + \xi_0 (x_a - x_b) + \xi_2 (x_c - x_b)$$
$$x_p = \xi_0 x_a + (1 - \xi_0 - \xi_2) x_b + \xi_2 x_c$$
$$x_p = \xi_0 x_a + \xi_1 x_b + \xi_2 x_c \tag{3.22}$$

Analogously, it can be found that

$$y_p = \xi_0 y_a + \xi_1 y_b + \xi_2 y_c \tag{3.23}$$

Similar to the one-dimensional constrained numbers the cartesian coordinate numbers (x_p, y_p) of the point P in figure 3.2.4(b) are a linear combination of the cartesian coordinates $(x_a, y_a), (x_b, y_b), (x_c, y_c)$ of the points A, B, and C where the scalars ξ_0, ξ_1, ξ_2 are the two-dimensional constrained numbers of the point P in figure 3.2.4(a) and in the level $\ell = 1$.

3.2.3 Coordinates orthogonal to the sides of the triangle

So far the location of the point P was obtained by the construction of lines parallel to the sides of the triangle through the points f_1 parallel to BC and through g_1 parallel to AC. As a consequence of the linear dependency of the constrained coordinates the line through h_1 parallel to AB also intercepts the point P.

These same parallel lines can be constructed using the orthogonal distance from the point P to each side as shown in figure 3.2.5.

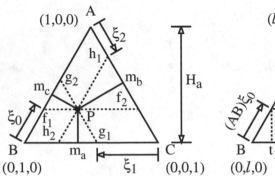

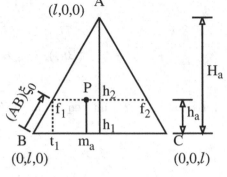

Figure 3.2.5 *Location of a two-dimensional constrained number in the level $\ell = 1$.*

Figure 3.2.6 *Relation between the orthogonal distances from the point P to the side BC the height with respect to the vertex A.*

The similarity of the triangles $\triangle Bf_1t_1$ and $\triangle Bah_1$, shown in figure 3.2.6, gives the following relation

$$\frac{\overline{AB}\xi_0}{\overline{AB}} = \frac{\overline{t_1 f_1}}{\overline{Ah_1}} = \frac{h_a}{H_a} \tag{3.24}$$

From the definition 3.0.13 of natural coordinates $\overline{AB} = \ell = 1$, therefore,

$$\xi_0 = \frac{h_a}{H_a} \tag{3.25}$$

Analogously, the other coordinate numbers are given by

$$\xi_1 = \frac{h_b}{H_b} \qquad \xi_2 = \frac{h_c}{H_c} \tag{3.26}$$

where h_b, H_b are orthogonal to the side AC and h_c, H_c are orthogonal to the side AB.

The constraint equation for the two-dimensional constrained numbers system gives the following relation

$$\xi_0 + \xi_1 + \xi_2 = \frac{h_a}{H_a} + \frac{h_b}{H_b} + \frac{h_c}{H_c} = 1 \tag{3.27}$$

which can be rewritten as

$$\frac{\overline{BC} \times h_a/2}{\overline{BC} \times H_a/2} + \frac{\overline{AB} \times h_b/2}{\overline{AB} \times H_b/2} + \frac{\overline{BA} \times h_c/2}{\overline{BA} \times H_c/2} = 1 \tag{3.28}$$

where

$$\frac{\overline{BC} \times H_a}{2} = \frac{\overline{AB} \times H_b}{2} = \frac{\overline{BA} \times H_c}{P2} = \text{area} (\triangle ABC) \qquad (3.29)$$

and

$$\frac{\overline{BC} \times H_a}{2} = \text{area} (\triangle BPC)$$

$$\frac{\overline{AB} \times H_b}{2} = \text{area} (\triangle APB) \qquad (3.30)$$

$$\frac{\overline{BA} \times H_c}{2} = \text{area} (\triangle BPA)$$

therefore, the constraint equation for the two-dimensional constrained numbers system can also be written as

$$\xi_0 = \frac{\text{area} (\triangle BPC)}{\text{area} (\triangle ABC)} \qquad \xi_1 = \frac{\text{area} (\triangle APB)}{\text{area} (\triangle ABC)} \qquad \xi_2 = \frac{\text{area} (\triangle BPA)}{\text{area} (\triangle ABC)}$$

$$(3.31)$$

for any triangle $\triangle ABC$.

The ratios shown in equation (3.31) are well known in finite element books such as Huebner [19].

3.3 Three-dimensional natural coordinate numbers

3.3.1 Location of a point

The level EFG is located with respect to the vertex B by the coordinate number ξ_0. The definition of the coordinate numbers in the triangle EFG in figure 3.3.2 are similar to those defined for the triangle $\triangle ABC$ in figure 3.2.1 and it is shown in the 3.3.1.

Table 3.3.1 *Definition of the natural coordinate numbers in the sublevel defined by the triangle $\triangle EFG$.*

figure 3.2.1		figure 3.3.1	
side	variable	side	variable
AB	ξ_0	BC	ξ_1
BC	ξ_1	CD	ξ_2
CA	ξ_2	DB	ξ_3

The analogy in table 3.3.1 permits the location of the points E, e_1, f_1, and g_1 as follows

1. ξ_0 is the natural coordinate number along the edge AB locating the point E,

2. ξ_1 is the natural coordinate number along the edge BC locating the point e_1,

3. ξ_2 is the natural coordinate number along the edge CD locating the point f_1,

4. ξ_3 is the natural coordinate number along the edge BD locating the point g_1.

The constraint equation (1.33) on page 9 for three-dimensional constrained natural numbers is given by

$$\xi_0 + \xi_1 + \xi_2 + \xi_3 = 1 \tag{3.32}$$

Multiplying equation (3.32) by the value ℓ of the level if follows

$$\ell\xi_0 + \ell\xi_1 + \ell\xi_2 + \ell\xi_3 = \ell \tag{3.33}$$

On the other hand, the sum of the constrained coordinate numbers equals the value of the level, that is

$$w_0 + w_1 + w_2 + w_3 = \ell \tag{3.34}$$

therefore,

$$w_0 = \ell\xi_0 \qquad w_1 = \ell\xi_1 \qquad w_2 = \ell\xi_2 \qquad w_3 = \ell\xi_3 \tag{3.35}$$

This result can also be obtained by similarity of two tetrahedrons, one with edges equal to 1 and the other with edges equal to ℓ. If all the edges of the tetrahedron are different then, instead of ℓ, it must be used the length of each edge, that is

$$w_0 = \overline{AB} \times \xi_0 \qquad w_1 = \overline{BC} \times \xi_1 \qquad w_2 = \overline{CD} \times \xi_2 \qquad w_3 = \overline{BD} \times \xi_3 \tag{3.36}$$

which is shown in figure 3.1.3

The distance $\overline{Fg_1}$ must not computed based on the side $\overline{EF}$ of the triangle $\triangle EFG$ in figure 3.3.2, but on the side $\overline{BC}$ of the triangle $\triangle BCD$, which is the edge of the tetrahedron. The same applies for all other coordinate numbers.

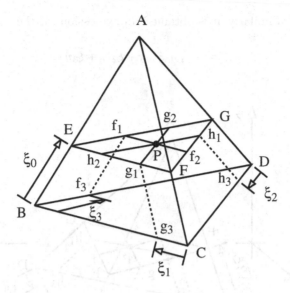

Figure 3.3.2 *Location of a three-dimensional constrained number in the level $\ell = 1$.*

3.3.2 Relation between the three-dimensional natural coordinate numbers and the three-dimensional cartesian coordinate numbers

The coordinate x_p is given by

$$x_p = x_b + (\overline{BE})_x + (\overline{Eh_2})_x + (\overline{h_2P})_x \tag{3.37}$$

As shown in figure 3.3.3 the projection of the line segments $\overline{BE}, \overline{Eh_2}$ and $\overline{h_2P}$ on the direction X is given by

$$(\overline{BE})_x = \overline{BA}\,\xi_0 = \xi_0\,(x_a - x_b)$$
$$(\overline{Eh_2})_x = \overline{BC}\,\xi_2 = \xi_2\,(x_c - x_b)$$
$$(\overline{h_2P})_x = \overline{BD}\,\xi_3 = \xi_3\,(x_d - x_b) \tag{3.38}$$

Introducing equation (3.38) into equation (3.37) it follows that

$$
\begin{aligned}
x_p &= x_b + \xi_0\,(x_a - x_b) + \xi_2\,(x_c - x_b) + \xi_3\,(x_d - x_b) \\
&= \xi_0 x_a + (1 - \xi_0 - \xi_2 - \xi_3)\,x_b + \xi_2 x_c + \xi_3 x_d \\
&= \xi_0 x_a + \xi_1 x_b + \xi_2 x_c + \xi_3 x_d
\end{aligned}
\tag{3.39}
$$

Equation (3.39) shows that the cartesian coordinate number x_p is a linear combination of the coordinates x_a, x_b, x_c, x_d or the vertices of the tetrahedron in figure 3.3.3. Similarly, it is obtained an expression for the computation of y_p as

$$y_p = \xi_0 y_a + \xi_1 y_b + \xi_2 y_c + \xi_3 y_d \qquad (3.40)$$

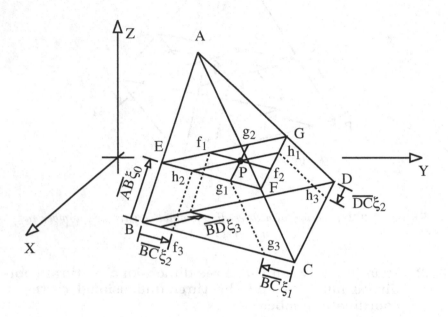

Figure 3.3.3 *Location of a three-dimensional constrained number in a generic tetrahedron.*

3.3.3   Coordinates orthogonal to the faces of the tetrahedron

Let $\overline{An}$ be a line segment orthogonal to the plane that contains the triangle $\triangle BCD$. Let $\triangle EFG$ be a triangle in a plane parallel to the face BCD of the tetrahedron. Let ξ_0 be the natural coordinate number locating the point E along the edge BA and on a plane containing the triangle $\triangle EFG$ and the point P.

For clarity, the triangle $\triangle Ban$ in figure 3.3.4(a) is reproduced in figure 3.3.4(b). The triangles $\triangle Bem$ and $\triangle Ban$ are similar, therefore,

$$\frac{\overline{BE}}{\overline{BA}} = \frac{h_a}{H_a} \qquad (3.41)$$

which gives

$$\frac{\overline{BA}\,\xi_0}{\overline{BA}} = \frac{h_a}{H_a} \qquad (3.42)$$

and then

$$\xi_0 = \frac{h_a}{H_a} \tag{3.43}$$

where ξ_0 is the ratio of the distance h_a of a plane containing the point P parallel to a face of the tetrahedron and the distance H_a from the vertex to that face.

Similarly,

$$\xi_1 = \frac{h_b}{H_b} \qquad \xi_2 = \frac{h_c}{H_c} \qquad \xi_3 = \frac{h_d}{H_d} \tag{3.44}$$

The natural coordinate numbers can be obtained as the ratio of the volumes of two tetrahedrons observing that

$$\xi_0 = \frac{(A_{bcd} \times h_b)/3}{(A_{bcd} \times H_b)/3} \tag{3.45}$$

Where A_{bcd} is the area of the triangle $\triangle BCD$ face of the tetrahedron. Recalling that the volume of a tetrahedron is $(base \times height)/3$ the equation (3.45) can be rewritten as the ratio of volumes as

$$\xi_0 = \frac{V_{bcdp}}{V_{bcda}} \tag{3.46}$$

where V_{bcdp} denotes the volume of the tetrahedron whose vertices are $BCDP$ and V_{bcda} denotes the volume of the tetrahedron whose vertices are $BCDA$.

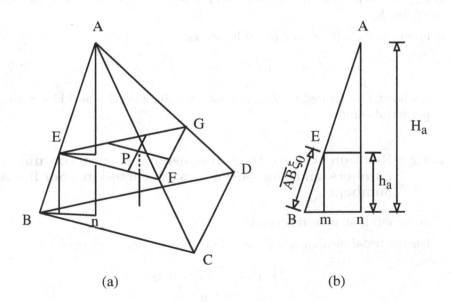

(a)   (b)

Figure 3.3.4 *Distance h_a between the planes EFG and BCD.*

3.4 δ-dimensional natural coordinate numbers

3.4.1   Location of a point

The constraint equation (1.33) on page 9 for a δ-dimensional constrained numbers and underlying set $\mathbb{Z}^+$ where $|w_i| = w_i$ is

$$\sum_{j=0}^{\delta} w_j = \ell \in \mathbb{Z}^+ \tag{3.47}$$

Dividing (3.47) by ℓ

$$\sum_{j=0}^{\delta} \frac{w_j}{\ell} = 1 \tag{3.48}$$

Denoting $\xi_i = w_i/\ell$ equation (3.48) yields the constraint equation for the natural numbers in a three-dimensional constrained space as

$$\sum_{j=0}^{\delta} \xi_j = 1 \tag{3.49}$$

Let ξ_j be the natural coordinate number along the hyper edge $\overline{AB}_j$ locating the point E_j.

From equation (3.48) and (3.49) it follows

$$w_j = \overline{AB}_j \xi_j \qquad j = 0, 1, \ldots, \delta \tag{3.50}$$

and a point P is located by the coordinate numbers $(w_0, \ldots, w_\delta)$ in a generic hyper-tetrahedron.

3.4.2   Relation between the δ-dimensional natural coordinate numbers and the δ-dimensional cartesian coordinate numbers

To establish an induction consider

For the two-dimensional constrained number system

$$x_p = x_b + \overline{AB} \times \xi_0 + \overline{BC} \times \xi_2$$
$$x_b + (x_a - x_b)\xi_0 + (x_c - x_b)\xi_2$$
$$\xi_0 x_a + \xi_1 x_b + \xi_2 x_c \tag{3.51}$$

as shown in figure 3.4.1(a). The same expression for coordinate number x_p can be obtained starting from the vertex A as shown in figure 3.4.1(b) giving

$$x_p = x_a + \overline{AB} \times \xi_1 + \overline{AC} \times \xi_2$$
$$x_a + (x_b - x_a)\,\xi_1 + (x_c - x_a)\,\xi_2$$
$$\xi_0 x_a + \xi_1 x_b + \xi_2 x_c \qquad (3.52)$$

or starting from the vertex C as shown in figure 3.4.1(c) giving

$$x_p = x_c + \overline{AB} \times \xi_0 + \overline{BC} \times \xi_1$$
$$x_c + (x_a - x_c)\,\xi_0 + (x_b - x_c)\,\xi_1$$
$$\xi_0 x_a + \xi_1 x_b + \xi_2 x_c \qquad (3.53)$$

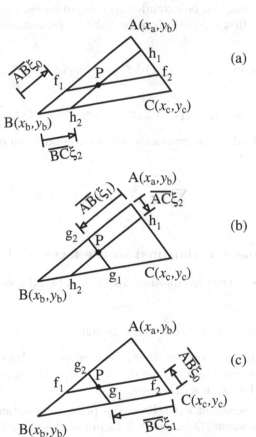

Figure 3.4.1 *Computation of the cartesian coordinate numbers of the point P in a two-dimensional constrained number system.*

The computation of the cartesian coordinate number x_p in three-dimensional constrained number system requires one more coordinate with respect to the two-dimensional case, that is, the first ξ_0 locating the sublevel represented by the triangle $\triangle EFG$ in figure 3.3.3. Let ξ_1, ξ_2, and ξ_3 be the coordinate numbers locating the point P in the sublevel analogous the the coordinate numbers ξ_0, ξ_1, and ξ_2 in figures 3.4.1(a)(b) and (c). Evaluating the expression for x_p starting from the vertex B it follows

$$x_p = x_b + \xi_0 \left(x_a - x_b\right) + \xi_2 \left(x_c - x_b\right) + \xi_3 \left(x_d - x_b\right) \qquad (3.54)$$

The natural coordinate number ξ_1 is not present in the equation (3.54) because it is redundant. As was shown for the two-dimensional case in figures 3.4.1(a)(b) and (c) if the reference vertex changes, the redundant variable also changes such that the final result is the same.

The above reasoning can be extended to a δ-dimensional by always adding the location of the first sublevel given by the natural constrained number ξ_0 as follows

$$x_p = x_1 + \xi_0 \left(x_0 - x_1\right) + \xi_2 \left(x_2 - x_1\right) + \xi_3 \left(x_3 - x_1\right) + \cdots + \xi_\delta \left(x_\delta - x_1\right) \quad (3.55)$$

which gives

$$x_p = \xi_0 x_0 + \xi_1 x_1 + \xi_2 x_2 + \xi_3 x_3 + \cdots + \xi_\delta x_\delta \qquad (3.56)$$

Similarly it is obtained an expression for the computation of all coordinate numbers of the point P as

$$(x_p)_j = \sum_{i=0}^{\delta} \xi_i x_j \qquad j = 0, 1, \ldots, \delta \qquad (3.57)$$

3.4.3   Coordinates orthogonal to the faces of the δ-hedron

To generalize the results in equations (3.43) and (3.44) consider the following

1. For a three-dimensional coordinate system

   (a) A 4-hedron (tetrahedron) can be seen as the closed region formed by the intersection of four noncoincident $2D$-planes whose region is called *solid* in a $3D$-space.

   (b) Denote *vertex* of a 4-hedron the point intersection of each three noncoincident $2D$-planes that form the closed region. Each vertex is considered as a $0D$-plane and their number is given by $C(4, 3) = 4$.

   Denote *edge* of a 4-hedron the line segment between two vertices. Each edge is contained in the line ($1D$-plane) intersection of two noncoincident $2D$-planes. The number of edges is given by $C(4, 2) = 6$.

Denote *face* of a 4-hedron the closed region on each 2D-plane formed by the intersection with the other three 2D-planes. Since each face is a closed region in a 2D-plane formed by the three 1D-plane (lines), which are the intersection of the other three 2D-planes with this 2D-plane then each face is a 3-hedron, that is, a triangle. The number of faces is given by $C(4,1) = 4$. For example the triangle $\triangle ABC$ shown in figure 3.3.4(a). Each face is on one 2D-plane. The number of faces is given by $C(4,1) = 4$.

(c) Let A, B, C, and D be vertices of a 4-hedron (tetrahedron). The the line segments $\overline{AB}$, $\overline{AC}$, $\overline{AD}$, $\overline{BC}$, $\overline{CD}$, and $\overline{DA}$ are edges defining the closed region $ABCD$, which is a tetrahedron as shown in figure 3.3.4(a).

(d) Denote basis of the 4-hedron with respect to a vertex the 3-hedron (triangle) that do not contain this vertex. Then the triangle $\triangle BCD$ in figure 3.3.4(a) is a basis of the 4-hedron (tetrahedron) with respect to the vertex A. Therefore, a 4-hedron (tetrahedron) is a solid in a 3D space defined by four noncoincident 3-hedron (triangle).

(e) Let E, F, and G be a sublevel, which is a (3-hedron) (a triangle $\triangle EFG$) parallel to the basis 3-hedron (triangle $\triangle BCD$) and containing the point $P = (\xi_0, \xi_1, \xi_2, \xi_3)$. Let h_a denote the distance between the sublevel $\triangle EFG$ and the the basis $\triangle BCD$ as shown in figure 3.3.4(a).

(f) Let H_a denotes the distance from the vertex A to the basis $\triangle BCD$ (3-hedron) as shown in figure 3.3.4(b).

(g) it was already shown that the ratio h_a/H_a defines the natural coordinate number $\xi_0 = h_a/H_a$, which expresses the ratio of the length of two line segments. Since both 3-hedron have the same base, then

$$\xi_0 = \frac{h_a}{H_a} = \frac{\text{volume (3-hedron } PBCD)}{\text{volume (3-hedron } ABCD)}$$

$$= \frac{\text{area (2-hedron } \triangle BCD)h_a/3}{\text{area (2-hedron } \triangle BCD)H_a/3} \tag{3.58}$$

2. For a four-dimensional coordinate system

(a) A 5-hedron (pentahedron) can be seen as the closed region formed by the intersection of five noncoincident 3D-planes whose region is called *hyper-solid* in a 4D-space.

(b) Denote *vertex* of a 5-hedron the point intersection of each four noncoincident 3D-planes that form the closed region. Each vertex is considered as a 0D-plane and their number is given by $C(5,4) = 5$.

Denote *edge* of a 5-hedron the line segment between each two vertices. Each edge is contained in the line (1D-plane) intersection of three noncoincident $3D$-planes. The number of edges is given by $C(5,3) = 10$.

Denote *face* of a 5-hedron the closed region on each $3D$-plane formed by the intersection with the other four $3D$-planes. Since each face is a closed region in a $3D$-plane formed by the four $2D$-plane (planes), which are the intersection of the other four $3D$-planes with this $3D$-plane then each face is a 4-hedron, that is, a tetrahedron. The number of faces is given by $C(5,2) = 10$.

(c) Let A, B, C, D, and E be vertices of a 5-hedron (pentahedron). The line segments $\overline{AB}$, $\overline{AC}$, $\overline{AD}$, $\overline{AE}$, $\overline{BC}$, $\overline{BD}$, $\overline{BE}$, $\overline{CD}$, $\overline{CE}$, and $\overline{DE}$ are edges defining the closed region $ABCDE$, which is a pentahedron.

(d) Denote basis of the 5-hedron with respect to a vertex the 4-hedron (tetrahedron) that do not contain this vertex. Then the tetrahedron $\triangle BCDE$ is a basis of the 5-hedron (pentahedron) with respect to the vertex A. Therefore, a 5-hedron (pentahedron) is a solid in a $4D$ space defined by five noncoincident 4-hedron (tetrahedron).

(e) Let F, G, H, and I be a sublevel, which is 4-hedron (tetrahedron $FGHI$), parallel to the basis (4-hedron BCDE) and containing the point $P = (\xi_0, \xi_1, \xi_2, \xi_3, \xi_4)$. Let h_a denote the distance between the sublevel 4-hedron $FGHI$ and the the basis 4-hedron $BCDE$.

(f) Let H_a denotes the distance from the vertex A to the basis 4-hedron $BCDE$, that is, the tetrahedron $BCDE$.

(g) the ratio h_a/H_a defines the natural coordinate number $\xi_0 = h_a/H_a$, which expresses the ratio of the length of two line segments where h_a is the height of the 5-hedron $PBCDE$ and H_a is the height of the 5-hedron $ABCDE$. Since both 5-hedron have the same base, then

$$\xi_0 = \frac{h_a}{H_a} = \frac{\text{volume (5-hedron } PBCDE)}{\text{volume (5-hedron } ABCDE)} \tag{3.59}$$

3. For a δ-dimensional coordinate system

(a) A $\delta + 1$-hedron can be seen as the closed region formed by the intersection of $(\delta + 1)$ noncoincident $(\delta - 1)D$-planes whose region is called *hyper-solid* in a δD-space.

(b) Denote *vertex* of a $(\delta + 1)$-hedron the point intersection of each δ noncoincident $(\delta - 1)D$-planes that form the closed region. Each vertex is considered as a $0D$-plane and their number is given by $C((\delta + 1), \delta) = \delta + 1$.

Denote *edge* of a $(\delta + 1)$-hedron the line segment between each two vertices. Each edge is contained in the line ($1D$-plane) intersection of $(\delta - 1)$ noncoincident $(\delta - 1)D$-planes. The number of edges is given by $C((\delta + 1), (\delta - 1)) = (\delta + 1)\delta/2$.

Denote *face* of a $(\delta + 1)$-hedron the closed region on each $(\delta - 1)D$-plane formed by the intersection with the other δ planes of the type $(\delta - 1)D$-planes. Since each face is a closed region in a $(\delta - 1)D$-plane formed by $(\delta - 1)$ planes of the type $(\delta - 2)D$-plane, which are the intersection of the other δ planes of the type $(\delta - 1)D$-plane with this $(\delta - 1)D$-plane then each face is a $(\delta - 1)$-hedron and it is on the intersection of $(\delta - 2)$ planes of the type $(\delta - 1)D$-plane. The number of faces is given by $C((\delta + 1), (\delta - 2)) = (\delta - 1)\delta(\delta + 1)/6$.

(c) Let $A_0, A_1, \ldots, A_{\delta-1}$ be vertices of a $(\delta + 1)$-hedron. Then the line segments $\overline{A_0 A_1}, \ldots, \overline{A_{\delta-2} A_{\delta-1}}$ are edges defining the closed region $A_0 \cdots A_{\delta-1}$.

(d) Denote basis of the $(\delta + 1)$-hedron with respect to a vertex the δ-hedron that do not contain this vertex. Then the δ-hedron $A_1 \cdots A_{\delta-1}$ is a basis of the $(\delta + 1)$-hedron with respect to the vertex A_0. Therefore, a $(\delta + 1)$-hedron is a solid in a δD space defined by $(\delta + 1)$ noncoincident δ-hedron.

(e) Let $B_1, B_2, \ldots, B_{\delta-1}$ be a sublevel, which is a δ-hedron parallel to the basis $A_1, A_2, \ldots, A_{\delta-1}$ and containing the point $P = (\xi_0, \xi_1, \ldots, \xi_{\delta-1})$. Let h_a denote the distance between the sublevel $B_1, B_2, \ldots, B_{\delta-1}$ and the the basis $A_1, A_2, \ldots, A_{\delta-1}$.

(f) Let H_a denote the distance from a vertex A to the basis δ-hedron $A_1, A_2, \ldots, A_{\delta-1}$.

(g) The ratio h_a/H_a defines the natural coordinate number $\xi_0 = h_a/H_a$, which expresses the ratio of the length of two line segments where h_a is the height of the $(\delta + 1)$-hedron $P, B_1, B_2, \ldots, B_{\delta-1}$ and H_a is the height of the $(\delta + 1)$-hedron $A_0, A_1, A_2, \ldots, A_{\delta-1}$. Since both $\delta + 1$-hedron have the same base then

$$\xi_0 = \frac{h_a}{H_a} = \frac{\text{volume }((\delta + 1)\text{-hedron } P, B_1, B_2, \ldots, B_{\delta-1})}{\text{volume }((\delta + 1)\text{-hedron } A_0, A_1, A_2, \ldots, A_{\delta-1})} \quad (3.60)$$

4. For a two-dimensional coordinate system

A two-dimensional system can be obtained with $\delta = 2$ as shown below.

(a) A 3-hedron or triangle can be seen as the closed region formed by the intersection of three noncoincident $1D$ planes (lines) whose region is called *area* in a $2D$-space.

(b) Denote *vertex* of a 3-hedron the point intersection of each two noncoincident $1D$-planes (lines) that form the closed region. Each vertex is considered as a $0D$-plane and their number is given by $C(3, 2) = 3$.

Denote *edge* of a 3-hedron the line segment between two vertices. Each edge is contained in the line (1D-plane) intersection of 1 non-coincident 1D-planes, which is a line. The number of edges is given by $C(3, 1) = 3$. For case of triangles the edges are called *sides*.

The 3-hedron has no faces. The number $C(3, 0) = C(3, 3) = 1$ gives the number of closed regions in the 2D-space, which is one.

(c) Let A, B, and C be vertices of a 3-hedron. Then the line segments $\overline{AB}$, $\overline{BC}$, and $\overline{CA}$ are edges and define the closed region $\triangle ABC$, which is a triangle as shown in figure 3.4.1.

(d) Denote basis of the 3-hedron with respect to a vertex the 2-hedron that do not contain this vertex. Then the 1-hedron B, C the line segment $\overline{BC}$ is a basis of the 3-hedron with respect to the vertex A as shown in figure 3.2.6. Therefore, a 3-hedron (a triangle) is a solid in a 2D space (area) defined by 3 noncoincident 2-hedron, that is, by three line segments.

(e) Let (f_1, f_2) be a sublevel, which is a 2-hedron parallel to the basis (B, C) and containing the point $P = (\xi_0, \xi_1, \xi_2$. Let h_a denote the distance between the sublevel (f_1, f_2) and the basis (B, C) as shown in figure 3.2.6.

(f) Let H_a denote the distance from a vertex A to the basis 2-hedron (B, C).

(g) It was already shown that the ratio h_a / H_a defines the natural coordinate number $\xi_0 = h_a / H_a$, which expresses the ratio of the length of two line segments. Since both 2-hedron have the same base then

$$\xi_0 = \frac{h_a}{H_a} = \frac{\text{volume (2-hedron } PBC)}{\text{volume (2-hedron } ABC)}$$

$$= \frac{\text{length (1-hedron } \overline{BC})h_a/2}{\text{length (1-hedron } \overline{BC})H_a/2} = \frac{\text{area } \triangle PBC}{\text{area } \triangle ABC} \qquad (3.61)$$

For the applications in finite element computation it is not necessary to know the volumes associated with the natural coordinates but to know that they are equal to the ratio of height and distance of sublevel to apply the integration expressions along the area or volume. To obtain the cartesian coordinates of a point it is enough to know the natural coordinates of the cartesian coordinates of the vertices.

Chapter 4

Computation of the number of elements

4.1 Zero-dimensional constrained space

4.1.1 Underlying set $\mathbb{Z}$

4.1.1.1 Computation of the number of elements in a set L_ℓ^0 in the level ℓ

The sets L_ℓ^0 of the constrained numbers for the underling set $\mathbb{Z}$ as defined in equation (1.78) on page 26 are

$$L_\ell^0 = \{(\ell), (-\ell)\} \qquad \ell = 0, 1, 2, \ldots \tag{4.1}$$

with two elements per level. For the level $\ell = 0$ the set $L_0^0 = \{(0), (-0)\}$ is represented geometrically by a point on a line. This point is defined as the origin of the zero dimensional constrained number system. It is usual to represent the set $\{(0), (-0)\}$ by $\{(0)\}$ and here both notations will be used. The elements on the sets whose level is greater than zero, that is $(\ell > 0)$, are represented by two points, one defined as negative and the other as positive. They are not represented symmetrically with respect to the origin as it is usual to do for the integer numbers representation on a line. For example, the sets

$$L_0^0 = \{(0), (-0)\},$$
$$L_1^0 = \{(1), (-1)\},$$
$$L_2^0 = \{(2), (-2)\},$$
$$L_3^0 = \{(2), (-2)\} \tag{4.2}$$

are shown in figure 4.1.1 in a nonsymmetrical representation on the line.

Equations (1.81) on page 27 provide a natural definition for an ordering relation based on the levels. Figure 4.1.1 shows the first four sets L_0^0, L_1^0, L_2^0, and L_3^0 using the level ordering and these sets can be written ordered as $L_0^0 \prec L_1^0 \prec L_2^0 \prec L_3^0$. The definition of ordering relation for the constrained numbers is given in chapter 5 as well as a justification for its choice.

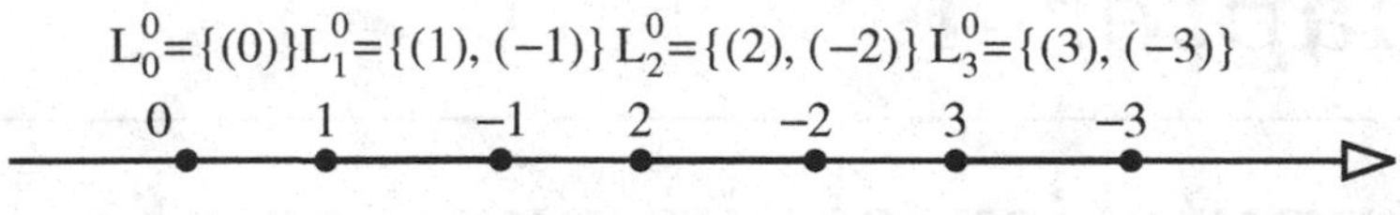

$$L_0^0 = \{(0)\} \quad L_1^0 = \{(1),\,(-1)\} \quad L_2^0 = \{(2),\,(-2)\} \quad L_3^0 = \{(3),\,(-3)\}$$

Figure 4.1.1 *Representation of ordered levels L_0^0, L_1^0, L_2^0, and L_3^0 in a zero-dimensional constrained number system.*

Notation 4.1.2

   n_ℓ^δ denotes the number of elements in the set L_ℓ^δ, that is, a set in the level ℓ of a δ-dimensional constrained space.

   $n^\delta(\ell)$ denotes the number of elements in the union of all sets from level zero until the level ℓ inclusive in a δ-dimensional constrained space.

   $C(\ell, m)$ denotes the combination of ℓ elements taken m at a time.

end of notation 4.1.2

The number of elements in the set L_ℓ^0 in the level ℓ is given by equation (2.61) on page 50 and for the $\mathbb{Z}$ it is

$$n_\ell^0 = \begin{cases} C(\ell, 0) = 1 & \text{if } \ell = 0 \\ 2C(\ell, 0) = 2 & \text{if } \ell > 0 \end{cases} \tag{4.3}$$

4.1.1.2   Computation of the number of elements in the set $L^0(\ell)$

Let $L^0(\ell)$ be the union of the subsets L_i^0 from level $i = 0$ until the level $i = \ell$ and whose notation is

$$L^0(\ell) = \bigcup_{i=0}^{\ell} L_i^0 \tag{4.4}$$

The number $n^0(\ell)$ of elements in the set $L^0(\ell)$ is obtained by adding the number n_i^0 of elements in each of the sets L_i^0 from $i = 0$ until $i = \ell$ inclusive, that is

$$n^0(\ell) = \sum_{i=0}^{\ell} n_i^0 \tag{4.5}$$

where the number of elements n_i^0 in each level i is given by equation (4.3). Since there are ℓ sets with two elements and only one set with one element then equation (4.5) must be split into two parts as

$$n^0(\ell) = n_0^0 + \sum_{i=1}^{\ell} n_i^0$$

$$= C(\ell, 0) + \sum_{i=1}^{\ell} 2C(\ell, 0)$$

$$= C(\ell, 0) + 2C(\ell, 1) = 1 + 2\ell \qquad \ell \geq 0 \qquad (4.6)$$

4.1.2 Underlying set $\mathbb{Z}^+$

4.1.2.1 Computation of the number of elements in the set L_ℓ^0 in the level ℓ.

Since the set $\mathbb{Z}^+$ contains no negative integers then the sets whose underlying set is $\mathbb{Z}^+$ are subsets of the sets whose underlying set is $\mathbb{Z}$. As an example, the sets in equations (1.81) on page 27 and (1.83) can be written as

$$\{(0), (1), (2), \ldots, (\ell), \ldots \} \subset \{(0), (1), (-1), (2), (-2), \ldots, (\ell), (-\ell), \ldots \} \tag{4.7}$$

The equation (1.83) on page 27 shows the set $\{(0), (1), (2), \ldots, (\ell), \ldots\}$ whose elements are union singleton sets. The first four of these elements as well as the associated sets are represented geometrically in figure 4.1.3.

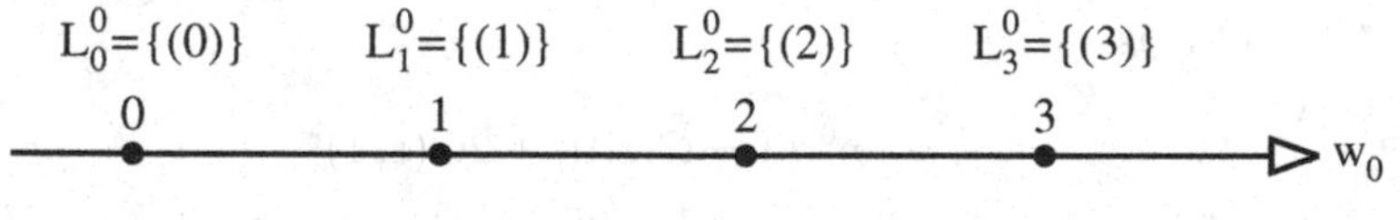

Figure 4.1.3 *Number of points in each level for the first four sets $L_i^0, i = 0, 1, 2, 3, 4$ in a zero-dimensional space.*

The number of elements in the level ℓ is denoted by

$$n_\ell^0 = C(\ell, 0) = 1 \tag{4.8}$$

4.1.2.2 Computation of the number of elements in the set $L^0(\ell)$

Analogous to equation (4.4) the set $L^0(\ell)$ is the union of all subsets L_i^0 from the level $i = 0$ until the level $i = \ell$, which can be represented as

$$L^0(\ell) = \bigcup_{i=0}^{\ell} L_i^0 \tag{4.9}$$

The number of elements $n^0(\ell)$ in the set $L^0(\ell)$ equals the sum of the number n_i^0 of elements of all its subsets L_i^0. Denote this sum by

$$n^0(\ell) = \sum_{i=0}^{\ell} n_i^0 \tag{4.10}$$

where n_i^0 is given by equation (4.8). Substituting

$$n^0(\ell) = \sum_{i=0}^{\ell} C(\ell, 0)$$
$$= C((\ell + 1), 1) = 1 + \ell \qquad \ell \geq 0 \tag{4.11}$$

4.1.3   Summary for the zero-dimensional space

1. If the underlying set is $\mathbb{Z}$

   (a) The number of elements in the set L_ℓ^0 in the level ℓ was obtained in equation (4.3) and it is

   $$n_\ell^0 = \begin{cases} C(\ell, 0) = 1 & \text{if} \quad \ell = 0 \\ 2C(\ell, 0) = 2 & \text{if} \quad \ell > 0 \end{cases} \tag{4.12}$$

   (b) The number of elements in the set $L^0(\ell)$, which include the elements from the level zero until the level ℓ was obtained in equation (4.6) and it is

   $$n^0(\ell) = C(\ell, 0) + 2C(\ell, 1)$$
   $$= 1 + 2\ell \tag{4.13}$$

2. If the underlying set is $\mathbb{Z}^+$

   (a) The number of elements in the set L_ℓ^0 in the level ℓ was obtained in equation (4.8) and it is

   $$n_\ell^0 = C(\ell, 0) = 1 \qquad \ell \geq 0 \tag{4.14}$$

   (b) The number of elements in the set $L^0(\ell)$, which include the elements from the level zero until the level ℓ was obtained in equation (4.11) and it is

   $$n^0(\ell) = C((\ell + 1), 1)$$
   $$= 1 + \ell \qquad \ell \geq 0 \tag{4.15}$$

4.2 One-dimensional constrained space

4.2.1 Underlying set $\mathbb{Z}$

4.2.1.1 Computation of the number of elements in the set L^1_ℓ in the level ℓ

The set L^1_ℓ is obtained from equation (2.15) with $\delta = 1$ and $\lambda_0 = \ell$ as

$$L^1_\ell = \bigcup_{w_0=\lambda_0}^{-\lambda_0} \{(w_0)\} \times L^0_{\lambda_1} \tag{4.16}$$

where $\lambda_1 = \ell - |w_0|$. The union in equation (4.16) can be divided into three parts, the positive values $w_0 = \lambda_0, \ldots, 1$, for $w_0 = 0$, and the negative values $w_0 = -1, \ldots, -\lambda_0$ as

$$L^1_\ell = \begin{cases} \{(0)\} \times L^0_{\lambda_0} & \lambda_0 = 0 \\ \left\{ \displaystyle\bigcup_{w_0=\lambda_0}^{1} \{(w_0)\} \times L^0_{\lambda_1} \right\} \cup \{\{(0)\} \times L^0_{\lambda_1}\} \cup \left\{ \displaystyle\bigcup_{w_0=-1}^{-\lambda_0} \{(w_0)\} \times L^0_{\lambda_1} \right\} \\ & \lambda_0 > 0 \end{cases} \tag{4.17}$$

where λ_1 assume the values $0, 1, \ldots, (\lambda_0 - 1), \lambda_0, (\lambda_0 - 1), \ldots, 0$ while $w_0 = \lambda_0, \ldots, -\lambda_1$. Since $\lambda_0 = \ell$ the values of λ_0 can also be written as $0, 1, \ldots, (\ell - 1), \ell, (\ell - 1), \ldots, 0$.

From the definition of sublevel $\lambda_1 = \ell - |w_0|$ then $|w_0| = \ell - \lambda_1$. The solutions of this equation for λ_1 are $\lambda_1 = \ell - w_0$ and $\lambda_1 = \ell + w_0$. The change of variables for the indices is obtained considering that

1. if $w_0 = \lambda_0$ then from $\lambda_1 = \ell - w_0$ it follows $\lambda_1 = 0$

2. if $w_0 = 1$ then from $\lambda_1 = \ell - w_0$ it follows $\lambda_1 = \ell - 1$

3. if $w_0 = -1$ then from $\lambda_1 = \ell + w_0$ it follows $\lambda_1 = \ell - 1$

4. if $w_0 = -\ell$ then from $\lambda_1 = \ell + w_0$ it follows $\lambda_1 = 0$

Substituting these values for λ_1 and w_0 into equation (4.17) it follows

$$L^1_\ell = \begin{cases} \{(0)\} \times L^0_{\lambda_0} & \lambda_0 = 0 \\ \left\{ \displaystyle\bigcup_{\lambda_1=0}^{\ell-1} \{(\ell - \lambda_1)\} \times L^0_{\lambda_1} \right\} \cup \{\{(0)\} \times L^0_{\lambda_1}\} \cup \left\{ \displaystyle\bigcup_{\lambda_1=\ell-1}^{0} \{-(\ell - \lambda_1)\} \times L^0_{\lambda_1} \right\} \\ & \lambda_0 > 0 \end{cases} \tag{4.18}$$

The number of elements in the set $\left\{ \bigcup_{w_0=\lambda_0}^{1} \{(w_0)\} \times L_{\lambda_1}^0 \right\}$ in equation (4.17) equals the number of elements in the set $L_{\lambda_1}^0$ since $\{(w_0)\}$ is a singleton factor in the cartesian product. The same is true for the set $\left\{ \bigcup_{w_0=-1}^{-\lambda_0} \{(w_0)\} \times L_{\lambda_1}^0 \right\}$. Then the number of elements in the set L_ℓ^1 at the level $\ell > 0$ can be computed as the sum of the elements in each union in the equation (4.17) as follows

$$n_\ell^1 = \left[\sum_{\lambda_1=0}^{\ell-1} n_{\lambda_1}^0 \right] + n_\ell^0 + \left[\sum_{\lambda_1=\ell-1}^{0} n_{\lambda_1}^0 \right] = 2 \left[\sum_{\lambda_1=0}^{\ell-1} n_{\lambda_1}^0 \right] + n_\ell^0 \qquad \ell > 0 \quad (4.19)$$

From (4.12) it follows that $n_0^0 = 1$ if $\lambda_1 = 0$ and $n_i^0 = 2$ if $\lambda_1 > 0$ where λ_1 is the level of the zero-dimensional set $L_{\lambda_1}^0$, therefore, λ_1 plays the role of ℓ in equation (4.12). Rewriting equation (4.19) to take into consideration the different amount of elements it gives

$$n_\ell^1 = 2 \left[n_0^0 + \sum_{\lambda_1=1}^{\ell-1} n_{\lambda_1}^0 \right] + n_\ell^0 = 2n_0^0 + n_\ell^0 + 2 \sum_{\lambda_1=1}^{\ell-1} n_{\lambda_1}^0 \qquad \ell > 0 \quad (4.20)$$

Replacing $n_0^0, n_{\lambda_1}^0$ by theirs values given given by equation (4.12) into (4.20) it allows the computation of the number of elements as

$$\begin{aligned} n_\ell^1 &= 2C(0,0) + 2C(\ell,0) + 2\sum_{i=1}^{\ell-1} C(i,0) \\ &= 4C(\ell,0) + 4C((\ell-1),1) \\ &= 4C(\ell,1) \\ &= 4 + 4(\ell-1) = 4\ell \end{aligned} \qquad (4.21)$$

where $C(0,0) = C(\ell,0)$. See section 4.7 for the combinatorics properties used here. The number of elements in the set L_ℓ^1 can be written as

$$n_\ell^1 = \begin{cases} C(\ell,0) = 1 & \text{if } \ell = 0 \\ 4\left[C(\ell,0) + C((\ell-1),1)\right] = 4C(\ell,1) = 4\ell & \text{if } \ell > 0 \end{cases} \qquad (4.22)$$

If $\ell = 0$ then there is only one point, namely $(0,0)$, which is the origin of the one-dimensional constrained system.

4.2.1.2   Computation of the number of elements in the set $L^1(\ell)$

The set $L^1(\ell)$ is obtained by the union

$$L^1(\ell) = \bigcup_{i=0}^{\ell} L_i^1 \qquad (4.23)$$

of the subsets $L_i^1, i = 0, \ldots, \ell$ whose number of elements is given by equation (4.22). The number of elements in $L^1(\ell)$ writes

$$n^1(\ell) = \sum_{i=0}^{\ell} n_i^1 \qquad (4.24)$$

Equation (4.22) gives one expression for the computation of n_0^1 and another for $n_i^1, i > 0$ where i plays the role of ℓ in equation (4.22). Considering these two cases the equation (4.24) must be written as

$$n^1(\ell) = n_0^1 + \sum_{i=1}^{\ell} n_i^1 \qquad (4.25)$$

The resulting equation after substituting n_0^1 and n_i^1 given by equation (4.22) into the equation (4.25) is

$$n^1(\ell) = n_0^1 + \sum_{i=1}^{\ell} n_i^1$$

$$= C(\ell, 0) + \sum_{i=1}^{\ell} 4C(i, 1) = C(\ell, 0) + 4C((\ell + 1), 2)$$

$$= 1 + 4\frac{\ell(\ell + 1)}{2!} = 1 + 2(\ell + 1)\ell \qquad (4.26)$$

Example 4.2.1 *Given an one-dimensional constrained system whose underlying set is $\mathbb{Z}$ find the number of elements in the level $\ell = 4$ and the number of elements from the origin until the level $\ell = 4$.*

1. Number of elements in the level $\ell = 4$.

   Equation (4.20) gives for $\ell > 0$ the value $n_4^1 = 4\ell = 4 \times 4 = 16$. This can be verified in figure 4.2.2 for $\lambda^0 = \ell = 4$.

2. The number of elements until the level ℓ is given by equation 4.26. For the level $\ell = 4$ the number is $n^1(4) = 1 + 2(\ell + 1)\ell = 1 + 2(4 + 1) \times 4 = 41$. From figure 4.2.2 it follows for the levels $\lambda^0 = \lambda_0 = \ell = 0, 1, 2, 3, 4$ that $n_0^1 = 1, n_1^1 = 4, n_2^1 = 8, n_3^1 = 12, n_4^1 = 16$, which gives $1 + 4 + 8 + 12 + 16 = 41$.

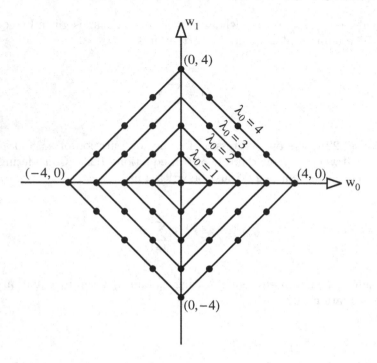

Figure 4.2.2 *Elements belonging to thee set $L^1(4)$ in a system whose underlying set is $\mathbb{Z}$.*

end of example 4.2.1

4.2.2   Underlying set $\mathbb{Z}^+$

4.2.2.1   Computation of the number of elements in the set L^1_ℓ in the level ℓ.

For the underlying set $\mathbb{Z}^+$ the equation (4.16) writes

$$L^1_\ell = \bigcup_{w_0=\lambda_0}^{0} \big\{ (w_0) \big\} \times L^0_{\lambda_1} \tag{4.27}$$

Similarly to the case where the underlying set is $\mathbb{Z}$, subsection 4.2.1, the sublevels λ_1 associated with $w_0 = \lambda_0, \ldots, 0$ are $\lambda_1 = \ell, (\ell-1), \ldots, 1, 0$. Then the set L^1_ℓ can be obtained as

$$L^1_\ell = \bigcup_{\lambda_1=0}^{\lambda_0} \big\{ (\lambda_0 - \lambda_1) \big\} \times L^0_{\lambda_1} \tag{4.28}$$

where $\lambda_0 = \ell$ and the sets $L_i^0, i = 0, \ldots, \ell$ contain one element each according to equation (4.14). The number of elements in the set L_ℓ^1 can be evaluated as

$$n_\ell^1 = \sum_{\lambda_1=0}^{\ell} n_{\lambda_1}^0 \qquad (4.29)$$

After the substitution of equation (4.14) into equation (4.29) it gives for each one-dimensional level the following number of elements

$$n_\ell^1 = \sum_{i=0}^{\ell} C(\lambda_1, 0) = C((\ell+1), 1)$$

$$= \frac{\ell+1}{1!} = \ell + 1$$

$$= \frac{\prod_{\lambda_1=1}^{1}(\ell + \lambda_1)}{1!} \qquad (4.30)$$

The number of points n_ℓ^1 can be written in terms of the coordinate numbers recalling the definition of level $\ell = |w_0| + |w_1| = w_0 + w_1$ for the nonnegative integers. The substitution of $\ell = w_0 + w_1$ into (4.30) gives the number of points (elements) per level in terms of the coordinate numbers as

$$n_\ell^1 = \frac{w_0 + w_1 + 1}{1!} \qquad (4.31)$$

Example 4.2.3 *Computation of the elements in the set L_4^1*

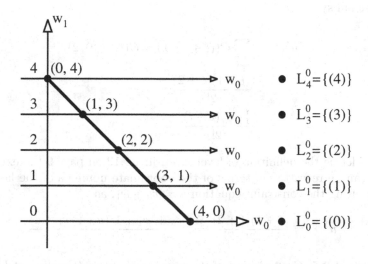

Figure 4.2.4 *Each point on the right represents a set in the zero-dimensional space to be associated to the corresponding w_0 coordinate.*

Figure 4.2.4 shows the number of points for the level $\ell = 4$, that is, $n_4^1 = 1 + 4 = 5$. Each point on the right of figure 4.2.4 represents a set in the zero-dimensional space whose sublevel is $\lambda_1 = \ell - w_0$, to be associated to the corresponding $w_0 = 4, 3, 2, 1, 0$ coordinate. Therefore, the sublevels, associated with w_0 are $\lambda_1 = 0, 1, 2, 3, 4$. There is no need to consider the absolute value of the coordinates since they are all positive by definition of the underlying set $\mathbb{Z}^+$.

end of example 4.2.3

4.2.2.2   Computation of the number of elements in the set $L^1(\ell)$

The set $L^1(\ell)$ is obtained by the union

$$L^1(\ell) = \bigcup_{i=0}^{\ell} L_i^1 \tag{4.32}$$

of the subsets $L_i^1, i = 0, \ldots, \ell$ whose number of elements is given by equation (4.30). The number of elements in $L^1(\ell)$ writes

$$n^1(\ell) = \sum_{i=0}^{\ell} n_i^1 \tag{4.33}$$

Substituting n_ℓ^0 given by equation (4.30) into equation (4.33) it gives the number of the elements from the level zero until the level ℓ for the one-dimensional system as

$$
\begin{aligned}
n^1(\ell) &= \sum_{i=0}^{\ell} C((\ell+1), 1) = C((\ell+2), 2) \\
&= \frac{(\ell+1)(\ell+2)}{2!} \\
&= \frac{\prod_{i=1}^{2}(\ell+i)}{2!}
\end{aligned}
\tag{4.34}
$$

Considering the definition of level given in 1.1.12 on page 9 the expression (4.34) can be rewritten in terms of the coordinate numbers of the last level, that is, using the constraint equation $\ell = w_0 + w_1$, as

$$n^1(\ell) = \frac{(\ell+1)(\ell+2)}{2!} = \frac{(w_0 + w_1 + 1)(w_0 + w_1 + 2)}{2!} \tag{4.35}$$

Example 4.2.5 *Computation of the number of elements in the set $L^1(4)$*

Figure 4.2.6 shows the total number the points $n^1(4)$ for the one-dimensional system in the level $\ell = 4$.

The number of elements in the set $L^1(4)$ is

$$n^1(4) = \frac{(4+1)(4+2)}{2!} = 15 \qquad (4.36)$$

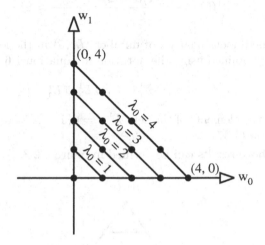

Figure 4.2.6 *Total number the points $n^1(4)$ in the set L_4^1.*

end of example 4.2.5

Example 4.2.7 *Evaluate the number of points in each level and the total number of points in the level $\ell = 5$ of a one-dimensional system with $\mathbb{Z}^+$ as the underlying set.*

1. The number of points per level, that is, the number of elements in the set L_i^1, is obtained using the equation (4.30). These number of points are $1, 2, 3, 4, 5, 6$ from level zero to level $\ell = 5$. The respective coordinate numbers are

$$L_0^1 = \{(0,0)\}$$
$$L_1^1 = \{(1,0),(0,1)\}$$
$$L_2^1 = \{(2,0),(1,1),(0,2)\}$$
$$L_3^1 = \{(3,0),(2,1),(1,2),(0,3)\}$$
$$L_4^1 = \{(4,0),(3,1),(2,2),(1,3),(0,4)\}$$
$$L_5^1 = \{(5,0),(4,1),(3,2),(2,3),(1,4),(0,5)\} \qquad (4.37)$$

The sets above contains zero-dimensional constrained elements with coordinates in a two-dimensional cartesian axes. It is analog to a line, which is is a one-dimensional geometric entity, on a plane which is a two-dimensional space.

2. Total number of points from the level zero until the level
$\ell = 5$, that is, the number of elements in the set $L^1(5)$, is
given by equation (4.34) on page 144 and it evaluates to

$$n^1(5) = \frac{(\ell+1)(\ell+2)}{2} = \frac{6 \times 7}{2} = 21 \qquad (4.38)$$

The one-dimensional set of numbers $L^1(5)$ in the level $\ell = 5$
can be obtained using the notation in equation 1.6 as

$$L^1(5) = L_0^1 \cup L_1^1 \cup L_2^1 \cup L_3^1 \cup L_4^1 \cup L_5^1 \qquad (4.39)$$

where the elements of the sets $L_i^1, i = 0, 1, \ldots, 5$ are given by
equation (4.37).

The above results can be verified in figure 4.2.8.

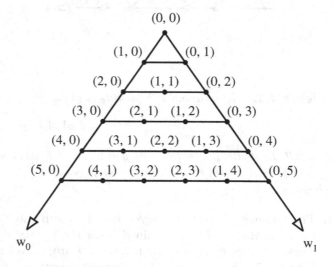

Figure 4.2.8 *Elements of the set $L^1(5)$, which contains
the elements from the level $\ell = 0$ until the level $\ell = 5$.*

end of example 4.2.7

The coordinate numbers belonging to the one-dimensional space whose underlying set is $\mathbb{Z}^+$ can be represented graphically by points in a triangle,
fiigure 4.2.8, and it is known as the *Pascal triangle*.

Example 4.2.9 *Compute the number of points in the set L_3^1 represented geometrically in a two-dimensional cartesian space shown in figure 4.2.10 and in the set L_4^1, which is represented in figure 4.2.11.*

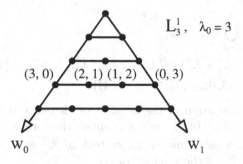

Figure 4.2.10 *Set $L_3^1 = \{(3,0),(2,1),(1,2),(0,3)\}$ of points.*

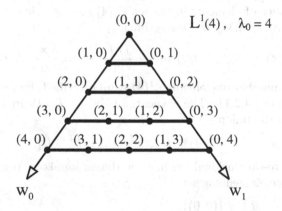

Figure 4.2.11 *The elements of the set $L^1(4)$ belong to a one-dimensional constrained space.*

1. Figure 4.2.10 shows a one-dimensional constrained set of numbers represented in a two-dimensional cartesian space. Both coordinate numbers assume the values $0, 1, 2, 3$ with $w_0 = 0$ on the w_0 axis and $w_1 = 0$ on the w_1 axis. Thus each point can be referred to by using the coordinate w_0 or w_1 along the line connecting the zero values of each coordinate. The points associate to these coordinates are identified by (w_0, w_1) with the additional property that $w_0 + w_1 = \lambda_0$, For example, the point $(2, 1)$ has the property that $\lambda_0 = 2 + 1 = 3$.

The elements in the set L_3^1 are points of one-dimensional constrained space written with coordinates in a two-dimensional cartesian space. They can be constructed by the union of the cartesiam product of zero-dimensional constrained numbers as

$$L_3^1 = \bigcup_{i=0}^{3} L_{3-i}^0 \times L_i^0$$

$$= \{L_3^0 \times L_0^0\} \cup \{L_2^0 \times L_1^0\} \cup \{L_1^0 \times L_2^0\} \cup \{L_0^0 \times L_3^0\}$$

$$= \{(3,0),(2,1),(1,2),(0,3)\} \tag{4.40}$$

The computation of the number of elements is made using the expression (4.11), which evaluate the total number of elements of a zero-dimensional system for $\mathbb{Z}^+$ as the underlying set. For $\ell = 3$ those numbers are

$$n^0(3) = C((\ell+1),1) = \ell+1 = C(4,1) = 4 \tag{4.41}$$

2. Figure 4.2.11 shows a one-dimensional constrained set of numbers represented in a two-dimensional cartesian space. The number of elements in the set $L^1(4)$ can be calculated using equation (4.34) and the result is

$$n^1(4) = C((\ell+2),2) = \frac{(\ell+1)(\ell+2)}{2!} = \frac{5 \times 6}{2} = 15 \tag{4.42}$$

The number of elements $n^1(4)$ can be verified, by inspection, in figure 4.2.11. The elements of the set $L^1(4)$ are obtained from the union

$$L^1(4) = L_0^1 \cup L_1^1 \cup L_2^1 \cup L_3^1 \cup L_4^1 \tag{4.43}$$

of zero-dimensional sets in a one-dimensional constrained space whose elements are

$$L_0^1 = \{(0,0)\},$$
$$L_1^1 = \{(1,0),(0,1)\},$$
$$L_2^1 = \{(2,0),(1,1),(0,2)\},$$
$$L_3^1 = \{(3,0),(2,1),(1,2),(0,3)\},$$
$$L_4^1 = \{(4,0),(3,1),(2,2),(1,3),(0,4)\}. \tag{4.44}$$

Substituting (4.44) into (4.43) it gives the set

$$L^1(4) = \{ (0,0),(1,0),(0,1),(2,0),(1,1),(0,2),$$
$$(3,0),(2,1),(1,2),(0,3),$$
$$(4,0),(3,1),(2,2),(1,3),(0,4) \} \tag{4.45}$$

end of example 4.2.9

4.2.3 Summary for the one-dimensional numbers

1. For the underlying set $\mathbb{Z}$ the number of elements is given by

   (a) From equation (4.22) the number of elements in the set L_ℓ^1 in the level $\ell \geq 0$ is

   $$n_\ell^1 = \begin{cases} C(\ell, 0) = 1 & \text{if} \quad \ell = 0 \\ 4C(\ell, 1) = 4\ell & \text{if} \quad \ell > 0 \end{cases} \qquad (4.46)$$

   (b) From equation (4.26) the number of elements in the sets $L^1(\ell)$ from the level zero through the level ℓ is given by

   $$\begin{aligned} n^1(\ell) &= C(\ell, 0) + 4C((\ell+1), 2) \\ &= 1 + 2(\ell+1)\ell \qquad \ell \geq 0 \end{aligned} \qquad (4.47)$$

2. If the underlying set is $\mathbb{Z}^+$ then the number of elements are given by

   (a) Number of elements in the level ℓ

   From equation (4.30) the number of elements in the set L_ℓ^1 in the level $\ell \geq 0$ is

   $$n_\ell^1 = C((\ell+1), 1) = \frac{\ell+1}{1!} \qquad \ell \geq 0 \qquad (4.48)$$

   (b) From equation (4.34) the number of elements in the sets $L^1(\ell)$ from the level zero through the level ℓ is given by

   $$\begin{aligned} n^1(\ell) &= C((\ell+2), 2) = \frac{(\ell+1)(\ell+2)}{2!} \qquad \ell \geq 0 \\ &= \frac{\prod_{i=1}^2 (\ell+i)}{2!} \end{aligned} \qquad (4.49)$$

4.3   Two-dimensional space

4.3.1   Underlying set $\mathbb{Z}$

4.3.1.1   Computation of the number of elements in the set L_ℓ^2 in the level ℓ for the underlying set $\mathbb{Z}$.

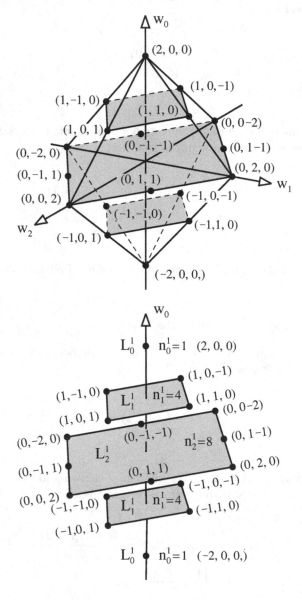

Figure 4.3.1 *Geometric representation of the set of numbers L_2^2 in the level $\ell = 2$ in a two-dimensional $\delta = 2$ constrained number system represented with $d = 3$ cartesian coordinates.*

The two-dimensional space can be constructed as a sequence of one-dimensional spaces for the case of $\ell = 2$ as show in figure 4.3.1. To each coordinate number $w_2 = \ell, \ldots, 0, \ldots, -\ell$ there is associated a one-dimensional set along the plane $w_0 \times w_1$ and in the sublevel $\lambda_1 = \ell - |w_0|$. The elements in the set L_ℓ^2 are obtained performing the cartesian product

$$L_\ell^2 = \bigcup_{w_0 = -\ell}^{\ell} \{(w_0)\} \times L_{\lambda_1}^1 \tag{4.50}$$

These one-dimensional spaces $L_{\lambda_1}^1$ are represented in figure 4.3.1 by a shaded rectangle. For example, the points $(0, 1, -1)$ and $(0, 2, 0)$ belong to the one-dimensional space in the plane $w_1 \times w_2$ defined by $(w_1, w_2) = (2, 0)$, $(w_1, w_1) = (0, 2)$, and the coordinate number $w_0 = 0$. Figure 4.3.1 shows at its right the one-dimensional spaces corresponding to each w_0 coordinate number. The coordinate numbers in the plane parallel to $w_1 \times w_2$ belong to the sublevel $\lambda_1 = \lambda^0 - |w_0|$.

The sublevels λ_1 associated to $w_0 = \ell, \ldots, -1, 0, 1, \ldots, -\ell$ are

$$\lambda_1 = 0, 1, \ldots, (\ell - 1), \ell, (\ell - 1), \ldots, 1, 0 \tag{4.51}$$

Then the set L_ℓ^2 can be obtained rewriting equation (4.18) with the indices transformation $\lambda_1 = \ell - |w_0|$ as

$$L_\ell^2 = \begin{cases} \{(0)\} \times L_{\lambda_0}^1 & \lambda_0 = 0 \\ \left\{ \displaystyle\bigcup_{\lambda_1=0}^{\ell-1} \{(\ell - \lambda_1)\} \times L_{\lambda_1}^1 \right\} \cup \{\{(0)\} \times L_{\lambda_1}^1\} \cup \\ \left\{ \displaystyle\bigcup_{\lambda_1=\ell-1}^{0} \{-(\ell - \lambda_1)\} \times L_{\lambda_1}^1 \right\} & \lambda_0 > 0 \end{cases} \tag{4.52}$$

Since the coordinate number w_0 assume the values $w_2 = \ell, \ldots, 0, \ldots, -\ell$ this implies that there are $(2\ell + 1)$ coordinate numbers along the axis w_0. The number of elements in the level ℓ can be written as

$$n_\ell^2 = \left[\sum_{\lambda_1=0}^{\ell-1} n_{\lambda_1}^1 \right] + n_\ell^1 + \left[\sum_{\lambda_1=\ell-1}^{0} n_{\lambda_1}^1 \right] \tag{4.53}$$

Among the subsets along the axis w_0 two have only one elements, they are the set L_0^1 associated to $w_0 = \ell$ and the set L_0^1 associated to $w_0 = -\ell$. The number of elements in the set L_0^1 is denoted as n_0^1 and given by equation (4.46). All other subsets have 4ℓ points. Rewriting equation (4.53) to consider the different number of elements in the sets it follows

$$n_\ell^2 = 2 \left[\sum_{\lambda_1=0}^{\ell-1} n_{\lambda_1}^1 \right] + n_\ell^1 = 2 \left[n_0^1 + \sum_{\lambda_1=1}^{\ell-1} n_{\lambda_1}^1 \right] + n_\ell^1 \tag{4.54}$$

Substituting n_0^1 and $n_{\lambda_1}^1$ given by equation (4.46) into equation (4.54)

$$n_\ell^2 = 2\left[C(\ell,0) + \sum_{\lambda_1=1}^{\ell-1} 4C(\lambda_1,1)\right] + 4C(\ell,1)$$

$$= 2\left[C(\ell,0) + 4C(\ell,2)\right] + 4C(\ell,1)$$

$$= 2 + 8\frac{\ell(\ell-1)}{2!} + 4\ell \tag{4.55}$$

$$= 2 + 4(\ell-1)\ell + 4\ell = 2 + 4\ell^2 \qquad \ell > 0$$

Therefore,

$$n_\ell^2 = \begin{cases} C(\ell,0) = 1 & \text{if } \ell = 0 \\ 2C(\ell,0) + 4C(\ell,1) + 8C(\ell,2) = 2 + 4\ell^2 & \text{if } \ell > 0 \end{cases} \tag{4.56}$$

4.3.1.2   Computation of the number of elements in the set $L^2(\ell)$ for the underlying set $\mathbb{Z}$

The set $L^2(\ell)$ is obtained by the union

$$L^2(\ell) = \bigcup_{i=0}^{\ell} L_i^2 \tag{4.57}$$

of the subsets $L_i^2, i = 0,\ldots,\ell$ whose number of elements is given by equation (4.56). The number of elements in $L^2(\ell)$ writes

$$n^2(\ell) = \sum_{i=0}^{\ell} n_i^2 \tag{4.58}$$

Equation (4.56) gives one expression for the computation of n_0^2 and another for $n_i^2, i > 0$. Considering this, the equation (4.58) must be rewritten as

$$n^2(\ell) = \begin{cases} n_0^2 & \ell = 0 \\ n_0^2 + \sum_{i=1}^{\ell} n_i^2 & \ell > 0 \end{cases} \tag{4.59}$$

Substituting n_0^2 and n_i^2 given by equation (4.56) into equation (4.59) it gives the total number of points $n^2(\ell)$ of a two-dimensional constrained system from level zero until the level ℓ is given by

$$n^2(\ell) = 1 + \sum_{i=1}^{\ell}(2 + 4i^2)$$

$$= 1 + (2 + 4 \times 1) + (2 + 4 \times 2^2) + (2 + 4 \times 3^2) + \cdots$$

$$= 1 + (2 + 2 + 2 + \cdots) + 4(1^2 \times 2^2 \times 3^2 + \cdots)$$

$$= 1 + 2\ell + 4\sum_{i=1}^{\ell} i^2 \qquad \ell > 0 \tag{4.60}$$

Considering that, $i = 0$ will not change the value of $\sum_{i=1}^{\ell} i^2$ the summation can be written starting from $i = 0$, which yields

$$n^2(\ell) = 1 + 2\ell + 4\sum_{i=0}^{\ell} i^2 \qquad \ell > 0 \tag{4.61}$$

The number of elements can also be evaluated using the algebra of combinations, that is,

$$n^2(\ell) = C(\ell, 0) + \sum_{i=1}^{\ell}(2C(i,0) + 4C(i,1) + 8C(i,2))$$

$$= C(\ell, 0) + 2C(\ell, 1) + 4C((\ell+1), 2) + 8C((\ell+1), 3)$$

$$= 1 + 2\ell + 4\left[\frac{\ell(\ell+1)}{2!} + 2\frac{(\ell-1)\ell(\ell+1)}{3!}\right] \tag{4.62}$$

$$= 1 + 2\ell + 4\left[\frac{3\ell(\ell+1)}{6} + 2\frac{(\ell-1)\ell(\ell+1)}{6}\right]$$

$$= 1 + 2\ell + 4\left[\frac{\ell(\ell+1)[3 + 2(\ell-1)]}{6}\right]$$

$$= 1 + 2\ell + 4\left[\frac{\ell(\ell+1)(2\ell+1)}{6}\right] \qquad \ell \geq 0$$

Example 4.3.2 *Given the set L_2^2 shown in figure 4.3.1 find the number of elements.*

The set L_2^2 belongs to a two-dimensional space and it is in the level $\ell = 2$. Substituting into expression (4.56)

$$n_2^2 = 2 + 4\ell^2 = 2 + 4 \times 2^2 = 18 \tag{4.63}$$

end of example 4.3.2

Example 4.3.3 *Find the number of elements in the set $L^2(2)$ if $\mathbb{Z}$ is the underlying set.*

The set $L^2(2)$ belongs to a two-dimensional space and its subsets are

$$L^2(2) = L_0^2 \cup L_1^2 \cup L_2^2 \tag{4.64}$$

The number of elements $n^2(2)$ in the set $L^2(2)$ is

$$n^2(2) = n_0^2 + n_1^2 + n_2^2 \tag{4.65}$$

where

$$
\begin{aligned}
n_0^2 &= 1 \\
n_1^2 &= 2 + 4\ell = 2 + 4 \times 1^2 = 6 \\
n_2^2 &= 2 + 4\ell = 2 + 4 \times 2^2 = 18
\end{aligned}
\tag{4.66}
$$

Substituting into (4.65)

$$n^2(2) = n_0^2 + n_1^2 + n_2^2 = 1 + 6 + 18 = 25 \tag{4.67}$$

The number of elements in the set L_1^2 in level is $\ell = 1$ is 6 and it can be seen in figure 4.3.4. For the set L_2^2 in level $\ell = 2$ the number of elements is 18 and it can be seen in figure 4.3.1.

The number of elements can also be obtained from equation (4.62) as

$$
\begin{aligned}
n^2(2) &= 1 + 2\ell + 4 \left[\frac{\ell(\ell+1)(2\ell+1)}{6} \right] \\
&= 1 + 4 + 4 \left[\frac{2(2+1)(2 \times 2 + 1)}{6} \right] = 1 + 4 + 20 = 25
\end{aligned}
\tag{4.68}
$$

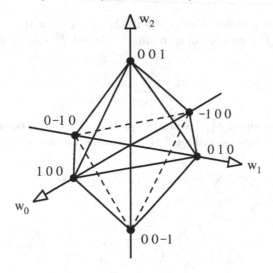

Figure 4.3.4 *Geometric representation of the set of numbers L_1^2 in the level $\ell = 1$ in a two-dimensional $\delta = 2$ constrained number system with $d = 3$ cartesian coordinates.*

end of example 4.3.3

4.3.2 Underlying set $\mathbb{Z}^+$

4.3.2.1 Computation of the number of elements in the set L_ℓ^2 for the underlying set $\mathbb{Z}^+$

For the underlying set $\mathbb{Z}^+$ the equation (4.27) writes

$$L_\ell^2 = \bigcup_{w_0 = \lambda_0}^{0} \{(w_0)\} \times L_{\lambda_1}^1 \tag{4.69}$$

Similarly, to the case where the underlying set is $\mathbb{Z}$, subsection 4.3.1, the sublevels λ_1 associated with $w_0 = \lambda_0, \ldots, 0$ are $\lambda_1 = \ell, (\ell - 1), \ldots, 1, 0$. Then the set L_ℓ^2 can be obtained as

$$L_\ell^2 = \bigcup_{\lambda_1 = 0}^{\lambda_0} \{(\lambda_0 - \lambda_1)\} \times L_{\lambda_1}^1 \tag{4.70}$$

where $\lambda_0 = \ell$. The number of elements in the sets $L_i^1, i = 0, \ldots, \ell$ in given by equation (4.48). The number of elements in the set L_ℓ^2 can be evaluated as

$$n_\ell^2 = \sum_{\lambda_1=0}^{\ell} n_{\lambda_1}^1 \tag{4.71}$$

After the substitution of $n_{\lambda_1}^1$ given by equation (4.48) into equation (4.70) it gives for each two-dimensional level the following number of elements

$$n_\ell^2 = \sum_{i=0}^{\ell} C((i+1), 1) = C((\ell+2), 2)$$

$$= \frac{(\ell+1)(\ell+2)}{2!}$$

$$= \frac{\prod_{i=1}^2 (\ell + i)}{2!} \qquad \ell > 0 \tag{4.72}$$

Note that $n_\ell^2 = n^1(\ell)$.

The number of elements in the set L_ℓ^2 is given by

$$n_\ell^2 = \begin{cases} C(\ell, 0) = 1 & \text{if } \ell = 0 \\ C((\ell+2), 2) = \dfrac{(\ell+1)(\ell+2)}{2!} & \text{if } \ell > 0 \end{cases} \tag{4.73}$$

Example 4.3.5

The elements in the set $L^2(5)$, which are numbers in a two-dimensional space, level $\ell = 5$ and underlying set $\mathbb{Z}^+$, are represented graphically by points in a pyramid such as that shown in figure 4.3.6.

Each triangle corresponds to the sets $L_{\lambda_0}^2$ in a two-dimensional constrained space in the levels $\lambda_0 = 0, 1, 2, 3, 4, 5$. Each line on the triangles represents the sets $L_{\lambda_1}^1$ in a one-dimensional constrained space, and each point on a line represents the sets $L_{\lambda_2}^0$ in a zero-dimensional constrained space.

For example, for the level $\lambda_0 = 5$ the zero-dimensional space have the first sublevel $\lambda_1 = 0, 1, 2, 3, 4, 5$. Along the first sublevel $\lambda_1 = 4$, there are the second sublevels $\lambda_2 = 0, 1, 2, 3, 4$. The point $w = (w_0, w_1, w_2) = (1, 2, 2)$ has level $\lambda_0 = 1+2+2 = 5$, first sublevel $\lambda_1 = 2 + 2 = 4$, and second sublevel $\lambda_2 = 2$. Therefore, $(w_0, w_1, w_2) = (1, 2, 2) \in L_5^2$ and $(w_0, w_1, w_2) = (1, 2, 2) \in L^2(5)$ since $L_5^2 \subset L^2(5)$. The first sublevel of w, which has the value $\lambda_1 = 4$, is $(w_1, w_2) = (2, 2) \in L_4^1$, and the second sublevel of w, denoted by $\lambda_2 = 2$, is $(w_2) = (2) \in L_2^0$.

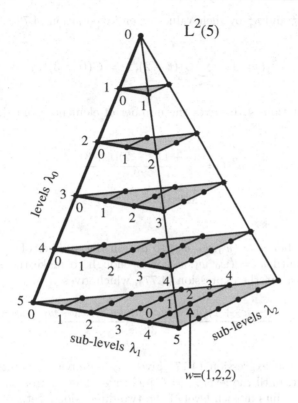

Figure 4.3.6 *Each horizontal triangle represents a one-dimensional sublevel. Each line on the triangle represents a zero-dimensional sublevel.*

end of example 4.3.5

4.3.2.2 Computation of the number of elements in the set $L^2(\ell)$ for the underlying set $\mathbb{Z}^+$

The set $L^2(\ell)$ is obtained by the union

$$L^2(\ell) = \bigcup_{i=0}^{\ell} L_i^2 \tag{4.74}$$

of the subsets $L_i^2, i = 0, \ldots, \ell$ whose number of elements is given by equation (4.72). The number of elements in $L^2(\ell)$ writes

$$n^2(\ell) = \sum_{i=0}^{\ell} n_i^2 = n_0^2 + \sum_{i=1}^{\ell} n_i^2 \tag{4.75}$$

Substituting n_0^2 and n_i^2 by their values given by equation (4.72) it follows

$$n^2(\ell) = 1 + \sum_{i=1}^{\ell} C((\ell+2), 2) = C((\ell+3), 3) \qquad (4.76)$$

Expanding equation (4.76) gives the number of elements from the level zero until the level ℓ as

$$n^2(\ell) = \frac{(\ell+1)(\ell+2)(\ell+3)}{3!}$$
$$= \frac{\prod_{i=1}^{3}(\ell+i)}{3!} \qquad \ell \geq 0 \qquad (4.77)$$

Given a number $w = (w_0, w_1, w_2)$ to obtain the number of elements from $w = (0,0,0)$ until $w = (w_0, w_1, w_2)$ it is enough to make the substitution $\ell = w_0 + w_1 + w_2$ in the expression (4.77), which gives

$$n^2(\ell) = \frac{(w_0 + w_1 + w_2 + 1)(w_0 + w_1 + w_2 + 2)(w_0 + w_1 + w_2 + 3)}{3!} \qquad (4.78)$$

Recall that the expression (4.77) gives the number of points of a two-dimensional set, until the level $\lambda_0 = \ell$, inclusive. The expression (4.72) gives the number of points in each level of the two-dimensional set.

Example 4.3.7

Figure 4.3.8 is the geometric representation of the set $L^2(3)$ of two-dimensional constrained numbers in the level $\ell = 3$ where each point presents an element of the set. The number of its points can be computed using the equation (4.77) as

$$n^2(3) = \frac{(\ell+1)(\ell+2)(\ell+3)}{3!}$$
$$= \frac{(3+1)(3+2)(3+3)}{3!} = \frac{4 \times 5 \times 6}{6} = 20$$

This same number can be obtained from figure 4.3.8, by inspection, that is, $n^2(3) = 1 + 3 + 6 + 10 = 20$. The number of elements in the set L_2^2 is given by

$$n_2^2 = \frac{(\ell+1)(\ell+2)}{2!} = \frac{(2+1)(2+2)}{2!} = \frac{3 \times 4}{2} = 6$$

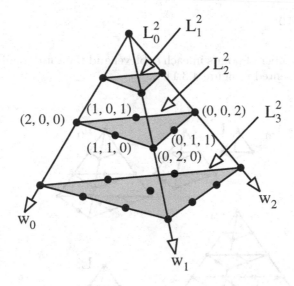

Figure 4.3.8 *Geometric representation of the set* $L^2(3)$
with $n^2(3) = 20$ *elements. The number of elements in the*
subset L^2_2 *is 6.*

end of example 4.3.7

Example 4.3.9

Figure 4.3.6 is the geometric representation of a two-dimensional
set of elements in the set $L^2(5)$. The number of its elements can
be computed from (4.77) as

$$n^2(5) = \frac{(\ell+1)(\ell+2)(\ell+3)}{3!}$$

$$= \frac{(5+1)(5+2)(5+3)}{3!} = \frac{6 \times 7 \times 8}{6} = 56$$

From figure 4.3.6, adding the number of points of each triangle
$n^2(5) = 1 + 3 + 6 + 10 + 15 + 21 = 56$. The number of elements in
the set n^2_5, whose level is $\lambda_0 = 5$, can be obtained from equation
(4.72) and it is given by

$$n^2_5 = \frac{(\ell+1)(\ell+2)}{2!} = \frac{(5+1)(5+2)}{2!} = \frac{6 \times 7}{2} = 21$$

end of example 4.3.9

Example 4.3.10

Evaluate the number of points in each sublevel and the total number of points in the set represented in figure 4.3.11

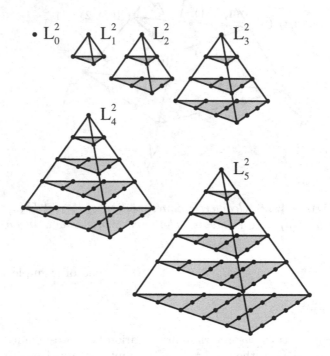

Figure 4.3.11 *Each pyramid represents a two-dimensional sublevel. Each horizontal triangle represents a one-dimensional sublevel. Each line on the triangle represents a zero dimensional sublevel.*

1. Number of elements in each sublevel

    The number of elements in each sublevel can be evaluated using equation (4.72) as follows

    (a) The number of elements in the set L_0^2, level $\lambda_0 = \ell = 0$ is

    $$n_0^2 = \frac{(0+1)(0+2)}{2!} = 1$$

    and the elements in the set are

    $$L_0^2 = \left\{ (0,0,0) \right\}$$

(b) The number of elements in the set L_1^2, level $\lambda_0 = \ell = 1$ is

$$n_1^2 = \frac{(1+1)(1+2)}{2!} = 3$$

and the elements in the set are

$$L_1^2 = \{(1,0,0),(0,1,0),(0,0,1)\}$$

(c) The number of elements in the set L_2^2, level $\lambda_0 = \ell = 2$ is

$$n_2^2 = \frac{(2+1)(2+2)}{2!} = 6$$

and the elements in the set are

$$L_2^2 = \{(2,0,0),(1,1,0),(1,0,1),(0,2,0),(0,1,1),(0,0,2)\}$$

(d) The number of elements in the set L_4^2, level $\lambda_0 = \ell = 3$ is

$$n_3^2 = \frac{(3+1)(3+2)}{2!} = 10$$

and the elements in the set are

$$L_3^2 = \{(3,0,0),(2,1,0),(2,0,1),(1,2,0),(1,1,1),(1,0,2),$$
$$(0,3,0),(0,2,1),(0,1,2),(0,0,3)\}$$

(e) The number of elements in the set L_4^2, level $\lambda_0 = \ell = 4$ is

$$n_4^2 = \frac{(4+1)(4+2)}{2!} = 15$$

and the elements in the set are

$$L_4^2 = \{(4,0,0),(3,1,0),(3,0,1),(2,2,0),(2,1,1),$$
$$(2,0,2),(1,3,0),(1,2,1),(1,1,2),(1,0,3),(0,4,1),$$
$$(0,3,1),(0,2,2),(0,1,3),(0,0,4)\}$$

(f) The number of elements in the set L_5^2, level $\lambda_0 = \ell = 5$ is

$$n_5^2 = \frac{(5+1)(5+2)}{2!} = 21$$

and the elements in the set are

$$L_4^2 = \{(5,0,0),(4,1,0),(4,0,1),(3,2,0),(3,1,1),(3,0,2),(2,3,0),$$
$$(2,2,1),(2,1,2),(2,0,3),(1,4,0),(1,3,1),(1,2,2),(1,1,3),$$
$$(1,0,4),(0,5,0),(0,4,1),(0,3,2),(0,2,3),(0,1,4),(0,0,5)\}$$

2. Total number of points from $\ell = 0$ until $\ell = 5$ is given by

$$n^2(5) = \frac{(\ell+1)(\ell+2)(\ell+3)}{3!} = \frac{(5+1)(5+2)(5+3)}{3!} = 56$$

which equals the partial sum

$$n^2(5) = n_0^2 + n_1^2 + n_2^2 + n_3^2 + n_4^2 + n_5^2$$
$$= 1 + 3 + 6 + 10 + 15 + 21 = 56$$

This sum contains the elements of the union

$$L^2(5) = L_0^2 \cup L_1^2 \cup L_2^2 \cup L_3^2 \cup L_4^2 \cup L_5^2$$

that is the definition of the set $L^2(5)$.

end of example 4.3.10

4.3.3   Summary for the two-dimensional numbers

1. For the underlying set $\mathbb{Z}$ the number of elements is given by

   (a) From equation (4.56) the number of elements in the set L_ℓ^2 in the level ℓ is

   $$n_\ell^2 = \begin{cases} C(\ell,0) = 1 & \text{if } \ell = 0 \\ 2C(\ell,0) + 4C(\ell,1) + 8C(\ell,2) = 2 + 4\ell^2 & \text{if } \ell > 0 \end{cases}$$
   $$(4.79)$$

   (b) From equation (4.62) the number of elements in the set $L^2(\ell)$ from the level zero through the level ℓ is given by

   $$n^2(\ell) = \begin{cases} C(\ell,0) = 1 \quad \text{if } \ell = 0 \\ \\ 1 + 2\ell + 4\displaystyle\sum_{i=0}^{\ell} i^2 = C(\ell,0) + 2C(\ell,1) + 4C((\ell+1),2) \\ \qquad\qquad + 8C((\ell+1),3) \\ \qquad = 1 + 2\ell + 4\left[\dfrac{\ell(\ell+1)(2\ell+1)}{3!}\right] \\ \qquad\qquad \text{if } \ell > 0 \end{cases}$$
   $$(4.80)$$

2. For the underlying set $\mathbb{Z}^+$ the number of elements is given by

(a) From equation (4.72) the number of elements in the set L_ℓ^2 in the level ℓ is

$$n_\ell^2 = C((\ell+2), 2) = \frac{(\ell+1)(\ell+2)}{2!} \quad \text{if} \quad \ell \geq 0 \qquad (4.81)$$

(b) From equation (4.77) the number of elements in the set $L^2(\ell)$ from the level zero through the level ℓ is given by

$$n^2(\ell) = C((\ell+3), 3) = \frac{(\ell+1)(\ell+2)(\ell+3)}{3!} \quad \text{if} \quad \ell \geq 0 \quad (4.82)$$

4.4   Three-dimensional constrained space

4.4.1   Underlying set $\mathbb{Z}$

4.4.1.1   Computation of the number of elements in the set L_ℓ^3 in the level ℓ for the underlying set $\mathbb{Z}$.

In accordance with the generation rules in subsection 2.2 on page 33 the coordinate number w_0 assumes the a sequence of values $w_0 = \lambda_0, \ldots, 0, \ldots, -\lambda_0$, which gives a total of $(2\lambda_0 + 1)$ coordinate numbers of the type w_0. Equation (2.15) associates to each w_0 a two-dimensional set in the level $\lambda_1 = \ell - |w_0| = \lambda_0 - |w_0|$, which for $\delta = 3$ writes

$$L_\ell^3 = \bigcup_{w_0=\lambda_0}^{-\lambda_0} \{(w_0)\} \times L_{\lambda_1}^3 \qquad (4.83)$$

The sublevels λ_1 associated to $w_0 = \lambda_0, \ldots, 1, 0, -1, \ldots, \lambda_0$ are $\lambda_1 = 0, 1, \ldots, (\ell - 1), \ell, (\ell - 1), \ldots, 1, 0$. Then the set L_ℓ^2 can be obtained rewriting equation (4.18) with the indices transformation $\lambda_1 = \ell - |w_0|$ as

$$L_\ell^3 = \begin{cases} \{(0)\} \times L_0^2 & \ell = 0 \\ \left[\displaystyle\bigcup_{\lambda_1=0}^{\ell-1} \{(\ell-\lambda_1)\} \times L_{\lambda_1}^2\right] \cup [\{(0)\} \times L_{\lambda_1}^2] \cup \left[\displaystyle\bigcup_{\lambda_1=\ell-1}^{0} \{-(\ell-\lambda_1)\} \times L_{\lambda_1}^2\right] \\ \hfill \ell > 0 \end{cases}$$

$$(4.84)$$

if $w_0 = 0$ then $\lambda_1 = \ell - |w_0| = \ell$, which is the value of the index λ_1 in the cartesian product $[\{(0)\} \times L_{\lambda_1}^2]$ in equation (4.84).

Then the number of elements in the level ℓ can be written as

$$n_\ell^3 = \begin{cases} n_0^2 & \ell = 0 \\ \left[\displaystyle\sum_{\lambda_1=0}^{\ell-1} n_{\lambda_1}^2 \right] + n_\ell^2 + \left[\displaystyle\sum_{\lambda_1=0}^{\ell-1} n_{\lambda_1}^2 \right] & \ell > 0 \end{cases} \qquad (4.85)$$

For $\ell > 0$ the number of elements can be evaluated according to

$$n_\ell^3 = \left[\sum_{\lambda_1=0}^{\ell-1} n_{\lambda_1}^2 \right] + n_\ell^2 + \left[\sum_{\lambda_1=0}^{\ell-1} n_{\lambda_1}^2 \right] = 2 \left[n_0^2 + \sum_{\lambda_1=1}^{\ell-1} n_{\lambda_1}^2 \right] + n_\ell^2 \qquad \ell > 0 \quad (4.86)$$

The substitution of $n_{\lambda_1}^2$ obtained in equation (4.79) into equation (4.86) gives an expression for the computation of n_ℓ^3 as

$$\begin{aligned} n_\ell^3 &= 2 \left[1 + \sum_{i=0}^{\ell-1} (2 + 4i^2) \right] + (2 + 4\ell^2) \\ &= 2 \left[1 + 2(\ell-1) + 4 \sum_{i=0}^{\ell-1} i^2 \right] + (2 + 4\ell^2) \\ &= 2 + 4(\ell-1) + 8 \sum_{i=0}^{\ell-1} i^2 + (2 + 4\ell^2) \\ &= 4 \left(\ell + \ell^2 + 2 \sum_{i=0}^{\ell-1} i^2 \right) \qquad \ell > 0 \end{aligned} \qquad (4.87)$$

Substituting into equation (4.87) the expression of the sum of squares given by equation (4.147) it follows

$$\begin{aligned} n_\ell^3 &= 4 \left[\ell + \ell^2 + 2 \sum_{i=0}^{\ell-1} i^2 \right] \\ &= 4 \left[\ell(\ell+1) + 2\frac{(\ell-1)\ell(\ell+1)}{3!} \right] = \frac{4\ell}{3} \left[3\ell(\ell+1) + (\ell-1)(\ell+1) \right] \\ &= \frac{4\ell}{3} \left[3\ell + 3 + 2\ell^2 - 3\ell + 1 \right] = \frac{4\ell}{3} \left[4 + 2\ell^2 \right] = \frac{8\ell(2 + \ell^2)}{3} \end{aligned} \qquad (4.88)$$

Using the algebra of combinations

$$n_\ell^3 = 2 \left[n_0^2 + \sum_{i=1}^{\ell-1} n_i^2 \right] + n_\ell^2$$

$$= 2 \left[C(\ell,0) + \sum_{i=1}^{\ell-1} (2C(i,0) + 4C(i,1) + 8C(i,2)) \right]$$
$$+ (2C(\ell,0) + 4C(\ell,1) + 8C(\ell,2))$$
$$= 2 \left[C(\ell,0) + 2C((\ell-1),1) + 4C(\ell,2) + 8C(\ell,3) \right]$$
$$+ (2C(\ell,0) + 4C(\ell,1) + 8C(\ell,2))$$
$$= 4C(\ell,0) + 4C((\ell-1),1) + 4C(\ell,1) + 16C(\ell,2) + 16C(\ell,3)$$
$$= 4 + 4(\ell-1) + 4\ell + 16\frac{\ell(\ell-1)}{2!} + 16\frac{\ell(\ell-1)(\ell-2)}{3!}$$
$$= 8\ell + \frac{8}{3}\ell(\ell-1)(\ell+1) = \frac{8}{3}[3\ell + \ell(\ell-1)(\ell+1)]$$
$$= \frac{8}{3}\ell(3 + \ell^2 - 1) = \frac{8\ell(\ell^2+2)}{3} \tag{4.89}$$

The number of elements in the set L_ℓ^3 in terms of combinations writes

$$n_\ell^3 = 4\left(C(\ell,0) + C((\ell-1),1) + C(\ell,1) + 4C(\ell,2) + 4C(\ell,3)\right)$$
$$= 8\left[C(\ell,1) + 2C(\ell,2) + 2C(\ell,3)\right] \tag{4.90}$$

4.4.1.2 Computation of the number of elements in the set $L^3(\ell)$ for the underlying set $\mathbb{Z}$

The set $L^3(\ell)$ is obtained by the union

$$L^3(\ell) = \bigcup_{i=0}^{\ell} L_i^3 \tag{4.91}$$

of the subsets $L_i^3, i = 0, \ldots, \ell$ whose number of elements is given by equation (4.88) or equation (4.90) if the combinatorial notation is used. The number of elements in $L^3(\ell)$ writes

$$n^3(\ell) = \sum_{i=0}^{\ell} n_i^3 \tag{4.92}$$

Equation (4.88) gives one expression for the computation of n_0^3 and another for $n_i^3, i > 0$. Considering these two cases, the equation (4.92) must be rewritten as

$$n^3(\ell) = \begin{cases} n_0^3 = 1 & \ell = 0 \\ n_0^3 + \sum_{i=1}^{\ell} n_i^3 & \ell > 0 \end{cases} \tag{4.93}$$

Substituting n_ℓ^3 obtained in equation (4.87) for $\ell > 0$ into equation (4.93) it gives the total number of points $n^3(\ell)$ of a two-dimensional constrained system from level $\lambda_0 = 0$ until the level $\lambda_0 = \ell$ as follows

$$n^3(\ell) = 1 + 4\sum_{i=1}^{\ell}\left[i + i^2 + 2\sum_{j=0}^{i-1}j^2\right] \tag{4.94}$$

If instead of replacing n_ℓ^3 obtained in equation (4.87), it is substituted by the value given by equation (4.90), which was obtained using the algebra of combinations it gives

$$
\begin{aligned}
n^3(\ell) &= 1 + 8\sum_{i=1}^{\ell}(C(\ell,1) + 2C(\ell,2) + 2C(\ell,3)) \\
&= 1 + 8\left[C((\ell+1),2) + 2C((\ell+1),3) + 2C((\ell+1),4)\right] \\
&= 1 + 8\left[\frac{(\ell+1)\ell}{2!} + 2\frac{(\ell+1)\ell(\ell-1)}{3!} + 2\frac{(\ell+1)\ell(\ell-1)(\ell-2)}{4!}\right] \\
&= 1 + \frac{2}{3}\left(6(\ell+1)\ell + 4(\ell+1)\ell(\ell-1) + (\ell+1)\ell(\ell-1)(\ell-2)\right) \\
&= 1 + \frac{2}{3}\ell(\ell+1)\left[6 + 4(\ell-1) + (\ell-2)(\ell-1)\right] \\
&= 1 + \frac{2}{3}\ell(\ell+1)\left[6 + 4\ell - 4 + \ell^2 - 3\ell + 2)\right] \\
&= 1 + \frac{2}{3}\ell(\ell+1)(\ell^2 + \ell + 4) \tag{4.95}
\end{aligned}
$$

Therefore,

$$n^3(\ell) = \begin{cases} 1 + 8\left[C((\ell+1),2) + 2C((\ell+1),3) + 2C((\ell+1),4)\right] \\ 1 + \dfrac{2}{3}\ell(\ell+1)(\ell^2+\ell+4) \end{cases} \quad \ell > 0 \tag{4.96}$$

4.4.2   Underlying set $\mathbb{Z}^+$

4.4.2.1   Computation of the number of elements in the set L_ℓ^3 in the level ℓ

The sublevel λ_1 associated to $w_0 = \lambda_0, \ldots, 0$ are $\lambda^1 = 0, 1, \ldots, (\ell-1), \ell$. Then the set L_ℓ^3 can be obtained as

$$L_\ell^3 = \bigcup_{w_0=\lambda_0}^{0}\{w_0\} \times L_{\lambda_1}^2 \tag{4.97}$$

Transforming the index as $\lambda_1 = \ell - |w_0|$ it follows

$$L_\ell^3 = \bigcup_{\lambda_1=0}^{\ell-1} \{(\lambda_0 - \lambda_1)\} \times L_{\lambda_1}^2 \tag{4.98}$$

Then the number of elements in the level ℓ can be written as

$$n_\ell^3 = \sum_{\lambda_1=0}^{\ell-1} n_{\lambda_1}^2 \tag{4.99}$$

where $n_{\lambda_1}^2$ is the number of elements in the set $L_{\lambda_1}^2$ and it is given by equation (4.81). Substituting it into equation (4.99) it follows

$$n_\ell^3 = \sum_{\lambda_1=0}^{\ell} C((\lambda_1 + 2), 2) = C((\ell + 3), 3) \tag{4.100}$$

This result can be proved using the property 4.7.4 in the section 4.7 making first the definition $j := \lambda_1 + 2$ and then modifying the indices as: for $\lambda_1 = 0$ then $j = 2$, for $\lambda_1 = \ell$ then $j = n = \ell + 2$. Substituting these indices into the equation in the property 4.7.4 it gives

$$\sum_{\lambda_1=0}^{\ell} C((\lambda_1 + 2), 2) = \sum_{j=2}^{\ell+2} C(j, 2) = C((\ell + 3), 3) \tag{4.101}$$

which completes the proof.

The expansion of the result obtained in equation (4.100) gives

$$n_\ell^3 = C((\ell + 3), 3) = \frac{(\ell + 1)(\ell + 2)(\ell + 3)}{3!}$$

$$= \frac{\prod_{i=1}^{3}(\ell + i)}{3!} \tag{4.102}$$

4.4.2.2 Computation of the number of elements in the set $L^3(\ell)$ for the underlying set $\mathbb{Z}^+$

The set $L^2(\ell)$ is obtained from the union

$$L^3(\ell) = \bigcup_{i=0}^{\ell} L_i^3 \tag{4.103}$$

of the subsets $L_i^1, i = 0, \ldots, \ell$. The number of elements in $L^3(\ell)$ writes

$$n^3(\ell) = \sum_{i=0}^{\ell} n_i^3 \tag{4.104}$$

where n_i^3 is given by equation (4.102). Then

$$n^3(\ell) = \sum_{i=1}^{\ell} C((\ell+3),3) = C((\ell+4),4) \tag{4.105}$$

whose proof is analogous to that in equation (4.101). Expanding the combination it gives

$$n^3(\ell) = C((\ell+4),4)$$
$$= \frac{(\ell+1)(\ell+2)(\ell+3)(\ell+4)}{4!}$$
$$= \frac{\prod_{i=1}^{4}(\ell+i)}{4!} \tag{4.106}$$

Given a number $w = (w_0, w_1, w_2, w_3)$ to obtain the number of points from $(0,0,0,0)$ until w it is enough to make the substitution $\ell = w_0 + w_1 + w_2 + w_3$ in the expression (4.106), which gives

$$n^3(\ell) = \tag{4.107}$$
$$\frac{(w_0 + w_1 + w_2 + w_3 + 1)(w_0 + w_1 + w_2 + w_3 + 2)(w_0 + w_1 + w_2 + w_3 + 3)}{4!}$$
$$\tag{4.108}$$

Example 4.4.1

The expression (4.106) gives the number of points in a three-dimensional set $L^3(\ell)$, from the level $\lambda_0 = 0$ until the level $\lambda_0 = \ell$, inclusive. For a three-dimensional set in the level $\ell = 3$ the number of elements are

$$n^3(3) = \frac{(3+1)(3+2)(3+3)(3+4)}{4!} = \frac{4 \times 5 \times 6 \times 7}{2 \times 3 \times 4} = 35$$

The expression (4.102) gives the number of points in each level of the three-dimensional set $L^3(\ell)$. Performing the computation for $\ell = 0, 1, 2, 3$ it follows

1. For level $\lambda_0 = \ell = 0$ there is only one element in the set L_0^3, that is $n_0^3 = 1$.

2. The number of points the set L_1^3 in the level $\lambda_0 = \ell = 1$ is

$$n_1^3 = \frac{(\ell+1)(\ell+2)(\ell+3)}{3!}$$
$$= \frac{(1+1)(1+2)(1+3)}{3!} = \frac{2 \times 3 \times 4}{6} = 4$$

which equals the number of elements in table 2.7.22.

3. The number of points in the set L_2^3 in the level $\lambda_0 = \lambda^0 = \ell = 2$ is

$$n_2^3 = \frac{(\ell+1)(\ell+2)(\ell+3)}{3!}$$

$$= \frac{(2+1)(2+2)(2+3)}{3!} = \frac{3 \times 4 \times 5}{6} = 10$$

which equals the number of elements in table 2.8.23.

4. The number of points the set L_3^3 in the level $\lambda_0 = \lambda^0 = \ell = 3$ is

$$n_3^3 = \frac{(\ell+1)(\ell+2)(\ell+3)}{3!}$$

$$= \frac{(3+1)(3+2)(3+3)}{3!} = \frac{4 \times 5 \times 6}{6} = 20$$

which equals the number of elements in table 2.9.10.

Adding the number of points in each level $n^3(3) = n_0^3 + n_1^3 + n_2^3 + n_3^3 = 1 + 4 + 10 + 20 = 35$.

end of example 4.4.1

4.4.3 Summary for the three-dimensional numbers

1. For the underlying set $\mathbb{Z}$ the number of elements is given by

(a) From equations (4.87), (4.89), and (4.90) the number of elements in the set L_ℓ^3 in the level ℓ is

$$n_\ell^3 = \begin{cases} C(\ell,0) = 1 & \text{if } \ell = 0 \\ 4\left(\ell + \ell^2 + 2\sum_{i=0}^{\ell-1} i^2\right) \\ \dfrac{8\ell(\ell^2+2)}{3} & \text{if } \ell > 0 \\ 4\left(2C(\ell,1) + 4C(\ell,2) + 4C(\ell,3)\right) \end{cases}$$

$$(4.109)$$

(b) From equations (4.94) and (4.95) the number of elements in the set $L^3(\ell)$ from the level zero through the level ℓ is given by

$$n^3(\ell) = \begin{cases} C(\ell,0) = 1 & \text{if } \ell = 0 \\ 1 + 8\left[C((\ell+1),2) + 2C((\ell+1),3) + 2C((\ell+1),4)\right] \\ & \text{if } \ell > 0 \\ 1 + \dfrac{2}{3}\ell(\ell+1)(\ell^2+\ell+4) \end{cases}$$

$$(4.110)$$

2. For the underlying set $\mathbb{Z}^+$ the number of elements is given by

   (a) From equation (4.102) the number of elements in the set L_ℓ^3 in the level ℓ is

   $$n_\ell^3 = C((\ell+3),3) = \frac{(\ell+1)(\ell+2)(\ell+3)}{3!} \qquad \ell \geq 0 \qquad (4.111)$$

   (b) From equation (4.106) the number of elements in the set $L^3(\ell)$ from the level zero through the level ℓ is given by

   $$n^3(\ell) = C((\ell+4),4) = \frac{(\ell+1)(\ell+2)(\ell+3)(\ell+4)}{4!} \qquad \ell \geq 0$$
   $$(4.112)$$

4.5 δ-dimensional constrained space

4.5.1   Underlying set $\mathbb{Z}^+$

4.5.1.1   Computation of the number of elements in the set L_ℓ^δ in the level ℓ

The generalization of the equation (4.111) gives

$$n_\ell^\delta = C((\ell+\delta),\delta) = \frac{(\ell+1)(\ell+2)\cdots(\ell+\delta)}{\delta!} \qquad \ell \geq 0 \qquad (4.113)$$

In a compact form it writes

$$n_\ell^\delta = \frac{\prod_{i=1}^{\delta}(\ell+i)}{\delta!} \qquad (4.114)$$

The number of elements in the set L_ℓ^δ in the level ℓ can also be obtained by the computation of the number of solutions of the equation

$$w_0 + w_1 + \cdots + w_\delta = \ell \qquad (4.115)$$

which from Rosen [31] is given by $C((n+r-1),r)$ r-combinations from a set with n elements when repetition of the elements are allowed. Applying it to equation (4.115) where $n = \ell+d-1 = \ell+\delta$ and $r = \ell$ it gives result obtained in equation (4.113).

4.5.1.2 Computation of the number of elements in the set $L^\delta(\ell)$

The generalization of the equation (4.112) gives

$$n^\delta(\ell) = C((\ell + \delta + 1), (\delta + 1)) = \frac{(\ell + 1)(\ell + 2) \cdots (\ell + \delta + 1)}{(\delta + 1)!} \qquad \ell \geq 0$$

(4.116)

In a compact form it writes

$$n^\delta(\ell) = \frac{\prod_{i=1}^{\delta+1}(\ell + i)}{(\delta + 1)!}$$

(4.117)

4.6 Summary

4.6.1 If the underlying set is $\mathbb{Z}$

4.6.1.1 Zero-dimensional system and underlying set $\mathbb{Z}$

1. The number of elements in the set L_ℓ^0 in the level ℓ was obtained in equation (4.3) and it is

$$n_\ell^0 = \begin{cases} C(\ell, 0) = 1 & \text{if} \quad \ell = 0 \\ 2C(\ell, 0) = 2 & \text{if} \quad \ell > 0 \end{cases}$$

(4.118)

2. The number of elements in the set $L^0(\ell)$, which include the elements from the level zero through the level ℓ was obtained in equation (4.6) and it is

$$n^0(\ell) = C(\ell, 0) + 2C(\ell, 1)$$
$$= 1 + 2\ell \qquad \ell \geq 0$$

(4.119)

4.6.1.2 One-dimensional system and underlying set $\mathbb{Z}$

1. The number of elements in the set L_ℓ^1 in the level ℓ was obtained in equation (4.22) and it is

$$n_\ell^1 = \begin{cases} C(\ell, 0) = 1 & \text{if} \quad \ell = 0 \\ 4\left[C(\ell, 0) + C_1((\ell - 1), 1)\right] = 4C(\ell, 1) = 4\ell & \text{if} \quad \ell > 0 \end{cases}$$

(4.120)

2. The number of elements in the set $L^1(\ell)$ from the level zero through the level ℓ was obtained in equation (4.26) and it is

$$n^1(\ell) = C(\ell, 0) + 4C((\ell+1), 2) = 1 + 2(\ell+1)\ell \qquad \ell \geq 0 \qquad (4.121)$$

4.6.1.3   Two-dimensional system and underlying set $\mathbb{Z}$

1. The number of elements in the set L_ℓ^2 in the level ℓ was obtained in equation (4.56) and it is

$$n_\ell^2 = \begin{cases} C(\ell, 0) = 1 & \text{if } \ell = 0 \\ 2\left[C(\ell, 0) + 2C(\ell, 1) + 4C(\ell, 2)\right] = 2 + 4\ell^2 & \text{if } \ell > 0 \end{cases}$$
$$(4.122)$$

2. The number of elements in the set $L^2(\ell)$ from the level zero through the level ℓ was obtained in equation (4.62) and it is

$$n^2(\ell) = \begin{cases} C(\ell, 0) = 1 & \text{if } \ell = 0 \\[2mm] 1 + 2\ell + 4\displaystyle\sum_{i=0}^{\ell} i^2 = C(\ell, 0) + 2C(\ell, 1) + 4C((\ell+1), 2) \\[2mm] \qquad\qquad + 8C((\ell+1), 3) & \text{if } \ell > 0 \\[2mm] \qquad = 1 + 2\ell + 4\left[\dfrac{\ell(\ell+1)(2\ell+1)}{3!}\right] \end{cases}$$
$$(4.123)$$

4.6.1.4   Three-dimensional system and underlying set $\mathbb{Z}$

1. The number of elements in the set L_ℓ^3 in the level ℓ was obtained in equations (4.87), (4.89), and (4.90) and it is

$$n_\ell^3 \begin{cases} = C(\ell, 0) = 1 & \text{if } \ell = 0 \\[2mm] \begin{cases} = 4\left(\ell + \ell^2 + 2\displaystyle\sum_{i=0}^{\ell-1} i^2\right) \\[2mm] = \dfrac{8\ell(\ell^2+2)}{3} \\[2mm] = 4\left[C(\ell, 0) + C((\ell-1), 1) + C(\ell, 1) + 4C(\ell, 2) + 4C(\ell, 3)\right] \\[2mm] = 8\left[C(\ell, 1) + 2C(\ell, 2) + 2C(\ell, 3)\right] \end{cases} & \text{if } \ell > 0 \end{cases}$$
$$(4.124)$$

2. The number of elements in the set $L^3(\ell)$, which includes the elements from the level zero through the level ℓ was obtained in equations (4.94)

and (4.95) and it is

$$
n^3(\ell)
\begin{cases}
= C(\ell, 0) = 1 & \text{if} \quad \ell = 0 \\[2mm]
= 1 + 4\displaystyle\sum_{i=1}^{\ell}\left[i + i^2 + 2\sum_{j=0}^{i-1} j^2\right] \\[4mm]
= 1 + \dfrac{4}{6}\ell(\ell+1)(\ell^2 + \ell + 4) & \text{if} \quad \ell > 0 \\[3mm]
= 1 + 4\big[C(\ell, 1) + C(\ell, 2) + C((\ell+1), 2) \\[1mm]
\qquad\quad + 4C((\ell+1), 3) + 4C((\ell+1), 4)\big]
\end{cases}
$$

$$(4.125)$$

4.6.2 If the underlying set is $\mathbb{Z}^+$

4.6.2.1 Zero-dimensional system and underlying set $\mathbb{Z}^+$

1. The number of elements in the set L_ℓ^0 at the level ℓ was obtained in equation (4.8) and it is

$$
n_\ell^0 = C(\ell, 0) = 1 \qquad \ell \geq 0 \tag{4.126}
$$

2. The number of elements in the set $L^0(\ell)$, which include the elements from the level zero through the level ℓ was obtained in equation (4.11) and it is

$$
\begin{aligned}
n^0(\ell) &= C((\ell+1), 1) \\
&= 1 + \ell \qquad \ell \geq 0
\end{aligned} \tag{4.127}
$$

4.6.2.2 One-dimensional system and underlying set $\mathbb{Z}^+$

1. The number of elements in the set L_ℓ^1 at the level ℓ was obtained in equation (4.30) and it is

$$
n_\ell^1 = C((\ell+1), 1) = \frac{\ell+1}{1!} \qquad \ell \geq 0 \tag{4.128}
$$

2. The number of elements in the set $L^1(\ell)$, which include the elements from the level zero through the level ℓ was obtained in equation (4.34) and it is

$$
\begin{aligned}
n^1(\ell) &= C((\ell+2), 2) = \frac{(\ell+1)(\ell+2)}{2!} \\
&= \frac{\prod_{i=1}^{2}(\ell+i)}{2!} \qquad \ell \geq 0
\end{aligned} \tag{4.129}
$$

4.6.2.3   Two-dimensional system and underlying set $\mathbb{Z}^+$

1. The number of elements in the set L_ℓ^2 at the level ℓ was obtained in equation (4.81) and it is

$$n_\ell^2 = C((\ell+2), 2) = \frac{(\ell+1)(\ell+2)}{2!} \qquad \ell \geq 0 \qquad (4.130)$$

2. The number of elements in the set $L^2(\ell)$, which include the elements from the level zero through the level ℓ was obtained in equation (4.82) and it is

$$n^2(\ell) = C((\ell+3), 3) = \frac{(\ell+1)(\ell+2)(\ell+3)}{3!} \qquad \ell \geq 0 \qquad (4.131)$$

4.6.2.4   Three-dimensional system and underlying set $\mathbb{Z}^+$

1. The number of elements in the set L_ℓ^3 at the level ℓ was obtained in equation (4.102) and it is

$$n_\ell^3 = C((\ell+3), 3) = \frac{(\ell+1)(\ell+2)(\ell+3)}{3!} \qquad \ell \geq 0 \qquad (4.132)$$

2. The number of elements in the set $L^3(\ell)$, which are those from the level zero through the level ℓ was obtained in equation (4.106) and it is

$$n^3(\ell) = 1 + C((\ell+4), 4) = \frac{(\ell+1)(\ell+2)(\ell+3)(\ell+4)}{4!} \qquad \ell \geq 0$$
$$(4.133)$$

4.6.2.5 δ-dimensional system and underlying set $\mathbb{Z}^+$

1. The number of elements in the set L_ℓ^δ in the level ℓ was obtained in equations (4.113) and (4.114) and it is

$$\begin{aligned} n_\ell^\delta &= C((\ell+\delta), \delta) \\ &= \frac{(\ell+1)(\ell+2)\cdots(\ell+\delta)}{\delta!} \qquad \ell \geq 0 \\ &= \frac{\prod_{i=1}^{\delta}(\ell+i)}{\delta!} \qquad\qquad (4.134) \end{aligned}$$

2. The number of elements in the set $L^\delta(\ell)$, which are those from the level zero through the level ℓ was obtained in equation (4.116) and (4.117)

and it is

$$n^\delta(\ell) = C((\ell + \delta + 1), (\delta + 1))$$
$$= \frac{(\ell + 1)(\ell + 2) \cdots (\ell + \delta + 1)}{(\delta + 1)!} \qquad \ell \geq 0$$
$$= \frac{\prod_{i=1}^{\delta+1}(\ell + i)}{(\delta + 1)!} \qquad (4.135)$$

4.7 Mathematical tools

4.7.1 Sum of the terms of an arithmetic progression

The sum of the terms of an arithmetic progression where a_n is the last term, a_1 is the first, and n is the ratio is given by

$$S = \frac{n(a_n + a_1)}{2} \qquad (4.136)$$

4.7.2 Sum of squares

The sum of the sequence of the square of the first m integers is given by

$$S_m = \sum_{i=1}^{m} i^2 = \frac{m(m+1)}{2!} + \frac{2(m-1)m(m+1)}{3!}$$
$$= \frac{m(m+1)(2m+1)}{3!}$$
$$= C_2^{m+1} + C_3^{m+1} \qquad (4.137)$$

Proof:

The sum

$$S_5 = \sum_{i=1}^{5} i^2 = 1^2 + 2^2 + 3^2 + 4^2 + 5^2 \qquad (4.138)$$

can be rewritten as

$$S_5 = 1^2 + (1+1)^2 + (1+1+1)^2 + (1+1+1+1)^2 + (1+1+1+1+1)^2 \qquad (4.139)$$

where

$$2^2 = (1+1)^2 = 1^2 + 1^2 + 2 = 2 \times 1 + 1 \times 2 \qquad (4.140)$$

$$3^2 = (1+1+1)^2 = (1^2 + 1^2 + 1^2) + (2+2+2) = 3 \times 1 + 3 \times 2 \qquad (4.141)$$

$$4^2 = (1+1+1+1)^2 = (1^2 + 1^2 + 1^2 + 1^2) + (2+2+2+2+2+2)$$
$$= 4 \times 1 + 6 \times 2 \qquad (4.142)$$

$$5^2 = (1+1+1+1+1)^2 = (1^2 + 1^2 + 1^2 + 1^2 + 1^2)$$
$$= +(2+2+2+2+2+2+2+2+2+2) = 5 \times 1 + 10 \times 2 \qquad (4.143)$$

The numbers $1, 3, 6, 10$ represent the number of elements n_ℓ^2 in the sets $L_\ell^2, \ell = 0, 1, 2, 3$ if the underlying set is $\mathbb{Z}^+$, which is given by (4.81) and it writes

$$n_\ell^2 = \frac{(\ell+1)(\ell+2)}{2!} \qquad (4.144)$$

The sum $1 + 3 + 6 + 10$ equals to the number of elements in the set $L^2(\ell)$, which is given by (4.82) and it writes

$$n^2(\ell) = \frac{(\ell+1)(\ell+2)(\ell+3)}{3!} \qquad (4.145)$$

The numbers $\ell = 1, 3, 6, 10$ correspond to $m = 2, 3, 4, 5$, therefore, $\ell = n - 2$. The sum of the squares until $m = 5$ can be written as

$$S_5 = 1 \times 1 + 0 \times 2 + 2 \times 1 + 1 \times 2 + 3 \times 1 + 3 \times 2 + 4 \times 1$$
$$+ 6 \times 2 + 5 \times 1 + 10 \times 2$$
$$= (1 + 2 + 3 + 4 + 5) + (0 + 1 + 3 + 6 + 10) \times 2$$
$$= \frac{5(5+1)}{2!} + \frac{2(5-1)5(5+1)}{3!} = 55 \qquad (4.146)$$

Generalizing, by induction,

$$S_m = \sum_{i=1}^{m} i^2 = \frac{m(m+1)}{2!} + \frac{2(m-1)m(m+1)}{3!}$$
$$= \frac{m(m+1)(2m+1)}{3!}$$
$$= C((m+1), 2) + C((m+1), 3) \qquad (4.147)$$

See, for example, Gradshteyn and Ryzhik [15] to compare the results.

4.7.3 Sum of cubes

Consider

$$n^3 = n^3$$
$$(n-1)^3 = n^3 + 3 \times n^2 \times 1 + 3 \times n \times 1^2 + 1$$
$$(n-2)^3 = n^3 + 3 \times n^2 \times 2 + 3 \times n \times 2^2 + 2^3$$
$$(n-3)^3 = n^3 + 3 \times n^2 \times 3 + 3 \times n \times 3^2 + 3^3 \qquad (4.148)$$

Then it can be written

$$n^3 + (n-1)^3 + (n-2)^3 = 4n^3 - 3n^2(1+2+3) + 3n(1^2+2^2+3^2) - (1^3+2^3+3^3) \qquad (4.149)$$

Define $S_n = n^3 + (n-1)^3 + (n-2)^3 + \cdots + (n-(n-1))$. Extending equation (4.149) until $(n-1)$ it follows

$$S_n = \sum_{j=0}^{n-1}(n-j)^3 = \sum_{j=1}^{n} j^3$$
$$= n^3 \sum_{j=0}^{n-1} j^0 - 3n^2 \sum_{j=0}^{n-1} j^1 + 3n \sum_{j=0}^{n-1} j^2 - \sum_{j=0}^{n-1} j^3$$
$$= n^3 \times n - 3n^2 \left(1 + 2 + \cdots + (n-1)\right) + 3n \left(1^2 + 2^2 + \cdots + (n-1)^2\right)$$
$$\quad - \left(1^3 + 2^3 + \cdots + (n-1)^3\right) - n^3 + n^3$$
$$= n^4 - 3n^2 \frac{n(n-1)}{2} + 3n \left[\frac{(n-1)n}{2} + \frac{(n-2)(n-1)n}{3}\right] - S_n + n^3$$
$$\qquad (4.150)$$

$$2S_n = \frac{2}{2}(n^4 + n^3) - \frac{3}{2}n^3(n-1) + \frac{1}{2}n^2(n-1)(2n-1)$$
$$= \frac{n^2}{2} \left[2n(n+1) - 3n(n-1) + (n-1)(2n-1)\right]$$
$$= \frac{n^2}{2} \left[n^2 + 2n + 1\right]$$
$$= \frac{n^2}{2}(n+1)^2 = \left[\frac{n(n+1)}{2}\right]^2 = (C((n+1),2))^2 \qquad (4.151)$$

$$S_n = \sum_{j=1}^{n} j^3 = \frac{n^2}{4}(n+1)^2 = \left(C_2^{n+1}\right)^2 \qquad (4.152)$$

See, for example Gradshteyn and Ryzhik [15] to compare the results.

4.7.4   Selected combinatorics properties

Property 4.7.1

$$C_2^m = \frac{m!}{(m-2)!2!} = \frac{(m-1)m}{2!}$$

$$C_3^m = \frac{m!}{(m-3)!3!} = \frac{(m-2)(m-1)m}{3!}$$

$$C_\delta^m = \frac{m!}{(m-\delta)!\delta!} = \frac{(m-\delta+1)\cdots(m-2)(m-1)m}{\delta!} \qquad (4.153)$$

$$\sum_{i=1}^{m} C(i,2) = C(m,3) \qquad (4.154)$$

use $\ell + 2 = m$, and n_ℓ^2 and $n^2(\ell)$ to prove it.

end of property 4.7.1

Property 4.7.2

$$S_n = \sum_{i=k}^{j=n} C(i,0) = C((n-k+1),1) \qquad k = 0,1 \qquad (4.155)$$

proof: using induction

1. First term

   The initial term satisfies the condition $S_0 = C(0,0) = C(1,1)$ since $S_0 = C(0,0) = 1 = C(1,1)$. For the second term

   $$S_1 = C(1,1) + C(1,0)$$

   $$= \frac{1!}{(1-1)!\,1!} + \frac{1!}{(1-0)!\,0!}$$

   $$= 1 + 1 = 2 = \frac{2!}{(2-1)!\,1!} = C(2,1) \qquad (4.156)$$

   therefore, it can be written as $S_1 = S_0 + C(1,0) = C(1,1) + C(1,0) = C(2,1)$.

2. Induction

   The term n being true implies that the term $n+1$ is also true, that is, $S_n = S_{n-1} + C(n,0) = C((n+1),1)$ is true. Assume

that $S_{n-1} = C(0,0) + C(1,0) + \cdots + C((n-1),0) = C(n,1)$
then

$$S_n = S_{n-1} + C(n,0) = C(n,1) + C(n,0)$$

$$= \frac{n!}{(n-1)!\,1!} + \frac{n!}{(n-0)!\,0!} = \frac{n!\,n}{(n-1)!\,n\,1!} + \frac{n!\,1}{(n-0)!\,0!\,1}$$

$$= \frac{n!\,(n+1)}{n!\,1!} = \frac{(n+1)!}{n!\,1!} = C((n+1),1) \tag{4.157}$$

end of property 4.7.2

Property 4.7.3

$$S_n = \sum_{i=k}^{i=n} C(i,1) = C((n+k),2) \qquad k = 0,1 \tag{4.158}$$

A first proof:

$$S_n = \sum_{i=k}^{i=n} C(i,1) = C(k,1) + C((k+1),1) + \cdots + C(n,1)$$

$$= k + (k+1) + \cdots + n$$

$$= \frac{(n+k)(n-k+1)}{2}$$

$$= \frac{(n+k)(n+k+1)}{2!} \frac{n-k+1}{n+k+1} \tag{4.159}$$

then

$$S_n = \begin{cases} \displaystyle\sum_{i=k}^{i=n} C(i,1) = C((n+1),2) & k = 0,1 \\[2mm] \dfrac{n-k+1}{n+k+1} C((n+k+1),2) & k \geq 0 \end{cases} \tag{4.160}$$

A second proof: using induction show that

$$S_n = \sum_{j=1}^{j=n} C(j,1) = C((n+1),2) \tag{4.161}$$

1. First term

   The initial term satisfies the condition $S_0 = C(1,1) = C(2,2) = 1$. For the second term

   $$S_1 = C(2,2) + C(2,1)$$
   $$= \frac{2!}{(2-2)!\,2!} + \frac{2!}{(2-1)!\,1!}$$
   $$= 1 + 2 = 3 = \frac{3!}{(3-2)!\,2!} = C(3,2) \qquad (4.162)$$

   therefore, it can be written $S_1 = S_0 + C(2,1) = C(2,2) + C(2,1) = C(3,2)$.

2. Induction

   The term n being true implies that the term $n+1$ is also true, that is, $S_n = S_{n-1} + C(n,1) = C((n+1),2)$ is true. Assume that $S_{n-1} = C(1,1) + C(2,1) + \cdots + C((n-1),1) = C(n,2)$ then

   $$S_n = S_{n-1} + C(n,1) = C(n,2) + C(n,1)$$
   $$= \frac{n!}{(n-2)!\,2!} + \frac{n!}{(n-1)!\,1!}$$
   $$= \frac{n!\,(n-1)}{(n-2)!\,2!\,(n-1)} + \frac{n!\,2}{(n-1)!\,1!\,2}$$
   $$= \frac{n!\,(n-1)}{(n-1)!\,2!} + \frac{n!\,2}{(n-1)!\,2!}$$
   $$= \frac{n!}{(n-1)!\,2!}\,(n-1+2) = \frac{n!(n+1)}{(n-1)!\,2!}$$
   $$= \frac{(n+1)!}{(n-1)!\,2!} = C((n+1),2) \qquad (4.163)$$

There is no pole at $n = 1$ since

$$C((n+1),2) = \frac{(n+1)!}{(n-1)!\,2!} = \frac{n(n+1)}{2} \qquad (4.164)$$

which gives for $n = 0$ the value $C(1,2) = 0$. The sum can be written as

$$S_n = \sum_{j=0}^{j=n} C(j,1) = \sum_{j=1}^{j=n} C(j,1) = C((n+1),2) \qquad (4.165)$$

end of property 4.7.3

Property 4.7.4

$$S_n = \sum_{j=2}^{j=n} C(j,2) = C((n+1),3) \qquad (4.166)$$

Proof:

1. First term

   The initial term satisfies the condition $S_0 = C(2,2) = C(3,3)$. For the second term it follows

   $$S_1 = S_0 + C(3,2) = C(3,3) + C(3,2) = C(4,3) \qquad (4.167)$$

   which can be seen from

   $$\begin{aligned} S_1 &= C(3,3) + C(3,2) \\ &= \frac{3!}{(3-3)!3!} + \frac{3!}{(3-2)!2!} \\ &= 1 + 3 = 4 = \frac{4!}{(4-3)!3!} = C(4,3) \qquad (4.168) \end{aligned}$$

2. Induction

   Assume that

   $$\begin{aligned} S_{n-1} &= C(2,2) + C(3,2) + C(4,2) + \cdots \\ &\quad + C((n-2),2) + C((n-1),2) \\ &= C(n,3) \qquad (4.169) \end{aligned}$$

   then

   $$\begin{aligned} S_n &= S_{n-1} + C(n,2) = C(n,3) + C(n,2) \\ &= \frac{n!}{(n-3)!\,3!} + \frac{n!}{(n-2)!\,2!} \\ &= \frac{n!\,(n-2)}{(n-3)!\,3!\,(n-2)} + \frac{n! \times 3}{(n-2)!\,2!\,3} \\ &= \frac{n!\,(n-2)}{(n-2)!\,3!} + \frac{3n!}{(n-2)!3!} \\ &= \frac{n!}{(n-2)!\,3!}[(n-2)+3] = \frac{n!\,(n+1)}{(n-2)!\,3!} = \frac{(n+1)!}{(n-2)!\,3!} \\ &= C((n+1),3) \qquad (4.170) \end{aligned}$$

where

$$C((n+1),3) = \frac{(n+1)!}{(n-2)!\,3!} = \frac{(n-1)n(n+1)}{3!} \qquad (4.171)$$

For $n = 0$ and $n = 1$ it follows $C(1,3) = C(2,3) = 0$, then it can be written that

$$S_n = \sum_{j=0}^{j=n} C(j,2) = \sum_{j=1}^{j=n} C(j,2) = \sum_{j=2}^{j=n} C(j,2) = C((n+1),3)$$

$$(4.172)$$

$$\text{end of property } 4.7.4$$

Property 4.7.5 *k-dimensional space*

Generalizing the above results

$$S_n = \sum_{j=k}^{j=n} C(j,k) = C((n+1),(k+1)) \qquad k = 0,1,2,\ldots,n$$

$$(4.173)$$

which can also be written as

$$S_n = \left[\sum_{j=k}^{j=n-1} C(j,k) \right] + C(n,k)$$

$$= C(n,(k+1)) + C(n,k)$$

$$= C((n+1),(k+1)) \qquad k = 0,1,2,\ldots,n \qquad (4.174)$$

$$\text{end of property } 4.7.5$$

Property 4.7.6

$$S_n = \sum_{j=0}^{j=n} C(n,j) = 2^n \qquad (4.175)$$

Proof:

1. First term
   Motivation

$$C(1,0) + C(1,1) = 2C(1,0) = 1 + 1 = 2^1$$
$$C(2,0) + C(2,1) + C(2,2) = 1 + 2 + 1 = 4 = 2^2$$
$$C(3,0) + C(3,1) + C(3,2) + C(3,3) = 1 + 3 + 3 + 1 = 8 = 2^3$$
$$C(4,0) + C(4,1) + C(4,2) + C(4,3) + C(4,4)$$
$$= 1 + 4 + 6 + 4 + 1 = 16 = 2^4$$
$$C(5,0) + C(5,1) + C(5,2) + C(5,3) + C(5,4) + C(5,5)$$
$$= 1 + 5 + 10 + 10 + 5 + 1 = 32 = 2^5 \qquad (4.176)$$

For the sake of clarity the following sums

$$S_4 = C(4,0) + C(4,1) + C(4,2) + C(4,3) + C(4,4) = 2^4$$
$$S_5 = C(5,0) + C(5,1) + C(5,2) + C(5,3) + C(5,4)$$
$$+ C(5,5) = 2^5 \qquad (4.177)$$

will be used as the initial step. Recalling equation (4.174) each term of the two sums above can be written as

$$C(5,0) = C(4,0) + 0$$
$$C(5,1) = C(4,1) + C(4,0)$$
$$C(5,2) = C(4,2) + C(4,1)$$
$$C(5,3) = C(4,3) + C(4,2)$$
$$C(5,4) = C(4,4) + C(4,3)$$
$$C(5,5) = 0 + C(4,4) \qquad (4.178)$$

which allows to write the sum S_5 as

$$S_5 = \sum_{j=0}^{5} C(5,j)$$
$$= \sum_{j=0}^{4} C(4,j) + \sum_{j=0}^{4} C(4,j)$$
$$= S_4 + S_4 = 2 \times 2^4 = 2^5 \qquad (4.179)$$

2. Induction

Assume that $S_{n-1} = 2^{n-1}$ then

$$S_n = \sum_{j=0}^{n} C(n,j)$$
$$= \sum_{j=0}^{n-1} C((n-1),j) + \sum_{j=0}^{n-1} C((n-1),j)$$
$$= S_{n-1} + S_{n-1} = 2 \times 2^{n-1} = 2^n \qquad (4.180)$$

end of property 4.7.6

4.7.4.1 Summary

$$\sum_{i=0}^{n} C(i,0) = C((n+1),1) \qquad \sum_{i=1}^{n} C(i,0) = C(n,1) \tag{4.181}$$

$$\sum_{i=0}^{n} C(i,1) = \sum_{i=1}^{n} C(i,1) = C((n+1),1) \tag{4.182}$$

$$\sum_{i=0}^{n} C(i,2) = \sum_{i=1}^{n} C(i,2) = \sum_{i=2}^{n} C(i,2) = C((n+1),3) \tag{4.183}$$

$$\sum_{i=1}^{n} (C(i,0) + C((i-1),1)) = C(n,1) + C(n,2)$$

$$= n + \frac{n(n-1)}{2!} = \frac{n(1+n-1)}{2}$$

$$= \frac{n(n+1)}{2!} = C((n+1),2) \tag{4.184}$$

Recalling that $C(n,i) = C(n,(n-i))$ it follows

$$S_n = \sum_{j=0}^{j=n} C(n,j) = 2^n \tag{4.185}$$

4.7.5 Pochhammer's symbol

The Pochhammer symbol $(\lambda)_n$, Srivastava and Manocha [33], is defined by

$$(\lambda)_n = \begin{cases} 1 & \text{if } n = 0 \\ (\lambda+0)(\lambda+1)\ldots(\lambda+n-1) & \text{if } n = 1,2,3,\ldots \end{cases} \tag{4.186}$$

The notation $(\lambda)_n$ contains part of the factorial product, and in particular it can contain all the factorial terms, that is, $(1)_n = n!$. It is also referred to as the factorial function.

In this work it uses either the Pochhammer symbol or the factorial notation

$$\frac{(\lambda+n-1)!}{(\lambda-1)!} \tag{4.187}$$

depending on the convenience. The expanded form

$$(\lambda+0)(\lambda+1)\ldots(\lambda+n-1) \tag{4.188}$$

will be used for computation.

Chapter 5

An ordering relation

5.1 Definition of order

5.1.1 Possible definitions

Definition 5.1.1 *Ordering*

1. First ordering level

   The highest level of ordering is based on the dimension of the set according to the equation (1.2) on page 4, that is,

   $$L^0 \prec L^1 \prec L^2 \prec \cdots \prec L^\delta \prec \cdots \tag{5.1}$$

   which corresponds to the first union in equation (1.1) on page 3 from $i = 0$ until $i = \delta$ were $\delta \in \mathbb{Z}^+$.

2. Second ordering level

   The subsets L_ℓ^δ of each set L^δ are ordered based on the value of the level ℓ to which the subset belongs, that is,

   $$L_0^\delta \prec L_1^\delta \prec L_2^\delta \prec \cdots \prec L_\ell^\delta \prec \cdots \tag{5.2}$$

   which corresponds to the second union in equation (1.1) from $\ell = 0$ until $i = \lambda$ were $\lambda \in \mathbb{G}^+$.

3. Third ordering level

   There are two possibilities for ordering at this level. Given a constrained number $(w_0, w_1, \ldots, w_\delta)$ either w_0 or w_δ can be defined as the least significant the coordinate number.

   (a) If w_0 is the least significant coordinate number.

       Consider the following polynomial of constant degree

       $$x^3 + x^2 y + x y^2 + y^3 \tag{5.3}$$

The exponents can be rewritten as

$$x^3 y^0 + x^2 y^1 + x^1 y^2 + x^0 y^3 \qquad (5.4)$$

and identified as the set of constrained numbers

$$L_3^1 = \{(3,0),(2,1),(1,2),(0,3)\} \qquad (5.5)$$

in the level $\ell = 3$.

If the polynomial in equation (5.4) will be written with the variables in the sequence $\{x, y\}$ with the power of the variable x ranging from the highest to the smallest then the power of the variables satisfies the following order

$$(3,0) \prec (2,1) \prec (1,2) \prec (0,3) \qquad (5.6)$$

which can be used as an ordering relation for the constrained numbers. That is, $(3,0)$ is the number with the smallest order since $w_1 = 0$ and $(0,3)$ is the number with the highest order because $w_1 = 3$. In this rule the first variable assume the values $w_0 = 3, 2, 1, 0$ while the second assume the values $w_1 = 0, 1, 2, 3$ such that the constraint equation $w_0 + w_1 = 3$ applies for each number. The construction rules 2.4.1 on page 40, defined in the section 2.4, for the underlying set $\mathbb{Z}^+$ is a generalization of the above ordering relation.

This is also the rule used in the construction of the Pascal triangle.

If the underlying set contains negative numbers the polynomial in the variables x, y of third degree writes

$$x^3 + x^2 y + \frac{x^2}{y} + xy^2 + \frac{x}{y^2} + y^3 + \frac{1}{y^3} + \frac{y^2}{x} + \frac{1}{xy^2} + \frac{y}{x^2} + \frac{1}{x^2 y} + \frac{1}{x^3} \quad (5.7)$$

The exponents can be rewritten as

$$
\begin{aligned}
& x^3 y^0 + x^2 y^1 + x^2 y^{-1} + x^1 y^2 + x^1 y^{-2} + x^0 y^3 \\
& + x^0 y^{-3} + x^{-1} y^2 + x^{-1} y^{-2} + x^{-2} y^1 \\
& + x^{-2} y^{-1} + x^{-3} y^0
\end{aligned}
\qquad (5.8)
$$

where the sequence of the exponents is

$$
\begin{aligned}
(3,0) \prec & (2,1) \prec (2,-1) \prec (1,2) \prec (1,-2) \\
& \prec (0,3) \prec (0,-3) \prec (-1,2) \prec (-1,-2) \\
& \prec (-2,1) \prec (-2,-1) \prec (-3,0)
\end{aligned}
\qquad (5.9)
$$

(b) If w_δ is the least significant coordinate number.

This is a rule similar to the one used to write a number abc in a base such as base 10. In this case c is the least significant digit. Under this rule a polynomial writes

$$y^3 + xy^2 + x^2y + x^3 \tag{5.10}$$

where the exponents follows the order

$$(0,3), (1,2), (2,1), (3,0) \tag{5.11}$$

Then $(0,3)$ is the number with the smallest order since $w_0 = 0$ and $(3,0)$ is the number with the highest order because $w_0 = 3$.

If the underlying set contains negative numbers the polynomial writes

$$\frac{1}{x^3} + \frac{1}{x^2y} + \frac{y}{x^2} + \frac{1}{xy^2} + \frac{y^2}{x} + \frac{1}{y^3} + y^3 + \frac{x}{y^2} + xy^2 + \frac{x^2}{y} + x^2y + x^3 \tag{5.12}$$

with the following order

$$(-3,0) \prec (-2,-1) \prec (-2,1) \prec (-1,-2) \prec (-1,2)$$
$$\prec (0,-3) \prec (0,3) \prec (1,-2) \prec (1,2)$$
$$\prec (2,-1) \prec (2,1) \prec (3,0) \tag{5.13}$$

for the exponents.

end of definition 5.1.1

5.1.2 Adopted definition

It was adopted the case "If w_δ is the least significant coordinate number."

The rules 2.2.1 on page 33 for the constrained number whose underlying set is $\mathbb{Z}$ is an extension of the rules for the underlying set $\mathbb{Z}^+$ to consider the negative numbers. The following was used as a guide line to select the ordering relation

1. the set $L^1(\ell)$ of constrained numbers for the underlying set $\mathbb{Z}^+$ must have the same ordering as the numbers in the Pascal Triangle.

2. the sets L_ℓ^δ for the underlying set $\mathbb{Z}^+$ must be subsets of the sets L_ℓ^δ if the underlying set is $\mathbb{Z}$

A consequence of this choice is that $(1) \prec (-1)$.

The rules 2.2.1 and 2.2.2 presented in the section 2.2 define the construction of a set constrained numbers following the above ordering for the underlying set $\mathbb{Z}$. The algorithm 2.2.3 generates the elements of the set L_ℓ^δ in order as well as the algorithm 2.3.1 for the elements of the set $L^\delta(\ell)$.

Similarly the rules 2.4.1 and 2.4.2 presented in the section 2.4 define the construction of a set constrained numbers following the above ordering for the underlying set $\mathbb{Z}^+$. The algorithms 2.4.3 and 2.4.4 generates the elements of the set L_ℓ^δ in order as well as the algorithms 2.5.1 and 2.5.2 for the elements of the set $L^\delta(\ell)$.

All the tables in chapter 2 showing the computation of the elements of a set exhibit the these elements following ordering defined above.

Notation 5.1.2 *Ordering*

The order of a δ-dimensional constrained number w with respect to the element with lowest order in the set $L^\delta(\ell)$, which is called the zero, will be denoted as $\mathcal{O}^\delta(w)$. The element with lowest order will be called the the zero element of the level ℓ.

The order of a δ-dimensional constrained number w with respect to the element with lowest order in the set L_ℓ^δ will be denoted as $\mathcal{O}_\ell^\delta(w)$. The element with lowest order will be called the first element of the level ℓ.

end of notation 5.1.2

Definition 5.1.3 *Minimum and maximum constrained numbers in a set*

Among the elements of a set the one that has the lowest order is called the minimum number and the one with the highest order is called maximum.

end of definition 5.1.3

Rule 5.1.4 *Minimum and maximum constrained numbers in a set*

Let $w = (w_0, w_1, \ldots, w_{\delta-1} w_\delta)$, where $w_i \in \mathbb{Z}$ or $w_i \in \mathbb{Z}^+$, be a constrained coordinate number.

1. The smallest number is defined as

$$\max w_0 \geq \max w_1 \geq \cdots \leq \max w_{\delta-1} \leq \max w_\delta$$

$$\text{such that} \quad \sum_{i=0}^{\delta} |\max w_i| = \ell \tag{5.14}$$

2. and the highest number is defined as

$$\min w_0 \leq \min w_1 \leq \cdots \leq \min w_{\delta-1} \leq \min w_\delta$$

$$\text{such that} \quad \sum_{i=0}^{\delta} |\min w_i| = \ell \tag{5.15}$$

end of rule 5.1.4

Example 5.1.5 *Let*

$$L^0(1) = \{(0), (1), (-1)\}$$

and

$$L^1(1) = \{(0,0), (1,0), (0,1), (0,-1), (-1,0)\}$$

which were obtained in example 1.1.10. Write the constrained numbers in order from the smallest to the highest.

1. Ordering by dimension

   By the definition 5.1.1 $L^0(1) \prec L^1(1)$, that is,

   $$\{(0), (1), (-1)\} \prec \{(0,0), (1,0), (0,1), (0,-1), (-1,0)\} \tag{5.16}$$

   since the dimension of $L^0(1)$ is smaller than that of $L^1(1)$, namely $0 < 1$.

2. Ordering the elements of $L^0(1)$

   (a) Ordering by subsets

   The subsets of the zero-dimensional space $L^0(1)$ are ordered as follows

   $$\{(0)\} \prec \{(1), (-1)\} \tag{5.17}$$

   where $L_0^0 = \{(0)\}$ and $L_1^0 = \{(1), (-1)\}$ satisfying the condition $L_0^0 \prec L_1^0$.

   (b) Ordering by level, that is, the elements of each subset
   For $L_0^0 = \{(0)\}$ there is nothing to do.
   For $L_1^0 = \{(1), (-1)\}$ they have the order $(1) \prec (-1)$.

   (c) The constrained numbers in $L^0(1)$ are ordered as

   $$(0) \prec (1) \prec (-1) \tag{5.18}$$

3. Ordering the elements of $L^1(1)$

   (a) Ordering by subsets

   The subsets of the one-dimensional space in the set $L^1(1)$ are ordered as follows

   $$\{(0,0)\} \prec \{(1,0), (0,1), (0,-1), (-1,0)\} \tag{5.19}$$

   where $L_0^1 = \{(0,0)\}$ and $L_1^1 = \{(1,0), (0,1), (0,-1), (-1,0)\}$ satisfying the condition $L_0^1 \prec L_1^1$.

(b) Ordering by level, that is, the elements of each subset
For $L_0^1 = \{(0,0)\}$ there is nothing to do.
The elements of the one-dimensional space in the set $L^1(1)$, level
$\ell = 1$, which are $\{(1,0), (0,1), (0,-1), (-1,0)\}$ have the following
order

$$(1,0) \prec (0,1) \prec (0,-1) \prec (-1,0) \qquad (5.20)$$

the coordinate w_0 of the minimum constrained number is $\max w_0 = \max(-1,0,1) = 1$. The maximum value for w_1 satisfying the condition $|w_0| + |w_1| = 1$ is $w_1 = 0$ and not $w_1 = 1$. Therefore, the minimum constrained number is $(1,0)$. A similar analysis will find that $(-1,0)$ is the maximum constrained number in this level.

(c) The constrained numbers in the set L^1 are ordered as

$$(0,0) \prec (0,-1) \prec (-1,0) \prec (1,0) \prec (0,1) \qquad (5.21)$$

4. Gathering all the above numbers, they obey the following order

$$(0) \prec (-1) \prec (1) \prec (0,0) \prec (0,-1) \prec (-1,0) \prec (1,0) \prec (0,1) \quad (5.22)$$

end of example 5.1.5

Example 5.1.6 *Write the elements in the set*

$$\{(-1,-1,-),(0,0,2),(1,1,0),(0,-1,1),(-1,0,-1),$$
$$(0,1,1),(2,0,0),(0,-2,0),(1,-1,0)\} \qquad (5.23)$$

in the level $\ell = 2$ in order from the minimum to the maximum.

The minimum number is the one that has the maximum w_0, which is $w_0 = 2$. Since $2 + w_1 + w_2 = 2$ then $w_1 = w_2 = 0$ and the number is $(2,0,0)$.

The next highest value for w_0 is $w_0 = 1$. There are two candidates for the second position $(1,1,0)$ and $(1,-1,0)$. The analysis of w_1 for both shows that $w_1 = 1$ is the maximum value, therefore, $(1,1,0) \prec (1,-1,0)$.

The numbers whose $w_0 = 0$ are $(0,0,2)$, $(0,-1,1)$, $(0,1,1)$, $(0,-2,0)$. The second coordinate has the following order from the maximum to the minimum $1,0,-1$ then the number have the following order $(0,1,1) \prec (0,0,2) \prec (0,-1,1) \prec (0,-2,0)$

A similar analysis for $w_0 = -1$ shows the following order order $(-1,0,-1) \prec (-1,-1,-)$

Writing all together it follows

$$(2,0,0) \prec (1,1,0) \prec (1,-1,0) \prec (0,1,1) \prec (0,0,2) \prec (0,-1,1) \prec$$
$$(0,-2,0) \prec (-1,0,-1) \prec (-1,-1,-) \qquad (5.24)$$

Compare result in equation (5.24) with the numbers in table 2.8.12.

end of example 5.1.6

5.2 Integer numbers as the underlying set

5.2.1 Zero-dimensional constrained space

The elements in the set L_ℓ^0 of a zero-dimensional constrained system in the level ℓ have the form $\{(\ell), (-\ell)\}$ then the number $w_0 = \ell$ has order zero and the number $w_0 = -\ell$ has order one since $\ell \prec -\ell$ by the ordering definition. Therefore, the order of the elements in the set L_ℓ^0 writes

$$\mathcal{O}^0(w_0) = \begin{cases} 0 & w_0 \geq 0 \\ 1 & w_0 < 0 \end{cases} \tag{5.25}$$

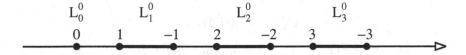

Figure 5.2.1 *Representation of ordered levels $L_0^0 \prec L_1^0 \prec L_2^0 \prec \cdots$ in a zero dimensional constrained number system.*

As an example, the constrained numbers in figure 5.2.1 are globally ordered as

$$\{(0)\} \prec \{(1) \prec (-1)\} \prec \{(2) \prec (-2)\} \prec \{(3) \prec (-3)\} \prec \cdots$$

where the braces define each set L_ℓ^0.

5.2.2 One-dimensional constrained space

Recall that the one-dimensional constrained system is denoted using two cartesian coordinates, namely (w_0, w_1). The order of a one-dimensional constrained system with $\mathbb{Z}$ as the underlying set is given by

1. The order with respect to the first element of the level is given by

$$\mathcal{O}_\ell^1(w_0, w_1) = \begin{cases} 0 & \text{if} \quad w_0 = \lambda_0 \\ 2(\lambda_0 - w_0) - 1 & \text{if} \quad w_0 \neq \lambda_0 \text{ and } w_1 \geq 0 \\ 2(\lambda_0 - w_0) & \text{if} \quad w_0 \neq \lambda_0 \text{ and } w_1 < 0 \end{cases} \tag{5.26}$$

2. The order with respect to the zero element of the level is given by

$$\mathcal{O}^1(\ell)(w_0, w_1) = \sum_{i=0}^{\lambda_0 - 1} n_i^1 + \mathcal{O}_\ell^1(w_0, w_1) \tag{5.27}$$

where $\lambda_0 = |w_0| + |w_1| = \ell$ is the level. The number of elements n_ℓ^1 in the the level ℓ of a one-dimensional constrained space with $\mathbb{Z}$ as the underlying set is given by equation (4.120).

Example 5.2.2 *Obtain the order of the elements in the set* L_1^1.

1. Computation of $\mathcal{O}^1(0, -1)$

   Since $w_1 = -1$ then the order is given by

   $$\mathcal{O}_1^0(w_0, w_1) = 2\,(\lambda_0 - w_0)$$

   which gives $2\,(1 - 0) = 2$. Therefore, $\mathcal{O}_1^0(0, -1) = 2$.

2. Computation of $\mathcal{O}_1^1(-1, 0)$

   Since $w_1 = 0$ then the order is given by

   $$\mathcal{O}_1^0(w_0, w_1) = 2\,(\lambda_0 - w_0) - 1$$

   which gives $2\,(1 - (-1)) - 1 = 3$. Therefore, $\mathcal{O}_1^0(-1, 0) = 3$.

3. Computation of $\mathcal{O}^1(1, 0)$

   Since $w_0 = \lambda_0 = 1$ then the order is given by

   $$\mathcal{O}_1^0(1, 0) = 0$$

4. Computation of $\mathcal{O}^1(0, 1)$

   Since $w_1 > 0$ then the order is given by

   $$\mathcal{O}_1^0(w_0, w_1) = 2\,(\lambda_0 - w_0) - 1$$

   which gives $2\,(1 - 0) - 1 = 1$. Therefore, $\mathcal{O}_1^0(0, 1) = 1$.

The above results can be compared with the values obtained in the table 2.7.6 on page 57, which is repeated in the table 5.2.3 with the addition of the respective orders.

Table 5.2.3 *Order of the one-dimensional numbers in the set* L_1^1.

w_0	w_1	numbers	$\mathcal{O}_1^1(w_0, w_1)$
1	0	$(1, 0)$	0
0	1	$(0, 1)$	1
0	-1	$(0, -1)$	2
-1	0	$(-1, 0)$	3

end of example 5.2.2

Example 5.2.4 *Find the order of the numbers*

$$(0, -2), (-1, -1), (-2, 0), \text{ and } (1, 1)$$

in the set $L^1(2)$.

1. Computation of $\mathcal{O}^1(2)(0, -2)$
   From equation (5.27)

$$\mathcal{O}^1(2)(0, -2) = \sum_{i=0}^{2-1} n_i^1 + \mathcal{O}_2^1(0, -2) \qquad (5.28)$$

   where $\quad n_0^1 = 1, \quad n_1^1 = 4\ell = 4 \times 1 = 4$
   and $\quad \mathcal{O}_2^1(0, -2) = 2(\lambda_0 - w_0) = 2(2 - 0) = 4$
   therefore, $\quad \mathcal{O}^1(2)(0, -2) = 1 + 4 + 4 = 9$.

2. Computation of $\mathcal{O}^1(-1, -1)$
   From equation (5.27)

$$\mathcal{O}^1(2)(-1, -1) = \sum_{i=0}^{2-1} n_i^1 + \mathcal{O}_2^1(-1, -1) \qquad (5.29)$$

   where $\quad n_0^1 = 1, \quad n_1^1 = 4\ell = 4 \times 1 = 4$
   and $\quad \mathcal{O}_2^1(-1, -1) = 2(\lambda_0 - w_0) = 2(2 - (-1)) = 6$
   therefore, $\quad \mathcal{O}^1(2)(-1, -1) = 1 + 4 + 6 = 11$.

3. Computation of $\mathcal{O}^1(-2, 0)$
   From equation (5.27)

$$\mathcal{O}^1(2)(-1, -1) = \sum_{i=0}^{2-1} n_i^1 + \mathcal{O}_2^1(-2, 0) \qquad (5.30)$$

   where $\quad n_0^1 = 1, \quad n_1^1 = 4\ell = 4 \times 1 = 4$
   and $\quad \mathcal{O}_2^1(-2, 0) = 2(\lambda_0 - w_0) - 1 = 2(2 - (-2)) - 1 = 7$
   therefore, $\quad \mathcal{O}^1(2)(-2, 0) = 1 + 4 + 7 = 12$.

4. Computation of $\mathcal{O}^1(1, 1)$
   From equation (5.27)

$$\mathcal{O}^1(2)(1, 1) = \sum_{i=0}^{2-1} n_i^1 + \mathcal{O}_2^1(1, 1) \qquad (5.31)$$

   where $\quad n_0^1 = 1, \quad n_1^1 = 4\ell = 4 \times 1 = 4$
   and $\quad \mathcal{O}_2^1(1, 1) = 2(\lambda_0 - w_0) - 1 = 2(2 - 1) - 1 = 1$
   therefore, $\quad \mathcal{O}^1(2)(1, 1) = 1 + 4 + 1 = 6$.

Table 5.2.5 *Order of the one-dimensional numbers in the set $L^1(2)$ in the level $\ell = 2$.*

w_0	w_1	numbers	$\mathcal{O}^1(w_0, w_1)$
0	0	$(0,0)$	0
1	0	$(1,0)$	1
0	1	$(0,1)$	2
0	-1	$(0,-1)$	3
-1	0	$(-1,0)$	4
2	0	$(2,0)$	5
1	1	$(1,1)$	6
1	-1	$(1,-1)$	7
0	2	$(0,2)$	8
0	-2	$(0,-2)$	9
-1	1	$(-1,1)$	10
-1	-1	$(-1,-1)$	11
-2	0	$(-2,0)$	12

Table 5.2.5 shows the order of the elements in the set $L^1(2)$. Compare this table with 2.7.16 on page 63 that was used for the computation of the constrained numbers in the set $L^1(2)$.

end of example 5.2.4

Example 5.2.6 *Order of the elements in the set L_3^1.*

Table 5.2.7 *Order of the elements in the one-dimensional set L_3^1 in the level $\ell = 3$.*

w_1	w_0	coordinates	$\mathcal{O}^1(w_0, w_1)$
3	0	$(3,0)$	0
2	1	$(2,1)$	1
2	-1	$(2,-1)$	2
1	2	$(1,2)$	3
1	-2	$(1,-2)$	4
0	3	$(0,3)$	5
0	-3	$(0,-3)$	6
-1	2	$(-1,2)$	7
-1	-2	$(-1,-2)$	8
-2	1	$(-2,1)$	9
-2	-1	$(-2,-1)$	10
-3	0	$(-3,0)$	11

Table 5.2.7 shows the computation of the elements in the set L_3^1 and their order. They can be rewritten according to their order as

$$(3,0) \prec (2,1) \prec (2,-1) \prec (1,2) \prec (1,-2) \prec (0,-3) \prec (0,3) \prec$$
$$(-1,2) \prec (-1,-2) \prec (-2,1) \prec (-2,-1) \prec (-3,0)$$

and they are represented geometrically in the figure 5.2.8.

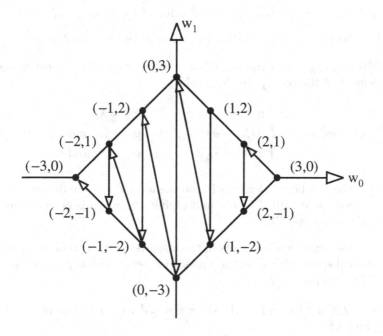

Figure 5.2.8 *Ordering from the smallest to the highest one-dimensional constrained number in the set L_3^1 and level $\lambda_0 = 3$*

end of example 5.2.6

5.2.3   Two-dimensional constrained space

The order of a two-dimensional constrained system with $\mathbb{Z}$ as the underlying set is given by

1. The order with respect to the first element of the level is given by

$$\mathcal{O}_\ell^2(w_0, w_1, w_2) = \begin{cases} 0 & \text{if } w_0 = \ell \\ \displaystyle\sum_{i=\lambda_0}^{w_0+1} n_{\lambda_0-|i|}^1 + \mathcal{O}_{\lambda_1}^1(w_1, w_2) & \text{if } w_0 < \ell \end{cases} \quad (5.32)$$

2. The order with respect to the zero element of the level is given by

$$\mathcal{O}^2(\ell)(w_0, w_1, w_2) = \sum_{i=0}^{\lambda_0 - 1} n_i^2 + \mathcal{O}_\ell^2(w_0, w_1, w_2) \tag{5.33}$$

where

1. $\lambda_0 = |w_0| + |w_1| + |w_2| = \ell$ is the value of the level

2. $\lambda_1 = |w_1| + |w_2| = \ell - |w_0|$ is the value of the first sublevel

3. $O_\ell^1(w_1, w_2)$ given by equation (5.26) is the order with respect to the first element of the set $L_{\lambda_1}^1$ in the level λ_1

$$\mathcal{O}_{\lambda_1}^1(w_1, w_2) = \begin{cases} 0 & \text{if} \quad w_1 = \lambda_1 \\ 2(\lambda_1 - w_1) - 1 & \text{if} \quad w_1 \neq \lambda_1 \text{ and } w_2 \geq 0 \\ 2(\lambda_1 - w_1) & \text{if} \quad w_1 \neq \lambda_1 \text{ and } w_2 < 0 \end{cases} \tag{5.34}$$

4. n_k^1 is the number of elements in the the level k of a one-dimensional constrained space with $\mathbb{Z}$ as the underlying set, which is given by equation (4.120) on page 171

5. n_k^2 is the number of elements in the the level k of a two-dimensional constrained space with $\mathbb{Z}$ as the underlying set, which is given by equation (4.122) on page 172

Tables 5.2.9, 5.2.10, and 5.2.12 show the order of the elements in the sets L_1^2, L_2^2, and L_3^2.

Table 5.2.9 *Order of the elements in a two-dimensional set L_1^2 in the level ℓ.*

w_0	w_1	w_2	numbers	$\mathcal{O}^2(w_0, w_1, w_2)$
1	0	0	$(1, 0, 0)$	0
0	1	0	$(0, 1, 0)$	1
0	0	1	$(0, 0, 1)$	2
0	0	-1	$(0, 0, -1)$	3
0	-1	0	$(0, -1, 0)$	4
-1	0	0	$(-1, 0, 0)$	5

The elements in the set L_1^2 can be rewritten according to their order as

$$(1, 0, 0) \prec (0, 1, 0) \prec (0, 0, 1) \prec (0, 0, -1) \prec (0, -1, 0) \prec (-1, 0, 0) \tag{5.35}$$

which is equivalent to

$$\{1 \times L_1^1\} \prec \{0 \times L_2^1\} \prec \{-1 \times L_1^1\} \tag{5.36}$$

Table 5.2.10 *Computation of the two-dimensional numbers in the set L_2^2.*

w_0	w_1	w_2	numbers	$\mathcal{O}^2(w_0, w_1, w_2)$
2	0	0	$(2,0,0)$	0
1	1	0	$(1,1,0)$	1
1	0	1	$(1,0,1)$	2
1	0	−1	$(1,0,-1)$	3
1	−1	0	$(1,-1,0)$	4
0	2	0	$(0,2,0)$	5
0	1	1	$(0,1,1)$	6
0	1	−1	$(0,1,-1)$	7
0	0	2	$(0,0,2)$	8
0	0	−2	$(0,0,-2)$	9
0	−1	1	$(0,-1,1)$	10
0	−1	−1	$(0,-1,-1)$	11
0	−2	0	$(0,-2,0)$	12
−1	1	0	$(-1,1,0)$	13
−1	−1	0	$(-1,-1,0)$	14
−1	0	1	$(-1,0,1)$	15
−1	0	−1	$(-1,0,-1)$	16
−2	0	0	$(-2,0,0)$	17

The elements in the set L_2^2 can be rewritten according to their order as

$$
\begin{aligned}
(2,0,0) \prec \\
(1,1,0) \prec (-1,0,-1) \prec (1,0,1) \prec (1,0,-1) \prec (1,-1,0) \prec \\
(0,2,0) \prec (0,1,1) \prec (0,1,-1) \prec (0,0,2) \prec (0,0,-2) \prec \\
(0,-1,1) \prec (0,-1,-1) \prec (0,-2,0) \prec \\
(-1,1,0) \prec (-1,-1,0) \prec (-1,0,1) \prec (-1,0,-1) \prec \\
(-2,0,0)
\end{aligned}
\tag{5.37}
$$

which is equivalent to

$$\{2 \times L_0^2\} \prec \{1 \times L_1^2\} \prec \{0 \times L_2^2\} \prec \{-1 \times L_1^2\} \prec \{-2 \times L_0^2\} \tag{5.38}$$

Example 5.2.11 *Find the order of the numbers* $(0, -2, -1)$, $(-1, -2, 0)$ *and* $(1, 1, 1)$ *in the set* L_3^2 *for* $\mathbb{Z}$ *as the underlying set.*

1. Computation of $\mathcal{O}^1(0, -2, -1)$

   From equation (5.32)

   $$\mathcal{O}_3^2(0, -2, -1) = \sum_{i=3}^{0+1} n_{\lambda_0 - |i|}^1 + \mathcal{O}_{\lambda_1}^1(w_1, w_2)$$
   $$= n_0^1 + n_1^1 + n_2^1 + \mathcal{O}_3^1(-2, -1) \tag{5.39}$$

   where $n_0^1 = 1$, $n_k^1 = 4k$,

   and

   $\mathcal{O}_3^1(-2, -1) = 2(\lambda_1 - w_1) = 2(|-2| + |-1| - (-1)) = 10.$

   Therefore, $\mathcal{O}_3^2(0, -2, -1) = 1 + 4 + 8 + 10 = 23$, which is the value shown in table 5.2.12.

2. Computation of $\mathcal{O}_3^1(-1, -2, 0)$

   From equation (5.32)

   $$\mathcal{O}_3^2(-1, -2, 0) = \sum_{i=3}^{-2+1} n_{3 - |i|}^1 + \mathcal{O}_{\lambda_1}^1(w_1, w_2)$$
   $$= n_0^1 + n_1^1 + n_2^1 + n_3^1 + n_2^1 + \mathcal{O}_2^1(-2, 0) \tag{5.40}$$

   where $n_0^1 = 1$, $n_k^1 = 4k$,

   and

   $\mathcal{O}_2^1(-2, 0) = 2(\lambda_1 - w_1) - 1 = 2(|-2| + |0| - 0) - 1 = 3.$

   Therefore, $\mathcal{O}_3^2(-1, -2, 0) = 1 + 4 + 8 + 12 + 8 + 3 = 36$, which is the value shown in table 5.2.12.

3. Computation of $\mathcal{O}_3^1(1, 1, 1)$

   From equation (5.32)

   $$\mathcal{O}_3^2(1, 1, 1) = \sum_{i=3}^{1+1} n_{3 - |i|}^1 + \mathcal{O}_{\lambda_1}^1(w_1, w_2) = n_0^1 + n_1^1 + \mathcal{O}_2^1(1, 1) \tag{5.41}$$

   where $n_0^1 = 1$, $n_k^1 = 4k$,

   and $\mathcal{O}_2^1(1, 1) = 2(\lambda_1 - w_1) - 1 = 2(|1| + |1| - 1) - 1 = 1.$

   Therefore, $\mathcal{O}_3^2(1, 1, 1) = 1 + 4 + 1 = 6$, which is the value shown in table 5.2.12.

   end of example 5.2.11

Table 5.2.12 *Order of the elements of the two-dimensional set L_3^2 in the level $\lambda_0 = 3$.*

w_0	w_1	w_2	numbers	$\mathcal{O}_3^2(w_0, w_1, w_2)$
3	0	0	$(3, 0, 0)$	0
2	1	0	$(2, 1, 0)$	1
2	0	1	$(2, 0, 1)$	2
2	0	-1	$(2, 0, -1)$	3
2	-1	0	$(2, -1, 0)$	4
1	2	0	$(1, 2, 0)$	5
1	1	1	$(1, 1, 1)$	6
1	1	-1	$(1, 1, -1)$	7
1	0	2	$(1, 0, 2)$	8
1	0	-2	$(1, 0, -2)$	9
1	-1	1	$(1, -1, 1)$	10
1	-1	-1	$(1, -1, -1)$	11
1	-2	0	$(0, -2, 0)$	12
0	3	0	$(0, 3, 0)$	13
0	2	1	$(0, 2, 1)$	14
0	2	-1	$(0, 2, -1)$	15
0	1	2	$(0, 1, 2)$	16
0	1	-2	$(0, 1, -2)$	17
0	0	3	$(0, 0, 3)$	18
0	0	-3	$(0, 0, -3)$	19
0	-1	2	$(0, -1, 2)$	20
0	-1	-2	$(0, -1, -2)$	21
0	-2	1	$(0, -2, 1)$	22
0	-2	-1	$(0, -2, -1)$	23
0	-3	0	$(0, -3, 0)$	24
-1	2	0	$(-1, 2, 1)$	25
-1	1	1	$(-1, 1, 1)$	26
-1	1	-1	$(-1, 1, -1)$	27
-1	0	2	$(-1, 0, 2)$	28
-1	0	-2	$(-1, 0, -2)$	29
-1	-1	1	$(-1, -1, 1)$	30
-1	-1	-1	$(-1, -1, -1)$	31
-1	-2	0	$(-1, -2, 0)$	32
-2	1	0	$(-2, 1, 0)$	33
-2	0	1	$(-2, 0, 1)$	34
-2	0	-1	$(-2, 0, -1)$	35
-2	-1	0	$(-2, -1, 0)$	36
-3	0	0	$(0, 0, 3)$	37

5.2.4   Three-dimensional constrained space

The order of a two-dimensional constrained system with $\mathbb{Z}$ as the underlying set is given by

1. The order with respect to the first element of the level is given by

$$\mathcal{O}_\ell^3(w_0, w_1, w_2, w_3) = \begin{cases} 0 & \text{if} \quad w_0 = \ell \\ \sum_{i=\lambda_0}^{w_0+1} n_{\lambda_0-|i|}^2 + \mathcal{O}_{\lambda_1}^2(w_1, w_2, w_3) & \text{if} \quad w_0 < \ell \end{cases}$$

(5.42)

The order $\mathcal{O}_{\lambda_1}^2(w_1, w_2, w_3)$ with respect to the first element of the set $L_{\lambda_1}^2$ in the level λ_1 can be evaluated using the equation (5.32) as follows

$$\mathcal{O}_{\lambda_1}^2(w_1, w_2, w_3) = \begin{cases} 0 & \text{if} \quad w_1 = \lambda_1 \\ \sum_{i=\lambda_1}^{w_1+1} n_{\lambda_1-|i|}^1 + \mathcal{O}_{\lambda_2}^1(w_2, w_3) & \text{if} \quad w_1 < \lambda_1 \end{cases}$$

(5.43)

and the order $\mathcal{O}_{\lambda_2}^1(w_2, w_3)$ with respect to the first element of the set $L_{\lambda_2}^1$ in the level λ_2 can be evaluated using the equation (5.26) as follows

$$\mathcal{O}_{\lambda_2}^1(w_2, w_3) = \begin{cases} 0 & \text{if} \quad w_2 = \lambda_2 \\ 2(\lambda_2 - w_2) - 1 & \text{if} \quad w_2 \neq \lambda_2 \text{ and } w_3 \geq 0 \\ 2(\lambda_2 - w_2) & \text{if} \quad w_2 \neq \lambda_2 \text{ and } w_3 < 0 \end{cases}$$

(5.44)

2. The order with respect to the zero element of the level is given by

$$\mathcal{O}^3(\ell)(w_0, w_1, w_2, w_3) = \sum_{i=0}^{\lambda_0-1} n_i^3 + \mathcal{O}_\ell^3(w_0, w_1, w_2, w_3)$$

(5.45)

where

1. $\lambda_0 = |w_0| + |w_1| + |w_2| + |w_3| = \ell$ is the value of the level

2. $\lambda_1 = |w_1| + |w_2| + |w_3| = \ell - |w_0|$ is the value of the first sublevel

3. $\lambda_2 = |w_2| + |w_3| = \ell - |w_0| - |w_1|$ is the value of the second sublevel

4. n_k^2 is the number of elements in the the level k of a one-dimensional constrained space with $\mathbb{Z}$ as the underlying set, which is given by equation (4.122) on page 172

5. n_k^3 is the number of elements in the the level k of a two-dimensional constrained space with $\mathbb{Z}$ as the underlying set, which is given by equation (4.124) on page 172

Table 5.2.13 *Order of the elements of the set L_1^3 in a three-dimensional constrained space in the level $\lambda_0 = 1$.*

w_0	w_1	w_2	w_3	coordinates	$\mathcal{O}^3(w_0, w_1, w_2, w_3)$
1	0	0	0	$(1, 0, 0, 0)$	0
0	1	0	0	$(0, 1, 0, 0)$	1
0	0	1	0	$(0, 0, 1, 0)$	2
0	0	0	1	$(-1, 0, 0, 1)$	3
0	0	0	-1	$(0, 0, 0, -1)$	4
0	0	-1	0	$(0, 0, -1, 0)$	5
0	-1	0	0	$(0, -1, 0, 0)$	6
-1	0	0	0	$(-1, 0, 0, 0)$	7

The elements in the set L_1^3 can be rewritten according to their order as

$(1, 0, 0, 0) \prec$
$(0, 1, 0, 0) \prec (0, 0, 1, 0) \prec (0, 0, 0, 1) \prec (0, 0, 0, -1) \prec (0, 0, -1, 0) \prec (0, -1, 0, 0) \prec$
$(-1, 0, 0, 0)$ \hfill (5.46)

which is equivalent to

$$\{1 \times L_1^2\} \prec \{0 \times L_2^2\} \prec \{-1 \times L_1^2\} \qquad (5.47)$$

Table 5.2.14 *Order of the elements of the set L_2^3 in a three-dimensional constrained space in the level $\lambda_0 = 2$.*

w_0	w_1	w_2	w_3	coordinates	$\mathcal{O}^3(w_0, w_1, w_2, w_3)$
2	0	0	0	$(2,0,0,0)$	0
1	1	0	0	$(1,1,0,0)$	1
1	0	1	0	$(1,0,1,0)$	2
1	0	0	1	$(1,0,0,1)$	3
1	0	0	-1	$(1,0,0,-1)$	4
1	0	-1	0	$(1,0,-1,0)$	5
1	-1	0	0	$(1,-1,0,0)$	6
0	2	0	0	$(0,2,0,0)$	7
0	1	1	0	$(0,1,1,0)$	8
0	1	0	1	$(0,1,0,1)$	9
0	1	0	-1	$(0,1,0,-1)$	10
0	1	-1	0	$(0,1,-1,0)$	11
0	0	2	0	$(0,0,2,0)$	12
0	0	1	1	$(0,0,1,1)$	13
0	0	1	-1	$(0,0,1,-1)$	14
0	0	0	2	$(0,0,0,2)$	15
0	0	0	-2	$(0,0,0,-2)$	16
0	0	-1	1	$(0,0,-1,1)$	17
0	0	-1	-1	$(0,0,-1,-1)$	18
0	0	-2	0	$(0,0,-2,0)$	19
0	-1	1	0	$(0,-1,1,0)$	20
0	-1	-1	0	$(0,-1,-1,0)$	21
0	-1	0	1	$(0,-1,0,1)$	22
0	-1	0	-1	$(0,-1,0,-1)$	23
0	-2	0	0	$(0,-2,0,0)$	24
-1	1	0	0	$(-1,1,0,0)$	25
-1	0	1	0	$(-1,0,1,0)$	26
-1	0	0	1	$(-1,0,0,1)$	27
-1	0	0	-1	$(-1,0,0,-1)$	28
-1	0	-1	0	$(-1,0,-1,0)$	29
-1	-1	0	0	$(-1,-1,0,0)$	30
-2	0	0	0	$(-2,0,0,0)$	31

Example 5.2.15 *Obtain the order of the elements* $(1,0,0,-1)$, $(0,-1,1,0)$, *and* $(-1,0,0,1)$ *in the set* L_2^3. *Compare the results with table 5.2.14 to verify the result.*

1. Computation of $\mathcal{O}_2^3(1,0,0,-1)$

   Since $w_0 \neq \lambda_0$ and $w_1 \neq \lambda_1$ the equations (5.42)–(5.43) can be written as one equation only as

   $$\mathcal{O}_\ell^3(w_0,w_1,w_2,w_3) = \sum_{i=\lambda_0}^{w_0+1} n_{\lambda_0-|i|}^2 + \sum_{j=\lambda_1}^{w_1+1} n_{\lambda_1-|j|}^1 + \mathcal{O}_{\lambda_2}^1(w_2,w_3) \quad (5.48)$$

   and in addition $w_3 \neq \lambda_2$ then $\mathcal{O}_{\lambda_2}^1(w_2,w_3)$ can be replaced by its expression given by equation (5.44) as

   $$\mathcal{O}_\ell^3(w_0,w_1,w_2,w_3) = \sum_{i=\lambda_0}^{w_0+1} n_{\lambda_0-|i|}^2 + \sum_{j=\lambda_1}^{w_1+1} n_{\lambda_1-|j|}^1 + 2(\lambda_2 - w_2) \quad (5.49)$$

   Substituting $w_0 = 1, w_1 = 0, w_2 = 0, w_3 = 1, \lambda_0 = 2, \lambda_1 = 1$, and $\lambda_2 = 1$ into equation (5.49) it gives the order as

   $$\mathcal{O}_2^3(1,0,0,-1) = \sum_{i=2}^{1+1} n_{2-|i|}^2 + \sum_{j=1}^{0+1} n_{1-|j|}^1 + 2(1-(-1))$$

   $$= n_0^2 + n_0^1 + 2(\lambda_2 - w_2) = 1 + 1 + 2 = 4 \quad (5.50)$$

   Compare this result with table 5.2.14.

2. Computation of $\mathcal{O}_2^3(0,-1,1,0)$

   Since $w_0 \neq \lambda_0$, $w_1 \neq \lambda_1$, and in $w_3 = \lambda_2$ then equations (5.42)–(5.44) can be written as one equation only as

   $$\mathcal{O}_\ell^3(w_0,w_1,w_2,w_3) = \sum_{i=\lambda_0}^{w_0+1} n_{\lambda_0-|i|}^2 + \sum_{j=\lambda_1}^{w_1+1} n_{\lambda_1-|j|}^1 \quad (5.51)$$

   Substituting $w_0 = 0, w_1 = -1, w_2 = 1, w_3 = 0, \lambda_0 = 2, \lambda_1 = 2$, and $\lambda_2 = 1$ into equation (5.51) it gives the order as

   $$\mathcal{O}_2^3(0,-1,1,0) = \sum_{i=2}^{0+1} n_{2-|i|}^2 + \sum_{j=2}^{-1+1} n_{1-|j|}^1$$

   $$= (n_0^2 + n_1^2) + (n_0^1 + n_1^1 + n_2^1)$$

   $$= (1+6) + (1+4+8) = 20 \quad (5.52)$$

   Compare this result with table 5.2.14.

3. Computation of $\mathcal{O}_2^3(-1, 0, 0, 1)$

Since $w_0 \neq \lambda_0$, $w_1 \neq \lambda_1$, and in $w_3 \neq \lambda_2$ then equations (5.42)–(5.44) can be written as one equation only as

$$\mathcal{O}_\ell^3(w_0, w_1, w_2, w_3) = \sum_{i=\lambda_0}^{w_0+1} n_{\lambda_0-|i|}^2 + \sum_{j=\lambda_1}^{w_1+1} n_{\lambda_1-|j|}^1 + 2(\lambda_2 - w_2) - 1 \quad (5.53)$$

Substituting $w_0 = -1, w_1 = 0, w_2 = 0, w_3 = 1$, $\lambda_0 = 2$, $\lambda_1 = 1$, and $\lambda_2 = 1$ into equation (5.53) it gives the order as

$$\begin{aligned}
\mathcal{O}_2^3(-1, 0, 0, 1) &= \sum_{i=2}^{-1+1} n_{2-|i|}^2 + \sum_{j=1}^{0+1} n_{1-|j|}^1 + 2(\lambda_2 - w_2) - 1 \\
&= (n_0^2 + n_1^2 + n_2^2) + n_0^1 + 2(1-0) - 1 \\
&= (1 + 6 + 18) + 1 + 1 = 27 \quad (5.54)
\end{aligned}$$

Compare this result with table 5.2.14.

end of example 5.2.15

5.3   Nonnegative integer numbers as the underlying set

5.3.1   Zero-dimensional constrained space

5.3.1.1   Order referred to the least significant number in the level

Each level of the zero-dimensional space contains only one element, therefore, its order is always zero, that is,

$$\mathcal{O}_0^0(w_0) = 0 \quad (5.55)$$

5.3.1.2   Order referred to the element zero of the space

The zero dimensional space is ordered by its levels. The sequence of levels equals the order, which are given by

$$\begin{aligned}
L_0^0 &= \{(0)\}, \quad L_1^0 = \{(1)\}, \quad L_2^0 = \{(2)\}, \\
L_3^0 &= \{(3)\}, \quad L_4^0 = \{(4)\}, \dots, L_\delta^0 = \{(\delta)\}, \dots
\end{aligned} \quad (5.56)$$

The number of elements in the set $L^0(\ell)$ is given by equation (4.127) on page 173 and it is

$$n^0(\ell) = 1 + \ell \quad (5.57)$$

and the order referred to the element (0) equals the number of elements minus one since the order of the first element has been defined as zero. Then

$$\mathcal{O}^1(w_0) = n^0(\ell) - 1 = \ell = w_0 \qquad (5.58)$$

5.3.2 One-dimensional constrained space

5.3.2.1 Order referred to the least significant number in the level

The least significant element of a one-dimensional space in the level ℓ is $(\ell, 0)$. The order with respect to this number is given by

$$\mathcal{O}^1_\ell(w_0, w_1) = n^0(w_1) - 1 = \lambda_1(w_0, w_1) = w_1 \qquad (5.59)$$

where $\lambda_1 = |w_1|$ is the first sublevel of (w_0, w_1).

5.3.2.2 Order referred to the element zero of the space

The order of the element (w_0, w_1) with respect to the origin $(0,0)$ can be obtained adding to the order of (w_0, w_1) with respect to the first element of the level $\ell = |w_0| + |w_1|$ the number of elements from $(0,0)$ until the previous level, which is $\lambda_0 - 1$. Then the order will be given by

$$\mathcal{O}^1(w_0, w_1) = n^1(\lambda_0 - 1) + \mathcal{O}^1_\ell(w_0, w_1) \qquad (5.60)$$

where $n^1(\lambda_0 - 1)$ is the number of elements from the level zero, which is $\lambda_0(0,0)$ until the previous level corresponding to the number (w_0, w_1), which is $[\lambda_0(w_0, w_1) - 1]$.

Making the substitution $\ell = \lambda_0 - 1$ in equation (4.129) it follows

$$n^1(\ell) = n^1(\lambda_0 - 1) = \frac{\lambda_0(\lambda_0 + 1)}{2!} \qquad (5.61)$$

where $\lambda_0 = \lambda_0(w_0, w_1)$. From equation (5.59)

$$\mathcal{O}^1_\ell(w_0, w_1) = \mathcal{O}^0(w_1) = \lambda_1 \qquad (5.62)$$

Substituting into (5.60)

$$\mathcal{O}^1(w_0, w_1) = \frac{\lambda_0(\lambda_0 + 1)}{2!} + \frac{\lambda_1}{1!} \qquad (5.63)$$

where $\lambda_0(w_0, w_1) = |w_0| + |w_1| = w_0 + w_1$ and $\lambda_1(w_0, w_1) = |w_1| = w_1$, since the coordinate numbers are nonnegative for the underlying set $\mathbb{Z}^+$.

In terms of the coordinate numbers the equation (5.63) yields

$$\mathcal{O}^1(w_0, w_1) = \frac{(w_0 + w_1)(w_0 + w_1 + 1)}{2} + \frac{w_1}{1} \qquad (5.64)$$

Example 5.3.1 *Evaluate* $\mathcal{O}_5^1(3,2)$ *and* $\mathcal{O}^1(3,2)$.

1. Order referred to the least significant number in the level

   Substituting $w_1 = 2$ into equation (5.59) it gives

   $$\mathcal{O}_\ell^1(w_0, w_1) = n^0(w_1) - 1 = w_1 = 2 \qquad (5.65)$$

   See figure 5.3.2 to verify this result.

2. Order referred to the element zero of the space

   The order is obtained by substituting $\lambda_0 = 3 + 2 = 5$ and $\lambda_1 = 2$ into equation (5.63), that is,

   $$\mathcal{O}^1(w_0, w_1) = \frac{\lambda_0(\lambda_0 + 1)}{2!} + \frac{\lambda_1}{1!} = \frac{5(5+1)}{2!} + \frac{2}{1!} = 15 + 2 = 17 \quad (5.66)$$

   Compare this result with table order of $(3,2)$ in the table 5.3.4 and figure 5.3.2.

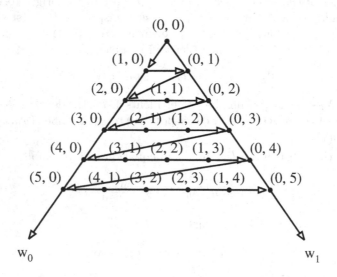

Figure 5.3.2 *The arrows show the order of the elements of the set* L_5^1.

end of example 5.3.1

Example 5.3.3 *Evaluate $\mathcal{O}^1(5,0)$.*

$\lambda_0(w_0, w_1) = |w_0| + |w_1| = 5 + 0 = 5$. $\lambda_1(w_0, w_1) = |w_1| = 0$. The order of $(5,0)$ is given by

$$\mathcal{O}^1(5,0) = \frac{\lambda_0(\lambda_0 + 1)}{2!} + \frac{\lambda_1}{1!} = \frac{5 \times 6}{2!} + \frac{0}{1!} = 15 + 0 = 15 \qquad (5.67)$$

Compare this result with the order of $(5,0)$ in the table 5.3.4.

Table 5.3.4 *Order of the elements in the set $L^1(5)$.*

(w_0, w_1)	$\mathcal{O}$	(w_0, w_1)	$\mathcal{O}$	(w_0, w_1)	$\mathcal{O}$
$(0,0)$	0	$(3,0)$	6	$(4,0)$	10
$(1,0)$	1	$(2,1)$	7	$(3,1)$	11
$(0,1)$	2	$(1,2)$	8	$(2,2)$	12
$(2,0)$	3	$(0,3)$	9	$(1,3)$	13
$(1,1)$	4			$(0,4)$	14
$(0,2)$	5				

(w_0, w_1)	$\mathcal{O}$
$(5,0)$	15
$(4,1)$	16
$(3,2)$	17
$(2,3)$	18
$(1,4)$	19
$(0,5)$	20

end of example 5.3.3

5.3.3   Two-dimensional space

5.3.3.1   Order referred to the least significant number in the level

The least significant element of a two-dimensional space in the level ℓ is $(\ell, 0, 0)$. Each level of a two-dimensional constrained space is a one-dimensional constrained space, therefore, the expression of compute the order with respect to the origin of the level is the same expression that is used to obtain the order of a one-dimensional constrained space referred to the point $(0,0)$, which is given by equation (5.63). Replacing (w_0, w_1) by (w_1, w_2) in equation (5.63) it gives the expression

$$\mathcal{O}_\ell^2(w_0, w_1, w_2) = n^1(\lambda_1 - 1) + \mathcal{O}^1(\lambda_2) = \frac{\lambda_1(\lambda_1 + 1)}{2!} + \frac{\lambda_2}{1!} \qquad (5.68)$$

for the computation of the order where the first sublevel is given by $\lambda_1 = \lambda_1(w_1, w_2) = w_1 + w_2$ and the second sublevel by $\lambda_2 = \lambda_2(w_2) = w_2$.

5.3.3.2   Order referred to the element zero of the space

The order of w is given by adding of the number of elements until the previous level to the order in the first dual sublevel. That is,

$$\mathcal{O}^2(w_0, w_1, w_2) = n^2(\lambda_0 - 1) + \mathcal{O}_\ell^2(w_0, w_1, w_2) \tag{5.69}$$

repeating here the equation (4.77) on page 158 as

$$n^2(\ell) = \frac{(\ell + 1)(\ell + 2)(\ell + 3)}{3!} \tag{5.70}$$

and performing the substitution $\ell = \lambda_0 - 1$, equation (5.70) becomes

$$n^2(\lambda_0 - 1) = \frac{\lambda_0(\lambda_0 + 1)(\lambda_0 + 2)}{6} \tag{5.71}$$

Substituting $n^2(\lambda_0 - 1)$ given by (5.71) and $\mathcal{O}_\ell^2(w_0, w_1, w_2)$ given by (5.68) into equation (5.69)

$$\mathcal{O}^2(w_0, w_1, w_2) = \frac{\lambda_0(\lambda_0 + 1)(\lambda_0 + 2)}{6} + \frac{\lambda_1(\lambda_1 + 1)}{2} + \frac{\lambda_2}{1} \tag{5.72}$$

Where

$\lambda_0(w_0, w_1, w_2) = |w_0| + |w_1| + |w_2|,$

$\lambda_1(w_0, w_1, w_2) = |w_1| + |w_2|,$ and

$\lambda_2(w_0, w_1, w_2) = |w_2|.$

In terms of the coordinate numbers the equation (5.72) writes

$$\mathcal{O}^2(w_0, w_1, w_2) = \frac{(w_0 + w_1 + w_2)(w_0 + w_1 + w_2 + 1)(w_0 + w_1 + w_2 + 2)}{6}$$
$$+ \frac{(w_1 + w_2)(w_1 + w_2 + 1)}{2} + \frac{w_2}{1} \tag{5.73}$$

Recall that all coordinate numbers are nonnegative for $\mathbb{Z}^+$ as the underlying set then the identity $|w_i| = w_i$, which permits to write the expressions for the level and sublevels without the absolute value for the coordinate numbers.

Rewriting (5.69) in an appropriate form preparing for generalization by induction it follows

$$\mathcal{O}^2(w_0, w_1, w_2) = \frac{(\lambda_0 + 0)(\lambda_0 + 1)(\lambda_0 + 2)}{3!} + \frac{(\lambda_1 + 0)(\lambda_1 + 1)}{2!} + \frac{(\lambda_2 + 0)}{1!} \tag{5.74}$$

Writing (5.74) in terms of combinations given in subsection 4.7.4 on page 178 it follows

$$\mathcal{O}^2(w_0, w_1, w_2) = C((\lambda_0 + 2), 3) + C((\lambda_1 + 1), 2) + C((\lambda_2 + 0), 1) \quad (5.75)$$

and writing (5.74) in product form

$$\mathcal{O}^2(w_0, w_1, w_2) = \frac{\prod_{j=0}^2 (\lambda_0 + j)}{3!} + \frac{\prod_{j=0}^1 (\lambda_1 + j)}{2!} + \frac{\prod_{j=0}^0 (\lambda_2 + j)}{1!} \quad (5.76)$$

5.3.4 Generalization for the constrained numbers

5.3.4.1 Order referred to the first element of the level

In terms of combinations

$$\mathcal{O}^\delta(w_0, w_1, w_2, ..., w_\delta) = \sum_{i=\delta-1}^{i=0} C((\lambda_{\delta-i} + i), (i+1)) \quad (5.77)$$

5.3.4.2 Order referred to the element zero of the space

Equations (5.75) and (5.76) suggest the following generalization

1. In terms of combinations

$$\mathcal{O}^\delta(w_0, w_1, w_2, ..., w_\delta) = \sum_{i=\delta}^{i=0} C((\lambda_{\delta-i} + i), (i+1)) \quad (5.78)$$

2. In product form

$$\mathcal{O}^\delta(w_0, w_1, w_2, ..., w_\delta) = \frac{\prod_{j=0}^\delta (\lambda_0 + j)}{(\delta + 1)!} + \frac{\prod_{j=0}^{\delta-1} (\lambda_1 + j)}{\delta!} + \cdots$$
$$\cdots + \frac{\prod_{j=0}^{\delta-2} (\lambda_2 + j)}{(\delta - 1)!} + \frac{\prod_{j=0}^2 (\lambda_{\delta-2} + j)}{3!}$$
$$+ \frac{\prod_{j=0}^1 (\lambda_{\delta-1} + j)}{2!} + \frac{\prod_{j=0}^0 (\lambda_\delta + j)}{1!} \quad (5.79)$$

which can be written in summation form as

$$\mathcal{O}^\delta(w) = \mathcal{O}(w_0, w_1, w_2, ..., w_\delta) = \sum_{s=0}^\delta \frac{\prod_{j=0}^s (\lambda_{\delta-s} + j)}{(s + 1)!} \quad (5.80)$$

5.4   Properties of the ordering function

Property 5.4.1

As consequence of definition 5.1.1

1. the first element of the set L_ℓ^δ in a level ℓ, that is, the smallest, is denoted as for the underlying set $\mathbb{Z}$

$$(\ell, 0, 0, \ldots, 0)$$

and the last element of the set L_ℓ^δ, that is, the one with highest value, is denoted as
$$(-\ell, 0, 0, \ldots, 0)$$

2. the first element of the set L_ℓ^δ in a level ℓ for the underlying set $\mathbb{Z}^+$, that is, the smallest value, is denoted as

$$(\ell, 0, \ldots, 0)$$

and the last element of the set L_ℓ^δ on the same level for the underlying set $\mathbb{Z}^+$, that is, the one with highest value, is denoted as

$$(0, 0, 0, \ldots, \ell)$$

end of property 5.4.1

Property 5.4.2 *Interval of level ℓ.*

1. for the underlying set $\mathbb{Z}$

The interval I containing all elements w in the level ℓ is given by

$$I_q = \Big[(\ell, 0, \ldots, 0), \, (-\ell, 0, \ldots, 0) \Big] \qquad (5.81)$$

2. for the underlying set $\mathbb{Z}$

The interval I containing all elements w in the level ℓ is given by

$$I_q = \Big[(\ell, 0, 0, \ldots, 0), \, (0, 0, \ldots, \ell) \Big] \qquad (5.82)$$

end of property 5.4.2

Notation 5.4.3 *Closed interval.*

Denote $[u, v]$ a closed interval where $u = (u_0, u_1, \ldots, u_\delta)$, $w = (w_0, w_1, \ldots, w_\delta)$, and $v = (v_0, v_1, \ldots, v_\delta)$ are constrained numbers such that $\mathcal{O}^\delta(u) \leq \mathcal{O}^\delta(w) \leq \mathcal{O}^\delta(v)$.

end of notation 5.4.3

Property 5.4.4 *Number of elements in an interval*

The number of elements in the closed interval $[u, v]$ is given by

$$n[u, v] = \mathcal{O}^\delta(v) - \mathcal{O}^\delta(u) + 1 \tag{5.83}$$

If the interval equals the the set of elements of a level, that is, L_ℓ^δ, then the number of elements is given by n_ℓ^δ in section 4.6 on page 171.

end of property 5.4.4

5.5 Summary

5.5.1 Summary for the underlying set is $\mathbb{Z}$

5.5.1.1 Zero-dimensional constrained space

The order of the zero-dimensional constrained numbers was obtained in equation (5.25) and it is

$$\mathcal{O}^0(w_0) = \begin{cases} 0 & w_0 \leq 0 \\ 1 & w_0 > 0 \end{cases} \tag{5.84}$$

5.5.1.2 One-dimensional constrained space

a- Order referred to the first element of the level

Order referred to the first element of the level was obtained in equation (5.26) and it is

$$\mathcal{O}_\ell^1(w_0, w_1) = \begin{cases} 0 & \text{if } w_0 = \lambda_0 \\ 2(\lambda_0 - w_0) - 1 & \text{if } w_0 \neq \lambda_0 \text{ and } w_1 \geq 0 \\ 2(\lambda_0 - w_0) & \text{if } w_0 \neq \lambda_0 \text{ and } w_1 < 0 \end{cases} \tag{5.85}$$

b- Order referred to the element zero of the level

The order with respect to the zero element of the level was obtained in equation (5.27) and it is

$$\mathcal{O}^1(\ell)(w_0, w_1) = \sum_{i=0}^{\lambda_0 - 1} n_i^1 + \mathcal{O}_\ell^1(w_0, w_1) \tag{5.86}$$

5.5.1.3   Two-dimensional constrained space

a- Order referred to the first element of the level

Order referred to the first element of the level was obtained in equation (5.32) and it is

$$\mathcal{O}_\ell^2(w_0, w_1, w_2) = \begin{cases} 0 & \text{if} \quad w_0 = \ell \\ \displaystyle\sum_{i=\lambda_0}^{w_0 + 1} n_{\lambda_0 - |i|}^1 + \mathcal{O}_{\lambda_1}^1(w_1, w_2) & \text{if} \quad w_0 < \ell \end{cases} \tag{5.87}$$

And the order $\mathcal{O}_{\lambda_2}^1(w_2, w_3)$ with respect to the first element of the set $L_{\lambda_2}^1$ in the level λ_2 can be evaluated using the equation (5.85) as follows

$$\mathcal{O}_{\lambda_1}^1(w_1, w_2) = \begin{cases} 0 & \text{if} \quad w_1 = \lambda_1 \\ 2(\lambda_1 - w_1) - 1 & \text{if} \quad w_1 \neq \lambda_1 \text{ and } w_2 \geq 0 \\ 2(\lambda_1 - w_1) & \text{if} \quad w_1 \neq \lambda_1 \text{ and } w_2 < 0 \end{cases} \tag{5.88}$$

b- Order referred to the element zero of the level

Order referred to the first element of the level was obtained in equation (5.33) and it is

$$\mathcal{O}^2(\ell)(w_0, w_1, w_2) = \sum_{i=0}^{\lambda_0 - 1} n_i^2 + \mathcal{O}_\ell^2(w_0, w_1, w_2) \tag{5.89}$$

5.5.1.4   Three-dimensional constrained space

a- Order referred to the first element of the level

Order referred to the first element of the level was obtained in equation (5.42) and it is

$$\mathcal{O}_\ell^3(w_0, w_1, w_2, w_3) = \begin{cases} 0 & \text{if} \quad w_0 = \ell \\ \displaystyle\sum_{i=\lambda_0}^{w_0 + 1} n_{\lambda_0 - |i|}^2 + \mathcal{O}_{\lambda_1}^2(w_1, w_2, w_3) & \text{if} \quad w_0 < \ell \end{cases} \tag{5.90}$$

The order $\mathcal{O}^2_{\lambda_1}(w_1, w_2, w_3)$ with respect to the first element of the set $L^2_{\lambda_1}$ in the level λ_1 can be evaluated using the equation (5.87) as follows

$$
\mathcal{O}^2_{\lambda_1}(w_1, w_2, w_3) = \begin{cases} 0 & \text{if} \quad w_1 = \lambda_1 \\ \displaystyle\sum_{i=\lambda_1}^{w_1+1} n^1_{\lambda_1-|i|} + \mathcal{O}^1_{\lambda_2}(w_2, w_3) & \text{if} \quad w_1 < \lambda_1 \end{cases} \tag{5.91}
$$

and the order $\mathcal{O}^1_{\lambda_2}(w_2, w_3)$ with respect to the first element of the set $L^1_{\lambda_2}$ in the level λ_2 can be evaluated using the equation (5.85) as follows

$$
\mathcal{O}^1_{\lambda_2}(w_2, w_3) = \begin{cases} 0 & \text{if} \quad w_2 = \lambda_2 \\ 2(\lambda_2 - w_2) - 1 & \text{if} \quad w_2 \neq \lambda_2 \text{ and } w_3 \geq 0 \\ 2(\lambda_2 - w_2) & \text{if} \quad w_2 \neq \lambda_2 \text{ and } w_3 < 0 \end{cases} \tag{5.92}
$$

b- Order referred to the element zero of the level

The order with respect to the zero element of the level was obtained in equation (5.45) and it is

$$
\mathcal{O}^3(\ell)(w_0, w_1, w_2, w_3) = \sum_{i=0}^{\lambda_0-1} n^3_i + \mathcal{O}^3_\ell(w_0, w_1, w_2, w_3) \tag{5.93}
$$

5.5.1.5 δ-dimensional constrained space

The expressions for the computation of the first element and the zero element of the level ℓ in a δ-dimensional space can be obtained by induction from the previous cases and they are:

a- Order referred to the first element of the level

Order referred to the first element of the level was obtained in equation (5.42) and it is

$$
\mathcal{O}^\delta_\ell(w_0, \ldots w_\delta) = \begin{cases} 0 & \text{if} \quad w_0 = \lambda_0 \\ \displaystyle\sum_{i=\lambda_0}^{w_0+1} n^{\delta-1}_{\lambda_0-|i|} + \mathcal{O}^{\delta-1}_{\lambda_1}(w_1, \ldots w_\delta) & \text{if} \quad w_0 < \lambda_0 \end{cases} \tag{5.94}
$$

where $\lambda_0 = \ell$. The order $\mathcal{O}^{\delta-k}_{\lambda_k}(w_k, \ldots w_\delta)$ with respect to the first element of the set $L^{\delta-k}_{\lambda_k}$ in the level λ_k can be evaluated using the equation (5.87) as follows

$$
\mathcal{O}^{\delta-k}_{\lambda_k}(w_k, \ldots w_\delta) = \begin{cases} 0 & \text{if} \quad w_k = \lambda_k \\ \displaystyle\sum_{i=\lambda_k}^{w_k+1} n^{\delta-k-1}_{\lambda_k-|i|} + \mathcal{O}^{\delta-k-1}_{\lambda_{k+1}}(w_{k+1}, \ldots, w_\delta) & \text{if} \quad w_k < \lambda_k \end{cases}
$$

$$
k = 1, \ldots, \delta - 2 \tag{5.95}
$$

and the order $\mathcal{O}^1_{\lambda_{\delta-1}}(w_{\delta-1}, w_\delta)$ with respect to the first element of the set $L^1_{\lambda_{\delta-1}}$ in the level $\lambda_{\delta-1}$ can be evaluated using the equation (5.85) as follows

$$\mathcal{O}^1_{\lambda_{\delta-1}}(w_{\delta-1}, w_\delta) = \begin{cases} 0 & \text{if } w_{\delta-1} = \lambda_{\delta-1} \\ 2\left(\lambda_{\delta-1} - w_{\delta-1}\right) - 1 & \text{if } w_{\delta-1} \neq \lambda_{\delta-1} \text{ and } w_\delta \geq 0 \\ 2\left(\lambda_{\delta-1} - w_{\delta-1}\right) & \text{if } w_{\delta-1} \neq \lambda_{\delta-1} \text{ and } w_\delta < 0 \end{cases} \tag{5.96}$$

b- Order referred to the element zero of the level

The order with respect to the zero element of the level was obtained in equation (5.45) and it is

$$\mathcal{O}^\delta(\ell)(w_0, \ldots w_\delta) = \sum_{i=0}^{\lambda_0-1} n_i^\delta + \mathcal{O}^\delta_\ell(w_0, \ldots w_\delta) \tag{5.97}$$

5.5.2 Summary for the underlying set is $\mathbb{Z}^+$

5.5.2.1 Zero-dimensional constrained space

a- Order referred to the first element of the level

Order referred to the first element of the level was obtained in equation (5.55) and it is

$$\mathcal{O}^0_0(w_0) = 0 \tag{5.98}$$

b- Order referred to the element zero of the level

The order of the zero-dimensional constrained numbers referred to the element zero of the space was obtained in equation (5.58) and it is

$$\mathcal{O}^1(w_0) = n^0(\ell) - 1 = \ell = \lambda_0 \tag{5.99}$$

where $\lambda_0 = |w_0|$ is the level of (w_0).

5.5.2.2 One-dimensional constrained space

a- Order referred to the first element of the level

Order referred to the first element of the level was obtained in equation (5.59) and it is

$$\mathcal{O}^1_\ell(w_0, w_1) = n^0(w_1) - 1 = \lambda_1 \tag{5.100}$$

where $\lambda_1 = \lambda_1(w_0, w_1) = |w_1|$ is the first sublevel of (w_0, w_1).

b- Order referred to the element zero of the level

The order of the one-dimensional constrained numbers referred to the element zero of the space was obtained in equation (5.63) and it is

$$\mathcal{O}^1(w_0, w_1) = \frac{\lambda_0(\lambda_0 + 1)}{2!} + \frac{\lambda_1}{1!} \tag{5.101}$$

5.5.2.3 Two-dimensional constrained space

a- Order referred to the first element of the level

Order referred to the first element of the level was obtained in equation (5.68) and it is

$$\mathcal{O}_\ell^2(w_0, w_1, w_2) = n^1(\lambda_1 - 1) + \mathcal{O}^1(\lambda_2) = \frac{\lambda_1(\lambda_1 + 1)}{2!} + \frac{\lambda_2}{1!} \qquad (5.102)$$

where $\lambda_1 = \lambda_1(w_0, w_1, w_2) = |w_1| + |w_2|$ is the first sublevel of (w_0, w_1, w_2) and $\lambda_2 = \lambda_1(w_0, w_1, w_2) = |w_2|$ is the second sublevel of (w_0, w_1, w_2).

b- Order referred to the element zero of the level

The order of the two-dimensional constrained numbers referred to the element zero of the space was obtained in equation (5.72) and it is

$$\mathcal{O}^2(w_0, w_1, w_2) = \frac{\lambda_0(\lambda_0 + 1)(\lambda_0 + 2)}{6} + \frac{\lambda_1(\lambda_1 + 1)}{2} + \frac{\lambda_2}{1} \qquad (5.103)$$

5.5.2.4 Three-dimensional constrained space

a- Order referred to the first element of the level

The order of the three-dimensional constrained numbers referred to the least significant number in the level was obtained in equation (5.77) with $\delta = 3$ and it is

$$\mathcal{O}_\ell^3(w_0, w_1, w_2, w_3) = \frac{\lambda_1(\lambda_1 + 1)(\lambda_1 + 2)}{3!} + \frac{\lambda_2(\lambda_2 + 1)}{2!} + \frac{\lambda_3}{1!} \qquad (5.104)$$

b- Order referred to the element zero of the level

The order of the three-dimensional constrained numbers referred to the element zero of the space was obtained in equation (5.78) with $\delta = 3$ and it is

$$\mathcal{O}^3(w_0, w_1, w_2, w_3) = \frac{\lambda_0(\lambda_0 + 1)(\lambda_0 + 2)(\lambda_0 + 3)}{4!}$$
$$+ \frac{\lambda_1(\lambda_1 + 1)(\lambda_1 + 2)}{3!} + \frac{\lambda_2(\lambda_2 + 1)}{2!} + \frac{\lambda_3}{1!} \qquad (5.105)$$

5.5.2.5 δ-dimensional constrained space

a- Order referred to the first element of the level

The order of the δ-dimensional constrained numbers referred to the least significant number in the level was obtained in equation (5.77) and it is

$$\mathcal{O}^\delta(w_0, w_1, w_2, ..., w_\delta) = \sum_{i=\delta-1}^{i=0} C((\lambda_{\delta-1} + i), (i+1)) \qquad (5.106)$$

b- Order referred to the element zero of the level

The order of the δ-dimensional constrained numbers referred to the element zero of the space was obtained in equation (5.78) and it is

$$\mathcal{O}^\delta(w_0, w_1, w_2, ..., w_\delta) = \sum_{i=\delta}^{i=0} C((\lambda_{\delta-1} + i), (i + 1)) \tag{5.107}$$

Chapter 6

Application to symbolic computation of derivatives

Application to symbolic computation of derivatives of the product of functions of several variables

---

6.1 Polynomial notation

6.1.1 Examples

A 3-dimensional complete polynomial of degree 3

$$
\begin{aligned}
f(x, y, z) = a_0 \\
+ a_1 x + a_2 y + a_3 z \\
+ a_4 x^2 + a_5 xy + a_6 xz + a_7 y^2 + a_8 yz + a_9 z^2 \\
+ a_{10} x^3 + a_{11} x^2 y + a_{12} x^2 z + a_{13} xy^2 + a_{14} xyz + a_{15} xz^2 + a_{16} y^3 \\
+ a_{17} y^2 z + a_{18} yz^2 + a_{19} z^3
\end{aligned}
\tag{6.1}
$$

will be used to introduce the notation and then the generalization of it.

Define

$$
\begin{aligned}
w &:= (w_0, w_1, w_2) \\
x &:= (x_0, x_1, x_2) = (x, y, z) \\
x^w &:= (x_0^{w_0}, x_1^{w_1}, x_2^{w_2}) \\
a_w &:= a_{(w_0, w_1, w_2)}
\end{aligned}
\tag{6.2}
$$

For example, the terms a_0, a_5, and a_{14} of the polynomial in equation (6.1) can be rewritten as

$$
\begin{aligned}
a_0 &= a_{(0,0,0)} x_0^0 x_1^0 x_2^0 \quad \text{where} \quad w = (0, 0, 0) \\
a_5 &= a_{(1,1,0)} x_0^1 x_1^1 x_2^0 \quad \text{where} \quad w = (1, 1, 0) \\
a_{14} &= a_{(1,1,1)} x_0^1 x_1^1 x_2^1 \quad \text{where} \quad w = (1, 1, 1)
\end{aligned}
\tag{6.3}
$$

The notation can be written in a more compact form as

$$a_w x^w = a_{(w_0,w_1,w_2)} x_0^{w_0} x_1^{w_1} x_2^{w_2} \tag{6.4}$$

With the above notation the 3-dimensional polynomial of degree 3 in equation (6.1) can be rewritten as

$$
\begin{aligned}
f(x,y,z) = {}& a_{(0,0,0)} x^0 y^0 z^0 \\
& + a_{(1,0,0)} x^1 y^0 z^0 + a_{(0,1,0)} x^0 y^1 z^0 + a_{(0,0,1)} x^0 y^0 z^1 \\
& + a_{(2,0,0)} x^2 y^0 z^0 + a_{(1,1,0)} x^1 y^1 z^0 + a_{(1,0,1)} x^1 y^0 z^1 \\
& + a_{(0,2,0)} x^0 y^2 z^0 + a_{(0,1,1)} x^0 y^1 z^1 + a_{(0,0,2)} x^0 y^0 z^2 \\
& + a_{(3,0,0)} x^3 y^0 z^0 + a_{(2,1,0)} x^2 y^1 z^0 + a_{(2,0,1)} x^2 y^0 z^1 \\
& + a_{(1,2,0)} x^1 y^2 z^0 + a_{(1,1,1)} x^1 y^1 z^1 + a_{(1,0,2)} x^1 y^0 z^2 \\
& + a_{(0,3,0)} x^0 y^3 z^0 + a_{(0,2,1)} x^0 y^2 z^1 \\
& + + a_{(0,1,2)} x^0 y^1 z^2 + a_{(0,0,3)} x^0 y^0 z^3
\end{aligned} \tag{6.5}
$$

Replacing the notation xyz by $x_0 x_1 x_2$ it follows

$$
\begin{aligned}
f(x,y,z) = {}& a_{(0,0,0)} x_0^0 x_1^0 x_2^0 \\
& + a_{(1,0,0)} x_0^1 x_1^0 x_2^0 + a_{(0,1,0)} x_0^0 x_1^1 x_2^0 + a_{(0,0,1)} x_0^0 x_1^0 x_2^1 \\
& + a_{(2,0,0)} x_0^2 x_1^0 x_2^0 + a_{(1,1,0)} x_0^1 x_1^1 x_2^0 + a_{(1,0,1)} x_0^1 x_1^0 x_2^1 \\
& + a_{(0,2,0)} x_0^0 x_1^2 x_2^0 + a_{(0,1,1)} x_0^0 x_1^1 x_2^1 + a_{(0,0,2)} x_0^0 x_1^0 x_2^2 \\
& + a_{(3,0,0)} x_0^3 x_1^0 x_2^0 + a_{(2,1,0)} x_0^2 x_1^1 x_2^0 + a_{(2,0,1)} x_0^2 x_1^0 x_2^1 \\
& + a_{(1,2,0)} x_0^1 x_1^2 x_2^0 + a_{(1,1,1)} x_0^1 x_1^1 x_2^1 + a_{(1,0,2)} x_0^1 x_1^0 x_2^2 \\
& + a_{(0,3,0)} x_0^0 x_1^3 x_2^0 + a_{(0,2,1)} x_0^0 x_1^2 x_2^1 + a_{(0,1,2)} x_0^0 x_1^1 x_2^2 \\
& + a_{(0,0,3)} x_0^0 x_1^0 x_2^3
\end{aligned} \tag{6.6}
$$

The powers of the terms of this polynomial can be obtained from table 2.8.25 for $L^2(3)$ in chapter 2.

A complete polynomial of degree $q = 3$ whose powers of each variable belongs to the set $\mathbb{Z}$ writes

$$
\begin{aligned}
f(x,y) = {}& a_{(0,0)} x_0^0 x_1^0 \\
& + a_{(1,0)} x_0^1 x_1^0 + a_{(0,1)} x_0^0 x_1^1 + a_{(0,-1)} x_0^0 x_1^{-1} + a_{(-1,0)} x_0^{-1} x_1^0 \\
& + a_{(2,0)} x_0^2 x_1^0 + a_{(1,1)} x_0^1 x_1^1 + a_{(1,-1)} x_0^1 x_1^{-1} + a_{(0,2)} x_0^0 x_1^2 + a_{(0,-2)} x_0^0 x_1^{-2} \\
& + a_{(-1,1)} x_0^{-1} x_1^1 + a_{(-1,-1)} x_0^{-1} x_1^{-1} + a_{(-2,0)} x_0^{-2} x_1^0 \\
& + a_{(3,0)} x_0^3 x_1^0 + a_{(2,1)} x_0^2 x_1^1 + a_{(2,-1)} x_0^2 x_1^{-1} + a_{(1,2)} x_0^1 x_1^2 + a_{(1,-2)} x_1^{-2} x_1^1 \\
& + a_{(0,3)} x_0^0 x_1^3 + a_{(0,-3)} x_0^0 x_1^{-3} + a_{(-1,2)} x_1^{-1} x_1^2 + a_{(-1,-2)} x_0^{-1} x_1^{-2} \\
& + a_{(-2,1)} x_0^{-2} x_1^1 + a_{(-2,-1)} x_0^{-2} x_1^{-1} + a_{(-3,0)} x_0^{-3} x_1^0
\end{aligned} \tag{6.7}
$$

The powers of the terms of this polynomial can be obtained from table 2.7.24 on page 67 for $L^1(3)$ in chapter 2.

6.1.2 Notation for a complete polynomial

6.1.2.1 Notation if the power of each variable in the term belongs to the set $\mathbb{Z}^+$

The algorithms 2.5.1 and 2.5.2 on page 48 presented in section 2.4 for the computation of the elements in the set $L^\delta(\ell)$ generates each coordinate number in an ordered form. A complete polynomial of degree $q = \ell$ contain all terms of degree zero, all terms of degree one, until all terms of degree q. In the case of the polynomial each generated term is added to the previous until the last therefore, it is enough to substitute the **for** loop by the summation symbol, that is, to replace

$$\textbf{for } \lambda_k = 0 \textbf{ to } \lambda_k = \lambda_{k-1} \textbf{ do} \qquad \text{by} \qquad \sum_{\lambda_k=0}^{\lambda_k=\lambda_{k-1}} \qquad (6.8)$$

, which gives the expression

$$f(x) = \sum_{\lambda_0=0}^{\lambda_0=q} \sum_{\lambda_1=0}^{\lambda_1=\lambda_0} \cdots \sum_{\lambda_{\delta-1}=0}^{\lambda_{\delta-1}=\lambda_{\delta-2}} a_{(w_0,w_1,\ldots,w_\delta)} x_0^{w_0} x_1^{w_1} \cdots x_\delta^{w_\delta} \qquad (6.9)$$

where each set of the summation index generates one constrained number w given by

$$w_0 = \lambda_0 - \lambda_1 \qquad w_1 = \lambda_1 - \lambda_2 \quad \cdots \quad w_{\delta-2} = \lambda_{\delta-2} - \lambda_{\delta-1}$$
$$w_{\delta-1} = \lambda_{\delta-1} - \lambda_\delta \qquad w_\delta = \lambda_\delta \qquad (6.10)$$

therefore, $w \in L^\delta(q)$.

If it is used the algorithm based on the coordinate numbers instead of the sublevels a polynomial is generated as follows

$$f(x) = \sum_{\lambda_0=0}^{\lambda_0=q} \sum_{w_0=\lambda_0}^{w_0=0} \cdots \sum_{w_{\delta-1}=\lambda_{\delta-1}}^{w_{\delta-1}=0} a_{(w_0,w_1,\ldots,w_\delta)} x_0^{w_0} x_1^{w_1} \cdots x_\delta^{w_\delta} \qquad (6.11)$$

where each set of the summation index generates one sublevel λ_i given by

$$\lambda_1 = \lambda_0 - w_0 \qquad \lambda_2 = \lambda_1 - w_1 \quad \cdots \quad \lambda_{k+1} = \lambda_k - w_k \quad \cdots$$
$$\cdots \quad \lambda_{\delta-1} = \lambda_{\delta-2} - w_{\delta-2} \qquad \lambda_\delta = \lambda_{\delta-1} - w_{\delta-1} \qquad w_\delta = \lambda_\delta \qquad (6.12)$$

and $w \in L^\delta(\ell)$.

Defining $x^w = x_0^{w_0} x_1^{w_1} \cdots x_\delta^{w_\delta}$ equation (6.11) can be rewritten as

$$f(x) = \sum_{\lambda_0=0}^{\lambda_0=q} \sum_{w_0=\lambda_0}^{w_0=0} \cdots \sum_{w_{\delta-1}=\lambda_{\delta-1}}^{w_{\delta-1}=0} a_{(w)} x^w \qquad (6.13)$$

6.1.2.2 Notation if the power of each variable in the term belongs to the set $\mathbb{Z}$

The algorithm 2.3.1 on page 38 presented in section 2.2 for the computation of the elements in the set $L^\delta(\ell)$ generates each coordinate number in an ordered form. Substituting the **for** loop by the summation symbol gives the expression

$$f(x) = \sum_{\lambda_0=0}^{\lambda_0=q} \sum_{w_0=\lambda_0}^{w_0=-\lambda_0} \cdots \sum_{w_{\delta-1}=\lambda_{\delta-1}}^{w_{\delta-1}=-\lambda_{\delta-1}} a_{(w_0,w_1,\ldots,w_\delta)} x_0^{w_0} x_1^{w_1} \cdots x_\delta^{w_\delta} \qquad (6.14)$$

where each set of the summation index generates one sublevel λ_i given by

$$\lambda_1 = \lambda_0 - |w_0| \qquad \lambda_2 = \lambda_1 - |w_1| \qquad \cdots \qquad \lambda_{k+1} = \lambda_k - |w_k| \qquad \cdots$$
$$\cdots \qquad \lambda_{\delta-1} = \lambda_{\delta-2} - |w_{\delta-2}| \qquad \lambda_\delta = \lambda_{\delta-1} - |w_{\delta-1}| \qquad w_\delta = \pm\lambda_\delta \qquad (6.15)$$

and $w \in L^\delta(\ell)$.

6.1.3 Examples

Example 6.1.1 *Polynomial representation*

Using the notation in equation (6.9) to represent the polynomial in equation (6.6) for the underlying set $\mathbb{Z}^+$ it follows

$$f(x_0, x_1, x_2) = \sum_{\lambda_0=0}^{3} \sum_{\lambda_1=0}^{\lambda_0} \sum_{\lambda_2=0}^{\lambda_1} a_{(w_0,w_1,w_2)} x_0^{w_0} x_1^{w_1} x_2^{w_2}$$

where $q = 3$, $w_0 = \lambda_0 - \lambda_1$, $w_1 = \lambda_1 - \lambda_2$, and $w_2 = \lambda_2$.

The expression in equation (6.14) can be used to represent the polynomial in equation (6.7) for the underlying set $\mathbb{Z}$, which gives

$$f(x_0, x_1) = \sum_{\lambda_0=0}^{3} \sum_{w_0=\lambda_0}^{-\lambda_0} a_{(w_0,w_1)} x_0^{w_0} x_1^{w_1}$$

where $\lambda_1 = \lambda_0 - w_0$, and $w_1 = \pm\lambda_1$.

end of example 6.1.1

Example 6.1.2 *Order of the term* $x^2z = x_0^2x_1^0x_2^1$ *in the polynomial* (6.6).

The term $x_0^2x_1^1x_2^0$ in the polynomial (6.6) has coordinate numbers $w = (2,1,0)$ and its order, from equation (5.103), is

$$\mathcal{O}_3^2(2,1,0) = \frac{\lambda_0(\lambda_0+1)(\lambda_0+2)}{3!} + \frac{\lambda_1(\lambda_1+1)}{2!} + \frac{\lambda_2}{1!}$$

$$= \frac{3 \times 4 \times 5}{3!} + \frac{1 \times 2}{2!} + \frac{0}{1!} = 10 + 1 = 11$$

where $\lambda_0 = 2+1+0 = 3$, $\lambda_1 = 1+0$, and $\lambda_2 = 0$.

The term $x_0^2x_1^1$ in the polynomial represented by the equation (6.7) has coordinate numbers $w = (2,1)$ and its order, from equation (5.86), is

$$\mathcal{O}_3^1(2,1) = \sum_{i=0}^{2} n_i^1 + \mathcal{O}_3^1(2,1) = n_0^1 + n_1^1 + n_2^1 + 2(\lambda_0 - w_0) - 1$$

$$= 1 + 4 + 8 + 1 = 14$$

end of example 6.1.2

Example 6.1.3 *Number of terms in the polynomial*

The number of terms in $f(x,y,z) = f(x_0,x_1,x_2)$ is given by the number of elements in the set $L^2(3)$ according to equation (4.131), that is,

$$n^2(3) = \frac{(\ell+1)(\ell+2)(\ell+3)}{3!}$$

$$= \frac{(3+1)(3+2)(3+3)}{3!} = \frac{4 \times 5 \times 6}{3!} = 20$$

where ℓ equals to the degree of the polynomial. See equation (6.1) and figure 6.1.5 to verify this result.

The number of terms in $f(x,y) = f(x_0,x_1)$ is given by the number of elements in the set $L^1(3)$ according to equation (4.121) for the underlying set $\mathbb{Z}$. The number of terms is

$$n^1(3) = 1 + 2(\ell+1)\ell = 1 + 2(3+1)3 = 25$$

See equation (6.7) to verify this result.

end of example 6.1.3

Example 6.1.4 *Terms in the level $\lambda_0 = 3$*

The terms in the level $\ell = 3$ of a two-dimensional constrained numbers whose underlying set is $\mathbb{Z}^+$ are the ones whose powers are the elements of the set L_3^2, which are shown in the table 2.8.23 and they are

$$x_0^3 x_1^0 x_2^0, \quad x_0^2 x_1^1 x_2^0, \quad x_0^2 x_1^0 x_2^1, \quad x_0^1 x_1^2 x_2^0, \quad x_0^1 x_1^1 x_2^1, \quad x_0^1 x_1^0 x_2^2,$$
$$x_0^0 x_1^3 x_2^0, \quad x_0^0 x_1^2 x_2^1, \quad x_0^0 x_1^1 x_2^2, \quad x_0^0 x_1^0 x_2^3$$

The number of terms is given by equation (4.130) as

$$n_2^2 = \frac{(\ell+1)(\ell+2)}{2!} = \frac{(3+1)(3+2)}{2!} = 10$$

The terms in the level $\ell = 3$ of a one-dimensional constrained numbers whose underlying set is $\mathbb{Z}$ are the ones whose powers are the elements of the set L_3^1, which are shown in the table 2.7.20 and they are

$$x_0^3 x_1^0, \quad x_0^2 x_1^1, \quad x_0^2 x_1^{-1}, \quad x_0^1 x_1^2, \quad x_0^1 x_1^{-2}, \quad x_0^0 x_1^3, \quad x_0^0 x_1^{-3},$$
$$x_0^{-1} x_1^2, \quad x_0^{-1} x_1^{-2}, \quad x_0^{-2} x_1^1, \quad x_0^{-2} x_1^{-1}, \quad x_0^{-3} x_1^0$$

The number of terms is given by equation (4.120) as

$$n_2^1 = 4\ell = 4 \times 3 = 12$$

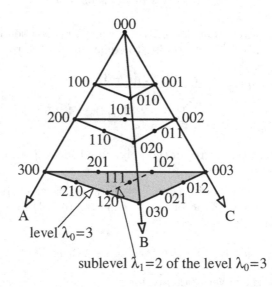

Figure 6.1.5 *The points in the pyramid represents the elements in the set L_3^2. The points in the triangle defined by the vertices (300)-(030)-(003) are the elements of the set L_3^2.*

end of example 6.1.4

6.1.4 Notation for the terms in the level ℓ of complete polynomials

6.1.4.1 Notation if the power of each variable in the term belongs to the set $\mathbb{Z}^+$

The exponents of the terms in the same level are the elements of the set L_ℓ^δ, which can be obtained using the algorithm 2.4.4 on page 45. Therefore, the terms in the level ℓ can be denoted as

$$f_\ell(x) = \overset{\lambda_1=\ell}{\underset{\lambda_1=0}{\sum}} \cdots \overset{\lambda_{\delta-1}=\lambda_{\delta-2}}{\underset{\lambda_{\delta-1}=0}{\sum}} a_{(w_0,w_1,\ldots,w_\delta)} x_0^{w_0} x_1^{w_1} \cdots x_\delta^{w_\delta} \tag{6.16}$$

For example, the terms degree $\ell = 3$ in the polynomial denoted in equation (6.5) writes

$$\begin{aligned}
f_3(x,y,z) &= \overset{3}{\underset{\lambda_1=0}{\sum}} \overset{\lambda_1}{\underset{\lambda_2=0}{\sum}} a_{(w_0,w_1,w_2)} x_0^{w_0} x_1^{w_1}, x_\delta^{w_2} \\
&= a_{(3,0,0)} x_0^3 x_1^0 x_2^0 + a_{(2,1,0)} x_0^2 x_1^1 x_2^0 + a_{(2,0,1)} x_0^2 x_1^0 x_2^1 \\
&\quad + a_{(1,2,0)} x_0^1 x_1^2 x_2^0 + a_{(1,1,1)} x_0^1 x_1^1 x_2^1 \\
&\quad + a_{(1,0,2)} x_0^1 x_1^0 x_2^2 + a_{(0,3,0)} x_0^0 x_1^3 x_2^0 + a_{(0,2,1)} x_0^0 x_1^2 x_2^1 \\
&\quad + a_{(0,1,2)} x_0^0 x_1^1 x_2^2 + a_{(0,0,3)} x_0^0 x_1^0 x_2^3
\end{aligned} \tag{6.17}$$

where $w_0 = 3 - \lambda_1$, $w_1 = \lambda_1 - \lambda_2$, and $w_2 = \lambda_2$ and $u \in L_2^1$. The elements of the set L_2^1 can also be found in the table 2.8.23 on page 88.

6.1.4.2 Notation if the power of each variable in the term belongs to the set $\mathbb{Z}$

Analogously to the case for the set $\mathbb{Z}^+$ the terms in the level ℓ of a polynomial whose exponents of the variables are elements of the set $\mathbb{Z}$ writes

$$f_\ell(x) = \overset{w_0=-\ell}{\underset{w_0=\ell}{\sum}} \cdots \overset{w_{\delta-1}=-\lambda_{\delta-1}}{\underset{w_{\delta-1}=\lambda_{\delta-1}}{\sum}} a_{(w_0,w_1,\ldots,w_\delta)} x_0^{w_0} x_1^{w_1} \cdots x_\delta^{w_\delta} \tag{6.18}$$

6.2   Partial derivative notation

The traditional mathematical notation for the partial derivatives of order $v_0, v_1, \ldots, v_{d-1}$ of a complete polynomial $f(x_0, x_1, \ldots, x_{d-1})$ of degree q with respect to d variables $x_0, x_1, \ldots, x_{d-1}$ is

$$\frac{\partial^{v_0}}{\partial x_0^{v_0}} \frac{\partial^{v_1}}{\partial x_1^{v_1}} \cdots \frac{\partial^{v_\delta}}{\partial x_{d-1}^{v_\delta}} f(x_0, x_1, \ldots, x_{d-1}) \tag{6.19}$$

where d denotes the number of coordinate numbers, which equals the number of variables in the polynomial and $\delta = d - 1$ denotes the dimension of the constrained space.

Using the constrained number notation the partial derivatives written in equation (6.19) is referred as *the derivatives of order v of a complete polynomial $f(x)$ of degree q with respect to d variables $x_i, i = 0, 1, \ldots, d - 1$* and the partial derivative is denoted as

$$f^{(v)}(x) \qquad \text{where} \quad v = (v_0, v_1, \ldots, v_\delta) \quad \text{and} \quad x = (x_0, x_1, \ldots, x_\delta) \tag{6.20}$$

The elements of the set of all partial derivatives of $f(x)$ from order zero until order q is denoted as

$$f^{(v)}(x) \qquad v = (0, 0, \ldots, 0), \ \ldots, \ (v_0, v_1, \ldots, v_{d-1}), \ \ldots, \ (0, 0, \ldots, q) \tag{6.21}$$

using the ordering concept for the constrained number system given in section 5.1 on page 185. An alternative notation is

$$f^{(v)}(x) \qquad \mathcal{O}^\delta(0, 0, \ldots, 0) \leq \mathcal{O}^\delta(v) \leq \mathcal{O}^\delta(0, 0, \ldots, q) \tag{6.22}$$

where $\mathcal{O}^\delta(0, 0, \ldots, 0)$ is the order of derivative given by the first element v of the set $L^\delta(q)$, which is the function itself and $\mathcal{O}^\delta(0, 0, \ldots, q)$ is the order of derivative given by the last element of that set.

Theorem 6.2.1 *Level of a term of a complete polynomial after the derivation*

Let

1. $\delta = d - 1,$ $\quad w = (w_0, w_1, \ldots, w_\delta),$ $\quad u = (u_0, u_1, \ldots, u_\delta),$ and $v_\delta)$

2. $f(x)$, where $x = (x_0, x_1, \ldots, x_\delta)$, be a complete polynomial of degree q in d variables.

Define the levels $\lambda_0(u) = \sum_{j=0}^{\delta} u_j$, $\lambda_0(v) = \sum_{j=0}^{\delta} v_j$, and $\lambda_0(w) = \sum_{j=0}^{\delta} w_j$.

The derivative of order v transforms the terms of $f(x)$ in the level $\lambda_0(w)$ into terms in the level $\lambda_0(u) = \lambda_0(w) - \lambda_0(v)$.

If $\lambda_0(w) - \lambda_0(v) < 0$, then the respective term vanishes.

Proof:

Let $x_0^{w_0} x_1^{w_1} \ldots x_\delta^{w_\delta}$ be the product of the variables in a generic term of the polynomial $f(x)$ in the level $\lambda_0(w)$. Let $x_0^{u_0} x_1^{u_1} \ldots x_\delta^{u_\delta}$ be the derivative of order v the product $x_0^{w_0} x_1^{w_1} \ldots x_\delta^{w_\delta}$ where $u_0 = w_0 - v_0, u_1 = w_1 - v_1, \ldots, u_\delta = w_\delta - v_\delta$ and u belongs to the level $\lambda_0(u)$. Then the level $\lambda_0(u)$ can be evaluated as

$$\lambda_0(u) = \sum_{j=0}^{\delta} u_j = \sum_{j=0}^{\delta} [w_j - v_j]$$

$$= \left[\sum_{j=0}^{\delta} w_j \right] - \left[\sum_{j=0}^{\delta} v_j \right] = \lambda_0(w) - \lambda_0(v) \qquad (6.23)$$

If $\lambda_0(w) - \lambda_0(v) < 0$, then the terms vanishes since it represents the derivative of one or more constant factors in the term, that is, of some x_i^0.

end of theorem 6.2.1

6.3   Derivative of order v of a polynomial

Theorem 6.3.1 *Polynomial derivative*

The derivative of order $v = (v_0, v_1, \ldots v_\delta)$ of a complete polynomial of degree q is a complete polynomial of degree $m = q - \lambda_0(v)$ where $w_k \in \mathbb{Z}^+$, $k = 0, \ldots, \delta$.

Proof:

Let $L^\delta(q)$ denote the set of exponents of the terms of the original polynomial and $L^\delta(m)$ denote the set of the exponents after the derivative operation.

Let $v = (v_0, 0, \ldots, 0)$ be the order of derivative with respect to w_0 and denote the exponents of the terms of the polynomial after the derivative by $u = (u_0, w_1, \ldots, w_\delta)$ where $u_0 = w_0 - v_0$.

The set $L^\delta(q)$ contains all subsets from the level $\lambda_0 = 0$ until $\lambda_0 = q$ and its elements can be generated by the rule 2.4.1 for the underlying set $\mathbb{Z}^+$.

Substituting $L_{\lambda_0}^{\delta}$ given by equation (2.38) on page 40 into equation (2.58) on page 46 it gives

$$L^{\delta}(q) = \bigcup_{\lambda_0=0}^{\lambda_0=q} \bigcup_{\lambda_1=0}^{\lambda_0} \{(\lambda_0 - \lambda_1)\} \times L_{\lambda_1}^{\delta-1} \qquad \delta > 0, \quad q > 0 \qquad (6.24)$$

where $\lambda_0 - \lambda_1 = w_0$. After the derivative all terms whose exponents are such that $u_0 = w_0 - v_0 < 0$, that is, such that $u_0 = \lambda_0 - \lambda_1 < 0$ will vanish. Together, all sets $L_{\lambda_1}^{\delta-1}$ such that $\lambda_0 - \lambda_1 < 0$, which correspond to the constrained numbers $\{(\lambda_0 - \lambda_1)\} \times L_{\lambda_1}^{\delta-1}$ will be removed from the original set. The values of $\lambda_0 = 0, \ldots, \ell$ transform into $\lambda_0 = 0 - v_0, \ldots, v_0 - v_0, \ldots, q - v_0$, that is, $\lambda_0 = 0, \ldots, m$ where $m = q - v_0$. The new set is given by

$$L^{\delta}(m) = \bigcup_{\lambda_0=0}^{\lambda_0=m} \bigcup_{\lambda_1=0}^{\lambda_0} \{(\lambda_0 - \lambda_1)\} \times L_{\lambda_1}^{\delta-1} \qquad \delta > 0, \quad m > 0 \qquad (6.25)$$

which is the algorithm to generate all the elements of the set $L^{\delta}(m)$ whose highest level is $\lambda_0 = m$, therefore, the derivative polynomial is complete and of degree m.

By corollary 2.10.5 on page 107 the permutation of the coordinate numbers do not change the elements of the set. Let $v = (v_0, v_1, \ldots, v_{\delta})$ be the order of derivative of a polynomial, since any coordinate number w_k can be moved to the first position and the elements of the new set, after performing the derivative of order v_k, will be given by equation (6.25). Equation (6.25) is the algorithm to generate all elements of the set, which means that the polynomial derivative is complete.

Since the coordinate numbers $w_k, k - 2, \ldots, \delta$ can be moved to the first position, this proves the theorem.

end of theorem 6.3.1

6.3.1   Polynomials whose powers are elements in the set $\mathbb{Z}^+$

Since the derivative of order $v = (v_0, v_1, \ldots, v_{\delta})$ is a complete polynomial of degree m, it can be written, using the notation in equation (6.9), as

$$f(x)^{(v)} = \sum_{r_0=0}^{r_0=m} \sum_{r_1=0}^{r_1=r_0} \cdots \sum_{r_{\delta-1}=0}^{r_{\delta-1}=r_{\delta-2}} a_{(u)} x_0^{u_0} x_1^{u_1} \cdots x_{\delta}^{u_{\delta}} \qquad (6.26)$$

where $a_{(u)} = a_{(u_0, \ldots, u_{\delta})}$ and

$$u_0 = r_0 - r_1 \quad u_1 = r_1 - r_2 \quad u_2 = r_2 - r_3 \quad \cdots$$
$$\cdots \quad u_{\delta-1} = r_{\delta-1} - r_{\delta} \quad u_{\delta} = r_{\delta} \qquad (6.27)$$

are elements of the set $L^\delta(m)$.

To obtain an expression for the derivative of order v consider the derivative of order v_0 of $x_0^{w_0}$, which is given by

$$\frac{d^{v_0}}{dx_0^{v_0}} x_0^{w_0} = (w_0 - 0)(w_0 - 1) \cdots [w_0 - (v_0 - 1)] x_0^{w_0 - v_0} \qquad (6.28)$$

Equating the exponents for x_0 in equations (6.26) and (6.28) it follows that $u_0 = w_0 - v_0$, which implies $w_0 = u_0 + v_0$. Therefore,

$$(w_0 - 0)(w_0 - 1) \cdots (w_0 - v_0 + 1) = (u_0 + v_0 - 0)(u_0 + v_0 - 1) \cdots (u_0 + 1) \quad (6.29)$$

The expression in equation (6.29) can be written in factorial form as

$$(u_0 + 1)(u_0 + 2) \cdots (u_0 + v_0 - 1)(u_0 + v_0) = \frac{(u_0 + v_0)!}{u_0!} \qquad (6.30)$$

or using the Pochhammer symbol in equation (4.186) on page 184 as

$$(u_0 + 1)(u_0 + 2) \cdots (u_0 + v_0 - 1)(u_0 + v_0) = (u_0 + 1)_{v_0} \qquad (6.31)$$

Recalling that $\lambda = u_0 + 1$ then $\lambda + n - 1 = u_0 + v_0$ gives $u_0 + 1 + n - 1 = u_0 + v_0$, which implies that $n = v_0$, thus $(\lambda)_n = (u_0 + 1)_{v_0}$ justifying the notation.

Introducing equation (6.30) into (6.28) the derivative can be written as

$$\frac{d^{v_0}}{dx_0^{v_0}} x_0^{w_0} = \frac{(u_0 + v_0)!}{u_0!} x_0^{u_0} \qquad (6.32)$$

Applying the result in equation (6.32) to all variables it permits to write the derivative of a generic term of the polynomial as

$$\left[a_{(w)} x_0^{w_0} x_1^{w_1} \cdots x_\delta^{w_\delta} \right]^{(v)} = a_{(w)} \left[\frac{(u_0 + v_0)!}{u_0!} \cdots \frac{(u_\delta + v_\delta)!}{u_\delta!} \right] \left[x_0^{u_0} \cdots x_\delta^{u_\delta} \right]$$

$$= a_{(u)} x_0^{u_0} x_1^{u_1} \cdots x_\delta^{u_\delta} \qquad (6.33)$$

where $a_{(w)} = a_{(w_0, \ldots, w_\delta)}$ and $a_{(u)} = a_{(u_0, \ldots, u_\delta)}$.

Equating the coefficients in equation (6.33), the coefficient $a_{(u)}$ of the derivative of order v can be obtained from the coefficient $a_{(w)}$ of the original polynomial as

$$a_{(u)} = a_{(w)} \left[\frac{(u_0 + v_0)!}{u_0!} \cdots \frac{(u_\delta + v_\delta)!}{u_\delta!} \right] \qquad (6.34)$$

and the exponents of the variables after the derivative are given by

$$u_0 = w_0 - v_0, \quad u_1 = w_1 - v_1, \quad \cdots u_\delta = w_\delta - v_\delta \qquad (6.35)$$

From equation (6.35) it follows that $w_0 = u_0 + v_0$, $w_1 = u_1 + v_1$, ..., $w_\delta = u_\delta + v_\delta$ then equation (6.26) can be written in terms of the derivative quantities as

$$f(x)^{(v)} = \sum_{r_0=0}^{r_0=m} \sum_{r_1=0}^{r_1=r_0} \cdots$$

$$\cdots \sum_{r_{\delta-1}=0}^{r_{\delta-1}=r_{\delta-2}} a_{(u+v)} \left[\frac{(u_0 + v_0)!}{u_0!} \cdots \frac{(u_\delta + v_\delta)!}{u_\delta!} \right] x_0^{u_0} x_1^{u_1} \cdots x_\delta^{u_\delta} \quad (6.36)$$

where according to the definition 1.1.18, on page 12, of addition of constrained numbers $u + v = (u_0 + v_0, \ldots, u_\delta + v_\delta)$ and so $a_{(u+v)} = a_{(u_0+v_0, \ldots, u_\delta+v_\delta)}$.

To compact the notation define

$$\frac{(u + v)!}{u!} := \frac{(u_0 + v_0)!}{u_0!} \cdots \frac{(u_\delta + v_\delta)!}{u_\delta!} = \prod_{k=0}^{\delta} \frac{(u_k + v_k)!}{u_k!} \quad (6.37)$$

then equation (6.36) rewrites

$$f(x)^{(v)} = \sum_{r_0=0}^{r_0=m} \sum_{r_1=0}^{r_1=r_0} \cdots \sum_{r_{\delta-1}=0}^{r_{\delta-1}=r_{\delta-2}} a_{(u+v)} \frac{(u + v)!}{u!} x_0^{u_0} x_1^{u_1} \cdots x_\delta^{u_\delta} \quad (6.38)$$

where the coordinate numbers in $u = (u_0, u_1, \ldots, u_\delta)$ are given by equation (6.27) and $v = (v_0, v_1, \ldots, v_\delta)$ is the order of derivative of the polynomial.

The comparison the polynomials denoted in equations (6.9), (6.26), and (6.38) gives a relation among the coefficient $a_{(w)}$ in equation (6.9) of the original polynomial, the coefficients $a_{(u)}$ in equation (6.26) whose equation is the notation for the polynomial derivative, and the coefficient $a_{(u+v)}$ in equation (6.38) for the computation of the polynomial derivative, as

$$a_{(u)} = a_{(w)} \frac{(u + v)!}{u!} = a_{(u+v)} \frac{(u + v)!}{u!} \quad (6.39)$$

6.3.2   Polynomial whose powers are elements in the set $\mathbb{Z}$

The polynomial derivative can be written, using the notation in equation (6.14), as

$$f(x)^{(v)} = \sum_{r_0=0}^{r_0=m} \sum_{u_0=r_0}^{u_0=-r_0} \cdots \sum_{u_{\delta-1}=r_{\delta-1}}^{-r_{\delta-1}} a_{(u)} x_0^{u_0} x_1^{u_1} \cdots x_\delta^{u_\delta} \quad (6.40)$$

where $a_{(u)} = a_{(u_0, \ldots, u_\delta)}$ and

$$u_0 = r_0 - r_1 \quad r_1 = r_0 - |u_0| \quad r_2 = r_1 - |u_1| \quad \cdots$$
$$r_{\delta-1} = r_{\delta-2} - |u_{\delta-2}| \quad r_\delta = r_{\delta-1} - |u_{\delta-1}| \quad u_\delta = \pm r_\delta \quad (6.41)$$

are elements of the set L_m^δ.

Since there are negative numbers, the Pochhammer symbol

$$(u_0 + 1)(u_0 + 2)\cdots(u_0 + v_0 - 1)(u_0 + v_0) = (u_0 + 1)_{v_0} \tag{6.42}$$

is more convenient, therefore, the equation (6.34) for the evaluation of the coefficients can be rewritten as

$$a_{(u)} = a_{(w)} \left[(u_0 + 1)_{v_0} \cdots (u_\delta + 1)_{v_\delta} \right] \tag{6.43}$$

where $a_{(w)} = a_{(w_0,\ldots,w_\delta)}$ and $a_{(u)} = a_{(u_0,\ldots,u_\delta)}$.

Define

$$(u + 1)_v := (u_0 + 1)_{v_0} \cdots (u_\delta + 1)_{v_\delta} \tag{6.44}$$

then the expression for the computation of the coefficients in equation (6.43) can be written in a compact form as

$$a_{(u)} = a_{(w)}(u + 1)_v \tag{6.45}$$

6.3.3 Algorithm to obtain the derivative of a polynomial whose powers are elements in the set $\mathbb{Z}^+$

A convenient method to obtain the polynomial derivative is

1. Compute the degree of the polynomial derivative $m = q - \lambda_0(v)$,

2. Obtain the elements of the set L_m^δ, which are the powers u_i of the variables x_i in each term of the polynomial derivative,

3. Evaluate the coefficients using the constrained numbers v and u.

6.3.4 Algorithm to obtain the derivative of a polynomial whose powers are elements in the set $\mathbb{Z}$

The previous algorithm cannot be used since the powers of the polynomial derivative, that is, u, are not elements of a constrained set. That is, the derivative operation is not closed for the powers of the terms as elements of a set of constrained numbers although the derivative can be obtained using similar rules.

1. Compute the power of the derivative term as $u = w - v$,

2. If for the variable x_i of a term $(w_i > 0$ and $v_i < 0)$ or $(w_i = 0$ and $v_i \neq 0)$ then the derivative of this term equals to zero,

3. Evaluate the coefficients using the equation. (6.45)

6.4 Partial integration notation

Notation 6.4.1 *Partial integration*

Let $w = (w_0, w_1, \ldots, w_\delta)$, $v = (v_0, v_1, \ldots, v_\delta)$. The traditional mathematical notation for the indefinite integral of a product of several variables is

$$\int \int \cdots \int a_w x_0^{w_0} x_1^{w_1} \cdots x_\delta^{w_\delta} \, dx_0 dx_1 \cdots dx_\delta \qquad (6.46)$$

and here it is denoted

$$[a_w x^w]^{(v)} := \left[a_{(w_0, w_1, \ldots, w_\delta)} x_0^{w_0} x_1^{w_1} \cdots x_\delta^{w_\delta} \right]^{(v_0, v_1, \ldots, v_\delta)}$$

$$:= \int_{(|v_0|)} \int_{(|v_1|)} \cdots \int_{(|v_\delta|)} a_w x_0^{w_0} x_1^{w_1} \cdots x_\delta^{w_\delta} \, dx_0 dx_1 \cdots dx_\delta \qquad (6.47)$$

where $v < 0$, that is, $v_0 < 0, v_1 < 0, \ldots, v_\delta < 0$ and it is called *partial integration*. The indices $|v_i|$ on each symbol of integration represents the order, which each variable x_i must be integrated.

end of notation 6.4.1

Extending the notation of partial integration in equation (6.47) to a polynomial, it can be written

$$f^{(v)}(x) \qquad \text{where} \quad v = (v_0, v_1, \ldots, v_\delta) < 0 \quad \text{and} \quad x = (x_0, x_1, \ldots, x_\delta) \qquad (6.48)$$

as *the integral of order v of a complete polynomial $f(x)$ of degree q with respect to d independent variables x* and it is denoted as

$$f^{(v)}(x) := \int_{(|v_0|)} \int_{(|v_1|)} \cdots \int_{(|v_\delta|)} f(x) \, dx_0 dx_1 \cdots dx_\delta \qquad v < 0 \qquad (6.49)$$

Analogously to the notation for the derivation of a polynomial the elements of the set $L^\delta(q)$ of all partial integrals of $f(x)$ from order zero until order q is denoted by

$$f^{(v)}(x) \qquad v = (0, 0, \ldots, 0), \quad \ldots, \quad (v_0, v_1, \ldots, v_\delta), \quad \ldots, \quad (0, 0, \ldots, q) \qquad (6.50)$$

using the ordering concept for the constrained number system given in section 5.1. An alternative notation is

$$f^{(v)}(x) \qquad \mathcal{O}^\delta(0, 0, \ldots, 0) \le \mathcal{O}^\delta(v) \le \mathcal{O}^\delta(0, 0, \ldots, q) \qquad (6.51)$$

where $\mathcal{O}^\delta(0, 0, \ldots, 0)$ is the order of integration given by the first element v of the set $L^\delta(q)$, which is the function itself and $\mathcal{O}^\delta(0, 0, \ldots, q)$ is the order of integration given by the last element of that set.

Theorem 6.4.2 *Level of the a term of a polynomial after the integration*

Let

1. $\delta = d - 1$, $w = (w_0, w_1, \ldots, w_\delta)$, $u = (u_0, u_1, \ldots, u_\delta)$, and
$v = (v_0, v_1, \ldots, v_\delta)$

2. $f(x)$, where $x = (x_0, x_1, \ldots, x_\delta)$, be a complete polynomial of degree q in d variables.

Define the levels $\lambda_0(u) = \sum_{j=0}^{\delta} u_j$, $\lambda_0(v) = \sum_{j=0}^{\delta} v_j$, and $\lambda_0(w) = \sum_{j=0}^{\delta} w_j$.

The integral of order v transforms the terms of $f(x)$ in the level $\lambda_0(w)$ into terms in the level $\lambda_0(v) = \lambda_0(w) - \lambda_0(u)$.

If $\lambda_0(w) - \lambda_0(u) < 0$, then $\lambda_0(v) := 0$ for $u, v, w \in \mathbb{Z}^+$, that is, if u, v, w belong to the underlying set $\mathbb{Z}^+$.

Proof:

Let $x_0^{w_0} x_1^{w_1} \ldots x_\delta^{w_\delta}$ be the product of the variables in a generic term of the polynomial $f(x)$ in the level $\lambda_0(w)$. The derivative of order v transforms the product $x_0^{w_0} x_1^{w_1} \ldots x_\delta^{w_\delta}$ into $x_0^{u_0} x_1^{u_1} \ldots x_\delta^{u_\delta}$ where $u_0 = w_0 - v_0, u_1 = w_1 - v_1, \ldots, u_\delta = w_\delta - v_\delta$ and u belongs to the level $\lambda_0(u)$. Therefore, the level $\lambda_0(u)$ can be evaluated as

$$\lambda_0(u) = \sum_{j=0}^{\delta} u_j = \sum_{j=0}^{\delta} [w_j - v_j]$$

$$= \left[\sum_{j=0}^{\delta} w_j\right] - \left[\sum_{j=0}^{\delta} v_j\right] = \lambda_0(w) + \lambda_0(v) \qquad v < 0 \qquad (6.52)$$

where $\lambda_0(v) = |v_0| + |v_1| + \cdots + |v_\delta|$, which justifies the plus sign in $\lambda_0(w) + \lambda_0(v)$.

end of theorem 6.2.1

6.4.1   Integration of order v of a polynomial

Let $f(x)^{(v)}$ denote the integral of order $v = (v_0, v_1, \ldots, v_\delta)$ of the polynomial $f(x)$ where $v < 0$, that is, $v_0 < 0, v_1 < 0, \ldots, v_\delta < 0$. Let q be the degree of the polynomial $f(x)$, which is also the level of the terms of highest power, for example, $\lambda_0(0, 0, \ldots, q) = q$. The highest level of the integral of the polynomial $f(x)^{(v)}$ is given by $m = q + \lambda_0(v)$ the lowest level is given by $n = \lambda_0(v)$.

6.4.1.1   Polynomial whose powers are in set $\mathbb{Z}^+$, which is the underlying set

The integration of a polynomial whose powers $w_i \in \mathbb{Z}^+$ can be performed similarly to the case of derivative.

To obtain an expression for the integration of order v consider the integral of order v_0 of $x_0^{w_0}$, which is given by

$$\int^{(v_0)} x_0^{w_0} dx_0 = (w_0 + 1)(w_0 + 2) \cdots (w_0 - v_0) \, x_0^{w_0 - v_0} \qquad (6.53)$$

Since $u_0 = w_0 - v_0$ and $w_0 = u_0 + v_0$ then

$$(w_0 + 1)(w_0 + 2) \cdots (w_0 - v_0) = (u_0 + v_0 + 1)(u_0 + v_0 + 2) \cdots (u_0) \quad (6.54)$$

The expression in equation (6.54) can be written in factorial form as

$$(u_0 + v_0 + 1)(u_0 + v_0 + 2) \cdots (u_0) = \frac{u_0!}{(u_0 + v_0)!} \qquad (6.55)$$

which, of course, is the inverse of the expression in equation (6.30) since the derivative of the integral must recover the original expression.

This allows one to write the coefficients using the Pochhammer symbol in equation (4.186), on page 184, as

$$(u_0 + v_0 + 1)(u_0 + v_0 + 2) \cdots (u_0) = \frac{u_0!}{(u_0 + v_0)!} = \frac{1}{(u_0 + 1)_{v_0}} \qquad (6.56)$$

or

$$(u_0 + v_0 + 1)(u_0 + v_0 + 2) \cdots (u_0) = (u_0 + v_0 + 1)_{1 - v_0} \qquad (6.57)$$

where since $\lambda = u_0 + v_0 + 1$, therefore, $\lambda + n - 1 = u_0$ gives $u_0 + v_0 + 1 + n - 1 = u_0$ and then $n = 1 - v_0$, thus $(\lambda)_n = (u_0 + v_0 + 1)_{1 - v_0}$, which justifies the notation.

Introducing equation (6.57) into (6.53) the integral can be written as

$$\int^{(v_0)} x_0^{w_0} dx_0 = [x_0^{w_0}]^{(v_0)} = \frac{1}{(u_0 + 1)_{v_0}} x_0^{u_0} \qquad (6.58)$$

where $v_0 < 0$.

Applying the result in equation (6.58) to all variables it permits to write the integral of a generic term of the polynomial as

$$\left[a_{(w)} x_0^{w_0} x_1^{w_1} \cdots x_\delta^{w_\delta} \right]^{(v)} = a_{(w)} \left[\frac{u_0!}{(u_0 + v_0)!} \cdots \frac{u_\delta!}{(u_\delta + v_\delta)!} \right] [x_0^{u_0} \cdots x_\delta^{u_\delta}]$$

$$= a_{(u)} x_0^{u_0} u_1^{u_1} \cdots u_\delta^{u_\delta} \qquad (6.59)$$

where $a_{(w)} = a_{(w_0, \dots, w_\delta)}$ and $a_{(u)} = a_{(u_0, \dots, u_\delta)}$.

Equating the coefficients in equation (6.59), the coefficient $a_{(u)}$ of the integral of order v can be obtained from the coefficient $a_{(w)}$ of the original polynomial as

$$a_{(u)} = a_{(w)} \left[\frac{u_0!}{(u_0 + v_0)!} \cdots \frac{u_\delta!}{(u_\delta + v_\delta)!} \right] \tag{6.60}$$

and the exponents of the variables after the integration are given by

$$u_0 = w_0 - v_0, \quad u_1 = w_1 - v_1, \quad \cdots u_\delta = w_\delta - v_\delta \tag{6.61}$$

The polynomial integral is not complete what can be seen from the equation (6.25). Let $v = (v_0, v_1, \ldots, v_\delta)$, the integration of order v_0 transforms $\lambda_0 = 0, \ldots, \ell$ into $\lambda_0 = v_0, \ldots, \ell + v_0$, which generates the set

$$L^\delta = \bigcup_{\lambda_0 = v_0}^{\lambda_0 = \ell + v_0} \bigcup_{\lambda_1 = 0}^{\lambda_0} \{(\lambda_0 - \lambda_1)\} \times L_{\lambda_1}^{\delta-1} \qquad \delta > 0, \quad v_0 > 0 \tag{6.62}$$

At least the elements whose first coordinate numbers are $w_0 = 0, \ldots, v_0 - 1$ are not present in the set and so the set L^δ is not complete.

6.5 Elements characterizing a polynomial

Degree
The degree q of the polynomial is given by the level of the last term.

Number of variables
The dimension of the constrained numbers equals to the number of variables of the polynomial minus one, that is, $\delta = d - 1$.

Number of terms
The number of terms is given by $n^\delta(q)$, which is the number of elements in the set $L^\delta(q)$, which can be obtained from equation (4.135) on page 175.

Last term
The value of the last term is obtained from the degree of the polynomial, which is $w = (0, 0, \cdots, q)$ if the underlying set is $\mathbb{Z}^+$ or $w = (-q, 0, \cdots, 0)$ if the underlying set is $\mathbb{Z}$.

Order of the last term

The order of the last term equals the number of terms less one, that is, $\mathcal{O}^\delta(q) = n^\delta(q) - 1$, the number of terms can be written as $n^\delta(q) = \mathcal{O}^\delta(q) + 1$.

Example 6.5.1 *Evaluate the derivative of the terms*

$$\left(a_{(2,0,1)}x_0^2x_1^0x_2^1 + a_{(0,0,3)}x_0^0x_1^0x_2^3\right)^{(1,0,1)}$$

in the polynomial (6.6).

Solution of the problem

1- Computation of the derivative of the first term

The derivative is defined by $v = (v_0, v_1, v_2) = (1, 0, 1)$ and the powers of the term is $w = (w_0, w_1, w_2) = (2, 0, 1)$. The exponents after performing the derivative operation are

$$u = (u_0, u_1, u_2) = w - v = (2, 0, 1) - (1, 0, 1)$$
$$= (2 - 1, 0 - 0, 1 - 1) = (1, 0, 0)$$

The coefficient generated by the powers of the variables after the derivative operation are obtained using

$$\frac{(u+v)!}{u!} = \frac{(u_0 + v_0)!}{(u_0)!} \frac{(u_1 + v_1)!}{(u_1)!} \frac{(u_2 + v_2)!}{(u_2)!}$$
$$= \frac{(1+1)!}{(1)!} \frac{(0+0)!}{(0)!} \frac{(0+1)!}{(0)!} = 2$$

Therefore,

$$\left(a_{(2,0,1)}x_0^2x_1^0x_2^1\right)^{(1,0,1)} = 2a_{(2,0,1)}x_0^1x_1^0x_2^0$$

2- Computation of the derivative of the second term

The derivative is defined by $v = (v_0, v_1, v_2) = (2, 0, 1)$ and the powers of the term is $w = (w_0, w_1, w_2) = (0, 0, 3)$. The exponents after performing the derivative operation are

$$u = (u_0, u_1, u_2) = w - v = (0, 0, 3) - (1, 0, 1)$$
$$= (0 - 1, 0 - 0, 3 - 1) = (-1, 0, 2)$$

If $w_i - v_i < 0$ then the derivative of $x_i^{w_i}$ of order v_i is zero. Therefore,

$$\left(a_{(0,0,3)}x_0^0x_1^0x_2^3\right)^{(1,0,1)} = 0$$

3- Solution

The solution is

$$\left(a_{(2,0,1)}x_0^2x_1^0x_2^1 + a_{(0,0,3)}x_0^0x_1^0x_2^3\right)^{(1,0,1)} = 0 + 2a_{(2,0,1)}x_0^1x_1^0x_2^0$$

end of example 6.5.1

Example 6.5.2 *Evaluate the derivative of order $v = (1,3)$ of the term $a_{(2,-1)}x_0^2x_1^{-1}$ in the polynomial (6.7), that is, evaluate*

$$\frac{\partial^1}{\partial x_0^1}\frac{\partial^3}{\partial x_1^3}\left(a_{(2,-1)}x_0^2x_1^{-1}\right)$$

Solution of the problem

The following constrained numbers can be identified $w = (2,-1)$ and $v = (1,3)$ thus the powers of the derivative term of the polynomial is given by $u = w - v = (2,-1) - (1,3) = (1,-4)$.

The contribution of the powers to the coefficient is given by

$$(u+1)_v = (u_0+1)_{v_0}(u_1+1)_{v_1}$$

where

$$(u_0+1)_{v_0} = (1+1) = 2$$
$$(u_1+1)_{v_1} = (-4+1)(-4+2)(-4+3) = -6$$

which gives

$$(u+1)_v = (u_0+1)_{v_0}(u_1+1)_{v_1} = 2(-6) = -12$$

then the derivative is

$$\frac{\partial^1}{\partial x_0^1}\frac{\partial^3}{\partial x_1^3}\left(a_{(2,-1)}x_0^2x_1^{-1}\right) = (-12)a_{(2,-1)}x_0^1x_1^{-4}$$

end of example 6.5.2

6.6   Order of derivative

The order of derivative of $f(x)$ is traditionally given by v. The order of v in the sequence of constrained numbers is given by $\mathcal{O}^\delta(v)$. As an example $f^{(3,0)}$ and $f^{(1,2)}$ are derivatives of order 3 under the usual sense and under the constrained number system concept they are of order

$$\mathcal{O}^1(3,0) = \frac{\lambda_0(\lambda_0 + 1)}{2!} + \frac{\lambda_1}{1!} = \frac{3(3+1)}{2!} + \frac{0}{1!} = 6 \tag{6.63}$$

and

$$\mathcal{O}^1(1,2) = \frac{\lambda_0(\lambda_0 + 1)}{2!} + \frac{\lambda_1}{1!} = \frac{3(3+1)}{2!} + \frac{2}{1!} = 8 \tag{6.64}$$

computed using equation (5.101), on page 214, for the one-dimensional constrained numbers.

6.7   Derivative of the product of functions in one variable

6.7.1   Derivative of the product of two functions using constrained multiplication

Let $P = A_0 A_1$ be the product of $t = 2$ functions of the same variable x. This product can be written as

$$P(x) = P^{(0)}(x) = [A_0 A_1]^{(0)} = A_0^{(0)} A_1^{(0)} \tag{6.65}$$

where the index (0) denotes derivative of order zero, which is the original function. Under this form it is a constrained number belonging to the set L_0^1 that is, $(0,0) \in L_0^1$.

The first derivative writes

$$P^{(1)}(x) = [A_0 A_1]^{(1)} = A_0^{(1)} A_1^{(0)} + A_0^{(0)} A_1^{(1)} \tag{6.66}$$

which can be considered as the terms in the level $\lambda_0 = 1$ of the expansion of a complete polynomial in the variables A_0 and A_1.

The indices representing the order of derivative of A_0 and A_1 in the expansion can be written as a pair, namely $(1,0)$ and $(0,1)$, which are the elements of the set L_1^1 of constrained numbers, that is,

$$L_1^1 = \{(1,0),(0,1)\} \tag{6.67}$$

The same result is obtained if the the constrained multiplication defined in 1.1.24, on page 14, is used, that is,

$$L_1^1 = L_0^1 \star L_1^1 = \{(0,0)\} \star \{(1,0),(0,1)\} = \{(1,0),(0,1)\} \tag{6.68}$$

The second derivative is given by

$$P^{(2)}(x) = [A_0 A_1]^{(2)} = A_0^{(2)} A_1^{(0)} + 2A_0^{(1)} A_1^{(1)} + A_0^{(0)} A_1^{(2)} \tag{6.69}$$

which can be considered as the terms in the level $\lambda_0 = 2$ of the expansion of a complete polynomial in the variables A_0 and A_1.

The constrained multiplication $L_1^1 \star L_1^1$ gives

$$\begin{aligned} L_1^1 \star L_1^1 &= \{(1,0),(0,1)\} \star \{(1,0),(0,1)\} \\ &= \{(2,0),(1,1),(1,1),(0,2)\} \end{aligned} \tag{6.70}$$

If a set has two equal elements, they are considered as only one, according to the definition of elements belonging to a set as is shown, for example, in Rosen [31]. Then equation (6.70) writes

$$L_1^1 \star L_1^1 = \{(2,0),(1,1),(0,2)\} \tag{6.71}$$

which is the set of constrained numbers in the level $\lambda_0 = 2$ and dimension $t - 1$, where λ_0 is the order of derivative and t the number of functions in the product.

Similarly, the order of derivative of the terms of the expansion of $[A_0 A_1]^{(3)}$ are the elements of the set

$$L_3^1 = L_2^1 \star L_1^1 = \{(3,0),(2,1),(1,2),(0,3)\} \tag{6.72}$$

where L_1^1 is playing the role of a derivative operator.

Given that the exponents are the elements of the set L_3^1 then the expression for the third derivative writes

$$P^{(3)}(x) = [A_0 A_1]^{(3)} = f_0 A_0^{(3)} A_1^{(0)} + f_1 A_0^{(2)} A_1^{(1)} + f_2 A_0^{(1)} A_1^{(2)} + f_3 A_0^{(0)} A_1^{(3)} \tag{6.73}$$

where f_i are the coefficients of the expansion of the derivative $[A_0 A_1]^{(3)}$.

By induction, the order of derivative of the terms of the expansion of $[A_0 A_1]^{(\lambda_0)}$ are the elements of the set

$$L_{\lambda_0}^1 = L_{\lambda_0 - 1}^1 \star L_1^1 \tag{6.74}$$

Using the notation in equation 6.16 for the terms of a complete polynomial in the level ℓ, which corresponds to terms of degree ℓ the derivative of order v of the product $A_0 A_1$ writes

$$P^{(v)}(x) = [A_0 A_1]^{(v)} = \sum_{\lambda_1 = 0}^{v} f_i x_0^{u_0} x_1^{u_1} \tag{6.75}$$

where

$$\ell = \lambda_0(v) = v$$
$$a_{(w_0, w_1, \ldots, w_\delta)} = f_i$$
$$\delta = d - 1 = 1$$
$$w_0 = u_0 = \lambda_0(v) - \lambda_1$$
$$w_1 = u_1 = \lambda_1 \qquad\qquad\qquad (6.76)$$

in equation 6.16.

The first sublevel of u is given by $\lambda_1(u) = u_1$ and the level is obtained from $\lambda_0(u) = u_0 + \lambda_1(u) = u_0 + u_1$.

6.7.2　Derivative of the product of t functions using constrained multiplication

The expression for the derivative operator given in equation (6.67) is also valid for the product of t functions as

$$L_1^{t-1} = \{(1, \ldots, 0), \cdots, (0, \ldots, 1)\} \qquad\qquad (6.77)$$

Then the order of derivative of the terms of the expansion of $[A_0 \cdots A_{t-1}]^{(\ell)}$ are the elements of the set

$$L_\ell^{t-1} = L_{\ell-1}^{t-1} \star L_1^{t-1} \qquad\qquad (6.78)$$

Using the notation 6.16 for the terms of a complete polynomial in the level ℓ, which corresponds to terms of degree ℓ the derivative of order v of the product of t functions writes

$$P^{(v)}(x) = [A_0 \cdots A_{t-1}]^{(v)} = \sum_{\lambda_1=0}^{v} \sum_{\lambda_2=0}^{\lambda_1} \cdots \sum_{\lambda_{t-1}=0}^{\lambda_{t-2}} f_i x_0^{u_0} x_1^{u_1} \cdots x_{t-1}^{u_{t-1}} \qquad (6.79)$$

where $u_i = \lambda_i - \lambda_{i+1}$ and $u_{t-1} = \lambda_{t-1}$. Since the elements of the set L_ℓ^{t-1} can be generated independently of the summation signs, the sublevels are given by $\lambda_i(u) = \sum_{j=i}^{t-1} u_j$ according to the definition of sublevel given in equation (1.35) on page 10.

6.7.3 Derivative of the product of two functions in one variable

Lemma 6.7.1

Let $A_0(x)$ and $A_1(x)$ be two functions of the independent variable x. The derivative of order v of the product of two functions according to the derivative rules and given by equation (6.75) expands as

$$\left[A_0(x) A_1(x) \right]^{(v)} = \sum_{\lambda_1=0}^{\lambda_0} F(\lambda_0, \lambda_1) A_0^{(u_0)}(x) A_1^{(u_1)}(x) \tag{6.80}$$

where the exponents of the factors $A_0(x), A_1(x)$ of the expansion are $u = (u_0, u_1)$ whose components are $u_0 = \lambda_0(v) - \lambda_1(v)$, and $u_1 = \lambda_1(v)$ and the constrained numbers u's are elements of the set $L_{\lambda_0(v)}^1$. The level of the constrained number u is given by $\lambda_0(u) = u_0 + u_1 = \lambda_0(v)$ and the first sublevel is given by $\lambda_1(u) = u_1$.

Then the coefficient of each term of the expansion is given by

$$F_v(u) = C(\lambda_0(u), \lambda_1(u)) \tag{6.81}$$

which is the combination of $\lambda_0(u)$ elements taken $\lambda_1(u)$ at a time.

Proof:

Define

$$\delta_1 = \delta_2 = \cdots = \delta_p = \frac{d}{dx}$$
$$\varepsilon_1 = \varepsilon_2 = \cdots = \varepsilon_r = \frac{d}{dx} \tag{6.82}$$

then

$$A_0^{(\lambda_0-\lambda_1)} A_1^{(\lambda_1)} = A_0^{(p)} A_1^{(r)} = (\delta_1 \delta_2 \cdots \delta_p) A_0 \, (\varepsilon_1 \varepsilon_2 \cdots \varepsilon_r) A_1 \tag{6.83}$$

There are p derivative operators of the type δ_j and r derivative operators of the type ε_k such that $p + r = \lambda_0 = \ell$. The total number of permutations is given by $\ell!$. Since there are p repetitions of δ_j and r of ε_k the total number of permutations with repetitions is given by

$$\alpha = \frac{\ell!}{p! \, r!} = \frac{\ell!}{(\ell - \lambda_1)! \, \lambda_1!} = \frac{\lambda_0!}{(\lambda_0 - \lambda_1)! \, \lambda_1!} \tag{6.84}$$

which represents the combination of λ_0 elements taken λ_1 at a time. Writing this result in combinatorics notation it follows

$$\alpha = C(\lambda_0, \lambda_1) \tag{6.85}$$

where α is the number of repetitions of the term $A_0^{(\lambda_0-\lambda_1)} A_1^{(\lambda_1)}$ in the expansion of the derivative of the product $A_0 A_1$ that is, α is the coefficient of the term $A_0^{(\lambda_0-\lambda_1)} A_1^{(\lambda_1)}$. This implies that

$$F_v(u) = C(\lambda_0, \lambda_1) = C(\lambda_0(u), \lambda_1(u)) = \frac{(\lambda_0)!}{(\lambda_0 - \lambda_1)! \lambda_1!} \qquad (6.86)$$

end of lemma 6.7.3

A second proof of the equation (6.80) can be performed using the *Binomial Theorem* whose proof can be found, for example, in Rosen [31].

Example 6.7.2 *Evaluate the second-order derivative of the product $A_0(x)A_1(x)$.*

From equation (6.80) for $v = (3)$

$$[A_0(x)\, A_1(x)]^{(3)} = \sum_{\lambda_1=0}^{\lambda_0} F(\lambda_0, \lambda_1) A_0^{(u_0)}(x)\, A_1^{(u_1)}(x) \qquad (6.87)$$

The set of the exponents for the expansion of the derivative of order $v = 3$ of the product of $t = 2$ functions is given by

$$L_v^{t-1} = L_3^1 = \{(3,0), (2,1), (1,2), (0,3)\} \qquad (6.88)$$

where $t - 1 = 2 - 1 = 1$ and $v = 3$.

The coefficients of the expansion are given by

$$\begin{aligned}
F_v(3,0) &= C\left(\lambda_0(3,0), \lambda_1(3,0)\right) = C(3,0) = 1\\
F_v(2,1) &= C\left(\lambda_0(2,1), \lambda_1(2,1)\right) = C(3,1) = 3\\
F_v(1,2) &= C\left(\lambda_0(1,2), \lambda_1(1,2)\right) = C(1,2) = 3\\
F_v(0,3) &= C\left(\lambda_0(0,3), \lambda_1(0,3)\right) = C(3,3) = 1
\end{aligned} \qquad (6.89)$$

which gives the following result

$$[A_0(x)\, A_1(x)]^{(3)} = A_0^{(3)} A_1^{(0)} + 3A_0^{(2)} A_1^{(1)} + 3A_0^{(1)} A_1^{(2)} + A_0^{(0)} A_1^{(3)} \qquad (6.90)$$

for the expansion of the third-order derivative of the product of two functions.

end of example 6.7.2

6.7.4 Derivative of the product of three functions in one variable

Lemma 6.7.3

Let $A_0(x)$, $A_1(x)$, and $A_2(x)$ be three functions of the independent variable x. The derivative of order v of the product of three functions according to the derivative rules and given by equation (6.79) expands as

$$\left[A_0(x)\, A_1(x)\, A_2(x) \right]^{(v)} =$$

$$\sum_{\lambda_1=0}^{\lambda_0(v)} \sum_{\lambda_2=0}^{\lambda_2=\lambda_1} F(\lambda_0, \lambda_1, \lambda_2)\, A_0^{(u_0)}(x)\, A_1^{(u_1)}(x)\, A_2^{(u_2)}(x) \qquad (6.91)$$

where the exponents of the factors $A_0(x), A_1(x), A_2(x)$ of the expansion are $u = (u_0, u_1, u_2)$ whose components are $u_0 = \lambda_0(v) - \lambda_1$, $u_1 = \lambda_1 - \lambda_2$, and $u_2 = \lambda_2$. The constrained numbers u's are elements of the set $L_{\lambda_0(v)}^2$. The level of the constrained number u is given by $\lambda_0(u) = u_0 + u_1 + u_2 = \lambda_0(v)$, the first sublevel is given by $\lambda_1(u) = u_1 + u_2$, and the second sublevel is $\lambda_2 = u_2$.

Then the coefficient of each term of the expansion is given by

$$F(\lambda_0, \lambda_1, \lambda_2) = C(\lambda_0, \lambda_1) C(\lambda_1, \lambda_2) \qquad (6.92)$$

which is the combination of λ_0 elements taken λ_1 at a time multiplied by the combination of λ_1 elements taken λ_2 at a time.

Proof:
Analogously to the lemma 6.7.1 define

$$\delta_1 = \delta_2 = \cdots = \delta_p = \frac{d}{dx}$$

$$\varepsilon_1 = \varepsilon_2 = \cdots = \varepsilon_q = \frac{d}{dx}$$

$$\sigma_1 = \sigma_2 = \cdots = \sigma_r = \frac{d}{dx} \qquad (6.93)$$

then

$$A_0^{(\lambda_0-\lambda_1)}\, A_1^{(\lambda_1-\lambda_2)}\, A_2^{(\lambda_2)}$$
$$= A_0^{(p)}\, A_1^{(q)}\, A_2^{(r)}$$
$$= (\delta_1\delta_2\cdots\delta_p)\, A_0\, (\varepsilon_1\varepsilon_2\cdots\varepsilon_q)\, A_1\, (\sigma_1\sigma_2\cdots\sigma_r)\, A_2 \qquad (6.94)$$

There are p derivative operators of the type δ_k, and q derivative operators of the type ε_m, and r derivative operators of the type σ_n such that $p + q + r = \lambda_0 = \ell$. The total number of permutations is given by $\ell!$. Since there are p

repetitions of δ_k, q repetitions of ε_m, and q repetitions of σ_n the total number of permutations with repetitions is given by

$$\alpha = \frac{\ell!}{p!\,q!\,r!}$$

$$= \frac{\ell!}{(\ell - \lambda_1)!\,(\lambda_1 - \lambda_2)!\,\lambda_2} = \frac{\lambda_0!}{(\lambda_0 - \lambda_1)!\,\lambda_1!}\,\frac{\lambda_1!}{(\lambda_1 - \lambda_2)!\,\lambda_2!}$$

$$= C\,(\lambda_0, \lambda_1)\ C\,(\lambda_1, \lambda_2) \tag{6.95}$$

Therefore,

$$F(\lambda_0, \lambda_1, \lambda_2) = C(\lambda_0, \lambda_1)C(\lambda_1, \lambda_2) \tag{6.96}$$

This completes the proof of the lemma.

end of lemma 6.7.3

Example 6.7.4 *Evaluate the second-order derivative of the product*

$$A_0(x)A_1(x)A_2(x)$$

Solution of the problem

1- Parameters

There are three functions, then $t = 3$ and so $t - 1 = 2$.
The order of derivative is two, therefore, $v = 2$.

2- Set of indices

The set of indices, which was obtained in the table 2.8.14, is

$$L_v^{t-1} = L_2^2 = \{(2,0,0), (1,1,0), (1,0,1), (0,2,0), (0,1,1), (0,0,2)\}$$

3- Coefficients

The coefficients are computed as

$$F\,(\lambda_0(2,0,0), \lambda_1(2,0,0), \lambda_2(2,0,0))$$
$$= C\,(\lambda_0(2,0,0), \lambda_1(2,0,0))\ C\,(\lambda_1(2,0,0), \lambda_2(2,0,0))$$
$$= C(2,0)C(0,0) = 1 \times 1 = 1$$

$$F\,(\lambda_0(1,1,0), \lambda_1(1,1,0), \lambda_2(1,1,0))$$
$$= C\,(\lambda_0(1,1,0), \lambda_1(1,1,0))\ C\,(\lambda_1(1,1,0), \lambda_2(1,1,0))$$
$$= C(2,1)C(1,0) = 2 \times 1 = 2$$

$$F\left(\lambda_0(1,0,1), \lambda_1(1,0,1), \lambda_2(1,0,1)\right)$$
$$= C\left(\lambda_0(1,0,1), \lambda_1(1,0,1)\right)\, C\left(\lambda_1(1,0,1), \lambda_2(1,0,1)\right)$$
$$= C(2,1)C(1,1) = 2 \times 1 = 2$$

$$F\left(\lambda_0(0,2,0), \lambda_1(0,2,0), \lambda_2(0,2,0)\right)$$
$$= C\left(\lambda_0(0,2,0), \lambda_1(0,2,0)\right)\, C\left(\lambda_1(0,2,0), \lambda_2(0,2,0)\right)$$
$$= C(2,2)C(2,0) = 1 \times 1 = 1$$

$$F\left(\lambda_0(0,1,1), \lambda_1(0,1,1), \lambda_2(0,1,1)\right)$$
$$= C\left(\lambda_0(0,1,1), \lambda_1(0,1,1)\right)\, C\left(\lambda_1(0,1,1), \lambda_2(0,1,1)\right)$$
$$= C(2,2)C(2,1) = 1 \times 2 = 2$$

$$F\left(\lambda_0(0,0,2), \lambda_1(0,0,2), \lambda_2(0,0,2)\right)$$
$$= C\left(\lambda_0(0,0,2), \lambda_1(0,0,2)\right)\, C\left(\lambda_1(0,0,2), \lambda_2(0,0,2)\right)$$
$$= C(2,2)C(2,2) = 1 \times 1 = 1$$

4- Solution

Which gives the following expansion

$$[A_0 A_1 A_2]^{(2)} = A_0^{(2)} A_1^{(0)} A_2^{(0)} + 2A_0^{(1)} A_1^{(1)} A_2^{(0)} + 2A_0^{(1)} A_1^{(0)} A_2^{(1)}$$
$$+ A_0^{(0)} A_1^{(2)} A_2^{(0)} + 2A_0^{(0)} A_1^{(1)} A_2^{(1)} + A_0^{(0)} A_1^{(0)} A_2^{(2)}$$

This result can be verified by working the derivative $[A_0 A_1 A_2]^{(2)}$ in the traditional way, that is, doing it by hand.

end of example 6.7.4

6.7.5   Derivative of the product of t functions in one variable

Theorem 6.7.5

Let $A_0(x)$, $A_1(x)$, ..., $A_{t-1}(x)$ be t functions of the independent variable x. The derivative of order v of the product of t functions according to the derivative rules and given by equation (6.79) expands as

$$\left[A_0(x)\, A_1(x) \ldots, A_{t-1}(x) \right]^{(v)}$$
$$= \sum_{\lambda_1=0}^{\lambda_1=\lambda_0} \sum_{\lambda_2=0}^{\lambda_2=\lambda_1} \cdots \sum_{\lambda_{t-1}=0}^{\lambda_{t-1}=\lambda_{t-2}} F(\lambda_0, \ldots, \lambda_{t-1}) A_0^{(u_0)}(x)\, A_1^{(u_1)}(x) \cdots$$
$$\cdots A_{t-2}^{(u_{t-2})}(x)\, A_{t-1}^{(u_{t-1})}(x) \tag{6.97}$$

where the exponents of the factors $A_0(x), \cdots, A_{t-1}(x)$ of the expansion are $u = (u_0, \ldots, u_{t-1})$ whose components are $u_i = \lambda_i - \lambda_{i+1}$. The constrained numbers u's are elements of the set L_v^{t-1}. The sublevels are given by $\lambda_i(u) = \sum_{j=i}^{t-1} u_j$ according to the definition of sublevel given in equation (1.35) on page 10.

Then the coefficient of each term of the expansion is given by

$$F(\lambda_0, \ldots, \lambda_{t-1}) = C(\lambda_0, \lambda_1)C(\lambda_1, \lambda_2) \cdots$$
$$\cdots C(\lambda_{t-3}, \lambda_{t-2})C(\lambda_{t-2}, \lambda_{t-1}) \qquad (6.98)$$

Proof:

The proof is a generalization of the lemma 6.7.3. The equation (6.94) can be rewritten as

$$A_0^{(\lambda_0-\lambda_1)} \cdots A_{t-2}^{(\lambda_{t-2}-\lambda_{t-1})} A_{t-1}^{(\lambda_{t-1})}$$
$$= A_0^{(p_0)} A_1^{(p_1)} \cdots A_{t-2}^{(p_{t-2})} A_{t-1}^{(p_{t-1})}$$
$$= (\delta_1 \cdots \delta_{p_0}) A_0 \cdots (\sigma_1 \cdots \sigma_{p_{t-1}}) A_{t-1} \qquad (6.99)$$

which gives

$$\alpha = \frac{\ell!}{p_0! \, p_1! \, \ldots, p_{t-2}!, p_{t-1}!}$$
$$= \frac{\ell!}{(\ell - \lambda_1)! \, (\lambda_1 - \lambda_2)! \cdots, (\lambda_{t-1} - \lambda_{t-2})! \, \lambda_{t-1}}$$
$$= \frac{\lambda_0!}{(\lambda_0 - \lambda_1)! \, \lambda_1!} \frac{\lambda_1!}{(\lambda_1 - \lambda_2)! \, \lambda_2!} \cdots$$
$$\cdots \frac{\lambda_{t-3}!}{(\lambda_{t-3} - \lambda_{t-2})! \, \lambda_{t-2}!} \frac{\lambda_{t-2}!}{(\lambda_{t-2} - \lambda_{t-1})! \, \lambda_{t-1}!}$$
$$= C(\lambda_0, \lambda_1)C(\lambda_1, \lambda_2) \cdots C(\lambda_{t-3}, \lambda_{t-2})C(\lambda_{t-2}, \lambda_{t-1}) \qquad (6.100)$$

Therefore,

$$F(\lambda_0, \ldots, \lambda_{t-1}) = C(\lambda_0, \lambda_1)C(\lambda_1, \lambda_2) \cdots$$
$$\cdots C(\lambda_{t-3}, \lambda_{t-2})C(\lambda_{t-2}, \lambda_{t-1}) \qquad (6.101)$$

This completes the proof of the lemma.

end of theorem 6.7.5

Example 6.7.6 *Evaluate the second-order derivative of the product*

$$A_0(x)A_1(x)A_2(x)A_3(x)$$

Solution of the problem

1- Parameters
There are three functions, then $t = 3$.
The order of derivative is two, therefore, $v = 2$.

2- Indices
The set of indices.

It can be generated using the rules in section 2.4 algorithm 2.4.4, which gives

$$L_v^{t-1} = L_3^2 = \{(2,0,0,0),(1,1,0,0),(1,0,1,0),(1,0,0,1),$$
$$(0,2,0,0),(0,1,1,0),$$
$$(0,1,0,1),(0,0,2,0),(0,0,1,1),(0,0,0,2)\}$$

3- Coefficients
The coefficients are obtained performing the computation

$$F(\lambda_0,\ldots,\lambda_3) = C(\lambda_0,\lambda_1) C(\lambda_1,\lambda_2) C(\lambda_2,\lambda_3)$$

which gives

for $(2,0,0,0)$	the coefficient is	$C(2,0)C(0,0)C(0,0) = 1$
for $(1,1,0,0)$	the coefficient is	$C(2,1)C(1,0)C(1,0) = 2$
for $(1,0,1,0)$	the coefficient is	$C(2,1)C(1,1)C(1,0) = 2$
for $(1,0,0,1)$	the coefficient is	$C(2,1)C(1,1)C(1,1) = 1$
for $(0,2,0,0)$	the coefficient is	$C(2,2)C(2,0)C(0,0) = 1$
for $(0,1,1,0)$	the coefficient is	$C(2,2)C(2,1)C(1,0) = 2$
for $(0,1,0,1)$	the coefficient is	$C(2,2)C(2,1)C(1,1) = 2$
for $(0,0,2,0)$	the coefficient is	$C(2,2)C(2,2)C(2,0) = 1$
for $(0,0,1,1)$	the coefficient is	$C(2,2)C(2,2)C(2,1) = 2$
for $(0,0,0,2)$	the coefficient is	$C(2,2)C(2,2)C(2,2) = 1$

4- Expansion

$$
\begin{aligned}
[A_0 A_1 A_2 A_3]^{(2)} = {} & A_0^{(2)} A_1^{(0)} A_2^{(0)} A_3^{(0)} + 2 A_0^{(1)} A_1^{(1)} A_2^{(0)} A_3^{(0)} \\
& + 2 A_0^{(1)} A_1^{(0)} A_2^{(1)} A_3^{(0)} + A_0^{(1)} A_1^{(0)} A_2^{(0)} A_3^{(1)} \\
& + A_0^{(0)} A_1^{(2)} A_2^{(0)} A_3^{(0)} + 2 A_0^{(0)} A_1^{(1)} A_2^{(1)} A_3^{(0)} \\
& + 2 A_0^{(0)} A_1^{(1)} A_2^{(0)} A_3^{(1)} + A_0^{(0)} A_1^{(0)} A_2^{(2)} A_3^{(0)} \\
& + 2 A_0^{(0)} A_1^{(0)} A_2^{(1)} A_3^{(1)} + A_0^{(0)} A_1^{(0)} A_2^{(0)} A_3^{(2)}
\end{aligned}
$$

end of example 6.7.6

6.8 Derivative of a product of functions of several variables

6.8.1 Derivative of a product of two functions of two variables

The indices corresponding to the order of derivative of each term of the expansion of the derivative of the product can be obtained as the cartesian product of the indices for the case of one variable and several functions. Each variable can be considered as a one-dimensional space, and several variables can be considered as the cartesian product of the respective one-dimensional space.

Let $[A_0(x_0, x_1) A_1(x_0, x_1)]^{(v_0, v_1)}$ be the derivative of order v_0 with respect to the variable x_0 and of order v_1 with respect to the variable x_1. The order of derivative with respect to x_0 of the terms in the expansion is given by the elements in the set $L_{v_0}^{t-1}$ of constrained numbers. With respect to the variable x_1 it is given by the elements in the set $L_{v_1}^{t-1}$, where t is the number of functions in the product. If the product of functions $A_0 A_1$ is derived with respect to x_0 and then with respect to x_1, the indices are the elements of the set

$$
L_v^{t-1} = L_{v_0 v_1}^{t-1} = L_{v_0}^{t-1} \times L_{v_1}^{t-1} \tag{6.102}
$$

Before developing the algorithm it will be presented an example.

Example 6.8.1 *Consider to obtain the derivative of*

$$
[A_0(x, y, z) A_1(x, y, z)]^{(v_0, v_1, v_2)} = \left[\left[\left[A_0(x, y, z) A_1(x, y, z) \right]^{(v_0)} \right]^{(v_1)} \right]^{(v_2)} \tag{6.103}
$$

To avoid a heavy notation the variables will be denoted as x, y, z instead of x_0, x_1, x_2.

Solution of the problem

1- Parameters

There are two functions, then $t = 2$, and there are three variables, then $d = 3$.

The order of derivative is $v = (v_0, v_1, v_2) = (2, 1, 1)$. That is, the product must be derived twice with respect to the variable x, and once with respect to the variables y and z.

2- Derivatives

Performing the derivatives in the traditional way it gives

$$
\begin{aligned}
[A_0 A_1]^{(2,1,1)} &= 1\, A_0^{(xxyz)} A_1 + 1\, A_0^{(xxy)} A_1^{(z)} + 1\, A_0^{(xxz)} A_1^{(y)} + 1\, A_0^{(xx)} A_1^{(yz)} \\
&+ 2\, A_0^{(xyz)} A_1^{(x)} + 2\, A_0^{(xy)} A_1^{(xz)} + 2\, A_0^{(xz)} A_1^{(xy)} + 2\, A_0^{(x)} A_1^{(xyz)} \\
&+ 1\, A_0^{(yz)} A_1^{(xx)} + 1\, A_0^{(y)} A_1^{(xxz)} + 1\, A_0^{(z)} A_1^{(xxy)} + 1\, A_0 A_1^{(xxyz)}
\end{aligned}
\tag{6.104}
$$

3- Coefficients

The coefficients of the terms and the respective order of derivative of the factors can be written as the following set

$$
\begin{aligned}
\big\{\, &[(1), (2, 1, 1), (0, 0, 0)], [(1), (2, 1, 0), (0, 0, 1)], [(1), (2, 0, 1), (0, 1, 0)], \\
&[(1), (2, 0, 0), (0, 1, 1)], [(2), (1, 1, 1), (1, 0, 0)], [(2), (1, 1, 0), (1, 0, 1)], \\
&[(2), (1, 0, 1), (1, 1, 0)], [(2), (1, 0, 0), (1, 1, 1)], [(1), (0, 1, 1), (2, 0, 0)], \\
&[(1), (0, 1, 0), (2, 0, 1)], [(1), (0, 0, 1), (2, 0, 1)], [(1), (0, 0, 0), (2, 1, 1)] \,\big\}
\end{aligned}
\tag{6.105}
$$

4- Generation of the order of derivative

1. Derivative with respect to x

   The derivative with respect to x is given by

$$
[A_0 A_1]^{(2)} = 1\, A_0^{(xx)} A_1 + 2\, A_0^{(x)} A_1^{(x)} + 1\, A_0 A_1^{(xx)}
\tag{6.106}
$$

   The order of derivative are the elements of the set

$$
L_2^1 = \{(2, 0), (1, 1), (0, 2)\}
\tag{6.107}
$$

The coefficients are obtained as follows

$$F_{v_0}(2,0) = C\left(\lambda_0(2,0), \lambda_1(2,0)\right) = C(2,0) = 1$$
$$F_{v_0}(1,1) = C\left(\lambda_0(1,1), \lambda_1(1,1)\right) = C(2,1) = 2$$
$$F_{v_0}(0,2) = C\left(\lambda_0(0,2), \lambda_1(0,2)\right) = C(2,2) = 1 \tag{6.108}$$

The set L_2^1 can be expanded with the introduction of the coefficients into their respective order of derivative as the first coordinate number as follows

$$\left(L_2^1\right)^+ = \{(1,2,0),(2,1,1),(1,0,2)\} \tag{6.109}$$

2. Derivative with respect to y

$$[A_0 A_1]^{(1)} = 1\,A_0^{(y)} A_1 + 1\,A_0 A_1^{(y)} \tag{6.110}$$

The order of derivative are the elements of the set

$$L_1^1 = \{(1,0),(0,1)\} \tag{6.111}$$

The coefficients are obtained as follows

$$F_{v_1}(1,0) = C\left(\lambda_0(1,0), \lambda_1(1,0)\right) = C(1,0) = 1$$
$$F_{v_1}(0,1) = C\left(\lambda_0(0,1), \lambda_1(0,1)\right) = C(0,1) = 1 \tag{6.112}$$

The set L_1^1 can be expanded with the introduction of the coefficients into their respective order of derivative as the first coordinate number as follows

$$\left(L_1^1\right)^+ = \{(1,1,0),(1,0,1)\} \tag{6.113}$$

3. Derivative with respect to z

Analogously for the derivative with respect to z it gives the expanded set as

$$\left(L_1^1\right)^+ = \{(1,1,0),(1,0,1)\} \tag{6.114}$$

5- Cartesian product

Performing the cartesian product $\left(L_2^1\right)^+ \times \left(L_1^1\right)^+ \times \left(L_1^1\right)^+$

$$\big\{(1,2,0),(2,1,1),(1,0,2)\big\} \times \big\{(1,1,0),(1,0,1)\big\} \times \big\{(1,1,0),(1,0,1)\big\}$$

$$= \big\{ \, [(1,1,1),(2,1,1),(0,0,0)],[(1,1,1),(2,1,0),(0,0,1)],$$

$$[(1,1,1),(2,0,1),(0,1,0)],[(1,1,1),(2,0,0),(0,1,1)],$$

$$[(2,1,1),(1,1,1),(1,0,0)],[(2,1,1),(1,1,0),(1,0,1)],$$

$$[(2,1,1),(1,0,1),(1,1,0)],[(2,1,1),(1,0,0),(1,1,1)],$$

$$[(1,1,1),(0,1,1),(2,0,0)],[(1,1,1),(0,1,0),(2,0,1)],$$

$$[(1,1,1),(0,0,1),(2,1,0)],[(1,1,1),(0,0,0),(2,1,1)] \, \big\}$$

$$(6.115)$$

Comparing equations (6.105) and (6.115) the coefficients can be obtained from the cartesian product as the product of the coordinate numbers of the first element in each bracket. This completes the algorithm.

To make the cartesian product clearer let

$$L_{v_0}^{t-1} = L_2^1 = \big\{ (u_{000}, u_{001}), (u_{010}, u_{011}), (u_{020}, u_{021}) \big\}$$
$$L_{v_1}^{t-1} = L_1^1 = \big\{ (u_{100}, u_{101}), (u_{110}, u_{111}) \big\}$$
$$L_{v_2}^{t-1} = L_1^1 = \big\{ (u_{200}, u_{201}), (u_{210}, u_{211}) \big\} \tag{6.116}$$

whose expanded sets writes

$$\big(L_{v_0}^{t-1}\big)^+ = \big(L_2^1\big)^+ = \big\{ \, (F_{v_0}(u_{00}), u_{000}, u_{001}),$$
$$(F_{v_0}(u_{01}), u_{010}, u_{011}),$$
$$(F_{v_0}(u_{02}), u_{020}, u_{021}) \, \big\}$$
$$\big(L_{v_1}^{t-1}\big)^+ = \big(L_1^1\big)^+ = \big\{ \, (F_{v_1}(u_{10}), u_{100}, u_{101}),$$
$$(F_{v_1}(u_{11}), u_{110}, u_{111}) \, \big\}$$
$$\big(L_{v_2}^{t-1}\big)^+ = \big(L_1^1\big)^+ = \big\{ \, (F_{v_2}(u_{20}), u_{200}, u_{201}),$$
$$(F_{v_2}(u_{21}), u_{210}, u_{211}) \, \big\} \tag{6.117}$$

where

$$(u_{00}) = (u_{000}, u_{001}) \qquad (u_{01}) = (u_{010}, u_{011})$$
$$(u_{02}) = (u_{020}, u_{021}) \qquad (u_{10}) = (u_{100}, u_{101})$$
$$(u_{11}) = (u_{110}, u_{111}) \qquad (u_{20}) = (u_{200}, u_{201})$$
$$(u_{21}) = (u_{210}, u_{211}) \tag{6.118}$$

The cartesian product of the sets in equation (6.116) writes

$$\left\{ \left(F_{v_0}(u_{00}), u_{000}, u_{001} \right), \left(F_{v_0}(u_{01}), u_{010}, u_{011} \right), \left(F_{v_0}(u_{02}), u_{020}, u_{021} \right) \right\} \times$$
$$\left\{ \left(F_{v_1}(u_{10}), u_{100}, u_{101} \right), \left(F_{v_1}(u_{11}), u_{110}, u_{111} \right) \right\} \times$$
$$\left\{ \left(F_{v_2}(u_{20}), u_{200}, u_{201} \right), \left(F_{v_2}(u_{21}), u_{210}, u_{211} \right) \right\} =$$

$$\Big\{ \Big[\left(F_{v_0}(u_{00}), u_{000}, u_{001} \right) \times \left(F_{v_1}(u_{10}), u_{100}, u_{101} \right) \times \left(F_{v_2}(u_{20}), u_{200}, u_{201} \right) \Big] \cup$$
$$\Big[\left(F_{v_0}(u_{00}), u_{000}, u_{001} \right) \times \left(F_{v_1}(u_{10}), u_{100}, u_{101} \right) \times \left(F_{v_2}(u_{21}), u_{210}, u_{211} \right) \Big] \cup$$
$$\Big[\left(F_{v_0}(u_{00}), u_{000}, u_{001} \right) \times \left(F_{v_1}(u_{11}), u_{110}, u_{111} \right) \times \left(F_{v_2}(u_{20}), u_{200}, u_{201} \right) \Big] \cup$$
$$\Big[\left(F_{v_0}(u_{00}), u_{000}, u_{001} \right) \times \left(F_{v_1}(u_{11}), u_{110}, u_{111} \right) \times \left(F_{v_2}(u_{21}), u_{210}, u_{211} \right) \Big] \cup$$

$$\Big[\left(F_{v_0}(u_{01}), u_{010}, u_{011} \right) \times \left(F_{v_1}(u_{10}), u_{100}, u_{101} \right) \times \left(F_{v_2}(u_{20}), u_{200}, u_{201} \right) \Big] \cup$$
$$\Big[\left(F_{v_0}(u_{01}), u_{010}, u_{011} \right) \times \left(F_{v_1}(u_{10}), u_{100}, u_{101} \right) \times \left(F_{v_2}(u_{21}), u_{210}, u_{211} \right) \Big] \cup$$
$$\Big[\left(F_{v_0}(u_{01}), u_{010}, u_{011} \right) \times \left(F_{v_1}(u_{11}), u_{110}, u_{111} \right) \times \left(F_{v_2}(u_{20}), u_{200}, u_{201} \right) \Big] \cup$$
$$\Big[\left(F_{v_0}(u_{01}), u_{010}, u_{011} \right) \times \left(F_{v_1}(u_{11}), u_{110}, u_{111} \right) \times \left(F_{v_2}(u_{21}), u_{210}, u_{211} \right) \Big] \cup$$

$$\Big[\left(F_{v_0}(u_{02}), u_{020}, u_{021} \right) \times \left(F_{v_1}(u_{10}), u_{100}, u_{101} \right) \times \left(F_{v_2}(u_{20}), u_{200}, u_{201} \right) \Big] \cup$$
$$\Big[\left(F_{v_0}(u_{02}), u_{020}, u_{021} \right) \times \left(F_{v_1}(u_{10}), u_{100}, u_{101} \right) \times \left(F_{v_2}(u_{21}), u_{210}, u_{211} \right) \Big] \cup$$
$$\Big[\left(F_{v_0}(u_{02}), u_{020}, u_{021} \right) \times \left(F_{v_1}(u_{11}), u_{110}, u_{111} \right) \times \left(F_{v_2}(u_{20}), u_{200}, u_{201} \right) \Big] \cup$$
$$\Big[\left(F_{v_0}(u_{02}), u_{020}, u_{021} \right) \times \left(F_{v_1}(u_{11}), u_{110}, u_{111} \right) \times \left(F_{v_2}(u_{20}), u_{200}, u_{201} \right) \Big] \Big\}$$

$$(6.119)$$

Expanding each bracket in equation (6.119) gives

$$\left\{ \left(F_{v_0}(u_{00}), u_{000}, u_{001} \right), \left(F_{v_0}(u_{01}), u_{010}, u_{011} \right), \left(F_{v_0}(u_{02}), u_{020}, u_{021} \right) \right\} \times$$
$$\left\{ \left(F_{v_1}(u_{10}), u_{100}, u_{101} \right), \left(F_{v_1}(u_{11}), u_{110}, u_{111} \right) \right\} \times$$
$$\left\{ \left(F_{v_2}(u_{20}), u_{200}, u_{201} \right), \left(F_{v_2}(u_{21}), u_{210}, u_{211} \right) \right\} =$$

$$\Big\{ \Big[\left(F_{v_0}(u_{00}), F_{v_1}(u_{10}), F_{v_2}(u_{20}) \right), \left(u_{000}, u_{100}, u_{200} \right), \left(u_{001}, u_{101}, u_{201} \right) \Big],$$
$$\Big[\left(F_{v_0}(u_{00}), F_{v_1}(u_{10}), F_{v_2}(u_{21}) \right), \left(u_{000}, u_{100}, u_{210} \right), \left(u_{001}, u_{101}, u_{211} \right) \Big],$$
$$\Big[\left(F_{v_0}(u_{00}), F_{v_1}(u_{11}), F_{v_2}(u_{20}) \right), \left(u_{000}, u_{110}, u_{200} \right), \left(u_{001}, u_{101}, u_{201} \right) \Big],$$
$$\Big[\left(F_{v_0}(u_{00}), F_{v_1}(u_{11}), F_{v_2}(u_{21}) \right), \left(u_{000}, u_{110}, u_{210} \right), \left(u_{001}, u_{111}, u_{211} \right) \Big],$$

$$\Big[\left(F_{v_0}(u_{01}), F_{v_1}(u_{10}), F_{v_2}(u_{20}) \right), \left(u_{010}, u_{100}, u_{200} \right), \left(u_{011}, u_{101}, u_{201} \right) \Big],$$
$$\Big[\left(F_{v_0}(u_{01}), F_{v_1}(u_{10}), F_{v_2}(u_{21}) \right), \left(u_{010}, u_{100}, u_{210} \right), \left(u_{011}, u_{101}, u_{211} \right) \Big],$$
$$\Big[\left(F_{v_0}(u_{01}), F_{v_1}(u_{11}), F_{v_2}(u_{20}) \right), \left(u_{010}, u_{110}, u_{200} \right), \left(u_{011}, u_{111}, u_{201} \right) \Big],$$
$$\Big[\left(F_{v_0}(u_{01}), F_{v_1}(u_{11}), F_{v_2}(u_{21}) \right), \left(u_{010}, u_{110}, u_{210} \right), \left(u_{011}, u_{111}, u_{211} \right) \Big],$$

$$\left[\left(F_{v_0}(u_{02}), F_{v_1}(u_{10}), F_{v_2}(u_{20})\right), \left(u_{020}, u_{100}, u_{200}\right), \left(u_{021}, u_{101}, u_{201}\right)\right],$$
$$\left[\left(F_{v_0}(u_{02}), F_{v_1}(u_{10}), F_{v_2}(u_{21})\right), \left(u_{020}, u_{100}, u_{210}\right), \left(u_{021}, u_{101}, u_{211}\right)\right],$$
$$\left[\left(F_{v_0}(u_{02}), F_{v_1}(u_{11}), F_{v_2}(u_{20})\right), \left(u_{020}, u_{110}, u_{200}\right), \left(u_{021}, u_{111}, u_{201}\right)\right],$$
$$\left.\left[\left(F_{v_0}(u_{02}), F_{v_1}(u_{11}), F_{v_2}(u_{21})\right), \left(u_{010}, u_{110}, u_{210}\right), \left(u_{011}, u_{111}, u_{211}\right)\right]\right\}$$
$$(6.120)$$

and the term of the expansion in equation (6.120) corresponding to the bracket

$$\left[\left(F_{v_0}(u_{01}), F_{v_1}(u_{10}), F_{v_2}(u_{20})\right), \left(u_{010}, u_{100}, u_{200}\right), \left(u_{011}, u_{101}, u_{201}\right)\right],$$
$$(6.121)$$

writes

$$\left(F_{v_0}(u_{01}) F_{v_1}(u_{10}) F_{v_2}(u_{20})\right) A_0^{(u_{010}, u_{100}, u_{200})} A_1^{(u_{011}, u_{101}, u_{201})}$$
$$(6.122)$$

end of example 6.8.1

Example 6.8.2 *Evaluate*

$$[A_0(x, y) A_1(x, y)]^{(2,3)}$$
$$(6.123)$$

Solution of the problem

1- Parameters

There are two functions, then $t = 2$, and two variables, then $d = 2$.
The order of derivative is $v = (v_0, v_1) = (2, 3)$.

2- Derivative with respect to x_0

Since $v_0 = 2$ and $t = 2$ the set indices is given by

$$L_{v_0}^{t-1} = L_2^1 = \{(2,0), (1,1), (0,2)\}$$
$$(6.124)$$

The contribution to the coefficients is given by

$$F_{v_0}(2,0) = C\left(\lambda_0(2,0), \lambda_1(2,0)\right) = C(2,0) = 1$$
$$F_{v_0}(1,1) = C\left(\lambda_0(1,1), \lambda_1(1,1)\right) = C(2,1) = 2$$
$$F_{v_0}(0,2) = C\left(\lambda_0(0,2), \lambda_1(0,2)\right) = C(2,2) = 1$$
$$(6.125)$$

Introducing the coefficients into the elements of the set L_2^1 in equation (6.124) it gives

$$\left(L_{v_0}^{t-1}\right)^+ = \left(L_2^1\right)^+ = \{(1,2,0), (2,1,1), (1,0,2)\}$$
$$(6.126)$$

3- Derivative with respect to x_1

Since $v_1 = 3$ and $t = 2$ the set indices is given by

$$L_{v_1}^{t-1} = L_3^1 = \{(3,0),(2,1),(1,2),(0,3)\} \tag{6.127}$$

The contribution to the coefficients is given by

$$\begin{aligned}
F_{v_1}(3,0) &= C\left(\lambda_0(3,0),\lambda_1(3,0)\right) = C(3,0) = 1, \\
F_{v_1}(1,2) &= C\left(\lambda_0(2,1),\lambda_1(2,1)\right) = C(3,1) = 3, \\
F_{v_1}(1,2) &= C\left(\lambda_0(1,2),\lambda_1(1,2)\right) = C(3,2) = 3, \\
F_{v_1}(0,3) &= C\left(\lambda_0(0,3),\lambda_1(0,3)\right) = C(3,3) = 1
\end{aligned} \tag{6.128}$$

Introducing the coefficients into the elements of the set L_3^1 in equation (6.127) it gives

$$\left(L_{v_1}^{t-1}\right)^+ = \left(L_3^1\right)^+ = \{(1,3,0),(3,2,1),(3,1,2),(1,0,3)\} \tag{6.129}$$

4- Cartesian product

The cartesian product gives

$$\begin{aligned}
\left(L_2^1\right)^+ \times \left(L_3^1\right)^+ = \big\{\, & (1,2,0),(2,1,1),(1,0,2) \,\big\} \times \\
& \big\{\, (1,3,0),(3,2,1),(3,1,2),(1,0,3) \,\big\} \\
= \big\{\, & [(1,1),(2,3),(0,0)],[(1,3),(2,2),(0,1)], \\
& [(1,3),(2,1),(0,2)]\,[(1,1),(1,0),(1,3)] \\
& [(2,1),(1,3),(1,0)],[(2,3),(1,2),(1,1)], \\
& [(2,3),(1,1),(1,2)],[(2,1),(1,0),(1,3)] \\
& [(1,1),(0,3),(2,0)],[(1,3),(0,2),(2,1)], \\
& [(1,3),(0,1),(2,2)],[(1,1),(0,0),(2,3)] \,\big\}
\end{aligned} \tag{6.130}$$

where each bracket contains the data for a term of the expansion. That is, the factors for the computation of the coefficient of the term and to the order of derivative of the product $A_0 A_1$. Therefore,

$$\begin{aligned}
[(1,1),(2,3),(0,0)] &\implies (1\times 1)A_0^{(2,3)}A_1^{(0,0)} = A_0^{(2,3)}A_1^{(0,0)} \\
[(1,3),(2,2),(0,1)] &\implies (1\times 3)A_0^{(2,2)}A_1^{(0,1)} = 3A_0^{(2,2)}A_1^{(0,1)} \\
[(1,3),(2,1),(0,2)] &\implies (1\times 3)A_0^{(2,1)}A_1^{(0,2)} = 3A_0^{(2,1)}A_1^{(0,2)} \\
[(1,1),(1,0),(1,3)] &\implies (1\times 1)A_0^{(1,0)}A_1^{(1,3)} = A_0^{(1,0)}A_1^{(1,3)},\dots,\ etc.
\end{aligned} \tag{6.131}$$

5- Solution

$$[A_0 A_1]^{(2,3)} = A_0^{(2,3)} A_1^{(0,0)} + 3A_0^{(2,2)} A_1^{(0,1)} + 3A_0^{(2,1)} A_1^{(0,2)} + A_0^{(2,0)} A_1^{(0,3)}$$
$$+ 2A_0^{(1,3)} A_1^{(1,0)} + 6A_0^{(1,2)} A_1^{(1,1)} + 6A_0^{(1,1)} A_1^{(1,2)} + 2A_0^{(1,0)} A_1^{(1,3)}$$
$$+ A_0^{(0,3)} A_1^{(2,0)} + 3A_0^{(0,2)} A_1^{(2,1)} + 3A_0^{(0,1)} A_1^{(2,2)} + A_0^{(0,0)} A_1^{(2,3)}$$

$$(6.132)$$

end of example 6.8.2

6.8.2 Derivative of the product of t functions of d variables

Algorithm 6.8.3 *Algorithm for the computation of the derivative of a product of functions*

1- A typical case

Consider to obtain the derivative of the expression

$$\big[A_0(x_0, x_1, \ldots, x_{d-1}) A_1(x_0, x_1, \ldots, x_{d-1}) \cdots$$
$$\cdots A_{t-1}(x_0, x_1, \ldots, x_{d-1}) \big]^{(v_0, v_1, \ldots, v_{d-1})} \qquad (6.133)$$

To compact the notation define the variables as $x := (x_0, x_1, \ldots, x_{d-1})$ and the order of derivative as $v := (v_0, v_1, \ldots, v_{d-1})$.

2- Parameters

There are t functions.

There are d variables.

The order of derivative is $v = (v_0, v_1, \ldots, v_{d-1})$. That is, the product must be derived v_i times with respect to the variable $x_i, i = 0, 1, \ldots, (d-1)$.

3- Generation of the order of derivative with respect to x_i

The order of derivative are the elements in the set

$$L_{v_i}^{t-1} = \big\{ (u_{(i,0)}), (u_{(i,1)}), \cdots, (u_{(i,n_i)}) \big\}$$
$$i = 0, 1, \ldots, (d-1) \qquad (6.134)$$

where $n_i = n_{v_i}^{t-1} - 1$. The number of elements in the set $L_{v_i}^{t-1}$ is denoted by $n_{v_i}^{t-1}$ and its computation is shown in chapter 4.

Each element in the set $L_{v_i}^{t-1}$ has $t-1$ coordinate numbers and they will be denoted as

$$(u_{(i,j)}) = \left(u_{(i,j,0)}, u_{(i,j,1)}, \ldots, u_{(i,j,t-1)} \right)$$
$$i = 0, 1, \ldots, (d-1) \qquad j = 0, 1, \ldots, n_i \tag{6.135}$$

The contribution of each element in equation (6.134) to the coefficients is computed as follows

$$F_{v_i}(u_{(i,j)}) = C\left(\lambda_0(u_{(i,j)}), \lambda_1(u_{(i,j)})\right) C\left(\lambda_1(u_{(i,j)}), \lambda_2(u_{(i,j)})\right) \cdots$$
$$\cdots C\left(\lambda_{t-2}(u_{(i,j)}), \lambda_{t-1}(u_{(i,j)})\right) \tag{6.136}$$

The set $L_{v_i}^{t-1}$ can be expanded with the introduction of the coefficients into their respective order of derivative as the first coordinate number as follows

$$\left(L_{v_i}^{t-1}\right)^+ = \Big\{ \left(F_{v_i}(u_{(i,0)}), u_{(i,0,0)}, u_{(i,0,1)}, \ldots, u_{(i,0,t-1)} \right),$$
$$\left(F_{v_i}(u_{(i,1)}), u_{(i,1,0)}, u_{(i,1,1)}, \ldots, u_{(i,1,t-1)} \right),$$
$$\cdots,$$
$$\left(F_{v_i}(u_{(i,j)}), u_{(i,j,0)}, u_{(i,j,1)}, \ldots, u_{(i,j,t-1)} \right),$$
$$\cdots,$$
$$\left(F_{v_{n_i}}(u_{(i,j)}), u_{(i,j,0)}, u_{(i,j,1)}, \ldots, u_{(i,j,t-1)} \right) \Big\} \tag{6.137}$$

4- Cartesian product

The cartesian product of the sets writes

$$\left(L_{v_0}^{t-1} \right)^+ \times \left(L_{v_1}^{t-1} \right)^+ \times \cdots \times \left(L_{v_{d-1}}^{t-1} \right)^+ \tag{6.138}$$

and it expands as

$$\left(L_{v_0}^{t-1} \right)^+ \times \left(L_{v_1}^{t-1} \right)^+ \times \cdots \times \left(L_{v_{d-1}}^{t-1} \right)^+ =$$
$$\bigcup_{k_0=0}^{n_0} \bigcup_{k_1=0}^{n_1} \bigcup_{k_1=0}^{n_2} \cdots \bigcup_{k_{d-2}=0}^{n_{d-2}} \bigcup_{k_{d-1}=0}^{n_{d-1}} \Big[u_{(0,k_0)} \times u_{(1,k_1)} \times \cdots$$
$$\cdots \times u_{(d-2,k_{d-2})} \times u_{(d-1,k_{d-1})} \Big] \tag{6.139}$$

where $n_i = n_{v_i}^{t-1} - 1$. The total number of unions is given by the product

$$n = \left(n_{v_0}^{t-1} \right) \left(n_{v_1}^{t-1} \right) \cdots \left(n_{v_{d-1}}^{t-1} \right) \tag{6.140}$$

Each $u_{(i,k_i)}$ contains $t + 1$ coordinate numbers where the first is the contribution to the coefficient, and the others are the contributions to the order of derivative. The element $u_{(i,k_i)}$ is denoted as

$$u_{(i,k_i)} = \left(F_{v_i}(u_{i,k_i}), u_{(i,k_i,0)}, u_{(i,k_i,1)}, \cdots, u_{(i,k_i,t-1)} \right) \tag{6.141}$$

The expansion of each bracket gives $t + 1$ elements generated as

- the first element contains the first coordinate number of each $u_{(i,k_i)}$.

- the second element contains the second coordinate number of each $u_{(i,k_i)}$, and so on.

therefore, the expansion of each bracket can be denoted as

$$\left[\left(F_{v_0}(u_{(0,k_0)}), F_{v_1}(u_{(1,k_1)}), \cdots, F_{v_{d-1}}(u_{(d-1,k_{d-1})}) \right), \right.$$
$$\left(u_{(0,k_0,0)}, u_{(1,k_1,0)}, \cdots, u_{(d-2,k_{d-2},0)}, u_{(d-1,k_{d-1},0)} \right),$$
$$\vdots$$
$$\left. \left(u_{(0,k_0,t-1)}, u_{(1,k_1,t-1)}, \cdots, u_{(d-2,k_{d-2},t-1)}, u_{(d-1,k_{d-1},t-1)} \right) \right] \tag{6.142}$$

This completes the algorithm.

end of algorithm 6.8.3

Example 6.8.4 *Expand the derivative of*

$$\frac{d}{dx_0} \frac{d^3}{dx_1^3} \frac{d^2}{dx_3^2} [A_0(x) A_1(x) A_2(x)] \tag{6.143}$$

where $x = (x_0, x_1, x_2, x_3)$.

Solution of the problem

1- Notation

The derivative can be written as

$$[A_0(x) A_1(x) A_2(x)]^{(1,3,0,2)} = [A_0 A_1 A_2]^{(1,3,0,2)} \tag{6.144}$$

2- Parameters

1. Number of functions

There are three functions $t = 3$ then the constrained numbers representing the derivatives are two-dimensional numbers, that is, $t - 1 = 3 - 1 = 2$. The set of constrained numbers has $t = 3$ coordinate numbers.

2. Number of variables

   There are four variables $d = 4$, which gives the dimension $d - 1 = 3$.

3. Order of derivative

   The order of derivative constrained number is given by $(v_0, v_1, v_2, v_3) = (1, 3, 0, 2)$.

3- Computation of the sets related to each variable and their respective coefficients

1. Contribution of the variable x_0.

   The parameters $t = 3$ and $v_0 = 1$ characterize the set as

   $$L_{v_0}^{t-1} = L_1^2 = \{(1,0,0), (0,1,0), (0,0,1)\} \tag{6.145}$$

   which can be obtained using the algorithm shown in section 2.4.4 on page 45.

   The contribution of each element in L_1^2 given by equation (6.145) to the coefficients is computed using the table 6.8.5 below

   Table 6.8.5 *Contributions of the elements of the set L_1^2, corresponding to the variable x_0, to the coefficients, where $C_{\lambda_1}^{\lambda_0} := C(\lambda_0, \lambda_1)$ and $C_{\lambda_2}^{\lambda_1} := C(\lambda_1, \lambda_2)$.*

coeff.	number	λ_0	λ_1	λ_2	$C_{\lambda_1}^{\lambda_0}$	$C_{\lambda_2}^{\lambda_1}$	$C_{\lambda_1}^{\lambda_0} C_{\lambda_2}^{\lambda_1}$
$F_{v_0}(1,0,0)$	(1, 0, 0)	1	0	0	$C_0^1 = 1$	$C_0^0 = 1$	$C_0^1 C_0^0 = 1$
$F_{v_0}(0,1,0)$	(0, 1, 0)	1	1	0	$C_1^1 = 1$	$C_0^1 = 1$	$C_1^1 C_0^1 = 1$
$F_{v_0}(0,0,1)$	(0, 0, 1)	1	1	1	$C_1^1 = 1$	$C_1^1 = 1$	$C_1^1 C_1^1 = 1$

   The expanded set is

   $$\left(L_1^2\right)^+ = \{(1,1,0,0), (1,0,1,0), (1,0,0,1)\} \tag{6.146}$$

2. Contribution of the variable x_1.

   The parameters $t = 3$ and $v_1 = 3$ characterize the set as

   $$L_{v_1}^{t-1} = L_3^2 = \{(3,0,0), (2,1,0), (2,0,1), (1,2,0), (1,1,1), (1,0,2),$$
   $$(0,3,0), (0,2,1), (0,1,2), (0,0,3)\} \tag{6.147}$$

   The contribution of each element in L_3^2 given by equation (6.147) to the coefficients is computed using the table 6.8.6 below

Table 6.8.6 *Contributions of the elements of the set L_3^2, corresponding to the variable x_1, to the coefficients, where $C_{\lambda_1}^{\lambda_0} := C(\lambda_0, \lambda_1)$ and $C_{\lambda_2}^{\lambda_1} := C(\lambda_1, \lambda_2)$.*

coeff.	number	λ_0	λ_1	λ_2	$C_{\lambda_1}^{\lambda_0}$	$C_{\lambda_2}^{\lambda_1}$	$C_{\lambda_1}^{\lambda_0} C_{\lambda_2}^{\lambda_1}$
$F_{v_1}(3,0,0)$	(3, 0, 0)	3	0	0	$C_0^3 = 1$	$C_0^0 = 1$	$C_0^3 C_0^0 = 1$
$F_{v_1}(2,1,0)$	(2, 1, 0)	3	1	1	$C_1^3 = 3$	$C_0^1 = 1$	$C_1^3 C_0^1 = 3$
$F_{v_1}(2,0,1)$	(2, 0, 1)	3	1	1	$C_1^3 = 3$	$C_1^1 = 1$	$C_1^3 C_1^1 = 3$
$F_{v_1}(1,2,0)$	(1, 2, 0)	3	2	0	$C_2^3 = 3$	$C_1^2 = 1$	$C_2^3 C_1^2 = 3$
$F_{v_1}(1,1,1)$	(1, 1, 1)	3	2	1	$C_2^3 = 3$	$C_1^2 = 2$	$C_2^3 C_1^2 = 6$
$F_{v_1}(1,0,2)$	(1, 0, 2)	3	2	2	$C_2^3 = 3$	$C_2^2 = 1$	$C_2^3 C_2^2 = 3$
$F_{v_1}(0,3,0)$	(0, 3, 0)	3	3	0	$C_3^3 = 1$	$C_0^2 = 1$	$C_3^3 C_2^2 = 1$
$F_{v_1}(0,2,1)$	(0, 2, 1)	3	3	1	$C_3^3 = 1$	$C_1^3 = 3$	$C_3^3 C_1^3 = 3$
$F_{v_1}(0,1,2)$	(0, 1, 2)	3	3	2	$C_3^3 = 1$	$C_2^3 = 3$	$C_3^3 C_2^3 = 3$
$F_{v_1}(0,0,3)$	(0, 0, 3)	3	3	3	$C_3^3 = 1$	$C_3^3 = 3$	$C_3^3 C_3^3 = 1$

The expanded set writes

$$(L_3^2)^+ = \{(1,3,0,0),(3,2,1,0),(3,2,0,1),(3,1,2,0),(6,1,1,1),$$
$$(3,1,0,2),(1,0,3,0),(3,0,2,1),(3,0,1,2),(1,0,0,3)\} \tag{6.148}$$

3. Contribution of the variable x_2.

   The parameters $t = 3$ and $v_2 = 0$ characterize the set as

$$L_{v_2}^{t-1} = L_0^2 = \{(0,0,0)\} \tag{6.149}$$

The contribution of each element in L_0^2 given by equation (6.149) to the coefficients is computed using the table 6.8.7 below

Table 6.8.7 *Contributions of the elements of the set L_0^2, corresponding to the variable x_2, to the coefficients.*

coeff.	number	λ_0	λ_1	λ_2	$C_{\lambda_1}^{\lambda_0}$	$C_{\lambda_2}^{\lambda_1}$	$C_{\lambda_1}^{\lambda_0} C_{\lambda_2}^{\lambda_1}$
$F_{v_2}(0,0,0)$	(0, 0, 0)	0	0	0	$C_0^0 = 1$	$C_0^0 = 1$	$C_0^0 C_0^0 = 1$

The expanded set writes

$$(L_0^2)^+ = \{(1,0,0,0)\} \tag{6.150}$$

4. Contribution of the variable x_3.

The parameters $t = 3$ and $v_3 = 2$ characterize the set as

$$L_{v_3}^{t-1} = L_2^2 = \{(2,0,0),(1,1,0),(1,0,1),(0,2,0),(0,1,1),(0,0,2)\}$$
(6.151)

The contribution of each element in L_2^2 given by equation (6.151) to the coefficients is computed using the table 6.8.8 below

Table 6.8.8 *Contributions of the elements of the set L_2^2, corresponding to the variable x_3, to the coefficients.*

coeff.	number	λ_0	λ_1	λ_2	$C_{\lambda_1}^{\lambda_0}$	$C_{\lambda_2}^{\lambda_1}$	$C_{\lambda_1}^{\lambda_0}C_{\lambda_2}^{\lambda_1}$
$F_{v_3}(2,0,0)$	$(2,0,0)$	2	0	0	$C_0^2=1$	$C_0^0=1$	$C_0^2C_0^0=1$
$F_{v_3}(1,1,0)$	$(1,1,0)$	2	1	0	$C_1^2=2$	$C_0^1=1$	$C_1^2C_0^1=2$
$F_{v_3}(1,0,1)$	$(1,0,1)$	2	1	1	$C_1^2=2$	$C_1^1=1$	$C_1^2C_1^1=2$
$F_{v_3}(0,2,0)$	$(0,2,0)$	2	2	0	$C_2^2=1$	$C_0^2=1$	$C_2^2C_1^2=1$
$F_{v_3}(0,1,1)$	$(0,1,1)$	2	1	1	$C_2^2=1$	$C_1^2=2$	$C_2^2C_1^2=2$
$F_{v_3}(0,0,2)$	$(0,0,2)$	2	2	2	$C_2^2=1$	$C_2^2=1$	$C_2^2C_2^2=1$

The expanded set writes

$$\left(L_2^2\right)^+ = \{(1,2,0,0),(2,1,1,0),(2,1,0,1),(1,0,2,0),$$
$$(2,0,1,1),(1,0,0,2)\}$$
(6.152)

4- Computation of the order of derivative of each term of the expansion

The cartesian product is given by

$$\left(L_1^2\right)^+ \times \left(L_3^2\right)^+ \times \left(L_0^2\right)^+ \times \left(L_3^2\right)^+ =$$
$$\{(1,1,0,0) \times (1,3,0,0) \times (1,0,0,0) \times (1,2,0,0)\}$$
$$\cup \{(1,1,0,0) \times (1,3,0,0) \times (1,0,0,0) \times (2,1,1,0)\}$$
$$\cup \{(1,1,0,0) \times (1,3,0,0) \times (1,0,0,0) \times (2,1,0,1)\}$$
$$\cup \{(1,1,0,0) \times (1,3,0,0) \times (1,0,0,0) \times (1,0,2,0)\}$$
$$\cup \{(1,1,0,0) \times (1,3,0,0) \times (1,0,0,0) \times (2,0,1,1)\}$$
$$\cup \{(1,1,0,0) \times (1,3,0,0) \times (1,0,0,0) \times (1,0,0,2)\} \cup \cdots$$
(6.153)

the result is

$$\left(L_1^2\right)^+ \times \left(L_3^2\right)^+ \times \left(L_0^2\right)^+ \times \left(L_3^2\right)^+ =$$

$$[(1,1,1,1),(1,3,0,2),(0,0,0,0),(0,0,0,0)],$$
$$[(1,1,1,2),(1,3,0,1),(0,0,0,1),(0,0,0,0)],$$
$$[(1,1,1,2),(1,3,0,1),(0,0,0,0),(0,0,0,1)],$$
$$[(1,1,1,1),(1,3,0,0),(0,0,0,2),(0,0,0,0)],$$
$$[(1,1,1,2),(1,3,0,0),(0,0,0,1),(0,0,0,1)],$$
$$[(1,1,1,1),(1,3,0,0),(0,0,0,0),(0,0,0,2)],\cdots \qquad (6.154)$$

The number of terms of the expansion, which equals the number of elements of the set cartesian product, is given by

$$n = n_1^2 \times n_3^2 \times n_0^2 \times n_2^2 = 3 \times 10 \times 1 \times 6 = 180 \qquad (6.155)$$

and each term has $d = 4$ coordinate numbers.

That is, each bracket corresponds to a term of the expansion, for example,

$$[(1,1,1,2),(1,3,0,0),(0,0,0,1),(0,0,0,1)] \qquad (6.156)$$

The number of brackets equals to 180 that is, the number of terms in the expansion.

The number of elements in each bracket equals the number of functions in product plus one to take care of the coefficients. That is, it equals to $t+1 = 4$ since the number of factors equals to $t = 3$. The elements in equation (6.156) are

$$(1,1,1,2), \quad (1,3,0,0), \quad (0,0,0,1), \quad (0,0,0,1) \qquad (6.157)$$

The number of coordinates in each element equals to the number of variables of each factor in the product, which is $d = 4$. For example, $(1,1,1,2)$ has four coordinates for the computation of the coefficient of the term. The element $(1,3,0,0)$ has four coordinates corresponding to the order of derivative of the first factor, that is,

$$(1 \times 1 \times 1 \times 2)\, A_0^{(1,3,0,0)}\, A_1^{(0,0,0,1)}\, A_2^{(0,0,0,1)} = 2\, A_0^{(1,3,0,0)}\, A_1^{(0,0,0,1)}\, A_2^{(0,0,0,1)}$$
$$(6.158)$$

5- Computation of the coefficients

The coefficients are obtained by multiplying the coordinate numbers of the first element as follows

$$f(1,1,1,1) = 1 \times 1 \times 1 \times 1 = 1$$
$$f(1,1,1,2) = 1 \times 1 \times 1 \times 2 = 2 \qquad etc. \qquad (6.159)$$

6- Solution

The first six terms of the expansion are

$$[A_0 A_1 A_2]^{(1,3,0,2)} = A_0^{(1,3,0,2)} A_1^{(0,0,0,0)} A_2^{(0,0,0,0)} + 2A_0^{(1,3,0,1)} A_1^{(0,0,0,1)} A_2^{(0,0,0,0)}$$
$$+ 2A_0^{(1,3,0,1)} A_1^{(0,0,0,0)} A_2^{(0,0,0,1)} + A_0^{(1,3,0,0)} A_1^{(0,0,0,2)} A_2^{(0,0,0,0)}$$
$$+ 2A_0^{(1,3,0,0)} A_1^{(0,0,0,1)} A_2^{(0,0,0,1)} + A_0^{(1,3,0,0)} A_1^{(0,0,0,0)} A_2^{(0,0,0,2)}$$
$$+ \cdots$$

end of example 6.8.4

Example 6.8.9 *Expand the derivative of*

$$\frac{d}{dx} \frac{d^2}{dy^2} \frac{d}{dz} A_0(x, y, z)$$

Solution of the problem

1- Notation

The derivative can be written in more compact form as $[A_0]^{(1,2,1)}$.

2- Parameters

1. Number of functions

   There is one only function, then $t = 1$.

2. Number of variables

   There are $d = 3$ variables.

3. Order of derivative

   The order of derivative is $v = (1, 2, 1)$.

3- Computation of the sets relates to each variable

1. Contribution of the variable x.

   The parameters $t = 1$ and $v_0 = 1$ characterize the set $L_{v_0}^{t-1}$ as

   $$L_1^0 = \{(1)\} \tag{6.160}$$

The contribution of the variable x to the coefficients of the expansion is given, by definition, by

$$F_{v_0}(1) = 1 \qquad (6.161)$$

which gives the expanded set as

$$\left(L_1^0\right)^+ = \{(1,1)\} \qquad (6.162)$$

2. Contribution of the variable y

The parameters $t = 1$ and $v_1 = 2$ characterize the set $L_{v_1}^{t-1}$ as

$$L_2^0 = \{(2)\} \qquad (6.163)$$

The contribution of the variable y to the coefficients of the expansion is given, by definition, by

$$F_{v_1}(1) = 1 \qquad (6.164)$$

which gives the expanded set as

$$\left(L_2^0\right)^+ = \{(1,2)\} \qquad (6.165)$$

3. Contribution of the variable z

The parameters $t = 1$ and $v_2 = 1$ characterize the set $L_{v_2}^{t-1}$ as

$$L_1^0 = \{(1)\} \qquad (6.166)$$

The contribution of the variable z to the coefficients of the expansion is given, by definition, by

$$F_{v_2}(1) = 1 \qquad (6.167)$$

which gives the expanded set as

$$\left(L_1^0\right)^+ = \{(1,1)\} \qquad (6.168)$$

6- Computation of the order of derivative of each term of the expansion

The cartesian product is given by

$$\left(L_1^0\right)^+ \times \left(L_2^0\right)^+ \times \left(L_1^0\right)^+ = \{(1,1)\} \times \{(1,2)\} \times \{(1,1)\}$$
$$= \{(1,1,1), (1,2,1)\} \qquad (6.169)$$

7- Computation of the coefficients

The multiplication of the coordinate numbers of the first element is given by

$$1 \times 1 \times 1 = 1 \tag{6.170}$$

8- Write the expansion

$$[A_0]^{(1,2,1)} = 1\, A_0^{(1,2,1)} \tag{6.171}$$

that is, in this case there is no expansion.

end of example 6.8.9

Example 6.8.10 *Expand the derivative of*

$$\frac{d^3}{dx^3}\,[A_0(x)A_1(x)]$$

Solution of the problem

1- Notation

The derivative can be written in more compact form as $[A_0(x)A_1(x)]^{(3)}$.

2- Parameters

1. Number of functions

    There are two function, then $t = 2$.

2. Number of variables

    There is only one variable, then $d = 1$.

3. Order of derivative

    The order of derivative is $v = (3)$.

3- Computation of the sets relates to each variable

1. Contribution of the variable x.

    The parameters $t = 2$ and $v_0 = 3$ characterize the set $L_{v_0}^{t-1}$ as

    $$L_3^1 = \{(3,0),(2,1),(1,2),(0,3)\} \tag{6.172}$$

    The contribution of the variable x to the coefficients of the expansion is evaluated using the table 6.8.11

Table 6.8.11 *Contributions of the elements of the set L_3^1, corresponding to the variable x, to the coefficients.*

coeff.	number	λ_0	λ_1	$C_{\lambda_1}^{\lambda_0}$
$F_{v_0}(3,0)$	$(3,0)$	$3+0=3$	$0=0$	$C_0^3=1$
$F_{v_0}(2,1)$	$(2,1)$	$2+1=3$	$1=1$	$C_1^3=3$
$F_{v_0}(1,2)$	$(1,2)$	$1+2=3$	$2=2$	$C_2^3=3$
$F_{v_0}(0,3)$	$(0,3)$	$0+3=3$	$3=3$	$C_3^3=1$

The expanded set writes

$$\left(L_3^1\right)^+ = \{(1,3,0),(3,2,1),(3,1,2),(1,0,3)\} \tag{6.173}$$

4- The cartesian product

The cartesian product is given by

$$\left(L_3^1\right)^+ = [(1,3,0)],[(3,2,1)],[(3,1,2)],[(1,0,3)] \tag{6.174}$$

since there is only one factor in the cartesian product.

5- Computation of the coefficients

The coefficient of each term of the expansion is given by the first coordinate number in the respective bracket, which are $1, 3, 3$, and 1.

6- Write the expansion

$$[\dot{A}_0(x)A_1(x)]^{(3)} = A_0^{(3)}A_1^{(0)} + 3A_0^{(2)}A_1^{(1)} + 3A_0^{(1)}A_1^{(2)} + A_0^{(0)}A_1^{(3)}$$

end of example 6.8.10

6.9   Expansion of the power of a sum of functions

6.9.1   Expansion of the power of a sum of two functions

This is given by the Binomial theorem [31] writes

$$(x+y)^n = \sum_{j=0}^{n} C_j^n x^{n-j} y^j \tag{6.175}$$

This expression is the same as the equation (6.80)where the product is replaced by addition, and the derivative by the power is given by ℓ, that is,

$$(A_0(x) + A_1(x))^\ell = \sum_{r=0}^{r=n} C_r^\ell A_0^{(\ell-r)}(x) A_1^{(r)}(x) \tag{6.176}$$

6.9.2   Expansion of the power of a sum of t functions

As consequence of the above section, the algorithm using constrained numbers also applies to the power of the sum of functions.

Example 6.9.1 *Obtain the expansion of the square of the sum $A_0 + A_1 + A_2$ that is,*

$$(A_0 + A_1 + A_2)^p \tag{6.177}$$

where $p = 2$.

Solution of the problem

1- Parameters

1. Number of functions: $t = 3$.

2. Power to which it must be computed: $p = 2$.

3. Number of variables: do not apply.

2- Computation of the set of the powers

The set of constrained numbers whose elements are the power of each term of the expansion is given by

$$L_p^{t-1} = L_2^2 = \{(2,0,0), (1,1,0), (1,0,1), (0,2,0), (0,1,1), (0,0,2)\} \tag{6.178}$$

3- Computation of the coefficients

The coefficients are evaluated using the expression

$$F_p = C\left(\lambda_0(u_0), \lambda_1(u_0)\right)\ C\left(\lambda_0(u_1), \lambda_1(u_1)\right) \tag{6.179}$$

where $u \in L_2^2$. The detailed computation is shown in table 6.9.2.

Table 6.9.2 *Contributions of the elements of the set L_2^3 to the power of terms of the expansion, where $C_{\lambda_1}^{\lambda_0} := C(\lambda_0, \lambda_1)$ and $C_{\lambda_2}^{\lambda_1} := C(\lambda_1, \lambda_2)$.*

coeff.	number	λ_0	λ_1	λ_2	$C_{\lambda_1}^{\lambda_0}$	$C_{\lambda_2}^{\lambda_1}$	$C_{\lambda_1}^{\lambda_0} C_{\lambda_2}^{\lambda_1}$
$F_p(2,0,0)$	$(2,0,0)$	2	0	0	$C_0^2 = 1$	$C_0^0 = 1$	$C_0^2 C_0^0 = 1$
$F_p(1,1,0)$	$(1,1,0)$	2	1	0	$C_1^2 = 2$	$C_0^1 = 1$	$C_1^2 C_0^1 = 2$
$F_p(1,0,1)$	$(1,0,1)$	2	1	1	$C_1^2 = 2$	$C_1^1 = 1$	$C_1^2 C_1^1 = 2$
$F_p(0,2,0)$	$(0,2,0)$	2	2	0	$C_2^2 = 1$	$C_0^2 = 1$	$C_2^2 C_0^2 = 1$
$F_p(0,1,1)$	$(0,1,1)$	2	1	1	$C_1^2 = 2$	$C_1^1 = 1$	$C_1^2 C_1^1 = 2$
$F_p(0,0,2)$	$(0,0,2)$	2	2	2	$C_2^2 = 1$	$C_2^2 = 1$	$C_2^2 C_2^2 = 1$

The expanded set is given by

$$(L_2^2)^+ = \{(1,2,0,0), (2,1,1,0), (2,1,0,1), (1,0,2,0), (2,0,1,1), (1,0,0,2)\} \tag{6.180}$$

4- Solution

The expansion is given by

$$(A_0 + A_1 + A_2)^2 = A_0^2 + 2A_0A_1 + 2A_0A_2 + A_1^2 + 2A_1A_2 + A_2^2 \tag{6.181}$$

end of example 6.9.1

6.10 Computation of the derivatives of the form $1/(x-\beta)^s$

It will be considered here only the cases of the derivatives that will be used to obtain the Hermite Interpolating Polynomials.

6.10.1 One function and one variable

Let $P(x) = (x - \beta)^s$ be a function of a single variable x. Consider the expression

$$\frac{1}{P(x)} := \frac{1}{(x - \beta)^s} \tag{6.182}$$

If $(x + \beta)$ it is sufficient to take $\beta = -\beta$.

The derivative of order v of $1/P(x)$ with respect to x can be obtained observing that

$$\left[\frac{1}{P(x)}\right] = \left[\frac{1}{(x-\beta)^s}\right] = \left[(x-\beta)^{-s}\right]$$

$$\left[\frac{1}{P(x)}\right]^{(1)} = \frac{d}{dx}\left[(x-\beta)^{-s}\right] = (-s)\left[(x-\beta)^{-s-1}\right]$$

$$= (-1)(s+0)\left[(x-\beta)^{-(s+1)}\right]$$

$$\left[\frac{1}{P(x)}\right]^{(2)} = \frac{d^2}{dx^2}\left[(x-\beta)^{-s}\right] = (-s)[-(s+1)]\left[(x-\beta)^{-s-2}\right]$$

$$= (-1)^2(s+0)(s+1)\left[(x-\beta)^{-(s+2)}\right]$$

$$\vdots$$

$$\left[\frac{1}{P(x)}\right]^{(v)} = \frac{d^v}{dx^v}\left[(x-\beta)^{-s}\right]$$

$$= (-s)[-(s+1)]\cdots[-(s+v-1)]\left[(x-\beta)^{-s-v}\right]$$

$$= (-1)^v(s+0)(s+1)\cdots(s+v-1)\left[(x-\beta)^{-(s+v)}\right]$$

$$= (-1)^v(s)_v\left[(x-\beta)^{-(s+v)}\right] \tag{6.183}$$

where $(s)_v$ is the Pochhammer notation given in equation (4.186) on page 184. It can be evaluated in terms of factorial as

$$(s)_v = (s+0)(s+1)\cdots(s+v-1)$$

$$= \frac{1\times 2\times\cdots\times(s-1)(s+0)(s+1)\cdots(s+v-1)}{1\times 2\times\cdots\times(s-1)}$$

$$= \frac{(s+v-1)!}{(s-1)!} \tag{6.184}$$

It will be used either the Pochhammer notation or the factorial notation

$$\frac{b!}{(a-1)!} = (a+0)\times(a+1)\times\cdots\times(b-1)\times(b-0) \tag{6.185}$$

depending on the convenience for the deduction of the expressions.

Substituting (6.184) into (6.183) then the derivative of $[1/P(x)]^{(v)}$ can be written as

$$\left[\frac{1}{P(x)}\right]^{(v)} = \left[\frac{1}{(x-\beta)^s}\right]^{(v)}$$

$$= (-1)^v \frac{(s+v-1)!}{(s-1)!}(x-\beta)^{-(s+v)}$$

$$= \left[(-1)^v \frac{(s+v-1)!}{(s-1)!}\frac{1}{(x-\beta)^v}\right]\frac{1}{(x-\beta)^s} \tag{6.186}$$

Define

$$E^{(v)}(x) := (-1)^v \frac{(s+v-1)!}{(s-1)!}\frac{1}{(x-\beta)^v} \tag{6.187}$$

Observing that $(x-\beta)^v = (-1)^v(\beta-x)^v$ the equation (6.187) can be rewritten as

$$E^{(v)}(x) = \frac{(s+v-1)!}{(s-1)!}\frac{1}{(\beta-x)^v} \tag{6.188}$$

For some applications $s = \eta + 1$, which gives

$$E^{(v)}(x) = \frac{(\eta+v)!}{\eta!}\frac{1}{(\beta-x)^v} \tag{6.189}$$

The equation (6.186) can be rewritten as

$$\left[\frac{1}{P(x)}\right]^{(v)} = \frac{d^v}{dx^v}\left[\frac{1}{(x-\beta)^s}\right]$$

$$= E^{(v)}(x)\frac{1}{(x-\beta)^s}$$

$$= E^{(v)}(x)\left[\frac{1}{P(x)}\right] \tag{6.190}$$

and for some applications it particularizes to

$$\left[\frac{1}{P(x)}\right]^{(v)} = E^{(v)}(x)\frac{1}{(x-\beta)^s}$$

$$= \frac{(\eta+v)!}{\eta!}\frac{1}{(\beta-x)^v}\left[\frac{1}{P(x)}\right] \tag{6.191}$$

6.10.2 Product of t functions of the same variable x

Let

$$\frac{1}{P(x)} = \frac{1}{P_0(x)}\frac{1}{P_1(x)}\cdots\frac{1}{P_{t-1}(x)}$$

$$= \frac{1}{(x-\beta_0)^{s_0}}\frac{1}{(x-\beta_1)^{s_1}}\cdots\frac{1}{(x-\beta_{t-1})^{s_{t-1}}} \tag{6.192}$$

The derivative of order (v) can be obtained by expanding the product with the aid of the algorithm described in subsection 2.4, on page 40, and it can be written as

$$\left[\frac{1}{P(x)}\right]^{(v)} = \left[\prod_{i=0}^{i=t-1}\frac{1}{P_i(x)}\right]^{(v)}$$

$$= \sum_u F_u \left[\frac{1}{P_0(x)}\right]^{(u_0)}\left[\frac{1}{P_1(x)}\right]^{(u_1)}\cdots\left[\frac{1}{P_{t-1}(x)}\right]^{(u_{t-1})} \quad \forall u \in L_v^{t-1}$$

$$(6.193)$$

Each term of the expansion is evaluated using the expression in equation (6.190) as follows

$$\left[\frac{1}{P_i(x_i)}\right]^{(u_i)} = E^{(u_i)}(x)\left[\frac{1}{P_i(x)}\right] \tag{6.194}$$

where

$$E^{(u_i)}(x) := \frac{(s_i + u_i - 1)!}{(s_i - 1)!}\frac{1}{(\beta_i - x)^{u_i}} \tag{6.195}$$

is evaluated using the equation (6.189) for each s_i.

For some applications $s_i = \eta + 1$ and

$$E^{(u_i)}(x) = \frac{(\eta + u_i)!}{\eta!}\frac{1}{(\beta_i - x)^{u_i}} \tag{6.196}$$

Substituting equation (6.194) into equation (6.193) and factoring out $1/P_i(x)$ it follows

$$\left[\prod_{i=0}^{i=t-1}\frac{1}{P_i(x)}\right]^{(v)} =$$

$$\sum_u F_u\left[E_0^{(u_0)}(x)E_1^{(u_1)}(x)\cdots E_{t-1}^{(u_{t-1})}(x)\right]\left[\frac{1}{P_0(x)}\frac{1}{P_1(x)}\cdots\frac{1}{P_{t-1}(x)}\right] \tag{6.197}$$

Define

$$E^{(u)}(x) := E_0^{(u_0)}(x)E_1^{(u_1)}(x)\cdots E_{t-1}^{(u_{t-1})}(x)$$

$$:= \prod_{i=0}^{t-1}E_i^{(u_i)}(x) \tag{6.198}$$

where $E^{(u_i)}(x)$ is given by equation (6.195) or (6.196) and $u = (u_0, u_1, \ldots, u_{t-1}) \in L_v^{t-1}$.

The derivative in equation (6.197) writes

$$\left[\frac{1}{P(x)}\right]^{(v)} = \left[\prod_{i=0}^{i=t-1}\frac{1}{P_i(x)}\right]^{(v)} = \left[\sum_u F_u E^{(u)}(x)\right]\frac{1}{P(x)} \tag{6.199}$$

Define

$$E_F^{(v)}(x) := \sum_u F_u E^{(u)}(x) \qquad \forall u \in L_v^{t-1} \qquad (6.200)$$

where $E^{(u)}(x)$ is given by equation (6.198) and the coefficients F_u are obtained from the elements of the set of constrained numbers L_v^{t-1} using the theorem 6.7.5.

The equation (6.199) can be written in a compact form as

$$\left[\frac{1}{P(x)}\right]^{(v)} = E_F^{(v)}(x)\frac{1}{P(x)} \qquad (6.201)$$

Example 6.10.1 *Evaluate the derivative*

$$\left[\frac{1}{P(x)}\right]^{(3)} = \left[\frac{1}{(x-a)^{\eta+1}}\frac{1}{(x-b)^{\eta+1}}\right]^{(3)} \qquad (6.202)$$

Solution of the problem

1- Notation
Define

$$\frac{1}{P_0(x)} := \frac{1}{(x-a)^{\eta+1}} \qquad \frac{1}{P_1(x)} := \frac{1}{(x-b)^{\eta+1}} \qquad (6.203)$$

then equation (6.202) can be rewritten as

$$\left[\frac{1}{P(x)}\right]^{(3)} = \left[\frac{1}{P_0(x)}\frac{1}{P_1(x)}\right]^{(3)} \qquad (6.204)$$

which has a form similar to that of equation (6.199).

2- Identification of the parameters

1. Number of functions $t = 2$, which are $1/P_0(x)$ and $1/P_1(x)$.

2. Number of variables $d = 1$, which is x.

3. Order of derivative $v = 3$.

3- Computation of the set of the order of derivative of each term
To obtain the coefficients F_u and the terms $E^{(u)}(x)$ it is necessary first to obtain the elements u of the set $L_v^{t-1} = L_3^1$ with the aid of the algorithm

2.4.4, on page 45, or the algorithm 2.77, on page 54, which is particularized for the one-dimensional constrained numbers. The elements of the set L_3^1 are also exhibited in the table 2.7.22, on page 66, and it is shown below

$$L_v^{t-1} = L_3^1 = \{(3,0), (2,1), (1,2), (0,3)\} \tag{6.205}$$

The elements of the set L_3^1 represent the order of derivative of the expansion in equation (6.197).

4- Computation of the coefficients

The expression for the computation of the coefficients of the expansion was obtained in the subsection 6.7.3, equation (6.81), and it is

$$F_u = C(\lambda_0, \lambda_1) \tag{6.206}$$

The detailed computation of F_u is shown in the table 6.10.2.

Table 6.10.2 *Contributions of the elements of the set L_3^1 to the coefficients F_u of each term of the expansion.*

number u	λ_0	λ_1	$F_u = C(\lambda_0, \lambda_1)$
$(3,0)$	$3 + 0 = 3$	$0 = 0$	$C(3,0) = 1$
$(2,1)$	$2 + 1 = 3$	$1 = 1$	$C(3,1) = 3$
$(1,2)$	$1 + 2 = 3$	$2 = 2$	$C(3,2) = 3$
$(0,3)$	$0 + 3 = 3$	$3 = 3$	$C(3,3) = 1$

The coefficients F_u obtained in the fourth column are summarized below

$$F_u = \begin{bmatrix} 1 & 3 & 3 & 1 \end{bmatrix} \tag{6.207}$$

5- The expansion

The expansion of the derivative of the product shown in equation (6.201) writes, for this example, as

$$\left[\frac{1}{P(x)} \right]^{(3)} = E_F^{(3)}(x) \left[\frac{1}{P(x)} \right] \tag{6.208}$$

where $E_F^{(3)}(x)$ is given by equation (6.200), and u are the elements of the set L_3^1 displayed in equation (6.205). Substituting the values of u into equation (6.200) it follows that

$$E_F^{(3)}(x) = F_{(3,0)} E^{(3,0)}(x) + F_{(2,1)} E^{(2,1)}(x) + F_{(1,2)} E^{(1,2)}(x) + F_{(0,3)} E^{(0,3)}(x) \tag{6.209}$$

Expanding each $E^{(u_i)}(x)$ in equation (6.209) using equation (6.198) and replacing the values of F_u given by equation (6.207) it follows

$$E_F^{(3)}(x) = E_0^{(3)}(x)E_1^{(0)}(x) + 3E_0^{(2)}(x)E_1^{(1)}(x) + 3E_0^{(1)}(x)E_1^{(2)}(x)$$
$$+ E_0^{(0)}(x)E_1^{(3)}(x) \qquad (6.210)$$

where each $E_0^{(u_{i,0})}(x)$

$$E_0^{(3)}(x) = \frac{(\eta+1+3-1)!}{(\eta+1-1)!} \frac{1}{(a-x)^3} = \frac{(\eta+3)!}{\eta!} \frac{1}{(a-x)^3}$$

$$E_0^{(2)}(x) = \frac{(\eta+1+2-1)!}{(\eta+1-1)!} \frac{1}{(a-x)^2} = \frac{(\eta+2)!}{\eta!} \frac{1}{(a-x)^2}$$

$$E_0^{(1)}(x) = \frac{(\eta+1+1-1)!}{(\eta+1-1)!} \frac{1}{(a-x)^1} = \frac{(\eta+1)!}{\eta!} \frac{1}{(a-x)^1}$$

$$E_0^{(0)}(x) = \frac{(\eta+1+0-1)!}{(\eta+1-1)!} \frac{1}{(a-x)^0} = \frac{(\eta+0)!}{\eta!} \frac{1}{(a-x)^0} = 1 \qquad (6.211)$$

and analogously for $E_1^{(3)}(x) \cdots E_1^{(0)}(x)$.

6- Solution

The solution is obtained substituting $E_k^{(m)}(x)$ from equation (6.211) giving $E_F^{(3)}(x)$ then introduce equation (6.210) into equation (6.209) and the results into equation (6.208) as follows

$$\left[\frac{1}{P(x)}\right]^{(3)} = E_F^{(3)}(x)\left[\frac{1}{P(x)}\right]$$

$$= \left[F_{(3,0)}E^{(3,0)}(x) + F_{(2,1)}E^{(2,1)}(x) + F_{(1,2)}E^{(1,2)}(x)\right.$$
$$\left. + F_{(0,3)}E^{(0,3)}(x)\right] \frac{1}{P(x)}$$

$$= \left[E_0^{(3)}(x)E_1^{(0)}(x) + 3E_0^{(2)}(x)E_1^{(1)}(x) + 3E_0^{(1)}(x)E_1^{(2)}(x)\right.$$
$$\left. + E_0^{(0)}(x)E_1^{(3)}(x)\right] \frac{1}{P(x)}$$

$$= \left[\frac{(\eta+3)!}{\eta!} \frac{1}{(a-x)^3} \times \frac{(\eta+0)!}{\eta!} \frac{1}{(b-x)^0}\right.$$
$$+ 3\frac{(\eta+2)!}{\eta!} \frac{1}{(a-x)^2} \times \frac{(\eta+1)!}{\eta!} \frac{1}{(b-x)^1}$$
$$+ 3\frac{(\eta+1)!}{\eta!} \frac{1}{(a-x)^1} \times \frac{(\eta+2)!}{\eta!} \frac{1}{(b-x)^2}$$
$$\left. + \frac{(\eta+0)!}{\eta!} \frac{1}{(a-x)^0} \times \frac{(\eta+3)!}{\eta!} \frac{1}{(b-x)^3}\right] \frac{1}{P(x)} \qquad (6.212)$$

end of example 6.10.1

6.10.3   Computation of the derivatives of the product of d functions, which are product of k_j functions of the same variable x_j

The equation (6.192) in the subsection 6.10.2 can be generalized by defining a product of d functions of the form $1/P(x)$ as follows

$$\frac{1}{P_i(x)} = \frac{1}{P_{i,0}(x_0)} \frac{1}{P_{i,1}(x_1)} \cdots \frac{1}{P_{i,d-1}(x_{d-1})} \qquad (6.213)$$

where each $1/P_i(x_i)$ has the form

$$\frac{1}{P_{i,j}(x_i)} = \frac{1}{(x_j - \beta_{j,j,0})^{s_0}} \frac{1}{(x_j - \beta_{j,j,1})^{s_1}} \cdots \frac{1}{(x_j - \beta_{j,j,k})^{s_k}} \qquad (6.214)$$

For example, the $d = 4$ functions below

$$\frac{1}{P_0(x_0)} = \frac{1}{(x_0 - 3)^2} \frac{1}{(x_0 - 5)^1} \frac{1}{(x_0 - 7)^2}$$

$$\frac{1}{P_1(x_1)} = \frac{1}{(x_1 + 3)^3} \frac{1}{(x_1 + 5)^2}$$

$$\frac{1}{P_2(x_2)} = \frac{1}{(x_2 + 0)^4} \frac{1}{(x_2 + 1)^3} \frac{1}{(x_2 + 2)^3} \frac{1}{(x_2 + 3)^4}$$

$$\frac{1}{P_3(x_3)} = \frac{1}{(x_3 - 1)^1} \qquad (6.215)$$

The derivative of order $v = (v_0, \ldots, v_{d-1})$ of the product of d functions in equation (6.213) can be denoted as

$$\left[\frac{1}{P(x)} \right]^{(v)} = \left[\prod_{i=0}^{i=d-1} \frac{1}{P_i(x_i)} \right]^{(v_i)} \qquad (6.216)$$

Expanding the derivative in equation (6.216) it gives

$$\left[\frac{1}{P(x)} \right]^{(v)} = \left[\prod_{i=0}^{i=d-1} \frac{1}{P_i(x_i)} \right]^{(v_i)} = \prod_{i=0}^{i=d-1} \left[\frac{1}{P_i(x_i)} \right]^{(v_i)}$$

$$= \left[\frac{1}{P_0(x_0)} \right]^{(v_0)} \left[\frac{1}{P_1(x_1)} \right]^{(v_1)} \cdots \left[\frac{1}{P_{d-1}(x_{d-1})} \right]^{(v_{d-1})} \qquad (6.217)$$

where the derivative of order v_i of each factor $1/P_i(x_i)$ corresponds to the derivative of a product of k_i functions of the same variable x_i whose derivative is computed using the expression (6.201). The result is

$$\prod_{i=0}^{d-1}\left[\frac{1}{P_i(x)}\right]^{(v_i)} = \left[E_F^{(v_0)}(x_0)E_F^{(v_1)}(x_1)\cdots E_F^{(v_{d-1})}(x_{d-1})\right]\left[\frac{1}{P(x)}\right]$$

$$= \left[\prod_{i=0}^{d-1}E^{(v_i)}(x_i)\right]\frac{1}{P(x)} \qquad (6.218)$$

therefore,

$$\left[\frac{1}{P(x)}\right]^{(v)} = E_F^{(v)}(x)\frac{1}{P(x)} \qquad (6.219)$$

where

$$E^{(v)}(x) := \prod_{i=0}^{d-1}E^{(v_i)}(x_i) \qquad \text{where } v_i \in L_v^{d-1} \qquad (6.220)$$

The definition in equation (6.220) applies to $x = (x_0, \ldots, x_{d-1})$ variables while the definition in equation (6.201) applies to one variable x only.

Example 6.10.3 *Evaluate the derivative*

$$\left[\frac{1}{P(x)}\right]^{(1,3,0,2)} = \left[\frac{1}{P_0(x_0)}\frac{1}{P_1(x_1)}\frac{1}{P_2(x_2)}\frac{1}{P_3(x_3)}\right]^{(1,3,0,2)}$$

$$= \left[\frac{1}{P_0(x_0)}\right]^{(1)}\left[\frac{1}{P_1(x_1)}\right]^{(3)}\left[\frac{1}{P_2(x_2)}\right]^{(0)}\left[\frac{1}{P_3(x_3)}\right]^{(2)} \qquad (6.221)$$

where $P_i(x_i), i = 0,1,2,3$ are given by equation (6.215).

Solution of the problem

1- Parameters

1. Number of variables $d = 4$, one for each $P_i(x_i)$.

2. Number of functions in each $P_i(x_i)$.

   For x_0, there are $k_0 = 3$, for x_1, there are $k_1 = 2$, for x_2, there are $k_2 = 4$, for x_3, there are $k_3 = 1$.

3. Order of derivative $v = (v_0, v_1, v_2, v_3) = (1,3,0,2)$

2- Contribution of $P_0(x_0)$ to the expansion of the derivative

There are $t = 3$ functions of the same variable x_0 in the product, therefore, the derivative is performed using the equation (6.201) as follows

1. Equation

$$
\left[\frac{1}{P_0(x_0)}\right]^{(1)} = \left[\prod_{i=0}^{2} \frac{1}{P_{0,i}(x_0)}\right]^{(1)}
$$

$$
= \left[\frac{1}{(x_0 - 3)^2} \frac{1}{(x_0 - 5)^1} \frac{1}{(x_0 - 7)^1}\right]^{(1)} \tag{6.222}
$$

2. Set

$$
L_{v_0}^{k_0 - 1} = L_1^2 = \{(1,0,0), (0,1,0), (0,0,1)\} \tag{6.223}
$$

Denote by $u_i, i = 0, 1, 2$ the elements of L_1^2.

3. Coefficients

The coefficients are given by

$$
F_u = \begin{bmatrix} 1 & 1 & 1 \end{bmatrix} \tag{6.224}
$$

where $u \in L_1^2$. The detailed computation is shown in table 6.10.4.

Table 6.10.4 *Contributions of the elements of the set L_1^2, corresponding to the product of functions $P_0(x_0)$, to the coefficients.*

number u_i	λ_0	λ_1	λ_2	$C(\lambda_0, \lambda_1)$	$C(\lambda_1, \lambda_2)$
(1, 0, 0)	$1 + 0 + 0 = 1$	$0 + 0 = 0$	0	$C(1,0) = 1$	$C(0,0) = 1$
(0, 1, 0)	$0 + 1 + 0 = 1$	$1 + 0 = 1$	0	$C(1,1) = 1$	$C(1,0) = 1$
(0, 0, 1)	$0 + 0 + 1 = 1$	$0 + 1 = 1$	1	$C(1,1) = 1$	$C(1,1) = 1$

4. Computation of $E^{(u)}(x)$.

$$
E^{(u)}(x_0) = E_0^{(u_0)}(x_0) + E_1^{(u_1)}(x_0) + E_2^{(u_2)}(x_0)
$$

$$
= E_0^{(1,0,0)}(x_0) + E_1^{(0,1,0)}(x_0) + E_2^{(0,0,1)}(x_0) \tag{6.225}
$$

For $u_0 = (1,0,0) \in L_1^2$ if follows

$$E_0^{(u_0)}(x_0) = E_{0,0}^{(1)}(x_0)\, E_{0,1}^{(0)}(x_0)\, E_{0,2}^{(0)}(x_0)$$

$$= \frac{(s_0 + u_0 - 1)!}{(s_0 - 1)!}\, \frac{1}{(b_0 - x_0)^{u_0}} \times \frac{(s_1 + u_1 - 1)!}{(s_1 - 1)!}\, \frac{1}{(b_1 - x_0)^{u_1}}$$

$$\times \frac{(s_2 + u_2 - 1)!}{(s_2 - 1)!}\, \frac{1}{(b_2 - x_0)^{u_2}}$$

$$= \frac{(2 + 1 - 1)!}{(2 - 1)!}\, \frac{1}{(3 - x_0)^1} \times \frac{(2 + 0 - 1)!}{(1 - 1)!}\, \frac{1}{(5 - x_0)^0}$$

$$\times \frac{(2 + 0 - 1)!}{(2 - 1)!}\, \frac{1}{(7 - x_0)^0} = \frac{2}{(3 - x_0)} \qquad (6.226)$$

For $u_1 = (0,1,0) \in L_1^2$ if follows

$$E_1^{(u_1)}(x_0) = E_{1,0}^{(0)}(x_0)\, E_{1,0}^{(1)}(x_0)\, E_{1,2}^{(0)}(x_0)$$

$$= \frac{(s_0 + u_0 - 1)!}{(s_0 - 1)!}\, \frac{1}{(b_0 - x_0)^{u_0}} \times \frac{(s_1 + u_1 - 1)!}{(s_1 - 1)!}\, \frac{1}{(b_1 - x_0)^{u_1}}$$

$$\times \frac{(s_2 + u_2 - 1)!}{(s_2 - 1)!}\, \frac{1}{(b_2 - x_0)^{u_2}}$$

$$= \frac{(2 + 0 - 1)!}{(2 - 1)!}\, \frac{1}{(3 - x_0)^0} \times \frac{(1 + 1 - 1)!}{(1 - 1)!}\, \frac{1}{(5 - x_0)^1}$$

$$\times \frac{(2 + 0 - 1)!}{(2 - 1)!}\, \frac{1}{(7 - x_0)^0} = \frac{1}{(5 - x_0)} \qquad (6.227)$$

For $u_2 = (0,0,1) \in L_1^2$ if follows

$$E_2^{(u_2)}(x_0) = E_{2,0}^{(0)}(x_0)\, E_{2,1}^{(0)}(x_0)\, E_{2,2}^{(1)}(x_0)$$

$$= \frac{(s_0 + u_0 - 1)!}{(s_0 - 1)!}\, \frac{1}{(b_0 - x_0)^{u_0}} \times \frac{(s_1 + u_1 - 1)!}{(s_1 - 1)!}\, \frac{1}{(b_1 - x_0)^{u_1}}$$

$$\times \frac{(s_2 + u_2 - 1)!}{(s_2 - 1)!}\, \frac{1}{(b_2 - x_0)^{u_2}}$$

$$= \frac{(2 + 0 - 1)!}{(2 - 1)!}\, \frac{1}{(3 - x_0)^0} \times \frac{(1 + 0 - 1)!}{(1 - 1)!}\, \frac{1}{(5 - x_0)^0}$$

$$\times \frac{(2 + 1 - 1)!}{(2 - 1)!}\, \frac{1}{(7 - x_0)^1} = \frac{2}{(7 - x_0)} \qquad (6.228)$$

5. Solution

$$\left[\frac{1}{P_0(x_0)}\right]^{(1)} = \left[\frac{2}{(3 - x_0)} + \frac{1}{(5 - x_0)} + \frac{2}{(7 - x_0)}\right]\left[\frac{1}{P_0(x_0)}\right] \qquad (6.229)$$

3- Contribution of $P_1(x_1)$ to the expansion

There are $t = 2$ functions of the same variable x_1 in the product, therefore, the derivative is performed using the equation (6.201) as follows

1. Equation

$$\left[\frac{1}{P_1(x_1)}\right]^{(3)} = \left[\prod_{i=0}^{1}\frac{1}{P_{1,i}(x_1)}\right]^{(3)} = \left[\frac{1}{(x_1+3)^3}\frac{1}{(x_1+5)^2}\right]^{(3)} \quad (6.230)$$

2. Set

$$L_{v_1}^{k_1-1} = L_3^1 = \{(3,0),(2,1),(1,2),(0,3)\} \quad (6.231)$$

Denote by $u_i, i = 0,1,2,3$ the elements of L_3^1.

3. Coefficients

The coefficients are given by

$$F_u = \begin{bmatrix} 1 & 3 & 3 & 1 \end{bmatrix} \quad (6.232)$$

where $u \in L_3^1$. The detailed computation is shown in table 6.10.5.

Table 6.10.5 *Contributions of the elements of the set L_3^1, corresponding to the product of functions $P_1(x_1)$, to the coefficients.*

number u_i	λ_0	λ_1	$F_u = C(\lambda_0, \lambda_1)$
$(3,0)$	$3 + 0 = 3$	$0 = 0$	$C(3,0) = 1$
$(2,1)$	$2 + 1 = 3$	$1 = 1$	$C(3,1) = 3$
$(1,2)$	$1 + 2 = 3$	$2 = 2$	$C(3,2) = 3$
$(0,3)$	$0 + 3 = 3$	$3 = 3$	$C(3,3) = 1$

4. Computation of $E^{(u)}(x)$.

$$E^{(u)}(x_0) = F_{u_0}E_0^{(u_0)}(x_1) + F_{u_1}E_1^{(u_1)}(x_1) + F_{u_2}E_2^{(u_2)}(x_1)$$
$$+ F_{u_3}E_3^{(u_3)}(x_1)$$
$$= E_0^{(3,0)}(x_1) + 3E_1^{(2,1)}(x_1) + 3E_2^{(1,2)}(x_1) + E_3^{(0,3)}(x_1) \quad (6.233)$$

For $u_0 = (3,0) \in L_3^1$ if follows

$$
\begin{aligned}
E_0^{(u_0)}(x_1) &= E_{0,0}^{(1)}(x_1)\, E_{0,1}^{(0)}(x_1) \\
&= \frac{(s_0 + u_0 - 1)!}{(s_0 - 1)!}\, \frac{1}{(b_0 - x_0)^{u_0}} \times \frac{(s_1 + u_1 - 1)!}{(s_1 - 1)!}\, \frac{1}{(b_1 - x_0)^{u_1}} \\
&= \frac{(3 + 3 - 1)!}{(3 - 1)!}\, \frac{1}{(-3 - x_1)^3} \times \frac{(2 + 0 - 1)!}{(2 - 1)!}\, \frac{1}{(-5 - x_1)^0} \\
&= \frac{60}{(-3 - x_1)^3} \quad\quad\quad\quad\quad\quad\quad\quad\quad (6.234)
\end{aligned}
$$

For $u_1 = (2,1) \in L_3^1$ if follows

$$
\begin{aligned}
E_1^{(u_1)}(x_0) &= E_{1,0}^{(0)}(x_1)\, E_{1,0}^{(1)}(x_1) \\
&= \frac{(s_0 + u_0 - 1)!}{(s_0 - 1)!}\, \frac{1}{(b_0 - x_1)^{u_0}} \times \frac{(s_1 + u_1 - 1)!}{(s_1 - 1)!}\, \frac{1}{(b_1 - x_1)^{u_1}} \\
&= \frac{(3 + 2 - 1)!}{(3 - 1)!}\, \frac{1}{(3 - x_1)^2} \times \frac{(2 + 1 - 1)!}{(2 - 1)!}\, \frac{1}{(5 - x_1)^1} \\
&= \frac{12}{(-3 - x_1)^2}\, \frac{2}{(5 - x_1)^1} \quad\quad\quad\quad\quad\quad (6.235)
\end{aligned}
$$

For $u_2 = (1,2) \in L_3^1$ if follows

$$
\begin{aligned}
E_2^{(u_2)}(x_0) &= E_{2,0}^{(0)}(x_1)\, E_{2,0}^{(1)}(x_1) \\
&= \frac{(s_0 + u_0 - 1)!}{(s_0 - 1)!}\, \frac{1}{(b_0 - x_1)^{u_0}} \times \frac{(s_1 + u_1 - 1)!}{(s_1 - 1)!}\, \frac{1}{(b_1 - x_1)^{u_1}} \\
&= \frac{(3 + 1 - 1)!}{(3 - 1)!}\, \frac{1}{(3 - x_1)^1} \times \frac{(2 + 2 - 1)!}{(2 - 1)!}\, \frac{1}{(5 - x_1)^2} \\
&= \frac{3}{(-3 - x_1)^1}\, \frac{3}{(5 - x_1)^2} \quad\quad\quad\quad\quad\quad (6.236)
\end{aligned}
$$

For $u_3 = (0,3) \in L_3^1$ if follows

$$
\begin{aligned}
E_3^{(u_3)}(x_0) &= E_{3,0}^{(0)}(x_0)\, E_{3,0}^{(1)}(x_0) \\
&= \frac{(s_0 + u_0 - 1)!}{(s_0 - 1)!}\, \frac{1}{(b_0 - x_1)^{u_0}} \times \frac{(s_1 + u_1 - 1)!}{(s_1 - 1)!}\, \frac{1}{(b_1 - x_1)^{u_1}} \\
&= \frac{(3 + 0 - 1)!}{(3 - 1)!}\, \frac{1}{(3 - x_1)^0} \times \frac{(2 + 3 - 1)!}{(2 - 1)!}\, \frac{1}{(5 - x_1)^3} \\
&= \frac{1}{(-3 - x_1)^0}\, \frac{24}{(5 - x_1)^3} \quad\quad\quad\quad\quad\quad (6.237)
\end{aligned}
$$

5. Solution

$$
\left[\frac{1}{P_1(x_1)}\right]^{(3)} = \left[\frac{60}{(-3-x_1)^3} + \frac{12}{(-3-x_1)^2\,(-5-x_1)^1} + \frac{2}{}\right.
$$
$$
\left. + \frac{3}{(-3-x_1)^1\,(-5-x_1)^2} + \frac{3}{} + \frac{24}{(-5-x_1)^3}\right]\left[\frac{1}{P_1(x_1)}\right]
$$
$$
\tag{6.238}
$$

4- Contribution of $P_2(x_2)$ to the expansion

There are $t = 4$ functions of the same variable x_2 in the product, therefore, the derivative is performed using the equation (6.201) as follows

1. Equation

$$
\left[\frac{1}{P_2(x_2)}\right]^{(0)} = \left[\prod_{i=0}^{3}\frac{1}{P_{2,i}(x_2)}\right]^{(0)}
$$
$$
= \left[\frac{1}{(x_2+0)^4}\frac{1}{(x_2+1)^3}\frac{1}{(x_2+2)^3}\frac{1}{(x_2+3)^4}\right]^{(0)}
\tag{6.239}
$$

2. Set

$$
L_{v_2}^{k_2-1} = L_0^3 = \{(0,0,0,0)\}
\tag{6.240}
$$

Denote by u_0 the elements of L_0^3.

3. Coefficients

The coefficients are given by

$$
F_u = \begin{bmatrix} 1 \end{bmatrix}
\tag{6.241}
$$

where $u \in L_0^3$. The detailed computation is shown in table 6.10.6.

Table 6.10.6 *Contributions of the elements of the set L_0^3, corresponding to the product of functions $P_2(x_2)$, to the coefficients, where $C_{\lambda_1}^{\lambda_0} := C\,(\lambda_0, \lambda_1)$, $C_{\lambda_2}^{\lambda_1} := C\,(\lambda_1, \lambda_2)$, and $C_{\lambda_3}^{\lambda_2} := C\,(\lambda_2, \lambda_3)$.*

number u_i	λ_0	λ_1	λ_2	λ_3
$(0,0,0,0)$	$0+0+0+0=0$	$0+0+0=0$	$0+0=0$	$0=0$

number	$C_{\lambda_1}^{\lambda_0}$	$C_{\lambda_2}^{\lambda_1}$	$C_{\lambda_3}^{\lambda_2}$	$F_u = C_{\lambda_1}^{\lambda_0} C_{\lambda_2}^{\lambda_1} C_{\lambda_3}^{\lambda_2}$
$(0,0,0,0)$	$C_0^0 = 1$	$C_0^0 = 1$	$C_0^0 = 1$	$C_0^0 C_0^0 C_0^0 = 1$

4. Computation of $E^{(u)}(x)$.

$$E^{(u)}(x_2) = E_0^{(u)}(x_2) \tag{6.242}$$

For $u_0 = (0,0,0,0) \in L_0^3$ if follows

$$
\begin{aligned}
E_0^{(u)}(x_2) &= E_{0,0}^{(0)}(x_2) \\
&= \frac{(s_0 + u_0 - 1)!}{(s_0 - 1)!} \frac{1}{(b_0 - x_2)^{u_0}} \times \frac{(s_1 + u_1 - 1)!}{(s_1 - 1)!} \frac{1}{(b_1 - x_2)^{u_1}} \\
&\quad \times \frac{(s_2 + u_2 - 1)!}{(s_2 - 1)!} \frac{1}{(b_2 - x_2)^{u_2}} \times \frac{(s_3 + u_3 - 1)!}{(s_3 - 1)!} \frac{1}{(b_3 - x_2)^{u_3}} \\
&= \frac{(4 + 0 - 1)!}{(4 - 1)!} \frac{1}{(-0 - x_2)^0} \times \frac{(3 + 0 - 1)!}{(3 - 1)!} \frac{1}{(-1 - x_2)^0} \\
&\quad \times \frac{(3 + 0 - 1)!}{(3 - 1)!} \frac{1}{(-2 - x_2)^0} \times \frac{(4 + 0 - 1)!}{(4 - 1)!} \frac{1}{(-3 - x_2)^0} \\
&= 1 \tag{6.243}
\end{aligned}
$$

5. Solution

$$\left[\frac{1}{P_2(x_0)} \right]^{(0)} = 1 \times \left[\frac{1}{P_2(x_2)} \right] \tag{6.244}$$

5- Contribution of $P_3(x_3)$ to the expansion

There are $t = 1$ functions of the same variable x_3 in the product, therefore, the derivative is performed using the equation (6.201) as follows

1. Equation

$$\left[\frac{1}{P_3(x_3)} \right]^{(2)} = \left[\prod_{i=0}^0 \frac{1}{P_{3,i}(x_3)} \right]^{(2)} = \left[\frac{1}{(x_3 - 1)^1} \right]^{(2)} \tag{6.245}$$

2. Set

$$L_{v_3}^{k_3 - 1} = L_2^0 = \{(2)\} \tag{6.246}$$

Denote by u_0 the elements of L_2^0.

3. Coefficients

The coefficients are given by

$$F_u = \begin{bmatrix} 1 \end{bmatrix} \tag{6.247}$$

where $u \in L_2^0$.

4. Computation of $E^{(u)}(x)$.

$$E^{(u)}(x_3) = E_0^{(u)}(x_3) \tag{6.248}$$

For $u = (2) \in L_2^0$ if follows

$$E_0^{(u)}(x_3) = E_{0,0}^{(2)}(x_3)$$

$$= \frac{(s_0 + u_0 - 1)!}{(s_0 - 1)!} \frac{1}{(b_0 - x_3)^{u_0}}$$

$$= \frac{(1 + 2 - 1)!}{(1 - 1)!} \frac{1}{(1 - x_3)^2} = \frac{2}{(1 - x_3)^2} \tag{6.249}$$

5. Solution

$$\left[\frac{1}{P_3(x_3)} \right]^{(2)} = \left[\frac{2}{(1 - x_3)^2} \right] \left[\frac{1}{P_3(x_3)} \right] = \frac{2}{(x_3 - 1)^3} \tag{6.250}$$

6- Solution of the problem

$$\left[\frac{1}{P(x)} \right]^{(1,3,0,2)} = \left[\frac{1}{P_0(x_0)} \frac{1}{P_1(x_1)} \frac{1}{P_2(x_2)} \frac{1}{P_3(x_3)} \right]^{(1,3,0,2)}$$

$$= \left[\frac{1}{P_0(x_0)} \right]^{(1)} \left[\frac{1}{P_1(x_1)} \right]^{(3)} \left[\frac{1}{P_2(x_2)} \right]^{(0)} \left[\frac{1}{P_3(x_3)} \right]^{(2)}$$

$$= \left[\frac{2}{(3 - x_0)} + \frac{1}{(5 - x_0)} + \frac{2}{(7 - x_0)} \right]$$

$$\times \left[\frac{60}{(-3 - x_1)^3} + \frac{12}{(-3 - x_1)^2 (-5 - x_1)^1} \right.$$

$$\left. + \frac{3}{(-3 - x_1)^1 (-5 - x_1)^2} + \frac{3}{(-5 - x_1)^3} + \frac{24}{(-5 - x_1)^3} \right] \times 1 \times \left[\frac{2}{(1 - x_3)^2} \right]$$

$$\times \left[\frac{1}{P_0(x_0)} \right] \left[\frac{1}{P_1(x_1)} \right] \left[\frac{1}{P_2(x_0)} \right] \left[\frac{1}{P_3(x_3)} \right] \tag{6.251}$$

end of example 6.10.3

6.10.4   Summary

6.10.4.1   Derivative of order v of $t = 1$ function and $d = 1$ variable x

For one function $(t = 1)$ and one variable $(d = 1)$ equation (6.190) gives

$$\left[\frac{1}{P(x)} \right]^{(v)} = E^{(v)}(x) \frac{1}{(x - \beta)^s} = E^{(v)}(x) \left[\frac{1}{P(x)} \right] \tag{6.252}$$

where from equation (6.187)

$$E^{(v)}(x) = \frac{(s+v-1)!}{(s-1)!}\frac{1}{(\beta-x)^v} \tag{6.253}$$

For some applications $s = \eta + 1$ and equation (6.189) gives

$$E^{(v)}(x) = \frac{(\eta+v)!}{\eta!}\frac{1}{(\beta-x)^v} \tag{6.254}$$

6.10.4.2 Derivative of order t of the same variable variable x

From equation (6.201)

$$\left[\frac{1}{P(x)}\right]^{(v)} = E_F^{(v)}(x)\frac{1}{P(x)} \tag{6.255}$$

where $E_F^{(u)}(x)$ is given by equation (6.200) as

$$E_F^{(v)}(x) := \sum_u F_u E^{(u)}(x) \tag{6.256}$$

where $E^{(u)}(x)$ is given by equation (6.198) as

$$E^{(u)}(x) = \prod_{i=0}^{d-1} E^{(u_i)}(x) \tag{6.257}$$

The coefficients F_u are obtained from the elements of the set of constrained numbers L_v^{t-1} using the theorem 6.7.5.

Each $E^{(u_i)}(x)$ in equation (6.257) is given by (6.253) or (6.254).

6.10.4.3 Computation of the derivatives of the product of d functions, which are product of k_j functions of the same variable x_j

From the equation (6.213)

$$\frac{1}{P_i(x)} = \frac{1}{P_{i,0}(x_0)}\frac{1}{P_{i,1}(x_1)}\cdots\frac{1}{P_{i,d-1}(x_{d-1})} \tag{6.258}$$

where each $1/P_i(x_i)$ has the form

$$\frac{1}{P_{i,j}(x_i)} = \frac{1}{(x_j-\beta_{j,j,0})^{s_0}}\frac{1}{(x_j-\beta_{j,j,1})^{s_1}}\cdots\frac{1}{(x_j-\beta_{j,j,k})^{s_k}} \tag{6.259}$$

The general expression is given by (6.219)

$$\left[\frac{1}{P(x)}\right]^{(v)} = E_F^{(v)}(x)\frac{1}{P(x)} \tag{6.260}$$

where

$$E^{(v)}(x) := \prod_{i=0}^{d-1} E^{(v_i)}(x_i) \tag{6.261}$$

6.11 Computation of the derivatives of the form $1/(c_0 x_0 + c_1 x_1 + c_2)$

6.11.1 Derivative of one function

Define

$$\frac{1}{P(x)} := \frac{1}{(c_0 x_0 + c_1 x_1 + c_2)^s} \tag{6.262}$$

where $x = (x_0, x_1)$. For the majority of the applications $s = \eta + 1$.

The derivative of order $v = (v_0, v_1)$ writes

$$\left[\frac{1}{P(x)} \right]^{(v_0, v_1)} = \left[\frac{1}{(c_0 x_0 + c_1 x_1 + c_2)^s} \right]^{(v_0, v_1)} \tag{6.263}$$

Following the same procedure as in the subsection 6.10.1 for each variable it follows

$$\left[\frac{1}{P(x)} \right]^{(v_0, v_1)} = \left[\left[\frac{1}{P(x)} \right]^{(v_0)} \right]^{(v_1)} \tag{6.264}$$

where

$$\left[\frac{1}{P(x)} \right]^{(v_0)} = (-1)^{v_0} c_0^{v_0} (s+0) \cdots (s + v_0 - 1)(c_0 x_0 + c_1 x_1 + c_2)^{-(s+v_0)} \tag{6.265}$$

and

$$\left[\left[\frac{1}{P(x)} \right]^{(v_0)} \right]^{(v_1)}$$

$$= (-1)^{v_0 + v_1} c_0^{v_0} c_1^{v_1} (s+0) \cdots (s + v_0 + v_1 - 1)(c_0 x_0 + c_1 x_1 + c_2)^{-(s+v_0+v_1)}$$

$$= (-1)^{v_0 + v_1} c_0^{v_0} c_1^{v_1} \frac{(s + v_0 + v_1 - 1)!}{(s-1)!} (c_0 x_0 + c_1 x_1 + c_2)^{-(s+v_0+v_1)} \tag{6.266}$$

Define

$$E^{(v_0+v_1)}(x_0, x_1) := (-1)^{v_0+v_1} c_0^{v_0} c_1^{v_1} \frac{(s + v_0 + v_1 - 1)!}{(s-1)!} \frac{1}{(c_0 x_0 + c_1 x_1 + c_2)^{v_0+v_1}} \tag{6.267}$$

then

$$\left[\frac{1}{P(x)} \right]^{(v_0,v_1)} = E^{(v_0+v_1)}(x_0,x_1) \frac{1}{(c_0 x_0 + c_1 x_1 + c_2)^s}$$

$$= E^{(v_0+v_1)}(x) \left[\frac{1}{P(x)} \right] \tag{6.268}$$

6.11.2   Derivative of product of t functions

Define

$$\frac{1}{P(x)} := \frac{1}{(c_{0,0} x_0 + c_{0,1} x_1 + c_{0,2})^{s_0}} \cdots \frac{1}{(c_{t-1,0} x_0 + c_{t-1,1} x_1 + c_{t-1,2})^{s_{t-1}}} \tag{6.269}$$

and

$$\frac{1}{P_k(x)} := \frac{1}{(c_{k,0} x_0 + c_{k,1} x_1 + c_{k,2})^{s_k}} \tag{6.270}$$

then

$$\frac{1}{P(x)} = \prod_{k=0}^{t-1} \frac{1}{P_k(x)} \tag{6.271}$$

where $x = (x_0, x_1)$.

The derivative of order $v = (v_0, v_1)$ writes

$$\left[\frac{1}{P(x)} \right]^{(v_0,v_1)} = \left[\prod_{k=0}^{t-1} \frac{1}{P_k(x)} \right]^{(v_0,v_1)} \tag{6.272}$$

Equation (6.272) represents the product of t functions $1/P_k(x)$ in $d = 2$ variables (x_0, x_1). The algorithm for the derivative of this product of functions is found in subsection 6.8.2.

Part II

Hermite Interpolating Polynomials

Chapter 7

Multivariate Hermite Interpolating Polynomial

7.1 Definition

Definition 7.1.1 *Domain*

Let $\mathcal{S}$ be a set of $\sigma > d$ hypersurfaces in $\mathbb{F}^d$, where $\mathbb{F}$ is a field, defined as

$$\mathcal{S} = \{\varphi_i(x) \mid \varphi_i(x) = 0,\, i = 0,\ldots,\sigma - 1\} \qquad (7.1)$$

Let Ω be a bounded domain in $\mathbb{F}^d$ whose boundary $\partial\Omega$ is defined by a subset of $\mathcal{S}$ forming a closed region. The hypersurfaces that do not belong to the boundary subset are assumed to cross the region. Define as reference node e a point on $d \leq \rho < \sigma$ hypersurfaces then it can be written

$$\bigcap_{I \subset [0,\sigma-1]} \varphi_i(x) = \mathbb{F}^0 \qquad i \in I \qquad (7.2)$$

where I has ρ elements and, therefore, it is a proper subset of the index set $[0, \sigma - 1]$. A subset $\chi_e \subset \mathcal{S}$ of ρ hypersurfaces containing the reference node e can be denoted as

$$\chi_e = \left\{ \varphi_i(x) = 0 \in \mathcal{S} \,\middle|\, \varphi_i(\alpha) = 0 \ \forall\, i \in I \right\} \qquad (7.3)$$

where $\alpha = (\alpha_0, \alpha_1, \ldots, \alpha_\delta)$ are the coordinates numbers of the reference node e, which is on the intersection of, at least, d hypersurfaces since its dimension is zero.

The region of interest is assumed to have, at least, $d + 1$ nodes in a $\mathbb{F}^d$ space such that each node is at the intersection of at least d noncoincident boundaries and so they are able to form a closed space, which justifies the condition $\sigma \geq d + 1$. The coordinate numbers of the reference node will be denoted by $\alpha = (\alpha_0, \alpha_1, \ldots, \alpha_\delta)$.

For example, in a two-dimensional Cartesian space, a triangular region can be defined by three noncoincident curves intercepting at three different noncollinear points such as shown in figure 7.1.2.

287

In particular if the three curves are coincident the algorithm here proposed will not work if all points are on the curve since it assumes that at least one point does not belong to the curve. For the case of the circle shown if figure 7.1.2 it can be used polar coordinates and, for example, the center of the circle as the point outside the curve.

In a three-dimensional space a closed region can be defined by four noncoincident surfaces intercepting at four non-coplanar points. The surface can be an sphere defined by four different points.

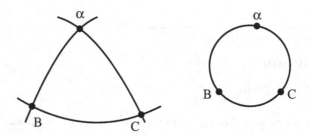

Figure 7.1.2 *Triangular region defined by three curves passing through three points and defined by three arcs of the same circle.*

end of definition 7.1.1

Definition 7.1.3 *Visibility of a reference node.*

Define as visibility of the reference node the subset of hypersurfaces $V_e \subset S$ complement of χ_e with respect to S, that is, $V_e := S - \chi_e$ is the set of hypersurfaces that do not contain the reference node. Let τ be the number of elements in V_e, that is, the number of hypersurfaces that do not contain the reference node.

The curve BC in figure 7.1.2 belongs to the visibility of the node e whose coordinate is α. That is, BC is the curve that the node α sees.

end of definition 7.1.3

Definition 7.1.4 *Multivariate Hermite Interpolating Polynomial*

Let

1. d be the dimension of a Cartesian space

2. $\delta = d - 1$ the dimension of a constrained number with d coordinate numbers

3. η a nonnegative integer

4. $\mathcal{O}^\delta(\mu) = \mathcal{O}^\delta(\mu_0, \mu_1, \ldots, \mu_\delta)$ represent the order of a constrained coordinate number $\mu = (\mu_0, \mu_1, \ldots, \mu_\delta) \in L^\delta(\eta)$ such that $\mathcal{O}^\delta(0, 0, \ldots, 0) \leq \mathcal{O}^\delta(\mu) \leq \mathcal{O}^\delta(0, 0, \ldots, \eta)$ (see property 5.4.2)

5. $\alpha = (\alpha_0, \alpha_1, \ldots, \alpha_\delta)$ be the coordinate numbers of the reference node e

6. $x = (x_0, x_1, \ldots, x_\delta)$ be a point in the domain Ω or boundary $\partial\Omega$

The set of functions $\Phi_{e,n}(x)$ satisfying the properties:

1. The derivatives of order $\mathcal{O}^\delta(\mu) = \mathcal{O}(n)$, where

$$\mathcal{O}^\delta(0, 0, \ldots, 0) \leq \mathcal{O}^\delta(n) \leq \mathcal{O}^\delta(0, 0, \ldots, \eta)$$

equal one when evaluated at the reference node e, that is,

$$\left. \Phi_{e,n}^{(\mu)}(x) \right|_{x=\alpha} = 1 \quad \text{if} \quad \mathcal{O}^\delta(\mu) = \mathcal{O}^\delta(n) \tag{7.4}$$

2. The derivatives of order $\mathcal{O}^\delta(\mu) \neq \mathcal{O}^\delta(n)$, when evaluated at the same reference node, equal zero

$$\left. \Phi_{e,n}^{(\mu)}(x) \right|_{x=\alpha} = 0 \quad \text{if} \quad \mathcal{O}^\delta(\mu) \neq \mathcal{O}^\delta(n) \tag{7.5}$$

where

$$\mathcal{O}^\delta(0, 0, \ldots, 0) \leq \mathcal{O}^\delta(\mu) \leq \mathcal{O}^\delta(0, 0, \ldots, 0, \eta)$$

and

$$\mathcal{O}^\delta(0, 0, \ldots, 0) \leq \mathcal{O}^\delta(n) \leq \mathcal{O}^\delta(0, 0, \ldots, 0, \eta)$$

3. The derivatives of order $\mathcal{O}^\delta(\mu)$ evaluated at points belonging to any of the τ hypersurfaces not containing the reference node equal to zero

$$\left. \Phi_{e,n}^{(\mu)}(x) \right|_{x=\beta_k} = 0 \quad k = 0, 1, \ldots, \tau - 1 \tag{7.6}$$

where

$$\mathcal{O}^\delta(0, 0, \ldots, 0) \leq \mathcal{O}^\delta(u) \leq \mathcal{O}^\delta(0, 0, \ldots, 0, \eta)$$

and β_k is a point on the hypersurface $\varphi_{e,k}(x) = 0$, $k = 0, \ldots, \tau - 1$, that is, $\varphi_{e,k}(\beta_k) = 0$, $k = 0, \ldots, \tau - 1$.

are called here the *d-dimensional or multivariate Hermite Interpolating Polynomial*. Note that η is not the degree of the polynomial but the order of derivative up to which the polynomial satisfies the conditions above.

The theorem 7.4.5 extends the above definitions to the case where 1 and 0 are the identity elements of the multiplication and addition, respectively, of a field $\mathbb{F}$.

end of definition 7.1.4

Notation 7.1.5

The notation $\Phi_{e,n}^{(\mu)}(x)$ identifies a function $\Phi(x)$. The subindex e identifies to which node the polynomial refers and the subindex n denotes the order of derivative for which the function has value one at the reference node. The upper-index μ tells the actual order of derivative of the function $\Phi(x)$. Therefore, when $\mu = n$ the value of the function $\Phi(x)$ is one at the reference node and zero if $\mu \neq n$.

end of notation 7.1.5

7.2   Polynomial construction

A polynomial with the above properties can be constructed using the following product

$$\Phi_{e,n}(x) = P_e(x)\, Q_{e,n}(x)\, f_{e,n}(\overline{a}, x) \qquad (7.7)$$

where each factor satisfies, at least, one of the properties stated in definition 7.1.4 and $\overline{a}$ represents the coefficients of the terms of the polynomial $f_{e,n}(\overline{a}, x)$. From the equation (7.4) it can be seen that at the reference node α all three factors in equation (7.7) do not vanish, that is, if $\mu = n = (0, \ldots, 0)$ then

$$\Phi_{e,0}^{(0)}(\alpha) = 1 \implies P_e(\alpha) \neq 0, \qquad Q_{e,0}(\alpha) \neq 0 \qquad f_{e,0}(\overline{a}, x) \neq 0 \qquad (7.8)$$

7.2.1   The polynomial $P_e(x)$

The polynomial $P_e(x)$ and its derivatives of order up to $(0, 0, \ldots, 0, \eta)$ must satisfy the following properties

Property 7.2.1

1. the polynomial $P_e(x)$ vanishes at any point that belongs at least to one of the hypersurfaces $\varphi_{e,k}(x) = 0$, $k = 0, \ldots, \tau - 1$, which does not contain the reference node, that is it must vanish at the hypersurfaces that belong to the visibility set $\mathcal{V}_e$

2. the polynomial $P_e(x)$ does not vanish when evaluated at the hypersurfaces that contain the reference node e, which implies that it does not vanish at the reference node e intersection of these surfaces. This statement can be written as

$$P_e(x)^{(\mu)}\Big|_{x=\alpha} \neq 0 \qquad \mu = (0, \ldots, 0), \ldots, (0, \ldots, \eta) \qquad (7.9)$$

where α is the coordinate number of the reference node e.

A polynomial satisfying the above properties can be written as the product

$$P_e(x) = \prod_{k=0}^{\tau-1} \varphi_{e,k}(x) \tag{7.10}$$

where $\varphi_{e,k}(x) \in V_e$.

end of property 7.2.1

Theorem 7.2.2 *Condition for the vanishing of $P_e(x)$ at V_e*

The polynomial $P_e(x)$ defined by

$$P_e(x) = \prod_{k=0}^{\tau-1} [\varphi_{e,k}(x)]^{\eta+1} \tag{7.11}$$

satisfies the condition described in the property 7.2.1.

Proof:

Without loss of generality consider $\eta = 2$ and $\tau = 3$, then

$$P_e(x) = \prod_{k=0}^{k=2} [\varphi_{e,k}(x)]^3 = [\varphi_{e,0}(x)\varphi_{e,1}(x)\varphi_{e,2}(x)]^3 \tag{7.12}$$

The first derivative is

$$\frac{d}{dx}P_e(x) = 3\left[\varphi_{e,0}(x)\varphi_{e,1}(x)\varphi_{e,2}(x)\right]^2 \left[\varphi_{e,0}(x)\varphi_{e,1}(x)\varphi_{e,2}(x)\right]^{(1)} \tag{7.13}$$

and the second derivative, which is of order $\eta = 2$, is

$$\frac{d^2}{dx^2}P_e(x) = 6\left[\varphi_{e,0}(x)\varphi_{e,1}(x)\varphi_{e,2}(x)\right]\left[\varphi_{e,0}(x)\varphi_{e,1}(x)\varphi_{e,2}(x)\right]^{(1)}$$
$$+ 3\left[\varphi_{e,0}(x)\varphi_{e,1}(x)\varphi_{e,2}(x)\right]^2 \left[\varphi_{e,0}(x)\varphi_{e,1}(x)\varphi_{e,2}(x)\right]^{(2)} \tag{7.14}$$

where

$$\left[\varphi_{e,0}(x)\varphi_{e,1}(x)\varphi_{e,2}(x)\right] = 0 \quad \text{and} \quad \left[\varphi_{e,0}(x)\varphi_{e,1}(x)\varphi_{e,2}(x)\right]^2 = 0$$

since

$$\varphi_{e,0}(x) = \varphi_{e,2}(x) = \varphi_{e,2}(x) = 0.$$

Each term has always a factor of the form $\varphi_{e,k}(x)^m$ where $0 < m \leq \eta+1$, which guarantees the vanishing of $P_e(x)$ and their derivatives of order up to η at each hypersurface $\varphi_{ej}(x) = 0$ in the visibility set of the reference node e.

Define

$$P_{e,k}(x) := [\varphi_{e,k}(x)]^{\eta+1} \tag{7.15}$$

then equation (7.11) can be rewritten as

$$P_e(x) = \prod_{k=0}^{\tau-1} P_{e,k}(x) \tag{7.16}$$

end of theorem 7.2.2

Notation 7.2.3 *Power and order of derivative*

The derivative is denoted by an upper index between parenthesis as, for example, the second order derivative of $\phi(x)$ is denoted as $\phi(x)^{(2)}$, that is,

$$\phi(x)^{(2)} := \frac{d^2}{dx^2} \phi(x) \tag{7.17}$$

The power is denoted by an upper index without parenthesis as, for example, the square of the function $\phi(x)$ is denoted as $\phi(x)^2$, that is,

$$\phi(x)^2 := \phi(x) \times \phi(x) \tag{7.18}$$

end of notation 7.2.3

The polynomial $P_e(x)$ given by (7.29), generated with respect to the reference node, also applies to a region divided into finite elements. In this case τ represents the number of hypersurfaces in the visibility set $\mathcal{V}_e$ for all finite elements of the domain.

The degree of $P_e(x)$ according to the equation (7.11) is given by

$$p_p = t(\eta + 1) \tag{7.19}$$

where t equals to the sum of the degrees of $\varphi_{e,k}(x)$ for $k = 0, 1, \ldots, \tau$.

Example 7.2.4 *Hyperpolyhedron region*

For the region shown in figure 7.2.5 the following can be identified

1. the dimension of the cartesian space is $d = 2$

2. the reference node e has the coordinate $\alpha = a_3 = (-5, 2)$

3. there are $\rho = 2$ hypersurfaces containing the reference node e, which are the lines through the points a_3, a_2, and a_3, a_4

4. there are $\tau = 3$ hypersurfaces in visibility set $\mathcal{V}_e$ of the reference node e. They are $\varphi_{3,0}, \varphi_{3,1}$, and $\varphi_{3,2}$, which allows to write the polynomial $P_e(x)$ as

$$P_e(x) = \prod_{k=0}^{k=3} P_{3,k}(x) = P_{3,0}(x)P_{3,1}(x)P_{3,2}(x)P_{3,3}(x) \tag{7.20}$$

where

$$P_{3,0}(x) = [\varphi_{3,0}(x_0, x_1)]^{\eta+1} = (3x_0 + 10x_1 + 64)^{\eta+1} \qquad (7.21)$$

$$P_{3,1}(x) = [\varphi_{3,1}(x_0, x_1)]^{\eta+1} = (-5x_0 + 6x_1 + 52)^{\eta+1} \qquad (7.22)$$

$$P_{3,2}(x) = [\varphi_{3,2}(x_0, x_1)]^{\eta+1} = (6x_0 + 5x_1 - 38)^{\eta+1} \qquad (7.23)$$

The introduction of equations (7.21)–(7.23) into equation (7.20) it gives

$$P_3(x) = (3x_0 + 10x_1 + 64)^{\eta+1} (-5x_0 + 6x_1 + 52)^{\eta+1} (6x_0 + 5x_1 - 38)^{\eta+1} \qquad (7.24)$$

which is an expression for the computation of $P_3(x)$ whose degree is $p_p = 3(\eta + 1)$.

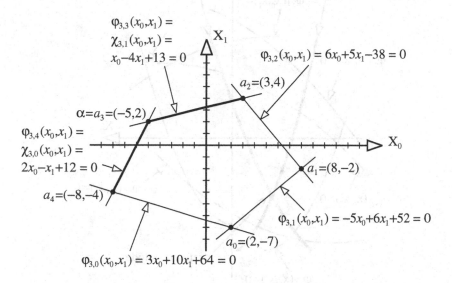

Figure 7.2.5 *Region defined by a set of lines connecting the vertices* $a_j, j = 0, 1, 2, 3, 4$. *The point* $\alpha = a_3 = (-5, 2)$ *represents the reference node* e.

end of example 7.2.4

Example 7.2.6 *Hypersurface region*

For the region shown in figure 7.2.7 it can be identified that:

1. the dimension of the cartesian space is $d = 2$.

2. the reference node e has coordinate α.

3. there are $\rho = 2$ hypersurfaces containing the reference node α, which are $\chi_{e,0}(x) = 0$ and $\chi_{e,1}(x) = 0$.

4. there are $\tau = 6$ hypersurfaces in visibility set $\mathcal{V}_e$ of the reference node e whose equations are $\varphi_{i,0}(x) = 0, i = 0, \ldots, 5$.

5. The boundary $\partial\Omega$ of the domain Ω in figure 7.2.7 is formed by the curves $\varphi_{e,0}(x_0, x_1) = 0$, $\varphi_{e,1}(x_0, x_1) = 0$, $\varphi_{e,3}(x_0, x_1) = 0$, $\varphi_{e,4}(x_0, x_1) = 0$, and $\varphi_{e,5}(x_0, x_1) = 0$.

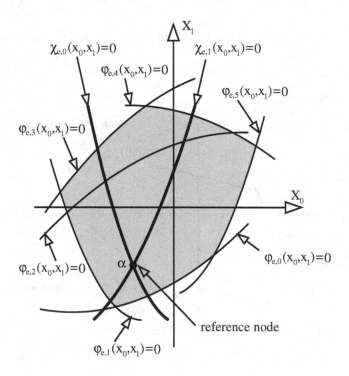

Figure 7.2.7 *Two-dimensional region divided into a finite number of subregions (finite elements) by a set of curves, where $\rho = 2$, $\tau = 6$, and $d = 2$. The point α represents a reference node e.*

The polynomial $P_3(x)$ writes

$$P_3(x) = \prod_{k=0}^{k=5} P_{e,k}(x)$$

$$= P_{e,0}(x)P_{e,1}(x)P_{e,2}(x)P_{e,3}(x)P_{e,4}(x)P_{e,5}(x)$$

$$= \left[\varphi_{e,0}(x_0, x_1)\varphi_{e,1}(x_0, x_1)\varphi_{e,2}(x_0, x_1)\varphi_{e,3}(x_0, x_1)\varphi_{e,4}(x_0, x_1)\varphi_{e,5}(x_0, x_1) \right]^{\eta+1}$$

$$(7.25)$$

If the reference node α changes then the numbering of the hypersurfaces may change since the visibility changes. No restriction is made with respect to the shape of the functions $\varphi_{e,k}(x)$, therefore, the functions $\Phi_{e,n}(x)$ include the concept of sub-, iso-, and superparametric elements.

end of example 7.2.6

7.2.1.1 Particularization of $P_e(x)$ for a rectangular hyperparallelopiped

1- Equation of a plane orthogonal to a coordinate axis

The equation of a hyperplane orthogonal the the axis X_i, and perpendicular to the line α, β_i, where β_i is a point on the plane can be obtained from the condition that the vector AB is orthogonal to the axis X_i whose unit direction vector is $i = (0, 0, \ldots, 1, \ldots, 0)$, as shown in figure 7.2.8.

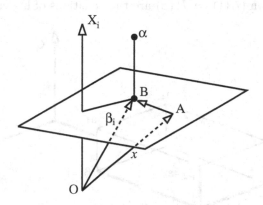

Figure 7.2.8 *Plane orthogonal to the axis X_i and containing the point β_i.*

This condition, in vector equation, implies that the scalar product of the two vectors must be equal to zero, that is,

$$(x - \beta_i) \bullet i = 0 \tag{7.26}$$

where $OA = x$ and $OB = \beta_i$. Expanding the coordinates

$$(x_0 - \beta_{i,0}, x_1 - \beta_{i,1}, \ldots, x_i - \beta_{i,i}, \ldots, x_{d-1} - \beta_{i,d-1}) \bullet (0, 0, \ldots, 1, \ldots, 0) = 0 \tag{7.27}$$

which gives the equations

$$x_i - \beta_{i,i} = 0 \qquad i = 0, 1, \ldots, d-1 \tag{7.28}$$

Notation 7.2.9 *Coordinate numbers of the points in the visibility set $\mathcal{V}_e$ for a rectangular hyperparallelopiped*

Let $\beta_i = (\beta_{i,0}, \beta_{i,1}, \ldots, \beta_{i,d-1})$ denote a node along the line segment parallel to the coordinate axis X_i and linking the reference node α to the node β_i. The set of all nodes in a rectangular hyperparallelopiped with this property is denoted as β. The line segment $\overline{\alpha, \beta_i}$ is the orthogonal segment defining the plane i in the visibility of the reference node α.

In the case of figure 7.2.10 the set β is identified as $\beta = \{\beta_0, \beta_1, \beta_2\} \in \mathcal{V}_e$ where, for example, the line segment $\overline{\alpha, \beta_1}$ defines the plane 1 that contains the nodes a_0, a_1, a_4, a_5 and which plane belongs to the visibility of $\alpha = a_2$.

end of notation 7.2.9

2- Rectangular hyperparallelopiped

For a rectangular hyperparallelopiped, Behr and Jungst [4], with edges parallel to the coordinate axes (faces orthogonal to the coordinate axes) the factors in equation (7.11) for $P_e(x)$ are the equations of hyperplanes.

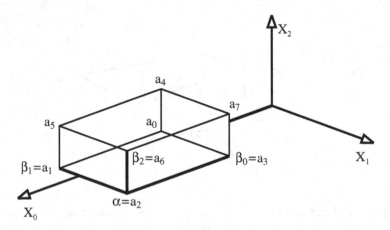

Figure 7.2.10 *Rectangular parallelopiped with edges parallel to the coordinate axes.*

The advantages of a rectangular hyperparallelopiped region as the one shown in figure 7.2.10 are

1. the only coordinate number of $\beta_{i,j}$ that is, different from the coordinates numbers $(\alpha_0, \alpha_1, \ldots, \alpha_{d-1})$ of α is $\beta_{i,i}$. For example, if $i = 1$ then $\alpha_0 = \beta_{1,0}$ and $\alpha_2 = \beta_{1,2}$ but $\alpha_1 \neq \beta_{1,1}$.

2. the hypersurfaces $\varphi_{e,k}(x)$ are hyperplanes, that is,

$$\varphi_{e,i}(x) = x_i - \beta_{i,i} \qquad i = 0, 1, \ldots, d-1 \qquad (7.29)$$

which gives

$$P_{e,i}(x) = (x_i - \beta_{i,i})^{\eta+1} \tag{7.30}$$

This permits to write the polynomial $P_e(x)$ for a hyperparallelopiped region as

$$P_e(x) = \prod_{i=0}^{d-1} (x_i - \beta_{i,i})^{\eta+1} \tag{7.31}$$

whose degree is $p_p = d(\eta + 1)$.

3. the number of hyperplanes satisfying the condition (7.28) equals the number d of coordinate axes.

4. the equation (7.29) represents the hyperplanes that are in the visibility set $\mathcal{V}_e$ of the reference node whose coordinate is given by α.

Example 7.2.11 *Polynomial $P_e(x)$ for the parallelopiped exhibited in figure 7.2.10.*

The node β_0 is on the line β_0-α parallel to the coordinate axis X_0, similarly the node β_1 is on the line β_1-α parallel to the coordinate axis X_1, and the node β_2 is on the line β_2-α parallel to the coordinate axis X_2. The polynomial is given by

$$P_e(x) = (x_0 - \beta_{0,0})^{\eta+1} (x_0 - \beta_{1,1})^{\eta+1} (x_0 - \beta_{2,2})^{\eta+1}$$

and its degree is $p_p = 3(\eta + 1)$.

end of example 7.2.11

If the hyperparallelopiped is symmetric with respect to the coordinate axes as shown in figure 7.2.12 then $\alpha_i = -\beta_{i,i}$. That is, the reference node α and the node β_i along the line segment α-β_i parallel to the axis X_i have the same absolute value. All other coordinate numbers related to both nodes are equal.

3- Rectangular hyperparallelopiped symmetric with respect to the coordinate axes

A hyper-parallelopiped that is, symmetric with respect to the coordinate axes as shown in figure 7.2.12 satisfies the condition

$$\beta_{i,i} = -\alpha_i \qquad i = 0, 1, \ldots, d-1 \tag{7.32}$$

and the equation (7.31) writes

$$P_e(x) = \prod_{i=0}^{d-1} (x_i + \alpha_i)^{\eta+1} \tag{7.33}$$

The value of $P_e(x)$ at the reference node is obtained for $x = \alpha$ and it is

$$P_e(\alpha) = \prod_{i=0}^{d-1} (2\alpha_i)^{\eta+1} \tag{7.34}$$

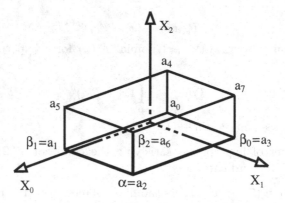

Figure 7.2.12 *Rectangular parallelopiped symmetric with respect to the coordinate axes.*

4- Normalized hypercube symmetric with respect to the coordinate axes

Normalization here is understood as the transformation of the interval $[a, b]$ along each the coordinate axis to the interval to $[-1, 1]$. Thus the normalized domain defined by a hypercube such that

$$\Omega = [-1, 1] \times \cdots \times [-1, 1] \tag{7.35}$$

which is symmetric with respect to the coordinate axes, as shown in figure 7.2.13.

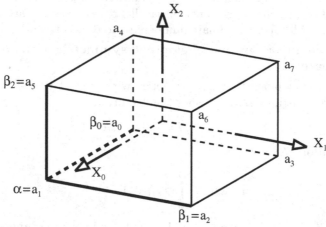

Figure 7.2.13 *Hyper-cube symmetric with respect to the coordinate axes.*

Example 7.2.14 *Write the polynomial $P_e(x)$ for the three-dimensional space shown in figure 7.2.13 where the reference node e is $\alpha = (1, -1, -1)$*

From equation (7.33) it follows

$$P_e(x) = (x_0 - \beta_{0,0})^{\eta+1} (x_1 - \beta_{1,1})^{\eta+1} (x_2 - \beta_{2,2})^{\eta+1}$$
$$= (x_0 + 1)^{\eta+1} (x_1 - 1)^{\eta+1} (x_2 - 1)^{\eta+1}$$

where $\beta_{0,0} = -1$, $\beta_{1,1} = 1$, and $\beta_{2,2} = -1$.

end of example 7.2.14

5- Symmetric hyperparallelepiped domain divided into a set of finite elements

Figure 7.2.15 shows a two-dimensional domain $\Omega = [-a, a] \times [-b, b]$ symmetric with respect to the coordinate axes and divided into finite subdomains called finite elements.

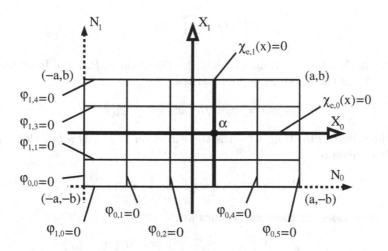

Figure 7.2.15 *Two-dimensional finite element mesh for the computation of the Hermite Interpolating Polynomials.*

For the region shown in figure 7.2.15 it can be identified that:

1. the dimension of the cartesian space is $d = 2$.

2. the reference node is a_e.

3. there are $\rho = 2$ hypersurfaces containing the reference node α, which are $\chi_{e,0}(x) = \varphi_{1,2} = 0$ and $\chi_{e,1}(x) = \varphi_{0,3} = 0$, which they are two lines in the case of figure 7.2.15.

4. there are $\tau = 9$ lines in visibility set $\mathcal{V}_e$ of the reference node α. Five of these lines are vertical and they are $\varphi_{0,0} = 0$, $\varphi_{0,1} = 0$, $\varphi_{0,2} = 0$, $\varphi_{0,4} = 0$, and $\varphi_{0,5} = 0$. There are four horizontal lines, which are $\varphi_{1,0} = 0$, $\varphi_{1,1} = 0$, $\varphi_{1,3} = 0$, and $\varphi_{1,4} = 0$.

5. The boundary $\partial\Omega$ of the domain Ω is formed by the lines $\varphi_{0,0} = 0$, $\varphi_{1,0} = 0$, $\varphi_{0,5} = 0$, and $\varphi_{1,4} = 0$.

The polynomial $P_e(x)$, related to the reference node α, as shown in figure 7.2.15, that is, the polynomial that vanishes at the boundary of all elements in the domain, except at the boundaries containing the reference node α, is given by

$$P_e(x) = [\varphi_{0,0}\,\varphi_{0,1}\,\varphi_{0,2}\,\varphi_{0,4}\,\varphi_{0,5}]^{\eta+1} \times [\varphi_{1,0}\,\varphi_{1,1}\,\varphi_{1,3}\,\varphi_{1,4}]^{\eta+1} \qquad (7.36)$$

A hyperparallelepided domain as shown in figure 7.2.15 permits do write the polynomial $P_e(x)$ as a product of polynomials along each direction as

$$P_e(x) = \prod_{k=0}^{d-1} P_{e,k}(x) \qquad (7.37)$$

For the horizontal direction, which is the direction 0, $P_{e,0}$ is a function of x_0 only and, for the above example it writes

$$P_{e,0}(x_0) = [\varphi_{0,0}\,\varphi_{0,1}\,\varphi_{0,2}\,\varphi_{0,4}\,\varphi_{0,5}\,]^{\eta+1} \qquad (7.38)$$

For the vertical direction, which is the direction 1, $P_{e,1}$ is a function x_1 only and it is given by

$$P_{e,1}(x_1) = [\varphi_{1,0}\,\varphi_{1,1}\,\varphi_{1,3}\varphi_{1,4}\,\varphi_{1,5}]^{\eta+1} \qquad (7.39)$$

The equations for the vertical lines are

$$\varphi_{0,i}(x_0) = x_0 - \beta_{0,i} = 0 \quad i = 0,1,2,4,5 \qquad (7.40)$$

and for the horizontal lines they are

$$\varphi_{1,j}(x_1) = x_0 - \beta_{1,j} = 0 \quad j = 0,1,3,4 \qquad (7.41)$$

and the polynomial $P_e(x)$ can be written as

$$P_e(x) = P_{e,0}(x_0)P_{e,1}(x_1) \qquad (7.42)$$

where

$$P_{e,0}(x_0) = \prod_{\substack{i=0 \\ i\neq 3}}^{i=5} (x_0 - \beta_{0,i})^{\eta+1} \qquad (7.43)$$

and

$$P_{e,1}(x_1) = \prod_{\substack{j=0 \\ i \neq 2}}^{j=4} (x_1 - \beta_{1,j})^{\eta+1}$$
(7.44)

6- Circular domain

For the region shown in figure 7.2.16(a) it can be identified that:

1. there are two polar coordinates but only the radius is used since the angle can take any value, therefore, $d = 1$.

2. the reference node A has coordinate $\rho = 0$.

3. the hypersurfaces containing the reference node are circles whose radius is zero. They can be denoted as $\chi_A(\rho, \alpha)|_{\rho=0} = 0$.

4. there are $\tau = 2$ hypersurfaces in visibility set V_e of the reference node A whose equations are $\varphi_B(\rho, \alpha) = 0$ and $\varphi_C(\rho, \alpha) = 0$.

5. the boundary $\partial\Omega$ of the domain $\Omega_i, i = 0, 1$ of the elements in figure 7.2.16 is formed by the curves $\varphi_B(\rho, \alpha) = 0$, and $\varphi_C(\rho, \alpha) = 0$

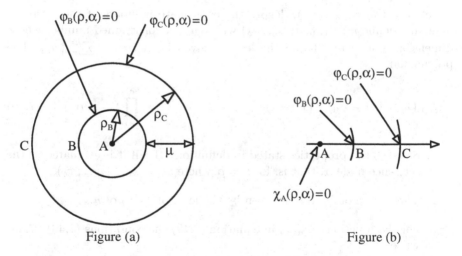

Figure (a)   Figure (b)

Figure 7.2.16 *Two circular finite element regions defined by two concentric circles.*

The polynomial $P_A(\rho, \alpha)$ writes

$$P_A(\rho, \alpha) = \varphi_B(\rho, \alpha)^{\eta+1} \varphi_C(\rho, \alpha)^{\eta+1}$$
$$= (\rho - \rho_B)^{\eta+1} (\rho - \rho_C)^{\eta+1}$$
(7.45)

Figure 7.2.16(b) shows the hypersurface $\chi_A(\rho,\alpha)|_{\rho=0} = 0$ through the reference node A as well as the hypersurfaces in the visibility of the node A, which are $\varphi_B(\rho,\alpha) = 0$ and $\varphi_C(\rho,\alpha) = 0$.

7.2.2   The polynomial $Q_{e,n}(x)$

The property stated in equation (7.5) can be constructed defining a polynomial $Q_{e,n}(x)$ such that the conditions

$$Q_{e,n}^{(\mu)}(x)\big|_{x=\alpha} = 0 \qquad \text{if} \quad \mathcal{O}^\delta(\mu) < \mathcal{O}^\delta(n) \qquad (7.46)$$

$$Q_{e,n}^{(\mu)}(x)\big|_{x=\alpha} = K \neq 0 \qquad \text{if} \quad \mathcal{O}^\delta(\mu) = \mathcal{O}^\delta(n) \qquad (7.47)$$

$$Q_{e,n}^{(\mu)}(x)\big|_{x=\alpha} = 0 \qquad \text{if} \quad \mathcal{O}^\delta(\mu) > \mathcal{O}^\delta(n) \qquad (7.48)$$

where K is a nonzero constant. These conditions hold at the reference node α $\forall \mu \in [(0,\dots,0),\dots,(0,\dots,\eta)]$. See 5.4.3 for the notation of closed interval of constrained numbers. If $\mathcal{O}^\delta(n) = 0$ then it can be chosen $Q_{e,n}(x) = K = 1$, which it will satisfy the conditions defined in equations (7.46)–(7.48).

Theorem 7.2.17

Let $\alpha = (\alpha_0, \alpha_1, \dots, \alpha_\delta)$, denote the coordinate numbers of the reference node and define $n := (n_0, n_1, \dots, n_\delta)$ where n is a constrained number whose dimension is $\delta = d - 1$ and the level is given by $\lambda_0(n) = \sum_{k=0}^{k=\delta} n_k$. The polynomial

$$Q_{e,n}(x) = (x_0 - \alpha_0)^{n_0}(x_1 - \alpha_1)^{n_1}\dots(x_\delta - \alpha_\delta)^{n_\delta} = \prod_{k=0}^{\delta}(x_k - \alpha_k)^{n_k} \quad (7.49)$$

1. Satisfies the properties stated in definition 7.1.4 if it is evaluated at the reference node α, that is, for $x = \alpha$ where $x := (x_0, x_1, \dots, x_\delta)$

2. The degree of $Q_{e,n}(x)$ is given by the level of $\lambda_0(n)$ of n.

3. For the function $Q_{e,n}(x)$ in equation (7.49) the conditions (7.46)–(7.48) write

$$Q_{e,n}^{(u)}(x)\big|_{x=\alpha} = 0 \qquad \text{if} \quad \mathcal{O}^\delta(\mu) < \mathcal{O}^\delta(n) \qquad (7.50)$$

$$Q_{e,n}^{(\mu)}(x)\big|_{x=\alpha} = i! \qquad \text{if} \quad \mathcal{O}^\delta(\mu) = \mathcal{O}^\delta(n) \qquad (7.51)$$

$$Q_{e,n}^{(\mu)}(x)\big|_{x=\alpha} = 0 \qquad \text{if} \quad \mathcal{O}^\delta(\mu) > \mathcal{O}^\delta(n) \qquad (7.52)$$

Proof:

1. If $\mathcal{O}^\delta(0) \leq \mathcal{O}^\delta(\mu) < \mathcal{O}^\delta(n)$ the derivatives of order $\mathcal{O}^\delta(\mu)$ of the polynomial defined in equation (7.49) are identically zero. For those derivatives there exists in the product terms of the form $(x_k - \alpha_{e,k})^{n_k - \mu_k}$ where $n_k - \mu_k > 0$ and when it is evaluated at $x = \alpha$ it will give $(\alpha_{e,k} - \alpha_{e,k})^{n_k - \mu_k} = 0$ and this will make the product in equation (7.49) to be zero.

2. If $\mathcal{O}^\delta(\mu) = \mathcal{O}^\delta(n)$ the derivative equals the $n! = n_0! \, n_1! \, \ldots \, n_k! \, \ldots \, n_\delta!$ where $n_k!$ is the the factorial of the exponent of each variable x_k, which is also a constant value.

3. If $\mathcal{O}^\delta(0, 0, \ldots, \eta) \geq \mathcal{O}^\delta(\mu) > \mathcal{O}^\delta(n)$ then the order of derivative of $Q_{e,n}^{(\mu)}(x)$ can be written as $\mathcal{O}^\delta(\mu) = \mathcal{O}^\delta(n + \xi)$ where $\xi = (\xi_0, \xi_1, \ldots, \xi_\delta)$, that is,

$$
\begin{aligned}
[Q_{e,n}(x)]^{(\mu)} &= [Q_{e,n}(x)]^{(n+\xi)} \\
&= \left[[Q_{e,n}(x)]^{(n)} \right]^{(\xi)} \\
&= [n!\kappa]^{(\xi)} = 0 \qquad \mathcal{O}^\delta(n) < \mathcal{O}^\delta(n + \xi) \leq \mathcal{O}^\delta(0, 0, \ldots, \eta)
\end{aligned}
$$
(7.53)

which equals to the derivative of order $\mathcal{O}^\delta(\xi)$ of the constant $i!\kappa$ and thus equals to zero.

Since $\mu = \mathcal{O}^\delta(0, \ldots, 0), \ldots, \mathcal{O}^\delta(0, \ldots, \eta)$ then the degree of $Q_{e,n}(x)$ when using the equation (7.49) goes from $\lambda_0(0, \ldots, 0) = 0$ until $\lambda_0(0, \ldots, \eta) = \eta$.

end of theorem 7.2.17

7.2.3 The polynomial $f_{e,n}(\bar{a}, x)$

The property in equation (7.4) will be obtained evaluating the coefficients of the polynomial $f_{e,n}(\bar{a}, x)$ such that, at the reference node,

$$
\Phi_{e,n}^{(\mu)}(x)\big|_{x=\alpha} = P_e(x)\, Q_{e,n}(x)\, f_{e,n}(\bar{a}, x)\big|_{x=\alpha} = 0 \qquad \text{if} \quad \mathcal{O}^\delta(\mu) < \mathcal{O}^\delta(n)
$$
(7.54)

$$
\Phi_{e,n}^{(\mu)}(x)\big|_{x=\alpha} = P_e(x)\, Q_{e,n}(x)\, f_{e,n}(\bar{a}, x)\big|_{x=\alpha} = 1 \qquad \text{if} \quad \mathcal{O}^\delta(\mu) = \mathcal{O}^\delta(n)
$$
(7.55)

$$
\Phi_{e,n}^{(\mu)}(x)\big|_{x=\alpha} = P_e(x)\, Q_{e,n}(x)\, f_{e,n}(\bar{a}, x)\big|_{x=\alpha} = 0 \qquad \text{if} \quad \mathcal{O}^\delta(\mu) > \mathcal{O}^\delta(n)
$$
(7.56)

hold. No conditions are imposed upon the roots of $f_{e,n}(\bar{a}, x)$, which may be complex, real, or both.

Theorem 7.2.18

The minimum degree of the polynomial $Q_{e,n}(x)f_{e,n}(\overline{a},x)$ is η.

Proof:

The only equation that can be used to evaluate the coefficients of $f_{e,n}(\overline{a},x)$ is (7.55) since $Q_{e,n}(x)$ guarantees the vanishing of equations (7.54) and (7.56). The derivative of order $\mathcal{O}^{\delta}(n)$ of the equation (7.55) gives

$$\Phi_{e,n}^{(n)}(x)\big|_{x=\alpha} = n!P_e(x)f_{e,n}(\overline{a},x)\big|_{x=\alpha} = 1 \qquad (7.57)$$

where $P_e(\alpha) \neq 0$ and $f_{e,n}(\overline{a},\alpha) \neq 0)$. The function $n!P_e(x)f_{e,n}(\overline{a},x)$ must also satisfy the conditions in equation (7.56) for $\mu = \mathcal{O}^{\delta}(n)+1,\ldots,(0,\ldots,\eta)$ in addition to the condition in equation (7.57) for $u = \mathcal{O}^{\delta}(n)$. The derivative of order r of the equation (7.57) satisfying equation (7.56) writes

$$n!P_e(x)^{(r)}f_{e,n}(\overline{a},x) + n!P_e(x)f_{e,n}^{(r)}(\overline{a},x)\bigg|_{x=\alpha} = 0$$
$$\text{where} \quad \mathcal{O}^{\delta}(r) = \left[\mathcal{O}^{\delta}(0)+1\right],\ldots,\left[\mathcal{O}^{\delta}(0,0,\ldots,\eta) - \mathcal{O}^{\delta}(i)\right] \qquad (7.58)$$

Since $P_e(x)^{(r)}\big|_{x=\alpha} \neq 0$ by the property 7.2.1, $f_{e,n}(\overline{a},x)\big|_{x=\alpha} \neq 0$ by equation (7.57), therefore

$$P_e(x)^{(r)}f_{e,n}(\overline{a},x)\big|_{x=\alpha} \neq 0 \qquad (7.59)$$

which implies that

$$P_e(x)f_{e,n}^{(r)}(\overline{a},x)\big|_{x=\alpha} \neq 0 \qquad (7.60)$$

and then

$$f_{e,n}^{(r)}(\overline{a},x)\big|_{x=\alpha} \neq 0 \qquad \mathcal{O}^{\delta}(r) = \mathcal{O}^{\delta}(0),\ldots,\left[\mathcal{O}^{\delta}(0,0,\ldots,\eta) - \mathcal{O}^{\delta}(n)\right] \quad (7.61)$$

To satisfy the equation (7.61) the derivative $f_{e,n}^{(r)}$ of order

$$\left[(0,0,\ldots,\eta) - \mathcal{O}^{\delta}(n)\right] \qquad (7.62)$$

must be, at least, a constant different from zero. Recalling that the highest level λ_0 of a polynomial equals to its degree, subsection 6.5, it can be written that

$$\lambda_0\left[(0,0,\ldots,\eta) - (n)\right] = \lambda_0(0,0,\ldots,\eta) - \lambda_0(n) = \eta - \lambda_0(n) \qquad (7.63)$$

where $\lambda_0\left[(0,0,\ldots,\eta) - (n)\right]$ is the order of the last term of the complete polynomial $f_{e,n}(\overline{a},x)$. This gives the relation

$$\left[\text{degree } Q_{e,n}(x)f_{e,n}(\overline{a},x)\right] = \left[\text{degree } Q_{e,n}(x)\right] + \left[\text{degree } f_{e,n}(\overline{a},x)\right]$$
$$= \lambda_0(n) + \eta - \lambda_0(n) = \eta \qquad (7.64)$$

where η is the minimum degree for which the polynomial

$$Q_{e,n}(x)f_{e,n}(\bar{a}, x) \qquad (7.65)$$

still satisfies the conditions in equations (7.54)–(7.56).

end of theorem 7.2.18

Corollary 7.2.19

The minimum degree of $f_{e,n}(\bar{a}, x)$ is given by

$$q = \eta - \lambda_0(n) \qquad (7.66)$$

where the degree of $Q_{e,n}(x)$ is given by $\lambda_0(n)$, that is, by the value of the level of n.

Proof:
This a consequence of the equation (7.64).

end of corollary 7.2.19

Corollary 7.2.20

All Hermite Interpolating Polynomials $\Phi_{e,n}(x)$ related to a reference node k whose coordinate numbers are given by

$$\alpha = (\alpha_0, \alpha_1, \ldots, \alpha_{d-1}) \qquad (7.67)$$

have the same degree.

Proof:
From the construction of $\Phi_{e,n}(x)$ given by equation (7.7) it follows

$$[\text{degree } \Phi_{e,n}(x)] = [\text{degree } P_e(x)] + [\text{degree } Q_{e,n}(x)f_{e,n}(\bar{a}, x)] \qquad (7.68)$$

The degree of $P_e(x)$ is given by

1. If the domain is a hyperparallelopiped

$$\text{degree } [P_e(x)] = d(\eta + 1) \qquad (7.69)$$

   where d is the dimension of the cartesian space, that is, the number of coordinate axes.

2. If the domain is divided in finite elements.

$$\text{degree } [P_e(x)] = t(\eta + 1) \qquad (7.70)$$

   where $t = \sum_i t_i$ and t_i is the number of hyperplanes orthogonal to the coordinate axis x_i.

Since the degree of $Q_{e,n}(x)f(\bar{a}, x)$ is constant, say k, whose minimum value is $k = \eta$, then

$$p = \text{degree } [\Phi_{e,n}(x)] = t(\eta + 1) + k \qquad (7.71)$$

where t and $\eta + 1$ are constants, therefore, p is a constant for all derivatives of order u and all functions of order i in equations (7.54–7.56) that is, for all polynomials $\Phi_{e,n}(x)$ related to same reference node.

end of corollary 7.2.20

A consequence of Corollary 7.2.20 is that the Multivariate Hermite Interpolating Polynomial, whose domain is a rectangular hyperparallelopiped, must be characterized by the (d, η, k) where d, the number of variables and η the maximum order of derivative requested for continuity, and k is the degree of the product $Q_{e,n}(x)f(\bar{a}, x)$. The degree of the polynomial must not be used as characterization since it contains implicitly more than one information.

In the case of a hyperparallelopiped divided into finite elements it may be characterized by (t, η, k).

Corollary 7.2.21

Let $\mu = (0, 0, \ldots, 0), \ldots, (0, 0, \ldots, \eta)$ be the order of the derivatives of a polynomial $\Phi_{e,n}(x)$ given by equation (7.7) and satisfying the properties defined in equations (7.4)–(7.6). The polynomial $f_{e,n}(\bar{a}, x)$ and its derivatives of order

$$\mathcal{O}^{\delta}(v) = (0, 0, \ldots, 0), \ldots, (0, 0, \ldots, q) \qquad (7.72)$$

do not vanish at the reference node $x = \alpha$ where q is given by corollary 7.2.19.

Proof:

From equation (7.66) in corollary 7.2.19 $q = \eta - \lambda_0(n)$ and with the aid of the equation (7.63) it follows

$$\mathcal{O}^{\delta}(q) = \mathcal{O}^{\delta}[(0, 0, \ldots, \eta) - (n)] = [\mathcal{O}^{\delta}(0, 0, \ldots, \eta) - \mathcal{O}^{\delta}(n)] \qquad (7.73)$$

replacing $[\mathcal{O}^{\delta}(0, 0, \ldots, \eta) - \mathcal{O}^{\delta}(n)]$ by $\mathcal{O}^{\delta}(q)$ and r by v into equation (7.61) it gives

$$f_{e,n}^{(v)}(\bar{a}, x)\big|_{x=\alpha} \neq 0 \qquad \mathcal{O}^{\delta}(v) = \mathcal{O}^{\delta}(0), \ldots, \mathcal{O}^{\delta}(q) \qquad (7.74)$$

end of corollary 7.2.21

Corollary 7.2.22

All coefficients of a complete polynomial $f_{e,n}(\bar{a}, x)$ do not vanish

Proof:

According to the corollary 7.2.21 $f_{e,n}(\bar{a}, x)$ and all its derivatives do not vanish at the reference node $x = \alpha$ and so $f_{e,n}^{(q)}(\bar{a}, x) \neq 0$. That is, the coefficients of the terms of highest degree are not zero.

Since $f_{e,n}^{(q-1)}(\bar{a}, x) \neq 0$ then it can be written that

$$f_{e,n}^{(q-1)}(x) = \bar{a}_{q-1} + \bar{a}_q x \big|_{x=\alpha} \neq 0 \qquad (7.75)$$

where equation (7.75) must be true for any α, therefore, $\bar{a}_q \alpha \neq 0$ and so all the coefficients $\bar{a}_{q-1}$ do not vanish.

Continuing until the terms in the level zero it can be seen that the coefficients at all levels are not zero.

end of corollary 7.2.22

7.3 Computation of the coefficients of $f_{e,n}(\bar{a}, x)$

7.3.1 Hypersurface region

The computation of the coefficients of $f_{e,n}(\bar{a}, x)$ for a hypersurface region is performed solving the set of simultaneous linear algebraic equations given by equations (7.55) and (7.56), respectively, for x evaluated at the reference node α, that is, solving

$$\left[Q_{e,n}(x)^{(n)} P_e(x) f_{e,n}(\bar{a}, x) \right] \bigg|_{x=\alpha_e} = 1 \qquad n = (n_0, n_1, \dots, n_{d-1}) \qquad (7.76)$$

$$\left[Q_{e,n}(x)^{(n)} P_e(x) f_{e,n}(\bar{a}, x) \right]^{(v)} \bigg|_{x=\alpha_e} = 0 \qquad (0, 0, \dots, 0) < v \leq (0, 0, \dots, q)$$

$$(7.77)$$

for the coefficients $\bar{a}$ where the degree of the polynomial $f_{e,n}(\bar{a}, x)$ is given by the corollary 7.2.19 as $q = \eta - \lambda_0(n)$.

The constrained numbers v were defined in the equation (7.74), corollary 7.2.21, as $\mathcal{O}^\delta(v) = \mathcal{O}^\delta(0), \dots, \mathcal{O}^\delta(q)$ therefore, they are numbers in the range $(0, 0, \dots, 0) < v \leq (0, 0, \dots, q)$, which implies that

$$v = (v_0, v_1, \dots, v_k, \dots, v_{d-1}) \in L^\delta(q) \qquad (7.78)$$

The number of polynomials related to a reference node equals to the number of elements in the set $L^\delta(q)$, which is denoted as $n^\delta(q)$ and whose expression is found in chapter 4.

For the particular case of a rectangular hyperparallelopiped the coefficients of $f_{e,n}(\bar{a}, x)$ can be obtained by the procedure below.

7.3.2   Construction of the system of equations

As it was shown in equation (7.57) for each $n = (n_0, n_1, \ldots, n_{d-1})$ that is, for each derivative of order $\mathcal{O}^\delta(\mu) = \mathcal{O}^\delta(n)$, there is an associated polynomial $f_{e,n}(\overline{a}, x)$ given by

$$n! P_e(x) f_{e,n}(\overline{a}, x) \Big|_{x=\alpha} = 1 \tag{7.79}$$

where $(0, 0, \ldots, 0) \leq (n_0, n_1, \ldots, n_{d-1}) \leq (0, 0, \ldots, 0, \eta)$.

The number of coefficients $\overline{a}$ of the polynomial $f_{e,n}(\overline{a}, x)$ equals the number of its terms, which equals the number of elements $n^\delta(q)$ in the set $L^\delta(q)$ and it can be evaluated using equation (4.135) on page 175. The computation of the coefficients $\overline{a}$ is done using the condition given by equation (7.56). Instead of deriving equation (7.55) or (7.79) it is more convenient to solve the equation (7.79) for $f_{e,n}(\overline{a}, x)$ and then perform the derivatives.

It is known that the terms of a polynomial are linearly independent, Lange [25, pages 116–117] and that for a complete polynomial it is necessary $n^\delta(q)$ conditions to evaluate the coefficients $\overline{a}$ uniquely. Solving the equation (7.79) for $f_{e,n}(\overline{a}, x)$ and considering equation (7.74) it follows that

$$[f_{e,n}(\overline{a}, x)]^{(v)} \Big|_{x=\alpha} = \left[\frac{1}{n! P_e(x)} \right]^{(v)} \Big|_{x=\alpha} \neq 0 \qquad \mathcal{O}^\delta(0), \ldots, \mathcal{O}^\delta(q) \tag{7.80}$$

substituting $\mathcal{O}^\delta(0), \ldots, \mathcal{O}^\delta(q)$ by $(0, 0, \ldots, 0) \leq v \leq (0, 0, \ldots, q)$ in equation (7.80) it follows that

$$[f_{e,n}(\overline{a}, x)]^{(v)} \Big|_{x=\alpha} = \left[\frac{1}{n! P_e(x)} \right]^{(v)} \Big|_{x=\alpha} \neq 0 \qquad (0, 0, \ldots, 0) \leq v \leq (0, 0, \ldots, q) \tag{7.81}$$

Equation (7.81) expresses a set of linear algebraic equations in triangular form relating the coefficients of the polynomial $f_{e,n}(\overline{a}, x)$. The computation of the coefficients $\overline{a}$ of $f_{e,n}(\overline{a}, x)$ for each n must be done performing the derivatives in the order $v = (0, 0, \ldots, q), \ldots, (0, 0, \ldots, 0)$. The derivative of order $(0, 0, \ldots, q)$ permits the evaluation of the coefficient of the term of highest position. The terms in lower positions are obtained with the aid of the terms in higher positions as it will be shown soon.

7.3.3   Computation of the term $[1/P_e(x)]^{(v)}$

The derivative of $1/P_e(x)$ in equation (7.81) can be written as

$$\left[\frac{1}{P_e(x)} \right]^{(v)} \Big|_{x=\alpha} = E_F^{(v)}(x) \frac{1}{P_e(x)} \Big|_{x=\alpha} \tag{7.82}$$

where $E_F^{(v)}(x)$ depends on the domain. Section 6.10, on page 265, shows the computation of the derivatives of $1/P_e(x)$ for the usual cases.

The expansion of the derivative in equation (7.82) can be obtained using the equations (6.217) and (6.218), on page 273, as follows

$$
\left[\frac{1}{P(x)}\right]^{(v)} = \left[\frac{1}{P_0(x_0)}\right]^{(v_0)} \left[\frac{1}{P_1(x_1)}\right]^{(v_1)} \cdots \left[\frac{1}{P_i(x_i)}\right]^{(v_i)} \cdots \left[\frac{1}{P_{d-1}(x_{d-1})}\right]^{(v_{d-1})}
$$

$$
= \left[E_F^{(v_0)}(x_0)E_F^{(v_1)}(x_1)\cdots E_F^{(v_{d-1})}(x_{d-1})\right]\left[\frac{1}{P(x)}\right] \tag{7.83}
$$

Comparing equations (7.83) and equation (7.82) it permits to write $E_F^{(v)}(x)$ as

$$
E_F^{(v)}(x) = E_F^{(v_0)}(x_0)E_F^{(v_1)}(x_1)\cdots E_F^{(v_i)}(x_i)\cdots E_F^{(v_{d-1})}(x_{d-1}) \tag{7.84}
$$

For the particular case of a rectangular hyperparallelopiped it gives the factors $E_F^{(v_i)}(x)$ are obtained from the equation (6.196), on page 268, as follows

$$
E_F^{(v_i)}(x) = \frac{(\eta + v_i)!}{\eta!}\frac{1}{(\beta_{i,i} - x_i)^{v_i}} \tag{7.85}
$$

and, when evaluated at the reference node e it writes

$$
E_F^{(v_i)}(\alpha_i) = \frac{(\eta + v_i)!}{\eta!}\frac{1}{(\beta_{i,i} - \alpha_i)^{v_i}} \tag{7.86}
$$

for each coordinate number $x_i = \alpha_i$, where $\beta_{i,i}$ are the coordinate numbers of the points at the visibility of the reference node e according to the notation 7.2.9.

Introducing $E_F^{(v_i)}$ given by equation (7.86) into equation (7.84) it gives

$$
E_F^{(v)}(x)\Big|_{x=\alpha} = \prod_{i=0}^{d-1} E_F^{(v_i)}(x_i)\Big|_{x_i=\alpha_i}
$$

$$
= \prod_{i=0}^{d-1} \frac{(\eta + v_i)!}{\eta!}\frac{1}{(\beta_{i,i} - x_i)^{v_i}}\Big|_{x_i=\alpha_i} \tag{7.87}
$$

7.3.4 Computation of $[f_{e,n}(\overline{a}, x)]^{(v)}$

The procedure below is valid either for a hyperparallelopiped divided into finite elements or consisting of one element only.

The minimum degree q of the polynomial $f_{e,n}(\overline{a}, x)$ is given by the corollary 7.2.19, which can be written as

$$
q \geq \eta - \lambda_0(n) \tag{7.88}
$$

From the notation for a complete polynomial given in section 6.1.2.1 equation (6.13), on page 219, the polynomial $f_{e,n}(\bar{a}, x)$ can be written as

$$f_{e,n}(\bar{a}, x) = \sum_{\lambda_0=0}^{\lambda_0=q} \sum_{\lambda_1=0}^{\lambda_1=\lambda_0} \cdots \sum_{\lambda_{\delta-1}=0}^{\lambda_{\delta-1}=\lambda_{\delta-2}} \bar{a}_w \, x_0^{w_0} x_1^{w_1} \cdots x_\delta^{w_\delta} \tag{7.89}$$

and is called the original polynomial of degree q that is, the polynomial before the application of the derivative operator.

According to the theorem 6.3.1, on page 225, the derivative of a complete polynomial whose exponents are elements of the set $\mathbb{Z}^+$ is a complete polynomial whose notation is given in equation (6.26) and it writes

$$[f_{e,n}(\bar{a}, x)]^{(v)} = \sum_{r_0=0}^{r_0=m} \sum_{r_1=0}^{r_1=r_0} \cdots \sum_{r_{\delta-1}=0}^{r_{\delta-1}=r_{\delta-2}} \bar{a}_{(u)} \, x_0^{u_0} u_1^{u_1} \cdots u_\delta^{u_\delta} \tag{7.90}$$

whose degree is

$$m = q - \lambda_0(v) \tag{7.91}$$

and the derivative is evaluated using the equation (6.38) as follows

$$[f_{e,n}(\bar{a}, x)]^{(v)} = \sum_{r_0=0}^{r_0=m} \sum_{r_1=0}^{r_1=r_0} \cdots \sum_{r_{\delta-1}=0}^{r_{\delta-1}=r_{\delta-2}} \bar{a}_{(u+v)} \frac{(u+v)!}{u!} x_0^{u_0} x_1^{u_1} \cdots x_\delta^{u_\delta} \tag{7.92}$$

where $u \in L^{d-1}(m)$. Since $\delta = d - 1$ it can also be written as $u \in L^\delta(m)$.

The equation (7.92) can be split into two parts. The first part contains only the terms of power zero, that is, the terms that belong to the level $\lambda_0(u) = 0$ for which $u_0 = 0, u_1 = 0, \ldots, u_\delta = 0$. The second part gathers all other terms of the polynomial, that is, from level $\lambda_0(u) = 1$ until $\lambda_0(u) = m$. The separation into two parts can be written as

$$[f_{e,n}(\bar{a}, x)]^{(v)} = \bar{a}_v \, (v_0! v_1! \cdots v_\delta!) + \sum_{r_0=1}^{r_0=m} \sum_{r_1=0}^{r_1=r_0} \cdots$$

$$\cdots \sum_{r_{\delta-1}=0}^{r_{\delta-1}=r_{\delta-2}} \left[\bar{a}_{(u+v)} \frac{(u+v)!}{u!} \right] x_0^{u_0} x_1^{u_1} \cdots x_\delta^{u_\delta} \tag{7.93}$$

where the first term in the righthand side of equation (7.93) is obtained as the particular case of the coefficients in equation (7.92) for $u = 0$ as

$$\bar{a}_{(v)} \, (v_0! v_1! \cdots v_\delta!) = \left[\bar{a}_{(u+v)} \frac{(u+v)!}{u!} \right]_{u=0} \tag{7.94}$$

The coefficients $\bar{a}_{(v)}$ and $\bar{a}_{(u+v)}$, in equation (7.93), belong to the original polynomial $f_{e,n}(\bar{a}, x)$ in equation (7.89). The coefficients $\bar{a}_{(v)}$ belongs to the level $\lambda_0(w) = \lambda_0(v)$ and the coefficients $\overline{a}_{(u+v)}$, where $u + v = w$, belong to the levels from $\lambda_0(w) = \lambda_0(v) + 1$ until $\lambda_0(w) = q$.

The index r_0 starts at one in the first summation since the term of power zero, which are terms in the level $\lambda(u) = 0$, have been removed from the summation and, therefore, must not be considered.

Define

$$v! := v_0! v_1! \cdots v_\delta! \tag{7.95}$$

$$\overline{S}_{(v)}(\overline{a}_{(u+v)}, x) := 0 \qquad \text{if} \quad \lambda_0(v) = q \tag{7.96}$$

$$\overline{S}_{(v)}(\overline{a}_{(u+v)}, x) := \sum_{r_0=1}^{r_0=m} \sum_{r_1=0}^{r_1=r_0} \cdots$$

$$\cdots \sum_{r_{\delta-1}=0}^{r_{\delta-1}=r_{\delta-2}} \overline{a}_{(u+v)} \frac{(u+v)!}{u!} x_0^{u_0} x_1^{u_1} \cdots x_\delta^{u_\delta}$$

$$\text{if} \quad 0 \le \lambda_0(v) < q \tag{7.97}$$

Substituting equations (7.95)– (7.97) into equation (7.93) the derivative $[f_{e,n}(\overline{a}, x)]^{(v)}$ can be written as

$$[f_{e,n}(\overline{a}, x)]^{(v)} = v! \overline{a}_{(v)} + \overline{S}_{(v)}(\overline{a}_{(u+v)}, x) \tag{7.98}$$

which is the equation that will be used to evaluate the coefficients $\overline{a}_{(v)}$. Since $[f_{e,n}(\overline{a}, x)]^{(v)}$ is a complete polynomial of degree m, then $\overline{S}_{(v)}(\overline{a}_{(u+v)}, x)$ is a complete polynomial without the constant term.

7.3.5 Solution of the system of equations

Introducing $[f_{e,n}(\overline{a}, x)]^{(v)}$ given by equation (7.98) into equation (7.81) it follows

$$v! \overline{a}_v + \overline{S}_v(\overline{a}_{(u+v)}, x) \bigg|_{x=\alpha} = \left[\frac{1}{n! P_e(x)}\right]^{(v)} \bigg|_{x=\alpha} \qquad (0, 0, \ldots, 0) \le v \le (0, 0, \ldots, q) \tag{7.99}$$

The solution of equation (7.99) for $\overline{a}_v$ yields

$$\overline{a}_{(v)} = \frac{1}{v!} \left[\left[\frac{1}{n! P_e(x)}\right]^{(v)} - \overline{S}_v(\overline{a}_{(u+v)}, x) \right]_{x=\alpha} \tag{7.100}$$

where the coefficient $\overline{a}_v$ must be obtained from the sequence of derivatives given by $\lambda_0(v) = q, q - 1, \ldots, 0$. That is, the coefficients $\overline{a}_{(v)}$ are evaluated, level by level, from the highest level of the polynomial, which is $\lambda_0(v) = q$, to the lowest level $\lambda_0(v) = 0$, which corresponds to the constant term.

7.3.5.1　Computation of the coefficients in the level $\lambda_0(v) = q$

For the level $\lambda_0(v) = q$ the term $\overline{S}_v(\overline{a}_{(u+v)}, x)$ equals zero in equation (7.100) and the coefficients $\overline{a}_{(v)}$ are then given by

$$\overline{a}_{(v)} = \frac{1}{v!} \left[\frac{1}{n! P_e(x)} \right]^{(v)} \bigg|_{x=\alpha} \tag{7.101}$$

Introducing the derivative given by equation (7.82) into equation (7.101) it gives

$$\overline{a}_{(v)} = \frac{1}{v!} \left[E_F^{(v)}(\alpha) \frac{1}{n! P_e(\alpha)} \right] = \frac{1}{n! P_e(\alpha)} \frac{E_F^{(v)}(\alpha)}{v!} \tag{7.102}$$

where $v \in L_{\lambda_0(v)}^{\delta} = L_q^{d-1} \subset L^{\delta}(q)$. Define

$$a_{(v)} = \frac{E_F^{(v)}(\alpha)}{v!} \tag{7.103}$$

where $E_F^{(v)}(\alpha)$ depends on the type of mesh chosen. Then equation (7.102) can be rewritten as

$$\overline{a}_{(v)} = \frac{1}{n! P_e(\alpha)} a_{(v)} \tag{7.104}$$

or

$$\overline{a}_{(u+v)} \big|_{u=0} = \frac{1}{n! P_e(\alpha)} a_{(u+v)} \big|_{u=0} \tag{7.105}$$

and equation (7.98) becomes

$$[f_{e,n}(a, x)]^{(v)} = v! a_{(v)} \tag{7.106}$$

7.3.5.2　Computation of the coefficients in the levels $\lambda_0(v) < q$

From theorem 6.3.1, on page 225, the degree of the derivative of order v of a complete polynomial is given by

$$m = q - \lambda_0(v) \tag{7.107}$$

For the level $\lambda_0(v) = q - 1$ the upper limit of the first summation in

$$\overline{S}_v(\overline{a}_{(u+v)}, x) = \sum_{r_0=1}^{r_0=m} \sum_{r_1=0}^{r_1=r_0} \cdots \sum_{r_{\delta-1}=0}^{r_{\delta-1}=r_{\delta-2}} \overline{a}_{(u+v)} \frac{(u+v)!}{u!} x_0^{u_0} x_1^{u_1} \cdots x_{\delta}^{u_{\delta}} \tag{7.108}$$

is $m = q - \lambda_0(v) = q - q - 1 = 1$. Then $\overline{S}_v(\overline{a}_{(u+v)}, x)$ contains the coefficients of elements, which are in the level $\lambda_0(w) = \lambda_0(u + v) = q$ of the original polynomial and it is given by equation (7.89), where $u \in L^{\delta}(m) - L_0^{\delta} = L_1^{\delta} - L_0^{\delta}$ where $\delta = d - 1$.

If $m = q - \lambda_0(v) = q - q - 2 = 2$ then $\overline{S}_v(\overline{a}_{(u+v)}, x)$ contains the coefficients of elements, which are in the levels $\lambda_0(w) = \lambda_0(u + v) = q$ and $\lambda_0(w) = \lambda_0(u + v) = q - 1$ where $u \in L^\delta(m) - L_0^\delta = L_2^\delta - L_0^\delta$, etc. until $\lambda_0(v) = 0$, which corresponds to $m = q$. Therefore, m goes from $m = 1$ until $m = q$ for the computation of all the coefficients of the polynomial $f_{e,i}(\overline{a}, x)$. Note that v is a variable, which can be seen from (7.81).

Introducing $\overline{a}_{u+v} = \overline{a}_w$, which are coefficients of the original polynomial $f_{e,i}(\overline{a}, x)$ on the level $\lambda_0(v) = q$ and are given by equation (7.105) for $u \neq 0$, into equation (7.108) it yields

$$\overline{S}_v(\overline{a}_{(u+v)}, x) = \sum_{r_0=1}^{r_0=m} \sum_{r_1=0}^{r_1=r_0} \cdots \sum_{r_{\delta-1}=0}^{r_{\delta-1}=r_{\delta-2}} \frac{1}{n! P_e(\alpha)} a_{(u+v)} \frac{(u+v)!}{u!} x_0^{u_0} x_1^{u_1} \cdots x_\delta^{u_\delta}$$

(7.109)

Factoring out $1/(n! P_e(\alpha))$, which is a constant

$$\overline{S}_v(\overline{a}_{(u+v)}, x) = \frac{1}{n! P_e(\alpha)} \sum_{r_0=1}^{r_0=m} \sum_{r_1=0}^{r_1=r_0} \cdots \sum_{r_{\delta-1}=0}^{r_{\delta-1}=r_{\delta-2}} a_{(u+v)} \frac{(u+v)!}{u!} x_0^{u_0} x_1^{u_1} \cdots x_\delta^{u_\delta}$$

(7.110)

Writing in compact form

$$\overline{S}_v(\overline{a}_{(u+v)}, x) = \frac{1}{n! P_e(\alpha)} S_v(a_{(u+v)}, x)$$

(7.111)

where

$$S_v(a_{(u+v)}, \alpha) = \sum_{r_0=1}^{r_0=m} \sum_{r_1=0}^{r_1=r_0} \cdots \sum_{r_{\delta-1}=0}^{r_{\delta-1}=r_{\delta-2}} a_{(u+v)} \frac{(u+v)!}{u!} \alpha_0^{u_0} \alpha_1^{u_1} \cdots \alpha_\delta^{u_\delta}$$

(7.112)

The substitution of $\overline{S}_v(\overline{a}_{(u+v)}, x)$ given by equation (7.111) into equation (7.100) allows to evaluate the coefficient $\overline{a}_{(v)}$ as

$$\overline{a}_{(v)} = \frac{1}{v!} \left[\frac{E_F^{(v)}(\alpha)}{n! P_e(\alpha)} - \frac{1}{n! P_e(\alpha)} S_v(a_{(u+v)}, \alpha) \right]$$

(7.113)

Factoring out $1/(n! P_e(\alpha))$ it becomes

$$\overline{a}_{(v)} = \frac{1}{n! P_e(\alpha)} \frac{1}{v!} \left[E_F^{(v)}(\alpha) - S_v(a_{(u+v)}, \alpha) \right]$$

(7.114)

The above expression shows that the coefficients $a_{(v)}$ on the level $r_0 = \lambda_0(v)$ can be obtained in terms of the coefficients $a_{(u+v)}$ from the level $r_0 = q$ until the level $r_0 = \lambda_0(v) + 1$. This establishes a recurrence for the computation of the coefficients $a_{(v)}$ in terms of $a_{(u+v)}$.

Define
$$a_{(v)} := \frac{1}{v!} \left[E_F^{(v)}(\alpha) - S_v \left(a_{(u+v)}, \alpha \right) \right]$$
(7.115)

then equation (7.114) writes

$$\bar{a}_{(v)} = \frac{1}{n! P_e(\alpha)} a_{(v)}$$
(7.116)

Summarizing, the coefficients of the polynomial $f_{e,n}(\bar{a}, x)$ can be evaluated using the equation (7.102) or equation (7.114) as follows

$$\bar{a}_{(v)} = \frac{1}{n! P_e(\alpha)} \frac{E_F^{(v)}(\alpha)}{v!} \qquad \text{if} \qquad \lambda_0(v) = q$$
(7.117)

$$\bar{a}_{(v)} = \frac{1}{n! P_e(\alpha)} \frac{1}{v!} \left[E_F^{(v)}(\alpha) - S_v \left(a_{(u+v)}, \alpha \right) \right]$$
$$\text{if} \qquad \lambda_0(v) = q-1, q-2, \ldots, 2, 1, 0$$
(7.118)

where

$$\bar{a}_{(v)} = \frac{1}{n! P_e(\alpha)} a_{(v)}$$
(7.119)

for both cases. Substituting $\bar{a}_{(v)}$ given by equation (7.119) and $\bar{S}_v(\bar{a}_{(u+v)}, x)$ given by equation (7.111) into equation (7.98) it gives

$$\left[f_{e,n} \left(\frac{1}{n! P_e(\alpha)} a, x \right) \right]^{(v)} = v! \left[\frac{1}{n! P_e(\alpha)} a_{(v)} \right] + \frac{1}{n! P_e(\alpha)} S_v(a_{(u+v)}, x)$$
(7.120)

Cancelling out the factor $1/(n! P_e(\alpha))$ equation (7.120) becomes

$$[f_{e,n}(a, x)]^{(v)} = v! a_{(v)} + S_v(a_{(u+v)}, x)$$
(7.121)

where the coefficients of the polynomial $f_{e,n}(a, x)$ can be evaluated using

$$a_{(v)} = \frac{E_F^{(v)}(\alpha)}{v!} \qquad \text{if} \qquad \lambda_0(v) = q$$
(7.122)

$$a_{(v)} = \frac{1}{v!} \left[E_F^{(v)}(\alpha) - S_v \left(a_{(u+v)}, \alpha \right) \right]$$
$$\text{if} \qquad \lambda_0(v) = q-1, q-2, \ldots, 2, 1, 0$$
(7.123)

The factor $E_F^{(v)}(\alpha)$ is computed using the appropriate case in section 6.10 and $S_{(v)}$ by equation (7.112). The degree of the polynomial derivative $S_v \left(a_{(u+v)}, \alpha \right)$ is given by $m = q - \lambda_0(v)$ and the exponents $u = (u_0, \ldots, u_{d-1})$ of the variables $x = (x_0, \ldots, x_{d-1})$ in each term are obtained as

$$u \in \left(L^\delta(m) - L_0^\delta \right)$$
(7.124)

Comparing equations (7.120) and (7.121) it gives the following relation

$$f_{e,n}(\bar{a}, x) = \frac{1}{n! P_e(\alpha)} f_{e,n}(a, x) \tag{7.125}$$

which, when substituted into the equation (7.7) gives the following expression

$$\Phi_{e,n}(x) = \frac{1}{n! P_e(\alpha)} P_e(x) Q_{e,n}(x) f_{e,n}(a, x) \tag{7.126}$$

for the generation of the Hermite Interpolating Polynomials.

Taking the derivative $\mathcal{O}(u) = \mathcal{O}(n)$ of the expression in equation (7.126) it allows to rewrite the equation (7.55) as

$$\Phi_{e,n}^{(n)}(\alpha) = P_e(x)\, n!\, f_{e,i}(\bar{a}, x)\Big|_{x=\alpha} = 1$$

$$= \frac{1}{n!\, P_e(\alpha)} P_e(x)\, n!\, f_{e,n}(a, x)\Big|_{x=\alpha} = 1$$

$$= \frac{1}{P_e(\alpha)} P_e(x)\, f_{e,n}(a, x)\Big|_{x=\alpha} = 1 \tag{7.127}$$

Equation (7.127) implies that

$$f_{e,n}(a, x)\Big|_{x=\alpha} = 1 \tag{7.128}$$

7.4 Properties of the Hermite Interpolating Polynomials

7.4.1 Computation of the coefficients of $f_{h,n}(g, x)$ in terms of the coefficients of $f_{e,n}(a, x)$ if the reference nodes e and h differ only for one coordinate number

Theorem 7.4.1

Let

$$\Phi_{e,n}(x) = \frac{1}{n!\, P_e(\alpha)} P_e(x) Q_{e,n}(x) f_{e,n}(a, x) \tag{7.129}$$

and

$$\Phi_{h,n}(x) = \frac{1}{n!\, P_h(\alpha)} P_h(x) Q_{h,n}(x) f_{h,n}(g, x) \tag{7.130}$$

be two sets of Hermite Interpolating Polynomials defined on the same hypercube domain symmetric with respect to the origin of the reference system as shown in figure 7.4.2 below. The first set is related to the reference node e whose coordinate numbers are $\alpha = (\alpha_0, \ldots, \alpha_k, \ldots, \alpha_{d-1})$ and the second set related to the reference node h whose coordinate numbers are $\gamma = (\alpha_0, \ldots, -\alpha_k, \ldots, \alpha_{d-1})$.

The coefficients g_v of the polynomial $f_{h,n}(g,x)$ can be obtained in terms of the coefficients a_v of the polynomial $f_{e,n}(a,x)$ using the following transformations

$$g_v = (-1)^{v_k} a_v \tag{7.131}$$

where (7.124) and the constant factor $P_h(\gamma)$ can be obtained from $P_e(\alpha)$ as follows

$$P_h(\gamma) = (-1)^{n+1} P_e(\alpha) \tag{7.132}$$

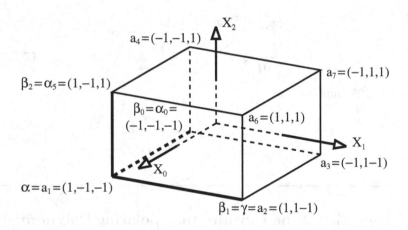

Figure 7.4.2 *Cube symmetric with respect to the coordinate axes.*

Figure 7.4.2 shows

1. The reference node $\alpha = a_1 = (1, -1, -1)$ and its visibility set

$$\beta = \{\beta_{0,0}, \beta_{1,1}, \beta_{2,2}\} = \{a_{0,0}, a_{2,1}, a_{5,2}\} = \{-1, 1, 1\}$$

2. The reference node $\gamma = a_2 = (1, 1, -1)$ and its visibility set

$$\delta = \{\delta_{0,0}, \delta_{1,1}, \delta_{2,2}\} = \{a_{3,0}, a_{1,1}, a_{6,2}\} = \{-1, -1, 1\}$$

where $\{\delta_{0,0}, \delta_{1,1}, \delta_{2,2}\} = \{\beta_{0,0}, -\beta_{1,1}, \beta_{2,2}\}$, that is, $\delta = \{\beta_{0,0}, -\beta_{1,1}, \beta_{2,2}\}$ for the coordinate number in the position $k = 1$, which is a consequence of the hypothesis $\gamma_k = -\alpha_k$.

Proof of $g_v = (-1)^{v_k} a_v$:

By hypothesis α and γ have the same coordinate numbers except for the coordinate k where $\gamma_k = -\alpha_k$ as consequence for β and δ it follows that $\delta_{k,k} = -\beta_{k,k}$.

The coefficients of the polynomial $f_{e,n}(a, x)$ are obtained using the equation (7.122) or the equation (7.123), in both cases it is necessary to evaluate $E_F^{(v)}(\alpha)$ as well as $E_F^{(v)}(\gamma)$ and for the case (7.123) is also necessary to evaluate the terms $S_v\left(a_{(u+v)}, \alpha\right)$ and $S_v\left(a_{(u+v)}, \gamma\right)$

1- Relation between $E_F^{(v)}(\alpha)$ and $E_F^{(v)}(\gamma)$

The term $E_F^{(v)}(\alpha)$ is given by the equation (7.87) as follows

$$E_F^{(v)}(x) \Big|_{x=\alpha} \prod_{i=0}^{d-1} \frac{(\eta + v_i)!}{\eta!} \frac{1}{(\beta_{i,i} - x_i)^{v_i}} \Big|_{x_i = \alpha_i} \qquad (7.133)$$

If $x_i = \gamma_i$ in equation (7.133) then

$$E_F^v(\gamma) = \frac{(\eta + v_0)!}{\eta!} \frac{1}{(\delta_{0,0} - \gamma_0)^{v_0}} \times \cdots \times \frac{(\eta + v_k)!}{\eta!} \frac{1}{(\delta_{k,k} - \gamma_k)^{v_k}} \times \cdots$$
$$\cdots \times \frac{(\eta + v_{d-1})!}{\eta!} \frac{1}{(\delta_{d-1,d-1} - \gamma_{d-1})^{v_{d-1}}} \qquad (7.134)$$

For the coordinate $i = 0, \ldots, d - 1$ in a rectangular hyperparallelopiped equation (7.32) gives $\delta_{i,i} = -\gamma_i$, then $\delta_{i,i} - \gamma_i = (-\gamma_i) - \gamma_i = -2\gamma_i$. Substituting into equation (7.134) it gives

$$E_F^v(\gamma) = \frac{(\eta + v_0)!}{\eta!} \frac{1}{(-2\gamma_0)^{v_0}} \times \cdots \times \frac{(\eta + v_k)!}{\eta!} \frac{1}{(-2\gamma_k)^{v_k}} \times \cdots$$
$$\cdots \times \frac{(\eta + v_{d-1})!}{\eta!} \frac{1}{(-2\gamma_{d-1})^{v_{d-1}}} \qquad (7.135)$$

If $x_i = \alpha_i$ in equation (7.133) then

$$E_F^v(\alpha) = \frac{(\eta + v_0)!}{\eta!} \frac{1}{(\beta_{0,0} - \alpha_0)^{v_0}} \times \cdots \times \frac{(\eta + v_k)!}{\eta!} \frac{1}{(\beta_{k,k} - \alpha_k)^{v_k}} \times \cdots$$
$$\cdots \times \frac{(\eta + v_{d-1})!}{\eta!} \frac{1}{(\beta_{d-1,d-1} - \alpha_{d-1})^{v_{d-1}}} \qquad (7.136)$$

For the coordinate $i = 0, \ldots, d - 1$ in a rectangular hyper-parallelopiped equation (7.32) gives $\beta_{i,i} = -\alpha_i$, then $\beta_{i,i} - \alpha_i = (-\alpha_i) - \alpha_i = -2\alpha_i$. Substituting into equation (7.136) it gives

$$
E_F^v(\alpha) = \frac{(\eta + v_0)!}{\eta!} \frac{1}{(-2\alpha_0)^{v_0}} \times \cdots \times \frac{(\eta + v_k)!}{\eta!} \frac{1}{(-2\alpha_k)^{v_k}} \times \cdots
$$

$$
\cdots \times \frac{(\eta + v_{d-1})!}{\eta!} \frac{1}{(-2\alpha_{d-1})^{v_{d-1}}} \tag{7.137}
$$

Since $\gamma_i = \alpha_i$ for $i = 0, \ldots, d - 1$ except for $i = k$ where $\gamma_k = -\alpha_k$ by hypothesis, then all the factors in equations (7.135) and (7.137) with exception of the factor $i = k$ for which

$$
\frac{(\eta + v_k)!}{\eta!} \frac{1}{(-2\gamma_k)^{v_k}} = (-1)^{v_k} \frac{(\eta + v_k)!}{\eta!} \frac{1}{(-2\alpha_k)^{v_k}} \tag{7.138}
$$

and then

$$
E_F^{(v)}(\gamma) = (-1)^{v_k} E_F^{(v)}(\alpha) \tag{7.139}
$$

2- Relation between the coefficients in the level $\lambda_0(v) = q$.

Equation (7.122) yields

$$
a_v = \frac{E_F^{(v)}}{v!}(\alpha) \qquad \text{and} \qquad g_v = \frac{E_F^{(v)}}{v!}(\gamma) \tag{7.140}
$$

Dividing both sides of equation (7.139) by $v!$ and considering g_v and g_v given by equation (7.140) it gives

$$
\frac{E_F^{(v)}}{v!}(\gamma) = (-1)^{v_k} \frac{E_F^{(v)}}{v!}(\alpha) \qquad \Longrightarrow \qquad g_v = (-1)^{v_k} g_a \tag{7.141}
$$

3- Relation between $S_v\left(a_{(u+v)}, \alpha\right)$ and $S_v\left(a_{(u+v)}, \gamma\right)$

The term $S_v\left(a_{(u+v)}, x\right)$ is given by equation (7.112) as

$$
S_v(a_{(u+v)}, \alpha) = \sum_{r_0=1}^{r_0=m} \sum_{r_1=0}^{r_1=r_0} \cdots \sum_{r_{\delta-1}=0}^{r_{\delta-1}=r_{\delta-2}} a_{(u+v)} \frac{(u+v)!}{u!} \alpha_0^{u_0} \alpha_1^{u_1} \cdots \alpha_\delta^{u_\delta} \tag{7.142}
$$

By hypothesis $\gamma_k = -\alpha_k$ then the relation between the product of the variables writes

$$
\gamma_0^{u_0} \gamma_1^{u_1} \cdots \gamma_k^{u_k} \cdots \gamma_{d-1}^{u_{d-1}} = \alpha_0^{u_0} \alpha_1^{u_1} \cdots (-\alpha_k)^{u_k} \cdots \alpha_{d-1}^{u_{d-1}}
$$

$$
= (-1)^{u_k} \alpha_0^{u_0} \alpha_1^{u_1} \cdots \alpha_k^{u_k} \cdots \alpha_{d-1}^{u_{d-1}} \tag{7.143}
$$

Consider the computation of the coefficients in the level $\lambda_0(v) = q - 1$, then the coefficients $a_{(u+v)}$ and $g_{(u+v)}$ were obtained using the expression (7.140) and so they satisfy the relation

$$g_{(u+v)} = (-1)^{u_k+v_k} a_{(u+v)} \qquad (7.144)$$

then the relation between a term in the expression (7.142) for the computation a_v and g_v writes

$$g_{(u+v)} \frac{(u+v)!}{u!} \gamma_0^{u_0} \gamma_1^{u_1} \cdots \gamma_\delta^{u_\delta} = (-1)^{u_k+v_k} a_{(u+v)} \frac{(u+v)!}{u!} (-1)^{v_k} \alpha_0^{u_0} \alpha_1^{u_1} \cdots \alpha_\delta^{u_\delta}$$

$$= (-1)^{2u_k+v_k} a_{(u+v)} \frac{(u+v)!}{u!} \alpha_0^{u_0} \alpha_1^{u_1} \cdots \alpha_\delta^{u_\delta}$$

$$= (-1)^{v_k} a_{(u+v)} \frac{(u+v)!}{u!} \alpha_0^{u_0} \alpha_1^{u_1} \cdots \alpha_\delta^{u_\delta}$$

$$(7.145)$$

the application of the above result to the equation (7.142) gives the following relation

$$S_v \left(g_{(u+v)}, \gamma \right) = (-1)^{v_k} S_v \left(a_{(u+v)}, \alpha \right) \qquad (7.146)$$

Substituting the results in equation (7.139) and equation (7.146) into the equation (7.123) it gives the relation

$$\frac{1}{v!} \left[E_F^{(v)}(\gamma) - S_v \left(g_{(u+v)}, \gamma \right) \right] = (-1)^{v_k} \frac{1}{v!} \left[E_F^{(v)}(\alpha) - S_v \left(a_{(u+v)}, \alpha \right) \right] \qquad (7.147)$$

which yields

$$g_v = (-1)^{v_k} a_v \qquad (7.148)$$

The result in equation (7.148) for $\lambda_0(v) = q - 1$ permits to extend the same relation for $\lambda_0(v) = q - 1, \ldots, 0$, which proves the expression (7.131).

Proof of $P_h(\gamma) = (-1)^{\eta+1} P_e(\alpha)$:

Consider

$$P_h(\delta) = \left[(\gamma_0 - \delta_{0,0})(\gamma_1 - \delta_{1,1}) \cdots (\gamma_k - \delta_{k,k}) \cdots (\gamma_{d-1} - \delta_{d-1,d-1}) \right]^{\eta+1} \qquad (7.149)$$

and

$$P_e(\alpha) = \left[(\alpha_0 - \beta_{0,0})(\alpha_1 - \beta_{1,1}) \cdots (\alpha_k - \beta_{k,k}) \cdots (\alpha_{d-1} - \beta_{d-1,d-1}) \right]^{\eta+1} \qquad (7.150)$$

For a rectangular hyperparallelopiped equation (7.32) gives $\gamma_i - \delta_{i,i} = 2\gamma_i$ and $\alpha_i - \beta_{i,i} = 2\alpha_i$. By hypothesis $\gamma_k = -\alpha_k$, then the following relation between $P_g(\delta)$ and $P_a(\alpha)$

$$
\begin{aligned}
P_h(\delta) &= \Big[(2\gamma_0) \times \ldots \times (2\gamma_k) \times \ldots \times (2\gamma_{d-1}) \Big]^{\eta+1} \\
&= \Big[(2\alpha_0) \times \ldots \times (-1)(2\alpha_k) \times \ldots \times (2\alpha_{d-1}) \Big]^{\eta+1} \\
&= (-1)^{\eta+1} P_e(\alpha)
\end{aligned}
\tag{7.151}
$$

end of theorem 7.4.1

7.4.2　Computation of the coefficients of $f_{h,n}(g,x)$ in terms of the coefficients of $f_{e,n}(a,x)$ if the reference nodes e and h differ for several coordinate number

Corollary 7.4.3

Let

$$
\Phi_{e,n}(x) = \frac{1}{P_e(\alpha)} P_e(x) Q_{e,n}(x) f_{e,n}(a, x)
\tag{7.152}
$$

and

$$
\Phi_{h,n}(x) = \frac{1}{P_h(\alpha)} P_h(x) Q_{h,n}(x) f_{h,n}(g, x)
\tag{7.153}
$$

be two sets of Hermite Interpolating Polynomials defined on the same hypercube domain symmetric with respect to the origin of the reference system as shown in figure 7.4.2 below. The first set is related to the reference node a whose coordinate numbers are $\alpha = (\alpha_0, \ldots, \alpha_k, \ldots, \alpha_{d-1})$ and the second set related to the reference node h whose coordinate numbers are $\gamma = (\alpha_0, \ldots, -\alpha_k, \ldots, \alpha_{d-1})$.

Assume that there are $p \leq d$ coordinate numbers such that $\gamma_p = -\alpha_p$ and the rest satisfy the condition $\gamma_j = \alpha_j$. As consequence $\delta_p = -\beta_p$ and the rest satisfy the condition $\delta_j = \beta_j$.

The coefficients g_t of the polynomial $f_{g,n}(x)$ can be obtained in terms of the coefficients a_s of the polynomial $f_{e,n}(x)$ using the following transformations

$$
g_v = \big[(-1)^{v_0} (-1)^{v_1} \cdots (-1)^{v_{p-1}} \big] a_v
\tag{7.154}
$$

and

$$
P_h(\gamma) = \big[(-1)^{\eta+1} \cdots (-1)^{\eta+1} \big] P_e(\alpha)
\tag{7.155}
$$

where there are p factors of the form $(-1)^{\eta+1}$.

Proof:

It is enough to apply the theorem 7.4.1 several times.

end of corollary 7.4.3

Example 7.4.4

As an example let $d = 1$, $\eta = 4$, and $i_0 = 1$. For $\alpha = -4$ and $\beta = 4$ the Hermite Interpolating Polynomial is given by

$$\Phi_{a,1}(x) = \frac{1}{-8^5}(x-4)^5(x+4)\frac{1}{8^3}(5952 + 2960x + 540x^2 + 35x^3) \quad (7.156)$$

where

$$a_0 = 5952 \qquad a_1 = 2960 \qquad a_2 = 540 \qquad a_3 = 35 \qquad (7.157)$$

Equation (7.131) applies for each coefficient that is,

$$g_v = (-1)^{v_k} a_v \qquad (7.158)$$

where $(-1)^{v_k}$ refers to the k-th variable in the terms. For this example there is only one variable, therefore, the index $k = 0$ and then $v = (v_0)$. For $\gamma = -\alpha = 4$ and $\delta = -\beta = -4$ it follows

$$g_0 = (-1)^0 a_0 = 5952 \qquad g_1 = (-1)^1 a_1 = -2960$$
$$g_2 = (-1)^2 a_2 = 540 \qquad g_3 = (-1)^3 a_3 = -35 \qquad (7.159)$$

and

$$P_g(\gamma) = (-1)^{(v_0)} P_a(\alpha) = (-1)^{4+1}\frac{1}{-8^5} \qquad (7.160)$$

giving the polynomial

$$\Phi_{g,1}(x) = \frac{1}{8^5}(x+4)^5(x-4)\frac{1}{8^3}(5952 - 2960x + 540x^2 - 35x^3) \quad (7.161)$$

end of example 7.4.4

7.4.3   Relation between the polynomials $\Psi_{e,n}^{(n)}(x)|_{x=\alpha}$ and $\Phi_{e,n}^{(n)}(x)|_{x=\alpha}$ if $\Psi_{e,n}^{(n)}(x)|_{x=\alpha} = \varphi \Phi_{e,n}^{(n)}(x)|_{x=\alpha}$

Theorem 7.4.5

Assume $\varphi \in \mathbb{F}$. Let $\Psi_{e,n}(x)$ and $\Phi_{e,n}(x)$ be polynomials such that

$$\Psi_{e,n}^{(n)}(x)\Big|_{x=\alpha} = \frac{1}{n!\,P_e(\alpha)}P_e(x)\mathcal{Q}_{e,n}(x)\mathcal{F}_{e,n}(g,x)\Big|_{x=\alpha} = \varphi \qquad (7.162)$$

and

$$\Phi_{e,n}^{(n)}(x)\Big|_{x=\alpha} = \frac{1}{n!\,P_e(\alpha)}P_e(x)\mathcal{Q}_{e,n}(x)f_{e,n}(a,x)\Big|_{x=\alpha} = 1 \qquad (7.163)$$

over the same domain Ω, related to the same reference node e, and same maximum derivative η, then

1. The coefficients of the polynomials $\mathcal{F}_{e,n}(g, x)$ and $f_{e,n}(a, x)$ are related as

$$g_{(v)} = \varphi a_{(v)} \quad \forall v \in L^{d-1}(q) \tag{7.164}$$

2. The polynomials $\mathcal{F}_{e,n}(g, x)$ and $f_{e,n}(a, x)$ are related as

$$\mathcal{F}_{e,n}(g, x) = \varphi f_{e,n}(a, x) \quad \forall x \in \Omega \quad \text{and} \quad \forall n \in L^{d-1}(\eta) \tag{7.165}$$

3. The polynomials $\Psi_{e,i}^{(i)}(x)$ and $\Phi_{e,i}^{(i)}(x)$ are related as

$$\Psi_{e,n}^{(n)}(x) = \varphi \Phi_{e,n}^{(n)}(x) \quad \forall x \in \Omega \quad \text{and} \quad \forall n \in L^{d-1}(\eta) \tag{7.166}$$

Proof:

If the domain Ω, the reference node e, and the degree of derivative η are the same then

$$P_e(\alpha) = P_e(\alpha) \qquad P_e(x) = P_e(x) \qquad Q_{e,n}(x) = Q_{e,n}(x) \tag{7.167}$$

Therefore, it is sufficient to prove the equation (7.164) and the others two will be true as a consequence.

The derivative of order $\mathcal{O}(n)$ of $\Psi_{e,n}(x)$ in equation (7.162) and $\Phi_{e,n}(x)$ in equation (7.163) according to equation (7.128) implies that

$$\mathcal{F}_{e,n}(g, x)\Big|_{x=\alpha} = \varphi \quad \text{and} \quad f_{e,n}(a, x)\Big|_{x=\alpha} = 1 \tag{7.168}$$

Since

$$f_{e,n}(a, x)\Big|_{x=\alpha} = 1 \implies \varphi f_{e,n}(a, x)\Big|_{x=\alpha} = \varphi = \mathcal{F}_{e,n}(g, x)\Big|_{x=\alpha} \tag{7.169}$$

then

$$\varphi f_{e,n}(a, x)\Big|_{x=\alpha} = f_{e,n}(\varphi a, x)\Big|_{x=\alpha} = \mathcal{F}_{e,n}(g, x)\Big|_{x=\alpha} \tag{7.170}$$

The two polynomials $f_{e,n}(\varphi a, x)$ and $\mathcal{F}_{e,n}(g, x)$ will be identical if the equality

$$f_{e,n}(\varphi a, x) = \mathcal{F}_{e,n}(g, x) \tag{7.171}$$

is true for all $x \in \Omega$, which requires the equality of all the coefficients, that is,

$$\varphi a_{(v)} = g_{(v)} \quad \forall v \in L^{d-1}(q) \tag{7.172}$$

and this completes the proof.

end of theorem 7.4.5

7.4.4 Relation between $[1/(n!P_e(x))]^{(\nu)}$ and $[1/(n!P_e(x)]^{(\mu)}$ if μ is a permutation of ν

Lemma 7.4.6

Let $\mu = (\mu_0, \mu_1, \ldots, \mu_{d-1})$ be a permutation of $\nu = (\nu_0, \nu_1, \ldots, \nu_{d-1})$ then the equality

$$\left[\frac{1}{n!P_e(x)}\right]^{(\nu)} = \left(E_F^{(\nu/\mu)}(x)\right)\left[\frac{1}{n!P_e(x)}\right]^{(\mu)} \tag{7.173}$$

at the same reference node of a hyper-parallelopiped domain.

Proof:

The derivative of order ν of $1/P_e(x)$, recalling equation (7.82), is given by

$$\left[\frac{1}{P_e(x)}\right]^{(\nu)} = \frac{E_F^{(\nu)}(x)}{P_e(x)} \tag{7.174}$$

and the derivative of order μ is given by

$$\left[\frac{1}{P_e(x)}\right]^{(\mu)} = \frac{E_F^{(\mu)}(x)}{P_e(x)} \tag{7.175}$$

where $E_F^{(\nu)}(x)$ and $E_F^{(\mu)}(x)$ are given by equation (7.84).

Dividing equation (7.174) by equation (7.175) and rearranging the factors

$$\frac{[1/P_e(x)]^{(\nu)}}{[1/P_e(x)]^{(\mu)}} = \frac{E_F^{(\nu)}(x)}{P_e(x)}\frac{P_e(x)}{E_e^{(\mu)}(x)} = E_F^{(\nu/\mu)}(x) \tag{7.176}$$

where

$$E_F^{(\nu/\mu)}(x) := \frac{E_F^{(\nu)}(x)}{E_F^{(\mu)}(x)} = \frac{\prod_{j=0}^{d-1}(x_j - \beta_{j,j})^{\mu_j}}{\prod_{j=0}^{d-1}(x_j - \beta_{j,j})^{\nu_j}} = \prod_{j=0}^{d-1}(x_j - \beta_{j,j})^{\mu_j - \nu_j} \tag{7.177}$$

Since restriction is made with respect to the values of x then equation (7.175) is true for all $x = (x_0, x_1, \ldots, x_{d-1})$. This completes the proof.

end of lemma 7.4.6

7.4.5   Equality of the coefficients of the terms of the polynomial $f_{e,n}(a,x)$ that are in the same level and which indices are permutations

Theorem 7.4.7

Let

1. $\mu = (\mu_0, \mu_1, \ldots, \mu_{d-1})$, and $\nu = (\nu_0, \nu_1, \ldots, \nu_{d-1})$ be constrained coordinate numbers such that $\mu, \nu \in L^{\delta}(q)$

2. ν be a permutation of μ denoted as $\nu = \mathcal{P}(\mu)$, as shown in Espence and Eynden [12]

3. $a_{(\mu)}$ and $a_{(\nu)}$ be two coefficients of the terms

$$a_{(\mu)}x_0^{\mu_0} x_0^{\mu_1} \cdots x_0^{\mu_{d-1}} \quad \text{and} \quad a_{(\nu)}x_0^{\nu_0} x_0^{\nu_1} \cdots x_0^{\nu_{d-1}} \tag{7.178}$$

of the polynomial $f_{e,n}(a,x)$.

4. $v = (v_0, v_1, \ldots, v_{\delta})$ and $z = (z_0, z_1, \ldots, z_{\delta})$ be two constrained numbers in the set L_n^{δ} where $n = 0, 1, \ldots, q$.

Then

1. For each coefficient $a_{(v)}$ in the level $n = \lambda_0(v)$ there exists one and only one coefficient $a_{(z)}$ such that $z = \mathcal{P}(v)$ where $v, z \in L_n^{\delta}$.

2. For the reference node $\alpha = (1, \ldots, 1)$ or $\alpha = (-1, \ldots, -1)$ and for a hypercube domain symmetric with respect to the origin of the reference coordinate system, the coefficients of the terms that are in the same level of the polynomial $f_{e,n}(a,x)$ are equal, that is,

$$a_{\mu} = a_{\nu} \quad \text{if} \quad \nu = \mathcal{P}(\mu) \in L_{\lambda_0(\mu)}^{\delta} \tag{7.179}$$

Proof of part 1:

By hypothesis the indices of the coefficients $a_{(v)}$ and $a_{(z)}$ belong to L_n^{δ}, which is a complete set and, therefore, it contains all combinations of the coordinate numbers $(v_0, v_1, \ldots, v_{\delta})$ of the number v that satisfy the constraint equation. Since a permutation $\mathcal{P}$ of a set T is a bijective function from T to T, Hungerford [20, pp. 151], then given a $v, z \in L_n^{\delta}$ the number z satisfying the permutation $z = \mathcal{P}(v)$ is unique.

Proof of part 2:

1. Coefficients $v, z \in L_q^{\delta}$

Let $v \in L_q^{\delta}$ be the order of derivative to obtain the coefficient $a_{(v)}$ and $z \in L_q^{\delta}$ the order of derivative to compute the coefficient $a_{(z)}$.

The coefficients $a_{(v)}$ and $a_{(z)}$ in the level q are evaluated at the reference node e, whose coordinate is α, using the equation (7.122) as

$$a_{(v)} = \frac{E_F^{(v)}(\alpha)}{v!} \quad \text{and} \quad a_{(z)} = \frac{E_F^{(z)}(\alpha)}{z!} \tag{7.180}$$

Dividing $a_{(v)}$ by $a_{(z)}$ and using the Lemma 7.4.6 it follows the relation

$$\frac{a_{(v)}}{a_{(z)}} = \frac{E_F^v(\alpha) \, z!}{E_F^z(\alpha) \, v!} = E_F^{v/z} \tag{7.181}$$

between the two coefficients.

Since $z = \mathcal{P}(v)$ it implies that $v! = z!$ and then the equation (7.181) transforms into

$$a_{(v)} = \left(E_F^{v/z}\right) a_{(z)} \tag{7.182}$$

From the definition of $E_F^{(v)}$ in equations (7.84) and equation (7.86) it can be written that

$$E_F^v = \frac{(\eta + v_0)!}{\eta!} \times \cdots \times \frac{(\eta + v_\delta)!}{\eta!} \times \frac{1}{(\beta_{0,0} - \alpha_0)^{v_0}} \times \cdots \times \frac{1}{(\beta_{\delta,\delta} - \alpha_\delta)^{v_\delta}} \tag{7.183}$$

and

$$E_F^z = \frac{(\eta + z_0)!}{\eta!} \times \cdots \times \frac{(\eta + z_\delta)!}{\eta!} \times \frac{1}{(\beta_{0,0} - \alpha_0)^{z_0}} \times \cdots \times \frac{1}{(\beta_{\delta,\delta} - \alpha_\delta)^{z_\delta}} \tag{7.184}$$

Since $z = \mathcal{P}(v)$ then

$$\frac{(\eta + z_0)!}{\eta!} \times \cdots \times \frac{(\eta + z_\delta)!}{\eta!} = \frac{(\eta + z_0)!}{\eta!} \times \cdots \times \frac{(\eta + z_\delta)!}{\eta!} \tag{7.185}$$

In the case of a hypercube symmetric with respect to the origin of the reference system the equation (7.34) shows that $\alpha_j - \beta_{j,j} = 2, j = 0, \ldots, d - 1$, which gives the equality

$$\frac{1}{(\beta_{0,0} - \alpha_0)^{v_0}} \times \cdots \times \frac{1}{(\beta_{\delta,\delta} - \alpha_\delta)^{v_\delta}} = \frac{1}{(\beta_{0,0} - \alpha_0)^{z_0}} \times \cdots \times \frac{1}{(\beta_{\delta,\delta} - \alpha_\delta)^{z_\delta}} \tag{7.186}$$

and, therefore,

$$E_F^{v/z} = 1 \qquad \Longrightarrow \qquad E_F^v = E_F^z \qquad (7.187)$$

which reduces the expression in equation (7.182) to

$$a_{(v)} = a_{(z)} \qquad (7.188)$$

2. Coefficients $v, z \in L_{q-1}^\delta$

Let $v \in L_{q-1}^\delta$ be the order of derivative to obtain the coefficient $a_{(v)}$ and $z \in L_{q-1}^\delta$ the order of derivative to compute the coefficients $z \in L_{q-1}^\delta$.

The coefficients $a_{(v)}$ and $a_{(z)}$ in the level $q-1$ are evaluated at the reference node e, whose coordinate is α, using the equation (7.123) as

$$a_{(v)} = \frac{1}{v!} \left[E_F^{(v)}(\alpha) - S_v \left(a_{(u+v)}, \alpha \right) \right] \qquad (7.189)$$

and

$$a_{(z)} = \frac{1}{z!} \left[E_F^{(z)}(\alpha) - S_z \left(a_{(r+z)}, \alpha \right) \right] \qquad (7.190)$$

where $m = q - \lambda_0(v) = 1$, $u, r \in (L^\delta(m) - L_0^\delta) = (L^\delta(1) - L_0^\delta)$, $u \neq r$, $\mu = u + v$ and $\nu = r + z$

Solving equation (7.121) for $S_v(a_{(u+v)}, x)$ it gives

$$S_v(a_{(u+v)}, x) = \left[f_{e,n} \left(a_{(u+v)}, x \right) \right]^{(v)} - v! a_{(v)} \qquad (7.191)$$

where $v \in L_{q-1}^\delta$ and $u \in L_{q-\lambda_0(v)}^\delta = L^\delta(q) - \{(v)\}$ and

$$(u + v) \in \left(L_{q-\lambda_0(v)}^\delta + \{(v)\} \right) \subset L^\delta(q) \qquad (7.192)$$

Similarly

$$S_v(a_{(r+z)}, x) = \left[f_{e,n} \left(a_{(r+z)}, x \right) \right]^{(z)} - z! a_{(z)} \qquad (7.193)$$

where $z \in L_{q-1}^\delta$ and $r \in L_{q-\lambda_0(z)}^\delta = L^\delta(q) - \{(z)\}$ and

$$(r + z) \in \left(L_{q-\lambda_0(z)}^\delta + \{(z)\} \right) \subset L^\delta(q) \qquad (7.194)$$

By the theorem 2.10.10, on page 110, it follows that $(r + z) = \mathcal{P}(u + v)$. Since $(r + z), (u + v) \in L_q^\delta$ then the coefficients $a_{(r+z)}$ and $a_{(u+v)}$ are equal.

Observing that $L_{q-\lambda_0(v)}^\delta = L_{q-\lambda_0(z)}^\delta$ then $\lambda_0(u) = \lambda_0(r)$, therefore,

$$\alpha_0^{u_0} \alpha_1^{u_1} \cdots \alpha_\delta^{u_\delta} = \alpha_0^{r_0} \alpha_1^{r_1} \cdots \alpha_\delta^{r_\delta} \qquad (7.195)$$

either $\alpha_i = 1$ or $\alpha_i = -1$ and

$$\frac{(u+z)!}{u!} = \frac{(r+v)!}{r!} \tag{7.196}$$

since $(r+v) = \mathcal{P}(u+z)$, as consequence

$$S_v\left(a_{(u+v)}, \alpha\right) = S_z\left(a_{(r+z)}, \alpha\right) \tag{7.197}$$

Recalling that $E_F^v = E_F^z$ if $z = \mathcal{P}(v)$ then

$$a_{(v)} = a_{(z)} \tag{7.198}$$

3. Induction

Assume that the coefficients $a_{(v)} = a_{(z)}$ at the level n if $z = (v)$. The coefficients of the level $n-1$ are evaluated using the coefficients at the previous levels, which are equal if $z = (v)$, which implies the equalities

$$E_F^v = E_F^z \qquad \text{and} \qquad S_v\left(a_{(u+v)}, \alpha\right) = S_z\left(a_{(r+z)}, \alpha\right) \tag{7.199}$$

recalling that $v! = z!$ if $z = (v)$ then

$$a_{(v)} = a_{(z)} \tag{7.200}$$

end of theorem 7.4.7

Example 7.4.8

The polynomial $\Phi_{e,n}^{(i_0, i_1)}$ of degree $q = 3$ related to the node $a_e = (-1, -1)$ is given by

$$\begin{aligned}
\Phi_{a,10}(x,y) = {} & \frac{1}{1!0!} \frac{1}{(-2)^{10}} \frac{1}{2^3} (x_0 - 1)^5 (x_1 - 1)^5 (x_0 + 1) \left(378\, x_0^0 x_1^0 \right. \\
& + 460\, x_0^1 x_1^0 + 460\, x_0^0 x_1^1 + 210\, x_0^2 x_1^0 + 350\, x_0^1 x_1^1 + 210\, x_0^0 x_1^2 \\
& \left. + 35\, x_0^3 x_1^0 + 75\, x_0^2 x_1^1 + 75\, x_0^1 x_1^2 + 35\, x_0^0 x_1^3 \right)
\end{aligned} \tag{7.201}$$

At the level $q = 3$ the number $z = (0,3)$ is a permutation of $v = (3,0)$ and their coefficients are $a_{(3,0)} = a_{(0,3)} = 35$.

For the level $q-1 = 2$ for the computation of the coefficient $a_{(2,0)}$ the order of derivative is $v = (2,0)$, which gives

$$\left[f_{(e,n)}(a, x) \right]^{(2,0)} = 210\, x_0^0 x_1^0 + 35\, x_0^1 x_1^0 + 75\, x_0^0 x_1^1 \tag{7.202}$$

from where

$$S_v\left(a_{(u+v)}, \alpha\right) = 35\, x_0^1 x_1^0 + 75\, x_0^0 x_1^1 \tag{7.203}$$

and for the computation of the coefficient $a_{(0,2)}$ the order of derivative is $z = (0,2)$, which gives

$$\left[f_{(e,n}(a,x)\right]^{(0,2)} = 210\, x_0^0 x_1^0 + 75\, x_0^1 x_1^0 + 35\, x_0^0 x_1^1 \qquad (7.204)$$

from where

$$S_z\left(a_{(r+z)},\alpha\right) = 75\, x_0^1 x_1^0 + 35\, x_0^0 x_1^1 \qquad (7.205)$$

when evaluated at the reference node $\alpha = (-1,-1)$ it gives

$$S_v\left(a_{(u+v)},\alpha\right) = -35 - 75$$
$$S_z\left(a_{(r+z)},\alpha\right) = -75 - 35 \qquad (7.206)$$

showing the equality.

end of example 7.4.8

Example 7.4.9

The polynomial $\Phi_{e,n}^{(i_0,i_1,i_2)}$ of degree $q = 3$ related to the node $a_e = (-1,-1,-1)$ is given by

$$\Phi_{a,000}(x_0,x_1,x_2) = \frac{1}{0!0!0!}\left[\frac{1}{2^4}\frac{1}{2^4}\frac{1}{2^4}\right](x_0-1)^4(x_1-1)^4(x_2-1)^4\frac{1}{2^1}$$

$$\times \Big(\; 144\, x_0^0 x_1^0 x_2^0 + 121\, x_0^1 x_1^0 x_2^0 + 121\, x_0^0 x_1^1 x_2^0 + 121\, x_0^0 x_1^0 x_2^1$$
$$+ 40\, x_0^2 x_1^0 x_2^0 + 64\, x_0^1 x_1^1 x_2^0 + 64\, x_0^1 x_1^0 x_2^1 + 40\, x_0^0 x_1^2 x_2^0 + 64\, x_0^0 x_1^1 x_2^1 + 40\, x_0^0 x_1^0 x_2^2$$
$$+ 5\, x_0^3 x_1^0 x_2^0 + 10\, x_0^2 x_1^1 x_2^0 + 10\, x_0^2 x_1^0 x_2^1 + 10\, x_0^1 x_1^2 x_2^0 + 16\, x_0^1 x_1^1 x_2^1$$
$$+ 10\, x_0^1 x_1^0 x_2^2 + 5\, x_0^0 x_1^3 x_2^0 + 10\, x_0^0 x_1^2 x_2^1 + 10\, x_0^0 x_1^1 x_2^2 + 5\, x_0^0 x_1^0 x_2^3 \;\Big) \qquad (7.207)$$

At the level $q = 3$ the number

$$(2,1,0),(2,0,1),(1,2,0),(1,0,2),(0,2,1),(0,1,2) \qquad (7.208)$$

are a permutation of $v = (2,1,0)$ and their coefficients are $a_{(3,0)} = a_{(0,3)} = 35$.

$$a_{(2,1,0)} = a_{(2,0,1)} = a_{(1,2,0)} = a_{(1,0,2)} = a_{(0,2,1)} = a_{(0,1,2)} = 10 \qquad (7.209)$$

For the level $q - 1 = 2$ for the computation of the coefficient $a_{(1,1,0)}$ the order of derivative is $v = (1,1,0)$, which gives

$$\left[f_{(e,n}(a,x)\right]^{(1,1,0)} = 64\, x_0^0 x_1^0 x_2^0 + 10\, x_0^1 x_1^0 x_2^0 + 10\, x_0^0 x_1^1 x_2^0 + 16\, x_0^0 x_1^0 x_2^1 \quad (7.210)$$

from where

$$S_v\left(a_{(u+v)}, \alpha\right) = 10\, x_0^1 x_1^0 x_2^0 + 10\, x_0^0 x_1^1 x_2^0 + 16\, x_0^0 x_1^0 x_2^1 \qquad (7.211)$$

and for the computation of the coefficient $a_{(0,1,1)}$ the order of derivative is $z = (0, 1, 1)$, which gives

$$\left[f_{(e,n)}(a, x)\right]^{(0,1,1)} = 64\, x_0^0 x_1^1 x_2^2 + 16\, x_0^0 x_1^0 x_2^2 + 10\, x_0^0 x_1^1 x_2^0 + 10\, x_0^0 x_1^0 x_2^1 \quad (7.212)$$

from where

$$S_z\left(a_{(r+z)}, \alpha\right) = 16\, x_0^1 x_1^0 x_2^0 + 10\, x_0^0 x_1^1 x_2^0 + 10\, x_0^0 x_1^0 x_2^1 \qquad (7.213)$$

when evaluated at the reference node $\alpha = (-1, -1, -1)$ it gives

$$S_v\left(a_{(u+v)}, \alpha\right) = -10 - 10 - 16$$
$$S_z\left(a_{(r+z)}, \alpha\right) = -16 - 10 - 10 \qquad (7.214)$$

showing the equality.

end of example 7.4.9

Lemma 7.4.10

Let

$$\frac{P_e(x)}{P_e(-2\alpha)} = \frac{(x - \alpha)^{\eta+1}}{(-\alpha - \alpha)^{\eta+1}} = \left(\frac{x - \alpha}{-2\alpha}\right)^{\eta+1} \qquad (7.215)$$

and

$$\frac{P_\varepsilon(x)}{P_\epsilon(2\alpha)} = \frac{(x + \alpha)^{\eta+1}}{(\alpha + \alpha)^{\eta+1}} = \left(\frac{x + \alpha}{2\alpha}\right)^{\eta+1} \qquad (7.216)$$

be polynomials related to the nodes e and ϵ whose coordinates are $e = -\alpha$ and $\varepsilon = \alpha$, respectively. According to the notation used $\beta_e = \alpha$ and $\beta_\varepsilon = -\alpha$. Then

1. Part 1

$$\frac{P_\epsilon(x)}{P_\epsilon(2\alpha)} = \frac{P_e(-x)}{P_e(-2\alpha)} \qquad (7.217)$$

2. Part 2

$$\left[\frac{P_\varepsilon(x)}{P_\varepsilon(2\alpha)}\right]^{(u)} = (-1)^u \left[\frac{P_e(-x)}{P_e(-2\alpha)}\right]^{(u)} \qquad (7.218)$$

Proof of part 1:

$$\frac{P_e(-x)}{P_e(-2\alpha)} = \frac{((-x) - \alpha)^{\eta+1}}{(-\alpha - \alpha)^{\eta+1}} = \left(\frac{x + \alpha}{2\alpha}\right)^{\eta+1} = \frac{P_\epsilon(x)}{P_\epsilon(2\alpha)} \qquad (7.219)$$

Proof of part 2:

$$\frac{1}{P_e(-2\alpha)}\left[P_e(-x)\right]^{(u)} = \frac{(\eta+1)!}{(\eta+1-u)!}\left(\frac{(-x)-\alpha}{-2\alpha}\right)^{\eta+1}((-)x-\alpha)^{-u}$$

$$= \frac{(\eta+1)!}{(\eta+1-u)!}\left(\frac{x+\alpha}{2\alpha}\right)^{\eta+1}(x+\alpha)^{-u}(-1)^{-u}$$

$$= (-1)^{-u}\frac{1}{P_\varepsilon(2\alpha)}\left[P_\varepsilon(x)\right]^{(u)} \tag{7.220}$$

thus

$$\left[\frac{P_\varepsilon(x)}{P_\varepsilon(2\alpha)}\right]^{(u)} = (-1)^u\left[\frac{P_\varepsilon(-x)}{P_e(-2\alpha)}\right]^{(u)} \tag{7.221}$$

and the proofs are completed.

If there are several variables, equations (7.217) and (7.218) holds for each variable. The "and only if" part can be proved in a similar way.

end of lemma 7.216

Lemma 7.4.11

Let

$$Q_e(x) = (x+\alpha)^i \tag{7.222}$$

$$Q_\varepsilon(x) = (x-\alpha)^i \tag{7.223}$$

be the polynomials $Q_e(x)$ and $Q_\varepsilon(x)$ defined according to the theorem 7.2.17 and related to the nodes e and ϵ whose coordinates are $-\alpha$ and α, respectively. According to the notation used $\beta_e = \alpha$ and $\beta_\epsilon = -\alpha$. Then

$$Q_\varepsilon(x) = (-1)^i Q_e(-x) \tag{7.224}$$

Moreover

$$\left[Q_\varepsilon(x)\right]^{(u)} = (-1)^i(-1)^u\left[Q_e(-x)\right]^{(u)} \tag{7.225}$$

Proof:

Proof of the equation (7.224).

$$Q_e(-x) = ((-x)+\alpha)^i = (-1)^i(x-\alpha)^i = (-1)^i Q_\varepsilon(x) \tag{7.226}$$

Proof of the equation (7.225).

$$\left[Q_e(-x) \right]^{(u)} = \frac{(i_j)!}{(i_j - u)!} \left((-x)_j + \alpha_j \right)^{-u} Q_e(-x)$$

$$= \frac{(i_j)!}{(i_j - u)!} (x_j - \alpha_j)^{-u} (-1)^{-u}(-1)^i Q_\varepsilon(x)$$

$$= (-1)^{-u}(-1)^i \left[\frac{(i_j)!}{(i_j - u)!} (x_j - \alpha_j)^{-u} Q_\varepsilon(x) \right]$$

$$= (-1)^{-u}(-1)^i \left[Q_\varepsilon(x) \right]^{(u)} \tag{7.227}$$

which completes the proofs.

end of lemma 7.4.11

Lemma 7.4.12

Let

$$f_{e,n}(a, x) = \sum_{i=0}^{\eta} a_i x^i \tag{7.228}$$

and

$$f_\varepsilon(g, x) = \sum_{i=0}^{\eta} g_i x^i \tag{7.229}$$

be two polynomials whose coefficients a_i and g_i are given by equations (7.122) and (7.123). Then

$$f_{e,n}(a, -x) = f_{\epsilon,n}(g, x) \tag{7.230}$$

Moreover

$$\left[f_{\varepsilon,n}(g, x) \right]^{(u)} = (-1)^u \left[f_{e,n}(a, -x) \right]^{(u)} \tag{7.231}$$

Proof:

Proof of the equation (7.230).

$$f_{e,n}(a, -x) = \sum_{i=0}^{\eta} a_i(-x)^i = \sum_{i=0}^{\eta}(-1)^i a_i x^i = \sum_{i=0}^{\eta} g_i x^i = f_\epsilon(g, x) \tag{7.232}$$

Proof of the equation (7.231).

$$\left[f_{e,n}(a,-x) \right]^{(u)} = \sum_{i=0}^{i=\eta} a_i(-x)^i(-x)^{-u}$$

$$= \sum_{i=0}^{i=\eta} (-1)^i a_i x^i x^{-u}(-1)^{-u}$$

$$= \sum_{i=0}^{i=\eta} g_i x^i x^{-u}(-1)^{-u}$$

$$= \left[f_\epsilon(g,y)(-1)^{-u} \right]^{(u)} \tag{7.233}$$

then

$$\left[f_\epsilon(g,x) \right]^{(u)} = (-1)^u \left[f_e(a,-x) \right]^{(u)} \tag{7.234}$$

which completes the proofs.

end of lemma 7.4.12

7.4.6   Computation of the polynomials related to one node in terms of the polynomials related to a symmetric node

Theorem 7.4.13

Let $\Phi_{e,n}^{(v)}(a,x)$ and $\Phi_{\varepsilon,n}^{(v)}(g,x)$ be two Hermite Interpolating Polynomials related to the nodes e and ε in a hypercube domain symmetric with respect to the axes of the coordinate system and whose coordinates are $-\alpha$ and α, respectively. According to the notation used $\beta_e = \alpha$ and $\beta_\varepsilon = -\alpha$. Let a_i and g_i be the coefficients of $f_{e,n}(a,-x)$ and $f_{\varepsilon,n}(g,x)$ given by equations (7.122-7.123). Then

$$\Phi_{\varepsilon,n}^{(v)}(g,x) = (-1)^{\lambda_0(v)}(-1)^n \Phi_{e,n}^{(v)}(a,-x) \tag{7.235}$$

Proof:

Recalling equation (7.126)

$$\Phi_{\varepsilon,n}^{(v)}(x) = \left[\frac{1}{n! P_\varepsilon(\alpha)} P_\varepsilon(x) \, Q_{\varepsilon,n}(x) \, f_{\varepsilon,n}(g,x) \right]^{(v)} \tag{7.236}$$

which expands, according to the lemma 6.7.4, on page 241, as follows

$$\Phi_{\varepsilon,n}^{(v)}(x) = \sum_{\lambda_1=0}^{\lambda_0(v)} \sum_{\lambda_2=0}^{\lambda_2=\lambda_1} F(\lambda_0,\lambda_1,\lambda_2) \left[\frac{P_\varepsilon(x)}{P_\varepsilon(\alpha)} \right]^{(u_0)} \left[Q_{\varepsilon,n}(x) \right]^{(u_1)} \left[f_{\varepsilon,n}(g,x) \right]^{(u_2)}$$

$$\tag{7.237}$$

using the lemmas 7.216, 7.4.11, and 7.4.12 and recalling that $u_0 = \lambda_0(v) - \lambda_1$, $u_1 = \lambda_1 - \lambda_2$, and $u_2 = \lambda_2$ it follows

$$\Phi_{\varepsilon,n}^{(v)}(x) = \sum_{\lambda_1=0}^{\lambda_0(v)} \sum_{\lambda_2=0}^{\lambda_2=\lambda_1} F(\lambda_0, \lambda_1, \lambda_2) \left[(-1)^{(u_0)} \frac{P_e(x)}{P_e(-\alpha)} \right]^{(u_0)}$$
$$\times (-1)^n (-1)^{(u_1)} \left[Q_{e,n}(x) \right]^{(u_1)} (-1)^{(u_2)} \left[f_{e,n}(a,x) \right]^{(u_2)} \quad (7.238)$$

Performing the product of the (-1) factors it is found that

$$(-1)^{(u_0)} \times (-1)^{(u_1)} \times (-1)^{(u_2)} = (-1)^{(\lambda_0-\lambda_1)}(-1)^{(\lambda_2-\lambda_1)}(-1)^{(\lambda_2)}$$
$$= (-1)^{(\lambda_0)} \quad (7.239)$$

introducing the result in equation (7.239) into equation (7.238) it gives

$$\Phi_{\varepsilon,n}^{(v)}(x) = (-1)^{(\lambda_0)}(-1)^n \sum_{\lambda_1=0}^{\lambda_0(v)} \sum_{\lambda_2=0}^{\lambda_2=\lambda_1} F(\lambda_0, \lambda_1, \lambda_2) \left[\frac{P_e(x)}{P_e(-\alpha)} \right]^{(u_0)}$$
$$\times \left[Q_{e,n}(x) \right]^{(u_1)} \left[f_{e,n}(a,x) \right]^{(u_2)}$$
$$= (-1)^{(\lambda_0)}(-1)^n \Phi_{e,n}^{(v)}(-x) \quad (7.240)$$

which completes the proof.

end of theorem 7.4.13

These theorem show that, for a hypercube symmetric with respect to the coordinate axes of the reference system, given the Hermite Interpolating Polynomials related to a node, the polynomials related to the other nodes are obtained by appropriate sign manipulation. This reduces the computational time for the evaluation of that polynomials related to all nodes of a hypercubic domain.

Chapter 8

Generation of the Hermite Interpolating Polynomials

8.1 Polynomials of minimum degree

The expression for the generation of the Hermite Interpolating Polynomials is given by equation (7.126) as

$$\Phi_{e,n}(x) = \frac{1}{n!} \frac{P_e(x)}{P_e(\alpha)} Q_{e,n}(x) f_{e,n}(a, x) \tag{8.1}$$

The minimum degree of the polynomial $f_{e,n}(\bar{a}, x)$, according to the corollary 7.2.19 on page 305, is given by $q = \eta - \lambda_0(n)$, where $\lambda_0(n)$ is the degree of the polynomial $Q_{e,n}(x)$.

Without loss of generality consider a three-dimensional space domain.

1. For $\lambda_0(n) = \lambda_0(0,0,0) = 0$ then $Q_{e,n}(x) = 1, \forall x \in \Omega$ and the equation (8.1) takes the form

$$\Phi_{e,0}(x) = \frac{1}{(0,0,0)!} \frac{P_e(x)}{P_e(\alpha)} f_{e,0}(a, x) \tag{8.2}$$

and the degree of $f_{e,0}(a, x)$ is given by $q = \eta - \lambda_0(0,0,0) = \eta$. Where

$$Q_{e,0}(x) = (x_0 - \alpha_0)^0 (x_0 - \alpha_1)^0 (x_2 - \alpha_2)^0 = 1 \tag{8.3}$$

is the simplest form that $Q_{e,0}(x)$ can assume.

2. For $\lambda_0(n) = \lambda_0(1,0,0) = 1$ then

$$\frac{\partial}{\partial x_0} Q_{e,1}(x) \bigg|_{x_0 = \alpha_0} = 1 \tag{8.4}$$

where $\alpha = (\alpha_0, \alpha_1, \alpha_2)$ is the coordinate of the reference node and $x = (x_0, x_1, x_2)$ is a variable that take values in the domain Ω. If

$$Q_{e,1}(x) = (x_0 - \alpha_0)^1 (x_0 - \alpha_1)^0 (x_2 - \alpha_2)^0 = (x_0 - \alpha_0) \tag{8.5}$$

then the equation (8.1) takes the form

$$\Phi_{e,1}(x) = \frac{1}{(1,0,0)!} \frac{P_e(x)}{P_e(\alpha)} (x_0 - \alpha_0) f_{e,1}(a, x) \tag{8.6}$$

and the degree of $f_{e,1}(a, x)$ is given by $q = \eta - \lambda_0(1, 0, 0) = \eta - 1$.

3. For $\lambda_0(n) = \lambda_0(0, 1, 0) = 1$ then

$$\frac{\partial}{\partial x_1} Q_{e,2}(x) \bigg|_{x_1 = \alpha_1} = 1 \tag{8.7}$$

If

$$Q_{e,2}(x) = (x_0 - \alpha_0)^0 (x_0 - \alpha_1)^1 (x_2 - \alpha_2)^0 = (x_1 - \alpha_1) \tag{8.8}$$

then the equation (8.1) takes the form

$$\Phi_{e,2}(x) = \frac{1}{(0,1,0)!} \frac{P_e(x)}{P_e(\alpha)} (x_1 - \alpha_1) f_{e,2}(a, x) \tag{8.9}$$

and the degree of $f_{e,2}(a, x)$ is given by $q = \eta - \lambda_0(0, 1, 0) = \eta - 1$.

The computation of the coefficients of $f_{e,n}(a, x)$ is a function of $P_e(x)/P_e(\alpha)$ but not of $Q_{e,n}(x)$, therefore, the polynomials $f_{e,k}(a, x)$ such that $\lambda_0(k) = 1$ are the same.

4. For $\lambda_0(n) = \lambda_0(0, 0, 1) = 1$ then

$$\Phi_{e,3}(x) = \frac{1}{(0,0,1)!} \frac{P_e(x)}{P_e(\alpha)} (x_2 - \alpha_2) f_{e,3}(a, x) \tag{8.10}$$

where

$$Q_{e,3}(x) = (x_0 - \alpha_0)^0 (x_0 - \alpha_1)^0 (x_2 - \alpha_2)^1 = (x_2 - \alpha_2) \tag{8.11}$$

The degree of $f_{e,3}(a, x)$ is given by $q = \eta - \lambda_0(0, 0, 1) = \eta - 1$.

Since $\lambda_0(1, 0, 0) = \lambda_0(0, 1, 0) = \lambda_0(0, 0, 1) = 1$ then $f_{e,1}(a, x) = f_{e,2}(a, x) = f_{e,3}(a, x)$.

5. For $\lambda_0(n) = \lambda_0(2, 0, 0) = 2$ then

$$\frac{\partial^2}{\partial x_0^2} Q_{e,4}(x) \bigg|_{x_0 = \alpha_0} = 1 \tag{8.12}$$

If

$$Q_{e,4}(x) = (x_0 - \alpha_0)^2 (x_0 - \alpha_1)^0 (x_2 - \alpha_2)^0 = (x_0 - \alpha_0)^2 \tag{8.13}$$

then the equation (8.1) takes the form

$$\Phi_{e,4}(x) = \frac{1}{(2,0,0)!} \frac{P_e(x)}{P_e(\alpha)} (x_0 - \alpha_0)^2 f_{e,4}(a, x) \tag{8.14}$$

The degree of $f_{e,4}(a, x)$ is given by $q = \eta - \lambda_0(2, 0, \ldots, 0) = \eta - 2$.

6. For $\lambda_0(n) = \lambda_0(1,1,0,\ldots,0) = 2$ then

$$\left. \frac{\partial^2}{\partial x_0^2} Q_{e,5}(x) \right|_{x_0=\alpha_0} = 1 \tag{8.15}$$

If

$$Q_{e,5}(x) = (x_0 - \alpha_0)^1 (x_0 - \alpha_1)^1 (x_2 - \alpha_2)^0 = (x_0 - \alpha_0)(x_1 - \alpha_1) \tag{8.16}$$

then the equation (8.1) takes the form

$$\Phi_{e,5}(x) = \frac{1}{(2,0,0)!} \frac{P_e(x)}{P_e(\alpha)} (x_0 - \alpha_0)(x_1 - \alpha_1) f_{e,5}(a, x) \tag{8.17}$$

The degree of $f_{e,5}(a, x)$ is given by $q = \eta - \lambda_0(2,0,0) = \eta - 2$.
Since $\lambda_0(2,0,0) = \lambda_0(1,1,0) = \cdots = 2$ then $f_{e,4}(a, x) = f_{e,5}(a, x) = \cdots$.

7. and so on.

The computation of the coefficients of the polynomials $f_{e,n}(a, x)$ is shown in section 7.3.

8.2 The degree of $f_{e,n}(a, x)$ is kept constant

8.2.1 The degree of $f_{e,n}(a, x)$ is η

That is, all polynomials will be generated using $f_{e,0}(a, x)$. Therefore, the equation (8.1) takes the form

$$\Phi_{e,n}(x) = \frac{1}{n!} \frac{P_e(x)}{P_e(\alpha)} Q_{e,n}(x) f_{e,0}(a, x) \tag{8.18}$$

and each polynomial, for a three-dimensional space domain writes

1. For $\lambda_0(n) = \lambda_0(0,0,0) = 0$

$$\Phi_{e,0}(x) = \frac{1}{(0,0,0)!} \frac{P_e(x)}{P_e(\alpha)} f_{e,0}(a, x) \tag{8.19}$$

which is the same polynomial as the one shown in equation (8.2).

2. For $\lambda_0(n) = \lambda_0(1,0,0) = 1$ the polynomial $Q_{e,1}(x)$ is given by equation (8.5) then

$$\Phi_{e,1}(x) = \frac{1}{(1,0,0)!} \frac{P_e(x)}{P_e(\alpha)} (x_0 - \alpha_0) f_{e,0}(a, x) \tag{8.20}$$

The polynomial $f_{e,0}(a, x)$ is equivalent to the polynomial $f_{e,1}(a, x)$ obtained for $q = \eta - \lambda_0(n) + 1$, that is, with q greater that the minimum value, which is given by $\eta - \lambda_0(n)$.

3. For $\lambda_0(n) = \lambda_0(0, 1, 0) = 1$ the polynomial $Q_{e,2}(x)$ is given by equation (8.8) then

$$\Phi_{e,2}(x) = \frac{1}{(0, 1, 0)!} \frac{P_e(x)}{P_e(\alpha)} (x_1 - \alpha_1) f_{e,0}(a, x) \tag{8.21}$$

4. For $\lambda_0(n) = \lambda_0(0, 0, 1) = 1$ the polynomial $Q_{e,3}(x)$ is given by equation (8.11) then

$$\Phi_{e,3}(x) = \frac{1}{(0, 0, 1)!} \frac{P_e(x)}{P_e(\alpha)} (x_2 - \alpha_2) f_{e,0}(a, x) \tag{8.22}$$

Recall that $f_{e,1}(a, x) = f_{e,2}(a, x) = f_{e,3}(a, x)$ since $\lambda_0(1, 0, 0) = \lambda_0(0, 1, 0) = \lambda_0(0, 0, 1) = 1$.

5. For $\lambda_0(n) = \lambda_0(2, 0, 0) = 2$ the polynomial $Q_{e,4}(x)$ is given by equation (8.13) then

$$\Phi_{e,4}(x) = \frac{1}{(2, 0, 0)!} \frac{P_e(x)}{P_e(\alpha)} (x_0 - \alpha_0)^2 f_{e,0}(a, x) \tag{8.23}$$

The polynomial $f_{e,0}(a, x)$ is equivalent to the polynomial $f_{e,4}(a, x)$ obtained for $q = \eta - \lambda_0(n) + 2$, that is, with q greater that the minimum value, which is given by $\eta - \lambda_0(n)$.

6. For $\lambda_0(n) = \lambda_0(1, 1, 0) = 2$ the polynomial $Q_{e,5}(x)$ is given by equation (8.16) then

$$\Phi_{e,5}(x) = \frac{1}{(1, 2, 0)!} \frac{P_e(x)}{P_e(\alpha)} (x_0 - \alpha_0)(x_1 - \alpha_1) f_{e,0}(a, x) \tag{8.24}$$

7. and so on ...

The polynomials can also be constructed evaluating $f_{e,0}(a, x)$ and $f_{e,1}(a, x)$ and generate the other polynomials with $f_{e,1}(a, x)$.

Chapter 9

Hermite Interpolating Polynomials: the classical and present approaches

9.1 Classical approach, one variable

Let f be a polynomial in the variable x and of degree 5. Let $f^{(1)}$ and $f^{(2)}$ be the derivatives of order one and two of f, that is,

$$f^{(0)} = a_0 + a_1 x + a_2 x^2 + a_3 x^3 + a_4 x^4 + a_5 x^5 \tag{9.1}$$

$$f^{(1)} = a_1 + 2a_2 x + 3a_3 x^2 + 4a_4 x^3 + 5a_5 x^4 \tag{9.2}$$

$$f^{(2)} = 2a_2 + 6a_3 x + 12a_4 x^2 + 20a_5 x^3 \tag{9.3}$$

Let $f_a^{(0)}$, $f_a^{(1)}$, and $f_a^{(2)}$, denote the value of the function and its derivatives of order one and two at the reference node $a = (-1)$ whose coordinate is $x = -1$.

$$f_a^{(0)} = a_0 - a_1 + a_2 - a_3 + a_4 - a_5$$

$$f_a^{(1)} = a_1 - 2a_2 + 3a_3 - 4a_4 + 5a_5$$

$$f_a^{(2)} = 2a_2 - 6a_3 + 12a_4 - 20a_5$$

Let $f_b^{(0)}$, $f_b^{(1)}$, and $f_b^{(2)}$, denote the value of the function and its derivatives of order one and two at the reference node $b = (1)$ whose coordinate is $x = 1$.

$$f_b^{(0)} = a_0 + a_1 + a_2 + a_3 + a_4 + a_5$$

$$f_b^{(1)} = a_1 + 2a_2 + 3a_3 + 4a_4 + 5a_5$$

$$f_b^{(2)} = 2a_2 + 6a_3 + 12a_4 + 20a_5$$

Writing in matrix form

$$
\begin{bmatrix}
1 & -1 & 1 & -1 & 1 & -1 \\
1 & 1 & 1 & 1 & 1 & 1 \\
0 & 1 & -2 & 3 & -4 & 5 \\
0 & 1 & 2 & 3 & 4 & 5 \\
0 & 0 & 2 & -6 & 12 & -20 \\
0 & 0 & 2 & 6 & 12 & 20
\end{bmatrix}
\begin{bmatrix}
a_0 \\
a_1 \\
a_2 \\
a_3 \\
a_4 \\
a_5
\end{bmatrix}
=
\begin{bmatrix}
f_a^{(0)} \\
f_b^{(0)} \\
f_a^{(1)} \\
f_b^{(1)} \\
f_a^{(2)} \\
f_b^{(2)}
\end{bmatrix}
\tag{9.4}
$$

The solution is

$$
a_0 = \frac{1}{16}\left[8f_a^{(0)} + 8f_b^{(0)} + 5f_a^{(1)} - 5f_b^{(1)} + f_a^{(2)} + f_b^{(2)}\right]
\tag{9.5}
$$

$$
a_1 = \frac{1}{16}\left[-15f_a^{(0)} + 15f_b^{(0)} - 7f_a^{(1)} - 7f_b^{(1)} - f_a^{(2)} + f_b^{(2)}\right]
\tag{9.6}
$$

$$
a_2 = \frac{1}{8}\left[-3f_a^{(1)} + 3f_b^{(1)} - f_a^{(2)} - f_b^{(2)}\right]
\tag{9.7}
$$

$$
a_3 = \frac{1}{8}\left[5f_a^{(0)} - 5f_b^{(0)} + 5f_a^{(1)} + 5f_b^{(1)} + f_a^{(2)} - f_b^{(2)}\right]
\tag{9.8}
$$

$$
a_4 = \frac{1}{16}\left[f_a^{(1)} - f_b^{(1)} + f_a^{(2)} + f_b^{(2)}\right]
\tag{9.9}
$$

$$
a_5 = \frac{1}{16}\left[-3f_a^{(0)} + 3f_b^{(0)} - 3f_a^{(1)} - 3f_b^{(1)} - f_a^{(2)} + f_b^{(2)}\right]
\tag{9.10}
$$

Substituting the solution into equation (9.1) and factoring with respect to the value of the function and their derivatives at the points $x = -1$ and $x = 1$ it follows

$$
\begin{aligned}
f = {} & \left(\frac{1}{2} - \frac{15}{16}x + \frac{5}{8}x^3 - \frac{3}{16}x^5\right)f_a^{(0)} + \left(\frac{1}{2} + \frac{15}{16}x - \frac{5}{8}x^3 + \frac{3}{16}x^5\right)f_b^{(0)} \\
& + \left(\frac{5}{16} - \frac{7}{16}x - \frac{3}{8}x^2 + \frac{5}{8}x^3 + \frac{1}{16}x^4 - \frac{3}{16}x^5\right)f_a^{(1)} \\
& + \left(-\frac{5}{16} - \frac{7}{16}x + \frac{3}{8}x^2 + \frac{5}{8}x^3 - \frac{1}{16}x^4 - \frac{3}{16}x^5\right)f_b^{(1)} \\
& + \left(\frac{1}{16} - \frac{1}{16}x - \frac{1}{8}x^2 + \frac{1}{8}x^3 + \frac{1}{16}x^4 - \frac{1}{16}x^5\right)f_a^{(2)} \\
& + \left(\frac{1}{16} + \frac{1}{16}x - \frac{1}{8}x^2 - \frac{1}{8}x^3 + \frac{1}{16}x^4 + \frac{1}{16}x^5\right)f_b^{(2)}
\end{aligned}
\tag{9.11}
$$

The polynomials related to the reference node $a = (-1)$ are

$$\Phi_{a,0}(x) = \frac{1}{2} - \frac{15}{16}x + \frac{5}{8}x^3 - \frac{3}{16}x^5$$

$$\Phi_{a,1}(x) = \frac{5}{16} - \frac{7}{16}x - \frac{3}{8}x^2 + \frac{5}{8}x^3 + \frac{1}{16}x^4 - \frac{3}{16}x^5$$

$$\Phi_{a,2}(x) = \frac{1}{16} - \frac{1}{16}x - \frac{1}{8}x^2 + \frac{1}{8}x^3 + \frac{1}{16}x^4 - \frac{1}{16}x^5 \qquad (9.12)$$

The polynomials related to the reference node $b = (1)$ are

$$\Phi_{b,0}(x) = \frac{1}{2} + \frac{15}{16}x - \frac{5}{8}x^3 + \frac{3}{16}x^5$$

$$\Phi_{b,1}(x) = -\frac{5}{16} - \frac{7}{16}x + \frac{3}{8}x^2 + \frac{5}{8}x^3 - \frac{1}{16}x^4 - \frac{3}{16}x^5$$

$$\Phi_{b,2}(x) = \frac{1}{16} + \frac{1}{16}x - \frac{1}{8}x^2 - \frac{1}{8}x^3 + \frac{1}{16}x^4 + \frac{1}{16}x^5 \qquad (9.13)$$

Factoring the polynomials related to the reference node $a = (-1)$

$$\Phi_{a,0}(x) = \frac{1}{16}(x-1)^3\left(-3x^2 - 9x - 8\right)$$

$$\Phi_{a,1}(x) = \frac{1}{16}(x-1)^3(x+1)(-3x-5)$$

$$\Phi_{a,2}(x) = -\frac{1}{16}(x-1)^3(x+1)^2 \qquad (9.14)$$

Factoring the polynomials related to the reference node $b = (1)$

$$\Phi_{b,0}(x) = \frac{1}{16}(x+1)^3\left(3x^2 - 9x + 8\right)$$

$$\Phi_{b,1}(x) = \frac{1}{16}(x+1)^3(x-1)(-3x+5)$$

$$\Phi_{b,2}(x) = \frac{1}{16}(x+1)^3(x-1)^2 \qquad (9.15)$$

9.2 Computation of a one-variable polynomial

9.2.1 Generating expression

The generating form of the Hermite interpolating polynomial is given by equation (7.126), on page 315, which writes

$$\Phi_{e,i}(x) = \frac{1}{i!\,P_e(\alpha)}P_e(x)Q_e(x)f_{e,0}(x) \qquad (9.16)$$

The polynomial to be obtained and its derivatives of order up to $\eta = 2$ have $d = 1$ variable in the domain $[-1, 1]$ for each node in the domain, which are called the reference node.

Since there are two nodes then

1. the polynomials to be obtained related to the reference node $a = -1$ are: $\Phi_{a,0}(x)$, $\Phi_{a,1}(x)$, and $\Phi_{a,2}(x)$. They have the property that

$$\Phi_{a,0}^{(0)}(x)|_{x=-1} = 1, \quad \Phi_{a,1}^{(1)}(x)|_{x=-1} = 1, \quad \text{and} \quad \Phi_{a,2}^{(2)}(x)|_{x=-1} = 1 \tag{9.17}$$

2. the polynomials to be obtained related to the reference node $b = 1$ are: $\Phi_{b,0}(x)$, $\Phi_{b,1}(x)$, and $\Phi_{b,2}(x)$. They have the property that $\Phi_{b,0}^{(0)}(x)|_{x=1} = 1$, $\Phi_{b,1}^{(1)}(x)|_{x=1} = 1$, and $\Phi_{b,2}^{(2)}(x)|_{x=1} = 1$

9.2.2 Generation of the polynomials related to the reference node $a = -1$

9.2.2.1 Construction of the factor $P_a(x)$ and $P_a(\alpha)$

For a normalized domain symmetric with respect to the origin, that is $x \in [-1, 1]$, the polynomial $P_a(x)$ is given by equation (7.33) as

$$P_a(x) = \prod_{i=0}^{d-1} (x_i + \alpha_i)^{\eta+1} = (x + a)^{\eta+1} = (x - 1)^3 \tag{9.18}$$

Since $d = 1$, $\eta = 2$, and $a = -1$ the equation (9.18) gives

$$P_a(x) = (x + a)^{2+1} = (x - 1)^3 \tag{9.19}$$

The factor $P_a(x) = (x-1)^3$ guarantees that the polynomial and its derivatives of order up to $\eta = 2$ will vanish at the boundary $b = 1$, that is, at elements in the set $\mathcal{V}_e$ in the visibility of the reference node $a = -1$.

The factors $P_a(x)$ and $P_a(a)$ are the same for the polynomial and its derivatives related to the reference node $\alpha = a = -1$, that is, for the polynomials $\Phi_{a,0}(x)$, $\Phi_{a,1}(x)$, and $\Phi_{a,2}(x)$. The factor $P_a(a)$ is given by

$$P_a(\alpha)|_{a=-1} = P_e(a + a)^3 = (-1 - 1)^3 = -8 \tag{9.20}$$

9.2.3 Construction of the polynomial $\Phi_{a,0}(x)$

9.2.3.1 Properties of the polynomial $\Phi_{a,0}(x)$

Computation of the polynomial that equals one at the reference node $\alpha = a = -1$ and its first and second derivatives vanish at the same node. The value of this polynomial and its derivatives until order 2 equal to zero at the node $b = 1$. The polynomial with these properties is denoted as

$$\Phi_{a,i}(x) = \frac{1}{i! \, P_a(\alpha)} P_a(x) Q_a(x) f_{a1}(x) \tag{9.21}$$

where $i = 0$, $a_e = \alpha = (-1)$, and $\beta = b = 1$.

9.2.3.2 Construction of the polynomial $Q_{e,i}(x) = Q_{a,0}(x)$

The notation that the polynomial $\Phi_{a,0}(x)$ equals one at the reference node $a = -1$ and its derivatives vanish at the same node is given by $i = 0$. This condition is obtained by the factor $Q_e(x)$ whose expression is shown in equation (7.49) in the theorem 7.2.17, and it is

$$Q_{e,i}(x) = (x_0 - \alpha_0)^{i_0} (x_1 - \alpha_1)^{i_1} \ldots (x_\delta - \alpha_\delta)^{i_\delta} \tag{9.22}$$

therefore,

$$Q_{a,0}(x) = (x - a)^{i_0} = (x - a)^i = (x + 1)^0 = 1 \tag{9.23}$$

9.2.3.3 Construction of the polynomial $f_{a,0}(x)$

The minimum degree of $f_{a,0}(x)$ is given by equation (7.66) in the corollary 7.2.19, on page 305, as

$$q \geq \eta - \lambda_0(i) = 2 - \lambda_0(0) = 2 - 0 = 2 \tag{9.24}$$

and $q = 2$ it gives a polynomial of the form

$$f_{a,0}(x) = a_0 + a_1 x + a_2 x^2 \tag{9.25}$$

The evaluation of the coefficients a_0, a_1, and a_2 must be performed in the reverse order, that is, a_2, a_1, a_0. This is performed taking $\lambda_0(v) = q, q - 1, \ldots, 0$, that is, $\lambda_0(v) = 2, 1, 0$.

9.2.3.4 Computation of the coefficients in the level $\lambda_0(v) = 2$

There is only one coefficient in the level $\lambda_0(v) = 2$, which is a_2 and it can be evaluated using the equation (7.122), which writes

$$a_{(v)} = \frac{1}{v!} \left[E_F^{(v)}(\alpha) \right] \tag{9.26}$$

where $\alpha = a = -1$ and

$$v \in L_{\lambda_0(v)}^\delta = L_{\lambda_0(v)}^{d-1} = L_2^0 = \{(2)\} \tag{9.27}$$

The factor $E_F^{(v)}(\alpha)$ is given by equation (6.189) for the case of $d = 1$ variable as

$$E_F^{(v)}(x)\Big|_{x=\alpha} = E^{(v)}(x)\Big|_{x=\alpha} = \frac{(\eta+v)!}{\eta!}\frac{1}{(\beta-x)^v}\Big|_{x=\alpha} \tag{9.28}$$

which for $\beta = b = 1$, $\alpha = a = -1$, and $v = 2$ it evaluates as

$$E_F^{(2)}(-1) = \frac{(2+2)!}{2!}\frac{1}{(1-(-1))^2} = \frac{4!}{2!}\frac{1}{2^2} = 3 \tag{9.29}$$

Introducing $E_F^{(2)}(-1)$ given by equation (9.29) into equation (9.26) the coefficient a_2 evaluates

$$a_2 = \frac{1}{2!}\left[E_F^{(2)}(-1)\right] = \frac{1}{2!}[3] = \frac{3}{2} \tag{9.30}$$

9.2.3.5   computation of the coefficients in the level $\lambda_0(v) = q - 1 = 1$

The coefficients in the level $\lambda_0(v) = 1$ can be evaluated using the equation (7.123), on page 314, which writes

$$a_{(v)} = \frac{1}{v!}\left[E_F^{(v)}(\alpha) - S_v\left(a_{(u+v)},\alpha\right)\right] \tag{9.31}$$

where $E_F^{(v)}(\alpha)$ is given by equation (9.28) and $S_v\left(a_{(u+v)},\alpha\right)$ is given by equation (7.112) as

$$S_v(a_{(u+v)},\alpha) = \sum_{r_0=1}^{r_0=m}\sum_{r_1=0}^{r_1=r_0}\cdots\sum_{r_{\delta-1}=0}^{r_{\delta-1}=r_{\delta-2}} a_{(u+v)}\frac{(u+v)!}{u!}\alpha_0^{u_0}\alpha_1^{u_1}\cdots\alpha_\delta^{u_\delta} \tag{9.32}$$

1. The numbers v are elements of the set $L_{\lambda_0(v)}^{d-1} = L_1^0 = \{1\}$

2. The degree of the polynomial $S_v(a_{(u+v)},\alpha)$ is given by $m = q - \lambda_0(v) = 2 - 1 = 1$

3. The numbers u are elements of the set $\left(L^\delta(m) - L_0^\delta\right) = L^0(1) - L_1^0 = \{0,1\} - \{0\} = \{1\}$. Since the set $\{1\}$ has only one element then the polynomial $S_v(a_{(u+v)},\alpha)$ has only one term.

a- Computation of $E_F^{(v)}(\alpha)$ for $\lambda_0(v) = 1$

Rewriting the equation (9.28) with $x = \alpha = a = -1$, $\eta = 2$, $v \in \{1\}$, and $\beta = b = 1$ it follows

$$E_F^{(v)}(\alpha) = \frac{(\eta+v)!}{\eta!}\frac{1}{(\beta-\alpha)^v} = \frac{(2+1)!}{2!}\frac{1}{(1-(-1))^1} = \frac{3!}{2!}\frac{1}{2^1} = \frac{3}{2} \tag{9.33}$$

b- Computation of $S_v(a_{(u+v)}, \alpha)$ for $\lambda_0(v) = 1$

With $v \in \{1\}$, $\alpha = a = -1$, and $u \in \{1\}$ the equation (9.32) becomes

$$S_v(a_{(u+v)}, \alpha) = a_{(u+v)} \frac{(u+v)!}{u!} \alpha^u = a_{(1+1)} \frac{(1+1)!}{1!} (\alpha)^1 = \frac{3}{2} \frac{2!}{1!}(-1) = -3 \tag{9.34}$$

c- Computation of the term a_1

Substituting $E_F^{(v)}(\alpha)$ given by equation (9.33) and $S_v(a_{(u+v)}, \alpha)$ given by equation (9.34) into equation (9.31) it gives

$$
\begin{aligned}
a_1 &= \frac{1}{v!}\left[E_F^{(v)}(\alpha) - S_v(a_{(u+v)}, \alpha) \right] \\
&= \frac{1}{1!}\left[E_F^{(1)}(-1) - S_v(a_{(1+1)}, (-1)) \right] = \frac{1}{1!}\left[\frac{3}{2} - (-3) \right] = \frac{9}{2} \tag{9.35}
\end{aligned}
$$

9.2.3.6 Computation of the coefficients in the level $\lambda_0(v) = q - 2 = 0$

1. The numbers v are elements of the set $L_{\lambda_0(v)}^{d-1} = L_0^0 = \{0\}$

2. The degree of the polynomial $S_v(a_{(u+v)}, \alpha)$ is given by $m = q - \lambda_0(v) = 2 - 0 = 2$

3. The numbers u are elements of the set $\left(L^\delta(m) - L_0^\delta \right) = L^0(2) - L_1^0 = \{0, 1, 2\} - \{0\} = \{1, 2\}$. Since the set $\{1, 2\}$ has two element then the polynomial $S_v(a_{(u+v)}, \alpha)$ has two term.

a- Computation of $E_F^{(v)}(\alpha)$ for $\lambda_0(v) = 0$

Rewriting the equation (9.28) with $x = \alpha = a = -1$, $\eta = 2$, $v \in \{0\}$, and $\beta = b = 1$ it follows

$$E_F^{(v)}(\alpha) = \frac{(\eta + v)!}{\eta!} \frac{1}{(\beta - \alpha)^v} = \frac{(2+0)!}{2!} \frac{1}{(1-(-1))^0} = \frac{2!}{2!} \frac{1}{2^0} = 1 \tag{9.36}$$

b- Computation of $S_v(a_{(u+v)}, \alpha)$ for $\lambda_0(v) = 1$

With $v \in \{0\}$, $\alpha = a = -1$, and $u \in \{1, 2\}$ the equation (9.32) becomes

$$
\begin{aligned}
S_v(a_{(u+v)}, \alpha) &= a_{(u_0+v_0)} \frac{(u_0+v_0)!}{u_0!} \alpha^{u_0} + a_{(u_1+v_0)} \frac{(u_1+v_0)!}{u_1!} \alpha^{u_1} \\
&= a_{(1+0)} \frac{(1+0)!}{1!} (\alpha)^1 + a_{(2+0)} \frac{(2+0)!}{2!} (\alpha)^2 \\
&= \frac{9}{2} \frac{1!}{1!}(-1)^1 + \frac{3}{2} \frac{2!}{2!}(-1)^2 = -\frac{9}{2} + \frac{3}{2} = -3 \tag{9.37}
\end{aligned}
$$

c- computation of the term a_0

Substituting $E_F^{(v)}(\alpha)$ given by equation (9.36) and $S_v(a_{(u+v)}, \alpha)$ given by equation (9.37) into equation (9.31) it gives

$$
\begin{aligned}
a_0 &= \frac{1}{v!}\left[E_F^{(v)}(\alpha) - S_v(a_{u+v}, \alpha)\right] \\
&= \frac{1}{0!}\left\{E_F^{(0)}(-1) - [S_0(a_{1+0}, (-1)) + S_0(a_{2+0}, (-1))]\right\} = \frac{1}{0!}[1 - (-3)] = 4
\end{aligned}
\tag{9.38}
$$

9.2.3.7 Polynomial $f_{a,0}(x)$

Introducing the coefficients a_0 given by equation (9.38), a_1 given by equation (9.35), and a_2 given by equation (9.30) into the equation (9.24) it gives the polynomial $f_{a,0}(x)$ as

$$
f_{a,0}(x) = a_0 + a_1 x + a_2 x^2 = \frac{1}{2}\left(8 + 9x + 3x^2\right)
\tag{9.39}
$$

9.2.3.8 Polynomial solution $\Phi_{a,0}(x)$

Substituting $P_a(x)$ given by equation (9.18), $P_a(\alpha)$ given by equation (9.20), $Q_{a,0}(x)$ given by equation (9.23), and $f_{a,0}(x)$ given by the equation (9.39) into the equation (9.16) with $i = 0$ it gives the polynomial for $\Phi_{a,0}(x)$ as

$$
\begin{aligned}
\Phi_{a,0}(x) &= \frac{1}{i!\, P_a(\alpha)}\, P_a(x)\, Q_a(x)\, f_{a2}(x) \\
&= \frac{1}{0!\,(-8)}\,(x-1)^3\,(x+1)^0\,\frac{1}{2}\left(8 + 9x + 3x^2\right) \tag{9.40} \\
&= -\frac{1}{8}(x-1)^3\,\frac{1}{2}\left(8 + 9x + 3x^2\right) \tag{9.41} \\
&= \frac{1}{16}(x-1)^3\left(-8 - 9x - 3x^2\right) \tag{9.42}
\end{aligned}
$$

9.2.4 Construction of the polynomial $\Phi_{a,1}(x)$

9.2.4.1 Properties of the polynomial $\Phi_{a,1}(x)$

Computation of the polynomial whose first derivative equals one at the node $\alpha = a = -1$ and the polynomial and its second derivative equal to zero at the same node. The value of this polynomial and its derivatives until order 2 equal to zero at the node $b = 1$. The polynomial with these properties is denoted as

$$
\Phi_{a,i}(x) = \frac{1}{i!\, P_a(\alpha)} P_a(x) Q_a(x) f_{a1}(x)
\tag{9.43}
$$

where $i = 1$, $a_e = \alpha = -1$, and $\beta = b = 1$.

9.2.4.2 Construction of the polynomial $Q_{e,i}(x) = Q_{a,1}(x)$

The notation that the first derivative of the polynomial $\Phi_{a,1}(x)$ equals one at the reference node $a = -1$ and the polynomial and it second derivatives vanish at the same node is given by $i = 1$. This condition is obtained by the factor $Q_e(x)$ whose expression is shown in equation (7.49) in the theorem 7.2.17, on page 302, and it is

$$Q_{e,i}(x) = (x_0 - \alpha_0)^{i_0} (x_1 - \alpha_1)^{i_1} \dots (x_\delta - \alpha_\delta)^{i_\delta} \tag{9.44}$$

therefore,

$$Q_{a,1}(x) = (x - a)^{i_1} = (x - a)^i = (x + 1)^1 = (x + 1) \tag{9.45}$$

9.2.4.3 Construction of the polynomial $f_{a,1}(x)$

The minimum degree of $f_{a,1}(x)$ is given by equation (7.66) in the corollary 7.2.19, on page 305, as

$$q \geq \eta - \lambda_0(i) = 2 - \lambda_0(1) = 2 - 1 = 1 \tag{9.46}$$

and $q = 1$ it gives a polynomial of the form

$$f_{a,1}(x) = a_0 + a_1 x \tag{9.47}$$

The evaluation of the coefficients a_0, and a_1 must be performed in the reverse order, that is, a_1, a_0. This is performed taking $\lambda_0(v) = q, q - 1, \dots, 0$, that is, $\lambda_0(v) = 1, 0$.

9.2.4.4 computation of the coefficients in the level $\lambda_0(v) = 1$

There is only one coefficient in the level $\lambda_0(v) = 1$, which is a_1 and it can be evaluated using the equation (7.122), which writes

$$a_{(v)} = \frac{1}{v!} \left[E_F^{(v)}(\alpha) \right] \tag{9.48}$$

where $\alpha = a = -1$ and

$$v \in L_{\lambda_0(v)}^\delta = L_{\lambda_0(v)}^{d-1} = L_1^0 = \{(1)\} \tag{9.49}$$

The factor $E_F^{(v)}(\alpha)$ is given by equation (6.189) for the case of $d = 1$ variable as

$$E_F^{(v)}(x)\Big|_{x=\alpha} = E^{(v)}(x)\Big|_{x=\alpha} = \frac{(\eta + v)!}{\eta!} \frac{1}{(\beta - x)^v}\Big|_{x=\alpha} \tag{9.50}$$

which for $\beta = b = 1$, $\alpha = a = -1$, and $v = 1$ it evaluates as

$$E_F^{(1)}(-1) = \frac{(2+1)!}{2!} \frac{1}{(1 - (-1))^1} = \frac{3!}{2!} \frac{1}{2^1} = \frac{3}{2} \tag{9.51}$$

Introducing $E_F^{(1)}(-1)$ given by equation (9.51) into equation (9.48) the coefficient a_1 evaluates

$$a_1 = \frac{1}{1!} \left[E_F^{(1)}(-1) \right] = \frac{1}{1!} \begin{bmatrix} 3 \\ 2 \end{bmatrix} = \frac{3}{2} \tag{9.52}$$

9.2.4.5 Computation of the coefficients in the level $\lambda_0(v) = q - 1 = 0$

1. The numbers v are elements of the set $L_{\lambda_0(v)}^{d-1} = L_0^0 = \{0\}$

2. The degree of the polynomial $S_v(a_{(u+v)}, \alpha)$ is given by $m = q - \lambda_0(v) = 1 - 0 = 1$

3. The numbers u are elements of the set $\left(L^\delta(m) - L_0^\delta \right) = L^0(1) - L_1^0 = \{0, 1\} - \{0\} = \{1\}$. Since the set $\{1\}$ has only one element then the polynomial $S_v(a_{(u+v)}, \alpha)$ has only one term.

a- Computation of $E_F^{(v)}(\alpha)$ for $\lambda_0(v) = 0$

Rewriting the equation (9.53) with $x = \alpha = a = -1$, $\eta = 2$, $v \in \{0\}$, and $\beta = b = 1$ it follows

$$E_F^{(v)}(\alpha) = \frac{(\eta + v)!}{\eta!} \frac{1}{(\beta - \alpha)^v} = \frac{(2+0)!}{2!} \frac{1}{(1-(-1))^0} = \frac{2!}{2!} \frac{1}{1} = 1 \tag{9.53}$$

b- Computation of $S_v(a_{(u+v)}, \alpha)$ for $\lambda_0(v) = 1$

With $v \in \{0\}$, $\alpha = a = -1$, and $u \in \{1\}$ the equation (9.32) becomes

$$S_v(a_{(u+v)}, \alpha) = a_{(u+v)} \frac{(u+v)!}{u!} \alpha^u = a_{(1+0)} \frac{(1+0)!}{1!} (\alpha)^1 = \frac{3}{2} \frac{1!}{1!} (-1)^1 = -\frac{3}{2} \tag{9.54}$$

c- computation of the term a_1

Substituting $E_F^{(v)}(\alpha)$ given by equation (9.33) and $S_v(a_{(u+v)}, \alpha)$ given by equation (9.54) into equation (9.31) it gives

$$a_0 = \frac{1}{v!} \left[E_F^{(v)}(\alpha) - S_v(a_{(u+v)}, \alpha) \right]$$

$$= \frac{1}{0!} \left[E_F^{(0)}(-1) - S_0(a_{(1+0)}, (-1)) \right] = \frac{1}{0!} \left[1 - \left(-\frac{3}{2} \right) \right] = \frac{5}{2} \tag{9.55}$$

9.2.4.6 Polynomial $f_{a,1}(x)$

Introducing the coefficients a_0 given by equation (9.55), and a_1 given by equation (9.52), into the equation (9.46) it gives the polynomial $f_{a,1}(x)$ as

$$f_{a,1}(x) = a_0 + a_1 x = \frac{1}{2}(5 + 3x) \tag{9.56}$$

9.2.4.7 Polynomial solution $\Phi_{a,1}(x)$

Substituting $P_a(x)$ given by equation (9.18), $P_a(\alpha)$ given by equation (9.20), $Q_{a,0}(x)$ given by equation (9.23), and $f_{a,0}(x)$ given by the equation (9.56) into the equation (9.16) with $i = 1$ it gives the polynomial for $\Phi_{a,1}(x)$ as

$$\Phi_{a,1}(x) = \frac{1}{i!\, P_a(\alpha)} P_a(x) Q_a(x) f_{a2}(x)$$

$$= \frac{1}{1!\,(-8)} (x-1)^3 (x+1)^1 \, \frac{1}{2} (5+3x) \tag{9.57}$$

$$= -\frac{1}{16} (x-1)^3 (x+1)^1 (5+3x) \tag{9.58}$$

$$= \frac{1}{16} (x-1)^3 (x+1) (-5 - 3x) \tag{9.59}$$

9.2.5 Construction of the polynomial $\Phi_{a,2}(x)$

9.2.5.1 Properties of the polynomial $\Phi_{a,2}(x)$

Computation of the polynomial whose second derivative equals one at the node $\alpha = a = -1$ and the polynomial and its first derivative equal to zero at the same node. The value of this polynomial and its derivatives until order 2 equal to zero at the node $b = 1$. The polynomial with these properties is denoted as

$$\Phi_{a,i}(x) = \frac{1}{i!\, P_a(\alpha)} P_a(x) Q_a(x) f_{a1}(x) \tag{9.60}$$

where $i = 2$, $a_e = \alpha = -1$, and $\beta = b = 1$.

9.2.5.2 Construction of the polynomial $Q_{e,i}(x) = Q_{a,2}(x)$

The notation that the first derivative of the polynomial $\Phi_{a,2}(x)$ equals one at the reference node $a = -1$ and the polynomial and it second derivatives vanish at the same node is given by $i = 2$. This condition is obtained by the factor $Q_e(x)$ whose expression is shown in equation (7.49) in the theorem 7.2.17, on page 302, and it is

$$Q_{e,i}(x) = (x_0 - \alpha_0)^{i_0} (x_1 - \alpha_1)^{i_1} \ldots (x_\delta - \alpha_\delta)^{i_\delta} \tag{9.61}$$

therefore,

$$Q_{a,2}(x) = (x-a)^{i_1} = (x-a)^i = (x+1)^2 \tag{9.62}$$

9.2.5.3 Construction of the polynomial $f_{a,2}(x)$

The minimum degree of $f_{a,2}(x)$ is given by equation (7.66) in the corollary 7.2.19, on page 305, as

$$q \geq \eta - \lambda_0(i) = 2 - \lambda_0(2) = 2 - 2 = 0 \tag{9.63}$$

and $q = 0$ it gives a polynomial of the form

$$f_{a,2}(x) = a_0 \tag{9.64}$$

The evaluation of the coefficients a_0 must be performed in the reverse order, that is, a_0. This is performed taking $\lambda_0(v) = q, q-1, \ldots, 0$, that is, $\lambda_0(v) = 0$.

9.2.5.4   Computation of the coefficients in the level $\lambda_0(v) = 0$

There is only one coefficient in the level $\lambda_0(v) = 0$, which is a_0 and it can be evaluated using the equation (7.122), on page 314, which writes

$$a_{(v)} = \frac{1}{v!} \left[E_F^{(v)}(\alpha) \right] \tag{9.65}$$

where $\alpha = a = -1$ and

$$v \in L_{\lambda_0(v)}^{\delta} = L_{\lambda_0(v)}^{d-1} = L_0^0 = \{(0)\} \tag{9.66}$$

The factor $E_F^{(v)}(\alpha)$ is given by equation (6.189) for the case of $d = 1$ variable as

$$E_F^{(v)}(x) \Big|_{x=\alpha} = E^{(v)}(x) \Big|_{x=\alpha} = \frac{(\eta + v)!}{\eta!} \frac{1}{(\beta - x)^v} \Big|_{x=\alpha} \tag{9.67}$$

which for $\beta = b = 1$, $\alpha = a = -1$, and $v = 0$ it evaluates as

$$E_F^{(0)}(-1) = \frac{(2+0)!}{2!} \frac{1}{(1-(-1))^0} = \frac{2!}{2!} \frac{1}{2^0} = 1 \tag{9.68}$$

Introducing $E_F^{(0)}(-1)$ given by equation (9.68) into equation (9.65) the coefficient a_0 evaluates

$$a_1 = \frac{1}{0!} \left[E_F^{(0)}(-1) \right] = \frac{1}{0!} [1] = 1 \tag{9.69}$$

9.2.5.5   Polynomial $f_{a,2}(x)$

Introducing the coefficients a_0 given by equation (9.55), into the equation (9.63) it gives the polynomial $f_{a,2}(x)$ as

$$f_{a,2}(x) = a_0 = 1 \tag{9.70}$$

9.2.5.6   Polynomial solution $\Phi_{a,2}(x)$

Substituting $P_a(x)$ given by equation (9.18), $P_a(\alpha)$ given by equation (9.20), $Q_{a,0}(x)$ given by equation (9.61), and $f_{a,0}(x)$ given by the equation (9.70) into the equation (9.16) with $i = 2$ it gives the polynomial for $\Phi_{a,2}(x)$ as

$$\Phi_{a,2}(x) = \frac{1}{i!\,P_a(\alpha)} P_a(x) Q_a(x) f_{a2}(x)$$

$$= \frac{1}{2!\,(-8)} (x-1)^3 (x+1)^2 \,(1) \tag{9.71}$$

$$= -\frac{1}{16} (x-1)^3 (x+1)^2 \tag{9.72}$$

9.2.6 Generation of the polynomials related to the reference node $b = 1$

There is no need in performing the computation of the polynomials related to the reference node $b = 1$ it is enough to apply the theorem 7.4.1, on page 315, to the coefficients of the polynomials and make the appropriate corrections to the other factors.

9.2.7 Polynomial $\Phi_{b,0}(x)$

Computation of the polynomial whose value equals one at the node $b_e = (1)$ and whose derivatives until order 2 equal to zero at the same node. The value of this polynomial and its derivatives until order 2 equal to zero at the node $a_e = (-1)$.

Using the same notation as in the theorem 7.4.1 the data related to the node a are $\alpha = (\alpha_0) = (-1)$, $\beta = (\beta_0) = (1)$, and related to the node b are $\gamma = (-\alpha_0) = (1)$ and $\delta = (\delta_0) = (-\beta) = (-1)$.

The polynomial $\Phi_{a,0}(x)$ for the node a is given by equation (9.40) and it is

$$\Phi_{ai}(x) = \frac{1}{2^3} (x-1)^3 \, \frac{1}{2} \left(-8 - 9x - 3x^2 \right) \tag{9.73}$$

Theorem 7.4.1 gives the following relation for the coefficients of the polynomial $f_{a,i}(x)$ and $f_{b,i}(x)$ in the equation (7.131)

$$b_v = (-1)^{v_k} a_v \tag{9.74}$$

then

$$b_2 = (-1)^2 a_2 = (-1)^2 (-3) = -3 \tag{9.75}$$

$$b_1 = (-1)^1 a_1 = (-1)^1 (-9) = 9 \tag{9.76}$$

$$b_0 = (-1)^0 a_0 = (-1)^0 (-8) = -8 \tag{9.77}$$

therefore,

$$f_{b,i}(x) = 8 + 9x - 3x^2 \tag{9.78}$$

The constant factor changes according to the equation (7.132) in the theorem 7.4.1 as

$$P_b(\gamma) = (-1)^{\eta+1} P_a(\alpha) \tag{9.79}$$

$$P_b(1) = (-1)^{2+1} P_a(-1) = -\frac{1}{2^3} \tag{9.80}$$

Substituting the results in equations (9.79) and (9.75) into equation (9.73) it gives

$$\Phi_{bi}(x) = \frac{1}{16}(x-1)^3 \left(8 - 9x + 3x^2\right) \tag{9.81}$$

9.2.8   Polynomial $\Phi_{b,1}(x)$

The polynomial $\Phi_{a,1}(x)$ for the node a is given by equation (9.57) and it is

$$\Phi_{a,i}(x) = -\frac{1}{2^3}(x-1)^3(x+1)^1 \frac{1}{2}(5+3x) \tag{9.82}$$

The transformations of the coefficients of the polynomial $f_{a,1}(x)$ are given by

$$b_1 = (-1)^1 a_1 = -3 \tag{9.83}$$

$$b_0 = (-1)^0 a_0 = 5 \tag{9.84}$$

and the transformation of the constant factor is

$$P_b(\gamma) = (-1)^{\eta+1} P_a(\alpha) \tag{9.85}$$

$$P_b(1) = (-1)^{2+1} P_a(-1) = -\frac{1}{2^3} \tag{9.86}$$

Substituting

$$\Phi_{b,i}(x) = \frac{1}{16}(x-1)^3(x+1)^1 (5-3x) \tag{9.87}$$

9.2.9   Polynomial $\Phi_{b,2}(x)$

The polynomial $\Phi_{a,1}(x)$ for the node a is given by equation (9.71) and it is

$$\Phi_{a,i}(x) = -\frac{1}{2^3}(x-1)^3(x+1)^2 \frac{1}{2} \tag{9.88}$$

and the constant factor is

$$P_b(\gamma) = (-1)^{\eta+1} P_a(\alpha) \tag{9.89}$$

$$P_b(1) = (-1)^{2+1} P_a(-1) = -\frac{1}{2^3} \tag{9.90}$$

Substituting

$$\Phi_{b,i}(x) = \frac{1}{16}(x-1)^3 (x+1)^2 \tag{9.91}$$

9.3 Comparison between the two techniques

9.3.1 Classical approach

1. Requires the construction and solution of an algebraic system of linear equations.

2. Requires the algebraic manipulation of the solution to construct the polynomials.

3. To obtain a canonical form it is necessary to factor the resultant polynomials.

4. The complexity of the solution increases with the increasing degree of the polynomials and/or the increasing of the number of variables.

9.3.2 The present approach

1. There is no need for the construction and solution of an algebraic system of linear equations.

2. The algorithm gives the polynomials in their canonical form

3. There is no need for any algebraic manipulation. The algorithm is developed in such that all algebraic manipulations are performed implicitly.

4. Increasing the degree and/or number of variables has small effect in the increasing of the complexity of the solution.

5. The algorithm is suitable for computer programming for any number of variables and/or any degree of the polynomials since it does not require any algebraic manipulation.

6. The computational work is reduced:

   (a) the use of several properties permits to obtain almost half of the polynomials in terms of the others for the case of a domain symmetric with respect to the coordinate axis

   (b) the use of coefficient properties permits to obtain almost half of the coefficients as a functions of the others

7. The solution of nonlinear problems is reduced to the solution of a non-linear system of algebraic equations whose solution may not be easy to find.

Chapter 10

Normalized symmetric square domain

10.1 Computation of a two-variable polynomial

Computation of the Hermite Interpolating Polynomial $\Phi_{e,n}(x)$ related to the reference nodes $a_0 = (-1,-1)$, $a_1 = (1,-1)$, $a_2 = (-1,1)$, and $a_3 = (1,1)$. The domain is a square symmetric with respect to the coordinate axes and defined by $\Omega = [-1,1] \times [-1,1]$. The maximum order of derivative is $\eta = 4$ and the number of variables is $d = 2$. It will be performed a detailed computation for the polynomial related to the reference node $a_0 = (-1,-1)$. The polynomials related to the others can be derived using the theorem 7.4.7 on page 324.

10.2 Generation of the polynomial $\Phi_{a_0,00}(x_0, x_1)$ related to the reference node $a_0 = (-1,-1)$

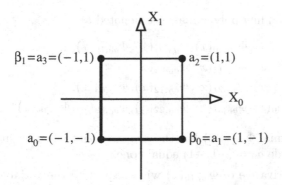

Figure 10.2.1 *Reference node a_0 in a symmetric normalized domain.*

10.2.1   Elements common to all polynomials related to the node $a_0 = (-1, -1)$

a- Characterization of the polynomials

1. Type of domain: a unit square symmetric with respect to the coordinate axes.

2. Number of variables: $d = 2$.

3. Maximum order of derivative: $\eta = 4$

4. Reference node: $a_0 = \alpha = (-1, -1)$

5. Nodes in the visibility of the reference node

   The elements of the visibility set V of the reference node a_0 are the planes $a_1 - a_2$ and $a_2 - a_3$. These planes are identified by the orthogonal distance from them to the axis to, which they are parallel. Their expression can be obtained from the equation (7.28), on page 295, where

$$\beta_0 = a_1 = (\beta_{0,0}, \beta_{0,1}) = (1, -1) \qquad \text{and}$$
$$\beta_1 = a_3 = (\beta_{1,0}, \beta_{1,1}) = (-1, 1) \tag{10.1}$$

6. The order of derivative of the Hermite Interpolating Polynomials related to the reference node are the constrained numbers in the set $L^\delta(\eta) = L^{d-1}(\eta) = L^1(4)$, which are

$$L^1(4) = \Big\{ (0,0), (1,0), (0,1), (2,0), (1,1), (0,2),$$
$$(3,0), (2,1), (1,2), (0,3), (4,0), (3,1), (2,2), (1,3), (0,4) \Big\} \tag{10.2}$$

The corresponding polynomials are denoted as

$$\Phi_{a_0,00}(x), \Phi_{a_0,10}(x), \Phi_{a_0,01}(x), \Phi_{a_0,10}(x),$$
$$\Phi_{a_0,20}(x), \Phi_{a_0,11}(x), \Phi_{a_0,02}(x),$$
$$\Phi_{a_0,30}(x), \Phi_{a_0,21}(x), \Phi_{a_0,12}(x), \Phi_{a_0,03}(x),$$
$$\Phi_{a_0,40}(x), \Phi_{a_0,31}(x), \Phi_{a_0,22}(x), \Phi_{a_0,13}(x), \Phi_{a_0,04}(x) \tag{10.3}$$

The polynomial $\Phi_{a_0,00}(x)$ has the property that when evaluated at the reference node $a_0 = (-1, -1)$ equals one.

The first derivative of $\Phi_{a_0,10}(x)$ with respect to the variable x_0 equals one at the reference node $a_0 = (-1, -1)$.

The cross derivative of $\Phi_{a_0,11}(x)$ with respect to the variables x_0 and x_1 equals one at the reference node $a_0 = (-1, -1)$, etc.

7. Number of polynomials related to a reference node

The number $n^\delta(\eta)$ of polynomials is given by the number of elements in the set $L^\delta(\eta) = L^1(4)$ whose expression is given by (4.49), on page 149, in the chapter 4

$$n^1(4) = \frac{(\eta+1)(\eta+2)}{2!} = \frac{5 \times 6}{2} = 15 \tag{10.4}$$

b- Common factors

Factors that are common to all polynomials related to the reference node $a_0 = (-1, -1)$

The polynomial in equation (7.31), on page 297, writes

$$\begin{aligned}
P_{a_0}(x) &= [\varphi_{a_0,0}(x_0, x_1)]^{\eta+1} [\varphi_{a_0,1}(x_0, x_1)]^{\eta+1} \\
&= (x_0 - \beta_{0,0})^{\eta+1}(x_1 - \beta_{1,1})^{\eta+1} \\
&= (x_0 - 1)^5 (x_1 - 1)^5
\end{aligned} \tag{10.5}$$

The constant term $P_{a_0}(\alpha)$ is given by

$$\begin{aligned}
P_{a_0}(x)\Big|_{x=\alpha} &= P_{a_0}(\alpha_0, \alpha_1) \\
&= (\alpha_0 - \beta_{0,0})^{\eta+1}(\alpha_1 - \beta_{1,1})^{\eta+1} \\
&= (-1-1)^5(-1-1)^5 = (-2)^{10} = 2^{10}
\end{aligned} \tag{10.6}$$

where $\alpha = a_0 = (-1, -1)$.

c- Expression for the polynomial $Q_{a_0,n}(x)$

The expression for the polynomial $Q_{a_0,n}(x)$ in equation (7.49), on page 302, for two variables writes

$$Q_{a_0,n}(x) = (x_0 - \alpha_0)^{n_0} (x_1 - \alpha_1)^{n_1} \tag{10.7}$$

d- Expression for the polynomial $\Phi_{a_0,n}(x)$

The expression for the polynomial is given by equation (7.126), on page 315, which writes

$$\Phi_{a_0,n}(x) = \frac{1}{n! P_{a_0}(\alpha)} P_{a_0}(x) Q_{a_0,n}(x) f_{a_0,n}(a, x) \tag{10.8}$$

where $n \in L^2(4)$.

10.2.2   Generation of the polynomial $\Phi_{a_0,00}(x_0, x_1)$

a- Properties of the polynomial $\Phi_{a_0,00}(x_0, x_1)$

1. Properties of the polynomial $\Phi_{a_0,00}(x_0, x_1)$

   (a) it equals one when evaluated at the reference node $a_0 = (-1, -1)$, that is,

   $$\Phi_{a_0,00}(x) \Big|_{x=a_0} = 1 \qquad (10.9)$$

   (b) it vanishes at the boundaries that do not intercept the reference node, which are the lines a_1-a_2 and a_2-a_3 where $a_1 = (1, -1)$, $a_2 = (1, 1)$, and $a_3 = (-1, 1)$. Therefore, the property

   $$\Phi_{a_0,00}(x) \Big|_{x=a_1} = \Phi_{a_0,00}(x) \Big|_{x=a_2} = \Phi_{a_0,00}(x) \Big|_{x=a_3} = 0 \quad (10.10)$$

2. Properties of the derivatives of the polynomial $\Phi_{a_0,00}(x_0, x_1)$

   (a) it has its derivatives until order $\eta = 4$ equal to zero at the reference node $a_0 = (-1, -1)$

   $$\Phi_{a_0,00}^{(k)}(x) \Big|_{x=a_0} = 0 \qquad k = 1, 2, 3, 4 \qquad (10.11)$$

   (b) it has the values of the derivatives until order $\eta = 4$ equal to zero at the boundaries that do not intercept the reference node, that is

   $$\Phi_{a_0,00}^{(k)}(x) \Big|_{x=a_1} = 0 \qquad \Phi_{a_0,00}^{(k)}(x) \Big|_{x=a_2} = 0$$

   $$\Phi_{a_0,00}^{(k)}(x) \Big|_{x=a_3} = 0 \qquad k = 1, 2, 3, 4 \qquad (10.12)$$

b- Construction of the constant term $n! P_{a_0}(\alpha)$

The factor $P_{a_0}(\alpha)$ was obtained in equation (10.6) and only $n!$ in equation (10.8) changes with the polynomial then, for $n = (0, 0)$, it follows that

$$n! P_{a_0}(\alpha) = (n_0! n_1!) P_{a_0}(\alpha) = (0! 0!)(-2)^{10} = 2^{10} \qquad (10.13)$$

c- Construction of the polynomial $Q_{a_0,n}(x)$

From equation (10.7) it follows, for $n = (0, 0)$, that

$$\begin{aligned} Q_{a_0,n}(x) &= (x_0 - (-1))^{n_0} (x_1 - (-1))^{n_1} \\ &= (x_0 + 1)^{n_0} (x_1 + 1)^{n_1} \\ &= (x_0 + 1)^0 (x_1 + 1)^0 = 1 \end{aligned} \qquad (10.14)$$

d- Expression for the polynomial $\Phi_{a_0,00}(x_0, x_1)$

Substituting $P_{a_0}(x)$ given by equation (10.5), $n!P_{a_0}(\alpha)$ given by equation (10.13), and $Q_{a_0,n}(x)$ given by equation (10.14) into equation (10.8) the polynomial $\Phi_{e,n}(x)$ writes

$$\Phi_{a_0,00}(x) = \frac{1}{n!P_{a_0}(\alpha)} P_{a_0}(x) \, Q_{a_0,n}(x) \, f_{a_0,n}(a, x)$$

$$= \frac{1}{2^{10}} \left[(x_0 - 1)^5 (x_2 - 1)^5 \right] (1) f_{a_0,00}(x_0, x_1) \tag{10.15}$$

10.2.3 Computation of the coefficients of the polynomial $f_{a_0,00}(x)$

a- minimum degree of the polynomial $f_{a_0,00}(a, x)$

From corollary 7.2.19, on page 305, it follows

$$q \geq \eta - \lambda_0(n) = 4 - \lambda_0(0,0) = 4 - 0 = 4 \tag{10.16}$$

where $\lambda_0(n)$ is the value of the level of the constrained number $n = (0,0)$, which is given by the definition 1.1.12, on page 9, and computed as

$$\sum_{j=0}^{j=d-1} n_j = \sum_{j=0}^{j=1} n_j = n_0 + n_1 = 0 + 0 = 0 \tag{10.17}$$

Then the equation (10.16) reduces to

$$q \geq \eta = 4 \tag{10.18}$$

A polynomial in two variables x_0, x_1 and of degree four writes

$$f_{a_0,00}(x) = a_{0,0}x_0^0 x_1^0 + a_{1,0}x_0^1 x_1^0 + a_{0,1}x_0^0 x_1^1 + a_{2,0}x_0^2 x_1^0 + a_{1,1}x_0^1 x_1^1 + a_{0,2}x_0^0 x_1^2$$
$$+ a_{3,0}x_0^3 x_1^0 + a_{2,1}x_0^2 x_1^1 + a_{1,2}x_0^1 x_1^2 + a_{0,3}x_0^0 x_1^3$$
$$+ a_{4,0}x_0^4 x_1^0 + a_{3,1}x_0^3 x_1^1 + a_{2,2}x_0^2 x_1^2 + a_{1,3}x_0^1 x_1^3 + a_{0,4}x_0^0 x_1^4 \tag{10.19}$$

The notation $a_{(w)}$, used later on, refers to the coefficients of the above polynomial.

b- Computation of the coefficients that belong to the terms in the level $\lambda_0(v) = q - 0 = 4$

1. Coefficients to be evaluated, which belong to the level $\lambda_0(v) = 4$.

   The indices of the coefficients of the terms of a polynomial elements v in the level $\lambda_0(v) = q = 4$ with two coordinates are the elements of the set L_q^δ of constrained numbers in the one-dimensional space given by

$L_4^1 = \{(4,0), (3,1), (2,2), (1,3), (0,4)\}$, which means that $v \in L_4^1$. The elements of the set L_ℓ^δ can obtained using the algorithm described in subsection 2.4.4. A particular case of this algorithm for computation of the elements of the set L_q^1 of a one-dimensional constrained numbers is shown in the equation (2.80), on page 55. Therefore, the coefficients to be evaluated in this level are $a_{(4,0)}$, $a_{(3,1)}$, $a_{(2,1)}$, $a_{(1,3)}$, $a_{(0,4)}$ as can be seen from the equation (10.19).

The domain for this problem is a square symmetric with respect to the coordinate axes as shown in figure 10.2.1, on page 355. This allows the use of the theorem 7.4.7, on page 324, with respect to the permutation of the indices and evaluate three out of $a_{(4,0)}$, $a_{(3,1)}$, $a_{(2,2)}$, $a_{(1,3)}$, and $a_{(0,4)}$. Knowing that $a_{(4,0)} = a_{(0,4)}$ and $a_{(3,1)} = a_{(1,3)}$ one can choose $a_{(4,0)}, a_{(3,1)}$, and $a_{(2,2)}$ as the coefficients to be calculated.

2. Choice of the equation to be used for the computation of the coefficients.

   The coefficients of $f_{e,n}(a, x)$ are obtained using equation (7.122) or (7.123), on page 314, which is

   $$a_{(v)} = \frac{1}{v!} E_F^{(v)}(\alpha) = \frac{1}{v_0! \, v_1!} E_F^{(v)}(\alpha) \qquad \forall v \in L_{\lambda_0(v)}^\delta \tag{10.20}$$

   where

   $$v \in L_{\lambda_0(v)}^\delta = L_4^1 = \{(4,0), (3,1), (2,2), (1,3), (0,4)\} \tag{10.21}$$

   From (7.87),

   $$E_F^{(v)}(\alpha) = E_F^{(v_0)}(\alpha) E_F^{(v_1)}(\alpha)$$
   $$= \left[\frac{(\eta + v_0)!}{\eta!} \frac{1}{(\beta_{00} - \alpha_0)^{v_0}} \right] \left[\frac{(\eta + v_1)!}{\eta!} \frac{1}{(\beta_{11} - \alpha_1)^{v_1}} \right] \tag{10.22}$$

   Equation (10.1) on page 356 shows that $\beta_0 = (\beta_{00}, \beta_{01}) = (1, -1)$ and $\beta_1 = (\beta_{10}, \beta_{11}) = (-1, 1)$, therefore, $\beta_{00} = 1$ and $\beta_{11} = 1$. The reference node is $\alpha = (\alpha_0, \alpha_1) = (-1, -1)$, thus $\alpha_0 = -1$ and $\alpha_1 = -1$.

3. Computation of the coefficients $a_{(w)} = a_{(u+v)}$ using equation (10.21).

   For the level $\lambda_0(v) = q = 4$ the values of v are the elements of the set L_4^1 and $u = (0,0)$, therefore, for $v = (4,0)$ it follows

   $$a_{(4,0)} = \frac{1}{4! \, 0!} \left[\frac{(4+4)!}{4!} \frac{1}{(1+1)^4} \right] \left[\frac{(4+0)!}{4!} \frac{1}{(1+1)^0} \right] = \frac{70}{2^4} \tag{10.23}$$

   where $w = (v+u) = (4,0)+(0,0) = (4,0)$, which identifies the coefficient as being $a_{(w)} = a_{(v+u)} = a_{(4,0)}$.

For $v = (3, 1)$ it follows

$$a_{(3,1)} = \frac{1}{3!1!} \left[\frac{(4+3)!}{4!} \frac{1}{(1+1)^3} \right] \left[\frac{(4+1)!}{4!} \frac{1}{(1+1)^1} \right] = \frac{175}{2^4} \quad (10.24)$$

where $w = (v + u) = (3, 1) + (0, 0) = (3, 1)$.

For $v = (2, 2)$ it follows

$$a_{(2,2)} = \frac{1}{2!2!} \left[\frac{(4+2)!}{4!} \frac{1}{(1+1)^2} \right] \left[\frac{(4+2)!}{4!} \frac{1}{(1+1)^2} \right] = \frac{225}{2^4} \quad (10.25)$$

where $w = (v + u) = (2, 2) + (0, 0) = (2, 2)$. The remaining coefficients are given by

$$a_{(1,3)} = a_{(3,1)} = \frac{175}{2^4} \quad \text{and} \quad a_{(0,4)} = a_{(4,0)} = \frac{70}{2^4} \quad (10.26)$$

with the aid of theorem 7.4.7, on page 324, for the equality of the coefficients if the indices are permutation or each other.

c- Computation of the coefficients that belong to the terms in the level $\lambda_0(v) = q - 1 = 3$

1. Coefficients to be evaluated, belong to the level $\lambda_0 = 3$.

   The elements in the level $\lambda_0(v) = 3$ with two coordinates are the constrained numbers in the one-dimensional space given by

   $$L_{\lambda_0(v)}^{\delta} = L_3^1 = \{ (3, 0), (2, 1), (1, 2), (0, 3) \} \quad (10.27)$$

   The elements of the set L_3^1 can obtained using the algorithm described in subsection 2.4.4, on page 45. Therefore, the coefficients $a_{(w)} = a_{(v)}$ to be evaluated in this level are $a_{(3,0)}$, $a_{(2,1)}$, $a_{(1,2)}$, and $a_{(0,3)}$.

   Using the theorem 7.4.7 with respect to the permutation of the indices it follows that $a_{(3,0)} = a_{(0,3)}$ and $a_{(2,1)} = a_{(1,2)}$.

2. Equation for the computation of the coefficients.

   The coefficients are obtained using equation (7.123), on page 314, which is

   $$a_{(v)} = \frac{1}{v!} \left[E_F^{(v)}(\alpha) - S_v \left(a_{u+v}, \alpha \right) \right] \qquad \lambda_0(v) = 3 \quad (10.28)$$

3. Computation of the term $S_v \left(a_{u+v}, \alpha \right)$

   The degree of the polynomial $S_v \left(a_{u+v}, \alpha \right)$ is given by $m = q - \lambda_0(v) = 4 - 3 = 1$ where $\lambda_0(v) = v_0 + v_1 = 3 + 0 = 3$. The exponents of the variables in each term of $S_v(a_{u+v}, \alpha)$ are the elements of the set shown in equation (7.124), on page 314, which is

$$u \in (L^\delta(m) - L_0^\delta) = (L^1(1) - L_0^1)$$
$$= \{(0,0), (1,0), (0,1)\} - \{(0,0)\}$$
$$= \{(1,0), (0,1)\} \tag{10.29}$$

which permits to obtain the indices of the coefficients of the terms of the polynomial $S_v(a_{u+v}, \alpha)$ as follows

(a) For $v = (3,0)$

If $u = (1,0)$ then $v+u = (3,0)+(1,0) = (4,0)$ then $a_{(v+u)} = a_{(4,0)}$, which is one of the coefficients obtained for the level $\lambda_0(v) = q = 4$.

If $u = (0,1)$ then $v+u = (3,0)+(0,1) = (3,1)$ then $a_{(v+u)} = a_{(3,1)}$, which is one of the coefficients obtained for the level $\lambda_0(v) = q = 4$.

(b) For $v = (2,1)$

If $u = (1,0)$ then $v+u = (2,1)+(1,0) = (3,1)$ then $a_{(v+u)} = a_{(3,1)}$, which is one of the coefficients obtained for the level $\lambda_0(v) = q = 4$.

If $u = (0,1)$ then $v+u = (2,1)+(0,1) = (2,1)$ then $a_{(v+u)} = a_{(2,2)}$, which is one of the coefficients obtained for the level $\lambda_0(v) = q = 4$.

The expression for $S_v(a_{u+v}, \alpha)$ is given by equation (7.112). For $d = 2$ and $m = 1$ it writes

$$S_v(a_{u+v}, \alpha) = \sum_{r_0=1}^{r_0=m} \sum_{r_1=0}^{r_1=r_0} a_{(v+u)} \frac{(v+u)!}{u!} \alpha_0^{u_0} \alpha_1^{u_1}$$

$$= \sum_{r_0=1}^{r_0=1} \sum_{r_1=0}^{r_1=r_0} a_{(v+u)} \frac{(v+u)!}{u!} \alpha_0^{u_0} \alpha_1^{u_1}$$

$$= a_{(v+(1,0))} \frac{(v+(1,0))!}{(1,0)!} \alpha_0^1 \alpha_1^0 + a_{(v+(0,1))} \frac{(v+(0,1))!}{(0,1)!} \alpha_0^0 \alpha_1^1 \tag{10.30}$$

Evaluating the polynomial $S_v(a_{u+v}, \alpha)$ at the reference node $\alpha = (-1, -1)$ for $a_{(v)} = a_{(3,0)}$ and $a_{(v)} = a_{(2,1)}$ it follows

(a) For $v = (3,0)$ the equation (10.30) expands as

$$S_{(3,0)}(a_{u+v}, \alpha) = a_{(4,0)} \frac{(4,0)!}{(1,0)!} (-1)^1 (-1)^0 + a_{(3,1)} \frac{(3,1)!}{(0,1)!} (-1)^0 (-1)^1 \tag{10.31}$$

where $(4,0)! = 4!0! = 24$, $(3,1)! = 3!1! = 6$, $(1,0)! = 1!0! = 1$, and $(0,1)! = 0!1! = 1$. Substituting into equation (10.31) it gives

$$S_{(3,0)}(a_{u+v}, \alpha) = -\frac{70}{2^4} \frac{24}{1} - \frac{175}{2^4} \frac{6}{1}$$

$$= -\frac{1}{2^4}(1680 + 1050) = -\frac{2730}{2^4} \tag{10.32}$$

(b) For $v = (2,1)$ the equation (10.30) expands as

$$S_{(2,1)}(a_{u+v}, \alpha) = a_{(3,1)} \frac{(3,1)!}{(1,0)!} (-1)^1 (-1)^0 + a_{(2,2)} \frac{(2,2)!}{(0,1)!} (-1)^0 (-1)^1$$

(10.33)

where $(3,1)! = 3!1! = 6$, $(2,2)! = 2!2! = 4$, $(1,0)! = 1!0! = 1$, and $(0,1)! = 0!1! = 1$. Substituting into equation (10.33) it gives

$$S_{(2,1)}(a_{u+v}, \alpha) = -\frac{175}{2^4} \frac{6}{1} - \frac{225}{2^4} \frac{4}{1}$$

$$= -\frac{1}{2^4} (1050 + 900) = -\frac{1950}{2^4}$$

(10.34)

4. Computation of the term $E_F^{(v)}(\alpha)$

The expression for $E_F^{(v)}(\alpha)$ is given by equation (7.87), on page 309, writes

$$E_F^{(v)}(\alpha) = E_F^{(v_0)}(\alpha) E_F^{(v_1)}(\alpha)$$

$$= \left[\frac{(\eta + v_0)!}{\eta!} \frac{1}{(\beta_{0,0} - \alpha_0)^{v_0}} \right] \left[\frac{(\eta + v_1)!}{\eta!} \frac{1}{(\beta_{1,1} - \alpha_1)^{v_1}} \right]$$

(10.35)

Evaluating the polynomial $E_F^v(\alpha)$ at the reference node $\alpha = (-1,-1)$ for $a_{(v)} = a_{(3,0)}$ and $a_{(v)} = a_{(2,1)}$ it follows

(a) For $v = (3,0)$ the equation (10.35) writes

$$E_F^{(3,0)}(\alpha) = E_F^{(3)}(-1,-1) E_F^{(0)}(-1,-1)$$

$$= \left[\frac{(4+3)!}{4!} \frac{1}{(1-(-1))^3} \right] \left[\frac{(4+0)!}{4!} \frac{1}{(1-(-1))^0} \right]$$

$$= \left[\frac{5 \times 6 \times 7}{(1+1)^3} \right] \times [1] = \frac{420}{2^4}$$

(10.36)

where $\eta = 4$, $\alpha = (-1,-1)$, and from equation (10.1) it follows $\beta_{0,0} = 1$, $\beta_{1,1} = 1$

(b) For $v = (2,1)$ the equation (10.35) writes

$$E_F^{(2,1)}(\alpha) = E_F^{(2)}(-1,-1) E_F^{(1)}(-1,-1)$$

$$= \left[\frac{(4+2)!}{4!} \frac{1}{(1-(-1))^2} \right] \left[\frac{(4+1)!}{1!} \frac{1}{(1-(-1))^1} \right]$$

$$= \left[\frac{5 \times 6}{2^2} \right] \times \left[\frac{5}{2} \right] = \frac{300}{2^4}$$

(10.37)

where $\eta = 4$, $\alpha = (-1,-1)$, and from equation (10.1) it follows $\beta_{0,0} = 1$, $\beta_{1,1} = 1$

5. Computation of the coefficients

(a) coefficient $a_{(3,0)}$ for which $v = (3,0)$

Substituting $420/2^4$ given by equation (10.36) and $S_v\,(a_{v+u}, \alpha)$ given by equation (10.32) into equation (10.28) it gives

$$a_{(v)} = \frac{1}{(v)!}\left[E_F^{(v)}(\alpha) - S_v\,(a_{v+u}, \alpha)\right]$$

$$a_{(3,0)} = \frac{1}{(3,0)!}\left[E_F^{(3,0)}(-1,-1) - S_{(3,0)}\,(a_{u+v}, (-1,-1))\right]$$

$$= \frac{1}{(3,0)!}\left[\frac{420}{2^4} - \frac{-2730}{2^4}\right] = \frac{3150}{6 \times 2^4} = \frac{525}{2^4} \qquad (10.38)$$

(b) coefficient $a_{(2,1)}$ for which $v = (2,1)$

Substituting $300/2^4$ given by equation (10.37) and $S_v\,(a_{v+u}, \alpha)$ given by equation (10.34) into equation (10.28) it gives

$$a_{(v)} = \frac{1}{(v)!}\left[E_F^{(v)}(\alpha) - S_v\,(a_{v+u}, \alpha)\right]$$

$$a_{(2,1)} = \frac{1}{(2,1)!}\left[E_F^{(2,1)}(-1,-1) - S_{(2,1)}\,(a_{u+v}, (-1,-1))\right]$$

$$= \frac{1}{(2,1)!}\left[\frac{300}{2^4} - \frac{-1950}{2^4}\right] = \frac{2250}{2 \times 2^4} = \frac{1125}{2^4} \qquad (10.39)$$

6. Computation of the coefficient $a_v = a_{(1,2)}$

Using the theorem 7.4.7, on page 324, with respect to the permutation of the indices it follows

$$a_{(1,2)} = a_{(2,1)} = \frac{1125}{2^4} \qquad (10.40)$$

7. Computation of the coefficient $a_v = a_{(0,3)}$.

$$a_{(0,3)} = a_{(3,0)} = \frac{525}{2^4} \qquad (10.41)$$

d- Computation of the coefficients that belong to the terms in the level $\lambda_0(v) = q - 2 = 2$

1. Coefficients to be evaluated, which belong to the level $\lambda_0(v) = 2$.

The elements in the level $\lambda_0(v) = 2$ with two coordinates, which are the constrained numbers in the one-dimensional space, are

$$L_{\lambda_0(v)}^{\delta} = L_2^1 = \{(2,0), (1,1), (0,2)\} \qquad (10.42)$$

The elements of the set L_2^1 can obtained using the algorithm described in subsection 2.4.4, on page 45. Therefore, the coefficients $a_{(w)} = a_{(v)}$ to be evaluated in this level are $a_{(2,0)}$, $a_{(1,1)}$, and $a_{(0,2)}$.

Using the theorem 7.4.7, on page 324, with respect to the permutation of the indices it follows that $a_{(2,0)} = a_{(0,2)}$.

2. Equation for the computation of the coefficients.

The coefficients are obtained using equation (7.123), which is

$$a_{(v)} = \frac{1}{v!} \left[E_F^{(v)}(\alpha) - S_v \left(a_{u+v}, \alpha \right) \right] \qquad \lambda_0(v) = 2 \qquad (10.43)$$

3. Computation of the term $S_v \left(a_{u+v}, \alpha \right)$

The degree of the polynomial $S_v \left(a_{u+v}, \alpha \right)$ is given by $m = q - \lambda_0(v) = 4 - 2 = 2$ where $\lambda_0(v) = v_0 + v_1 = 2 + 0 = 2$. The exponents of the variables in each term of $S_v \left(a_{u+v}, \alpha \right)$ are the elements of the set shown in equation (7.124), which is

$$u \in \left(L^\delta(m) - L_0^\delta \right) = \left(L^1(1) - L_0^1 \right)$$
$$= \left\{ (0,0), (1,0), (0,1), (2,0), (1,1), (0,2) \right\} - \left\{ (0,0) \right\}$$
$$= \left\{ (1,0), (0,1), (2,0), (1,1), (0,2) \right\} \qquad (10.44)$$

which permits to obtain the indices of the coefficients of the terms of the polynomial $S_v \left(a_{u+v}, \alpha \right)$ as follows

(a) For $v = (2,0)$

If $u = (1,0)$ then $v + u = (2,0) + (1,0) = (3,0)$, which gives $a_{(v+u)} = a_{(3,0)}$.
If $u = (0,1)$ then $v + u = (2,0) + (0,1) = (2,1)$, which gives $a_{(v+u)} = a_{(2,1)}$.
If $u = (2,0)$ then $v + u = (2,0) + (2,0) = (4,0)$, which gives $a_{(v+u)} = a_{(4,0)}$.
If $u = (1,1)$ then $v + u = (2,0) + (1,1) = (3,1)$, which gives $a_{(v+u)} = a_{(3,1)}$.
If $u = (0,2)$ then $v + u = (2,0) + (0,2) = (2,2)$, which gives $a_{(v+u)} = a_{(2,2)}$.
Where the indices of the coefficients $a_{(3,0)}$ and $a_{(2,1)}$ belong to the level $\lambda_0(v) = q - 1 = 3$. The indices of $a_{(4,0)}$, $a_{(3,1)}$, and $a_{(2,2)}$ belong to the level $\lambda_0(v) = q - 0 = 4$.

(b) For $v = (1,1)$

If $u = (1,0)$ then $v + u = (1,1) + (1,0) = (2,1)$, which gives $a_{(v+u)} = a_{(2,1)}$.
If $u = (0,1)$ then $v + u = (1,1) + (0,1) = (1,2)$, which gives $a_{(v+u)} = a_{(1,2)}$.

If $u = (2,0)$ then $v + u = (1,1) + (2,0) = (3,1)$, which gives
$a_{(v+u)} = a_{(3,1)}$.
If $u = (1,1)$ then $v + u = (1,1) + (1,1) = (2,2)$, which gives
$a_{(v+u)} = a_{(2,2)}$.
If $u = (0,2)$ then $v + u = (1,1) + (0,2) = (1,3)$, which gives
$a_{(v+u)} = a_{(1,3)}$.
Where the indices of the coefficients $a_{(2,1)}$ and $a_{(1,2)}$ belong to the
level $\lambda_0(v) = q - 1 = 3$. The indices of $a_{(3,1)}$, $a_{(2,2)}$, and $a_{(1,3)}$
belong to the level $\lambda_0(v) = q - 0 = 4$.

The expression for $S_v(a_{u+v}, \alpha)$ is given by equation (7.112). For $d = 2$
and $m = 2$ it writes

$$S_v(a_{u+v}, \alpha) = \sum_{r_0=1}^{r_0=m} \sum_{r_1=0}^{r_1=r_0} a_{(v+u)} \frac{(v+u)!}{u!} \alpha_0^{u_0} \alpha_1^{u_1}$$

$$= \sum_{r_0=1}^{r_0=2} \sum_{r_1=0}^{r_1=r_0} a_{(v+u)} \frac{(v+u)!}{u!} \alpha_0^{u_0} \alpha_1^{u_1}$$

$$= a_{(v+(1,0))} \frac{(v+(1,0))!}{(1,0)!} \alpha_0^1 \alpha_1^0 + a_{(v+(0,1))} \frac{(v+(0,1))!}{(0,1)!} \alpha_0^0 \alpha_1^1$$

$$+ a_{(v+(2,0))} \frac{(v+(2,0))!}{(2,0)!} \alpha_0^2 \alpha_1^0 + a_{(v+(1,1))} \frac{(v+(1,1))!}{(1,1)!} \alpha_0^1 \alpha_1^1$$

$$+ a_{(v+(0,2))} \frac{(v+(0,2))!}{(0,2)!} \alpha_0^0 \alpha_1^2 \tag{10.45}$$

The evaluation of the terms of the polynomial $S_v(a_{u+v}, \alpha)$, in equation
(10.45), at the reference node $\alpha = (-1,-1)$ for $a_{(v)} = a_{(2,0)}$ is shown in
the table 10.2.2 below

Table 10.2.2 *Computation of the terms for $S_{(2,0)}(a_{(u+v)}, \alpha)$.*

v	u	$(u+v)$	$a_{(u+v)}$	$(u+v)!/u!$	α^{u_0}	α^{u_1}	value
$(2,0)$	$(1,0)$	$(3,0)$	$252/2^4$	$(3,0)!/(1,0)! = 6$	$(-1)^1$	$(-1)^0$	$-3150/2^4$
	$(0,1)$	$(2,1)$	$1125/2^4$	$(2,1)!/(0,1)! = 2$	$(-1)^0$	$(-1)^1$	$-2250/2^4$
	$(2,0)$	$(4,0)$	$70/2^4$	$(4,0)!/(2,0)! = 12$	$(-1)^2$	$(-1)^0$	$840/2^4$
	$(1,1)$	$(3,1)$	$175/2^4$	$(3,1)!/(1,1)! = 6$	$(-1)^1$	$(-1)^1$	$1050/2^4$
	$(0,2)$	$(2,2)$	$225/2^4$	$(2,2)!/(0,2)! = 2$	$(-1)^2$	$(-1)^2$	$450/2^4$

Substituting into equation the values in the table 10.2.2 into the equation
(10.45) it gives

$$S_{(2,0)}(a_{u+v}, \alpha) = -\frac{3150}{2^4} - \frac{2250}{2^4} + \frac{840}{2^4} + \frac{1050}{2^4} + \frac{450}{2^4} = -\frac{3060}{2^4} \tag{10.46}$$

The evaluation of the terms of the polynomial $S_v(a_{u+v}, \alpha)$, in equation (10.45), at the reference node $\alpha = (-1, -1)$ for $a_{(v)} = a_{(1,1)}$ is shown in the table 10.2.3 below

Table 10.2.3 *Computation of the terms for* $S_{(1,1)}(a_{(u+v)}, \alpha)$.

v	u	$(u+v)$	$a_{(u+v)}$	$(u+v)!/u!$	α^{u_0}	α^{u_1}	value
$(1,1)$	$(1,0)$	$(2,1)$	$1125/2^4$	$(2,1)!/(1,0)! = 2$	$(-1)^1$	$(-1)^0$	$-2250/2^4$
	$(0,1)$	$(1,2)$	$1125/2^4$	$(1,2)!/(0,1)! = 2$	$(-1)^0$	$(-1)^1$	$-2250/2^4$
	$(2,0)$	$(3,1)$	$175/2^4$	$(3,1)!/(2,0)! = 3$	$(-1)^2$	$(-1)^0$	$525/2^4$
	$(1,1)$	$(2,2)$	$225/2^4$	$(2,2)!/(1,1)! = 4$	$(-1)^1$	$(-1)^1$	$900/2^4$
	$(0,2)$	$(1,3)$	$175/2^4$	$(1,3)!/(0,2)! = 3$	$(-1)^2$	$(-1)^2$	$525/2^4$

Substituting into equation the values in the table 10.2.3 into the equation (10.45) it gives

$$S_{(1,1)}(a_{u+v}, \alpha) = -\frac{2250}{2^4} - \frac{2250}{2^4} + \frac{525}{2^4} + \frac{900}{2^4} + \frac{525}{2^4} = -\frac{2550}{2^4} \quad (10.47)$$

4. Computation of the term $E_F^v(\alpha)$

The expression for $E_F^v(\alpha)$ is given by equation (7.87) writes

$$E_F^{(v)}(\alpha) = E_F^{(v_0)}(\alpha) E_F^{(v_1)}(\alpha)$$

$$= \left[\frac{(\eta + v_0)!}{\eta!} \frac{1}{(\beta_{0,0} - \alpha_0)^{v_0}} \right] \left[\frac{(\eta + v_1)!}{\eta!} \frac{1}{(\beta_{1,1} - \alpha_1)^{v_1}} \right] \quad (10.48)$$

Evaluating the polynomial $E_F^v(\alpha)$ at the reference node $\alpha = (-1, -1)$ for $a_{(v)} = a_{(2,0)}$ and $a_{(v)} = a_{(1,1)}$ it follows

(a) For $v = (v_0, v_1) = (2, 0)$ the equation (10.48) writes

$$E_F^{(2,0)}(\alpha) = E_F^{(2)}(-1, -1) E_F^{(0)}(-1, -1)$$

$$= \left[\frac{(4+2)!}{4!} \frac{1}{(1 - (-1))^2} \right] \left[\frac{(4+0)!}{4!} \frac{1}{(1 - (-1))^0} \right]$$

$$= \left[\frac{5 \times 6}{(1+1)^2} \right] \times [1] = \frac{120}{2^4} \quad (10.49)$$

where $\eta = 4$, $\alpha = (-1, -1)$, and from equation (10.1) it follows $\beta_{0,0} = 1$, $\beta_{1,1} = 1$

(b) For $v = (v_0, v_1) = (1, 1)$ the equation (10.48) writes

$$E_F^{(1,1)}(\alpha) = E_F^{(1)}(-1, -1) E_F^{(1)}(-1, -1)$$

$$= \left[\frac{(4+1)!}{4!} \frac{1}{(1 - (-1))^1} \right] \left[\frac{(4+1)!}{1!} \frac{1}{(1 - (-1))^1} \right]$$

$$= \left[\frac{5}{2^1} \right] \times \left[\frac{5}{2^1} \right] = \frac{25}{2^2} = \frac{100}{2^4} \quad (10.50)$$

where $\eta = 4$, $\alpha = (-1, -1)$, and from equation (10.1) it follows $\beta_{0,0} = 1$, $\beta_{1,1} = 1$

5. Computation of the coefficients

(a) coefficient $a_{(2,0)}$ for which $v = (2, 0)$

Substituting $E_F^{(2,0)}(\alpha)$ given by equation (10.49) and $S_{(2,0)}(a_{v+u}, \alpha)$ given by equation (10.46) into equation (10.43) it gives

$$a_{(v)} = \frac{1}{(v)!} \left[E_F^{(v)}(\alpha) - S_v(a_{v+u}, \alpha) \right]$$

$$a_{(2,0)} = \frac{1}{(2,0)!} \left[E_F^{(2,0)}(-1, -1) - S_{(2,0)}(a_{u+v}, (-1, -1)) \right]$$

$$= \frac{1}{(2,0)!} \left[\frac{120}{2^4} - \left(-\frac{3060}{2^4} \right) \right] = \frac{3180}{2 \times 2^4} = \frac{1590}{2^4} \qquad (10.51)$$

(b) coefficient $a_{(1,1)}$ for which $v = (1, 1)$

Substituting $E_F^{(1,1)}(\alpha)$ given by equation (10.50) and $S_{(1,1)}(a_{v+u}, \alpha)$ given by equation (10.47) into equation (10.43) it gives

$$a_{(v)} = \frac{1}{(v)!} \left[E_F^{(v)}(\alpha) - S_v(a_{v+u}, \alpha) \right]$$

$$a_{(1,1)} = \frac{1}{(1,1)!} \left[E_F^{(1,1)}(-1, -1) - S_{(1,1)}(a_{u+v}, (-1, -1)) \right]$$

$$= \frac{1}{(1,1)!} \left[\frac{100}{2^4} - \frac{-2550}{2^4} \right] = \frac{2650}{2^4} \qquad (10.52)$$

6. Computation of the coefficient $a_v = a_{(0,2)}$.

Using the theorem 7.4.7, on page 324, with respect to the permutation of the indices it follows

$$a_{(0,2)} = a_{(2,0)} = \frac{1590}{2^4} \qquad (10.53)$$

e- Computation of the coefficients that belong to the terms in the level $\lambda_0(v) = q - 3 = 1$

1. Coefficients to be evaluated, which belong to the level $\lambda_0(v) = 1$.

The elements in the level $\lambda_0(v) = 1$ with two coordinates are the constrained numbers in the one-dimensional space are

$$L_{\lambda_0(v)}^{\delta} = L_1^1 = \{ (1, 0), (0, 1) \} \qquad (10.54)$$

The elements of the set L_1^1 can obtained using the algorithm described in subsection 2.4.4, on page 45. Therefore, the coefficients $a_{(w)} = a_{(v)}$ to be evaluated in this level are $a_{(1,0)}$, and $a_{(0,1)}$.

Using the theorem 7.4.7 with respect to the permutation of the indices it follows that $a_{(1,0)} = a_{(0,1)}$.

2. Equation for the computation of the coefficients.

The coefficients are obtained using equation (7.123), which is

$$a_{(v)} = \frac{1}{v!} \left[E_F^{(v)}(\alpha) - S_v (a_{u+v}, \alpha) \right] \qquad \lambda_0(v) = 1 \qquad (10.55)$$

3. Computation of the term $S_v (a_{u+v}, \alpha)$

The degree of the polynomial $S_v (a_{u+v}, \alpha)$ is given by $m = q - \lambda_0(v) = 4 - 1 = 3$ where $\lambda_0(v) = v_0 + v_1 = 1 + 0 = 1$. The exponents of the variables in each term of $S_v (a_{u+v}, \alpha)$ are the elements of the set shown in equation (7.124), which is

$$\begin{aligned}
\left(L^\delta(m) - L_0^\delta \right) = &\left(L^1(1) - L_0^1 \right) \\
= &\{ (0,0), (1,0), (0,1), (2,0), (1,1), (0,2), (3,0), \\
&(2,1), (1,2), (0,3) \} - \{ (0,0) \} \\
= &\{ (1,0), (0,1), (2,0), (1,1), (0,2), (3,0), \\
&(2,1), (1,2), (0,3) \} \qquad (10.56)
\end{aligned}$$

which permits to obtain the indices of the coefficients of the terms of the polynomial $S_v (a_{u+v}, \alpha)$ for $v = (1,0)$ as follows

If $u = (1,0)$ then $v + u = (1,0) + (1,0) = (2,0)$, which gives $a_{(v+u)} = a_{(2,0)}$.

If $u = (0,1)$ then $v + u = (1,0) + (0,1) = (1,1)$, which gives $a_{(v+u)} = a_{(1,1)}$.

If $u = (2,0)$ then $v + u = (1,0) + (2,0) = (3,0)$, which gives $a_{(v+u)} = a_{(3,0)}$.

If $u = (1,1)$ then $v + u = (1,0) + (1,1) = (2,1)$, which gives $a_{(v+u)} = a_{(2,1)}$.

If $u = (0,2)$ then $v + u = (1,0) + (0,2) = (1,2)$, which gives $a_{(v+u)} = a_{(1,2)}$.

If $u = (3,0)$ then $v + u = (1,0) + (3,0) = (4,0)$, which gives $a_{(v+u)} = a_{(4,0)}$.

If $u = (2,1)$ then $v + u = (1,0) + (2,1) = (3,1)$, which gives $a_{(v+u)} = a_{(3,1)}$.

If $u = (1,2)$ then $v + u = (1,0) + (1,2) = (2,2)$, which gives $a_{(v+u)} = a_{(2,2)}$.

If $u = (0,3)$ then $v + u = (1,0) + (0,3) = (1,3)$, which gives
$a_{(v+u)} = a_{(1,3)}$.

Where the indices of the coefficients $a_{(2,0)}$ and $a_{(1,1)}$ belong to the level $\lambda_0(v) = q - 2 = 2$. The indices of $a_{(3,0)}$, $a_{(2,1)}$, and $a_{(1,2)}$ belong to the level $\lambda_0(v) = q - 1 = 3$. The indices of $a_{(4,0)}$, $a_{(3,1)}$, $a_{(2,2)}$, and $a_{(1,3)}$ belong to the level $\lambda_0(v) = q - 0 = 4$.

The expression for $S_v(a_{u+v}, \alpha)$ is given by equation (7.112). For $d = 2$ and $m = 3$ it writes

$$
S_v(a_{u+v}, \alpha) = \sum_{r_0=1}^{r_0=m} \sum_{r_1=0}^{r_1=r_0} a_{(v+u)} \frac{(v+u)!}{u!} \alpha_0^{u_0} \alpha_1^{u_1}
$$

$$
= \sum_{r_0=1}^{r_0=3} \sum_{r_1=0}^{r_1=r_0} a_{(v+u)} \frac{(v+u)!}{u!} \alpha_0^{u_0} \alpha_1^{u_1}
$$

$$
= a_{(v+(1,0))} \frac{(v+(1,0))!}{(1,0)!} \alpha_0^1 \alpha_1^0 + a_{(v+(0,1))} \frac{(v+(0,1))!}{(0,1)!} \alpha_0^0 \alpha_1^1
$$

$$
+ a_{(v+(2,0))} \frac{(v+(2,0))!}{(2,0)!} \alpha_0^2 \alpha_1^0 + a_{(v+(1,1))} \frac{(v+(1,1))!}{(1,1)!} \alpha_0^1 \alpha_1^1
$$

$$
+ a_{(v+(0,2))} \frac{(v+(0,2))!}{(0,2)!} \alpha_0^0 \alpha_1^2
$$

$$
+ a_{(v+(3,0))} \frac{(v+(3,0))!}{(3,0)!} \alpha_0^3 \alpha_1^0 + a_{(v+(1,1))} \frac{(v+(2,1))!}{(2,1)!} \alpha_0^2 \alpha_1^1
$$

$$
+ a_{(v+(1,2))} \frac{(v+(1,2))!}{(1,2)!} \alpha_0^1 \alpha_1^2 + a_{(v+(0,3))} \frac{(v+(0,3))!}{(0,3)!} \alpha_0^0 \alpha_1^3
$$

$$(10.57)$$

The evaluation of the terms of the polynomial $S_v(a_{u+v}, \alpha)$, in equation (10.57), at the reference node $\alpha = (-1, -1)$ for $a_{(v)} = a_{(1,0)}$ is shown in the table 10.2.4 below

Table 10.2.4 *Computation of the terms for $S_{(1,0)}(a_{(u+v)}, \alpha)$.*

v	u	$(u+v)$	$a_{(u+v)}$	$(u+v)!/u!$	α^{u_0}	α^{u_1}	value
$(1,0)$	$(1,0)$	$(2,0)$	$1590/2^4$	$(2,0)!/(1,0)! = 2$	$(-1)^1$	$(-1)^0$	$-3180/2^4$
	$(0,1)$	$(1,1)$	$2650/2^4$	$(1,1)!/(0,1)! = 1$	$(-1)^0$	$(-1)^1$	$-2650/2^4$
	$(2,0)$	$(3,0)$	$525/2^4$	$(3,0)!/(2,0)! = 3$	$(-1)^2$	$(-1)^0$	$1575/2^4$
	$(1,1)$	$(2,1)$	$1125/2^4$	$(2,1)!/(1,1)! = 2$	$(-1)^1$	$(-1)^1$	$2250/2^4$
	$(0,2)$	$(1,2)$	$1125/2^4$	$(1,2)!/(0,2)! = 1$	$(-1)^0$	$(-1)^2$	$1125/2^4$
	$(3,0)$	$(4,0)$	$70/2^4$	$(4,0)!/(3,0)! = 4$	$(-1)^3$	$(-1)^0$	$-280/2^4$
	$(2,1)$	$(3,1)$	$175/2^4$	$(3,1)!/(2,1)! = 3$	$(-1)^2$	$(-1)^1$	$-525/2^4$
	$(1,2)$	$(2,2)$	$225/2^4$	$(2,2)!/(1,2)! = 2$	$(-1)^1$	$(-1)^2$	$-450/2^4$
	$(0,3)$	$(1,3)$	$175/2^4$	$(1,3)!/(0,3)! = 1$	$(-1)^0$	$(-1)^3$	$-175/2^4$

Substituting into equation (10.57) it gives

$$S_{(1,0)}(a_{u+v}, \alpha) = -\frac{3180}{2^4} - \frac{2650}{2^4} + \frac{1575}{2^4} + \frac{2250}{2^4} + \frac{1125}{2^4}$$
$$-\frac{280}{2^4} - \frac{525}{2^4} - \frac{450}{2^4} - \frac{175}{2^4} = -\frac{2310}{2^4} \tag{10.58}$$

4. Computation of the term $E_F^v(\alpha)$

The expression for $E_F^v(\alpha)$ is given by equation (7.87), on page 309, writes

$$E_F^{(v)}(\alpha) = E_F^{(v_0)}(\alpha)E_F^{(v_1)}(\alpha)$$
$$= \left[\frac{(\eta + v_0)!}{\eta!} \frac{1}{(\beta_{0,0} - \alpha_0)^{v_0}}\right]\left[\frac{(\eta + v_1)!}{\eta!} \frac{1}{(\beta_{1,1} - \alpha_1)^{v_1}}\right] \tag{10.59}$$

Evaluating the polynomial $E_F^v(\alpha)$ at the reference node $\alpha = (-1, -1)$ for $a_{(v)} = a_{(1,0)}$ it follows

For $v = (v_0, v_1) = (1, 0)$ the equation (10.59) writes

$$E_F^{(1,0)}(\alpha) = E_F^{(1)}(-1, -1)E_F^{(0)}(-1, -1)$$
$$= \left[\frac{(4+1)!}{4!} \frac{1}{(1-(-1))^1}\right]\left[\frac{(4+0)!}{4!} \frac{1}{(1-(-1))^0}\right]$$
$$= \left[\frac{5}{(1+1)^1}\right] \times [1] = \frac{5}{2} = \frac{40}{2^4} \tag{10.60}$$

where $\eta = 4$, $\alpha = (-1, -1)$, and from equation (10.1) it follows $\beta_{0,0} = 1$, $\beta_{1,1} = 1$

5. Computation of the coefficient $a_{(1,0)}$ for which $v = (1, 0)$

Substituting $E_F^{(1,0)}(\alpha)$ given by equation (10.60) and $S_{(1,0)}(a_{v+u}, \alpha)$ given by equation (10.58) into equation (10.55) it gives

$$a_{(v)} = \frac{1}{(v)!}\left[E_F^{(v)}(\alpha) - S_v(a_{v+u}, \alpha)\right]$$
$$a_{(1,0)} = \frac{1}{(1,0)!}\left[E_F^{(1,0)}(-1, -1) - S_{(1,0)}(a_{u+v}, (-1, -1))\right]$$
$$= \frac{1}{(1,0)!}\left[\frac{40}{2^4} - \left(-\frac{2310}{2^4}\right)\right] = \frac{2350}{2^4} \tag{10.61}$$

6. Computation of the coefficient $a_v = a_{(0,1)}$.

Using the theorem 7.4.7, on page 324, with respect to the permutation of the indices it follows

$$a_{(0,1)} = a_{(1,0)} = \frac{2350}{2^4} \tag{10.62}$$

f- Computation of the coefficients that belong to the terms in the level $\lambda_0(v) = q - 4 = 0$

1. Coefficients to be evaluated, which belong to the level $\lambda_0(v) = 0$.

   The elements in the level $\lambda_0(v) = 0$ with two coordinates are the constrained numbers in the one-dimensional space are

   $$L^\delta_{\lambda_0(v)} = L^1_1 = \{(0,0)\} \tag{10.63}$$

   The elements of the set L^1_0 can obtained using the algorithm described in subsection 2.4.4, on page 45. Therefore, the coefficients $a_{(w)} = a_{(v)}$ to be evaluated in this level is $a_{(0,0)}$.

2. Equation for the computation of the coefficients.

   The coefficients are obtained using equation (7.123), on page 314, which is

   $$a_v = \frac{1}{v!} [E^v_F(\alpha) - S_v(a_{u+v}, \alpha)] \qquad \lambda_0(v) = 0 \tag{10.64}$$

3. Computation of the term $S_v(a_{u+v}, \alpha)$

   The degree of the polynomial $S_v(a_{u+v}, \alpha)$ is given by $m = q - \lambda_0(v) = 4 - 0 = 4$ where $\lambda_0(v) = v_0 + v_1 = 0 + 0 = 0$. The exponents of the variables in each term of $S_v(a_{u+v}, \alpha)$ are the elements of the set shown in equation (7.124), on page 314, which is

   $$\begin{aligned}
   (L^\delta(m) - L^\delta_0) &= (L^1(1) - L^1_0) \\
   &= \{(0,0), (1,0), (0,1), (2,0), (1,1), (0,2), (3,0), \\
   &\qquad (2,1), (1,2), (0,3), \\
   &\qquad (4,0), (3,1), (2,2), (1,3), (0,4)\} - \{(0,0)\} \\
   &= \{(1,0), (0,1), (2,0), (1,1), (0,2), (3,0), \\
   &\qquad (2,1), (1,2), (0,3), \\
   &\qquad (4,0), (3,1), (2,2), (1,3), (0,4)\} \tag{10.65}
   \end{aligned}$$

   which permits to obtain the indices of the coefficients of the terms of the polynomial $S_v(a_{u+v}, \alpha)$ for $v = (0,0)$ as follows

   If $u = (1,0)$ then $v + u = (0,0) + (1,0) = (1,0)$, which gives $a_{(v+u)} = a_{(1,0)}$.

   If $u = (0,1)$ then $v + u = (0,0) + (0,1) = (0,1)$, which gives $a_{(v+u)} = a_{(0,1)}$.

   If $u = (2,0)$ then $v + u = (0,0) + (2,0) = (2,0)$, which gives $a_{(v+u)} = a_{(2,0)}$.

   If $u = (1,1)$ then $v + u = (0,0) + (1,1) = (1,1)$, which gives $a_{(v+u)} = a_{(1,1)}$.

If $u = (0, 2)$ then $v + u = (0, 0) + (0, 2) = (0, 2)$, which gives $a_{(v+u)} = a_{(0,2)}$.

If $u = (3, 0)$ then $v + u = (0, 0) + (3, 0) = (3, 0)$, which gives $a_{(v+u)} = a_{(3,0)}$.

If $u = (2, 1)$ then $v + u = (0, 0) + (2, 1) = (2, 1)$, which gives $a_{(v+u)} = a_{(2,1)}$.

If $u = (1, 2)$ then $v + u = (0, 0) + (1, 2) = (1, 2)$, which gives $a_{(v+u)} = a_{(1,2)}$.

If $u = (0, 3)$ then $v + u = (0, 0) + (0, 3) = (0, 3)$, which gives $a_{(v+u)} = a_{(0,3)}$.

If $u = (4, 0)$ then $v + u = (0, 0) + (3, 0) = (4, 0)$, which gives $a_{(v+u)} = a_{(4,0)}$.

If $u = (3, 1)$ then $v + u = (0, 0) + (2, 1) = (3, 1)$, which gives $a_{(v+u)} = a_{(3,1)}$.

If $u = (2, 2)$ then $v + u = (0, 0) + (1, 2) = (2, 2)$, which gives $a_{(v+u)} = a_{(2,2)}$.

If $u = (1, 3)$ then $v + u = (0, 0) + (0, 3) = (1, 3)$, which gives $a_{(v+u)} = a_{(1,3)}$.

If $u = (0, 4)$ then $v + u = (0, 0) + (0, 3) = (0, 4)$, which gives $a_{(v+u)} = a_{(0,4)}$.

Where the indices of the coefficients $a_{(1,0)}$ and $a_{(0,1)}$ belong to the level $\lambda_0(v) = q - 3 = 1$. The indices of $a_{(2,0)}$, $a_{(1,1)}$, and $a_{(0,2)}$ belong to the level $\lambda_0(v) = q - 2 = 2$. The indices of $a_{(3,0)}$, $a_{(2,1)}$, $a_{(1,2)}$, and $a_{(0,3)}$ belong to the level $\lambda_0(v) = q - 1 = 3$. The indices of $a_{(4,0)}$, $a_{(3,1)}$, $a_{(2,2)}$, $a_{(1,3)}$, and $a_{(0,4)}$ belong to the level $\lambda_0(v) = q - 0 = 4$.

The expression for $S_v(a_{u+v}, \alpha)$ is given by equation (7.112), on page

313. For $d = 2$ and $m = 4$ it writes

$$S_v(a_{u+v}, \alpha) = \sum_{r_0=1}^{r_0=m} \sum_{r_1=0}^{r_1=r_0} a_{(v+u)} \frac{(v+u)!}{u!} \alpha_0^{u_0} \alpha_1^{u_1}$$

$$= \sum_{r_0=1}^{r_0=4} \sum_{r_1=0}^{r_1=r_0} a_{(v+u)} \frac{(v+u)!}{u!} \alpha_0^{u_0} \alpha_1^{u_1}$$

$$= a_{(v+(1,0))} \frac{(v+(1,0))!}{(1,0)!} \alpha_0^1 \alpha_1^0 + a_{(v+(0,1))} \frac{(v+(0,1))!}{(0,1)!} \alpha_0^0 \alpha_1^1$$

$$+ a_{(v+(2,0))} \frac{(v+(2,0))!}{(2,0)!} \alpha_0^2 \alpha_1^0 + a_{(v+(1,1))} \frac{(v+(1,1))!}{(1,1)!} \alpha_0^1 \alpha_1^1$$

$$+ a_{(v+(0,2))} \frac{(v+(0,2))!}{(0,2)!} \alpha_0^0 \alpha_1^2$$

$$+ a_{(v+(3,0))} \frac{(v+(3,0))!}{(3,0)!} \alpha_0^3 \alpha_1^0 + a_{(v+(1,1))} \frac{(v+(2,1))!}{(2,1)!} \alpha_0^2 \alpha_1^1$$

$$+ a_{(v+(1,2))} \frac{(v+(1,2))!}{(1,2)!} \alpha_0^1 \alpha_1^2 + a_{(v+(0,3))} \frac{(v+(0,3))!}{(0,3)!} \alpha_0^0 \alpha_1^3$$

$$+ a_{(v+(4,0))} \frac{(v+(4,0))!}{(4,0)!} \alpha_0^4 \alpha_1^0 + a_{(v+(3,1))} \frac{(v+(3,1))!}{(3,1)!} \alpha_0^3 \alpha_1^1$$

$$+ a_{(v+(2,2))} \frac{(v+(2,2))!}{(2,2)!} \alpha_0^2 \alpha_1^2 + a_{(v+(1,3))} \frac{(v+(1,3))!}{(1,3)!} \alpha_0^1 \alpha_1^3$$

$$+ a_{(v+(0,4))} \frac{(v+(0,4))!}{(0,4)!} \alpha_0^0 \alpha_1^4 \tag{10.66}$$

The evaluation of the terms of the polynomial $S_v(a_{u+v}, \alpha)$, in equation (10.66), at the reference node $\alpha = (-1, -1)$ for $a_{(v)} = a_{(1,0)}$ is shown in the table 10.2.5.

Substituting the values for the coefficients obtained in the table 10.2.5 into equation (10.66) it gives

$$S_{(1,0)}(a_{u+v}, \alpha) = -\frac{2350}{2^4} - \frac{2350}{2^4}$$
$$+ \frac{1590}{2^4} + \frac{2650}{2^4} + \frac{1590}{2^4}$$
$$- \frac{525}{2^4} - \frac{1125}{2^4} - \frac{1125}{2^4} - \frac{525}{2^4}$$
$$+ \frac{70}{2^4} + \frac{175}{2^4} + \frac{225}{2^4} + \frac{175}{2^4} + \frac{70}{2^4} = -\frac{8000}{2^4} + \frac{6545}{2^4}$$
$$- \frac{1455}{2^4} \tag{10.67}$$

Table 10.2.5 *Computation of the terms for* $S_{(0,0)}\left(a_{(u+v)}, \alpha\right)$

v	u	$(u+v)$	$a_{(u+v)}$	$(u+v)!/u!$	α^{u_0}	α^{u_1}	value
$(0,0)$	$(1,0)$	$(1,0)$	$2350/2^4$	$(1,0)!/(1,0)! = 1$	$(-1)^1$	$(-1)^0$	$-2350/2^4$
	$(0,1)$	$(0,1)$	$2350/2^4$	$(0,1)!/(0,1)! = 1$	$(-1)^0$	$(-1)^1$	$-2350/2^4$
	$(2,0)$	$(2,0)$	$1590/2^4$	$(2,0)!/(2,0)! = 1$	$(-1)^2$	$(-1)^0$	$1590/2^4$
	$(1,1)$	$(1,1)$	$2650/2^4$	$(2,1)!/(1,1)! = 1$	$(-1)^1$	$(-1)^1$	$2650/2^4$
	$(0,2)$	$(0,2)$	$1590/2^4$	$(1,2)!/(0,2)! = 1$	$(-1)^0$	$(-1)^2$	$1590/2^4$
	$(3,0)$	$(3,0)$	$525/2^4$	$(4,0)!/(3,0)! = 1$	$(-1)^3$	$(-1)^0$	$-525/2^4$
	$(2,1)$	$(2,1)$	$1125/2^4$	$(3,1)!/(2,1)! = 1$	$(-1)^2$	$(-1)^1$	$-1125/2^4$
	$(1,2)$	$(1,2)$	$1125/2^4$	$(2,2)!/(1,2)! = 1$	$(-1)^1$	$(-1)^2$	$-1125/2^4$
	$(0,3)$	$(0,3)$	$525/2^4$	$(1,3)!/(0,3)! = 1$	$(-1)^0$	$(-1)^3$	$-525/2^4$
	$(4,0)$	$(4,0)$	$70/2^4$	$(1,3)!/(0,3)! = 1$	$(-1)^4$	$(-1)^0$	$70/2^4$
	$(3,1)$	$(3,1)$	$175/2^4$	$(1,3)!/(0,3)! = 1$	$(-1)^3$	$(-1)^1$	$175/2^4$
	$(2,2)$	$(2,2)$	$225/2^4$	$(1,3)!/(0,3)! = 1$	$(-1)^2$	$(-1)^2$	$225/2^4$
	$(1,3)$	$(1,3)$	$175/2^4$	$(1,3)!/(0,3)! = 1$	$(-1)^1$	$(-1)^3$	$175/2^4$
	$(0,4)$	$(0,4)$	$70/2^4$	$(1,3)!/(0,3)! = 1$	$(-1)^0$	$(-1)^4$	$70/2^4$

4. Computation of the term $E_F^v(\alpha)$

The expression for $E_F^v(\alpha)$ is given by equation (7.87), on page 309, and it writes

$$E_F^{(v)}(\alpha) = E_F^{(v_0)}(\alpha)E_F^{(v_1)}(\alpha)$$
$$= \left[\frac{(\eta + v_0)!}{\eta!}\frac{1}{(\beta_{0,0} - \alpha_0)^{v_0}}\right]\left[\frac{(\eta + v_1)!}{\eta!}\frac{1}{(\beta_{1,1} - \alpha_1)^{v_1}}\right] \quad (10.68)$$

Evaluating the polynomial $E_F^v(\alpha)$ at the reference node $\alpha = (-1, -1)$ for $a_{(v)} = a_{(0,0)}$ it follows

For $v = (v_0, v_1) = (0, 0)$ the equation (10.68) writes

$$E_F^{(0,0)}(\alpha) = E_F^{(0)}(-1, -1)E_F^{(0)}(-1, -1)$$
$$= \left[\frac{(4+0)!}{4!}\frac{1}{(1-(-1))^0}\right]\left[\frac{(4+0)!}{4!}\frac{1}{(1-(-1))^0}\right]$$
$$= [1] \times [1] = \frac{16}{2^4} \quad (10.69)$$

where $\eta = 4$, $\alpha = (-1, -1)$, and from equation (10.1) it follows $\beta_{0,0} = 1$, $\beta_{1,1} = 1$

5. Computation of the coefficient $a_{(0,0)}$ for which $v = (0,0)$

Substituting $E_F^{(0,0)}(\alpha)$ given by equation (10.69) and $S_{(1,0)}(a_{v+u}, \alpha)$ given by equation (10.67) into equation (10.64) it gives

$$a_{(v)} = \frac{1}{(v)!} \left[E_F^{(v)}(\alpha) - S_v(a_{v+u}, \alpha) \right]$$

$$a_{(0,0)} = \frac{1}{(0,0)!} \left[E_F^{(0,0)}(-1,-1) - S_{(0,0)}(a_{u+v}, (-1,-1)) \right]$$

$$= \frac{1}{(0,0)!} \left[\frac{16}{2^4} - \left(-\frac{1455}{2^4} \right) \right] = \frac{1471}{2^4} \tag{10.70}$$

10.3 Generation of the polynomial $\Phi_{a_0,10}(x_0, x_1)$ related to the reference node $a_0 = (-1,-1)$

10.3.1 Expression for the polynomial $\Phi_{a_0,10}(x_0, x_1)$

$$\Phi_{a_0,10}(x_0, x_1) = \frac{1}{n!} \frac{P_{a_0}(x)}{P_{a_0}(\alpha)} Q_{a_0,n}(x) f_{a_0,n}(x) \tag{10.71}$$

where $n = (1,0)$ is the order of derivative for which the polynomial equals one at the reference node. The coordinate numbers of the reference node a_0 are $\alpha = (-1,-1)$.

The polynomial $P_{a_0}(x)$ is given by

$$P_{a_0}(x) = (x_0 - 1)^5 (x_1 - 1)^5 \tag{10.72}$$

The constant factor $P_{a_0}(\alpha)$ is given by

$$P_{a_0}(\alpha) = (-1 - 1)^5 (-1 - 1)^5 = 2^{10} \tag{10.73}$$

The polynomial factor $Q_{a_0,n}(x)$ is given by

$$Q_{a_0,n}(x) = (x_0 + 1)^{n_0} (x_1 + 1)^{n_1} = (x_0 + 1)^1 (x_1 + 1)^0 \tag{10.74}$$

10.3.2 Computation of the polynomial $f_{a_0,n}(x)$

a- Minimum degree of the polynomial $f_{a_0,10}(x)$
Minimum degree

$$q = \eta - \lambda_0(n) = 4 - \lambda_0(1,0) = 3 \tag{10.75}$$

where $\lambda_0(n)$ is the value of the level of the constrained number $n = (1,0)$, which is given by the definition 1.1.12, on page 9, and computed as

$$\sum_{j=0}^{j=d-1} n_j = \sum_{j=0}^{j=1} n_j = n_0 + n_1 = 1 + 0 = 1 \tag{10.76}$$

Then the equation (10.16) reduces to

$$q \geq \eta - 1 = 3 \tag{10.77}$$

b- Computation of the coefficients that belong to the terms in the level $\lambda_0(v) = q - 0 = 3$

1. The indices v of the coefficients are the constrained numbers in the set

$$L^\delta_{\lambda_0(v)} = L^1_3 = \{(3,0),(2,1),(1,2),(0,3)\} \tag{10.78}$$

2. Computation of the coefficient $a_{(0,3)}$

$$a_{(0,3)} = \frac{1}{n!}E_F^{v_0}(\alpha)E_F^{v_1}(\alpha) = \frac{1}{0!3!}E_F^0(\alpha)E_F^3(\alpha) \tag{10.79}$$

where

$$E_F^0(\alpha) = \frac{(\eta + v_0)!}{\eta!}\frac{1}{(\beta_{00} - \alpha_0)^{v_0}} = \frac{(4+0)!}{4!}\frac{1}{2^0} = 1$$
$$E_F^1(\alpha) = \frac{(\eta + v_1)!}{\eta!}\frac{1}{(\beta_{11} - \alpha_1)^{v_1}} = \frac{(4+3)!}{4!}\frac{1}{2^3} = \frac{210}{2^3} \tag{10.80}$$

then

$$a_{(0,3)} = \frac{1}{0!3!} \times 1 \times \frac{210}{2^3} = \frac{35}{2^3} \tag{10.81}$$

3. Computation of the coefficient $a_{(1,2)}$

$$a_{(1,2)} = \frac{1}{n!}E_F^{v_0}(\alpha)E_F^{v_1}(\alpha) = \frac{1}{1!2!}E_F^0(\alpha)E_F^3(\alpha) \tag{10.82}$$

where

$$E_F^1(\alpha) = \frac{(\eta + v_0)!}{\eta!}\frac{1}{(\beta_{00} - \alpha_0)^{v_0}} = \frac{(4+1)!}{4!}\frac{1}{2^1} = \frac{5}{2}$$
$$E_F^2(\alpha) = \frac{(\eta + v_1)!}{\eta!}\frac{1}{(\beta_{11} - \alpha_1)^{v_1}} = \frac{(4+2)!}{4!}\frac{1}{2^2} = \frac{30}{2^2} \tag{10.83}$$

then

$$a_{(1,2)} = \frac{1}{1!2!} \times \frac{5}{2} \times \frac{30}{2^2} = \frac{75}{2^3} \tag{10.84}$$

4. Computation of the coefficient $a_{(2,1)}$ and $a_{(3,0)}$

   The theorem 7.4.7, on page 324, says that for a normalized region symmetric with respect to the coordinate axes the coefficients in the same level whose indices are permutations are equal, therefore,

$$a_{(3,0)} = a_{(0,3)} = \frac{35}{2^3}$$

$$a_{(2,1)} = a_{(1,2)} = \frac{75}{2^3} \qquad (10.85)$$

c- Computation of the coefficients that belong to the terms in the level $\lambda_0(v) = q - 1 = 2$

1. The indices v of the coefficients are the constrained numbers in the set

$$L^{\delta}_{\lambda_0(v)} = L^1_2 = \{(2,0), (1,1), (0,2)\} \qquad (10.86)$$

2. The coefficients are obtained using equation (7.123), on page 314, whose equation is

$$a_{(v)} = \frac{1}{v!} \left[E_F^{(v)}(\alpha) - S_v (a_{v+u}, \alpha) \right] \qquad \lambda_0(v) = 2 \qquad (10.87)$$

3. Degree of the term $S_v (a_{u+v}, \alpha)$

   $m = q - \lambda_0(v) = 3 - 2 = 1$

   The exponents of the variables in each term of $S_v (a_{u+v}, \alpha)$ are the elements of the set shown in equation (7.124), on page 314, which is

$$u \in \left(L^{\delta}(m) - L^{\delta}_0 \right) = \left(L^1(1) - L^1_0 \right)$$
$$= \{(0,0), (1,0), (0,1)\} - \{(0,0)\}$$
$$= \{(1,0), (0,1)\} \qquad (10.88)$$

   and it gives, for $u \in \{(1,0), (0,1)\}$ the following expression

$$S_v (a_{v+u}, \alpha) = a_{(v+(1,0))} \frac{(v+(1,0))!}{(1,0)!} \alpha_0^1 \alpha_1^0 + a_{(v+(0,1))} \frac{(v+(0,1))!}{(0,1)!} \alpha_0^0 \alpha_1^1$$

$$(10.89)$$

4. Computation of the coefficient $a_{(0,2)}$ where $v = (0,2)$.

$$E_F^v(\alpha) = E_F^{v_0}(\alpha) E_F^{v_1}(\alpha) = E_F^0(\alpha) E_F^2(\alpha) \qquad (10.90)$$

where

$$E_F^0(\alpha) = \frac{(\eta + v_0)!}{\eta!} \frac{1}{(\beta_{00} - \alpha_0)^{v_0}} = \frac{(4+0)!}{4!} \frac{1}{2^0} = 1$$

$$E_F^2(\alpha) = \frac{(\eta + v_1)!}{\eta!} \frac{1}{(\beta_{11} - \alpha_1)^{v_1}} = \frac{(4+2)!}{4!} \frac{1}{2^2} = \frac{30}{2^2} \tag{10.91}$$

then

$$E_F^{(0,2)}(\alpha) = E_F^0(\alpha) E_F^2(\alpha) = 1 \times \frac{30}{2^2} = \frac{30}{2^2} \tag{10.92}$$

and

$$S_{(0,2)}(a_{v+u}, \alpha) = a_{((0,2)+(1,0))} \frac{((0,2)+(1,0))!}{(1,0)!} \alpha_0^1 \alpha_1^0$$

$$+ a_{((0,2)+(0,1))} \frac{((0,2)+(0,1))!}{(0,1)!} \alpha_0^0 \alpha_1^1$$

$$= a_{(1,2)} \frac{(1,2)!}{(1,0)!} \alpha_0^1 \alpha_1^0 + a_{(0,3)} \frac{(0,3)!}{(0,1)!} \alpha_0^0 \alpha_1^1$$

$$= \frac{75}{2^3} \frac{1!2!}{1!0!} (-1)^1 (-1)^0 + \frac{35}{2^3} \frac{0!3!}{0!1!} (-1)^0 (-1)^1 = -\frac{360}{2^3} \tag{10.93}$$

Substituting the values given by equations (10.92) and (10.93) into equation (10.87) it gives

$$a_{(0,2)} = \frac{1}{0!2!} \left[E_F^{(0,2)}(-1,-1) - S_{(0,2)}(a_{v+u}, \alpha) \right]$$

$$= \frac{1}{0!2!} \left[\frac{30}{2^2} + \frac{360}{2^3} \right] = \frac{210}{2^3} \tag{10.94}$$

5. Computation of the coefficient $a_{(1,1)}$ where $v = (1,1)$.

$$E_F^v(\alpha) = E_F^{v_0}(\alpha) E_F^{v_1}(\alpha) = E_F^1(\alpha) E_F^1(\alpha) \tag{10.95}$$

where

$$E_F^1(\alpha) = \frac{(\eta + v_0)!}{\eta!} \frac{1}{(\beta_{00} - \alpha_0)^{v_0}} = \frac{(4+1)!}{4!} \frac{1}{2^1} = \frac{5}{2}$$

$$E_F^1(\alpha) = \frac{(\eta + v_1)!}{\eta!} \frac{1}{(\beta_{11} - \alpha_1)^{v_1}} = \frac{(4+1)!}{4!} \frac{1}{2^1} = \frac{5}{2} \tag{10.96}$$

then

$$E_F^{(1,1)}(\alpha) = E_F^1(\alpha) E_F^1(\alpha) = \frac{5}{2} \times \frac{5}{2} = \frac{25}{2^2} \tag{10.97}$$

and

$$S_{(1,1)}(a_{v+u}, \alpha) = a_{((1,1)+(1,0))} \frac{((1,1)+(1,0))!}{(1,0)!} \alpha_0^1 \alpha_1^0$$

$$+ a_{((1,1)+(0,1))} \frac{((1,1)+(0,1))!}{(0,1)!} \alpha_0^0 \alpha_1^1$$

$$= a_{(2,1)} \frac{(2,1)!}{(1,0)!} \alpha_0^1 \alpha_1^0 + a_{(1,2)} \frac{(1,2)!}{(0,1)!} \alpha_0^0 \alpha_1^1$$

$$= \frac{75}{2^3} \frac{2!1!}{1!0!}(-1)^1(-1)^0 + \frac{75}{2^3} \frac{1!2!}{0!1!}(-1)^0(-1)^1 = -\frac{300}{2^3}$$

$$\text{(10.98)}$$

Substituting the values given by equations (10.97) and (10.98) into equation (10.87) it gives

$$a_{(1,1)} = \frac{1}{1!1!} \left[E_F^{(1,1)}(-1,-1) - S_{(1,1)}(a_{v+u}, \alpha) \right]$$

$$= \frac{1}{1!1!} \left[\frac{25}{2^2} + \frac{300}{2^3} \right] = \frac{350}{2^3} \qquad \text{(10.99)}$$

6. Computation of the coefficient $a_{(2,0)}$

The theorem 7.4.7, on page 324, says that for a normalized region symmetric with respect to the coordinate axes the coefficients in the same level whose indices are permutations are equal, therefore,

$$a_{(2,0)} = a_{(0,2)} = \frac{210}{2^3} \qquad \text{(10.100)}$$

d- Computation of the coefficients that belong to the terms in the level $\lambda_0(v) = q - 2 = 1$

1. The indices v of the coefficients are the constrained numbers in the set

$$L_{\lambda_0(v)}^{\delta} = L_1^1 = \{(1,0),(0,1)\} \qquad \text{(10.101)}$$

2. The coefficients are obtained using equation (7.123), on page 314, which is

$$a_{(v)} = \frac{1}{v!} \left[E_F^{(v)}(\alpha) - S_v(a_{v+u}, \alpha) \right] \qquad \lambda_0(v) = 1 \qquad \text{(10.102)}$$

3. Degree of the term $S_v(a_{u+v}, \alpha)$

$$m = q - \lambda_0(v) = 3 - 1 = 2$$

The exponents of the variables in each term of $S_v (a_{u+v}, \alpha)$ are the elements of the set shown in equation (7.124), on page 314, which is

$$u \in \left(L^\delta(m) - L_0^\delta\right) = \left(L^1(2) - L_0^1\right)$$
$$= \{ (0,0), (1,0), (0,1), (2,0), (1,1), (0,2) \} - \{ (0,0) \}$$
$$= \{ (1,0), (0,1), (2,0), (1,1), (0,2) \} \qquad (10.103)$$

and it gives, for $u \in \{(1,0), (0,1), (2,0), (1,1), (0,2)\}$ the following expression

$$
\begin{aligned}
S_v (a_{v+u}, \alpha) = &\, a_{(v+(1,0))} \frac{(v + (1,0))!}{(1,0)!} \alpha_0^1 \alpha_1^0 + a_{(v+(0,1))} \frac{(v + (0,1))!}{(0,1)!} \alpha_0^0 \alpha_1^1 \\
&+ a_{(v+(2,0))} \frac{(v + (2,0))!}{(2,0)!} \alpha_0^2 \alpha_1^0 + a_{(v+(1,1))} \frac{(v + (1,1))!}{(1,1)!} \alpha_0^1 \alpha_1^1 \\
&+ a_{(v+(0,2))} \frac{(v + (0,2))!}{(0,2)!} \alpha_0^0 \alpha_1^2 \qquad (10.104)
\end{aligned}
$$

4. Computation of the coefficient $a_{(0,1)}$ where $v = (0,1)$.

$$E_F^v(\alpha) = E_F^{v_0}(\alpha) E_F^{v_1}(\alpha) = E_F^0(\alpha) E_F^1(\alpha) \qquad (10.105)$$

where

$$
\begin{aligned}
E_F^0(\alpha) &= \frac{(\eta + v_0)!}{\eta!} \frac{1}{(\beta_{00} - \alpha_0)^{v_0}} = \frac{(4+0)!}{4!} \frac{1}{2^0} = 1 \\
E_F^1(\alpha) &= \frac{(\eta + v_1)!}{\eta!} \frac{1}{(\beta_{11} - \alpha_1)^{v_1}} = \frac{(4+1)!}{4!} \frac{1}{2^1} = \frac{5}{2} \qquad (10.106)
\end{aligned}
$$

then

$$E_F^{(0,1)}(\alpha) = E_F^0(\alpha) E_F^1(\alpha) = 1 \times \frac{5}{2} = \frac{5}{2} \qquad (10.107)$$

and

$$S_{(0,1)}\left(a_{v+u}, \alpha\right) = a_{((0,1)+(1,0))} \frac{((0,1)+(1,0))!}{(1,0)!} \alpha_0^1 \alpha_1^0$$

$$+ a_{((0,1)+(0,1))} \frac{((0,1)+(0,1))!}{(0,1)!} \alpha_0^0 \alpha_1^1$$

$$+ a_{((0,1)+(2,0))} \frac{((0,1)+(2,0))!}{(2,0)!} \alpha_0^2 \alpha_1^0$$

$$+ a_{((0,1)+(1,1))} \frac{((0,1)+(1,1))!}{(1,1)!} \alpha_0^1 \alpha_1^1$$

$$+ a_{((0,1)+(0,2))} \frac{((0,1)+(0,2))!}{(0,2)!} \alpha_0^0 \alpha_1^2$$

$$= a_{(1,1)} \frac{(1,1)!}{(1,0)!} \alpha_0^1 \alpha_1^0$$

$$+ a_{(0,2)} \frac{(0,2)!}{(0,1)!} \alpha_0^0 \alpha_1^1 + a_{(2,1)} \frac{(2,1)!}{(2,0)!} \alpha_0^2 \alpha_1^0$$

$$+ a_{(1,2)} \frac{(1,2)!}{(1,1)!} \alpha_0^1 \alpha_1^1 + a_{(0,3)} \frac{(0,3)!}{(0,2)!} \alpha_0^0 \alpha_1^2$$

$$= \frac{350}{2^3} \frac{1!1!}{1!0!}(-1)^1(-1)^0 + \frac{210}{2^3} \frac{0!2!}{0!1!}(-1)^0(-1)^1$$

$$+ \frac{75}{2^3} \frac{2!1!}{2!0!}(-1)^2(-1)^0 + \frac{75}{2^3} \frac{2!1!}{1!1!}(-1)^1(-1)^1$$

$$+ \frac{35}{2^3} \frac{0!3!}{0!2!}(-1)^0(-1)^2$$

$$= \frac{1}{2^3}\left(-350 - 420 + 75 + 150 + 105\right)$$

$$= -\frac{440}{2^3} \tag{10.108}$$

Substituting the values given by equations (10.107) and (10.108) into equation (10.102) it gives

$$a_{(0,1)} = \frac{1}{0!1!}\left[E_F^{(0,1)}(-1,-1) - S_{(0,1)}\left(a_{u+v}, \alpha\right)\right]$$

$$= \frac{1}{0!1!}\left[\frac{5}{2} + \frac{440}{2^3}\right] = \frac{460}{2^3} \tag{10.109}$$

5. Computation of the coefficient $a_{(1,0)}$ where $v = (1,0)$.

The theorem 7.4.7, on page 324, says that for a normalized region symmetric with respect to the coordinate axes the coefficients in the same level whose indices are permutations are equal, therefore,

$$a_{(1,0)} = a_{(0,1)} = \frac{460}{2^3} \tag{10.110}$$

e- Computation of the coefficients that belong to the terms in the level $\lambda_0(v) = q - 3 = 0$

1. The indices v of the coefficients are the constrained numbers in the set

$$L^{\delta}_{\lambda_0(v)} = L^1_1 = \{(0,0)\} \tag{10.111}$$

2. The coefficients are obtained using equation (7.123), on page 314, which is

$$a_{(v)} = \frac{1}{v!} \left[E_F^{(v)}(\alpha) - S_v(a_{v+u}, \alpha) \right] \qquad \lambda_0(v) = 0 \tag{10.112}$$

3. Degree of the term $S_v(a_{u+v}, \alpha)$

$$m = q - \lambda_0(v) = 3 - 0 = 3$$

The exponents of the variables in each term of $S_v(a_{u+v}, \alpha)$ are the elements of the set shown in equation (7.124), which is

$$
\begin{aligned}
u \in \left(L^{\delta}(m) - L^{\delta}_0 \right) &= \left(L^1(3) - L^1_0 \right) \\
&= \{ (0,0), (1,0), (0,1), (2,0), (1,1), (0,2), (3,0), \\
&\qquad (2,1), (1,2), (0,3) \} - \{ (0,0) \} \\
&= \{ (1,0), (0,1), (2,0), (1,1), (0,2), (3,0), \\
&\qquad (2,1), (1,2), (0,3) \}
\end{aligned} \tag{10.113}
$$

and it gives, for

$$u \in \{(1,0), (0,1), (2,0), (1,1), (0,2), (3,0), (2,1), (1,2), (0,3)\} \tag{10.114}$$

the following expression

$$
\begin{aligned}
S_v(a_{v+u}, \alpha) = {}& a_{(v+(1,0))} \frac{(v+(1,0))!}{(1,0)!} \alpha_0^1 \alpha_1^0 + a_{(v+(0,1))} \frac{(v+(0,1))!}{(0,1)!} \alpha_0^0 \alpha_1^1 \\
&+ a_{(v+(2,0))} \frac{(v+(2,0))!}{(2,0)!} \alpha_0^2 \alpha_1^0 + a_{(v+(1,1))} \frac{(v+(1,1))!}{(1,1)!} \alpha_0^1 \alpha_1^1 \\
&+ a_{(v+(0,2))} \frac{(v+(0,2))!}{(0,2)!} \alpha_0^0 \alpha_1^2 \\
&+ a_{(v+(3,0))} \frac{(v+(3,0))!}{(3,0)!} \alpha_0^3 \alpha_1^0 + a_{(v+(2,1))} \frac{(v+(2,1))!}{(2,1)!} \alpha_0^2 \alpha_1^1 \\
&+ a_{(v+(1,2))} \frac{(v+(1,2))!}{(1,2)!} \alpha_0^1 \alpha_1^2 + a_{(v+(0,3))} \frac{(v+(0,3))!}{(0,3)!} \alpha_0^0 \alpha_1^3
\end{aligned} \tag{10.115}
$$

4. Computation of the coefficient $a_{(0,0)}$ where $v = (0,0)$.

$$E_F^v(\alpha) = E_F^{v_0}(\alpha) E_F^{v_1}(\alpha) = E_F^0(\alpha) E_F^0(\alpha) \tag{10.116}$$

where

$$E_F^0(\alpha) = \frac{(\eta + v_0)!}{\eta!} \frac{1}{(\beta_{00} - \alpha_0)^{v_0}} = \frac{(4+0)!}{4!} \frac{1}{2^0} = 1$$

$$E_F^0(\alpha) = \frac{(\eta + v_1)!}{\eta!} \frac{1}{(\beta_{11} - \alpha_1)^{v_1}} = \frac{(4+0)!}{4!} \frac{1}{2^0} = 1 \tag{10.117}$$

then

$$E_F^{(0,1)}(\alpha) = E_F^0(\alpha) E_F^0(\alpha) = 1 \times 1 = 1 \tag{10.118}$$

and

$$\begin{aligned}
S_{(0,0)}(a_{v+u}, \alpha) = {}& a_{((0,0)+(1,0))} \frac{((0,0)+(1,0))!}{(1,0)!} \alpha_0^1 \alpha_1^0 \\
&+ a_{((0,0)+(0,1))} \frac{((0,0)+(0,1))!}{(0,1)!} \alpha_0^0 \alpha_1^1 \\
&+ a_{((0,0)+(2,0))} \frac{((0,0)+(2,0))!}{(2,0)!} \alpha_0^2 \alpha_1^0 \\
&+ a_{((0,0)+(1,1))} \frac{((0,0)+(1,1))!}{(1,1)!} \alpha_0^1 \alpha_1^1 \\
&+ a_{((0,0)+(0,2))} \frac{((0,0)+(0,2))!}{(0,2)!} \alpha_0^0 \alpha_1^2 \\
&+ a_{((0,0)+(3,0))} \frac{((0,0)+(3,0))!}{(3,0)!} \alpha_0^3 \alpha_1^0 \\
&+ a_{((0,0)+(2,1))} \frac{((0,0)+(2,1))!}{(2,1)!} \alpha_0^2 \alpha_1^1 \\
&+ a_{((0,0)+(1,2))} \frac{((0,0)+(1,2))!}{(1,2)!} \alpha_0^1 \alpha_1^2 \\
&+ a_{((0,0)+(0,3))} \frac{((0,0)+(0,3))!}{(0,3)!} \alpha_0^0 \alpha_1^3 \tag{10.119}
\end{aligned}$$

$$S_{(0,0)}(a_{v+u}, \alpha) = a_{(1,0)}\frac{(1,0)!}{(1,0)!}\alpha_0^1\alpha_1^0 + a_{(0,1)}\frac{(0,1)!}{(0,1)!}\alpha_0^0\alpha_1^1$$

$$+ a_{(2,0)}\frac{(2,0)!}{(2,0)!}\alpha_0^2\alpha_1^0 + a_{(1,1)}\frac{(1,1)!}{(1,1)!}\alpha_0^1\alpha_1^1$$

$$+ a_{(0,2)}\frac{(0,2)!}{(0,2)!}\alpha_0^0\alpha_1^2$$

$$+ a_{(3,0)}\frac{(3,0)!}{(3,0)!}\alpha_0^3\alpha_1^0 + a_{(2,1)}\frac{(2,1)!}{(2,1)!}\alpha_0^2\alpha_1^1$$

$$+ a_{(1,2)}\frac{(1,2)!}{(1,2)!}\alpha_0^1\alpha_1^2 + a_{(0,3)}\frac{(0,3)!}{(0,3)!}\alpha_0^0\alpha_1^3 \qquad (10.120)$$

$$S_{(0,0)}(a_{v+u}, \alpha) = \frac{460}{2^3}\frac{1!0!}{1!0!}(-1)^1(-1)^0 + \frac{460}{2^3}\frac{0!1!}{0!1!}(-1)^0(-1)^1$$

$$+ \frac{210}{2^3}\frac{2!0!}{2!0!}(-1)^2(-1)^0$$

$$+ \frac{350}{2^3}\frac{1!1!}{1!1!}(-1)^1(-1)^1 + \frac{210}{2^3}\frac{0!2!}{0!2!}(-1)^0(-1)^2$$

$$+ \frac{35}{2^3}\frac{3!0!}{3!0!}(-1)^3(-1)^0 + \frac{75}{2^3}\frac{2!1!}{2!1!}(-1)^2(-1)^1$$

$$+ \frac{75}{2^3}\frac{1!2!}{1!2!}(-1)^1(-1)^2 + \frac{35}{2^3}\frac{0!3!}{0!2!}(-1)^0(-1)^2$$

$$= \frac{1}{2^3}(-460 - 460 + 210 + 350 + 210 - 35$$

$$-75 - 75 - 35)$$

$$= \frac{1}{2^3}(-1140 + 770) = -\frac{370}{2^3} \qquad (10.121)$$

Substituting the values given by equations (10.118) and (10.121) into equation (10.112) it gives

$$a_{(0,0)} = \frac{1}{0!0!}\left[E_F^{(0,0)}(-1,-1) - S_{(0,0)}(a_{u+v}, \alpha)\right]$$

$$= \frac{1}{0!0!}\left[1 + \frac{370}{2^3}\right] = \frac{378}{2^3} \qquad (10.122)$$

10.3.3 Polynomial $\Phi_{a_0,10}(x_0, x_1)$

$$\Phi_{a,10}(x,y) = \frac{1}{1!0!}\frac{1}{(-2)^{10}}\frac{1}{2^3}(x-1)^5(y-1)^5(x+1)\left(378 + 460x_0 + 460x_1\right.$$

$$+ 210x_0^2 + 350x_0x_1 + 210x_1^2 + 35x_0^3 + 75x_0^2x_1 + 75x_0x_1^2 + 35x_1^3\left.\right) \qquad (10.123)$$

All the other polynomials are obtained using the same algorithm.

10.3.4   Summary of the polynomials $\Phi_{a_0,n}$ related to the reference node $a_0 = (-1, -1)$

For $d = 2$, $\eta = 4$, $\alpha = a_0 = (-1, -1)$, $\beta_0 = (\beta_{00}, \beta_{01}) = a_1 = (1, -1)$, and $\beta_1 = (\beta_{10}, \beta_{11}) = a_3 = (-1, 1)$, the Hermite Interpolating Polynomials writes

$$\Phi_{a,00}(x_0, x_1) = \frac{1}{0!0!} \frac{1}{(-2)^5} \frac{1}{(-2)^5} \frac{1}{2^4}(x_0 - 1)^5 (x_1 - 1)^5$$

$$\times \Big(1471 + 2350x_0 + 2350x_1 + 1590x_0^2 + 2650x_0x_1$$

$$+ 1590x_1^2 + 525x_0^3 + 1125x_0^2x_1 + 1125x_0x_1^2 + 525x_1^3$$

$$+ 70x_0^4 + 175x_0^3x_1 + 225x_0^2x_1^2 + 175x_0x_1^3 + 70x_1^4 \Big) \quad (10.124)$$

$$\Phi_{a,10}(x_0, x_1) = \frac{1}{1!0!} \frac{1}{(-2)^{10}} \frac{1}{2^3}(x_0 - 1)^5 (x_1 - 1)^5 (x_0 + 1) \Big(378 + 460x_0$$

$$+ 460x_1 + 210x_0^2 + 350x_0x_1 + 210x_1^2 + 35x_0^3$$

$$+ 75x_0^2x_1 + 75x_0x_1^2 + 35x_1^3 \Big) \quad (10.125)$$

$$\Phi_{a,01}(x_0, x_1) = \frac{1}{1!0!} \frac{1}{(-2)^{10}} \frac{1}{2^3}(x_0 - 1)^5 (x_1 - 1)^5 (x_1 + 1) \Big(378 + 460x_0$$

$$+ 460x_1 + 210x_0^2 + 350x_0x_1 + 210x_1^2 + 35x_0^3 + 75x_0^2x_1$$

$$+ +75x_0x_1^2 + 35x_1^3 \Big) \quad (10.126)$$

$$\Phi_{a,20}(x_0, x_1) = \frac{1}{2!0!} \frac{1}{(-2)^{10}} \frac{1}{2^2}(x_0 - 1)^5 (x_1 - 1)^5 (x_0 + 1)^2 \Big(79 + 65x_0$$

$$+ 65x_1 + 15x_0^2 + 25x_0x_1 + 15x_1^2 \Big) \quad (10.127)$$

$$\Phi_{a,11}(x_0, x_1) = \frac{1}{1!1!} \frac{1}{(-2)^{10}} \frac{1}{2^2}(x_0 - 1)^5 (x_1 - 1)^5 (x_0 + 1)(x_1 + 1) \Big(79 + 65x_0$$

$$+ 65x_1 + 15x_0^2 + 25x_0x_1 + 15x_1^2 \Big) \quad (10.128)$$

$$\Phi_{a,02}(x_0, x_1) = \frac{1}{0!2!} \frac{1}{(-2)^{10}} \frac{1}{2^2}(x_0 - 1)^5 (x_1 - 1)^5 (x_1 + 1)^2 \Big(79 + 65x_0$$

$$+ 65x_1 + 15x_0^2 + 25x_0x_1 + 15x_1^2 \Big) \quad (10.129)$$

$$\Phi_{a,30}(x_0, x_1) = \frac{1}{3!0!} \frac{1}{(-2)^{10}} \frac{1}{2^1} (x_0 - 1)^5 (x_1 - 1)^5 (x_0 + 1)^3 \left(12 + 5x_0 + 5x_1 \right)$$

(10.130)

$$\Phi_{a,21}(x_0, x_1) = \frac{1}{2!1!} \frac{1}{(-2)^{10}} \frac{1}{2^1} (x_0 - 1)^5 (x_1 - 1)^5 (x_0 + 1)^2 (x_1 + 1) \left(12 + 5x_0 + 5x_1 \right)$$

(10.131)

$$\Phi_{a,12}(x_0, x_1) = \frac{1}{1!2!} \frac{1}{(-2)^{10}} \frac{1}{2^1} (x_0 - 1)^5 (x_1 - 1)^5 (x_0 + 1)(x_1 + 1)^2 \left(12 + 5x_0 + 5x_1 \right)$$

(10.132)

$$\Phi_{a,03}(x_0, x_1) = \frac{1}{0!3!} \frac{1}{(-2)^{10}} \frac{1}{2^1} (x_0 - 1)^5 (x_1 - 1)^5 (x_1 + 1)^3 \left(12 + 5x_0 + 5x_1 \right)$$

(10.133)

$$\Phi_{a,40}(x_0, x_1) = \frac{1}{4!0!} \frac{1}{(-2)^{10}} (x_0 - 1)^5 (x_1 - 1)^5 (x_0 + 1)^4 \frac{1}{2^0} \left(1 \right)$$

(10.134)

$$\Phi_{a,31}(x_0, x_1) = \frac{1}{3!1!} \frac{1}{(-2)^{10}} (x_0 - 1)^5 (x_1 - 1)^5 (x_0 + 1)^3 (x_1 + 1) \frac{1}{2^0} \left(1 \right)$$

(10.135)

$$\Phi_{a,22}(x_0, x_1) = \frac{1}{2!2!} \frac{1}{(-2)^{10}} (x_0 - 1)^5 (x_1 - 1)^5 (x_0 + 1)^2 (x_1 + 1)^2 \frac{1}{2^0} \left(1 \right)$$

(10.136)

$$\Phi_{a,13}(x_0, x_1) = \frac{1}{1!3!} \frac{1}{(-2)^{10}} (x_0 - 1)^5 (x_1 - 1)^5 (x_0 + 1)^1 (x_1 + 1)^3 \frac{1}{2^0} \left(1 \right)$$

(10.137)

$$\Phi_{a,04}(x_0, x_1) = \frac{1}{0!4!} \frac{1}{(-2)^{10}} (x_0 - 1)^5 (x_1 - 1)^5 (x_1 + 1)^4 \frac{1}{2^0} \left(1 \right)$$

(10.138)

10.4   Generation of the polynomials $\Phi_{a_1,n}$ related to the reference node $(1,-1)$

10.4.1   Generation of the polynomials $\Phi_{a_1,n}$

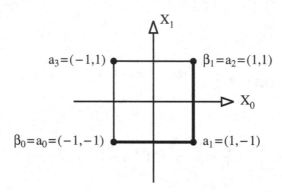

Figure 10.4.1 *Reference node a_1 in a symmetric normalized domain.*

The coordinate numbers of the node a_0 and the coordinate numbers of the node a_1 in figure 10.4.1 differ only by the first coordinate number. The theorem 7.4.1, on page 315, can be used to obtain the polynomials related to the reference node $(1,-1)$ for $\eta = 4$ knowing the polynomials related to the reference node $a_0 = (-1,-1)$.

The computation of the factor $P_{a_1}(a_1)$ can be performed using equation (7.132), on page 316, that is,

$$P_h(\gamma) = (-1)^{n+1} P_e(\alpha) \tag{10.139}$$

where $\gamma = a_1 = (1,-1)$ and $\alpha = a_0 = (-1,-1)$. Substituting these values into equation 10.139 it gives

$$P_{a_1}(a_1) = (-1)^5 P_{a_0}(a_0) = -P_{a_0}(-1,-1) \tag{10.140}$$

The coefficients of the terms of the polynomial $f_{h,n}(g,x) = f_{a_1,n}(g,x)$ relate to the reference node a_1 can be obtained from the coefficients of the polynomial $f_{e,n}(a,x) = f_{a_0}(a,x)$ using the equation (7.131), on page 316, and for the first variable it writes

$$g_{(v_0,v_1)} = (-1)^{v_0} a_{(v_0,v_1)} \tag{10.141}$$

The table 10.4.2 shows the computation of the coefficients of the polynomial $f_{a_1,n}(g,x)$ in terms of the coefficients of the polynomial $f_{a_0}(a,x)$ as follows

Table 10.4.2 *Computation of coefficients a_g of the polynomial $f_{a_1,n}(g,x)$ in terms of the coefficients a_e of the polynomial $f_{a_0,n}(a,x)$.*

a_e	v_0	v_1	$(-1)^{v_0}$	a_g
1471	0	0	$(-1)^0 = 1$	1471
2350	1	0	$(-1)^1 = -1$	-2350
2350	0	1	$(-1)^0 = 1$	2350
1590	2	0	$(-1)^2 = 1$	1590
2650	1	1	$(-1)^1 = -1$	-2650
1590	0	2	$(-1)^0 = 1$	1590
525	3	0	$(-1)^3 = -1$	-525
1125	2	1	$(-1)^2 = 1$	1125
1125	1	2	$(-1)^1 = -1$	-1125
525	0	3	$(-1)^0 = 1$	525
70	4	0	$(-1)^4 = 1$	70
175	3	1	$(-1)^3 = -1$	-175
225	2	2	$(-1)^2 = 1$	225
175	1	3	$(-1)^1 = -1$	-175
70	0	4	$(-1)^0 = 1$	70

Using the same algorithm to obtain all polynomials related to the reference nodes a_1 in terms of the polynomials related to the reference node a_0 it gives the result below.

10.4.2 Summary of the polynomials $\Phi_{a_1,n}$ related to the reference node $a_1 = (1, -1)$

For $d = 2$, $\eta = 4$, $\alpha = a_1 = (1, -1)$, $\beta_0 = (\beta_{00}, \beta_{01}) = a_0 = (-1, -1)$, and $\beta_1 = (\beta_{10}, \beta_{11}) = a_2 = (1, 1)$, the Hermite Interpolating Polynomials writes

$$
\begin{aligned}
\Phi_{a_1,00}(x_0, x_1) = {} & \frac{1}{0!0!} \left[\frac{1}{(2)^5} \frac{1}{(-2)^5} \right] (x_0 + 1)^5 (x_1 - 1)^5 \frac{1}{2^4} \Big(1471 - 2350 x_0 \\
& + 2350 x_1 + 1590 x_0^2 - 2650 x_0 x_1 + 1590 x_1^2 - 525 x_0^3 \\
& + 1125 x_0^2 x_1 - 1125 x_0 x_1^2 + 525 x_1^3 + 70 x_0^4 \\
& + 175 x_0^3 x_1 + 225 x_0^2 x_1^2 - 175 x_0 x_1^3 + 70 x_1^4 \Big)
\end{aligned}
$$

(10.142)

$$\Phi_{a_1,10}(x_0, x_1) = \frac{1}{1!0!} \left[\frac{1}{(-2)^5} \frac{1}{(2)^5} \right] (x_0 + 1)^5(x_1 - 1)^5(x_0 - 1)\frac{1}{2^3} \Big(378$$
$$- 460x_0 + 460x_1 + 210x_0^2 - 350x_0x_1 + 210x_1^2 - 35x_0^3$$
$$+ 75x_0^2x_1 - 75x_0x_1^2 + 35x_1^3 \Big) \tag{10.143}$$

$$\Phi_{a_1,01}(x_0, x_1) = \frac{1}{0!1!} \left[\frac{1}{(-2)^5} \frac{1}{(2)^5} \right] (x_0 + 1)^5(x_1 - 1)^5(x_1 + 1)\frac{1}{2^3} \Big(378$$
$$- 460x_0 + 460x_1 + 210x_0^2 - 350x_0v + 210x_1^2 - 35x_0^3$$
$$+ 75x_0^2x_1 - 75x_0x_1^2 + 35x_1^3 \Big) \tag{10.144}$$

$$\Phi_{a_1,20}(x_0, x_1) = \frac{1}{2!0!} \left[\frac{1}{(-2)^5} \frac{1}{(2)^5} \right] (x_0 + 1)^5(x_1 - 1)^5(x_0 - 1)^2$$
$$\times \frac{1}{2^2} \Big(79 - 65x_0 + 65x_1 + 15x_0^2 - 25x_0x_1 + 15x_1^2 \Big) \quad (10.145)$$

$$\Phi_{a_1,11}(x_0, x_1) = \frac{1}{1!1!} \left[\frac{1}{(-2)^5} \frac{1}{(2)^5} \right] (x_0 + 1)^5(x_1 - 1)^5(x_0 - 1)(x_1 + 1)$$
$$\times \frac{1}{2^2} \Big(79 - 65x_0 + 65x_1 + 15x_0^2 - 25x_0x_1 + 15x_1^2 \Big) \quad (10.146)$$

$$\Phi_{a_1,02}(x_0, x_1) = \frac{1}{0!2!} \left[\frac{1}{(-2)^5} \frac{1}{(2)^5} \right] (x_0 + 1)^5(x_1 - 1)^5(x_1 + 1)^2$$
$$\times \frac{1}{2^2} \Big(79 - 65x_0 + 65x_1 + 15x_0^2 - 25x_0x_1 + 15x_1^2 \Big) \quad (10.147)$$

$$\Phi_{a_1,30}(x_0, x_1) = \frac{1}{3!0!} \left[\frac{1}{(-2)^5} \frac{1}{(2)^5} \right] (x_0+1)^5(x_1-1)^5(x_0-1)^3\frac{1}{2^1} \Big(12-5x_0+5x_1 \Big) \tag{10.148}$$

$$\Phi_{a_1,21}(x_0, x_1)$$
$$= \frac{1}{2!1!} \left[\frac{1}{(-2)^5} \frac{1}{(2)^5} \right] (x_0 + 1)^5(x_1 - 1)^5(x_0 - 1)^2(x_1 + 1)\frac{1}{2^1} \Big(12 - 5x_0 + 5x_1 \Big) \tag{10.149}$$

$$\Phi_{a_1,12}(x_0, x_1)$$
$$= \frac{1}{1!2!} \left[\frac{1}{(-2)^5} \frac{1}{(2)^5} \right] (x_0 + 1)^5(x_1 - 1)^5(x_0 - 1)(x_1 + 1)^2\frac{1}{2^1} \Big(12 - 5x_0 + 5x_1 \Big) \tag{10.150}$$

$$\Phi_{a_1,03}(x_0, x_1)$$

$$= \frac{1}{0!3!} \left[\frac{1}{(-2)^5} \frac{1}{(2)^5} \right] (x_0+1)^5(x_1-1)^5(x_1+1)^3 \frac{1}{2^1} \left(12 - 5x_0 + 5x_1 \right)$$

$$\text{(10.151)}$$

$$\Phi_{a_1,40}(x_0, x_1) = \frac{1}{4!0!} \left[\frac{1}{(-2)^5} \frac{1}{(2)^5} \right] (x_0+1)^5(x_1-1)^5(x_0-1)^4 \frac{1}{2^0} \left(1 \right)$$

$$\text{(10.152)}$$

$$\Phi_{a_1,31}(x_0, x_1) = \frac{1}{3!1!} \left[\frac{1}{(-2)^5} \frac{1}{(2)^5} \right] (x_0+1)^5(x_1-1)^5(x_0-1)^3(x_1+1) \frac{1}{2^0} \left(1 \right)$$

$$\text{(10.153)}$$

$$\Phi_{a_1,22}(x_0, x_1) = \frac{1}{2!2!} \left[\frac{1}{(-2)^5} \frac{1}{(2)^5} \right] (x_0+1)^5(x_1-1)^5(x_0-1)^2(x_1+1)^2 \frac{1}{2^0} \left(1 \right)$$

$$\text{(10.154)}$$

$$\Phi_{a_1,13}(x_0, x_1) = \frac{1}{1!3!} \left[\frac{1}{(-2)^5} \frac{1}{(2)^5} \right] (x_0+1)^5(x_1-1)^5(x_0-1)^1(x_1+1)^3 \frac{1}{2^0} \left(1 \right)$$

$$\text{(10.155)}$$

$$\Phi_{a_1,04}(x_0, x_1) = \frac{1}{0!4!} \left[\frac{1}{(-2)^5} \frac{1}{(2)^5} \right] (x_0+1)^5(x_1-1)^5(x_1+1)^4 \frac{1}{2^0} \left(1 \right)$$

$$\text{(10.156)}$$

10.5 Generation of the polynomials $\Phi_{a_2,n}$ related to the reference node $(1,1)$

10.5.1 Generation of the polynomials

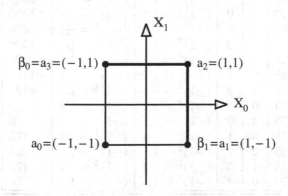

Figure 10.5.1 *Reference node a_2 in a symmetric normalized domain.*

The coordinate numbers of the node a_0 and the coordinate numbers of the node a_2 in figure 10.5.1, on page 391, have equal absolute values and opposite sings. The corollary 7.4.3, on page 320, can be used to obtain the polynomials related to the reference node $a_2 = (1,1)$ for $\eta = 4$ knowing the polynomials related to the reference node $a_0 = (-1,-1)$.

The computation of the factor $P_{a_2}(a_2)$ using equation (7.155), on page 320, which for two variables it writes

$$P_h(\gamma) = \left[(-1)^{\eta+1}(-1)^{\eta+1}\right] P_e(\alpha) \tag{10.157}$$

where $\gamma = a_2 = (1,1)$ and $\alpha = a_0 = (-1,-1)$. Substituting into equation 10.157 it gives

$$P_{a_2}(a_2) = (-1)^5(-1)^5 P_{a_0}(a_0) = P_{a_0}(-1,-1) \tag{10.158}$$

The coefficients of the terms of the polynomial $f_{h,n}(g,x) = f_{a_2,n}(g,x)$ relate to the reference node a_2 can be obtained from the coefficients of the polynomial $f_{e,n}(a,x) = f_{a_0}(a,x)$ using the equation (7.154) and for two variables it writes

$$g_{(v)} = \left[(-1)^{v_0}(-1)^{v_1}\right] a_{(v)} \tag{10.159}$$

The table 10.5.2 shows the computation of the coefficients of the polynomial $f_{a_1,n}(g,x)$ in terms of the coefficients of the polynomial $f_{a_0}(a,x)$ as follows

Table 10.5.2 *Computation of coefficients a_g of the polynomial $f_{a_2,n}(g,x)$ in terms of the coefficients a_e of the polynomial $f_{a_0,n}(a,x)$.*

a_e	v_0	v_1	$(-1)^{v_0}$	$(-1)^{v_1}$	a_g
1471	0	0	$(-1)^0 = 1$	$(-1)^0 = 1$	1471
2350	1	0	$(-1)^1 = -1$	$(-1)^0 = 1$	-2350
2350	0	1	$(-1)^0 = 1$	$(-1)^1 = -1$	-2350
1590	2	0	$(-1)^2 = 1$	$(-1)^0 = 1$	1590
2650	1	1	$(-1)^1 = -1$	$(-1)^1 = -1$	2650
1590	0	2	$(-1)^0 = 1$	$(-1)^2 = 1$	1590
525	3	0	$(-1)^3 = -1$	$(-1)^0 = 1$	-525
1125	2	1	$(-1)^2 = 1$	$(-1)^1 = -1$	-1125
1125	1	2	$(-1)^1 = -1$	$(-1)^2 = 1$	-1125
525	0	3	$(-1)^0 = 1$	$(-1)^3 = -1$	-525
70	4	0	$(-1)^4 = 1$	$(-1)^0 = 1$	70
175	3	1	$(-1)^3 = -1$	$(-1)^1 = -1$	175
225	2	2	$(-1)^2 = 1$	$(-1)^2 = 1$	225
175	1	3	$(-1)^1 = -1$	$(-1)^3 = -1$	175
70	0	4	$(-1)^0 = 1$	$(-1)^4 = 1$	70

The remaining $f_{a_2,n}(g,x)$ polynomials are obtained analogously.

10.5.2 Summary of the polynomials $\Phi_{a_2,n}$ related to the reference node $a_2 = (1,1)$

For $d = 2$, $\eta = 4$, $\alpha = a_2 = (1,1)$, $\beta_0 = (\beta_{00}, \beta_{01}) = a_3 = (-1,1)$, and $\beta_1 = (\beta_{10}, \beta_{11}) = a_1 = (1,-1)$, the Hermite Interpolating Polynomials writes

$$\Phi_{a_2,00}(x_0, x_1) = \frac{1}{0!0!} \left[\frac{1}{(2)^5} \frac{1}{(2)^5} \right] (x_0 + 1)^5 (x_1 + 1)^5 \frac{1}{2^4} \Big(1471 - 2350x_0$$
$$- 2350x_1 + 1590x_0^2 + 2650x_0 x_1 + 1590x_1^2 - 525x_0^3$$
$$- 1125x_0^2 x_1 - 1125x_0 x_1^2 - 525x_1^3 + 70x_0^4 + 175x_0^3 x_1$$
$$+ 225x_0^2 x_1^2 + 175x_0 x_1^3 + 70x_1^4 \Big) \tag{10.160}$$

$$\Phi_{a_2,10}(x_0, x_1) = \frac{1}{1!0!} \left[\frac{1}{(2)^5} \frac{1}{(2)^5} \right] (x_0 + 1)^5 (x_1 + 1)^5 (x_0 - 1) \frac{1}{2^3} \Big(378 - 460x_0$$
$$- 460x_1 + 210x_0^2 + 350x_0 x_1 + 210x_1^2 - 35x_0^3$$
$$- 75x_0^2 x_1 - 75x_0 x_1^2 - 35x_1^3 \Big) \tag{10.161}$$

$$\Phi_{a_2,01}(x_0, x_1) = \frac{1}{0!1!} \left[\frac{1}{(2)^5} \frac{1}{(2)^5} \right] (x_0 + 1)^5 (x_1 + 1)^5 (x_1 - 1) \frac{1}{2^3} \Big(378 - 460x_0$$
$$- 460x_1 + 210x_0^2 + 350x_0 x_1 + 210x_1^2 - 35x_0^3$$
$$- 75x_0^2 x_1 - 75x_0 x_1^2 - 35x_1^3 \Big) \tag{10.162}$$

$$\Phi_{a_2,20}(x_0, x_1) = \frac{1}{2!0!} \left[\frac{1}{(2)^5} \frac{1}{(2)^5} \right] (x_0 + 1)^5 (x_1 + 1)^5 (x_0 - 1)^2$$
$$\times \frac{1}{2^2} \Big(79 - 65x_0 - 65x_1 + 15x_0^2 + 25x_0 x_1 + 15x_1^2 \Big) \tag{10.163}$$

$$\Phi_{a_2,11}(x_0, x_1) = \frac{1}{1!1!} \left[\frac{1}{(2)^5} \frac{1}{(2)^5} \right] (x_0 + 1)^5 (x_1 + 1)^5 (x_0 - 1)(x_1 - 1)$$
$$\times \frac{1}{2^2} \Big(79 - 65x_0 - 65x_1 + 15x_0^2 + 25x_0 x_1 + 15x_1^2 \Big) \tag{10.164}$$

$$\Phi_{a_2,02}(x_0, x_1) = \frac{1}{0!2!} \left[\frac{1}{(2)^5} \frac{1}{(2)^5} \right] (x_0 + 1)^5 (x_1 + 1)^5 (x_1 - 1)^2$$
$$\times \frac{1}{2^2} \Big(79 - 65x_0 - 65x_1 + 15x_0^2 + 25x_0 x_1 + 15x_1^2 \Big) \tag{10.165}$$

$$\Phi_{a_2,30}(x,x_1) = \frac{1}{3!0!} \left[\frac{1}{(2)^5} \frac{1}{(2)^5} \right] (x_0+1)^5(x_1+1)^5(x_0-1)^3 \frac{1}{2^1} \left(12-5x_0-5x_1 \right)$$
(10.166)

$$\Phi_{a_2,21}(x_0,x_1)$$
$$= \frac{1}{2!1!} \left[\frac{1}{(2)^5} \frac{1}{(2)^5} \right] (x_0+1)^5(x_1+1)^5(x_0-1)^2(x_1-1) \frac{1}{2^1} \left(12-5x_0-5x_1 \right)$$
(10.167)

$$\Phi_{a_2,12}(x,x_1)$$
$$= \frac{1}{1!2!} \left[\frac{1}{(2)^5} \frac{1}{(2)^5} \right] (x_0+1)^5(x_1+1)^5(x_0-1)(x_1-1)^2 \frac{1}{2^1} \left(12-5x_0-5x_1 \right)$$
(10.168)

$$\Phi_{a_2,03}(x_0,x_1)$$
$$= \frac{1}{0!3!} \left[\frac{1}{(2)^5} \frac{1}{(2)^5} \right] (x_0+1)^5(x_1+1)^5(x_1-1)^3 \frac{1}{2^1} \left(12-5x_0-5x_1 \right)$$
(10.169)

$$\Phi_{a_2,40}(x_0,x_1) = \frac{1}{4!0!} \left[\frac{1}{(2)^5} \frac{1}{(2)^5} \right] (x_0+1)^5(x_1+1)^5(x_0-1)^4 \frac{1}{2^0} \left(1 \right)$$
(10.170)

$$\Phi_{a_2,31}(x_0,x_1) = \frac{1}{3!1!} \left[\frac{1}{(2)^5} \frac{1}{(2)^5} \right] (x_0+1)^5(x_1+1)^5(x_0-1)^3(x_1-1) \frac{1}{2^0} \left(1 \right)$$
(10.171)

$$\Phi_{a_2,22}(x_0,y) = \frac{1}{2!2!} \left[\frac{1}{(2)^5} \frac{1}{(2)^5} \right] (x_0+1)^5(x_1+1)^5(x_0-1)^2(x_1-1)^2 \frac{1}{2^0} \left(1 \right)$$
(10.172)

$$\Phi_{a_2,13}(x,x_1) = \frac{1}{1!3!} \left[\frac{1}{(2)^5} \frac{1}{(2)^5} \right] (x_0+1)^5(x_1+1)^5(x_0-1)^1(x_1-1)^3 \frac{1}{2^0} \left(1 \right)$$
(10.173)

$$\Phi_{a_2,04}(x_0,x_1) = \frac{1}{0!4!} \left[\frac{1}{(2)^5} \frac{1}{(2)^5} \right] (x_0+1)^5(x_1+1)^5(x_1-1)^4 \frac{1}{2^0} \left(1 \right) \quad (10.174)$$

10.6 Generation of the polynomials $\Phi_{a_3,n}$ related to the reference node $a_3 = (-1,1)$

10.6.1 Generation of the polynomials

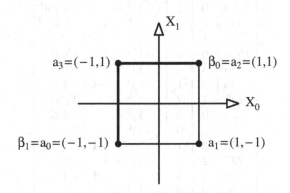

Figure 10.6.1 *Reference node a_3 in a symmetric normalized domain.*

The coordinate numbers of the node a_0 and the coordinate numbers of the node a_2 in figure 10.5.1 have equal absolute values and opposite sings. The corollary 7.4.3, on page 320, can be used to obtain the polynomials related to the reference node $a_3 = (-1,1)$ for $\eta = 4$ knowing the polynomials related to the reference node $a_0 = (-1,-1)$.

The computation of the factor $P_{a_3}(a_3)$ using equation (7.155), on page 320, which writes for the first variable

$$P_h(\gamma) = \left[(-1)^{\eta+1}\right] P_e(\alpha) \tag{10.175}$$

where $\gamma = a_3 = (-1,1)$ and $\alpha = a_0 = (-1,-1)$. Substituting into equation 10.175 it gives

$$P_{a_3}(a_3) = (-1)^5 P_{a_0}(a_0) = -P_{a_0}(-1,-1) \tag{10.176}$$

The coefficients of the terms of the polynomial $f_{h,n}(g,x) = f_{a_3,n}(g,x)$ relate to the reference node a_3 can be obtained from the coefficients of the polynomial $f_{e,n}(a,x) = f_{a_0}(a,x)$ using the equation (7.154) and for two variables it writes

$$g_{(v)} = \left[(-1)^{v_1}\right] a_{(v)} \tag{10.177}$$

The table 10.6.2 shows the computation of the coefficients of the polynomial $f_{a_3,n}(g,x)$ in terms of the coefficients of the polynomial $f_{a_0}(a,x)$ as follows

Table 10.6.2 *Computation of coefficients a_g of the polynomial $f_{a_3,n}(g,x)$ in terms of the coefficients a_e of the polynomial $f_{a_0,n}(a,x)$.*

a_e	v_0	v_1	$(-1)^{v_1}$	a_g
1471	0	0	$(-1)^0 = 1$	1471
2350	1	0	$(-1)^0 = 1$	2350
2350	0	1	$(-1)^1 = -1$	-2350
1590	2	0	$(-1)^0 = 1$	1590
2650	1	1	$(-1)^1 = -1$	-2650
1590	0	2	$(-1)^2 = 1$	1590
525	3	0	$(-1)^0 = 1$	525
1125	2	1	$(-1)^1 = -1$	-1125
1125	1	2	$(-1)^2 = 1$	1125
525	0	3	$(-1)^3 = -1$	-525
70	4	0	$(-1)^0 = 1$	70
175	3	1	$(-1)^1 = -1$	-175
225	2	2	$(-1)^2 = 1$	225
175	1	3	$(-1)^3 = -1$	-175
70	0	4	$(-1)^4 = 1$	70

The remaining $f_{a_3,n}(g,x)$ polynomials are obtained analogously.

10.6.2 Summary of the polynomials $\Phi_{a_3,n}$ related to the reference node $a_3 = (-1,1)$

For $d = 2$, $\eta = 4$, $\alpha = a_3 = (-1,1)$, $\beta_0 = (\beta_{00}, \beta_{01}) = a_2 = (1,1)$, and $\beta_1 = (\beta_{10}, \beta_{11}) = a_0 = (-1,-1)$, the Hermite Interpolating Polynomials writes

$$\Phi_{a_3,00}(x_0, x_1) = \frac{1}{0!0!} \left[\frac{1}{(2)^5} \frac{1}{(-2)^5} \right] (x_0 - 1)^5 (x_1 + 1)^5 \frac{1}{2^4} \left(1471 + 2350x_0 \right.$$
$$- 2350x_1 + 1590x_0^2 - 2650x_0x_1 + 1590x_1^2 + 525x_0^3 - 1125x_0^2x_1$$
$$+ 1125x_0x_1^2 - 525x_1^3 + 70x_0^4 - 175x_0^3x_1 + 225x_0^2x_1^2$$
$$\left. - 175x_0x_1^3 + 70x_1^4 \right) \tag{10.178}$$

$$\Phi_{a_3,10}(x_0, x_1) = \frac{1}{1!0!} \left[\frac{1}{(2)^5} \frac{1}{(-2)^5} \right] (x_0 - 1)^5 (x_1 + 1)^5 (x_0 + 1) \frac{1}{2^3} \left(378 \right.$$
$$+ 460x_0 - 460x_1 + 210x_0^2 - 350x_0x_1 + 210x_1^2 + 35x_0^3$$
$$\left. - 75x_0^2y + 75x_0x_1^2 - 35x_1^3 \right) \tag{10.179}$$

$$\Phi_{a3,01}(x_0, x_1) = \frac{1}{0!1!} \left[\frac{1}{(2)^5} \frac{1}{(-2)^5} \right] (x_0 - 1)^5 (x_1 + 1)^5 (x_1 - 1) \frac{1}{2^3} \Big(378$$
$$+ 460x_0 - 460x_1 + 210x_0^2 - 350x_0 x_1 + 210x_1^2 + 35x_0^3$$
$$- 75x_0^2 x_1 + 75x_0 x_1^2 - 35x_1^3 \Big) \qquad (10.180)$$

$$\Phi_{a3,20}(x_0, x_1) = \frac{1}{2!0!} \left[\frac{1}{(2)^5} \frac{1}{(-2)^5} \right] (x_0 - 1)^5 (x_1 + 1)^5 (x_0 + 1)^2$$
$$\times \frac{1}{2^2} \Big(79 + 65x_0 - 65x_1 + 15x_0^2 - 25x_0 x_1 + 15x_1^2 \Big) \quad (10.181)$$

$$\Phi_{a3,11}(x_0, x_1) = \frac{1}{1!1!} \left[\frac{1}{(2)^5} \frac{1}{(-2)^5} \right] (x_0 - 1)^5 (y + 1)^5 (x_0 + 1)(x_1 - 1)$$
$$\times \frac{1}{2^2} \Big(79 + 65x_0 - 65x_1 + 15x_0^2 - 25x_0 x_1 + 15x_1^2 \Big) \quad (10.182)$$

$$\Phi_{a3,02}(x_0, x_1) = \frac{1}{0!2!} \left[\frac{1}{(2)^5} \frac{1}{(-2)^5} \right] (x_0 - 1)^5 (x_1 + 1)^5 (x_1 - 1)^2$$
$$\times \frac{1}{2^2} \Big(79 + 65x_0 - 65x_1 + 15x_0^2 - 25x_0 x_1 + 15x_1^2 \Big) \quad (10.183)$$

$$\Phi_{a3,30}(x_0, x_1)$$
$$= \frac{1}{3!0!} \left[\frac{1}{(2)^5} \frac{1}{(-2)^5} \right] (x_0 - 1)^5 (x_1 + 1)^5 (x_0 + 1)^3 \frac{1}{2^1} \Big(12 + 5x_0 - 5x_1 \Big) \qquad (10.184)$$

$$\Phi_{a3,21}(x_0, x_1)$$
$$= \frac{1}{2!1!} \left[\frac{1}{(2)^5} \frac{1}{(-2)^5} \right] (x_0 - 1)^5 (x_1 + 1)^5 (x_0 + 1)^2 (x_1 - 1) \frac{1}{2^1} \Big(12 + 5x_0$$
$$- 5x_1 \Big) \qquad (10.185)$$

$$\Phi_{a3,12}(x_0, x_1)$$
$$= \frac{1}{1!2!} \left[\frac{1}{(2)^5} \frac{1}{(-2)^5} \right] (x_0 - 1)^5 (x_1 + 1)^5 (x_0 + 1)(x_1 - 1)^2 \frac{1}{2^1} \Big(12 + 5x_0 - 5y \Big) \qquad (10.186)$$

$$\Phi_{a3,03}(x_0, x_1)$$
$$= \frac{1}{0!3!} \left[\frac{1}{(2)^5} \frac{1}{(-2)^5} \right] (x_0 - 1)^5 (x_1 + 1)^5 (x_1 - 1)^3 \frac{1}{2^1} \Big(12 + 5x_0 - 5x_1 \Big) \qquad (10.187)$$

$$\Phi_{a_3,40}(x_0,x_1) = \frac{1}{4!0!}\left[\frac{1}{(2)^5}\frac{1}{(-2)^5}\right](x_0-1)^5(x_1+1)^5(x_0+1)^4\frac{1}{2^0}\begin{pmatrix}1\end{pmatrix}$$
$$(10.188)$$

$$\Phi_{a_3,31}(x_0,x_1) = \frac{1}{3!1!}\left[\frac{1}{(2)^5}\frac{1}{(-2)^5}\right](x_0-1)^5(x_1+1)^5(x_0+1)^3(x_1-1)\frac{1}{2^0}\begin{pmatrix}1\end{pmatrix}$$
$$(10.189)$$

$$\Phi_{a_3,22}(x_0,x_1) = \frac{1}{2!2!}\left[\frac{1}{(2)^5}\frac{1}{(-2)^5}\right](x_0-1)^5(x_1+1)^5(x_0+1)^2(x_1-1)^2\frac{1}{2^0}\begin{pmatrix}1\end{pmatrix}$$
$$(10.190)$$

$$\Phi_{a_3,13}(x_0,x_1) = \frac{1}{1!3!}\left[\frac{1}{(2)^5}\frac{1}{(-2)^5}\right](x_0-1)^5(x_1+1)^5(x_0+1)^1(x_1-1)^3\frac{1}{2^0}\begin{pmatrix}1\end{pmatrix}$$
$$(10.191)$$

$$\Phi_{a_3,04}(x_0,x_1) = \frac{1}{0!4!}\left[\frac{1}{(2)^5}\frac{1}{(-2)^5}\right](x_0-1)^5(x_1+1)^5(x_1-1)^4\frac{1}{2^0}\begin{pmatrix}1\end{pmatrix}$$
$$(10.192)$$

10.7 Generation of the polynomial $\Phi_{a_0,00}(x_0,x_1)$ related to the reference node $a_0 = (-1,-1)$ with $q = 6$

10.7.1 Computation of the coefficients of the polynomial $f_{a_0,00}(x)$

a- Degree of the polynomial $f_{a_0,00}(x)$

From corollary 7.2.19, on page 305, one has

$$q \geq \eta - \lambda_0(n) = 4 - \lambda_0(0,0) = 4 - 0 = 4 \qquad (10.193)$$

degree $q = 6 > 4$

$$
\begin{aligned}
f_{a_0,00}(x) = {} & a_{0,0}x_0^0x_1^0 + a_{1,0}x_0^1x_1^0 + a_{0,1}x_0^0x_1^1 + a_{2,0}x_0^2x_1^0 + a_{1,1}x_0^1x_1^1 + a_{0,2}x_0^0x_1^2 \\
& + a_{3,0}x_0^3x_1^0 + a_{2,1}x_0^2x_1^1 + a_{1,2}x_0^1x_1^2 + a_{0,3}x_0^0x_1^3 \\
& + a_{4,0}x_0^4x_1^0 + a_{3,1}x_0^3x_1^1 + a_{2,2}x_0^2x_1^2 + a_{1,3}x_0^1x_1^3 + a_{0,4}x_0^0x_1^4 \\
& + a_{5,0}x_0^5x_1^0 + a_{4,1}x_0^4x_1^1 + a_{3,2}x_0^3x_1^2 + a_{2,3}x_0^2x_1^3 + a_{1,4}x_0^1x_1^4 + a_{0,5}x_0^0x_1^5 \\
& + a_{6,0}x_0^6x_1^0 + a_{4,1}x_0^5x_1^1 + a_{4,2}x_0^4x_1^2 + a_{3,3}x_0^3x_1^3 + a_{2,4}x_0^2x_1^4 + a_{1,5}x_0^1x_1^5 \\
& + a_{0,6}x_0^0x_1^6
\end{aligned}
$$
$$(10.194)$$

The notation $a_{(w)}$, used later on, refers to the coefficients of the above polynomial.

b- Computation of the coefficients that belong to the terms in the level $\lambda_0(v) = q - 0 = 6$

1. Coefficients to be evaluated, which belong to the level $\lambda_0(v) = 6$.

$$v \in L_q^\delta = L_6^1 = \ cc(6,0), (5,1), (4,2), (3,3), (2,4), (1,5), (0,6) \}$$

$$(10.195)$$

2. Choice of the equation to be used for the computation of the coefficients.

   The coefficients of $f_{e,n}(x)$ are obtained using equation (7.122) or (7.123), on page 314, which is

$$a_v = \frac{1}{v!} E_F^v(\alpha) = \frac{1}{v_0! \, v_1!} E_F^v(\alpha) \qquad \forall v \in L_q^\delta \qquad (10.196)$$

   where $v \in L_q^\delta = L_4^1 = \{ (6,0), (5,1), (4,2), (3,3), (2,4), (1,5), (0,6) \}$. From (7.87),

$$E_F^{(v)}(\alpha) = E_F^{(v_0)}(\alpha) E_F^{(v_1)}(\alpha)$$
$$= \left[\frac{(\eta + v_0)!}{\eta!} \frac{1}{(\beta_{00} - \alpha_0)^{v_0}} \right] \left[\frac{(\eta + v_1)!}{\eta!} \frac{1}{(\beta_{11} - \alpha_1)^{v_1}} \right] \quad (10.197)$$

3. Computation of the coefficients $a_{(w)} = a_{(u+v)}$ using equation (10.196).

   For the level $\lambda_0(v) = q = 6$ the values of v are the elements of the set L_6^1 and $u = (0,0)$ therefore, for $v = (6,0)$ it follows

$$a_{(6,0)} = \frac{1}{6!0!} \left[\frac{(4+6)!}{4!} \frac{1}{(1+1)^6} \right] \left[\frac{(4+0)!}{4!} \frac{1}{(1+1)^0} \right] = \frac{151200}{6! \times 2^6} = \frac{210}{2^6}$$

$$(10.198)$$

   For $v = (5,1)$ it follows

$$a_{(5,1)} = \frac{1}{5!1!} \left[\frac{(4+5)!}{4!} \frac{1}{(1+1)^5} \right] \left[\frac{(4+1)!}{4!} \frac{1}{(1+1)^1} \right]$$
$$= \frac{15120}{5! \times 2^5} \times \frac{5}{2^1} = \frac{126}{2^5} \times \frac{5}{2^1} = \frac{630}{2^6} \qquad (10.199)$$

   where $w = (v + u) = (5,1) + (0,0) = (5,1)$.

   For $v = (4,2)$ it follows

$$a_{(4,2)} = \frac{1}{4!2!} \left[\frac{(4+4)!}{4!} \frac{1}{(1+1)^4} \right] \left[\frac{(4+2)!}{4!} \frac{1}{(1+1)^2} \right]$$
$$= \frac{1680}{4! \times 2^4} \times \frac{30}{2! \times 2^2} = \frac{70}{2^4} \times \frac{15}{2^2} = \frac{1050}{2^6} \qquad (10.200)$$

   where $w = (v + u) = (4,2) + (0,0) = (4,2)$.

For $v = (3, 3)$ it follows

$$a_{(3,3)} = \frac{1}{3!3!} \left[\frac{(4+3)!}{4!} \frac{1}{(1+1)^3} \right] \left[\frac{(4+3)!}{4!} \frac{1}{(1+1)^3} \right]$$

$$= \frac{210}{3! \times 2^3} \times \frac{210}{3! \times 2^3} = \frac{35}{2^3} \times \frac{35}{2^3} = \frac{1225}{2^6} \tag{10.201}$$

where $w = (v + u) = (3, 3) + (0, 0) = (3, 3)$.

The remaining coefficients are given by

$$a_{(0,6)} = a_{(6,0)} = \frac{210}{2^6}$$

$$a_{(1,5)} = a_{(5,1)} = \frac{630}{2^6}$$

$$a_{(2,4)} = a_{(4,2)} = \frac{1050}{2^6} \tag{10.202}$$

with the aid of theorem 7.4.7.

c- Computation of the coefficients that belong to the terms in the level $\lambda_0(v) = q - 1 = 5$

1. The indices v of the coefficients are the constrained numbers in the set

$$L^{\delta}_{\lambda_0(v)} = L^{1}_5 = \{ (5, 0), (4, 1), (3, 2), (2, 3), (1, 4), (0, 5) \} \tag{10.203}$$

2. The coefficients are obtained using equation (7.123), on page 314, which is

$$a_{(v)} = \frac{1}{v!} \left[E^{(v)}_F(\alpha) - S_v(a_{v+u}, \alpha) \right] \qquad \lambda_0(v) = 5 \tag{10.204}$$

3. Degree of the term $S_v(a_{u+v}, \alpha)$

$m = q - \lambda_0(v) = 6 - 5 = 1$

The exponents of the variables in each term of $S_v(a_{u+v}, \alpha)$ are the elements of the set shown in equation (7.124), on page 314, which is

$$\left(L^{\delta}(m) - L^{\delta}_0 \right) = \left(L^1(1) - L^1_0 \right) = \{(0,0), (1,0), (0,1)\} - \{(0,0)\}$$
$$= \{(1,0), (0,1)\} \tag{10.205}$$

and it gives, for $u \in \{(1,0), (0,1)\}$ the following expression

$$S_v(a_{v+u}, \alpha) = a_{(v+(1,0))} \frac{(v + (1,0))!}{(1,0)!} \alpha_0^1 \alpha_1^0 + a_{(v+(0,1))} \frac{(v + (0,1))!}{(0,1)!} \alpha_0^0 \alpha_1^1 \tag{10.206}$$

4. Computation of the coefficient $a_{(0,5)}$ where $v = (0,5)$.

$$E_F^v(\alpha) = E_F^{v_0}(\alpha)E_F^{v_1}(\alpha) = E_F^0(\alpha)E_F^5(\alpha) \tag{10.207}$$

where

$$E_F^0(\alpha) = \frac{(\eta + v_0)!}{\eta!} \frac{1}{(\beta_{00} - \alpha_0)^{v_0}} = \frac{(4+0)!}{4!} \frac{1}{2^0} = 1$$

$$E_F^5(\alpha) = \frac{(\eta + v_1)!}{\eta!} \frac{1}{(\beta_{11} - \alpha_1)^{v_1}} = \frac{(4+5)!}{4!} \frac{1}{2^5} = \frac{15120}{2^5} \tag{10.208}$$

then

$$E_F^{(0,5)}(\alpha) = E_F^0(\alpha)E_F^5(\alpha) = 1 \times \frac{15120}{2^5} = \frac{30240}{2^6} \tag{10.209}$$

and

$$S_{(0,5)}(a_{v+u}, \alpha) = a_{((0,5)+(1,0))} \frac{((0,5)+(1,0))!}{(1,0)!} \alpha_0^1 \alpha_1^0$$

$$+ a_{((0,5)+(0,1))} \frac{((0,5)+(0,1))!}{(0,1)!} \alpha_0^0 \alpha_1^1$$

$$= a_{(1,5)} \frac{(1,5)!}{(1,0)!} \alpha_0^1 \alpha_1^0 + a_{(0,6)} \frac{(0,6)!}{(0,1)!} \alpha_0^0 \alpha_1^1$$

$$= \frac{630}{2^6} \frac{1!5!}{1!0!} (-1)^1 (-1)^0 + \frac{210}{2^6} \frac{0!6!}{0!1!} (-1)^0 (-1)^1$$

$$= -\frac{75600}{2^6} - \frac{151200}{2^6} = -\frac{226800}{2^6} \tag{10.210}$$

Substituting the values given by equations (10.209) and (10.210) into equation (10.204)

$$a_{(0,5)} = \frac{1}{0!5!} \left[E_F^{(0,5)}(-1,-1) - S_{(0,5)}(a_{v+u}, \alpha) \right]$$

$$= \frac{1}{0!5!} \left[\frac{30240}{2^6} + \frac{226800}{2^6} \right] = \frac{257040}{5! \times 2^6} = \frac{2142}{2^6} \tag{10.211}$$

5. Computation of the coefficient $a_{(1,4)}$ where $v = (1,4)$.

$$E_F^v(\alpha) = E_F^{v_0}(\alpha)E_F^{v_1}(\alpha) = E_F^1(\alpha)E_F^4(\alpha) \tag{10.212}$$

where

$$E_F^1(\alpha) = \frac{(\eta + v_0)!}{\eta!} \frac{1}{(\beta_{00} - \alpha_0)^{v_0}} = \frac{(4+1)!}{4!} \frac{1}{2^1} = \frac{5}{2}$$

$$E_F^4(\alpha) = \frac{(\eta + v_1)!}{\eta!} \frac{1}{(\beta_{11} - \alpha_1)^{v_1}} = \frac{(4+4)!}{4!} \frac{1}{2^4} = \frac{1680}{2^4} \tag{10.213}$$

then

$$E_F^{(1,4)}(\alpha) = E_F^1(\alpha)E_F^4(\alpha) = \frac{5}{2} \times \frac{1680}{2^4} = \frac{8400}{2^5} = \frac{16800}{2^6} \quad (10.214)$$

and

$$
\begin{aligned}
S_{(1,4)}\left(a_{v+u}, \alpha\right) &= a_{((1,4)+(1,0))} \frac{((1,4)+(1,0))!}{(1,0)!} \alpha_0^1 \alpha_1^0 \\
&\quad + a_{((1,4)+(0,1))} \frac{((1,4)+(0,1))!}{(0,1)!} \alpha_0^0 \alpha_1^1 \\
&= a_{(2,4)} \frac{(2,4)!}{(1,0)!} \alpha_0^1 \alpha_1^0 + a_{(1,5)} \frac{(1,5)!}{(0,1)!} \alpha_0^0 \alpha_1^1 \\
&= \frac{1050}{2^6} \frac{2!4!}{1!0!}(-1)^1(-1)^0 + \frac{630}{2^6} \frac{1!5!}{0!1!}(-1)^0(-1)^1 \\
&= -\frac{50400}{2^6} - \frac{75600}{2^6} = -\frac{126000}{2^6} \quad (10.215)
\end{aligned}
$$

Substituting the values given by equations (10.214) and (10.215) into equation (10.204) it gives

$$
\begin{aligned}
a_{(1,4)} &= \frac{1}{1!4!}\left[E_F^{(1,4)}(-1,-1) - S_{(1,4)}\left(a_{v+u}, \alpha\right)\right] \\
&= \frac{1}{1!4!}\left[\frac{16800}{2^6} + \frac{126000}{2^6}\right] = \frac{5950}{2^6} \quad (10.216)
\end{aligned}
$$

6. Computation of the coefficient $a_{(2,3)}$ where $v = (2,3)$.

$$E_F^v(\alpha) = E_F^{v_0}(\alpha)E_F^{v_1}(\alpha) = E_F^2(\alpha)E_F^3(\alpha) \quad (10.217)$$

where

$$
\begin{aligned}
E_F^2(\alpha) &= \frac{(\eta + v_0)!}{\eta!} \frac{1}{(\beta_{00} - \alpha_0)^{v_0}} = \frac{(4+2)!}{4!} \frac{1}{2^2} = \frac{30}{2^2} \\
E_F^3(\alpha) &= \frac{(\eta + v_1)!}{\eta!} \frac{1}{(\beta_{11} - \alpha_1)^{v_1}} = \frac{(4+3)!}{4!} \frac{1}{2^3} = \frac{210}{2^3} \quad (10.218)
\end{aligned}
$$

then

$$E_F^{(2,3)}(\alpha) = E_F^2(\alpha)E_F^3(\alpha) = \frac{30}{2^2} \times \frac{210}{2^3} = \frac{6300}{2^5} = \frac{12600}{2^6} \quad (10.219)$$

and

$$S_{(2,3)}(a_{v+u}, \alpha) = a_{((2,3)+(1,0))} \frac{((2,3)+(1,0))!}{(1,0)!} \alpha_0^1 \alpha_1^0$$

$$+ a_{((2,3)+(0,1))} \frac{((2,3)+(0,1))!}{(0,1)!} \alpha_0^0 \alpha_1^1$$

$$= a_{(3,3)} \frac{(3,3)!}{(1,0)!} \alpha_0^1 \alpha_1^0 + a_{(2,4)} \frac{(2,4)!}{(0,1)!} \alpha_0^0 \alpha_1^1$$

$$= \frac{1225}{2^6} \frac{3!3!}{1!0!} (-1)^1 (-1)^0 + \frac{1050}{2^6} \frac{2!4!}{0!1!} (-1)^0 (-1)^1$$

$$= -\frac{44100}{2^6} - \frac{50400}{2^6} = -\frac{94500}{2^6} \qquad (10.220)$$

Substituting the values given by equations (10.219) and (10.220) into equation (10.204) it gives

$$a_{(2,3)} = \frac{1}{2!3!} \left[E_F^{(2,3)}(-1,-1) - S_{(2,3)}(a_{v+u}, \alpha) \right]$$

$$= \frac{1}{2!3!} \left[\frac{12600}{2^6} + \frac{94500}{2^6} \right] = \frac{8925}{2^6} \qquad (10.221)$$

7. Computation of the coefficients $a_{(5,0)}$, $a_{(4,1)}$, and $a_{(3,2)}$

The theorem 7.4.7, on page 324, says that for a normalized region symmetric with respect to the coordinate axes the coefficients in the same level whose indices are permutations are equal, therefore,

$$a_{(5,0)} = a_{(0,5)} = \frac{2142}{2^6}$$

$$a_{(4,1)} = a_{(1,4)} = \frac{5950}{2^6}$$

$$a_{(3,2)} = a_{(2,3)} = \frac{8925}{2^6} \qquad (10.222)$$

d- Computation of the coefficients that belong to the terms in the level $\lambda_0(v) = q - 2 = 4$

1. The indices v of the coefficients are the constrained numbers in the set

$$L_{\lambda_0(v)}^{\delta} = L_4^1 = \{ (4,0), (3,1), (2,2), (1,3), (0,4) \} \qquad (10.223)$$

2. The coefficients are obtained using equation (7.123), on page 314, which is

$$a_{(v)} = \frac{1}{v!} \left[E_F^{(v)}(\alpha) - S_v(a_{v+u}, \alpha) \right] \qquad \lambda_0(v) = 4 \qquad (10.224)$$

3. Degree of the term $S_v\,(a_{u+v}, \alpha)$

$$m = q - \lambda_0(v) = 6 - 4 = 2$$

The exponents of the variables in each term of $S_v\,(a_{u+v}, \alpha)$ are the elements of the set shown in equation (7.124), on page 314, which is

$$
\begin{aligned}
\left(L^\delta(m) - L_0^\delta\right) &= \left(L^1(2) - L_0^1\right) \\
&= \left\{\,(0,0), (1,0), (0,1), (2,0), (1,1), (0,2)\,\right\} - \left\{\,(0,0)\,\right\} \\
&= \left\{\,(1,0), (0,1), (2,0), (1,1), (0,2)\,\right\} \tag{10.225}
\end{aligned}
$$

and it gives, for $u \in \{(1,0), (0,1), (2,0), (1,1), (0,2)\}$ the following expression

$$
\begin{aligned}
S_v\,(a_{v+u}, \alpha) &= a_{(v+(1,0))} \frac{(v+(1,0))!}{(1,0)!}\,\alpha_0^1\alpha_1^0 + a_{(v+(0,1))} \frac{(v+(0,1))!}{(0,1)!}\,\alpha_0^0\alpha_1^1 \\
&\quad + a_{(v+(2,0))} \frac{(v+(2,0))!}{(2,0)!}\,\alpha_0^2\alpha_1^0 + a_{(v+(1,1))} \frac{(v+(1,1))!}{(1,1)!}\,\alpha_0^1\alpha_1^1 \\
&\quad + a_{(v+(0,2))} \frac{(v+(0,2))!}{(0,2)!}\,\alpha_0^0\alpha_1^2 \tag{10.226}
\end{aligned}
$$

4. Computation of the coefficient $a_{(0,4)}$ where $v = (0,4)$.

$$E_F^v(\alpha) = E_F^{v_0}(\alpha)E_F^{v_1}(\alpha) = E_F^0(\alpha)E_F^4(\alpha) \tag{10.227}$$

where

$$
\begin{aligned}
E_F^0(\alpha) &= \frac{(\eta + v_0)!}{\eta!}\,\frac{1}{(\beta_{00} - \alpha_0)^{v_0}} = \frac{(4+0)!}{4!}\,\frac{1}{2^0} = 1 \\
E_F^4(\alpha) &= \frac{(\eta + v_1)!}{\eta!}\,\frac{1}{(\beta_{11} - \alpha_1)^{v_1}} = \frac{(4+4)!}{4!}\,\frac{1}{2^4} = \frac{1680}{2^4} \tag{10.228}
\end{aligned}
$$

then

$$E_F^{(0,4)}(\alpha) = E_F^0(\alpha)E_F^4(\alpha) = 1 \times \frac{1680}{2^4} = \frac{1680}{2^4} = \frac{6720}{2^6} \tag{10.229}$$

and

$$
\begin{aligned}
S_{(0,4)}\,(a_{v+u}, \alpha) &= a_{((0,4)+(1,0))} \frac{((0,4)+(1,0))!}{(1,0)!}\,\alpha_0^1\alpha_1^0 \\
&\quad + a_{((0,4)+(0,1))} \frac{((0,4)+(0,1))!}{(0,1)!}\,\alpha_0^0\alpha_1^1 \\
&\quad + a_{((0,4)+(2,0))} \frac{((0,4)+(2,0))!}{(2,0)!}\,\alpha_0^2\alpha_1^0 \\
&\quad + a_{((0,4)+(1,1))} \frac{((0,4)+(1,1))!}{(1,1)!}\,\alpha_0^1\alpha_1^1 \\
&\quad + a_{((0,4)+(0,2))} \frac{((0,4)+(0,2))!}{(0,2)!}\,\alpha_0^0\alpha_1^2 \tag{10.230}
\end{aligned}
$$

$$S_{(0,4)}(a_{v+u}, \alpha) = a_{(1,4)} \frac{(1,4)!}{(1,0)!} \alpha_0^1 \alpha_1^0 + a_{(0,5)} \frac{(0,5)!}{(0,1)!} \alpha_0^0 \alpha_1^1$$

$$+ a_{(2,4)} \frac{(2,4)!}{(2,0)!} \alpha_0^2 \alpha_1^0 + a_{(1,5)} \frac{(1,5)!}{(1,1)!} \alpha_0^1 \alpha_1^1$$

$$+ a_{(0,6)} \frac{(0,6)!}{(0,2)!} \alpha_0^0 \alpha_1^2$$

$$= \frac{5950}{2^6} \frac{1!4!}{1!0!} (-1)^1 (-1)^0 + \frac{2142}{2^6} \frac{0!5!}{0!1!} (-1)^1 (-1)^0$$

$$+ \frac{1050}{2^6} \frac{2!4!}{2!0!} (-1)^2 (-1)^0 + \frac{630}{2^6} \frac{1!5!}{1!1!} (-1)^1 (-1)^1$$

$$+ \frac{210}{2^6} \frac{0!6!}{0!2!} (-1)^0 (-1)^2$$

$$= \frac{1}{2^6} \left(-142800 - 257040 + 25200 + 75600 + 75600 \right)$$

$$= -\frac{223400}{2^6} \tag{10.231}$$

Substituting the values given by equations (10.229) and (10.231) into equation (10.224) it gives

$$a_{(0,4)} = \frac{1}{0!4!} \left[E_F^{(0,4)}(-1,-1) - S_{(0,4)}(a_{v+u}, \alpha) \right]$$

$$= \frac{1}{0!4!} \left[\frac{6720}{2^6} + \frac{223400}{2^6} \right] = \frac{230160}{4! \times 2^6} = \frac{9590}{2^6} \tag{10.232}$$

5. Computation of the coefficient $a_{(1,3)}$ where $v = (1,3)$.

$$E_F^v(\alpha) = E_F^{v_0}(\alpha) E_F^{v_1}(\alpha) = E_F^1(\alpha) E_F^3(\alpha) \tag{10.233}$$

where

$$E_F^1(\alpha) = \frac{(\eta + v_0)!}{\eta!} \frac{1}{(\beta_{00} - \alpha_0)^{v_0}} = \frac{(4+1)!}{4!} \frac{1}{2^1} = \frac{5}{2}$$

$$E_F^3(\alpha) = \frac{(\eta + v_1)!}{\eta!} \frac{1}{(\beta_{11} - \alpha_1)^{v_1}} = \frac{(4+3)!}{4!} \frac{1}{2^3} = \frac{210}{2^3} \tag{10.234}$$

then

$$E_F^{(1,3)}(\alpha) = E_F^1(\alpha) E_F^3(\alpha) = \frac{5}{2} \times \frac{210}{2^3} = \frac{1050}{2^4} = \frac{4200}{2^6} \tag{10.235}$$

and

$$S_{(1,3)}(a_{v+u}, \alpha) = a_{((1,3)+(1,0))} \frac{((1,3)+(1,0))!}{(1,0)!} \alpha_0^1 \alpha_1^0$$

$$+ a_{((1,3)+(0,1))} \frac{((1,3)+(0,1))!}{(0,1)!} \alpha_0^0 \alpha_1^1$$

$$+ a_{((1,3)+(2,0))} \frac{((1,3)+(2,0))!}{(2,0)!} \alpha_0^2 \alpha_1^0$$

$$+ a_{((1,3)+(1,1))} \frac{((1,3)+(1,1))!}{(1,1)!} \alpha_0^1 \alpha_1^1$$

$$+ a_{((1,3)+(0,2))} \frac{((1,3)+(0,2))!}{(0,2)!} \alpha_0^0 \alpha_1^2 \qquad (10.236)$$

$$S_{(1,3)}(a_{v+u}, \alpha) = a_{(2,3)} \frac{(2,3)!}{(1,0)!} \alpha_0^1 \alpha_1^0 + a_{(1,4)} \frac{(1,4)!}{(0,1)!} \alpha_0^0 \alpha_1^1$$

$$+ a_{(3,3)} \frac{(3,3)!}{(2,0)!} \alpha_0^2 \alpha_1^0 + a_{(2,4)} \frac{(2,4)!}{(1,1)!} \alpha_0^1 \alpha_1^1 +$$

$$+ a_{(1,5)} \frac{(1,5)!}{(0,2)!} \alpha_0^0 \alpha_1^2 \qquad (10.237)$$

$$S_{(1,3)}(a_{v+u}, \alpha) = \frac{8925}{2^6} \frac{2!3!}{1!0!}(-1)^1(-1)^0 + \frac{5950}{2^6} \frac{1!4!}{0!1!}(-1)^1(-1)^0$$

$$+ \frac{1225}{2^6} \frac{3!3!}{2!0!}(-1)^2(-1)^0 + \frac{1050}{2^6} \frac{2!4!}{1!1!}(-1)^1(-1)^1$$

$$+ \frac{630}{2^6} \frac{1!5!}{0!2!}(-1)^0(-1)^2$$

$$= \frac{1}{2^6}(-107100 - 142800 + 22050 + 50400 + 37800)$$

$$= -\frac{139650}{2^6}$$

Substituting the values given by equations (10.235) and (10.237) into equation (10.224) it gives

$$a_{(1,3)} = \frac{1}{1!3!} \left[E_F^{(1,3)}(-1,-1) - S_{(1,3)}(a_{v+u}, \alpha) \right]$$

$$= \frac{1}{1!3!} \left[\frac{4200}{2^6} + \frac{139650}{2^6} \right] = \frac{143850}{3! \times 2^6} = \frac{23975}{2^6} \qquad (10.238)$$

6. Computation of the coefficient $a_{(2,2)}$ where $v = (2,2)$.

$$E_F^v(\alpha) = E_F^{v_0}(\alpha) E_F^{v_1}(\alpha) = E_F^2(\alpha) E_F^2(\alpha) \qquad (10.239)$$

where

$$E_F^2(\alpha) = \frac{(\eta + v_0)!}{\eta!} \frac{1}{(\beta_{00} - \alpha_0)^{v_0}} = \frac{(4+2)!}{4!} \frac{1}{2^2} = \frac{30}{2^2}$$

$$E_F^2(\alpha) = \frac{(\eta + v_1)!}{\eta!} \frac{1}{(\beta_{11} - \alpha_1)^{v_1}} = \frac{(4+2)!}{4!} \frac{1}{2^2} = \frac{30}{2^2} \qquad (10.240)$$

then

$$E_F^{(2,2)}(\alpha) = E_F^2(\alpha) E_F^2(\alpha) = \frac{30}{2^2} \times \frac{30}{2^2} = \frac{900}{2^4} = \frac{3600}{2^6} \qquad (10.241)$$

and

$$S_{(2,2)}(a_{v+u}, \alpha) = a_{((2,2)+(1,0))} \frac{((2,2)+(1,0))!}{(1,0)!} \alpha_0^1 \alpha_1^0$$

$$+ a_{((2,2)+(0,1))} \frac{((2,2)+(0,1))!}{(0,1)!} \alpha_0^0 \alpha_1^1$$

$$+ a_{((2,2)+(2,0))} \frac{((2,2)+(2,0))!}{(2,0)!} \alpha_0^2 \alpha_1^0$$

$$+ a_{((2,2)+(1,1))} \frac{((2,2)+(1,1))!}{(1,1)!} \alpha_0^1 \alpha_1^1$$

$$+ a_{((2,2)+(0,2))} \frac{((2,2)+(0,2))!}{(0,2)!} \alpha_0^0 \alpha_1^2 \qquad (10.242)$$

$$S_{(2,2)}(a_{v+u}, \alpha) = a_{(3,2)} \frac{(3,2)!}{(1,0)!} \alpha_0^1 \alpha_1^0 + a_{(2,3)} \frac{(2,3)!}{(0,1)!} \alpha_0^0 \alpha_1^1$$

$$+ a_{(4,2)} \frac{(4,2)!}{(2,0)!} \alpha_0^2 \alpha_1^0 + a_{(3,3)} \frac{(3,3)!}{(1,1)!} \alpha_0^1 \alpha_1^1 +$$

$$+ a_{(2,4)} \frac{(2,4)!}{(0,2)!} \alpha_0^0 \alpha_1^2 \qquad (10.243)$$

$$S_{(2,2)}(a_{v+u}, \alpha) = \frac{8925}{2^6} \frac{3!2!}{1!0!} (-1)^1 (-1)^0 + \frac{8925}{2^6} \frac{2!3!}{0!1!} (-1)^1 (-1)^0$$

$$+ \frac{1050}{2^6} \frac{4!2!}{2!0!} (-1)^2 (-1)^0 + \frac{1225}{2^6} \frac{3!3!}{1!1!} (-1)^1 (-1)^1 +$$

$$+ \frac{1050}{2^6} \frac{2!4!}{0!2!} (-1)^0 (-1)^2$$

$$= \frac{1}{2^6} (-107100 - 107100 + 25200 + 44100 + 25200)$$

$$= -\frac{119700}{2^6}$$

Substituting the values given by equations (10.241) and (10.243) into equation (10.224) it gives

$$a_{(2,2)} = \frac{1}{2!2!}\left[E_F^{(2,2)}(-1,-1) - S_{(2,2)}(a_{v+u},\alpha)\right]$$

$$= \frac{1}{2!2!}\left[\frac{3600}{2^6} + \frac{119700}{2^6}\right] = \frac{30825}{2^6} \qquad (10.244)$$

7. Computation of the coefficients $a_{(4,0)}$, and $a_{(3,1)}$

The theorem 7.4.7, on page 324, says that for a normalized region symmetric with respect to the coordinate axes the coefficients in the same level whose indices are permutations are equal, therefore,

$$a_{(4,0)} = a_{(0,4)} = \frac{9590}{2^6}$$

$$a_{(3,1)} = a_{(1,3)} = \frac{23975}{2^6} \qquad (10.245)$$

e- Computation of the coefficients that belong to the terms in the level $\lambda_0(v) = q - 3 = 3$

1. The indices v of the coefficients are the constrained numbers in the set

$$L_{\lambda_0(v)}^{\delta} = L_3^{1} = \{(3,0),(2,1),(1,2),(0,3)\} \qquad (10.246)$$

2. The coefficients are obtained using equation (7.123), on page 314, repeated below

$$a_{(v)} = \frac{1}{v!}\left[E_F^{(v)}(\alpha) - S_v(a_{v+u},\alpha)\right] \qquad \lambda_0(v) = 3 \qquad (10.247)$$

3. Degree of the term $S_v(a_{u+v},\alpha)$

$$m = q - \lambda_0(v) = 6 - 3 = 3$$

The exponents of the variables in each term of $S_v(a_{u+v},\alpha)$ are the elements of the set shown in equation (7.124), on page 314, which is

$$\left(L^{\delta}(m) - L_0^{\delta}\right) = \left(L^1(2) - L_0^1\right)$$
$$= \{(0,0),(1,0),(0,1),(2,0),(1,1),(0,2),(3,0),$$
$$(2,1),(1,2),(0,3)\} - \{(0,0)\}$$
$$= \{(1,0),(0,1),(2,0),(1,1),(0,2),(3,0),$$
$$(2,1),(1,2),(0,3)\} \qquad (10.248)$$

and it gives, for

$$u \in \{(1,0),(0,1),(2,0),(1,1),(0,2),(3,0),(2,1),(1,2),(0,3)\}$$

the following expression

$$
\begin{aligned}
S_v\left(a_{v+u},\alpha\right) = {} & a_{(v+(1,0))}\frac{(v+(1,0))!}{(1,0)!}\,\alpha_0^1\alpha_1^0 + a_{(v+(0,1))}\frac{(v+(0,1))!}{(0,1)!}\,\alpha_0^0\alpha_1^1 \\
& + a_{(v+(2,0))}\frac{(v+(2,0))!}{(2,0)!}\,\alpha_0^2\alpha_1^0 + a_{(v+(1,1))}\frac{(v+(1,1))!}{(1,1)!}\,\alpha_0^1\alpha_1^1 \\
& + a_{(v+(0,2))}\frac{(v+(0,2))!}{(0,2)!}\,\alpha_0^0\alpha_1^2 + a_{(v+(3,0))}\frac{(v+(3,0))!}{(3,0)!}\,\alpha_0^3\alpha_1^0 \\
& + a_{(v+(2,1))}\frac{(v+(2,1))!}{(2,1)!}\,\alpha_0^2\alpha_1^1 + a_{(v+(1,2))}\frac{(v+(1,2))!}{(1,2)!}\,\alpha_0^1\alpha_1^2 \\
& + a_{(v+(0,3))}\frac{(v+(0,3))!}{(0,3)!}\,\alpha_0^0\alpha_1^3
\end{aligned}
$$

4. Computation of the coefficient $a_{(0,3)}$ where $v = (0,3)$.

$$
E_F^v(\alpha) = E_F^{v_0}(\alpha)E_F^{v_1}(\alpha) = E_F^0(\alpha)E_F^3(\alpha) \tag{10.249}
$$

where

$$
\begin{aligned}
E_F^0(\alpha) &= \frac{(\eta+v_0)!}{\eta!}\frac{1}{(\beta_{00}-\alpha_0)^{v_0}} = \frac{(4+0)!}{4!}\frac{1}{2^0} = 1 \\
E_F^3(\alpha) &= \frac{(\eta+v_1)!}{\eta!}\frac{1}{(\beta_{11}-\alpha_1)^{v_1}} = \frac{(4+3)!}{4!}\frac{1}{2^3} = \frac{210}{2^3}
\end{aligned} \tag{10.250}
$$

then

$$
E_F^{(0,3)}(\alpha) = E_F^0(\alpha)E_F^3(\alpha) = 1\times\frac{210}{2^3} = \frac{210}{2^3} = \frac{1680}{2^6} \tag{10.251}
$$

and

$$S_v\left(a_{v+u}, \alpha\right) = a_{((0,3)+(1,0))}\frac{((0,3)+(1,0))!}{(1,0)!}\,\alpha_0^1\alpha_1^0$$

$$+ a_{((0,3)+(0,1))}\frac{((0,3)+(0,1))!}{(0,1)!}\,\alpha_0^0\alpha_1^1$$

$$+ a_{((0,3)+(2,0))}\frac{((0,3)+(2,0))!}{(2,0)!}\,\alpha_0^2\alpha_1^0$$

$$+ a_{((0,3)+(1,1))}\frac{((0,3)+(1,1))!}{(1,1)!}\,\alpha_0^1\alpha_1^1$$

$$+ a_{((0,3)+(0,2))}\frac{((0,3)+(0,2))!}{(0,2)!}\,\alpha_0^0\alpha_1^2$$

$$+ a_{((0,3)+(3,0))}\frac{((0,3)+(3,0))!}{(2,0)!}\,\alpha_0^3\alpha_1^0$$

$$+ a_{((0,3)+(2,1))}\frac{((0,3)+(2,1))!}{(2,1)!}\,\alpha_0^2\alpha_1^1$$

$$+ a_{((0,3)+(1,2))}\frac{((0,3)+(1,2))!}{(1,2)!}\,\alpha_0^1\alpha_1^2$$

$$+ a_{((0,3)+(0,3))}\frac{((0,3)+(0,3))!}{(0,3)!}\,\alpha_0^0\alpha_1^3$$

$$S_{(0,3)}\left(a_{v+u}, \alpha\right) = a_{(1,3)}\frac{(1,3)!}{(1,0)!}\,\alpha_0^1\alpha_1^0 + a_{(0,4)}\frac{(0,4)!}{(0,1)!}\,\alpha_0^0\alpha_1^1$$

$$+ a_{(2,3)}\frac{(2,3)!}{(2,0)!}\,\alpha_0^2\alpha_1^0 + a_{(1,4)}\frac{(1,4)!}{(1,1)!}\,\alpha_0^1\alpha_1^1$$

$$+ a_{(0,5)}\frac{(0,5)!}{(0,2)!}\,\alpha_0^0\alpha_1^2 + a_{(3,3)}\frac{(3,3)!}{(3,0)!}\,\alpha_0^3\alpha_1^0$$

$$+ a_{(2,4)}\frac{(2,4)!}{(2,1)!}\,\alpha_0^2\alpha_1^1 + a_{(1,5)}\frac{(1,5)!}{(1,2)!}\,\alpha_0^1\alpha_1^2$$

$$+ a_{(0,6)}\frac{(0,6)!}{(0,3)!}\,\alpha_0^0\alpha_1^3$$

$$S_{(0,3)}(a_{v+u}, \alpha) = \frac{23975}{2^6} \frac{1!3!}{1!0!}(-1)^1(-1)^0 + \frac{9590}{2^6} \frac{0!4!}{0!1!}(-1)^0(-1)^1$$

$$+ \frac{8925}{2^6} \frac{2!3!}{2!0!}(-1)^2(-1)^0 + \frac{5950}{2^6} \frac{1!4!}{1!1!}(-1)^1(-1)^1$$

$$+ \frac{2142}{2^6} \frac{0!5!}{0!2!}(-1)^0(-1)^2 + \frac{1225}{2^6} \frac{3!3!}{3!0!}(-1)^3(-1)^0$$

$$+ \frac{1050}{2^6} \frac{2!4!}{2!1!}(-1)^2(-1)^1 + \frac{630}{2^6} \frac{1!5!}{1!2!}(-1)^1(-1)^2$$

$$+ \frac{210}{2^6} \frac{0!6!}{0!3!}(-1)^0(-1)^3$$

$$S_{(0,3)}(a_{v+u}, \alpha) = \frac{1}{2^6}\big(-143850 - 230160 + 53550 + 142800 + 128520$$

$$- 7350 - 25200 - 37800 - 25200\big)$$

$$= -\frac{144690}{2^6} \tag{10.252}$$

Substituting the values given by equations (10.251) and (10.252) into equation (10.247) it gives

$$a_{(0,3)} = \frac{1}{0!3!}\left[E_F^{(0,3)}(-1, -1) - S_{(0,3)}(a_{v+u}, \alpha)\right]$$

$$= \frac{1}{0!3!}\left[\frac{1680}{2^6} + \frac{144690}{2^6}\right] = \frac{146370}{3! \times 2^6} = \frac{24395}{2^6} \tag{10.253}$$

5. Computation of the coefficient $a_{(1,2)}$ where $v = (1,2)$.

$$E_F^v(\alpha) = E_F^{v_0}(\alpha)E_F^{v_1}(\alpha) = E_F^1(\alpha)E_F^2(\alpha) \tag{10.254}$$

where

$$E_F^1(\alpha) = \frac{(\eta + v_0)!}{\eta!}\frac{1}{(\beta_{00} - \alpha_0)^{v_0}} = \frac{(4+1)!}{4!}\frac{1}{2^1} = \frac{5}{2}$$

$$E_F^2(\alpha) = \frac{(\eta + v_1)!}{\eta!}\frac{1}{(\beta_{11} - \alpha_1)^{v_1}} = \frac{(4+2)!}{4!}\frac{1}{2^2} = \frac{30}{2^2} \tag{10.255}$$

then

$$E_F^{(1,2)}(\alpha) = E_F^1(\alpha)E_F^2(\alpha) = \frac{5}{2} \times \frac{30}{2^2} = \frac{150}{2^3} = \frac{1200}{2^6} \tag{10.256}$$

and

$$S_v\left(a_{v+u}, \alpha\right) = a_{((1,2)+(1,0))}\frac{((1,2)+(1,0))!}{(1,0)!}\,\alpha_0^1\alpha_1^0$$

$$+\,a_{((1,2)+(0,1))}\frac{((1,2)+(0,1))!}{(0,1)!}\,\alpha_0^0\alpha_1^1$$

$$+\,a_{((1,2)+(2,0))}\frac{((1,2)+(2,0))!}{(2,0)!}\,\alpha_0^2\alpha_1^0$$

$$+\,a_{((1,2)+(1,1))}\frac{((1,2)+(1,1))!}{(1,1)!}\,\alpha_0^1\alpha_1^1$$

$$+\,a_{((1,2)+(0,2))}\frac{((1,2)+(0,2))!}{(0,2)!}\,\alpha_0^0\alpha_1^2$$

$$+\,a_{((1,2)+(3,0))}\frac{((1,2)+(3,0))!}{(2,0)!}\,\alpha_0^3\alpha_1^0$$

$$+\,a_{((1,2)+(2,1))}\frac{((1,2)+(2,1))!}{(1,1)!}\,\alpha_0^2\alpha_1^1$$

$$+\,a_{((1,2)+(1,2))}\frac{((1,2)+(1,2))!}{(0,2)!}\,\alpha_0^1\alpha_1^2$$

$$+\,a_{((1,2)+(0,3))}\frac{((1,2)+(0,3))!}{(0,2)!}\,\alpha_0^0\alpha_1^3$$

$$S_{(1,2)}\left(a_{v+u}, \alpha\right) = a_{(2,2)}\frac{(2,2)!}{(1,0)!}\,\alpha_0^1\alpha_1^0 + a_{(1,3)}\frac{(1,3)!}{(0,1)!}\,\alpha_0^0\alpha_1^1$$

$$+\,a_{(3,2)}\frac{(3,2)!}{(2,0)!}\,\alpha_0^2\alpha_1^0 + a_{(2,3)}\frac{(2,3)!}{(1,1)!}\,\alpha_0^1\alpha_1^1$$

$$+\,a_{(1,4)}\frac{(1,4)!}{(0,2)!}\,\alpha_0^0\alpha_1^2 + a_{(4,2)}\frac{(4,2)!}{(3,0)!}\,\alpha_0^3\alpha_1^0$$

$$+\,a_{(3,3)}\frac{(3,3)!}{(2,1)!}\,\alpha_0^2\alpha_1^1 + a_{(2,4)}\frac{(2,4)!}{(1,2)!}\,\alpha_0^1\alpha_1^2$$

$$+\,a_{(1,5)}\frac{(1,5)!}{(0,3)!}\,\alpha_0^0\alpha_1^3$$

$$S_{(1,2)}(a_{v+u}, \alpha) = \frac{30825}{2^6} \frac{2!2!}{1!0!}(-1)^1(-1)^0 + \frac{23975}{2^6} \frac{1!3!}{0!1!}(-1)^0(-1)^1$$

$$+ \frac{8925}{2^6} \frac{3!2!}{2!0!}(-1)^2(-1)^0 + \frac{8925}{2^6} \frac{2!3!}{1!1!}(-1)^1(-1)^1$$

$$+ \frac{5950}{2^6} \frac{1!4!}{0!2!}(-1)^0(-1)^2 + \frac{1050}{2^6} \frac{4!2!}{3!0!}(-1)^3(-1)^0$$

$$+ \frac{1225}{2^6} \frac{3!3!}{2!1!}(-1)^2(-1)^1 + \frac{1050}{2^6} \frac{2!4!}{1!2!}(-1)^1(-1)^2$$

$$+ \frac{630}{2^6} \frac{1!5!}{0!3!}(-1)^0(-1)^3$$

$$S_{(1,2)}(a_{v+u}, \alpha) = \frac{1}{2^6}\Big(-123300 - 143850 + 53550 + 107100 + 71400$$

$$- 8400 - 22050 - 25200 - 12600 \Big)$$

$$= -\frac{103350}{2^6} \tag{10.257}$$

Substituting the values given by equations (10.256) and (10.257) into equation (10.247) it gives

$$a_{(1,2)} = \frac{1}{1!2!} \left[E_F^{(1,2)}(-1,-1) - S_{(1,2)}(a_{v+u}, \alpha) \right]$$

$$= \frac{1}{1!2!} \left[\frac{1200}{2^6} + \frac{103350}{2^6} \right] = \frac{104550}{2! \times 2^6} = \frac{52275}{2^6} \tag{10.258}$$

6. Computation of the coefficients $a_{(3,0)}$, and $a_{(2,1)}$

The theorem 7.4.7, on page 324, says that for a normalized region symmetric with respect to the coordinate axes the coefficients in the same level whose indices are permutations are equal, therefore,

$$a_{(3,0)} = a_{(0,3)} = \frac{24395}{2^6}$$

$$a_{(2,1)} = a_{(1,2)} = \frac{52275}{2^6} \tag{10.259}$$

f- Computation of the coefficients that belong to the terms in the level $\lambda_0(v) = q - 4 = 2$

1. The indices v of the coefficients are the constrained numbers in the set

$$L_{\lambda_0(v)}^\delta = L_2^1 = \{(2,0), (1,1), (0,2)\} \tag{10.260}$$

2. The coefficients are obtained using equation (7.123), on page 314, which
is

$$a_{(v)} = \frac{1}{v!} \left[E_F^{(v)}(\alpha) - S_v\left(a_{v+u}, \alpha\right) \right] \qquad \lambda_0(v) = 2 \qquad (10.261)$$

3. Degree of the term $S_v\left(a_{u+v}, \alpha\right)$

$$m = q - \lambda_0(v) = 6 - 2 = 4$$

The exponents of the variables in each term of $S_v\left(a_{u+v}, \alpha\right)$ are the
elements of the set shown in equation (7.124), on page 314, which is

$$\begin{aligned}
\left(L^\delta(m) - L_0^\delta\right) = \left(L^1(2) - L_0^1\right) & \\
= \big\{ & (0,0), (1,0), (0,1), (2,0), (1,1), (0,2), (3,0), \\
& (2,1), (1,2), (0,3), \\
& (4,0), (3,1), (2,2), (1,3), (0,4) \big\} - \big\{(0,0)\big\} \\
= \big\{ & (1,0), (0,1), (2,0), (1,1), (0,2), (3,0), \\
& (2,1), (1,2), (0,3), \\
& (4,0), (3,1), (2,2), (1,3), (0,4) \big\} \qquad (10.262)
\end{aligned}$$

and it gives, for

$$\begin{aligned}
u \in \big\{ & (1,0), (0,1), (2,0), (1,1), (0,2), (3,0), (2,1), (1,2), (0,3), \\
& (4,0), (3,1), (2,2), (1,3), (0,4) \big\} \qquad (10.263)
\end{aligned}$$

the following expression

$$\begin{aligned}
S_v\left(a_{v+u}, \alpha\right) = & \; a_{(v+(1,0))} \frac{(v+(1,0))!}{(1,0)!} \alpha_0^1 \alpha_1^0 + a_{(v+(0,1))} \frac{(v+(0,1))!}{(0,1)!} \alpha_0^0 \alpha_1^1 \\
& + a_{(v+(2,0))} \frac{(v+(2,0))!}{(2,0)!} \alpha_0^2 \alpha_1^0 + a_{(v+(1,1))} \frac{(v+(1,1))!}{(1,1)!} \alpha_0^1 \alpha_1^1 \\
& + a_{(v+(0,2))} \frac{(v+(0,2))!}{(0,2)!} \alpha_0^0 \alpha_1^2 + a_{(v+(3,0))} \frac{(v+(3,0))!}{(3,1)!} \alpha_0^3 \alpha_1^0 \\
& + a_{(v+(2,1))} \frac{(v+(2,1))!}{(2,1)!} \alpha_0^2 \alpha_1^1 + a_{(v+(1,2))} \frac{(v+(1,2))!}{(1,2)!} \alpha_0^1 \alpha_1^2 \\
& + a_{(v+(0,3))} \frac{(v+(0,3))!}{(0,3)!} \alpha_0^0 \alpha_1^3 + a_{(v+(4,0))} \frac{(v+(4,0))!}{(4,0)!} \alpha_0^4 \alpha_1^0 \\
& + a_{(v+(3,1))} \frac{(v+(3,1))!}{(3,1)!} \alpha_0^3 \alpha_1^1 + a_{(v+(2,2))} \frac{(v+(2,2))!}{(2,2)!} \alpha_0^2 \alpha_1^2 \\
& + a_{(v+(1,3))} \frac{(v+(1,3))!}{(1,3)!} \alpha_0^1 \alpha_1^3 + a_{(v+(0,4))} \frac{(v+(0,4))!}{(0,4)!} \alpha_0^0 \alpha_1^4
\end{aligned}$$

4. Computation of the coefficient $a_{(0,2)}$ where $v = (0, 2)$.

$$E_F^v(\alpha) = E_F^{v_0}(\alpha)E_F^{v_1}(\alpha) = E_F^0(\alpha)E_F^2(\alpha) \qquad (10.264)$$

where

$$E_F^0(\alpha) = \frac{(\eta + v_0)!}{\eta!}\frac{1}{(\beta_{00} - \alpha_0)^{v_0}} = \frac{(4+0)!}{4!}\frac{1}{2^0} = 1$$

$$E_F^2(\alpha) = \frac{(\eta + v_1)!}{\eta!}\frac{1}{(\beta_{11} - \alpha_1)^{v_1}} = \frac{(4+2)!}{4!}\frac{1}{2^2} = \frac{30}{2^2} \qquad (10.265)$$

then

$$E_F^{(0,2)}(\alpha) = E_F^0(\alpha)E_F^2(\alpha) = 1 \times \frac{30}{2^2} = \frac{30}{2^2} = \frac{480}{2^6} \qquad (10.266)$$

and

$S_{(0,2)}(a_{v+u}, \alpha)$

$$= a_{((0,2)+(1,0))}\frac{((0,2)+(1,0))!}{(1,0)!}\alpha_0^1\alpha_1^0 + a_{((0,2)+(0,1))}\frac{((0,2)+(0,1))!}{(0,1)!}\alpha_0^0\alpha_1^1$$

$$+ a_{((0,2)+(2,0))}\frac{((0,2)+(2,0))!}{(2,0)!}\alpha_0^2\alpha_1^0 + a_{((0,2)+(1,1))}\frac{((0,2)+(1,1))!}{(1,1)!}\alpha_0^1\alpha_1^1$$

$$+ a_{((0,2)+(0,2))}\frac{((0,2)+(0,2))!}{(0,2)!}\alpha_0^0\alpha_1^2 + a_{((0,2)+(3,0))}\frac{((0,2)+(3,0))!}{(3,1)!}\alpha_0^3\alpha_1^0$$

$$+ a_{((0,2)+(2,1))}\frac{((0,2)+(2,1))!}{(2,1)!}\alpha_0^2\alpha_1^1 + a_{((0,2)+(1,2))}\frac{((0,2)+(1,2))!}{(1,2)!}\alpha_0^1\alpha_1^2$$

$$+ a_{((0,2)+(0,3))}\frac{((0,2)+(0,3))!}{(0,3)!}\alpha_0^0\alpha_1^3 + a_{((0,2)+(4,0))}\frac{((0,2)+(4,0))!}{(4,0)!}\alpha_0^4\alpha_1^0$$

$$+ a_{((0,2)+(3,1))}\frac{((0,2)+(3,1))!}{(3,1)!}\alpha_0^3\alpha_1^1 + a_{((0,2)+(2,2))}\frac{((0,2)+(2,2))!}{(2,2)!}\alpha_0^2\alpha_1^2$$

$$+ a_{((0,2)+(1,3))}\frac{((0,2)+(1,3))!}{(1,3)!}\alpha_0^1\alpha_1^3 + a_{((0,2)+(0,4))}\frac{((0,2)+(0,4))!}{(0,4)!}\alpha_0^0\alpha_1^4$$

$$S_{(0,2)}(a_{v+u}, \alpha) = a_{(1,2)} \frac{(1,2)!}{(1,0)!} \alpha_0^1 \alpha_1^0 + a_{(0,3)} \frac{(0,3)!}{(0,1)!} \alpha_0^0 \alpha_1^1$$

$$+ a_{(2,2)} \frac{(2,2)!}{(2,0)!} \alpha_0^2 \alpha_1^0 + a_{(1,3)} \frac{(1,3)!}{(1,1)!} \alpha_0^1 \alpha_1^1$$

$$+ a_{(0,4)} \frac{(0,4)!}{(0,2)!} \alpha_0^0 \alpha_1^2 + a_{(3,2)} \frac{(3,2)!}{(3,0)!} \alpha_0^3 \alpha_1^0$$

$$+ a_{(2,3)} \frac{(2,3)!}{(2,1)!} \alpha_0^2 \alpha_1^1 + a_{(1,4)} \frac{(1,4)!}{(1,2)!} \alpha_0^1 \alpha_1^2$$

$$+ a_{(0,5)} \frac{(0,5)!}{(0,3)!} \alpha_0^0 \alpha_1^3 + a_{((4,2)} \frac{(4,2)!}{(4,0)!} \alpha_0^4 \alpha_1^0$$

$$+ a_{((3,3)} \frac{(3,3)!}{(3,1)!} \alpha_0^3 \alpha_1^1 + a_{((2,4)} \frac{(2,4)!}{(2,2)!} \alpha_0^2 \alpha_1^2$$

$$+ a_{((1,5)} \frac{(1,5))!}{(1,3)!} \alpha_0^1 \alpha_1^3 + a_{((0,6)} \frac{(0,6)!}{(0,4)!} \alpha_0^0 \alpha_1^4$$

$$S_{(0,2)}(a_{v+u}, \alpha) = \frac{52275}{2^6} \frac{1!2!}{1!0!}(-1)^1(-1)^0 + \frac{24395}{2^6} \frac{0!3!}{0!1!}(-1)^0(-1)^1$$

$$+ \frac{30825}{2^6} \frac{2!2!}{2!0!}(-1)^2(-1)^0 + \frac{23975}{2^6} \frac{1!3!}{1!1!}(-1)^1(-1)^1$$

$$+ \frac{9590}{2^6} \frac{0!4!}{0!2!}(-1)^0(-1)^2 + \frac{8925}{2^6} \frac{3!2!}{3!0!}(-1)^3(-1)^0$$

$$+ \frac{8925}{2^6} \frac{2!3!}{2!1!}(-1)^2(-1)^1 + \frac{5950}{2^6} \frac{1!4!}{1!2!}(-1)^1(-1)^2$$

$$+ \frac{2142}{2^6} \frac{0!5!}{0!3!}(-1)^0(-1)^3 + \frac{1050}{2^6} \frac{4!2!}{4!0!}(-1)^4(-1)^0$$

$$+ \frac{1225}{2^6} \frac{3!3!}{3!1!}(-1)^3(-1)^1 + \frac{1050}{2^6} \frac{2!4!}{2!2!}(-1)^2(-1)^2$$

$$+ \frac{630}{2^6} \frac{1!5!}{1!3!}(-1)^1(-1)^3 + \frac{210}{2^6} \frac{0!6!}{0!4!}(-1)^0(-1)^4$$

$$S_{(0,2)}(a_{v+u}, \alpha) = \frac{1}{2^6}\big(-104550 - 146370 + 61650 + 143850 + 115080$$

$$- 17850 - 53550 - 71400 - 42840$$

$$+ 2100 + 7350 + 12600 + 12600 + 6300 \big)$$

$$= -\frac{75030}{2^6} \tag{10.267}$$

Substituting the values given by equations (10.266) and (10.267) into equation (10.261) it gives

$$
\begin{aligned}
a_{(0,2)} &= \frac{1}{0!2!} \left[E_F^{(0,2)}(-1,-1) - S_{(0,2)}(a_{v+u}, \alpha) \right] \\
&= \frac{1}{0!2!} \left[\frac{480}{2^6} + \frac{75030}{2^6} \right] = \frac{75510}{2! \times 2^6} = \frac{37755}{2^6}
\end{aligned}
\tag{10.268}
$$

5. Computation of the coefficient $a_{(1,1)}$ where $v = (1,1)$.

$$
E_F^v(\alpha) = E_F^{v_0}(\alpha) E_F^{v_1}(\alpha) = E_F^1(\alpha) E_F^1(\alpha)
\tag{10.269}
$$

where

$$
\begin{aligned}
E_F^1(\alpha) &= \frac{(\eta + v_0)!}{\eta!} \frac{1}{(\beta_{00} - \alpha_0)^{v_0}} = \frac{(4+1)!}{4!} \frac{1}{2^1} = \frac{5}{2} \\
E_F^1(\alpha) &= \frac{(\eta + v_1)!}{\eta!} \frac{1}{(\beta_{11} - \alpha_1)^{v_1}} = \frac{(4+1)!}{4!} \frac{1}{2^1} = \frac{5}{2}
\end{aligned}
\tag{10.270}
$$

then

$$
E_F^{(1,1)}(\alpha) = E_F^1(\alpha) E_F^2(\alpha) = \frac{5}{2} \times \frac{5}{2} = \frac{25}{2^2} = \frac{400}{2^6}
\tag{10.271}
$$

and

$$
\begin{aligned}
&S_{(1,1)}(a_{v+u}, \alpha) \\
&= a_{((1,1)+(1,0))} \frac{((1,1)+(1,0))!}{(1,0)!} \alpha_0^1 \alpha_1^0 + a_{((1,1)+(0,1))} \frac{((1,1)+(0,1))!}{(0,1)!} \alpha_0^0 \alpha_1^1 \\
&+ a_{((1,1)+(2,0))} \frac{((1,1)+(2,0))!}{(2,0)!} \alpha_0^2 \alpha_1^0 + a_{((1,1)+(1,1))} \frac{((1,1)+(1,1))!}{(1,1)!} \alpha_0^1 \alpha_1^1 \\
&+ a_{((1,1)+(0,2))} \frac{((1,1)+(0,2))!}{(0,2)!} \alpha_0^0 \alpha_1^2 + a_{((1,1)+(3,0))} \frac{((1,1)+(3,0))!}{(3,1)!} \alpha_0^3 \alpha_1^0 \\
&+ a_{((1,1)+(2,1))} \frac{((1,1)+(2,1))!}{(2,1)!} \alpha_0^2 \alpha_1^1 + a_{((1,1)+(1,2))} \frac{((1,1)+(1,2))!}{(1,2)!} \alpha_0^1 \alpha_1^2 \\
&+ a_{((1,1)+(0,3))} \frac{((1,1)+(0,3))!}{(0,3)!} \alpha_0^0 \alpha_1^3 + a_{((1,1)+(4,0))} \frac{((1,1)+(4,0))!}{(4,0)!} \alpha_0^4 \alpha_1^0 \\
&+ a_{((1,1)+(3,1))} \frac{((1,1)+(3,1))!}{(3,1)!} \alpha_0^3 \alpha_1^1 + a_{((1,1)+(2,2))} \frac{((1,1)+(2,2))!}{(2,2)!} \alpha_0^2 \alpha_1^2 \\
&+ a_{((1,1)+(1,3))} \frac{((1,1)+(1,3))!}{(1,3)!} \alpha_0^1 \alpha_1^3 + a_{((1,1)+(0,4))} \frac{((1,1)+(0,4))!}{(0,4)!} \alpha_0^0 \alpha_1^4
\end{aligned}
$$

$$S_{(1,1)}\left(a_{v+u}, \alpha\right)$$

$$= a_{(2,1)} \frac{(2,1)!}{(1,0)!} \; \alpha_0^1 \alpha_1^0 + a_{(1,2)} \frac{(1,2)!}{(0,1)!} \; \alpha_0^0 \alpha_1^1 + a_{(3,1)} \frac{(3,1)!}{(2,0)!} \; \alpha_0^2 \alpha_1^0$$

$$+ a_{(2,2)} \frac{(2,2)!}{(1,1)!} \; \alpha_0^1 \alpha_1^1 + a_{(1,3)} \frac{(1,3)!}{(0,2)!} \; \alpha_0^0 \alpha_1^2 + a_{(4,1)} \frac{(4,1)!}{(3,0)!} \; \alpha_0^3 \alpha_1^0$$

$$+ a_{(3,2)} \frac{(3,2)!}{(2,1)!} \; \alpha_0^2 \alpha_1^1 + a_{(2,3)} \frac{(2,3)!}{(1,2)!} \; \alpha_0^1 \alpha_1^2 + a_{(1,4)} \frac{(1,4)!}{(0,3)!} \; \alpha_0^0 \alpha_1^3$$

$$+ a_{((5,1)} \frac{(5,1)!}{(4,0)!} \; \alpha_0^4 \alpha_1^0 + a_{((4,2)} \frac{(4,2)!}{(3,1)!} \; \alpha_0^3 \alpha_1^1 + a_{((3,3)} \frac{(3,3)!}{(2,2)!} \; \alpha_0^2 \alpha_1^2$$

$$+ a_{((2,4)} \frac{(2,4))!}{(1,3)!} \; \alpha_0^1 \alpha_1^3 + a_{((1,5)} \frac{(1,5)!}{(0,4)!} \; \alpha_0^0 \alpha_1^4$$

$$S_{(1,1)}\left(a_{v+u}, \alpha\right)$$

$$= \frac{52275}{2^6} \frac{2!1!}{1!0!} (-1)^1 (-1)^0 + \frac{52275}{2^6} \frac{1!2!}{0!1!} (-1)^1 (-1)^0 + \frac{23975}{2^6} \frac{3!1!}{2!0!} (-1)^2 (-1)^0$$

$$+ \frac{30825}{2^6} \frac{2!2!}{1!1!} (-1)^1 (-1)^1 + \frac{23975}{2^6} \frac{1!3!}{0!2!} (-1)^0 (-1)^2 + \frac{5950}{2^6} \frac{4!1!}{3!0!} (-1)^3 (-1)^0$$

$$+ \frac{8925}{2^6} \frac{3!2!}{2!1!} (-1)^2 (-1)^1 + \frac{8925}{2^6} \frac{2!3!}{1!2!} (-1)^1 (-1)^2 + \frac{5950}{2^6} \frac{1!4!}{0!3!} (-1)^0 (-1)^3$$

$$+ \frac{630}{2^6} \frac{5!1!}{4!0!} (-1)^4 (-1)^0 + \frac{1050}{2^6} \frac{4!2!}{3!1!} (-1)^3 (-1)^1 + \frac{1225}{2^6} \frac{3!3!}{2!2!} (-1)^2 (-1)^2$$

$$+ \frac{1050}{2^6} \frac{2!4!}{1!3!} (-1)^1 (-1)^3 + \frac{630}{2^6} \frac{1!5!}{0!4!} (-1)^0 (-1)^4$$

$$S_{(1,1)}\left(a_{v+u}, \alpha\right) = \frac{1}{2^6}\big(-104550 - 104550 + 71925 + 123300 + 71925$$

$$- 23800 - 53550 - 53550 - 23800$$

$$+ 3150 + 8400 + 11025 + 8400 + 3150 \big)$$

$$= -\frac{62525}{2^6} \tag{10.272}$$

Substituting the values given by equations (10.271) and (10.272) into equation (10.261) it gives

$$a_{(1,1)} = \frac{1}{1!1!} \left[E_F^{(1,1)}(-1,-1) - S_{(1,1)}\left(a_{v+u}, \alpha\right) \right]$$

$$= \frac{1}{1!1!} \left[\frac{400}{2^6} + \frac{62525}{2^6} \right] = \frac{62925}{2^6} \tag{10.273}$$

6. Computation of the coefficients $a_{(2,0)}$

The theorem 7.4.7, on page 324, says that for a normalized region symmetric with respect to the coordinate axes the coefficients in the same level whose indices are permutations are equal, therefore,

$$a_{(2,0)} = a_{(0,2)} = \frac{37755}{2^6} \tag{10.274}$$

g- Computation of the coefficients that belong to the terms in the level $\lambda_0(v) = q - 5 = 1$

1. The indices v of the coefficients are the constrained numbers in the set

$$L^{\delta}_{\lambda_0(v)} = L^1_1 = \big\{ (1,0), (0,1) \big\} \tag{10.275}$$

2. The coefficients are obtained using equation (7.123), on page 314, whose equation is repeated below

$$a_{(v)} = \frac{1}{v!} \left[E_F^{(v)}(\alpha) - S_v \left(a_{v+u}, \alpha \right) \right] \qquad \lambda_0(v) = 1 \tag{10.276}$$

3. Degree of the term $S_v \left(a_{u+v}, \alpha \right)$

$$m = q - \lambda_0(v) = 6 - 1 = 5$$

The exponents of the variables in each term of $S_v \left(a_{u+v}, \alpha \right)$ are the elements of the set shown in equation (7.124), on page 314, which is

$$\begin{aligned}
\left(L^{\delta}(m) - L^{\delta}_0 \right) &= \left(L^1(2) - L^1_0 \right) \\
&= \big\{ (0,0), (1,0), (0,1), (2,0), (1,1), (0,2), (3,0), \\
&\qquad (2,1), (1,2), (0,3), \\
&\qquad (4,0), (3,1), (2,2), (1,3), (0,4), (5,0), (4,1), \\
&\qquad (3,2), (2,3), (1,4), (0,5) \big\} - \{(0,0)\} \\
&= \big\{ (1,0), (0,1), (2,0), (1,1), (0,2), (3,0), (2,1), (1,2), (0,3), \\
&\qquad (4,0), (3,1), (2,2), (1,3), (0,4), (5,0), (4,1), \\
&\qquad (3,2), (2,3), (1,4), (0,5) \big\}
\end{aligned} \tag{10.277}$$

and it gives, for

$$\begin{aligned}
u \in \big\{ &(1,0), (0,1), (2,0), (1,1), (0,2), (3,0), (2,1), (1,2), (0,3), \\
&(4,0), (3,1), (2,2), (1,3), (0,4), (5,0), (4,1), \\
&(3,2), (2,3), (1,4), (0,5) \big\}
\end{aligned} \tag{10.278}$$

the following expression

$$
\begin{aligned}
S_v\left(a_{v+u}, \alpha\right) = {} & a_{(v+(1,0))} \frac{(v+(1,0))!}{(1,0)!}\, \alpha_0^1 \alpha_1^0 + a_{(v+(0,1))} \frac{(v+(0,1))!}{(0,1)!}\, \alpha_0^0 \alpha_1^1 \\
& + a_{(v+(2,0))} \frac{(v+(2,0))!}{(2,0)!}\, \alpha_0^2 \alpha_1^0 + a_{(v+(1,1))} \frac{(v+(1,1))!}{(1,1)!}\, \alpha_0^1 \alpha_1^1 \\
& + a_{(v+(0,2))} \frac{(v+(0,2))!}{(0,2)!}\, \alpha_0^0 \alpha_1^2 + a_{(v+(3,0))} \frac{(v+(3,0))!}{(3,1)!}\, \alpha_0^3 \alpha_1^0 \\
& + a_{(v+(2,1))} \frac{(v+(2,1))!}{(2,1)!}\, \alpha_0^2 \alpha_1^1 + a_{(v+(1,2))} \frac{(v+(1,2))!}{(1,2)!}\, \alpha_0^1 \alpha_1^2 \\
& + a_{(v+(0,3))} \frac{(v+(0,3))!}{(0,3)!}\, \alpha_0^0 \alpha_1^3 + a_{(v+(4,0))} \frac{(v+(4,0))!}{(4,0)!}\, \alpha_0^4 \alpha_1^0 \\
& + a_{(v+(3,1))} \frac{(v+(3,1))!}{(3,1)!}\, \alpha_0^3 \alpha_1^1 + a_{(v+(2,2))} \frac{(v+(2,2))!}{(2,2)!}\, \alpha_0^2 \alpha_1^2 \\
& + a_{(v+(1,3))} \frac{(v+(1,3))!}{(1,3)!}\, \alpha_0^1 \alpha_1^3 + a_{(v+(0,4))} \frac{(v+(0,4))!}{(0,4)!}\, \alpha_0^0 \alpha_1^4 \\
& + a_{(v+(5,0))} \frac{(v+(5,0))!}{(5,0)!}\, \alpha_0^5 \alpha_1^0 + a_{(v+(4,1))} \frac{(v+(4,1))!}{(4,1)!}\, \alpha_0^4 \alpha_1^1 \\
& + a_{(v+(3,2))} \frac{(v+(3,2))!}{(3,2)!}\, \alpha_0^3 \alpha_1^2 + a_{(v+(2,3))} \frac{(v+(2,3))!}{(2,3)!}\, \alpha_0^2 \alpha_1^3 \\
& + a_{(v+(1,4))} \frac{(v+(1,4))!}{(1,4)!}\, \alpha_0^1 \alpha_1^4 + a_{(v+(0,5))} \frac{(v+(0,5))!}{(0,5)!}\, \alpha_0^0 \alpha_1^5
\end{aligned}
$$

4. Computation of the coefficient $a_{(0,1)}$ where $v = (0,1)$.

$$
E_F^v(\alpha) = E_F^{v_0}(\alpha) E_F^{v_1}(\alpha) = E_F^0(\alpha) E_F^1(\alpha) \tag{10.279}
$$

where

$$
\begin{aligned}
E_F^0(\alpha) &= \frac{(\eta + v_0)!}{\eta!}\, \frac{1}{(\beta_{00} - \alpha_0)^{v_0}} = \frac{(4+0)!}{4!}\, \frac{1}{2^0} = 1 \\
E_F^1(\alpha) &= \frac{(\eta + v_1)!}{\eta!}\, \frac{1}{(\beta_{11} - \alpha_1)^{v_1}} = \frac{(4+1)!}{4!}\, \frac{1}{2^1} = \frac{5}{2}
\end{aligned} \tag{10.280}
$$

then

$$
E_F^{(0,1)}(\alpha) = E_F^0(\alpha) E_F^1(\alpha) = 1 \times \frac{5}{2} = \frac{5}{2} = \frac{160}{2^6} \tag{10.281}
$$

and

$$S_{(0,1)}(a_{v+u}, \alpha)$$

$$= a_{((0,1)+(1,0))} \frac{((0,1)+(1,0))!}{(1,0)!} \alpha_0^1 \alpha_1^0 + a_{((0,1)+(0,1))} \frac{((0,1)+(0,1))!}{(0,1)!} \alpha_0^0 \alpha_1^1$$

$$+ a_{((0,1)+(2,0))} \frac{((0,1)+(2,0))!}{(2,0)!} \alpha_0^2 \alpha_1^0 + a_{((0,1)+(1,1))} \frac{((0,1)+(1,1))!}{(1,1)!} \alpha_0^1 \alpha_1^1$$

$$+ a_{((0,1)+(0,2))} \frac{((0,1)+(0,2))!}{(0,2)!} \alpha_0^0 \alpha_1^2 + a_{((0,1)+(3,0))} \frac{((0,1)+(3,0))!}{(3,1)!} \alpha_0^3 \alpha_1^0$$

$$+ a_{((0,1)+(2,1))} \frac{((0,1)+(2,1))!}{(2,1)!} \alpha_0^2 \alpha_1^1 + a_{((0,1)+(1,2))} \frac{((0,1)+(1,2))!}{(1,2)!} \alpha_0^1 \alpha_1^2$$

$$+ a_{((0,1)+(0,3))} \frac{((0,1)+(0,3))!}{(0,3)!} \alpha_0^0 \alpha_1^3 a_{((0,1)+(4,0))} \frac{((0,1)+(4,0))!}{(4,0)!} \alpha_0^4 \alpha_1^0 +$$

$$+ a_{((0,1)+(3,1))} \frac{((0,1)+(3,1))!}{(3,1)!} \alpha_0^3 \alpha_1^1 + a_{((0,1)+(2,2))} \frac{((0,1)+(2,2))!}{(2,2)!} \alpha_0^2 \alpha_1^2$$

$$+ a_{((0,1)+(1,3))} \frac{((0,1)+(1,3))!}{(1,3)!} \alpha_0^1 \alpha_1^3 + a_{((0,1)+(0,4))} \frac{((0,1)+(0,4))!}{(0,4)!} \alpha_0^0 \alpha_1^4$$

$$+ a_{((0,1)+(5,0))} \frac{((0,1)+(5,0))!}{(4,0)!} \alpha_0^5 \alpha_1^0 + a_{((0,1)+(4,1))} \frac{((0,2)+(4,1))!}{(3,1)!} \alpha_0^4 \alpha_1^1$$

$$+ a_{((0,1)+(3,2))} \frac{((0,1)+(3,2))!}{(2,2)!} \alpha_0^3 \alpha_1^2 + a_{((0,1)+(2,3))} \frac{((0,1)+(2,3))!}{(1,3)!} \alpha_0^2 \alpha_1^3$$

$$+ a_{((0,1)+(1,4))} \frac{((0,1)+(1,4))!}{(0,4)!} \alpha_0^1 \alpha_1^4 + a_{((0,1)+(0,5))} \frac{((0,1)+(0,5))!}{(0,4)!} \alpha_0^0 \alpha_1^5$$

$$S_{(0,1)}(a_{v+u}, \alpha)$$

$$= a_{(1,1)} \frac{(1,1)!}{(1,0)!} \alpha_0^1 \alpha_1^0 + a_{(0,2)} \frac{(0,2)!}{(0,1)!} \alpha_0^0 \alpha_1^1 + a_{(2,1)} \frac{(2,1)!}{(2,0)!} \alpha_0^2 \alpha_1^0 +$$

$$+ a_{(1,2)} \frac{(1,2)!}{(1,1)!} \alpha_0^1 \alpha_1^1 + a_{(0,3)} \frac{(0,3)!}{(0,2)!} \alpha_0^0 \alpha_1^2 + a_{(3,1)} \frac{(3,1)!}{(3,0)!} \alpha_0^3 \alpha_1^0$$

$$+ a_{(2,2)} \frac{(2,2)!}{(2,1)!} \alpha_0^2 \alpha_1^1 + a_{(1,3)} \frac{(1,3)!}{(1,2)!} \alpha_0^1 \alpha_1^2 + a_{(0,4)} \frac{(0,4)!}{(0,3)!} \alpha_0^0 \alpha_1^3$$

$$+ a_{((4,1)} \frac{(4,1)!}{(4,0)!} \alpha_0^4 \alpha_1^0 + a_{((3,2)} \frac{(3,2)!}{(3,1)!} \alpha_0^3 \alpha_1^1 + a_{((2,3)} \frac{(2,3)!}{(2,2)!} \alpha_0^2 \alpha_1^2$$

$$+ a_{((1,4)} \frac{(1,4)!}{(1,3)!} \alpha_0^1 \alpha_1^3 + a_{((0,5)} \frac{(0,5)!}{(0,4)!} \alpha_0^0 \alpha_1^4 + a_{((5,1)} \frac{(5,1)!}{(5,0)!} \alpha_0^5 \alpha_1^0$$

$$+ a_{((4,2)} \frac{(4,2)!}{(4,1)!} \alpha_0^4 \alpha_1^1 + a_{((3,3)} \frac{(3,3)!}{(3,2)!} \alpha_0^3 \alpha_1^2 + a_{((2,4)} \frac{(2,4))!}{(2,3)!} \alpha_0^2 \alpha_1^3$$

$$+ a_{((1,5)} \frac{(1,5)!}{(1,4)!} \alpha_0^1 \alpha_1^4 + a_{((0,6)} \frac{(0,6)!}{(0,5)!} \alpha_0^0 \alpha_1^5$$

$S_{(0,1)}\,(a_{v+u},\alpha)$

$$= \frac{62925}{2^6}\frac{1!1!}{1!0!}(-1)^1(-1)^0 + \frac{37755}{2^6}\frac{0!2!}{0!1!}(-1)^0(-1)^1$$

$$+ \frac{52275}{2^6}\frac{2!1!}{2!0!}(-1)^2(-1)^0 + \frac{52275}{2^6}\frac{1!2!}{1!1!}(-1)^1(-1)^1$$

$$+ \frac{24395}{2^6}\frac{0!3!}{0!2!}(-1)^0(-1)^2 + \frac{23975}{2^6}\frac{3!1!}{3!0!}(-1)^3(-1)^0$$

$$+ \frac{30825}{2^6}\frac{2!2!}{2!1!}(-1)^2(-1)^1 + \frac{23975}{2^6}\frac{1!3!}{1!2!}(-1)^1(-1)^2$$

$$+ \frac{9590}{2^6}\frac{0!4!}{0!3!}(-1)^0(-1)^3 + \frac{5950}{2^6}\frac{4!1!}{4!0!}\,(-1)^4(-1)^0$$

$$+ \frac{8925}{2^6}\frac{3!2!}{3!1!}\,(-1)^3(-1)^1 + \frac{8925}{2^6}\frac{2!3!}{2!2!}\,(-1)^2(-1)^2$$

$$+ \frac{5950}{2^6}\frac{1!4!}{1!3!}\,(-1)^1(-1)^3 + \frac{2142}{2^6}\frac{0!5!}{0!4!}\,(-1)^0(-1)^4$$

$$+ \frac{630}{2^6}\frac{5!1!}{5!0!}\,(-1)^5(-1)^0 + \frac{1050}{2^6}\frac{4!2!}{4!1!}\,(-1)^4(-1)^1$$

$$+ \frac{1225}{2^6}\frac{3!3!}{3!2!}\,(-1)^3(-1)^2 + \frac{1050}{2^6}\frac{2!4!}{2!3!}\,(-1)^2(-1)^3$$

$$+ \frac{630}{2^6}\frac{1!5!}{1!4!}\,(-1)^1(-1)^4 + \frac{210}{2^6}\frac{0!6!}{0!5!}\,(-1)^0(-1)^5$$

$$S_{(0,1)}\,(a_{v+u},\alpha) = \frac{1}{2^6}\big(-62925 - 75510 + 52275 + 104550 + 73185$$

$$- 23975 - 61650 - 71925 - 38360 + 5950$$

$$+ 17850 + 26775 + 23800 + 10710$$

$$- 630 - 2100 - 3675 - 4200 - 3150 - 1260 \big)$$

$$= -\frac{34265}{2^6} \tag{10.282}$$

Substituting the values given by equations (10.281) and (10.282) into equation (10.276) it gives

$$a_{(0,1)} = \frac{1}{0!1!}\left[E_F^{(0,1)}(-1,-1) - S_{(0,1)}\,(a_{v+u},\alpha) \right]$$

$$= \frac{1}{0!1!}\left[\frac{160}{2^6} + \frac{34265}{2^6} \right] = \frac{34425}{2^6} = \frac{37755}{2^6} \tag{10.283}$$

5. Computation of the coefficients $a_{(1,0)}$

The theorem 7.4.7, on page 324, says that for a normalized region symmetric with respect to the coordinate axes the coefficients in the same level whose indices are permutations are equal, therefore,

$$a_{(1,0)} = a_{(0,1)} = \frac{34425}{2^6} \qquad (10.284)$$

h- Computation of the coefficients that belong to the terms in the level $\lambda_0(v) = q - 6 = 0$

1. The indices v of the coefficients are the constrained numbers in the set

$$L^{\delta}_{\lambda_0(v)} = L^1_0 = \{(0,0)\} \qquad (10.285)$$

2. The coefficients are obtained using equation (7.123), on page 314, whose equation is repeated below

$$a_{(v)} = \frac{1}{v!} \left[E_F^{(v)}(\alpha) - S_v\left(a_{v+u}, \alpha\right) \right] \qquad \lambda_0(v) = 0 \qquad (10.286)$$

3. Degree of the term $S_v\left(a_{u+v}, \alpha\right)$

$m = q - \lambda_0(v) = 6 - 0 = 6$

The exponents of the variables in each term of $S_v\left(a_{u+v}, \alpha\right)$ are the elements of the set shown in equation (7.124), on page 314, which is

$$
\begin{aligned}
\left(L^{\delta}(m) - L^{\delta}_0\right) &= \left(L^1(2) - L^1_0\right) \\
&= \{(0,0), (1,0), (0,1), (2,0), (1,1), (0,2), (3,0), \\
&\quad (2,1), (1,2), (0,3), \\
&\quad (4,0), (3,1), (2,2), (1,3), (0,4), (5,0), (4,1), \\
&\quad (3,2), (2,3), (1,4), (0,5), \\
&\quad (6,0), (5,1), (4,2), (3,3), (2,4), (1,5), (0,6)\} - \{(0,0)\} \\
&= \{(1,0), (0,1), (2,0), (1,1), (0,2), (3,0), \\
&\quad (2,1), (1,2), (0,3), \\
&\quad (4,0), (3,1), (2,2), (1,3), (0,4), (5,0), (4,1), \\
&\quad (3,2), (2,3), (1,4), (0,5), \\
&\quad (6,0), (5,1), (4,2), (3,3), (2,4), (1,5), (0,6)\} \quad (10.287)
\end{aligned}
$$

and it gives, for

$$
\begin{aligned}
u \in \{ &(1,0), (0,1), (2,0), (1,1), (0,2), (3,0), (2,1), (1,2), (0,3), \\
&(4,0), (3,1), (2,2), (1,3), (0,4), (5,0), (4,1), \\
&(3,2), (2,3), (1,4), (0,5), \\
&(6,0), (5,1), (4,2), (3,3), (2,4), (1,5), (0,6)\} \qquad (10.288)
\end{aligned}
$$

the following expression

$$S_v \left(a_{v+u}, \alpha\right) = a_{(v+(1,0))} \frac{(v+(1,0))!}{(1,0)!} \, \alpha_0^1 \alpha_1^0 + a_{(v+(0,1))} \frac{(v+(0,1))!}{(0,1)!} \, \alpha_0^0 \alpha_1^1$$

$$+ \, a_{(v+(2,0))} \frac{(v+(2,0))!}{(2,0)!} \, \alpha_0^2 \alpha_1^0 + a_{(v+(1,1))} \frac{(v+(1,1))!}{(1,1)!} \, \alpha_0^1 \alpha_1^1$$

$$+ \, a_{(v+(0,2))} \frac{(v+(0,2))!}{(0,2)!} \, \alpha_0^0 \alpha_1^2$$

$$+ \, a_{(v+(3,0))} \frac{(v+(3,0))!}{(3,1)!} \, \alpha_0^3 \alpha_1^0 + a_{(v+(2,1))} \frac{(v+(2,1))!}{(2,1)!} \, \alpha_0^2 \alpha_1^1$$

$$+ \, a_{(v+(1,2))} \frac{(v+(1,2))!}{(1,2)!} \, \alpha_0^1 \alpha_1^2 + a_{(v+(0,3))} \frac{(v+(0,3))!}{(0,3)!} \, \alpha_0^0 \alpha_1^3$$

$$+ \, a_{(v+(4,0))} \frac{(v+(4,0))!}{(4,0)!} \, \alpha_0^4 \alpha_1^0 + a_{(v+(3,1))} \frac{(v+(3,1))!}{(3,1)!} \, \alpha_0^3 \alpha_1^1$$

$$+ \, a_{(v+(2,2))} \frac{(v+(2,2))!}{(2,2)!} \, \alpha_0^2 \alpha_1^2 + a_{(v+(1,3))} \frac{(v+(1,3))!}{(1,3)!} \, \alpha_0^1 \alpha_1^3$$

$$+ \, a_{(v+(0,4))} \frac{(v+(0,4))!}{(0,4)!} \, \alpha_0^0 \alpha_1^4$$

$$+ \, a_{(v+(5,0))} \frac{(v+(5,0))!}{(5,0)!} \, \alpha_0^5 \alpha_1^0 + a_{(v+(4,1))} \frac{(v+(4,1))!}{(4,1)!} \, \alpha_0^4 \alpha_1^1$$

$$+ \, a_{(v+(3,2))} \frac{(v+(3,2))!}{(3,2)!} \, \alpha_0^3 \alpha_1^2 + a_{(v+(2,3))} \frac{(v+(2,3))!}{(2,3)!} \, \alpha_0^2 \alpha_1^3$$

$$+ \, a_{(v+(1,4))} \frac{(v+(1,4))!}{(1,3)!} \, \alpha_0^1 \alpha_1^4 + a_{(v+(0,5))} \frac{(v+(0,5))!}{(0,5)!} \, \alpha_0^0 \alpha_1^5$$

$$+ \, a_{(v+(6,0))} \frac{(v+(6,0))!}{(6,0)!} \, \alpha_0^6 \alpha_1^0 + a_{(v+(5,1))} \frac{(v+(5,1))!}{(5,1)!} \, \alpha_0^5 \alpha_1^1$$

$$+ \, a_{(v+(4,2))} \frac{(v+(4,2))!}{(4,2)!} \, \alpha_0^4 \alpha_1^2 + a_{(v+(3,3))} \frac{(v+(3,3))!}{(3,3)!} \, \alpha_0^3 \alpha_1^3$$

$$+ \, a_{(v+(2,4))} \frac{(v+(2,4))!}{(2,4)!} \, \alpha_0^2 \alpha_1^4 + a_{(v+(1,5))} \frac{(v+(1,5))!}{(1,5)!} \, \alpha_0^1 \alpha_1^5$$

$$+ \, a_{(v+(0,6))} \frac{(v+(0,6))!}{(0,6)!} \, \alpha_0^0 \alpha_1^6 \tag{10.289}$$

4. Computation of the coefficient $a_{(0,0)}$ where $v = (0,0)$.

$$E_F^v(\alpha) = E_F^{v_0}(\alpha)E_F^{v_1}(\alpha) = E_F^0(\alpha)E_F^0(\alpha) \qquad (10.290)$$

where

$$E_F^0(\alpha) = \frac{(\eta + v_0)!}{\eta!}\frac{1}{(\beta_{00} - \alpha_0)^{v_0}} = \frac{(4+0)!}{4!}\frac{1}{2^0} = 1$$

$$E_F^0(\alpha) = \frac{(\eta + v_1)!}{\eta!}\frac{1}{(\beta_{11} - \alpha_1)^{v_1}} = \frac{(4+0)!}{4!}\frac{1}{2^0} = 1 \qquad (10.291)$$

then

$$E_F^{(0,0)}(\alpha) = E_F^0(\alpha)E_F^0(\alpha) = 1 \times 1 = 1 = \frac{64}{2^6} \qquad (10.292)$$

and

$$S_{(0,0)}(a_{v+u}, \alpha)$$

$$= a_{((0,0)+(1,0))}\frac{((0,0)+(1,0))!}{(1,0)!}\,\alpha_0^1\alpha_1^0 + a_{((0,0)+(0,1))}\frac{((0,0)+(0,1))!}{(0,1)!}\,\alpha_0^0\alpha_1^1$$

$$+ a_{((0,0)+(2,0))}\frac{((0,0)+(2,0))!}{(2,0)!}\,\alpha_0^2\alpha_1^0 + a_{((0,0)+(1,1))}\frac{((0,0)+(1,1))!}{(1,1)!}\,\alpha_0^1\alpha_1^1$$

$$+ a_{((0,0)+(0,2))}\frac{((0,0)+(0,2))!}{(0,2)!}\,\alpha_0^0\alpha_1^2$$

$$+ a_{((0,0)+(3,0))}\frac{((0,0)+(3,0))!}{(3,1)!}\,\alpha_0^3\alpha_1^0 + a_{((0,0)+(2,1))}\frac{((0,0)+(2,1))!}{(2,1)!}\,\alpha_0^2\alpha_1^1$$

$$+ a_{((0,0)+(1,2))}\frac{((0,0)+(1,2))!}{(1,2)!}\,\alpha_0^1\alpha_1^2 + a_{((0,0)+(0,3))}\frac{((0,0)+(0,3))!}{(0,3)!}\,\alpha_0^0\alpha_1^3$$

$$+ a_{((0,0)+(4,0))}\frac{((0,0)+(4,0))!}{(4,0)!}\,\alpha_0^4\alpha_1^0 + a_{((0,0)+(3,1))}\frac{((0,0)+(3,1))!}{(3,1)!}\,\alpha_0^3\alpha_1^1$$

$$+ a_{((0,0)+(2,2))}\frac{((0,0)+(2,2))!}{(2,2)!}\,\alpha_0^2\alpha_1^2 + a_{((0,0)+(1,3))}\frac{((0,0)+(1,3))!}{(1,3)!}\,\alpha_0^1\alpha_1^3$$

$$+ a_{((0,0)+(0,4))}\frac{((0,0)+(0,4))!}{(0,4)!}\,\alpha_0^0\alpha_1^4$$

$$+ a_{((0,0)+(5,0))}\frac{((0,0)+(5,0))!}{(5,0)!}\,\alpha_0^5\alpha_1^0 + a_{((0,0)+(4,1))}\frac{((0,0)+(4,1))!}{(4,1)!}\,\alpha_0^4\alpha_1^1$$

$$+ a_{((0,0)+(3,2))}\frac{((0,0)+(3,2))!}{(3,2)!}\,\alpha_0^3\alpha_1^2 + a_{((0,0)+(2,3))}\frac{((0,0)+(2,3))!}{(2,3)!}\,\alpha_0^2\alpha_1^3$$

$$+ a_{((0,0)+(1,4))}\frac{((0,0)+(1,4))!}{(1,3)!}\,\alpha_0^1\alpha_1^4 + a_{((0,0)+(0,5))}\frac{((0,0)+(0,5))!}{(0,5)!}\,\alpha_0^0\alpha_1^5$$

$$+ a_{((0,0)+(6,0))} \frac{((0,0)+(6,0))!}{(6,0)!} \alpha_0^6 \alpha_1^0 + a_{((0,0)+(5,1))} \frac{((0,0)+(5,1))!}{(5,1)!} \alpha_0^5 \alpha_1^1$$

$$+ a_{((0,0)+(4,2))} \frac{((0,0)+(4,2))!}{(4,2)!} \alpha_0^4 \alpha_1^2 + a_{((0,0)+(3,3))} \frac{((0,0)+(3,3))!}{(3,3)!} \alpha_0^3 \alpha_1^3$$

$$+ a_{((0,0)+(2,4))} \frac{((0,0)+(2,4))!}{(2,4)!} \alpha_0^2 \alpha_1^4 + a_{((0,0)+(1,5))} \frac{((0,0)+(1,5))!}{(1,5)!} \alpha_0^1 \alpha_1^5$$

$$+ a_{((0,0)+(0,6))} \frac{((0,0)+(0,6))!}{(0,6)!} \alpha_0^0 \alpha_1^6 \qquad (10.293)$$

$$S_{(0,0)} (a_{v+u}, \alpha)$$

$$+ a_{(1,0)} \frac{(1,0)!}{(1,0)!} \alpha_0^1 \alpha_1^0 + a_{(0,1)} \frac{(0,1)!}{(0,1)!} \alpha_0^0 \alpha_1^1 + a_{(2,0)} \frac{(2,0)!}{(2,0)!} \alpha_0^2 \alpha_1^0$$

$$+ a_{(1,1)} \frac{((1,1)!}{(1,1)!} \alpha_0^1 \alpha_1^1 a_{(0,2)} \frac{(0,2)!}{(0,2)!} \alpha_0^0 \alpha_1^2 + a_{(3,0)} \frac{(3,0)!}{(3,1)!} \alpha_0^3 \alpha_1^0$$

$$+ a_{(2,1)} \frac{((2,1)!}{(2,1)!} \alpha_0^2 \alpha_1^1 + a_{(1,2)} \frac{(1,2)!}{(1,2)!} \alpha_0^1 \alpha_1^2 + a_{(0,3)} \frac{(0,3)!}{(0,3)!} \alpha_0^0 \alpha_1^3$$

$$+ a_{(4,0))} \frac{(4,0)!}{(4,0)!} \alpha_0^4 \alpha_1^0 + a_{(3,1))} \frac{(3,1)!}{(3,1)!} \alpha_0^3 \alpha_1^1 + a_{(2,2)} \frac{(2,2)!}{(2,2)!} \alpha_0^2 \alpha_1^2$$

$$+ a_{(1,3)} \frac{(1,3)!}{(1,3)!} \alpha_0^1 \alpha_1^3 + a_{(0,4)} \frac{(0,4)!}{(0,4)!} \alpha_0^0 \alpha_1^4 + a_{(5,0))} \frac{(5,0)!}{(5,0)!} \alpha_0^5 \alpha_1^0$$

$$+ a_{(4,1)} \frac{(4,1)!}{(4,1)!} \alpha_0^4 \alpha_1^1 + a_{(3,2))} \frac{(3,2)!}{(3,2)!} \alpha_0^3 \alpha_1^2 + a_{(2,3)} \frac{(2,3)!}{(2,3)!} \alpha_0^2 \alpha_1^3$$

$$+ a_{(1,4)} \frac{(1,4)!}{(1,3)!} \alpha_0^1 \alpha_1^4 + a_{(0,5)} \frac{(0,5)!}{(0,5)!} \alpha_0^0 \alpha_1^5 + a_{(6,0)} \frac{(6,0)!}{(6,0)!} \alpha_0^6 \alpha_1^0 +$$

$$+ a_{(5,1)} \frac{((5,1)!}{(5,1)!} \alpha_0^5 \alpha_1^1 + a_{(4,2)} \frac{(4,2)!}{(4,2)!} \alpha_0^4 \alpha_1^2 + a_{(3,3)} \frac{(3,3)!}{(3,3)!} \alpha_0^3 \alpha_1^3$$

$$+ a_{(2,4)} \frac{(2,4)!}{(2,4)!} \alpha_0^2 \alpha_1^4 + a_{(1,5)} \frac{(1,5)!}{(1,5)!} \alpha_0^1 \alpha_1^5 + a_{(0,6)} \frac{(0,6)!}{(0,6)!} \alpha_0^0 \alpha_1^6$$

$$(10.294)$$

$$S_{(0,0)} (a_{v+u}, \alpha)$$

$$= \frac{34425}{2^6} (-1)^1 (-1)^0 + \frac{34425}{2^6} (-1)^0 (-1)^1 + \frac{37755}{2^6} (-1)^2 (-1)^0$$

$$+ \frac{62925}{2^6} (-1)^1(-1)^1 + \frac{37755}{2^6} (-1)^0(-1)^2 + \frac{24395}{2^6} (-1)^3(-1)^0$$

$$+ \frac{52275}{2^6} (-1)^2(-1)^1 + \frac{52275}{2^6} (-1)^1(-1)^2 + \frac{24395}{2^6} (-1)^0(-1)^3$$

$$+ \frac{9590}{2^6} (-1)^4(-1)^0 + \frac{23975}{2^6} (-1)^3(-1)^1 + \frac{30825}{2^6} (-1)^2(-1)^2$$

$$+ \frac{23975}{2^6} (-1)^1(-1)^3 + \frac{9590}{2^6} (-1)^0(-1)^4 + \frac{2142}{2^6} (-1)^5(-1)^0$$

$$+ \frac{5950}{2^6} (-1)^4(-1)^1 + \frac{8925}{2^6} (-1)^3(-1)^2 + \frac{8925}{2^6} (-1)^2(-1)^3$$

$$+ \frac{5950}{2^6} (-1)^1(-1)^4 + \frac{2142}{2^6} (-1)^0(-1)^5 + \frac{210}{2^6} (-1)^6(-1)^0$$

$$+ \frac{630}{2^6} (-1)^5(-1)^1 + \frac{1050}{2^6} (-1)^4(-1)^2 + \frac{1225}{2^6} (-1)^3(-1)^3$$

$$+ \frac{1050}{2^6} (-1)^2(-1)^4 + \frac{630}{2^6} (-1)^1(-1)^5 + \frac{210}{2^6} (-1)^0(-1)^6$$

$$S_{(0,0)} (a_{v+u}, \alpha) = \frac{1}{2^6} \big(-68850 + 138435 - 153340$$

$$+ 97955 - 34034 + 5005 \big)$$

$$= -\frac{14829}{2^6} \tag{10.295}$$

Substituting the values given by equations (10.292) and (10.295) into equation (10.286) it gives

$$a_{(0,0)} = \frac{1}{0!0!} \left[E_F^{(0,0)}(-1,-1) - S_{(0,0)} (a_{v+u}, \alpha) \right]$$

$$= \frac{1}{0!0!} \left[\frac{64}{2^6} + \frac{14829}{2^6} \right] = \frac{14893}{2^6} \tag{10.296}$$

10.7.2 Summary of the polynomials $\Phi_{a_0,n}$ related to the reference node $a_0 = (-1,-1)$ and $q = 6$

Degree of the polynomial $f_{a_0,00}(x)$

From corollary 7.2.19 the minimum degree is given by

$$q \geq \eta - \lambda_0(n) = 4 - \lambda_0(0,0) = 4 - 0 = 4 \tag{10.297}$$

adding 2 to the degree it gives $q = 4 + 2 = 6$.

For $d = 2$, $\eta = 4$, $\alpha = a_0 = (-1, -1)$, $\beta_0 = (\beta_{00}, \beta_{01}) = a_1 = (1, -1)$, and $\beta_1 = (\beta_{10}, \beta_{11}) = a_3 = (-1, 1)$, the Hermite Interpolating Polynomials writes

$$\Phi_{a_0,00}(x_0, x_1) = \left[\frac{1}{0!0!} \frac{1}{2^{10}} \right] (x_0 - 1)^5 (x_1 - 1)^5 \frac{1}{(-2)^6}$$

$$\times \Big(14893 + 34425x_0 + 34425x_1 + 37755x_0^2 + 62925x_0x_1$$

$$+ 37755x_1^2 + 24395x_0^3 + 52275x_0^2x_1 + 52275x_0x_1^2 + 24395x_1^3$$

$$+ 9590x_0^4 + 23975x_0^3x_1 + 30825x_0^2x_1^2 + 23975x_0x_1^3 + 9590x_1^4$$

$$+ 2142x_0^5 + 5950x_0^4x_1 + 8925x_0^3x_1^2 + 8925x_0^2x_1^3 + 5950x_0^1x_1^4$$

$$+ 2142x_1^5 + 210x_0^6 + 630x_0^5x_1 + 1050x_0^4x_1^2 + 1225x_0^3x_1^3$$

$$+ 1050x_0^2x_1^4 + 630x_0^1x_1^5 + 210x_1^6 \Big) \tag{10.298}$$

Degree of the polynomials $f_{a_0,10}(x)$ and $f_{a_0,01}(x)$ is obtained from

$$q \geq \eta - \lambda_0(n) = 4 - \lambda_0(1, 0) = 4 - 1 = 3 \tag{10.299}$$

adding 2 to the degree it gives $q = 3 + 2 = 5$.

$$\Phi_{a_0,10}(x_0, x_1) = \left[\frac{1}{1!0!} \frac{1}{2^{10}} \right] (x_0 - 1)^5 (x_1 - 1)^5 (x_0 + 1) \frac{1}{2^5} \Big(4944 + 9705x_0$$

$$+ 9705x_1 + 8640x_0^2 + 14400x_0x_1 + 8640x_1^2 + 4235x_0^3$$

$$+ 9075x_0^2x_1 + 9075x_0x_1^2 + 4235x_1^3 + 1120x_0^4 + 2800x_0^3x_1$$

$$+ 3600x_0^2x_1^2 + 2800x_0x_1^3 + 1120x_1^4 + 126x_0^5 + 350x_0^4x_1$$

$$+ 525x_0^3x_1^2 + 525x_0^2x_1^3 + 350x_0^1x_1^4 + 126x_1^5 \Big) \tag{10.300}$$

$$\Phi_{a_0,01}(x_0, x_1) = \left[\frac{1}{0!1!} \frac{1}{2^{10}} \right] (x_0 - 1)^5 (x_1 - 1)^5 (x_1 + 1) \frac{1}{2^5} \Big(4944 + 9705x_0$$

$$+ 9705x_1 + 8640x_0^2 + 14400x_0x_1 + 8640x_1^2 + 4235x_0^3$$

$$+ 9075x_0^2x_1 + 9075x_0x_1^2 + 4235x_1^3 + 1120x_0^4$$

$$+ 2800x_0^3x_1 + 3600x_0^2x_1^2 + 2800x_0x_1^3 + 1120x_1^4 + 126x_0^5$$

$$+ 350x_0^4x_1 + 525x_0^3x_1^2 + 525x_0^2x_1^3 + 350x_0^1x_1^4 + 126x_1^5 \Big) \tag{10.301}$$

Degree of the polynomials $f_{a_0,20}(x)$, $f_{a_0,11}(x)$, and $f_{a_0,02}(x)$ is given by

$$q \geq \eta - \lambda_0(n) = 4 - \lambda_0(2, 0) = 4 - 2 = 2 \tag{10.302}$$

adding 2 to the degree it gives $q = 2 + 2 = 4$.

$$\Phi_{a_0,20}(x_0, x_1) = \left[\frac{1}{2!0!}\frac{1}{2^{10}}\right](x_0 - 1)^5(x_1 - 1)^5(x_0 + 1)^2\frac{1}{2^4}\Big(1471 + 2350x_0$$
$$+ 2350x_1 + 1590x_0^2 + 2650x_0x_1 + 1590x_1^2 + 525x_0^3$$
$$+ 1125x_0^2x_1 + 1125x_0x_1^2 + 525x_1^3 + 70x_0^4$$
$$+ 175x_0^3x_1 + 225x_0^2x_1^2 + 175x_0x_1^3 + 70x_1^4\Big) \qquad (10.303)$$

$$\Phi_{a_0,11}(x_0, x_1) = \left[\frac{1}{1!1!}\frac{1}{2^{10}}\right](x_0 - 1)^5(x_1 - 1)^5(x_0 + 1)(x_1 + 1)\frac{1}{2^4}\Big(1471$$
$$+ 2350x_0 + 2350x_1 + 1590x_0^2 + 2650x_0x_1 + 1590x_1^2$$
$$+ 525x_0^3 + 1125x_0^2x_1 + 1125x_0x_1^2 + 525x_1^3$$
$$+ 70x_0^4 + 175x_0^3x_1 + 225x_0^2x_1^2 + 175x_0x_1^3 + 70x_1^4\Big) \quad (10.304)$$

$$\Phi_{a_0,02}(x_0, x_1) = \left[\frac{1}{0!2!}\frac{1}{(-2)^{10}}\right](x_0 - 1)^5(x_1 - 1)^5(x_1 + 1)^2\frac{1}{2^4}\Big(1471$$
$$+ 2350x_0 + 2350x_1 + 1590x_0^2 + 2650x_0x_1 + 1590x_1^2$$
$$+ 525x_0^3 + 1125x_0^2x_1 + 1125x_0x_1^2 + 525x_1^3$$
$$+ 70x_0^4 + 175x_0^3x_1 + 225x_0^2x_1^2 + 175x_0x_1^3 + 70x_1^4\Big) \quad (10.305)$$

Degree of the polynomials $f_{a_0,30}(x)$, $f_{a_0,21}(x)$, $f_{a_0,12}(x)$, and $f_{a_0,03}(x)$ is given by

$$q \geq \eta - \lambda_0(n) = 4 - \lambda_0(3,0) = 4 - 3 = 1 \qquad (10.306)$$

adding 2 to the degree it gives $q = 1 + 2 = 3$.

$$\Phi_{a_0,30}(x_0, x_1) = \left[\frac{1}{3!0!}\frac{1}{(-2)^{10}}\right](x_0 - 1)^5(x_1 - 1)^5(x_0 + 1)^3\frac{1}{2^3}\Big(378$$
$$+ 460x_0 + 460x_1 + 210x_0^2 + 350x_0x_1 + 210x_1^2$$
$$+ 35x_0^3 + 75x_0^2x_1 + 75x_0x_1^2 + 35x_1^3\Big) \qquad (10.307)$$

$$\Phi_{a_0,21}(x_0, x_1) = \left[\frac{1}{2!1!}\frac{1}{(-2)^{10}}\right](x_0 - 1)^5(x_1 - 1)^5(x_0 + 1)^2(x_1 + 1)\frac{1}{2^3}\Big(378$$
$$+ 460x_0 + 460x_1 + 210x_0^2 + 350x_0x_1 + 210x_1^2$$
$$+ 35x_0^3 + 75x_0^2x_1 + 75x_0x_1^2 + 35x_1^3\Big) \qquad (10.308)$$

$$\Phi_{a_0,12}(x_0, x_1) = \left[\frac{1}{1!2!}\frac{1}{(-2)^{10}}\right](x_0 - 1)^5(x_1 - 1)^5(x_0 + 1)(x_1 + 1)^2\frac{1}{2^3}\left(378\right.$$
$$+ 460x_0 + 460x_1 + 210x_0^2 + 350x_0x_1 + 210x_1^2$$
$$\left. + 35x_0^3 + 75x_0^2x_1 + 75x_0x_1^2 + 35x_1^3\right) \tag{10.309}$$

$$\Phi_{a_0,03}(x_0, x_1) = \left[\frac{1}{0!3!}\frac{1}{(-2)^{10}}\right](x_0 - 1)^5(x_1 - 1)^5(x_1 + 1)^3\frac{1}{2^3}\left(378\right.$$
$$+ 460x_0 + 460x_1 + 210x_0^2 + 350x_0x_1 + 210x_1^2$$
$$\left. + 35x_0^3 + 75x_0^2x_1 + 75x_0x_1^2 + 35x_1^3\right) \tag{10.310}$$

Degree of the polynomials $f_{a_0,40}(x)$, $f_{a_0,31}(x)$, $f_{a_0,22}(x)$, $f_{a_0,13}(x)$, and $f_{a_0,04}(x)$ is given by

$$q \geq \eta - \lambda_0(n) = 4 - \lambda_0(4, 0) = 4 - 4 = 0 \tag{10.311}$$

adding 2 to the degree it gives $q = 0 + 2 = 2$.

$$\Phi_{a_0,40}(x_0, x_1) = \left[\frac{1}{4!0!}\frac{1}{(-2)^{10}}\right](x_0 - 1)^5(x_1 - 1)^5(x_0 + 1)^4\frac{1}{2^2}\left(79\right.$$
$$\left. + 65x_0 + 65x_1 + 15x_0^2 + 25x_0x_1 + 15x_1^2\right) \tag{10.312}$$

$$\Phi_{a_0,31}(x_0, x_1) = \left[\frac{1}{3!1!}\frac{1}{(-2)^{10}}\right](x_0 - 1)^5(x_1 - 1)^5(x_0 + 1)^3(x_1 + 1)\frac{1}{2^2}\left(79\right.$$
$$\left. + 65x_0 + 65x_1 + 15x_0^2 + 25x_0x_1 + 15x_1^2\right) \tag{10.313}$$

$$\Phi_{a_0,22}(x_0, x_1) = \left[\frac{1}{2!2!}\frac{1}{(-2)^{10}}\right](x_0 - 1)^5(x_1 - 1)^5(x_0 + 1)^2(x_1 + 1)^2\frac{1}{2^2}\left(79\right.$$
$$\left. + 65x_0 + 65x_1 + 15x_0^2 + 25x_0x_1 + 15x_1^2\right) \tag{10.314}$$

$$\Phi_{a_0,13}(x_0, x_1) = \left[\frac{1}{1!3!}\frac{1}{(-2)^{10}}\right](x_0 - 1)^5(x_1 - 1)^5(x_0 + 1)^1(x_1 + 1)^3\frac{1}{2^2}\left(79\right.$$
$$\left. + 65x_0 + 65x_1 + 15x_0^2 + 25x_0x_1 + 15x_1^2\right) \tag{10.315}$$

$$\Phi_{a_0,04}(x_0, x_1) = \left[\frac{1}{0!4!}\frac{1}{(-2)^{10}}\right](x_0 - 1)^5(x_1 - 1)^5(x_1 + 1)^4\frac{1}{2^2}\left(79\right.$$
$$\left. + 65x_0 + 65x_1 + 15x_0^2 + 25x_0x_1 + 15x_1^2\right) \tag{10.316}$$

Degree of the polynomials $f_{a_0,50}(x)$, $f_{a_0,41}(x)$, $f_{a_0,32}(x)$, $f_{a_0,23}(x)$, $f_{a_0,14}(x)$, and $f_{a_0,05}(x)$ is given by

$$q \geq \eta - \lambda_0(n) = 4 - \lambda_0(5,0) = 4 - 5 = -1 \qquad (10.317)$$

adding 2 to the degree it gives $q = -1 + 2 = 1$.

$\Phi_{a_0,50}(x_0, x_1)$

$$= \left[\frac{1}{5!0!} \frac{1}{(-2)^{10}} \right] (x_0 - 1)^5 (x_1 - 1)^5 (x_0 + 1)^5 \frac{1}{2^1} \left(12 + 5x_0 + 5x_1 \right) \quad (10.318)$$

$\Phi_{a_0,41}(x_0, x_1)$

$$= \left[\frac{1}{4!1!} \frac{1}{(-2)^{10}} \right] (x_0 - 1)^5 (x_1 - 1)^5 (x_0 + 1)^4 (x_1 + 1)^1 \frac{1}{2^1} \left(12 + 5x_0 + 5x_1 \right)$$
$$(10.319)$$

$\Phi_{a_0,32}(x_0, x_1)$

$$= \left[\frac{1}{3!2!} \frac{1}{(-2)^{10}} \right] (x_0 - 1)^5 (x_1 - 1)^5 (x_0 + 1)^3 (x_1 + 1)^2 \frac{1}{2^1} \left(12 + 5x_0 + 5x_1 \right)$$
$$(10.320)$$

$\Phi_{a_0,23}(x_0, x_1)$

$$= \left[\frac{1}{2!3!} \frac{1}{(-2)^{10}} \right] (x_0 - 1)^5 (x_1 - 1)^5 (x_0 + 1)^2 (x_1 + 1)^3 \frac{1}{2^1} \left(12 + 5x_0 + 5x_1 \right)$$
$$(10.321)$$

$\Phi_{a_0,14}(x_0, x_1)$

$$= \left[\frac{1}{1!4!} \frac{1}{(-2)^{10}} \right] (x_0 - 1)^5 (x_1 - 1)^5 (x_0 + 1)^1 (x_1 + 1)^4 \frac{1}{2^1} \left(12 + 5x_0 + 5x_1 \right)$$
$$(10.322)$$

$\Phi_{a_0,05}(x_0, x_1)$

$$= \left[\frac{1}{0!5!} \frac{1}{(-2)^{10}} \right] (x_0 - 1)^5 (x_1 - 1)^5 (x_0 + 1)^0 (x_1 + 1)^5 \frac{1}{2^1} \left(12 + 5x_0 + 5x_1 \right)$$
$$(10.323)$$

Degree of the polynomials $f_{a_0,60}(x)$, $f_{a_0,51}(x)$, $f_{a_0,42}(x)$, $f_{a_0,33}(x)$, $f_{a_0,24}(x)$, $f_{a_0,15}(x)$, and $f_{a_0,06}(x)$ is given by

$$q \geq \eta - \lambda_0(n) = 4 - \lambda_0(6,0) = 4 - 6 = -2 \qquad (10.324)$$

adding 2 to the degree it gives $q = -2 + 2 = 0$.

$$\Phi_{a_0,60}(x_0, x_1) = \left[\frac{1}{6!0!}\frac{1}{(-2)^{10}}\right](x_0 - 1)^5(x_1 - 1)^5(x_0 + 1)^6(x_1 + 1)^0\frac{1}{2^0}\left(1\right)$$

$$(10.325)$$

$$\Phi_{a_0,51}(x_0, x_1) = \left[\frac{1}{5!1!}\frac{1}{(-2)^{10}}\right](x_0 - 1)^5(x_1 - 1)^5(x_0 + 1)^5(x_1 + 1)^1\frac{1}{2^0}\left(1\right)$$

$$(10.326)$$

$$\Phi_{a_0,42}(x_0, x_1) = \left[\frac{1}{4!0!}\frac{1}{(-2)^{10}}\right](x_0 - 1)^5(x_1 - 1)^5(x_0 + 1)^4(x_1 + 1)^2\frac{1}{2^0}\left(1\right)$$

$$(10.327)$$

$$\Phi_{a_0,33}(x_0, x_1) = \left[\frac{1}{3!3!}\frac{1}{(-2)^{10}}\right](x_0 - 1)^5(x_1 - 1)^5(x_0 + 1)^3(x_1 + 1)^3\frac{1}{2^0}\left(1\right)$$

$$(10.328)$$

$$\Phi_{a_0,24}(x_0, x_1) = \left[\frac{1}{2!4!}\frac{1}{(-2)^{10}}\right](x_0 - 1)^5(x_1 - 1)^5(x_0 + 1)^2(x_1 + 1)^4\frac{1}{2^0}\left(1\right)$$

$$(10.329)$$

$$\Phi_{a_0,15}(x_0, x_1) = \left[\frac{1}{1!5!}\frac{1}{(-2)^{10}}\right](x_0 - 1)^5(x_1 - 1)^5(x_0 + 1)^1(x_1 + 1)^5\frac{1}{2^0}\left(1\right)$$

$$(10.330)$$

$$\Phi_{a_0,06}(x_0, x_1) = \left[\frac{1}{0!6!}\frac{1}{(-2)^{10}}\right](x_0 - 1)^5(x_1 - 1)^5(x_0 + 1)^0(x_1 + 1)^6\frac{1}{2^0}\left(1\right)$$

$$(10.331)$$

Chapter 11

Rectangular nonsymmetric domain

11.1 Generation of a two-variable polynomial in a rectangular nonsymmetric domain

Generation of the Hermite Interpolating Polynomial $\Phi_{e,n}(x)$ related to the node $e = 3$, where the coordinate numbers are given by $a_e = a_3 = (2,4)$. The domain is rectangle, not symmetric, with respect to the coordinate as shown in figure 11.2.1. The maximum derivative is $\eta = 4$ and the number of variables is $d = 2$. The reference node is denoted in the expressions as $\alpha = (2,4)$. The nodes in the visibility of $\alpha = a_3$ are the node 2 where $b_0 = (\beta_{0,0}, \beta_{0,1})$, node 4 where $b_1 = (\beta_{1,0}, \beta_{1,1})$, and node 1.

11.2 Generation of the polynomial $\Phi_{a_3,00}(x_0, x_1)$ related to the reference node $a_3 = (2,4)$

11.2.1 Information about the polynomial

1. Type of domain: a generic rectangle nonsymmetric with respect to the coordinate axes.

2. Number of variables: $d = 2$.

3. Maximum order of derivative: $\eta = 4$

4. Reference node: $a_3 = \alpha = (2,4)$

5. Nodes in the visibility of the reference node

   The elements of the visibility set $\mathcal{V}$ of the reference node a_3 are the planes $a_1 - a_2$ and $a_1 - a_4$. These planes are identified by the orthogonal distance from them to the axis to which they are parallel. Their expression can be obtained from the equation (7.28) where

$$\beta_0 = a_4 = (\beta_{0,0}, \beta_{0,1}) = (6,4) \quad \text{and} \quad \beta_1 = a_2 = (\beta_{1,0}, \beta_{1,1}) = (2,12)$$
$$(11.1)$$

6. The order of derivative of the Hermite Interpolating Polynomials related to the reference node are the constrained numbers in the set $L^\delta(\eta) = L^2(4)$, which are

$$L^2(4) = \Big\{ \ (0,0), (1,0), (0,1), (2,0), (1,1), (0,2),$$
$$(3,0), (2,1), (1,2), (0,3), (4,0), (3,1), (2,2), (1,3), (0,4) \ \Big\}$$
$$(11.2)$$

The corresponding polynomials are denoted as

$$\Phi_{a_3,00}(x), \Phi_{a_3,10}(x), \Phi_{a_3,01}(x), \Phi_{a_3,10}(x),$$
$$\Phi_{a_3,20}(x), \Phi_{a_3,11}(x), \Phi_{a_3,02}(x),$$
$$\Phi_{a_3,30}(x), \Phi_{a_3,21}(x), \Phi_{a_3,12}(x), \Phi_{a_3,03}(x),$$
$$\Phi_{a_3,40}(x), \Phi_{a_3,31}(x), \Phi_{a_3,22}(x), \Phi_{a_3,13}(x), \Phi_{a_3,04}(x) \qquad (11.3)$$

The polynomial $\Phi_{a_3,00}(x)$ has the property that when evaluated at the reference node $a_3 = (2,4)$ equals one.

The first derivative of $\Phi_{a_3,10}(x)$ with respect to the variable x_0 equals one at the reference node $a_3 = (2,4)$.

The cross derivative of $\Phi_{a_0,11}(x)$ with respect to the variables x_0 and x_1 equals one at the reference node $a_3 = (2,4)$, etc.

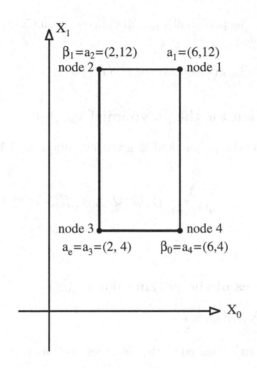

Figure 11.2.1 *Domain nonsymmetric with respect to the coordinate axes.*

11.2.2 Common factors related to the reference node $a_3 = (2, 4)$

Factors that are common to all polynomials related to the reference node $a_3 = (2, 4)$

The polynomial in equation (7.31), on page 297, writes

$$P_{a_3}(x) = P_{a_3}(x_0, x_1) = (x_0 - \beta_{0,0})^{\eta+1}(x_1 - \beta_{1,1})^{\eta+1}$$
$$= (x_0 - 6)^5(x_1 - 12)^5 \qquad (11.4)$$

The constant term $P_{a_3}(\alpha)$ is given by

$$P_{a_3}(x)\Big|_{x=\alpha} = P_{a_3}(\alpha) = P_{a_3}(\alpha_0, \alpha_1)$$
$$= (\alpha_0 - \beta_{0,0})^{\eta+1}(\alpha_1 - \beta_{1,1})^{\eta+1}$$
$$= (2 - 6)^5(4 - 12)^5 = (-4)^5(-8)^5 = 2^{25} \qquad (11.5)$$

where $\alpha = a_3 = (2, 4)$.

11.2.3   Expression for the polynomial $Q_{a_3,n}(x)$

The expression for the polynomial $Q_{a_3,n}(x)$ in equation (7.49), on page 302, for two variables writes

$$Q_{a_3,n}(x) = (x_0 - \alpha_0)^{n_0} (x_1 - \alpha_1)^{n_1} \tag{11.6}$$

11.2.4   Expression for the polynomial $\Phi_{a_3,n}(x)$

The expression for the polynomial is given by equation (7.126), on page 315, which rewrites

$$\Phi_{a_3,n}(x) = \frac{1}{n! P_{a_3}(\alpha)} P_{a_3}(x) Q_{a_3,n}(x) f_{a_3,n}(a,x) \tag{11.7}$$

where $n \in L^2(4)$.

11.2.5   Properties of the polynomial $\Phi_{a_3,00}(x_0, x_1)$

The polynomial $\Phi_{a_3,00}(x_0, x_1)$

1. Equals one when evaluated at the reference node $a_3 = (2, 4)$, that is,

$$\Phi_{a_3,00}(x) \Big|_{x=a_3} = 1 \tag{11.8}$$

2. Vanishes at the nodes different from the reference node, which are $a_1 = (6, 12)$, $a_2 = (2, 12)$, and $a_4 = (6, 4)$

$$\Phi_{a_3,00}(x) \Big|_{x=a_1} = \Phi_{a_3,00}(x) \Big|_{x=a_2} = \Phi_{a_3,00}(x) \Big|_{x=a_4} = 0 \tag{11.9}$$

3. Has its derivatives until order $\eta = 4$ equal to zero at the reference node $a_3 = (2, 4)$

$$\Phi_{a_3,00}^{(k)}(x) \Big|_{x=a_3} = 0 \qquad k = 1, 2, 3, 4 \tag{11.10}$$

4. Has its value as well as the values of the derivatives until order $\eta = 4$ equal to zero at the nodes $a_1 = (6, 12)$, $a_2 = (2, 12)$, and $a_4 = (6, 4)$

$$\Phi_{a_3,00}^{(k)}(x) \Big|_{x=a_1} = 0 \qquad \Phi_{a_3,00}^{(k)}(x) \Big|_{x=a_2} = 0$$

$$\Phi_{a_3,00}^{(k)}(x) \Big|_{x=a_4} = 0 \qquad k = 0, 1, 2, 3, 4 \tag{11.11}$$

11.2.6 Construction of the constant term $n!P_{a_3}(\alpha)$

The factor $P_{a_3}(\alpha)$ was obtained in equation (11.5) and only $n!$ changes with the polynomial then for $n = (0,0)$ it follows that

$$n!P_{a_3}(\alpha) = (n_0!n_1!)P_{a_3}(\alpha) = (0!0!)(2)^{25} = 2^{25} \qquad (11.12)$$

11.2.7 Construction of the polynomial $Q_{a_3,n}(x)$

From equation (11.6) it follows, for $n = (0,0)$, that

$$\begin{aligned} Q_{a_3,n}(x) &= (x_0 - (-1))^{n_0} (x_1 - (-1))^{n_1} \\ &= (x_0 + 1)^{n_0} (x_1 + 1)^{n_1} \\ &= (x_0 + 1)^0 (x_1 + 1)^0 = 1 \qquad (11.13) \end{aligned}$$

11.2.8 The polynomial $\Phi_{a_3,00}(x_0, x_1)$

The polynomial $\Phi_{e,n}(x)$ in equation (11.7) with $n = (0,0)$ writes

$$\begin{aligned} \Phi_{a_3,00}(x) &= \frac{1}{n!P_{a_3}(\alpha)} P_{a_3}(x)\, Q_{a_3,n}(x)\, f_{a_3,n}(a,x) \\ &= \frac{1}{2^{25}} \left[(x_0 - 6)^5 (x_2 - 12)^5 \right] (1)\, f_{a_3,00}(x_0, x_1) \qquad (11.14) \end{aligned}$$

11.2.9 Computation of the coefficients of the polynomial $f_{a_3,00}(x)$

a- Minimum degree of the polynomial $f_{a_0,00}(x)$

From corollary 7.2.19, on page 305, that is,

$$q \geq \eta - \lambda_0(n) = 4 - \lambda_0(0,0) = 4 - 0 = 4 \qquad (11.15)$$

where $\lambda_0(n)$ is the value of the level of the constrained number $n = (0,0)$, which is given by the definition 1.1.12 and computed as $\sum_{j=0}^{j=d-1} n_j = \sum_{j=0}^{j=1} n_j = n_0 + n_1 = 0 + 0 = 0$. A polynomial in two variables x_0, x_1 and of fourth order writes

$$\begin{aligned} f_{a_0,00}(x) &= a_{0,0}x_0^0 x_1^0 + a_{1,0}x_0^1 x_1^0 + a_{0,1}x_0^0 x_1^1 + a_{2,0}x_0^2 x_1^0 + a_{1,1}x_0^1 x_1^1 + a_{0,2}x_0^0 x_1^2 \\ &\quad + a_{3,0}x_0^3 x_1^0 + a_{2,1}x_0^2 x_1^1 + a_{1,2}x_0^1 x_1^2 + a_{0,3}x_0^0 x_1^3 \\ &\quad + a_{4,0}x_0^4 x_1^0 + a_{3,1}x_0^3 x_1^1 + a_{2,2}x_0^2 x_1^2 + a_{1,3}x_0^1 x_1^3 + a_{0,4}x_0^0 x_1^4 \qquad (11.16) \end{aligned}$$

The notation $a_{(w)}$, used later on, refers to the coefficients of the above polynomial.

b- Computation of the coefficients that belong to the terms in the level $\lambda_0(v) = q = 4$

1. Coefficients to be evaluated, which belong to the level $\lambda_0(v) = 4$.

   The indices of the coefficients of the terms of a polynomial elements v in the level $\lambda_0(v) = q = 4$ with two coordinates are the elements of the set $L^\delta_{\lambda_0(v)}$ of constrained numbers in the one-dimensional space given by

   $$L^\delta_{\lambda_0(v)} = L^1_4 = \{(4,0),(3,1),(2,2),(1,3),(0,4)\} \tag{11.17}$$

   which means that $v \in L^1_4$. The coefficients to be evaluated in this level are $a_{(4,0)}$, $a_{(3,1)}$, $a_{(2,1)}$, $a_{(1,3)}$, $a_{(0,4)}$.

   The domain for this problem is not symmetric with respect to the coordinate axes as shown in figure 11.2.1 therefore, the theorem 7.4.7, on page 324, with respect to the permutation of the indices do not apply.

2. Choice of the equation to be used for the computation of the coefficients.

   The coefficients of $f_{e,n}(x)$ are obtained using equation (7.122) or (7.123), on page 314, which is

   $$a_v = \frac{1}{v!} E^v_F(\alpha) = \frac{1}{v_0!\, v_1!} E^v_F(\alpha) \qquad \forall v \in L^\delta_{\lambda_0(v)} \tag{11.18}$$

   where $v \in L^\delta_{\lambda_0(v)} = L^1_4 = \{(4,0),(3,1),(2,2),(1,3),(0,4)\}$. From equation (7.87), on page 309, it follows

   $$E^{(v)}_F(\alpha) = E^{(v_0)}_F(\alpha) E^{(v_1)}_F(\alpha)$$
   $$= \left[\frac{(\eta + v_0)!}{\eta!}\,\frac{1}{(\beta_{00} - \alpha_0)^{v_0}}\right]\left[\frac{(\eta + v_1)!}{\eta!}\,\frac{1}{(\beta_{11} - \alpha_1)^{v_1}}\right] \tag{11.19}$$

3. Computation of the coefficients $a_{(w)} = a_{(u+v)}$ using equation (11.19).

   For the level $\lambda_0(v) = q = 4$ the values of v are the elements of the set L^1_4 and $u = (0,0)$, therefore, for $v = (4,0)$ it follows

   $$a_{(4,0)} = \frac{1}{4!0!}\left[\frac{(4+4)!}{4!}\,\frac{1}{(4)^4}\right]\left[\frac{(4+0)!}{4!}\,\frac{1}{(8)^0}\right] = \frac{35}{2^7} = \frac{560}{2^{11}} \tag{11.20}$$

   where $w = (v+u) = (4,0)+(0,0) = (4,0)$, which identifies the coefficient as being $a_{(w)} = a_{(v+u)} = a_{(4,0)}$.

   For $v = (3,1)$ it follows

   $$a_{(3,1)} = \frac{1}{3!1!}\left[\frac{(4+3)!}{4!}\,\frac{1}{(4)^3}\right]\left[\frac{(4+1)!}{4!}\,\frac{1}{(8)^1}\right] = \frac{175}{2^9} = \frac{700}{2^{11}} \tag{11.21}$$

   where $w = (v+u) = (3,1)+(0,0) = (3,1)$.

For $v = (2, 2)$ it follows

$$a_{(2,2)} = \frac{1}{2!2!} \left[\frac{(4+2)!}{4!} \frac{1}{(4)^2} \right] \left[\frac{(4+2)!}{4!} \frac{1}{(8)^2} \right] = \frac{225}{2^{10}} = \frac{450}{2^{11}} \qquad (11.22)$$

where $w = (v + u) = (2, 2) + (0, 0) = (2, 2)$.

For $v = (1, 3)$ it follows

$$a_{(1,3)} = \frac{1}{1!3!} \left[\frac{(4+1)!}{4!} \frac{1}{(4)^1} \right] \left[\frac{(4+3)!}{4!} \frac{1}{(8)^3} \right] = \frac{175}{2^{11}} \qquad (11.23)$$

where $w = (v + u) = (2, 2) + (0, 0) = (2, 2)$.

For $v = (0, 4)$ it follows

$$a_{(0,4)} = \frac{1}{0!4!} \left[\frac{(4+0)!}{4!} \frac{1}{(4)^0} \right] \left[\frac{(4+4)!}{4!} \frac{1}{(8)^4} \right] = \frac{35}{2^{11}} \qquad (11.24)$$

where $w = (v + u) = (2, 2) + (0, 0) = (2, 2)$.

b- Computation of the coefficients that belong to the terms in the level $\lambda_0(v) = q - 1 = 3$

1. Coefficients to be evaluated, which belong to the level $\lambda_0(v) = q - 1 = 3$.

   The elements in the level $\lambda_0(v) = 3$ with two coordinates are the constrained numbers in the one-dimensional space are

   $$L^{\delta}_{\lambda_0(v)} = L^1_3 = \{(3, 0), (2, 1), (1, 2), (0, 3)\} \qquad (11.25)$$

   The elements of the set L^1_3 can obtained using the algorithm described in subsection 2.4.4, on page 45. Therefore, the coefficients $a_{(w)} = a_{(v)}$ to be evaluated in this level are $a_{(3,0)}$, $a_{(2,1)}$, $a_{(1,2)}$, and $a_{(0,3)}$.

   The theorem 7.4.7, on page 324, do not apply if the domain is not symmetric with respect to the coordinate axes, which implies that all coefficients must be evaluated.

2. Equation for the computation of the coefficients.

   The coefficients are obtained using equation (7.123), on page 314, whose equation is repeated below

   $$a_v = \frac{1}{v!} \left[E^v_F(\alpha) - S_v (a_{u+v}, \alpha) \right] \qquad \lambda_0(v) = 3 \qquad (11.26)$$

3. Computation of the term $S_v(a_{u+v}, \alpha)$

The degree of the polynomial $S_v(a_{u+v}, \alpha)$ is given by $m = q - \lambda_0(v) = 4 - 3 = 1$ where $\lambda_0(v) = v_0 + v_1 = 3 + 0 = 3$. The exponents of the variables in each term of $S_v(a_{u+v}, \alpha)$ are the elements of the set shown in equation (7.124), on page 314, whose equation is

$$\left(L^\delta(m) - L_0^\delta\right) = \left(L^1(1) - L_0^1\right)$$
$$= \{(0,0), (1,0), (0,1)\} - \{(0,0)\}$$
$$= \{(1,0), (0,1)\} \tag{11.27}$$

which permits to obtain the indices of the coefficients of the terms of the polynomial $S_v(a_{u+v}, \alpha)$ as follows

(a) For $v = (3,0)$

If $u = (1,0)$ then $v+u = (3,0)+(1,0) = (4,0)$ then $a_{(v+u)} = a_{(4,0)}$, which is one of the coefficients obtained for the level $\lambda_0(v) = q = 4$.

If $u = (0,1)$ then $v+u = (3,0)+(0,1) = (3,1)$ then $a_{(v+u)} = a_{(3,1)}$, which is one of the coefficients obtained for the level $\lambda_0(v) = q = 4$.

(b) For $v = (2,1)$

If $u = (1,0)$ then $v+u = (2,1)+(1,0) = (3,1)$ then $a_{(v+u)} = a_{(3,1)}$, which is one of the coefficients obtained for the level $\lambda_0(v) = q = 4$.

If $u = (0,1)$ then $v+u = (2,1)+(0,1) = (2,1)$ then $a_{(v+u)} = a_{(2,2)}$, which is one of the coefficients obtained for the level $\lambda_0(v) = q = 4$.

The expression for $S_v(a_{u+v}, \alpha)$ is given by equation (7.112), on page 313. For $d = 2$ and $m = 1$ it writes

$$S_v(a_{u+v}, \alpha) = \sum_{r_0=0}^{r_0=m} \sum_{r_1=0}^{r_1=r_0} a_{(v+u)} \frac{(v+u)!}{u!} \alpha_0^{u_0} \alpha_1^{u_1}$$

$$= \sum_{r_0=0}^{r_0=1} \sum_{r_1=0}^{r_1=r_0} a_{(v+u)} \frac{(v+u)!}{u!} \alpha_0^{u_0} \alpha_1^{u_1}$$

$$= a_{(v+(1,0))} \frac{(v+(1,0))!}{(1,0)!} \alpha_0^1 \alpha_1^0 + a_{(v+(0,1))} \frac{(v+(0,1))!}{(0,1)!} \alpha_0^0 \alpha_1^1$$

$$\tag{11.28}$$

4. Computation of the term $E_F^v(\alpha)$

The expression for $E_F^v(\alpha)$ is given by equation (7.87) writes

$$E_F^{(v)}(\alpha) = E_F^{(v_0)}(\alpha) E_F^{(v_1)}(\alpha)$$

$$= \left[\frac{(\eta + v_0)!}{\eta!} \frac{1}{(\beta_{0,0} - \alpha_0)^{v_0}}\right] \left[\frac{(\eta + v_1)!}{\eta!} \frac{1}{(\beta_{1,1} - \alpha_1)^{v_1}}\right] \tag{11.29}$$

5. Computation of the coefficient $a_v = a_{3,0}$ where $v = (v_0, v_1) = (3, 0)$

(a) Computation of $E_F^{(3,0)}$.

Substituting $\eta = 4$, $v = (v_0, v_1) = (3, 0)$, $\alpha = (\alpha_0, \alpha_1) = (2, 4)$, into equation (11.29) it follows

$$E_F^{(3,0)}(\alpha) = \left[\frac{(4+3)!}{4!} \frac{1}{(4)^3} \right] \left[\frac{(4+0)!}{4!} \frac{1}{(8)^0} \right]$$

$$= \frac{5 \times 6 \times 7}{(4)^3} = \frac{105}{2^5} \qquad (11.30)$$

(b) Computation of $S_{3,0}$.

Since u belongs to the level $m = q - \lambda_0(v) = 4 - \lambda_0(3, 0) = 4 - 3 = 1$ then the constrained numbers u in the level $\lambda_0(v) = 1$ are elements of one-dimensional the set $L_1^1 = \{(1, 0), (0, 1)\}$.

Table 11.2.2 *Computation of the terms for the evaluation of* $S_{(3,0)} \left(a_{(u+v)}, \alpha \right)$.

v	u	$(u+v)$	$a_{(u+v)}$	$(u+v)!/u!$	α^{u_0}	α^{u_1}	value
$(3, 0)$	$(1, 0)$	$(4, 0)$	$35/2^7$	$(4,0)!/(1,0)! = 4!$	$(2)^1$	$(4)^0$	$105/2^3$
	$(0, 1)$	$(3, 1)$	$175/2^9$	$(3,1)!/(0,1)! = 3!$	$(2)^0$	$(4)^1$	$525/2^6$

Substituting the values for the terms obtained in the table 11.2.2 into the equation (11.28) it gives

$$S_{(3,0)} \left(a_{(u+v)}, \alpha \right) = \frac{105}{2^3} + \frac{525}{2^6} = \frac{840}{2^6} (840 + 525) = \frac{1365}{2^6}$$
$$(11.31)$$

(c) Computation of $a_{3,0}$.

Substituting v, $E_F^{(v)}(\alpha)$ given by equation (11.30) and $S_v \left(a_{u+v}, \alpha \right)$ given by (11.31) it follows

$$a_{(3,0)} = \frac{1}{v!} \left[E_F^{(3,0)}(\alpha) - S_{(3,0)} \left(a_{u+v}, \alpha \right) \right] = \frac{1}{3! \, 0!} \left[\frac{105}{2^5} - \frac{1365}{2^6} \right]$$

$$= -\frac{1}{3!} \frac{1155}{2^6} = -\frac{385}{2^7} = -\frac{6160}{2^{11}} \qquad (11.32)$$

6. Computation of the coefficient $a_v = a_{2,1}$ where $v = (v_0, v_1) = (2, 1)$

(a) Computation of $E_F^{(2,1)}$.

Substituting $\eta = 4$, $v = (v_0, v_1) = (2, 1)$, $\alpha = (\alpha_0, \alpha_1) = (2, 4)$, into equation (11.29) it follows

$$E_F^{(2,1)}(\alpha) = \left[\frac{(4+2)!}{4!} \frac{1}{(4)^2} \right] \left[\frac{(4+1)!}{4!} \frac{1}{(8)^1} \right] = \frac{5 \times 6}{4^2} \times \frac{5}{2^3} = \frac{75}{2^6} \tag{11.33}$$

(b) Computation of $S_{2,1}$.

For $v = (2, 1)$ it follows that $\lambda_0(v) = v_0 + v_1 = 2 + 1 = 3$ and $m = q - \lambda_0(v) = 4 - 3 = 1$.

Since u belongs to the level $m = 1$ then the constrained numbers u in the level $\lambda_0(v) = 1$ are elements of one-dimensional the set $L_1^1 = \{(1, 0), (0, 1)\}$.

Table 11.2.3 *Computation of the terms for the evaluation of* $S_{(2,1)}\left(a_{(u+v)}, \alpha\right)$.

v	u	$(u+v)$	$a_{(u+v)}$	$(u+v)!/u!$	α^{u_0}	α^{u_1}	value
$(2, 1)$	$(1, 0)$	$(3, 1)$	$175/2^9$	$(3, 1)!/(1, 0)! = 6$	$(2)^1$	$(4)^0$	$525/2^7$
	$(0, 1)$	$(2, 2)$	$225/2^{10}$	$(2, 2)!/(0, 1)! = 4$	$(2)^0$	$(4)^1$	$225/2^6$

Substituting the values for the terms obtained in the table 11.2.3 into the equation (11.28) it gives

$$S_{(2,1)}\left(a_{(u+v)}, \alpha\right) = \frac{525}{2^7} + \frac{225}{2^6} = \frac{1}{2^7}(525 + 450) = \frac{975}{2^7} \tag{11.34}$$

(c) Computation of $a_{2,1}$.

Substituting v, $E_F^{(v)}(\alpha)$ given by equation (11.32) and $S_v\left(a_{u+v}, \alpha\right)$ given by (11.34) it follows

$$a_{(2,1)} = \frac{1}{v!} \left[E_F^{(2,1)}(\alpha) - S_{(2,1)}\left(a_{u+v}, \alpha\right) \right] = \frac{1}{2! \, 1!} \left[\frac{75}{2^6} - \frac{975}{2^7} \right]$$

$$= \frac{1}{2! \, 1!} \left[\frac{150}{2^7} - \frac{975}{2^7} \right] = -\frac{825}{2 \times 2^7} = -\frac{825}{2^8} = -\frac{6600}{2^{11}} \tag{11.35}$$

7. Computation of the coefficient $a_v = a_{1,2}$ where $v = (v_0, v_1) = (1, 2)$

(a) Computation of $E_F^{(1,2)}$.

Substituting $\eta = 4$, $v = (v_0, v_1) = (1, 2)$, $\alpha = (\alpha_0, \alpha_1) = (2, 4)$, into equation (11.29) it follows

$$E_F^{(1,2)}(\alpha) = \left[\frac{(4+1)!}{4!} \frac{1}{(4)^1} \right] \left[\frac{(4+2)!}{4!} \frac{1}{(8)^2} \right] = \frac{5}{4} \times \frac{5 \times 6}{8^2} = \frac{75}{2^7} \tag{11.36}$$

(b) Computation of $S_{1,2}$.

For $v = (1, 2)$ it follows that $\lambda_0(v) = v_0 + v_1 = 1 + 2 = 3$ and $m = q - \lambda_0(v) = 4 - 3 = 1$.

Since u belongs to the level $m = 1$ then the constrained numbers u in the level $\lambda_0(v) = 1$ are elements of one-dimensional the set $L_1^1 = \{(1, 0), (0, 1)\}$.

Table 11.2.4 *Computation of the terms for the evaluation of* $S_{(1,2)}\left(a_{(u+v)}, \alpha\right)$.

v	u	$(u+v)$	$a_{(u+v)}$	$(u+v)!/u!$	α^{u_0}	α^{u_1}	value
$(1,2)$	$(1,0)$	$(2,2)$	$225/2^{10}$	$(2,2)!/(1,0)! = 4$	$(2)^1$	$(4)^0$	$225/2^7$
	$(0,1)$	$(1,3)$	$175/2^{11}$	$(1,3)!/(0,1)! = 6$	$(2)^0$	$(4)^1$	$525/2^8$

Substituting the values for the terms obtained in the table 11.2.4 into the equation (11.28) it gives

$$S_{(1,2)}\left(a_{(u+v)}, \alpha\right) = \frac{225}{2^7} + \frac{525}{2^8} = \frac{1}{2^8}(450 + 525) = \frac{975}{2^8} \quad (11.37)$$

(c) Computation of $a_{1,2}$.

Substituting v, $E_F^v(\alpha)$ and $S_v\left(a_{u+v}, \alpha\right)$ for its values into

$$\begin{aligned}
a_{(1,2)} &= \frac{1}{v!}\left[E_F^{(1,2)}(\alpha) - S_{(1,2)}\left(a_{u+v}, \alpha\right)\right] \\
&= \frac{1}{1!\,2!}\left[\frac{75}{2^7} - \frac{975}{2^8}\right] \\
&= \frac{1}{1!\,2!}\frac{1}{2^8}[150 - 975] \\
&= -\frac{825}{2 \times 2^8} = -\frac{825}{2^9} = -\frac{3300}{2^{11}} \quad (11.38)
\end{aligned}$$

8. Computation of the coefficient $a_v = a_{0,3}$ where $v = (v_0, v_1) = (0, 3)$

(a) Computation of $E_F^{(0,3)}$.

Substituting $\eta = 4$, $v = (v_0, v_1) = (0, 3)$, $\alpha = (\alpha_0, \alpha_1) = (2, 4)$, into equation (11.29) it follows

$$E_F^{(0,3)}(\alpha) = \left[\frac{(4+0)!}{4!}\frac{1}{(4)^0}\right]\left[\frac{(4+3)!}{4!}\frac{1}{(8)^3}\right] = \frac{5 \times 6 \times 7}{8^3} = \frac{105}{2^8} \quad (11.39)$$

(b) Computation of $S_{0,3}$.

For $v = (0, 3)$ it follows that $\lambda_0(v) = v_0 + v_1 = 0 + 3 = 3$ and $m = q - \lambda_0(v) = 4 - 3 = 1$.

Since u belongs to the level $m = 1$ then the constrained numbers u in the level $\lambda_0(v) = 1$ are elements of one-dimensional the set $L_1^1 = \{(1, 0), (0, 1)\}$.

Table 11.2.5 *Computation of the terms for the evaluation of* $S_{(0,3)}\left(a_{(u+v)}, \alpha\right)$.

v	u	$(u+v)$	$a_{(u+v)}$	$(u+v)!/u!$	α^{u_0}	α^{u_1}	value
$(0,3)$	$(1,0)$	$(1,3)$	$175/2^{11}$	$(1,3)!/(1,0)! = 6$	$(2)^1$	$(4)^0$	$525/2^9$
	$(0,1)$	$(0,4)$	$35/2^{11}$	$(0,4)!/(0,1)! = 24$	$(2)^0$	$(4)^1$	$105/2^6$

Substituting the values for the terms obtained in the table 11.2.5 into the equation (11.28) it gives

$$S_{(0,3)}\left(a_{(u+v)}, \alpha\right) = \frac{525}{2^9} + \frac{105}{2^6} = \frac{1}{2^9}(525 + 840) = \frac{1365}{2^9} \quad (11.40)$$

(c) Computation of $a_{0,3}$.

Substituting v, $E_F^v(\alpha)$ and $S_v\left(a_{u+v}, \alpha\right)$ for its values into

$$a_{(0,3)} = \frac{1}{v!}\left[E_F^{(0,3)}(\alpha) - S_{(0,3)}\left(a_{u+v}, \alpha\right)\right]$$

$$= \frac{1}{0!\,3!}\left[\frac{105}{2^8} - \frac{1365}{2^9}\right]$$

$$= \frac{1}{6}\frac{1}{2^9}[210 - 1365] = -\frac{385}{2^{10}} = -\frac{770}{2^{11}} \quad (11.41)$$

b- Computation of the coefficients that belong to the terms in the level $\lambda_0(v) = q - 2 = 2$

1. Coefficients to be evaluated, which belong to the level $\lambda_0(v) = 2$.

   The elements in the level $\lambda_0(v) = 2$ with two coordinates are the constrained numbers in the one-dimensional space are

$$L_{\lambda_0(v)}^\delta = L_2^1 = \{(2, 0), (1, 1), (0, 2)\} \quad (11.42)$$

The elements of the set L_2^1 can obtained using the algorithm described in subsection 2.4.4, on page 45. Therefore, the coefficients $a_{(w)} = a_{(v)}$ to be evaluated in this level are $a_{(2,0)}$, $a_{(1,1)}$, and $a_{(0,2)}$.

2. Equation for the computation of the coefficients.

The coefficients are obtained using equation (7.123), on page 314, which is repeated here as

$$a_v = \frac{1}{v!} \left[E_F^v(\alpha) - S_v \left(a_{u+v}, \alpha \right) \right] \qquad \lambda_0(v) = 2 \qquad (11.43)$$

3. Computation of the term $S_v \left(a_{u+v}, \alpha \right)$

The degree of the polynomial $S_v \left(a_{u+v}, \alpha \right)$ is given by $m = q - \lambda_0(v) = 4 - 2 = 2$ where $\lambda_0(v) = v_0 + v_1 = 2 + 0 = 2$. The exponents of the variables in each term of $S_v \left(a_{u+v}, \alpha \right)$ are the elements of the set shown in equation (7.124), which is

$$\begin{aligned}
\left(L^\delta(m) - L_0^\delta \right) &= \left(L^1(1) - L_0^1 \right) \\
&= \{(0,0), (1,0), (0,1), (2,0), (1,1), (0,2)\} - \{(0,0)\} \\
&= \{(1,0), (0,1), (2,0), (1,1), (0,2)\} \qquad (11.44)
\end{aligned}$$

which permits to obtain the indices of the coefficients of the terms of the polynomial $S_v \left(a_{u+v}, \alpha \right)$ as follows

(a) For $v = (2,0)$

If $u = (1,0)$ then $v + u = (2,0) + (1,0) = (3,0)$, which gives $a_{(v+u)} = a_{(3,0)}$.

If $u = (0,1)$ then $v + u = (2,0) + (0,1) = (2,1)$, which gives $a_{(v+u)} = a_{(2,1)}$.

If $u = (2,0)$ then $v + u = (2,0) + (2,0) = (4,0)$, which gives $a_{(v+u)} = a_{(4,0)}$.

If $u = (1,1)$ then $v + u = (2,0) + (1,1) = (3,1)$, which gives $a_{(v+u)} = a_{(3,1)}$.

If $u = (0,2)$ then $v + u = (2,0) + (0,2) = (2,2)$, which gives $a_{(v+u)} = a_{(2,2)}$.

Where the indices of the coefficients $a_{(3,0)}$ and $a_{(2,1)}$ belong to the level $\lambda_0(v) = q = 3$. And the indices of $a_{(4,0)}$, $a_{(3,1)}$, and $a_{(2,2)}$ belong to the level $\lambda_0(v) = q = 4$.

(b) For $v = (1,1)$

If $u = (1,0)$ then $v + u = (1,1) + (1,0) = (2,1)$, which gives $a_{(v+u)} = a_{(2,1)}$.

If $u = (0,1)$ then $v + u = (1,1) + (0,1) = (1,2)$, which gives $a_{(v+u)} = a_{(1,2)}$.

If $u = (2,0)$ then $v + u = (1,1) + (2,0) = (3,1)$, which gives $a_{(v+u)} = a_{(3,1)}$.

If $u = (1,1)$ then $v + u = (1,1) + (1,1) = (2,2)$, which gives $a_{(v+u)} = a_{(2,2)}$.

If $u = (0, 2)$ then $v + u = (1, 1) + (0, 2) = (1, 3)$, which gives $a_{(v+u)} = a_{(1,3)}$.

Where the indices of the coefficients $a_{(2,1)}$ and $a_{(1,2)}$ belong to the level $\lambda_0(v) = q = 3$. The indices of $a_{(3,1)}$, $a_{(2,2)}$, and $a_{(1,3)}$ belong to the level $\lambda_0(v) = q = 4$.

The expression for $S_v(a_{u+v}, \alpha)$ is given by equation (7.112), on page 313. For $d = 2$ and $m = 2$ it writes

$$S_v(a_{u+v}, \alpha) = \sum_{r_0=0}^{r_0=m} \sum_{r_1=0}^{r_1=r_0} a_{(v+u)} \frac{(v+u)!}{u!} \alpha_0^{u_0} \alpha_1^{u_1}$$

$$= \sum_{r_0=0}^{r_0=2} \sum_{r_1=0}^{r_1=r_0} a_{(v+u)} \frac{(v+u)!}{u!} \alpha_0^{u_0} \alpha_1^{u_1}$$

$$= a_{(v+(1,0))} \frac{(v+(1,0))!}{(1,0)!} \alpha_0^1 \alpha_1^0 + a_{(v+(0,1))} \frac{(v+(0,1))!}{(0,1)!} \alpha_0^0 \alpha_1^1$$

$$+ a_{(v+(2,0))} \frac{(v+(2,0))!}{(2,0)!} \alpha_0^2 \alpha_1^0 + a_{(v+(1,1))} \frac{(v+(1,1))!}{(1,1)!} \alpha_0^1 \alpha_1^1$$

$$+ a_{(v+(0,2))} \frac{(v+(0,2))!}{(0,2)!} \alpha_0^0 \alpha_1^2 \tag{11.45}$$

4. Computation of the term $E_F^v(\alpha)$

The expression for $E_F^v(\alpha)$ is given by equation (7.87) writes

$$E_F^{(v)}(\alpha) = E_F^{(v_0)}(\alpha) E_F^{(v_1)}(\alpha)$$

$$= \left[\frac{(\eta+v_0)!}{\eta!} \frac{1}{(\beta_{0,0} - \alpha_0)^{v_0}} \right] \left[\frac{(\eta+v_1)!}{\eta!} \frac{1}{(\beta_{1,1} - \alpha_1)^{v_1}} \right] \tag{11.46}$$

5. Computation of the coefficient $a_v = a_{(2,0)}$ where $v = (v_0, v_1) = (2, 0)$

(a) Computation of $E_F^{(2,0)}$.

Substituting $\eta = 4$, $v = (v_0, v_1) = (2, 0)$, $\alpha = (\alpha_0, \alpha_1) = (2, 4)$, into equation (11.46) it follows

$$E_F^{(2,0)}(\alpha) = \left[\frac{(4+2)!}{4!} \frac{1}{(4)^2} \right] \left[\frac{(4+0)!}{4!} \frac{1}{(8)^0} \right]$$

$$= \frac{5 \times 6}{4^2} = \frac{15}{2^3} \tag{11.47}$$

(b) Computation of $S_{(2,0)}$.

Since u belongs to the level $m = q - \lambda_0(v) = 4 - \lambda_0(2, 0) = 4 - 2 = 2$ then the constrained numbers u in the level $m = 1$ and $m = 2$ are elements of one-dimensional the set $L_1^1 \cup L_2^1$, which are $\{(1, 0), (0, 1), (2, 0), (1, 1), (0, 2)\}$.

Table 11.2.6 *Computation of the terms for the evaluation of* $S_{(2,0)}\left(a_{(u+v)}, \alpha\right).$

v	u	$(u+v)$	$a_{(u+v)}$	$(u+v)!/u!$	α^{u_0}	α^{u_1}	value
$(2,0)$	$(1,0)$	$(3,0)$	$-385/2^7$	$(3,0)!/(1,0)! = 6$	$(2)^1$	$(4)^0$	$-1155/2^5$
	$(0,1)$	$(2,1)$	$-825/2^8$	$(2,1)!/(0,1)! = 2$	$(2)^0$	$(4)^1$	$-825/2^5$
	$(2,0)$	$(4,0)$	$35/2^7$	$(4,0)!/(2,1)! = 12$	$(2)^2$	$(4)^0$	$105/2^3$
	$(1,1)$	$(3,1)$	$175/2^9$	$(3,1)!/(1,1)! = 6$	$(2)^1$	$(4)^1$	$525/2^5$
	$(0,2)$	$(2,2)$	$225/2^{10}$	$(2,2)!/(0,2)! = 2$	$(2)^0$	$(4)^2$	$225/2^5$

Substituting the values for the terms obtained in the table 11.2.6 into the equation (11.45) it gives

$$S_{(0,2)}\left(a_{(u+v)}, \alpha\right) = -\frac{1155}{2^5} - \frac{825}{2^5} + \frac{105}{2^3} + \frac{525}{2^5} + \frac{225}{2^5}$$

$$= \frac{1}{2^5}\left(-1155 - 825 + 420 + 525 + 225\right) = -\frac{810}{2^5}$$
(11.48)

(c) Computation of $a_{(2,0)}$.
Substituting v, $E_F^v(\alpha)$ and $S_v\left(a_{u+v}, \alpha\right)$ for its values into

$$a_{(2,0)} = \frac{1}{v!}\left[E_F^{(2,0)}(\alpha) - S_{(2,0)}\left(a_{u+v}, \alpha\right)\right] = \frac{1}{2!\,0!}\left[\frac{15}{2^3} - \frac{-810}{2^5}\right]$$
(11.49)

$$= \frac{1}{2}\frac{1}{2^5}\left(60 + 810\right) = \frac{870}{2^6} = \frac{27840}{2^{11}}$$
(11.50)

6. Computation of the coefficient $a_v = a_{(1,1)}$ where $v = (v_0, v_1) = (1,1)$

(a) Computation of $E_F^{(1,1)}$.
Substituting $\eta = 4$, $v = (v_0, v_1) = (1,1)$, $\alpha = (\alpha_0, \alpha_1) = (2,4)$, into equation (11.46) it follows

$$E_F^{(1,1)}(\alpha) = \left[\frac{(4+1)!}{4!}\frac{1}{(4)^1}\right]\left[\frac{(4+1)!}{4!}\frac{1}{(8)^1}\right]$$

$$= \frac{5 \times 5}{2^5} = \frac{25}{2^5}$$
(11.51)

(b) Computation of $S_{(1,1)}$.
Since u belongs to the level $m = q - \lambda_0(v) = 4 - \lambda_0(1,1) = 4 - 2 = 2$ then the constrained numbers u in the level $m = 1$ and $m = 2$ are elements of one-dimensional the set $L_1^1 \cup L_2^1$, which are $\{(1,0), (0,1), (2,0), (1,1), (0,2)\}$.

Table 11.2.7 *Computation of the terms for the evaluation of* $S_{(1,1)}\left(a_{(u+v)}, \alpha\right)$.

v	u	$(u+v)$	$a_{(u+v)}$	$(u+v)!/u!$	α^{u_0}	α^{u_1}	value
$(1,1)$	$(1,0)$	$(2,1)$	$-825/2^8$	$(2,1)!/(1,0)! = 2$	$(2)^1$	$(4)^0$	$-825/2^6$
	$(0,1)$	$(1,2)$	$-825/2^9$	$(1,2)!/(0,1)! = 2$	$(2)^0$	$(4)^1$	$-825/2^6$
	$(2,0)$	$(3,1)$	$175/2^9$	$(3,1)!/(2,0)! = 3$	$(2)^2$	$(4)^0$	$525/2^7$
	$(1,1)$	$(2,2)$	$225/2^{10}$	$(2,2)!/(1,1)! = 4$	$(2)^1$	$(4)^1$	$225/2^5$
	$(0,2)$	$(1,3)$	$175/2^{11}$	$(1,3)!/(0,2)! = 3$	$(2)^0$	$(4)^2$	$525/2^7$

Substituting the values for the terms obtained in the table 11.2.7 into the equation (11.45) it gives

$$S_{(1,1)}\left(a_{(u+v)}, \alpha\right) = -\frac{825}{2^6} - \frac{825}{2^6} + \frac{525}{2^7} + \frac{225}{2^5} + \frac{525}{2^7}$$

$$= \frac{1}{2^7}\left(-1650 - 1650 + 525 + 900 + 525\right) = -\frac{1350}{2^7} \tag{11.52}$$

(c) Computation of $a_{(1,1)}$.

Substituting v, $E_F^v(\alpha)$ and $S_v\left(a_{u+v}, \alpha\right)$ for its values into

$$a_{(1,1)} = \frac{1}{v!}\left[E_F^{(1,1)}(\alpha) - S_{(1,1)}\left(a_{u+v}, \alpha\right)\right] = \frac{1}{1!\,1!}\left[\frac{25}{2^5} - \frac{-1350}{2^7}\right] \tag{11.53}$$

$$= \frac{1}{2^7}\left(100 + 1350\right) = \frac{1450}{2^7} = \frac{23200}{2^{11}} \tag{11.54}$$

7. Computation of the coefficient $a_v = a_{(0,2)}$ where $v = (v_0, v_1) = (0, 2)$

(a) Computation of $E_F^{(0,2)}$.

Substituting $\eta = 4$, $v = (v_0, v_1) = (0, 2)$, $\alpha = (\alpha_0, \alpha_1) = (2, 4)$, into equation (11.46) it follows

$$E_F^{(0,2)}(\alpha) = \left[\frac{(4+0)!}{4!}\frac{1}{(4)^0}\right]\left[\frac{(4+2)!}{4!}\frac{1}{(8)^2}\right]$$

$$= \frac{5 \times 6}{2^6} = \frac{15}{2^5} \tag{11.55}$$

(b) Computation of $S_{(0,2)}$.

Since u belongs to the level $m = q - \lambda_0(v) = 4 - \lambda_0(0, 2) = 4 - 2 = 2$ then the constrained numbers u in the level $m = 1$ and $m = 2$ are elements of one-dimensional the set $L_1^1 \cup L_2^1$, which are $\{(1, 0), (0, 1), (2, 0), (1, 1), (0, 2)\}$.

Table 11.2.8 *Computation of the terms for the evaluation of* $S_{(0,2)}\left(a_{(u+v)}, \alpha\right)$.

v	u	$(u+v)$	$a_{(u+v)}$	$(u+v)!/u!$	α^{u_0}	α^{u_1}	value
$(0,2)$	$(1,0)$	$(1,2)$	$-825/2^9$	$(1,2)!/(1,0)! = 2$	$(2)^1$	$(4)^0$	$-825/2^7$
	$(0,1)$	$(0,3)$	$-385/2^{10}$	$(0,3)!/(0,1)! = 6$	$(2)^0$	$(4)^1$	$-1155/2^7$
	$(2,0)$	$(2,2)$	$225/2^{10}$	$(2,2)!/(2,0)! = 2$	$(2)^2$	$(4)^0$	$225/2^7$
	$(1,1)$	$(1,3)$	$175/2^{11}$	$(1,3)!/(1,1)! = 6$	$(2)^1$	$(4)^1$	$525/2^7$
	$(0,2)$	$(0,4)$	$35/2^{11}$	$(0,4)!/(0,2)! = 12$	$(2)^0$	$(4)^2$	$105/2^5$

Substituting the values for the terms obtained in the table 11.2.8 into the equation (11.45) it gives

$$S_{(0,2)}\left(a_{(u+v)}, \alpha\right) = -\frac{825}{2^7} - \frac{1155}{2^7} + \frac{225}{2^7} + \frac{525}{2^7} + \frac{105}{2^5}$$

$$= \frac{1}{2^7}\left(-825 - 1155 + 225 + 525 + 420\right) = -\frac{810}{2^7}$$

$$\tag{11.56}$$

(c) Computation of $a_{0,2}$.

Substituting v, $E_F^v(\alpha)$ and $S_v\left(a_{u+v}, \alpha\right)$ for its values into

$$a_{(0,2)} = \frac{1}{v!}\left[E_F^{(0,2)}(\alpha) - S_{(0,2)}\left(a_{u+v}, \alpha\right)\right] = \frac{1}{0!\,2!}\left[\frac{15}{2^5} - \frac{-810}{2^7}\right]$$

$$\tag{11.57}$$

$$= \frac{1}{2^7}\left(60 + 810\right) = \frac{870}{2^8} = \frac{6960}{2^{11}} \tag{11.58}$$

b- Computation of the coefficients that belong to the terms in the level $\lambda_0(v) = q - 3 = 1$

1. Coefficients to be evaluated, which belong to the level $\lambda_0(v) = 1$.

   The elements in the level $\lambda_0(v) = 1$ with two coordinates are the constrained numbers in the one-dimensional space are

   $$L_{\lambda_0(v)}^{\delta} = L_1^1 = \{(1,0), (0,1)\} \tag{11.59}$$

   The elements of the set L_1^1 can obtained using the algorithm described in subsection 2.4.4, on page 45. Therefore, the coefficients $a_{(w)} = a_{(v)}$ to be evaluated in this level are $a_{(1,0)}$, and $a_{(0,1)}$.

2. Equation for the computation of the coefficients.

   The coefficients are obtained using equation (7.123), on page 314, which is repeated below

   $$a_{(v)} = \frac{1}{v!}\left[E_F^{(v)}(\alpha) - S_v\left(a_{u+v}, \alpha\right)\right] \qquad \lambda_0(v) = 1 \tag{11.60}$$

3. Computation of the term $S_v(a_{u+v}, \alpha)$

The degree of the polynomial $S_v(a_{u+v}, \alpha)$ is given by $m = q - \lambda_0(v) = 4 - 1 = 3$ where $\lambda_0(v) = v_0 + v_1 = 1 + 0 = 1$. The exponents of the variables in each term of $S_v(a_{u+v}, \alpha)$ are the elements of the set shown in equation (7.124), on page 314, which is

$$\begin{aligned}
(L^\delta(m) - L_0^\delta) = (L^1(1) - L_0^1) &= L_1^1 \cup L_2^1 \cup L_3^1 \\
&= \{(0,0),(1,0),(0,1),(2,0),(1,1),(0,2),(3,0), \\
&\quad (2,1),(1,2),(0,3)\} - \{(0,0)\} \\
&= \{(1,0),(0,1),(2,0),(1,1),(0,2),(3,0), \\
&\quad (2,1),(1,2),(0,3)\}
\end{aligned} \tag{11.61}$$

which permits to obtain the indices of the coefficients of the terms of the polynomial $S_v(a_{u+v}, \alpha)$ for $v = (1,0)$ as follows

If $u = (1,0)$ then $v + u = (1,0) + (1,0) = (2,0)$, which gives $a_{(v+u)} = a_{(2,0)}$.

If $u = (0,1)$ then $v + u = (1,0) + (0,1) = (1,1)$, which gives $a_{(v+u)} = a_{(1,1)}$.

If $u = (2,0)$ then $v + u = (1,0) + (2,0) = (3,0)$, which gives $a_{(v+u)} = a_{(3,0)}$.

If $u = (1,1)$ then $v + u = (1,0) + (1,1) = (2,1)$, which gives $a_{(v+u)} = a_{(2,1)}$.

If $u = (0,2)$ then $v + u = (1,0) + (0,2) = (1,2)$, which gives $a_{(v+u)} = a_{(1,2)}$.

If $u = (3,0)$ then $v + u = (1,0) + (3,0) = (4,0)$, which gives $a_{(v+u)} = a_{(4,0)}$.

If $u = (2,1)$ then $v + u = (1,0) + (2,1) = (3,1)$, which gives $a_{(v+u)} = a_{(3,1)}$.

If $u = (1,2)$ then $v + u = (1,0) + (1,2) = (2,2)$, which gives $a_{(v+u)} = a_{(2,2)}$.

If $u = (0,3)$ then $v + u = (1,0) + (0,3) = (1,3)$, which gives $a_{(v+u)} = a_{(1,3)}$.

Where the indices of the coefficients $a_{(2,0)}$ and $a_{(1,1)}$ belong to the level $\lambda_0(v) = q = 2$. The indices of $a_{(3,0)}$, $a_{(2,1)}$, and $a_{(1,2)}$ belong to the level $\lambda_0(v) = q = 3$. The indices of $a_{(4,0)}$, $a_{(3,1)}$, $a_{(2,2)}$, and $a_{(1,3)}$ belong to the level $\lambda_0(v) = q = 4$.

The expression for $S_v(a_{u+v}, \alpha)$ is given by equation (7.112), on page

313. For $d = 2$ and $m = 3$ it writes

$$S_v(a_{u+v}, \alpha) = \sum_{r_0=0}^{r_0=m} \sum_{r_1=0}^{r_1=r_0} a_{(v+u)} \frac{(v+u)!}{u!} \alpha_0^{u_0} \alpha_1^{u_1}$$

$$= \sum_{r_0=0}^{r_0=3} \sum_{r_1=0}^{r_1=r_0} a_{(v+u)} \frac{(v+u)!}{u!} \alpha_0^{u_0} \alpha_1^{u_1}$$

$$= a_{(v+(1,0))} \frac{(v+(1,0))!}{(1,0)!} \alpha_0^1 \alpha_1^0 + a_{(v+(0,1))} \frac{(v+(0,1))!}{(0,1)!} \alpha_0^0 \alpha_1^1$$

$$+ a_{(v+(2,0))} \frac{(v+(2,0))!}{(2,0)!} \alpha_0^2 \alpha_1^0 + a_{(v+(1,1))} \frac{(v+(1,1))!}{(1,1)!} \alpha_0^1 \alpha_1^1$$

$$+ a_{(v+(0,2))} \frac{(v+(0,2))!}{(0,2)!} \alpha_0^0 \alpha_1^2$$

$$+ a_{(v+(3,0))} \frac{(v+(3,0))!}{(3,0)!} \alpha_0^3 \alpha_1^0 + a_{(v+(1,1))} \frac{(v+(2,1))!}{(2,1)!} \alpha_0^2 \alpha_1^1$$

$$+ a_{(v+(1,2))} \frac{(v+(1,2))!}{(1,2)!} \alpha_0^1 \alpha_1^2 + a_{(v+(0,3))} \frac{(v+(0,3))!}{(0,3)!} \alpha_0^0 \alpha_1^3$$

$$\tag{11.62}$$

4. Computation of the term $E_F^{(v)}(\alpha)$

The expression for $E_F^v(\alpha)$ is given by equation (7.87), on page 309, writes

$$E_F^{(v)}(\alpha) = E_F^{(v_0)}(\alpha) E_F^{(v_1)}(\alpha)$$

$$= \left[\frac{(\eta + v_0)!}{\eta!} \frac{1}{(\beta_{0,0} - \alpha_0)^{v_0}} \right] \left[\frac{(\eta + v_1)!}{\eta!} \frac{1}{(\beta_{1,1} - \alpha_1)^{v_1}} \right] \tag{11.63}$$

5. Computation of the coefficient $a_v = a_{1,0}$ where $v = (v_0, v_1) = (1,0)$

(a) Computation of $E_F^{(1,0)}$.

Substituting $\eta = 4$, $v = (v_0, v_1) = (1,0)$, $\alpha = (\alpha_0, \alpha_1) = (2,4)$, into equation (11.63) it follows

$$E_F^{(1,0)}(\alpha) = \left[\frac{(4+1)!}{4!} \frac{1}{(4)^1} \right] \left[\frac{(4+0)!}{4!} \frac{1}{(8)^0} \right] = \frac{5}{2^2} \tag{11.64}$$

(b) Computation of $S_{(1,0)}$.

Since u belongs to the level $m = q - \lambda_0(v) = 4 - \lambda_0(1,0) = 4 - 1 = 3$ then the constrained numbers u in the level $m = 1$, $m = 2$ and $m = 3$ are elements of one-dimensional the set $L_1^1 \cup L_2^1 \cup L_3^1$, which are $\{(1,0), (0,1), (2,0), (1,1), (0,2), (3,0), (2,1), (1,2), (0,3)\}$.

Table 11.2.9 *Computation of the terms for the evaluation of* $S_{(1,0)}\left(a_{(u+v)}, \alpha\right)$.

v	u	$(u+v)$	$a_{(u+v)}$	$(u+v)!/u!$	α^{u_0}	α^{u_1}	value
$(1,0)$	$(1,0)$	$(2,0)$	$870/2^6$	$(2,0)!/(1,0)! = 2$	$(2)^1$	$(4)^0$	$870/2^4$
	$(0,1)$	$(1,1)$	$1450/2^7$	$(1,1)!/(0,1)! = 1$	$(2)^0$	$(4)^1$	$1450/2^5$
	$(2,0)$	$(3,0)$	$-385/2^7$	$(3,0)!/(2,0)! = 3$	$(2)^2$	$(4)^0$	$-1155/2^5$
	$(1,1)$	$(2,1)$	$-825/2^8$	$(2,1)!/(1,1)! = 2$	$(2)^1$	$(4)^1$	$-825/2^4$
	$(0,2)$	$(1,2)$	$-825/2^9$	$(1,2)!/(0,2)! = 1$	$(2)^0$	$(4)^2$	$-825/2^5$
	$(3,0)$	$(4,0)$	$35/2^7$	$(4,0)!/(3,0)! = 4$	$(2)^3$	$(4)^0$	$35/2^2$
	$(2,1)$	$(3,1)$	$175/2^9$	$(3,1)!/(2,1)! = 3$	$(2)^2$	$(4)^1$	$525/2^5$
	$(1,2)$	$(2,2)$	$225/2^{10}$	$(2,2)!/(1,2)! = 2$	$(2)^1$	$(4)^2$	$225/2^4$
	$(0,3)$	$(1,3)$	$175/2^{11}$	$(1,3)!/(0,3)! = 1$	$(2)^0$	$(4)^3$	$175/2^5$

Substituting the values for the terms obtained in the table 11.2.9 into the equation (11.62) it gives

$$
\begin{aligned}
S_{(1,0)}\left(a_{(u+v)}, \alpha\right) &= \frac{870}{2^4} + \frac{1450}{2^5} - \frac{1155}{2^5} - \frac{825}{2^4} - \frac{825}{2^5} \\
&\quad + \frac{35}{2^2} + \frac{525}{2^5} + \frac{225}{2^4} + \frac{175}{2^5} \\
&= \frac{1}{2^5}\left(1740 + 1450 - 1155 - 1650 - 825\right. \\
&\qquad \left. + 280 + 525 + 450 + 175\right) \\
&= \frac{990}{2^5} = \frac{495}{2^4} \qquad\qquad (11.65)
\end{aligned}
$$

(c) Computation of $a_{1,0}$.

Substituting v, $E_F^{(v)}(\alpha)$ and $S_v\left(a_{u+v}, \alpha\right)$ for its values into

$$
a_{(1,0)} = \frac{1}{v!}\left[E_F^{(1,0)}(\alpha) - S_{(1,0)}\left(a_{u+v}, \alpha\right)\right] = \frac{1}{1!\,0!}\left[\frac{5}{2^2} - \frac{495}{2^4}\right] \tag{11.66}
$$

$$
= \frac{1}{2^2}(20 - 495) = -\frac{475}{2^4} = -\frac{60800}{2^{11}} \tag{11.67}
$$

6. Computation of the coefficient $a_v = a_{0,1}$ where $v = (v_0, v_1) = (0,1)$

(a) Computation of $E_F^{(0,1)}$.

Substituting $\eta = 4$, $v = (v_0, v_1) = (0,1)$, $\alpha = (\alpha_0, \alpha_1) = (2,4)$, into equation (11.63) it follows

$$
E_F^{(0,1)}(\alpha) = \left[\frac{(4+0)!}{4!}\frac{1}{(4)^0}\right]\left[\frac{(4+1)!}{4!}\frac{1}{(8)^1}\right] = \frac{5}{2^3} \tag{11.68}
$$

(b) Computation of $S_{(0,1)}$.

Since u belongs to the level $m = q - \lambda_0(v) = 4 - \lambda_0(0,1) = 4 - 1 = 3$ then the constrained numbers u in the level $m = 1$ and $m = 2$ are elements of one-dimensional the set $L_1^1 \cup L_2^1 \cup L_3^1$, which are $\{(1,0), (0,1), (2,0), (1,1), (0,2), (3,0), (2,1), (1,2), (0,3)\}$.

Table 11.2.10 *Computation of the terms for the evaluation of* $S_{(0,1)}\left(a_{(u+v)}, \alpha\right)$.

v	u	$(u+v)$	$a_{(u+v)}$	$(u+v)!/u!$	α^{u_0}	α^{u_1}	value
$(0,1)$	$(1,0)$	$(1,1)$	$1450/2^7$	$(1,1)!/(1,0)! = 1$	$(2)^1$	$(4)^0$	$1450/2^6$
	$(0,1)$	$(0,2)$	$870/2^8$	$(0,2)!/(0,1)! = 2$	$(2)^0$	$(4)^1$	$870/2^5$
	$(2,0)$	$(2,1)$	$-825/2^8$	$(2,1)!/(2,0)! = 1$	$(2)^2$	$(4)^0$	$-825/2^6$
	$(1,1)$	$(1,2)$	$-825/2^9$	$(1,2)!/(1,1)! = 2$	$(2)^1$	$(4)^1$	$-825/2^5$
	$(0,2)$	$(0,3)$	$-385/2^{10}$	$(0,3)!/(0,2)! = 3$	$(2)^0$	$(4)^2$	$-1155/2^6$
	$(3,0)$	$(3,1)$	$175/2^9$	$(3,1)!/(3,0)! = 1$	$(2)^3$	$(4)^0$	$175/2^6$
	$(2,1)$	$(2,2)$	$225/2^{10}$	$(2,2)!/(2,1)! = 2$	$(2)^2$	$(4)^1$	$225/2^5$
	$(1,2)$	$(1,3)$	$175/2^{11}$	$(1,3)!/(1,2)! = 3$	$(2)^1$	$(4)^2$	$525/2^6$
	$(0,3)$	$(0,4)$	$35/2^{11}$	$(0,4)!/(0,3)! = 4$	$(2)^0$	$(4)^3$	$35/2^3$

Substituting the values for the terms obtained in the table 11.2.10 into the equation (11.62) it gives

$$
\begin{aligned}
S_{(0,1)}\left(a_{(u+v)}, \alpha\right) &= \frac{1450}{2^6} + \frac{870}{2^5} - \frac{825}{2^6} - \frac{825}{2^5} - \frac{1155}{2^6} \\
&\quad + \frac{175}{2^6} + \frac{225}{2^5} + \frac{525}{2^6} + \frac{35}{2^3} \\
&= \frac{1}{2^6}\big(1450 + 1740 - 825 - 1650 - 1155 \\
&\quad + 175 + 450 + 525 + 280\big) \\
&= \frac{990}{2^6} = \frac{495}{2^5}
\end{aligned}
\tag{11.69}
$$

(c) Computation of $a_{0,1}$.

Substituting v, $E_F^{(v)}(\alpha)$ and $S_v\left(a_{u+v}, \alpha\right)$ for its values into

$$
a_{(0,1)} = \frac{1}{v!}\left[E_F^{(0,1)}(\alpha) - S_{(0,1)}\left(a_{u+v}, \alpha\right)\right] = \frac{1}{0!\,1!}\left[\frac{5}{2^3} - \frac{495}{2^5}\right]
\tag{11.70}
$$

$$
= \frac{1}{2^5}(20 - 495) = -\frac{475}{2^5} = -\frac{30400}{2^{11}}
\tag{11.71}
$$

b- Computation of the coefficients that belong to the terms in the level $\lambda_0(v) = q - 4 = 0$

1. Coefficients to be evaluated, which belong to the level $\lambda_0(v) = 0$.

   The elements in the level $\lambda_0(v) = 0$ with two coordinates are the constrained numbers in the one-dimensional space are

   $$L^\delta_{\lambda_0(v)} = L^1_1 = \{(0,0)\} \tag{11.72}$$

   The elements of the set L^1_0 can obtained using the algorithm described in subsection 2.4.4, on page 45. Therefore, the coefficients $a_{(w)} = a_{(v)}$ to be evaluated in this level is $a_{(0,0)}$.

2. Equation for the computation of the coefficients.

   The coefficients are obtained using equation (7.123), on page 314, which is

   $$a_{(v)} = \frac{1}{v!} \left[E^{(v)}_F(\alpha) - S_v\left(a_{u+v}, \alpha\right) \right] \qquad \lambda_0(v) = 0 \tag{11.73}$$

3. Computation of the term $S_v\left(a_{u+v}, \alpha\right)$

   The degree of the polynomial $S_v\left(a_{u+v}, \alpha\right)$ is given by $m = q - \lambda_0(v) = 4 - 0 = 4$ where $\lambda_0(v) = v_0 + v_1 = 0 + 0 = 0$. The exponents of the variables in each term of $S_v\left(a_{u+v}, \alpha\right)$ are the elements of the set shown in equation (7.124), on page 314, which is

   $$\begin{aligned}
   \left(L^\delta(m) - L^\delta_0\right) = &\left(L^1(1) - L^1_0\right) \\
   = &\big\{ (0,0), (1,0), (0,1), (2,0), (1,1), (0,2), (3,0), \\
   &\quad (2,1), (1,2), (0,3) \\
   &\quad (4,0), (3,1), (2,2), (1,3), (0,4) \big\} - \{(0,0)\} \\
   = &\big\{ (1,0), (0,1), (2,0), (1,1), (0,2), (3,0), \\
   &\quad (2,1), (1,2), (0,3), \\
   &\quad (4,0), (3,1), (2,2), (1,3), (0,4) \big\}
   \end{aligned} \tag{11.74}$$

   which permits to obtain the indices of the coefficients of the terms of the polynomial $S_v\left(a_{u+v}, \alpha\right)$ for $v = (0,0)$ as follows

   If $u = (1,0)$ then $v + u = (0,0) + (1,0) = (1,0)$, which gives $a_{(v+u)} = a_{(1,0)}$.

   If $u = (0,1)$ then $v + u = (0,0) + (0,1) = (0,1)$, which gives $a_{(v+u)} = a_{(0,1)}$.

   If $u = (2,0)$ then $v + u = (0,0) + (2,0) = (2,0)$, which gives $a_{(v+u)} = a_{(2,0)}$.

   If $u = (1,1)$ then $v + u = (0,0) + (1,1) = (1,1)$, which gives $a_{(v+u)} = a_{(1,1)}$.

If $u = (0, 2)$ then $v + u = (0, 0) + (0, 2) = (0, 2)$, which gives $a_{(v+u)} = a_{(0,2)}$.

If $u = (3, 0)$ then $v + u = (0, 0) + (3, 0) = (3, 0)$, which gives $a_{(v+u)} = a_{(3,0)}$.

If $u = (2, 1)$ then $v + u = (0, 0) + (2, 1) = (2, 1)$, which gives $a_{(v+u)} = a_{(2,1)}$.

If $u = (1, 2)$ then $v + u = (0, 0) + (1, 2) = (1, 2)$, which gives $a_{(v+u)} = a_{(1,2)}$.

If $u = (0, 3)$ then $v + u = (0, 0) + (0, 3) = (0, 3)$, which gives $a_{(v+u)} = a_{(0,3)}$.

If $u = (4, 0)$ then $v + u = (0, 0) + (4, 0) = (4, 0)$, which gives $a_{(v+u)} = a_{(4,0)}$.

If $u = (3, 1)$ then $v + u = (0, 0) + (3, 1) = (3, 1)$, which gives $a_{(v+u)} = a_{(3,1)}$.

If $u = (2, 2)$ then $v + u = (0, 0) + (2, 2) = (2, 2)$, which gives $a_{(v+u)} = a_{(2,2)}$.

If $u = (1, 3)$ then $v + u = (0, 0) + (1, 3) = (1, 3)$, which gives $a_{(v+u)} = a_{(1,3)}$.

If $u = (0, 4)$ then $v + u = (0, 0) + (0, 5) = (0, 3)$, which gives $a_{(v+u)} = a_{(0,4)}$.

Where the indices of the coefficients $a_{(1,0)}$ and $a_{(0,1)}$ belong to the level $\lambda_0(v) = q = 1$. The indices of $a_{(2,0)}$, $a_{(1,1)}$, and $a_{(0,2)}$ belong to the level $\lambda_0(v) = q = 2$. The indices of $a_{(3,0)}$, $a_{(2,1)}$, $a_{(1,2)}$, and $a_{(0,3)}$ belong to the level $\lambda_0(v) = q = 3$. The indices of $a_{(4,0)}$, $a_{(3,1)}$, $a_{(2,2)}$, $a_{(1,3)}$, and $a_{(0,4)}$ belong to the level $\lambda_0(v) = q = 4$.

The expression for $S_v(a_{u+v}, \alpha)$ is given by equation (7.112). For $d = 2$

and $m = 4$ it writes

$$S_v(a_{u+v}, \alpha) = \sum_{r_0=0}^{r_0=m} \sum_{r_1=0}^{r_1=r_0} a_{(v+u)} \frac{(v+u)!}{u!} \alpha_0^{u_0} \alpha_1^{u_1}$$

$$= \sum_{r_0=0}^{r_0=4} \sum_{r_1=0}^{r_1=r_0} a_{(v+u)} \frac{(v+u)!}{u!} \alpha_0^{u_0} \alpha_1^{u_1}$$

$$= a_{(v+(1,0))} \frac{(v+(1,0))!}{(1,0)!} \alpha_0^1 \alpha_1^0 + a_{(v+(0,1))} \frac{(v+(0,1))!}{(0,1)!} \alpha_0^0 \alpha_1^1$$

$$+ a_{(v+(2,0))} \frac{(v+(2,0))!}{(2,0)!} \alpha_0^2 \alpha_1^0 + a_{(v+(1,1))} \frac{(v+(1,1))!}{(1,1)!} \alpha_0^1 \alpha_1^1$$

$$+ a_{(v+(0,2))} \frac{(v+(0,2))!}{(0,2)!} \alpha_0^0 \alpha_1^2$$

$$+ a_{(v+(3,0))} \frac{(v+(3,0))!}{(3,0)!} \alpha_0^3 \alpha_1^0 + a_{(v+(1,1))} \frac{(v+(2,1))!}{(2,1)!} \alpha_0^2 \alpha_1^1$$

$$+ a_{(v+(1,2))} \frac{(v+(1,2))!}{(1,2)!} \alpha_0^1 \alpha_1^2 + a_{(v+(0,3))} \frac{(v+(0,3))!}{(0,3)!} \alpha_0^0 \alpha_1^3$$

$$+ a_{(v+(4,0))} \frac{(v+(4,0))!}{(4,0)!} \alpha_0^4 \alpha_1^0 + a_{(v+(3,1))} \frac{(v+(3,1))!}{(3,1)!} \alpha_0^3 \alpha_1^1$$

$$+ a_{(v+(2,2))} \frac{(v+(2,2))!}{(2,2)!} \alpha_0^2 \alpha_1^2 + a_{(v+(1,3))} \frac{(v+(1,3))!}{(1,3)!} \alpha_0^1 \alpha_1^3$$

$$+ a_{(v+(0,4))} \frac{(v+(0,4))!}{(0,4)!} \alpha_0^0 \alpha_1^4 \tag{11.75}$$

4. Computation of the term $E_F^{(v)}(\alpha)$

The expression for $E_F^v(\alpha)$ is given by equation (7.87) writes

$$E_F^{(v)}(\alpha) = E_F^{(v_0)}(\alpha) E_F^{(v_1)}(\alpha)$$

$$= \left[\frac{(\eta + v_0)!}{\eta!} \frac{1}{(\beta_{0,0} - \alpha_0)^{v_0}} \right] \left[\frac{(\eta + v_1)!}{\eta!} \frac{1}{(\beta_{1,1} - \alpha_1)^{v_1}} \right] \tag{11.76}$$

5. Computation of the coefficient $a_v = a_{0,0}$ where $v = (v_0, v_1) = (0,0)$

(a) Computation of $E_F^{(0,0)}$.

Substituting $\eta = 4$, $v = (v_0, v_1) = (0,0)$, $\alpha = (\alpha_0, \alpha_1) = (2,4)$, into equation (11.76) it follows

$$E_F^{(1,0)}(\alpha) = \left[\frac{(4+0)!}{4!} \frac{1}{(4)^0} \right] \left[\frac{(4+0)!}{4!} \frac{1}{(8)^0} \right] = 1 \tag{11.77}$$

(b) Computation of $S_{0,0}$.

Since u belongs to the level $m = q - \lambda_0(v) = 4 - \lambda_0(0,0) = 4 - 0 = 4$ then the constrained numbers u in the level $m = 1$, $m = 2$, $m = 3$, and $m = 4$ are elements of one-dimensional the set $L_1^1 \cup L_2^1 \cup L_3^1$, which are $\{(1,0), (0,1), (2,0), (1,1), (0,2), (3,0), (2,1), (1,2), (0,3), (4,0), (3,1), (2,2), (1,3), (0,4)\}$.

Table 11.2.11 *Computation of the terms for the evaluation of* $S_{(0,0)}\left(a_{(u+v)}, \alpha\right)$.

v	u	$(u+v)$	$a_{(u+v)}$	$(u+v)!/u!$	α^{u_0}	α^{u_1}	value
$(0,0)$	$(1,0)$	$(1,0)$	$-475/2^4$	$(1,0)!/(1,0)! = 1$	$(2)^1$	$(4)^0$	$-475/2^3$
	$(0,1)$	$(0,1)$	$-475/2^5$	$(0,1)!/(0,1)! = 1$	$(2)^0$	$(4)^1$	$-475/2^3$
	$(2,0)$	$(2,0)$	$870/2^6$	$(2,0)!/(2,0)! = 1$	$(2)^2$	$(4)^0$	$870/2^4$
	$(1,1)$	$(1,1)$	$1450/2^7$	$(1,1)!/(1,1)! = 1$	$(2)^1$	$(4)^1$	$1450/2^4$
	$(0,2)$	$(0,2)$	$870/2^8$	$(0,2)!/(0,2)! = 1$	$(2)^0$	$(4)^2$	$870/2^4$
	$(3,0)$	$(3,0)$	$-385/2^7$	$(3,0)!/(3,0)! = 1$	$(2)^3$	$(4)^0$	$-385/2^4$
	$(2,1)$	$(2,1)$	$-825/2^8$	$(2,1)!/(2,1)! = 1$	$(2)^2$	$(4)^1$	$-825/2^4$
	$(1,2)$	$(1,2)$	$-825/2^9$	$(1,2)!/(1,2)! = 1$	$(2)^1$	$(4)^2$	$-825/2^4$
	$(0,3)$	$(0,3)$	$-385/2^{10}$	$(0,3)!/(0,3)! = 1$	$(2)^0$	$(4)^3$	$-385/2^4$
	$(0,3)$	$(4,0)$	$35/2^7$	$(4,0)!/(4,0)! = 1$	$(2)^4$	$(4)^0$	$35/2^3$
	$(0,3)$	$(3,1)$	$175/2^9$	$(3,1)!/(3,1)! = 1$	$(2)^3$	$(4)^1$	$175/2^4$
	$(0,3)$	$(2,2)$	$225/2^{10}$	$(2,2)!/(2,2)! = 1$	$(2)^2$	$(4)^2$	$225/2^4$
	$(0,3)$	$(1,3)$	$175/2^{11}$	$(1,3)!/(1,3)! = 1$	$(2)^1$	$(4)^3$	$175/2^4$
	$(0,3)$	$(0,4)$	$35/2^{11}$	$(0,4)!/(0,4)! = 1$	$(2)^0$	$(4)^4$	$35/2^3$

Substituting the values for the terms obtained in the table 11.2.11 into the equation (11.75) it gives

$$
\begin{aligned}
S_{(0,0)}\left(a_{(u+v)}, \alpha\right) &= -\frac{475}{2^3} - \frac{475}{2^3} + \frac{870}{2^4} + \frac{1450}{2^4} + \frac{870}{2^4} - \frac{385}{2^4} \\
&\quad - \frac{825}{2^4} - \frac{385}{2^4} - \frac{385}{2^4} \\
&\quad + \frac{35}{2^3} + \frac{175}{2^4} + \frac{225}{2^4} + \frac{175}{2^4} + \frac{35}{2^3} \\
&= \frac{1}{2^4}\left(-950 - 950 + 870 + 1450 + 870\right. \\
&\quad - 385 - 825 - 825 - 385 \\
&\quad \left. + 70 + 175 + 225 + 175 + 70\right) \\
&= -\frac{415}{2^4} = \quad\quad\quad (11.78)
\end{aligned}
$$

(c) Computation of $a_{0,0}$.

Substituting $v = (0,0)$, $E_F^{(0,0)}(\alpha)$ and $S_{(0,0)}(a_{u+v}, \alpha)$ for its values into

$$a_{(0,0)} = \frac{1}{v!} \left[E_F^{(0,0)}(\alpha) - S_{(0,0)}(a_{u+v}, \alpha) \right] = \frac{1}{0!\,0!} \left[\frac{2^4}{2^4} - \frac{-415}{2^4} \right] \tag{11.79}$$

$$= \frac{1}{2^4} (16 + 415) = \frac{431}{2^4} = \frac{55168}{2^{11}} \tag{11.80}$$

11.3   Summary of the polynomials

11.3.1   Polynomials $\Phi_{e,i}$ related to the node $a_1 = (6, 12)$

Two dimensional $d = 2$ the Hermite Interpolating Polynomials, for maximum derivative of order $\eta = 4$, associate to the node a_1 whose coordinate numbers are $\alpha = (6, 12)$ and where $\beta_0 = (2, 12)$ and $\beta_1 = (6, 4)$.

$$\begin{aligned}
\Phi_{a,00}(x,y) = \frac{1}{0!\,0!} &\left[\frac{1}{(-4)^5} \frac{1}{(-8)^5} \right] (x_0 - 2)^5 (x_1 - 4)^5 \frac{1}{2^{11}} \Big(9219968 \\
&- 2896000x_0 - 1448000x_1 + 373440x_0^2 + 311200x_0x_1 \\
&+ 93360x_1^2 - 22960x_0^3 - 24600x_0^2x_1 - 12300x_0x_1^2 - 2870x_1^3 \\
&+ 560x_0^4 + 700x_0^3x_1 + 450x_0^2x_1^2 + 175x_0x_1^3 + 35x_1^4 \Big)
\end{aligned} \tag{11.81}$$

$$\begin{aligned}
\Phi_{a,10}(x,y) = \frac{1}{1!\,0!} &\left[\frac{1}{(-4)^5} \frac{1}{(-8)^5} \right] (x_0 - 2)^5 (x_1 - 4)^5 (x_0 - 6)^5 (x_1 - 12)^5 (x_0 - 2) \\
&\times \frac{1}{2^9} \Big(451712 - 106240x_0 - 53120x_1 + 9120x_0^2 + 7600x_0x_1 \\
&+ 2280x_1^2 - 280x_0^3 - 300x_0^2x_1 - 150x_0x_1^2 - 35x_1^3 \Big)
\end{aligned} \tag{11.82}$$

$$\begin{aligned}
\Phi_{a,01}(x,y) = \frac{1}{0!\,1!} &\left[\frac{1}{(-4)^5} \frac{1}{(-8)^5} \right] (x_0 - 2)^5 (x_1 - 4)^5 (x_0 - 6)^5 (x_1 - 12)^5 (x_1 - 4) \\
&\times \frac{1}{2^9} \Big(451712 - 106240x_0 - 53120x_1 + 9120x_0^2 + 7600x_0x_1 \\
&+ 2280x_1^2 - 280x_0^3 - 300x_0^2x_1 - 150x_0x_1^2 - 35x_1^3 \Big)
\end{aligned} \tag{11.83}$$

$$\Phi_{a,20}(x,y) = \frac{1}{2!0!} \left[\frac{1}{(-4)^5} \frac{1}{(-8)^5} \right] (x_0 - 2)^5 (x_1 - 4)^5 (x_0 - 6)^2$$
$$\times \frac{1}{2^6} \left(8944 - 1400x_0 - 700x_1 + 60x_0^2 + 50x_0x_1 + 15x_1^2 \right)$$
$$(11.84)$$

$$\Phi_{a,11}(x,y) = \frac{1}{1!1!} \left[\frac{1}{(-4)^5} \frac{1}{(-8)^5} \right] (x_0 - 2)^5 (x_1 - 4)^5 (x_0 - 6)(x_1 - 12)$$
$$\times \frac{1}{2^6} \left(8944 - 1400x_0 - 700x_1 + 60x_0^2 + 50x_0x_1 + 15x_1^2 \right)$$
$$(11.85)$$

$$\Phi_{a,02}(x,y) = \frac{1}{0!2!} \left[\frac{1}{(-4)^5} \frac{1}{(-8)^5} \right] (x_0 - 2)^5 (x_1 - 4)^5 (x_1 - 12)^2$$
$$\times \frac{1}{2^6} \left(8944 - 1400x_0 - 700x_1 + 60x_0^2 + 50x_0x_1 + 15x_1^2 \right)$$
$$(11.86)$$

$$\Phi_{a,30}(x,y) = \frac{1}{3!0!} \left[\frac{1}{(-4)^5} \frac{1}{(-8)^5} \right] (x_0 - 2)^5 (x_1 - 4)^5 (x_0 - 6)^3$$
$$\times \frac{1}{2^3} \left(128 - 10x_0 - 5x_1 \right)$$
$$(11.87)$$

$$\Phi_{a,21}(x,y) = \frac{1}{2!1!} \left[\frac{1}{(-4)^5} \frac{1}{(-8)^5} \right] (x_0 - 2)^5 (x_1 - 4)^5 (x_0 - 6)^2 (x_1 - 12)$$
$$\times \frac{1}{2^3} \left(128 - 10x_0 - 5x_1 \right)$$
$$(11.88)$$

$$\Phi_{a,12}(x,y) = \frac{1}{1!2!} \left[\frac{1}{(-4)^5} \frac{1}{(-8)^5} \right] (x_0 - 2)^5 (x_1 - 4)^5 (x_0 - 6)(x_1 - 12)^2$$
$$\times \frac{1}{2^3} \left(128 - 10x_0 - 5x_1 \right)$$
$$(11.89)$$

$$\Phi_{a,03}(x,y) = \frac{1}{0!3!} \left[\frac{1}{(-4)^5} \frac{1}{(-8)^5} \right] (x_0 - 2)^5 (x_1 - 12)^5 (x_1 - 4)^3$$
$$\times \frac{1}{2^3} \left(128 - 10x_0 - 5x_1 \right)$$
$$(11.90)$$

$$\Phi_{a,40}(x,y) = \frac{1}{4!0!} \left[\frac{1}{(-4)^5} \frac{1}{(-8)^5} \right] (x_0 - 2)^5 (x_1 - 12)^5 (x_0 - 6)^4 \frac{1}{2^0} \left(1 \right)$$
$$(11.91)$$

$$\Phi_{a,31}(x,y) = \frac{1}{3!1!} \left[\frac{1}{(-4)^5} \frac{1}{(-8)^5} \right] (x_0 - 2)^5 (x_1 - 4)^5 (x_0 - 6)^3 (x_1 - 12) \frac{1}{2^0} \left(1 \right)$$

$$(11.92)$$

$$\Phi_{a,22}(x,y) = \frac{1}{2!2!} \left[\frac{1}{(-4)^5} \frac{1}{(-8)^5} \right] (x_0 - 2)^5 (x_1 - 4)^5 (x_0 - 6)^2 (x_1 - 12)^2 \frac{1}{2^0} \left(1 \right)$$

$$(11.93)$$

$$\Phi_{a,13}(x,y) = \frac{1}{1!3!} \left[\frac{1}{(-4)^5} \frac{1}{(-8)^5} \right] (x_0 - 2)^5 (x_1 - 4)^5 (x_0 - 6)(x_1 - 12)^3 \frac{1}{2^0} \left(1 \right)$$

$$(11.94)$$

$$\Phi_{a,04}(x,y) = \frac{1}{0!4!} \left[\frac{1}{(-4)^5} \frac{1}{(-8)^5} \right] (x_0 - 2)^5 (x_1 - 4)^5 (x_1 - 12)^4 \frac{1}{2^0} \left(1 \right)$$

$$(11.95)$$

11.3.2 Polynomials $\Phi_{e,i}$ related to the node $a_2 = (2, 12)$

Two dimensional $d = 2$ the Hermite Interpolating Polynomials, for maximum derivative of order $\eta = 4$, associate to the node a_4 whose coordinate numbers are $\alpha = (2, 12)$ and where $\beta_0 = (6, 12)$ and $\beta_1 = (2, 4)$.

$$\Phi_{a,00}(x,y) = \frac{1}{0!0!} \left[\frac{1}{(-4)^5} \frac{1}{(-8)^5} \right] (x_0 - 6)^5 (x_1 - 4)^5 \frac{1}{2^{11}} \Big(490368$$
$$+ 182400x_0 - 174400x_1 + 37440x_0^2 - 52000x_0x_1$$
$$+ 23760x_1^2 + 5040x_0^3 - 7800x_0^2x_1 + 5100x_0x_1^2 - 1470x_1^3$$
$$+ 560x_0^4 - 700x_0^3x_1 + 450x_0^2x_1^2 - 175x_0x_1^4 + 35x_1^4 \Big) \qquad (11.96)$$

$$\Phi_{a,10}(x,y) = \frac{1}{1!0!} \left[\frac{1}{(-4)^5} \frac{1}{(-8)^5} \right] (x_0 - 6)^5 (x_1 - 4)^5 (x_0 - 2)$$
$$\times \frac{1}{2^9} \Big(42112 + 14080x_0 - 11520x_1 + 2400x_0^2 - 2800x_0x_1 + 1080x_1^2$$
$$+ 280x_0^3 - 300x_0^2x_1 + 150x_0x_1^2 - 35x_1^3 \Big) \qquad (11.97)$$

$$\Phi_{a,01}(x,y) = \frac{1}{0!1!} \left[\frac{1}{(-4)^5} \frac{1}{(-8)^5} \right] (x_0 - 6)^5 (x_1 - 4)^5 (x_1 - 12)$$
$$\times \frac{1}{2^9} \Big(42112 + 14080x_0 - 11520x_1 + 2400x_0^2 - 2800x_0x_1 + 1080x_1^2$$
$$+ 280x_0^3 - 300x_0^2x_1 + 150x_0x_1^2 - 35x_1^3 \Big) \qquad (11.98)$$

$$\Phi_{a,20}(x,y) = \frac{1}{2!0!} \left[\frac{1}{(-4)^5} \frac{1}{(-8)^5} \right] (x_0 - 6)^5 (x_1 - 4)^5 (x_0 - 2)^2$$

$$\times \frac{1}{2^6} \left(1584 + 440x_0 - 300x_1 + 60x_0^2 - 50x_0x_1 + 15x_1^2 \right) \quad (11.99)$$

$$\Phi_{a,11}(x,y) = \frac{1}{1!1!} \left[\frac{1}{(-4)^5} \frac{1}{(-8)^5} \right] (x_0 - 6)^5 (x_1 - 4)^5 (x_0 - 2)(x_1 - 12)$$

$$\times \frac{1}{2^6} \left(1584 + 440x_0 - 300x_1 + 60x_0^2 - 50x_0x_1 + 15x_1^2 \right)$$
$$(11.100)$$

$$\Phi_{a,02}(x,y) = \frac{1}{0!2!} \left[\frac{1}{(-4)^5} \frac{1}{(-8)^5} \right] (x_0 - 6)^5 (x_1 - 4)^5 (x_1 - 12)^2$$

$$\times \frac{1}{2^6} \left(1584 + 440x_0 - 300x_1 + 60x_0^2 - 50x_0x_1 + 15x_1^2 \right)$$
$$(11.101)$$

$$\Phi_{a,30}(x,y) = \frac{1}{3!0!} \left[\frac{1}{(-4)^5} \frac{1}{(-8)^5} \right] (x_0 - 6)^5 (x_1 - 4)^5 (x_0 - 2)^3$$

$$\times \frac{1}{2^3} \left(48 + 10x_0 - 5x_1 \right) \quad (11.102)$$

$$\Phi_{a,21}(x,y) = \frac{1}{2!1!} \left[\frac{1}{(-4)^5} \frac{1}{(-8)^5} \right] (x_0 - 6)^5 (x_1 - 4)^5 (x_0 - 2)^2 (x_1 - 12)$$

$$\times \frac{1}{2^3} \left(48 + 10x_0 - 5x_1 \right) \quad (11.103)$$

$$\Phi_{a,12}(x,y) = \frac{1}{1!2!} \left[\frac{1}{(-4)^5} \frac{1}{(-8)^5} \right] (x_0 - 6)^5 (x_1 - 4)^5 (x_0 - 2)(x_1 - 12)^2$$

$$\times \frac{1}{2^3} \left(48 + 10x_0 - 5x_1 \right) \quad (11.104)$$

$$\Phi_{a,03}(x,y) = \frac{1}{0!3!} \left[\frac{1}{(-4)^5} \frac{1}{(-8)^5} \right] (x_0 - 6)^5 (x_1 - 4)^5 (x_1 - 12)^3$$

$$\times \frac{1}{2^3} \left(48 + 10x_0 - 5x_1 \right) \quad (11.105)$$

$$\Phi_{a,40}(x,y) = \frac{1}{4!0!} \left[\frac{1}{(-4)^5} \frac{1}{(-8)^5} \right] (x_0 - 6)^5 (x_1 - 4)^5 (x_0 - 2)^4 \frac{1}{2^0} \left(1 \right)$$
$$(11.106)$$

$$\Phi_{a,31}(x,y) = \frac{1}{3!1!} \left[\frac{1}{(-4)^5} \frac{1}{(-8)^5} \right] (x_0 - 6)^5 (x_1 - 4)^5 (x_0 - 2)^3 (x_1 - 12) \frac{1}{2^0} \left(1 \right)$$

(11.107)

$$\Phi_{a,22}(x,y) = \frac{1}{2!2!} \left[\frac{1}{(-4)^5} \frac{1}{(-8)^5} \right] (x_0 - 6)^5 (x_1 - 4)^5 (x_0 - 2)^2 (x_1 - 12)^2 \frac{1}{2^0} \left(1 \right)$$

(11.108)

$$\Phi_{a,13}(x,y) = \frac{1}{1!3!} \left[\frac{1}{(-4)^5} \frac{1}{(-8)^5} \right] (x_0 - 6)^5 (x_1 - 4)^5 (x_0 - 2)(x_1 - 12)^3 \frac{1}{2^0} \left(1 \right)$$

(11.109)

$$\Phi_{a,04}(x,y) = \frac{1}{0!4!} \left[\frac{1}{(-4)^5} \frac{1}{(-8)^5} \right] (x_0 - 6)^5 (x_1 - 4)^5 (x_1 - 12)^4 \frac{1}{2^0} \left(1 \right)$$

(11.110)

11.3.3 Polynomials $\Phi_{e,n}$ related to the node $a_3 = (2,4)$

Two dimensional $d = 2$ the Hermite Interpolating Polynomials, for maximum derivative of order $\eta = 4$, associate to the node a_3 whose coordinate numbers are $\alpha = (2,4)$ and where $\beta_0 = (6,4)$ and $\beta_1 = (2,12)$.

$$\Phi_{a,00}(x,y) = \frac{1}{0!0!} \left[\frac{1}{(-4)^5} \frac{1}{(-8)^5} \right] (x_0 - 6)^5 (x_1 - 12)^5$$
$$\times \frac{1}{2^{11}} \left(55168 - 60800x_0 - 30400x_1 + 27840x_0^2 + 23200x_0x_1 \right.$$
$$+ 6960x_1^2 - 6160x_0^3 - 6600x_0^2x_1 - 3300x_0x_1^2 - 770x_1^3$$
$$\left. + 560x_0^4 + 700x_3^2x_1 + 450x_0^2x_1^2 + 175x_0x_1^3 + 35x_1^4 \right)$$

(11.111)

$$\Phi_{a,10}(x,y) = \frac{1}{1!0!} \left[\frac{1}{(-4)^5} \frac{1}{(-8)^5} \right] (x_0 - 6)^5 (x_1 - 12)^5 (x_0 - 2)$$
$$\times \frac{1}{2^9} \left(- 9088 + 7680x_0 + 3840x_1 - 2400x_0^2 - 2000x_0x_1 - 600x_1^2 \right.$$
$$\left. + 280x_0^3 + 300x_0^2x_1 + 150x_0x_1^2 + 35x_1^3 \right)$$

(11.112)

$$\Phi_{a,01}(x,y) = \frac{1}{0!1!} \left[\frac{1}{(-4)^5} \frac{1}{(-8)^5} \right] (x_0 - 6)^5 (x_1 - 12)^5 (x_1 - 4)$$
$$\times \frac{1}{2^9} \left(- 9088 + 7680x_0 + 3840x_1 - 2400x_0^2 - 2000x_0x_1 - 600x_1^2 \right.$$
$$\left. + 280x_0^3 + 300x_0^2x_1 + 150x_0x_1^2 + 35x_1^3 \right)$$

(11.113)

$$\Phi_{a,20}(x,y) = \frac{1}{2!0!} \left[\frac{1}{(-4)^5} \frac{1}{(-8)^5} \right] (x_0 - 6)^5 (x_1 - 12)^5 (x_0 - 2)^2$$
$$\times \frac{1}{2^6} \left(624 - 360x_0 - 180x_1 + 60x_0^2 + 50x_0x_1 + 15x_1^2 \right) \quad (11.114)$$

$$\Phi_{a,11}(x,y) = \frac{1}{1!1!} \left[\frac{1}{(-4)^5} \frac{1}{(-8)^5} \right] (x_0 - 6)^5 (x_1 - 12)^5 (x_0 - 2)(x_1 - 4)$$
$$\times \frac{1}{2^6} \left(624 - 360x_0 - 180x_1 + 60x_0^2 + 50x_0x_1 + 15x_1^2 \right) \quad (11.115)$$

$$\Phi_{a,02}(x,y) = \frac{1}{0!2!} \left[\frac{1}{(-4)^5} \frac{1}{(-8)^5} \right] (x_0 - 6)^5 (x_1 - 12)^5 (x_1 - 4)^2$$
$$\times \frac{1}{2^6} \left(624 - 360x_0 - 180x_1 + 60x_0^2 + 50x_0x_1 + 15x_1^2 \right) \quad (11.116)$$

$$\Phi_{a,30}(x,y) = \frac{1}{3!0!} \left[\frac{1}{(-4)^5} \frac{1}{(-8)^5} \right] (x_0 - 6)^5 (x_1 - 12)^5 (x_0 - 2)^3$$
$$\times \frac{1}{2^3} \left(-32 + 10x_0 + 5x_1 \right) \quad (11.117)$$

$$\Phi_{a,21}(x,y) = \frac{1}{2!1!} \left[\frac{1}{(-4)^5} \frac{1}{(-8)^5} \right] (x_0 - 6)^5 (x_1 - 12)^5 (x_0 - 2)^2 (x_1 - 4)$$
$$\times \frac{1}{2^3} \left(-32 + 10x_0 + 5x_1 \right) \quad (11.118)$$

$$\Phi_{a,12}(x,y) = \frac{1}{1!2!} \left[\frac{1}{(-4)^5} \frac{1}{(-8)^5} \right] (x_0 - 6)^5 (x_1 - 12)^5 (x_0 - 2)(x_1 - 4)^2$$
$$\times \frac{1}{2^3} \left(-32 + 10x_0 + 5x_1 \right) \quad (11.119)$$

$$\Phi_{a,03}(x,y) = \frac{1}{0!3!} \left[\frac{1}{(-4)^5} \frac{1}{(-8)^5} \right] (x_0 - 6)^5 (x_1 - 12)^5 (x_1 - 4)^3$$
$$\times \frac{1}{2^3} \left(-32 + 10x_0 + 5x_1 \right) \quad (11.120)$$

$$\Phi_{a,40}(x,y) = \frac{1}{4!0!} \left[\frac{1}{(-4)^5} \frac{1}{(-8)^5} \right] (x_0 - 6)^5 (x_1 - 12)^5 (x_0 - 2)^4 \frac{1}{2^0} \left(1 \right)$$
$$(11.121)$$

$$\Phi_{a,31}(x,y) = \frac{1}{3!1!} \left[\frac{1}{(-4)^5} \frac{1}{(-8)^5} \right] (x_0 - 6)^5 (x_1 - 12)^5 (x_0 - 2)^3 (x_1 - 4) \frac{0}{2^0} \left(1 \right)$$
$$(11.122)$$

$$\Phi_{a,22}(x,y) = \frac{1}{2!2!} \left[\frac{1}{(-4)^5} \frac{1}{(-8)^5} \right] (x_0 - 6)^5 (x_1 - 12)^5 (x_0 - 2)^2 (x_1 - 4)^2 \frac{1}{2^0} \left(1 \right)$$
$$(11.123)$$

$$\Phi_{a,13}(x,y) = \frac{1}{1!3!} \left[\frac{1}{(-4)^5} \frac{1}{(-8)^5} \right] (x_0 - 6)^5 (x_1 - 12)^5 (x_0 - 2)(x_1 - 4)^3 \frac{1}{2^0} \left(1 \right)$$
$$(11.124)$$

$$\Phi_{a,04}(x,y) = \frac{1}{0!4!} \left[\frac{1}{(-4)^5} \frac{1}{(-8)^5} \right] (x_0 - 6)^5 (x_1 - 12)^5 (x_1 - 4)^4 \frac{1}{2^0} \left(1 \right)$$
$$(11.125)$$

11.3.4   Polynomials $\Phi_{e,i}$ related to the node $a_4 = (6,4)$

Two dimensional $d = 2$ the Hermite Interpolating Polynomials, for maximum derivative of order $\eta = 4$, associate to the node a_4 whose coordinate numbers are $\alpha = (6,4)$ and where $\beta_0 = (2,4)$ and $\beta_1 = (6,12)$.

$$\Phi_{a,00}(x,y) = \frac{1}{0!0!} \left[\frac{1}{(-4)^5} \frac{1}{(-8)^5} \right] (x_0 - 2)^5 (x_1 - 12)^5$$
$$\times \frac{1}{2^{11}} \left(490368 - 348800x_0 + 91200x_1 + 95040x_0^2 - 52000x_0x_1 \right.$$
$$+ 9360x_1^2 - 11760x_0^3 + 10200x_0^2x_1 - 3900x_0x_1^2 + 630x_1^3$$
$$\left. + 560x_0^4 - 700x_0^3x_1 + 450x_0^2x_1^2 - 175x_0x_1^3 + 35x_1^4 \right) \qquad (11.126)$$

$$\Phi_{a,10}(x,y) = \frac{1}{1!0!} \left[\frac{1}{(-4)^5} \frac{1}{(-8)^5} \right] (x_0 - 2)^5 (x_1 - 12)^5 (x_0 - 6)$$
$$\times \frac{1}{2^9} \left(42112 - 23040x_0 + 7040x_1 + 4320x_0^2 - 2800x_0x_1 + 600x_1^2 \right.$$
$$\left. - 280x_0^3 + 300x_0^2x_1 - 150x_0x_1^2 + 35x_1^3 \right) \qquad (11.127)$$

$$\Phi_{a,01}(x,y) = \frac{1}{0!1!} \left[\frac{1}{(-4)^5} \frac{1}{(-8)^5} \right] (x_0 - 2)^5 (x_1 - 12)^5 (x_1 - 4)$$
$$\times \frac{1}{2^9} \left(42112 - 23040x_0 + 7040x_1 + 4320x_0^2 - 2800x_0x_1 + 600x_1^2 \right.$$
$$\left. - 280x_0^3 + 300x_0^2x_1 - 150x_0x_1^2 + 35x_1^3 \right) \qquad (11.128)$$

$$\Phi_{a,20}(x,y) = \frac{1}{2!0!} \left[\frac{1}{(-4)^5} \frac{1}{(-8)^5} \right] (x_0 - 2)^5 (x_1 - 12)^5 (x_0 - 6)^2$$

$$\times \frac{1}{2^6} \left(1584 - 600x_0 + 220x_1 + 60x_0^2 - 50x_0x_1 + 15x_1^2 \right)$$

$$(11.129)$$

$$\Phi_{a,11}(x,y) = \frac{1}{1!1!} \left[\frac{1}{(-4)^5} \frac{1}{(-8)^5} \right] (x_0 - 2)^5 (x_1 - 12)^5 (x_0 - 6)(x_1 - 4)$$

$$\times \frac{1}{2^6} \left(1584 - 600x_0 + 220x_1 + 60x_0^2 - 50x_0x_1 + 15x_1^2 \right)$$

$$(11.130)$$

$$\Phi_{a,02}(x,y) = \frac{1}{0!2!} \left[\frac{1}{(-4)^5} \frac{1}{(-8)^5} \right] (x_0 - 2)^5 (x_1 - 12)^5 (x_1 - 4)^2$$

$$\times \frac{1}{2^6} \left(1584 - 600x_0 + 220x_1 + 60x_0^2 - 50x_0x_1 + 15x_1^2 \right)$$

$$(11.131)$$

$$\Phi_{a,30}(x,y) = \frac{1}{3!0!} \left[\frac{1}{(-4)^5} \frac{1}{(-8)^5} \right] (x_0 - 2)^5 (x_1 - 12)^5 (x_0 - 6)^3$$

$$\times \frac{1}{2^3} \left(48 - 10x_0 + 5x_1 \right)$$

$$(11.132)$$

$$\Phi_{a,21}(x,y) = \frac{1}{2!1!} \left[\frac{1}{(-4)^5} \frac{1}{(-8)^5} \right] (x_0 - 2)^5 (x_1 - 12)^5 (x_0 - 6)^2 (x_1 - 4)$$

$$\times \frac{1}{2^3} \left(48 - 10x_0 + 5x_1 \right)$$

$$(11.133)$$

$$\Phi_{a,12}(x,y) = \frac{1}{1!2!} \left[\frac{1}{(-4)^5} \frac{1}{(-8)^5} \right] (x_0 - 2)^5 (x_1 - 12)^5 (x_0 - 6)(x_1 - 4)^2$$

$$\times \frac{1}{2^3} \left(48 - 10x_0 + 5x_1 \right)$$

$$(11.134)$$

$$\Phi_{a,03}(x,y) = \frac{1}{0!3!} \left[\frac{1}{(-4)^5} \frac{1}{(-8)^5} \right] (x_0 - 2)^5 (x_1 - 12)^5 (x_1 - 4)^3$$

$$\times \frac{1}{2^3} \left(48 - 10x_0 + 5x_1 \right)$$

$$(11.135)$$

$$\Phi_{a,40}(x,y) = \frac{1}{4!0!} \left[\frac{1}{(-4)^5} \frac{1}{(-8)^5} \right] (x_0 - 2)^5 (x_1 - 12)^5 (x_0 - 6)^4 \frac{1}{2^0} \left(1 \right)$$

$$(11.136)$$

$$\Phi_{a,31}(x,y) = \frac{1}{3!1!} \left[\frac{1}{(-4)^5} \frac{1}{(-8)^5} \right] (x_0 - 2)^5 (x_1 - 12)^5 (x_0 - 6)^3 (x_1 - 4) \frac{1}{2^0} \left(1 \right)$$

(11.137)

$$\Phi_{a,22}(x,y) = \frac{1}{2!2!} \left[\frac{1}{(-4)^5} \frac{1}{(-8)^5} \right] (x_0 - 2)^5 (x_1 - 12)^5 (x_0 - 6)^2 (x_1 - 4)^2 \frac{1}{2^0} \left(1 \right)$$

(11.138)

$$\Phi_{a,13}(x,y) = \frac{1}{1!3!} \left[\frac{1}{(-4)^5} \frac{1}{(-8)^5} \right] (x_0 - 2)^5 (x_1 - 12)^5 (x_0 - 6)(x_1 - 4)^3 \frac{1}{2^0} \left(1 \right)$$

(11.139)

$$\Phi_{a,04}(x,y) = \frac{1}{0!4!} \left[\frac{1}{(-4)^5} \frac{1}{(-8)^5} \right] (x_0 - 2)^5 (x_1 - 12)^5 (x_1 - 4)^4 \frac{1}{2^0} \left(1 \right)$$

(11.140)

Chapter 12

Generic domains

12.1 Expression for the polynomial

12.1.1 Information about the polynomial

1. Domain definition

   The domain is defined by lines through the points $a_0 = (2, -7)$, $a_1 = (8, -2)$, $a_2 = (3, 4)$, $a_3 = (-5, 2)$, $a_4 = (-8, -4)$, which lines are

   through the points a_4-a_0 the line is $\varphi_{a_3,0}(x_0, x_1) = 3x_0 + 10x_1 + 64 = 0$.

   through the points a_0-a_1 the line is $\varphi_{a_3,1}(x_0, x_1) = -5x_0 + 6x_1 + 52 = 0$.

   through the points a_1-a_2 the line is $\varphi_{a_3,2}(x_0, x_1) = 6x_0 + 5x_1 - 38 = 0$.

   through the points a_2-a_3 the line is $\varphi_{a_3,3}(x_0, x_1) = \chi_{a_3,0}(x) = x_0 - 4x_1 + 13 = 0$.

   through the points a_3-a_4 the line is $\varphi_{a_3,4}(x_0, x_1) = \chi_{a_3,1}(x) = 2x_0 - x_1 + 12 = 0$.

2. Number of variables: $d = 2$

3. Maximum order of derivative: $\eta = 4$

4. Reference node: $a_e = a_3$ whose coordinate numbers are denoted as $\alpha = (-5, 2)$

5. Nodes in the visibility of the reference node

   The elements of the visibility set $\mathcal{V}$ of the reference node a_3 are the lines $a_4 - a_0$, $a_0 - a_1$, and $a_1 - a_2$.

6. Order of derivative

   The order of derivative of the Hermite Interpolating Polynomials related to the reference node $a_e = (-5, 2)$ and the degree of derivative $\eta = 4$

are the constrained numbers in the set $L^\delta(\eta) = L^2(4)$, which are

$$L^2(4) = \Big\{ (0,0),(1,0),(0,1),(2,0),(1,1),(0,2),$$
$$(3,0),(2,1),(1,2),(0,3),(4,0),(3,1),(2,2),(1,3),(0,4) \Big\}$$
$$(12.1)$$

The corresponding polynomials are denoted as $Phi_{e,n}(x)$ where e represents the reference node a_3 and n are the coordinate numbers such that $n \in L^2(4)$. Therefore, the polynomials to be obtained are denoted as follows

$$\Phi_{a_3,00}(x), \Phi_{a_3,10}(x), \Phi_{a_3,01}(x), \Phi_{a_3,10}(x),$$
$$\Phi_{a_3,20}(x), \Phi_{a_3,11}(x), \Phi_{a_3,02}(x),$$
$$\Phi_{a_3,30}(x), \Phi_{a_3,21}(x), \Phi_{a_3,12}(x), \Phi_{a_3,03}(x),$$
$$\Phi_{a_3,40}(x), \Phi_{a_3,31}(x), \Phi_{a_3,22}(x), \Phi_{a_3,13}(x), \Phi_{a_3,04}(x) \qquad (12.2)$$

and defined in figure 12.1.1.

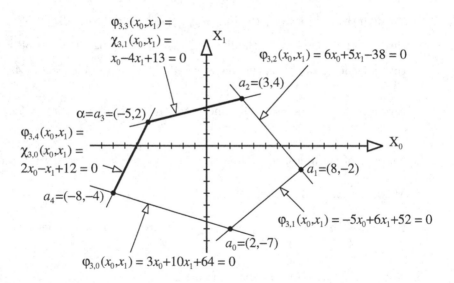

Figure 12.1.1 *Definition of a polygonal domain.*

12.1.2   Properties of the polynomial $\Phi_{a_3,00}(x_0, x_1)$

1. Properties of the polynomial $\Phi_{a_3,00}(x_0, x_1)$

(a) it equals one when evaluated at the reference node $a_3 = (-5, 2)$, that is,

$$\Phi_{a_3,00}(x)\Big|_{x=a_3} = 1 \qquad (12.3)$$

(b) it vanishes at the boundaries that do not intercept the reference node, which are: the boundary $\varphi_{a_3,0}(x_0, x_1)$ through $a_4 - a_0$, the boundary

$$\varphi_{a_3,1}(x_0, x_1) \quad \text{through} \quad a_0 - a_1 \qquad (12.4)$$

and the boundary

$$\varphi_{a_3,2}(x_0, x_1) \quad \text{through} \quad a_1 - a_2 \qquad (12.5)$$

where $a_4 = (-8, -4)$, $a_0 = (2, -7)$, $a_1 = (8, -2)$, and $a_2 = (3, 4)$. Therefore, the property

$$\Phi_{a_3,00}(x)\Big|_{x\in\varphi_{a_3,0}} = \Phi_{a_3,00}(x)\Big|_{x\in\varphi_{a_3,1}} = \Phi_{a_3,00}(x)\Big|_{x\in\varphi_{a_3,2}} = 0 \qquad (12.6)$$

2. Properties of the derivatives of the polynomial $\Phi_{a_3,00}(x_0, x_1)$

(a) it has its derivatives until order $\eta = 4$ equal to zero at the reference node $a_3 = (-5, 2)$, that is,

$$\Phi_{a_3,00}^{(k)}(x)\Big|_{x=a_3} = 0 \qquad k = 1, 2, 3, 4 \qquad (12.7)$$

(b) it has the values of the derivatives until order $\eta = 4$ equal to zero at the boundaries that do not intercept the reference node, that is

$$\Phi_{a_3,00}^{(k)}(x)\Big|_{x\in\varphi_{a_3,0}} = 0 \qquad \Phi_{a_3,00}^{(k)}(x)\Big|_{x\in\varphi_{a_3,1}} = 0$$

$$\Phi_{a_3,00}^{(k)}(x)\Big|_{x\in\varphi_{a_3,2}} = 0 \qquad k = 1, 2, 3, 4 \qquad (12.8)$$

12.2   Expression for the polynomial $\Phi_{e,n}(x)$

From the property defined in equation (12.3), the equation (7.7), on page 290, writes

$$\Phi_{e,n}(x)\Big|_{x=\alpha} = P_e(x)\, Q_{e,n}(x)\, f_{e,n}(\bar{a}, x)\Big|_{x=\alpha} \qquad (12.9)$$

The factor $P_e(x)$ can have several orders of magnitude due to the power $\eta + 1$. To reduce the order it is convenient to divide $P_e(x)$ by a factor of the same order of magnitude such as the value of $P_e(x)$ at $x = \alpha$, which is denoted as $P_e(\alpha)$. Introducing the constant factor $P_e(\alpha)$ into equation (12.9) it transforms into

$$\Phi_{e,n}(x) = \frac{P_e(x)}{P_e(\alpha)} \left[P_e(\alpha) Q_{e,n}(x) f_{e,n}(\bar{a}, x) \right] \tag{12.10}$$

The derivative of order n of the equation (12.10) writes

$$\left[\frac{P_e(x)}{P_e(\alpha)} \left[P_e(\alpha) Q_{e,n}(x) f_{e,n}(\bar{a}, x) \right] \right]^{(n)} \Bigg|_{x=\alpha} = 1 \tag{12.11}$$

Performing the derivative denoted in equation (12.11) it is obtained

$$\frac{P_e(x)}{P_e(\alpha)} \left[P_e(\alpha) n! \, f_{e,n}(\bar{a}, x) \right] \Bigg|_{x=\alpha} = 1 \tag{12.12}$$

Define

$$f_{e,n}(\bar{a}, x) := \frac{1}{n! \, P_e(\alpha)} f_{e,n}(a, x) \tag{12.13}$$

then the equation (12.12) can be rewritten as

$$\frac{P_e(x)}{P_e(\alpha)} f_{e,n}(a, x) \Bigg|_{x=\alpha} = 1 \tag{12.14}$$

which gives an expression for the computation of coefficients of the polynomial $f_{e,n}(a, x)$ as follows

$$f_{e,n}(a, x) \Bigg|_{x=\alpha} = \frac{P_e(\alpha)}{P_e(x)} \Bigg|_{x=\alpha} \tag{12.15}$$

Since $f_{e,n}(\bar{a}, x)$ and $f_{e,n}(a, x)$ are polynomials of the same degree in the same variables then the following relation

$$\bar{a} = \frac{1}{n! \, P_e(\alpha)} a \tag{12.16}$$

among their coefficients is true. This same result was obtained in equation (7.119), on page 314.

Substituting $f_{e,n}(\bar{a}, x)$ defined in equation (12.13) into equation (7.7), on page 290, it gives

$$\Phi_{e,n}(x) = \frac{1}{n!} \frac{1}{P_e(\alpha)} P_e(x) Q_{e,n}(x) f_{e,n}(a, x) \tag{12.17}$$

where $P_e(\alpha)$ can be any number. If

$$P_e(\alpha) = P_e(x) \Bigg|_{x=\alpha} \tag{12.18}$$

then

$$\frac{P_e(x)}{P_e(\alpha)}\bigg|_{x=\alpha} = 1 \qquad (12.19)$$

Introducing the result in equation (12.19) into equation (12.14) it gives the following property

$$f_{e,n}(a, x)\bigg|_{x=\alpha} = 1 \qquad \forall \ n \in L^\delta(\eta) \qquad (12.20)$$

Evaluating the expression in equation (12.17) at the reference point $x = \alpha$ and substituting the results obtained in equations (12.19) and (12.20) it follows the property

$$\Phi_{e,n}(x)\bigg|_{x=\alpha} = \frac{Q_{e,n}(x)}{n!}\bigg|_{x=\alpha} \qquad (12.21)$$

As shown in section 7.3 for the computation of the coefficients a of $f_{e,n}(a, x)$, the technique consists in the evaluation of the derivatives of terms of the equation (12.16) of order equal to the power of the highest term of the polynomial $f_{e,n}(a, x)$ to the lowest, that is, solving the system of equations

$$\left[f_{e,n}(a, x) \right]^{(v)}\bigg|_{x=\alpha} = \left[\frac{P_e(\alpha)}{n! \, P_x(x)} \right]^{(v)}\bigg|_{x=\alpha} \qquad (12.22)$$

Sometimes it is convenient to split $P_e(\alpha)$ according to the factors of $P_e(x)$, that is,

$$\frac{P_e(x)}{P_e(\alpha)} = \frac{\Phi_{e,0}(x)}{P_{e,0}(\alpha)} \cdots \frac{\Phi_{e,k}(x)}{P_{e,k}(\alpha)} \qquad (12.23)$$

where $P_{e,i}(\alpha) = \Phi_{e,i}(x)|_{x=\alpha}$.

The computation of the derivatives in equation (12.21) can be performed observing that

$$P_e(x) = \prod_k \Phi_{e,k}(x) \qquad (12.24)$$

and using the algorithm given in section 6.8.2 for derivative of the product of t functions and d variables.

In the general case it must be added the algorithm for the evaluation of the derivative of order i with respect to the variable j of each function, that is,

$$\frac{d_i}{dx_j^i}\Phi_{e,k}(x) = \left[\Phi_{e,k}(x_0, \ldots, x_j, \ldots, x_\delta) \right]^{(0,\ldots,i,\ldots,0)} \qquad (12.25)$$

where i occupies the j-th position.

12.2.1   Common factors related to the reference node $a_3 = (-5, 2)$

Variable factors Factors that are common to all polynomials related to the reference node $a_3 = (-5, 2)$

The polynomial in equation (7.11), on page 291, writes

$$P_{a_3}(x) = [\varphi_{a_3,0}(x_0, x_1)]^{\eta+1} [\varphi_{a_3,1}(x)]^{\eta+1} [\varphi_{a_3,2}(x_0, x_1)]^{\eta+1}$$

$$= [3x_0 + 10x_1 + 64]^{\eta+1} [-5x_0 + 6x_1 + 52]^{\eta+1} [6x_0 + 5x_1 - 38]^{\eta+1}$$
$$(12.26)$$

where $\eta = 4$, which gives

$$P_{a_3}(x) = [3x_0 + 10x_1 + 64]^5 [-5x_0 + 6x_1 + 52]^5 [6x_0 + 5x_1 - 38]^5$$
$$(12.27)$$

Constant factor The constant factor $P_{a_3}(\alpha)$ is the value of $P_{a_3}(x)$ at the reference node $\alpha = a_3 = (-5, 2)$, that is,

$$P_{a_3}(\alpha) \Big|_{x=\alpha}$$
$$= [3(-5) + 10(2) + 64]^5 [-5(-5) + 6(2) + 52]^5 [6(-5) + 5(2) - 38]^5$$
$$= 69^5 \times 89^5 \times (-58)^5 = 0.4876918749e2910^{29} \qquad (12.28)$$

12.2.2   Expression for the polynomial $Q_{a_3,n}(x)$

The expression for the polynomial $Q_{a_3,n}(x)$ in equation (7.49), on page 302, for two variables writes

$$Q_{a_3,n}(x) = (x_0 - \alpha_0)^{n_0} (x_1 - \alpha_1)^{n_1} \qquad (12.29)$$

12.2.3   Construction of the constant term $n! P_{a_3}(\alpha)$

The factor $P_{a_3}(\alpha)$ was obtained in equation (10.6), on page 357, and only $n!$ changes with the polynomial then for $n = (0, 0)$ it follows that

$$n! P_{a_3}(\alpha) = (n_0! n_1!) P_{a_3}(\alpha) = (0! 0!) 69^5 \times 89^5 \times (-58)^5 = 0.4876918749e2910^{29}$$
$$(12.30)$$

12.2.4   Construction of the polynomial $Q_{a_3,n}(x)$

For $n = (0, 0)$ it follows

$$Q_{a_3,n}(x) = (x_0 - (-1))^{n_0} (x_1 - (-1))^{n_1}$$
$$= (x_0 + 1)^{n_0} (x_1 + 1)^{n_1}$$
$$= (x_0 + 1)^0 (x_1 + 1)^0 = 1 \qquad (12.31)$$

12.2.5 The polynomial $\Phi_{a_3,00}(x_0, x_1)$

The polynomial $\Phi_{e,n}(x)$ in equation (12.17) with $n = (0,0)$ writes

$$\Phi_{a_3,00}(x) = \frac{1}{n!} P_{a_3}(x) Q_{3,n}(x) f_{a_3,n}(\bar{a}, x)$$
$$= [3x_0 + 10x_1 + 64]^5 [-5x_0 + 6x_1 + 52]^5$$
$$\times [6x_0 + 5x_1 - 38]^5 (1) f_{a_3,00}(\bar{a}, x_0, x_1) \qquad (12.32)$$

12.2.6 Computation of the coefficients of the polynomial $f_{a_3,00}(a, x)$

The minimum degree of the polynomial $f_{a_3,00}(a, x)$ is given, according to the corollary 7.2.19, on page 305, by

$$q \geq \eta - \lambda_0(n) = 4 - \lambda_0(0,0) = 4 - 0 = 4 \qquad (12.33)$$

a- Computation of the coefficients that belong to the terms in the level $\lambda_0(v) = q = 4$

The coefficients to be evaluated in this level are $a_{(4,0)}$, $a_{(3,1)}$, $a_{(2,1)}$, $a_{(1,3)}$, $a_{(0,4)}$ where the indices of the coefficients are elements of the set $L_4^1 = \{(4,0), (3,1), (2,2), (1,3), (0,4)\}$. The computation is performed using the equation (12.21)

$$[f_{e,n}(a, x)]^{(v)}\Big|_{x=\alpha_e} = \left[\frac{1}{n! P_e(x)}\right]^{(v)}\Bigg|_{x=\alpha} \qquad (0,0,\ldots,0) \leq v \leq (0,0,\ldots,q)$$
$$(12.34)$$

and the coefficients for $f_{e,n}(a, x)$ for the level $\lambda_0(v) = 4$ are

$$a_{(0,4)} = 0.2988386871^{-1} \qquad a_{(1,3)} = -0.4046531032^{-1}$$
$$a_{(2,2)} = 0.3916597322^{-1} \qquad a_{(3,1)} = -0.2264043092^{-1}$$
$$a_{(4,0)} = 0.1347747184^{-1} \qquad (12.35)$$

b- Computation of the coefficients that belong to the terms in the level $\lambda_0(v) = q - 1 = 3$

$$a_{(0,3)} = -0.5397242942 \qquad a_{(1,2)} = 0.7623398278$$
$$a_{(2,1)} = -0.5920813203 \qquad a_{(3,0)} = 0.3723429254 \qquad (12.36)$$

c- Computation of the coefficients that belong to the terms in the level $\lambda_0(v) = q - 2 = 2$

$$a_{(0,2)} = 4.421054465 \qquad a_{(1,1)} = -5.529026611$$
$$a_{(2,0)} = 4.119813776 \qquad (12.37)$$

d- Computation of the coefficients that belong to the terms in the level $\lambda_0(v) = q - 3 = 1$

$$a_{(0,1)} = -19.56536527 \qquad a_{(1,0)} = 22.04539678 \qquad (12.38)$$

e- Computation of the coefficients that belong to the terms in the level $\lambda_0(v) = q - 4 = 0$

$$a_{(0,0)} = 50.00252938 \qquad (12.39)$$

12.3 Triangular domain

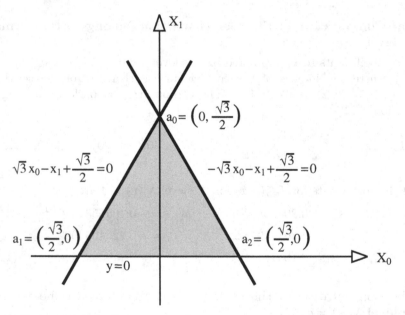

Figure 12.3.1 *Reference node $a_e = a_0$ in a triangular domain.*

12.3.1 Information about the polynomial

The data for the computation of the polynomials related to the triangular domain shown in figure 12.3.1 are

1. The domain is defined by lines through the points $a_0 = (0, \sqrt{3}/2)$, $a_1 = (-1/2, 0)$, $a_2 = (0, 1/2)$, which lines are

through the points a_2-a_3 the line is $\varphi_{a_0,0}(x_0, x_1) = \chi_{a_0,0}(x) = \sqrt{3}x_0 - x_1 + \sqrt{3}/2 = 0$.

through the points a_3-a_4 the line is $\varphi_{a_0,1}(x_0, x_1) = \chi_{a_0,1}(x) = -\sqrt{3}x_0 - x_1 + \sqrt{3}/2 = 0$.

through the points a_1-a_2 the line is $\varphi_{a_3,2}(x_0, x_1) = x_1 = 0$.

2. Number of variables: $d = 2$

3. Maximum order of derivative: $\eta = 4$

4. Reference node: $a_e = a_0 = \alpha = (0, \sqrt{3}/2)$

5. Nodes in the visibility of the reference node

   The elements of the visibility set $\mathcal{V}$ of the reference node a_3 is the line $a_1 - a_2$.

6. The order of derivative of the Hermite Interpolating Polynomials related to the reference node are the constrained numbers in the set $L^\delta(\eta) = L^2(4)$, which are

$$L^2(4) = \Big\{ (0,0), (1,0), (0,1), (2,0), (1,1), (0,2),$$
$$(3,0), (2,1), (1,2), (0,3), (4,0), (3,1), (2,2), (1,3), (0,4) \Big\}$$
$$(12.40)$$

The corresponding polynomials are denoted as

$$\Phi_{a_0,00}(x), \Phi_{a_0,10}(x), \Phi_{a_0,01}(x), \Phi_{a_0,10}(x),$$
$$\Phi_{a_0,20}(x), \Phi_{a_0,11}(x), \Phi_{a_0,02}(x),$$
$$\Phi_{a_0,30}(x), \Phi_{a_0,21}(x), \Phi_{a_0,12}(x), \Phi_{a_0,03}(x),$$
$$\Phi_{a_0,40}(x), \Phi_{a_0,31}(x), \Phi_{a_0,22}(x), \Phi_{a_0,13}(x), \Phi_{a_0,04}(x) \quad (12.41)$$

The polynomial $\Phi_{a_0,00}(x)$ has the property that when evaluated at the reference node $a_3 = (0, \sqrt{3}/2)$ equals one.

The first derivative of $\Phi_{a_0,10}(x)$ with respect to the variable x_0 equals one at the reference node $a_3 = (0, \sqrt{3}/2)$.

The cross derivative of $\Phi_{a_0,11}(x)$ with respect to the variables x_0 and x_1 equals one at the reference node $a_3 = (0, \sqrt{3}/2)$, etc.

12.3.2 Common factors related to the reference node $a_0 = (0, \sqrt{3}/2)$

Factors that are common to all polynomials related to the reference node $a_0 = (0, \sqrt{3}/2)$

The polynomial in equation (7.11), on page 291, writes

$$P_{a_0}(x) = [\varphi_{a_0,0}(x_0, x_1)]^{\eta+1} = [y]^{\eta+1} \tag{12.42}$$

where $\eta = 4$, which gives

$$P_{a_0}(x) = [y]^5 \tag{12.43}$$

The constant term $P_{a_0}(\alpha)$ is the value of $P_{a_0}(x)$ at the reference node $\alpha = a_0 = (0, \sqrt{3}/2))$, that is,

$$P_{a_0}(x)\Big|_{x=\alpha} = \left[\frac{\sqrt{3}}{2}\right]^5 \tag{12.44}$$

12.3.3   Expression for the polynomial $Q_{a_0,n}(x)$

The expression for the polynomial $Q_{a_0,n}(x)$ in equation (7.49), on page 302, for two variables writes

$$Q_{a_0,n}(x) = (x_0 - \alpha_0)^{n_0} (x_1 - \alpha_1)^{n_1} \tag{12.45}$$

12.3.4   Expression for the polynomial $\Phi_{a_0,n}(x)$

The expression for the polynomial is given by equation (7.126), on page 315, which writes

$$\Phi_{a_0,n}(x) = \frac{1}{n! P_{a_0}(\alpha)} P_{a_0}(x) Q_{a_0,n}(x) f_{a_0,n}(a, x) \tag{12.46}$$

where $n \in L^2(4)$.

12.3.5   Properties of the polynomial $\Phi_{a_0,00}(x_0, x_1)$

1. Properties of the polynomial $\Phi_{a_0,00}(x_0, x_1)$

   (a) it equals one when evaluated at the reference node $a_0 = (0, \sqrt{3}/2)$, that is,

   $$\Phi_{a_0,00}(x)\Big|_{x=a_0} = 1 \tag{12.47}$$

   (b) it vanishes at the boundaries that do not intercept the reference node, which is $y = 0$ through $a_1 - a_2$. Therefore, the property

   $$\Phi_{a_0,00}(x)\Big|_{x \in \varphi_{a_0,0}} = 0 \tag{12.48}$$

2. Properties of the derivatives of the polynomial $\Phi_{a_0,00}(x_0, x_1)$

   (a) it has its derivatives until order $\eta = 4$ equal to zero at the reference node $a_0 = (0, \sqrt{3}/2)$

$$\Phi_{a_0,00}^{(k)}(x)\Big|_{x=a_0} = 0 \qquad k = 1, 2, 3, 4 \qquad (12.49)$$

   (b) it has the values of the derivatives until order $\eta = 4$ equal to zero at the boundaries that do not intercept the reference node

$$\Phi_{a_0,00}^{(k)}(x)\Big|_{x \in \varphi_{a_0,0}} = 0 \qquad k = 1, 2, 3, 4 \qquad (12.50)$$

12.3.6 Construction of the constant term $n!P_{a_0}(\alpha)$

The factor $P_{a_0}(\alpha)$ was obtained in equation (12.44) and only $n!$ changes with the polynomial, then for $n = (0, 0)$ it follows that

$$n!P_{a_0}(\alpha) = (n_0!n_1!)P_{a_3}(\sqrt{3}/2) = (0!0!)\left(\frac{\sqrt{3}}{2}\right)^5 = \frac{9\sqrt{3}}{32} \qquad (12.51)$$

12.3.7 Construction of the polynomial $Q_{a_0,n}(x)$

For $n = (0, 0)$ it follows

$$\begin{aligned} Q_{a_0,n}(x) &= (x_0 - (-1))^{n_0} (x_1 - (-1))^{n_1} \\ &= (x_0 + 1)^{n_0} (x_1 + 1)^{n_1} \\ &= (x_0 + 1)^0 (x_1 + 1)^0 = 1 \end{aligned} \qquad (12.52)$$

12.3.8 The polynomial $\Phi_{a_0,00}(x_0, x_1)$

The polynomial $\Phi_{e,n}(x)$ in equation (12.17) with $n = (0, 0)$ writes

$$\begin{aligned} \Phi_{a_0,00}(x) &= \frac{1}{n!}P_{a_0}(x)\,Q_{0,n}(x)\,f_{a_0,n}(\bar{a}, x) \\ &= [y]^5\,(1)\,f_{a_0,00}(\bar{a}, x_0, x_1) \end{aligned} \qquad (12.53)$$

12.3.9 Computation of the coefficients of the polynomial $f_{a_0,00}(x)$

The minimum degree of the polynomial $f_{a_0,00}(x)$ is given, according to the corollary 7.2.19, on page 305, by

$$q \geq \eta - \lambda_0(n) = 4 - \lambda_0(0, 0) = 4 - 0 = 4 \qquad (12.54)$$

a- Computation of the coefficients that belong to the terms in the level $\lambda_0(v) = q = 4$

The coefficients to be evaluated in this level are $a_{(4,0)}$, $a_{(3,1)}$, $a_{(2,1)}$, $a_{(1,3)}$, $a_{(0,4)}$ where the indices of the coefficients are elements of the set $L_4^1 = \{(4,0), (3,1), (2,2), (1,3), (0,4)\}$. The computation is performed using the equation (7.81)

$$[f_{e,n}(\bar{a}, x)]^{(v)}\Big|_{x=\alpha_e} = \left[\frac{1}{n!P_e(x)}\right]^{(v)}\Big|_{x=\alpha} \neq 0 \qquad (0,0,\ldots,0) \leq v \leq (0,0,\ldots,q) \tag{12.55}$$

For $v = (0,4)$ it follows

$$a_{(0,4)} = \frac{1}{v!}\left[\frac{1}{n!P_e(x)}\right]^{(4,0)}\Big|_{x=(0,\sqrt{3}/2)} = \frac{35840\sqrt{3}}{243} \tag{12.56}$$

To reduce the order of magnitude of the coefficients close to 10^0 both sides of equation (12.55) both sides will be multiplied by $P_e(\alpha)$, which gives

$$[f_{e,n}(\bar{a}, x)]^{(v)}\Big|_{x=\alpha_e} P_e(\alpha) = \left[\frac{1}{n!P_e(x)}\right]^{(v)}\Big|_{x=\alpha} P_e(\alpha)$$
$$(0,0,\ldots,0) \leq v \leq (0,0,\ldots,q) \tag{12.57}$$

and the coefficients for $f_{e,n}(\bar{a}, x)$ are

$$a_{(0,4)} = \frac{1120}{9} \qquad a_{(1,3)} = 0 \qquad a_{(2,2)} = 0 \qquad a_{(3,1)} = 0 \qquad a_{(4,0)} = 0 \tag{12.58}$$

b- Computation of the coefficients that belong to the terms in the level $\lambda_0(v) = q - 1 = 3$

$$a_{(0,3)} = -280\sqrt{3} \qquad a_{(1,2)} = 0 \qquad a_{(2,1)} = 0 \qquad a_{(3,0)} = 0 \tag{12.59}$$

c- Computation of the coefficients that belong to the terms in the level $\lambda_0(v) = q - 2 = 2$

$$a_{(0,2)} = 720 \qquad a_{(1,1)} = 0 \qquad a_{(2,0)} = 0 \tag{12.60}$$

d- Computation of the coefficients that belong to the terms in the level $\lambda_0(v) = q - 3 = 1$

$$a_{(0,1)} = -280\sqrt{3} \qquad a_{(1,0)} = 0 \tag{12.61}$$

e- Computation of the coefficients that belong to the terms in the level $\lambda_0(v) = q - 4 = 0$

$$a_{(0,0)} = 126 \qquad (12.62)$$

The expression for the first Hermite Interpolating Polynomial writes

$$\Phi_{a_0,00}(x) = \frac{1}{n!} \frac{1}{P_{a_0}(\alpha)} P_{a_0}(x) Q_{a_0}(x) f_{a_0,n}(a, x)$$

$$= \frac{1}{0!} \frac{32}{9\sqrt{3}} [y]^5 (1) \left(126 - 280\sqrt{3}y + 720y^2 - 280\sqrt{3}y^3 + \frac{1120}{9}y^4 \right)$$

$$(12.63)$$

Chapter 13

Extensions of the constrained numbers

13.1 Comparison to the divided difference method

13.1.1 Comparison

The one-dimensional Hermite Interpolating Polynomial whose function equals one at the point $x = -1$, zero at the point $x = 1$, and all derivatives up to order 3 equal zero at $x = -1$ and $x = 1$ can be obtained using divided differences [22]. The table with the divided differences writes

Table 13.1.1 *Divided difference values.*

x_0	-1	1	0	0	0	$-1/16$	$4/32$	$-10/64$	$20/128$
x_1	-1	1	0	0	$-1/8$	$3/16$	$-6/32$	$10/64$	
x_2	-1	1	0	$-1/4$	$2/8$	$-3/16$	$4/32$		
x_3	-1	1	$-1/2$	$1/4$	$-1/8$	$1/16$			
x_4	1	0	0	0	0				
x_5	1	0	0	0					
x_6	1	0	0						
x_7	1	0							

The solution is written in the form

$$p(x) = \sum_{k=0}^{n} f[x_0, x_1, \dots, x_k] \prod_{j=0}^{k-1} (x - x_j) \tag{13.1}$$

yielding

$$p(x) = 1 - \frac{1}{16}(x+1)^4 + \frac{4}{32}(x+1)^4(x-1) - \frac{10}{64}(x+1)^4(x-1)^2 + \frac{20}{128}(x+1)^4(x-1)^3 \tag{13.2}$$

Performing the algebraic manipulation is its obtained

$$p(x) = 1 + \frac{1}{32}(x+1)^4 \left(-2 + 4(x-1) - 5(x-1)^2 + 5(x-1)^3\right)$$

$$= 1 + \frac{1}{32}(x+1)^4 \left(5x^3 - 20x^2 + 29x - 16\right)$$

$$= \frac{1}{32}(x-1)^4 \left(5x^3 + 20x^2 + 29x + 16\right) \tag{13.3}$$

which equals to the solution using the proposed algorithm. The solution using divided differences, although not difficult to be generated, yields a solution in a more complex form.

The complexity increases still more for the multidimensional case as reported in Capello, Gallopoulos, and Koç [5] or in Gasca and Maeztu [13].

13.1.2   Example

Determine a polynomial that takes these values:

$$p(1) = 2 \qquad p'(1) = 3 \qquad p(2) = 6 \qquad p'(2) = 7 \qquad p''(2) = 8 \tag{13.4}$$

This example is evaluated using extended Newton divided differences in [22], example 3.

The interval is $[1, 2]$ and this is a zero-dimensional problem. For the first node $\alpha = \alpha_0 = 1$ and $\beta = \beta_0 = 2$, and the order of the derivative is $\eta = 1$, thus

$$\Phi_{a0} = (x-2)^2 f(x) \tag{13.5}$$

The computation of $f(x)$ gives $f(x) = -2 + 4x$. For the first derivative it follows

$$\Phi_{a1} = (x-2)^2 (x-1) f(x) \tag{13.6}$$

where $f(x) = 3$.

For the second node $\alpha = \alpha_0 = 2$ and $\beta = \beta_0 = 1$ and the order of derivative is $\eta = 2$. Thus

$$\Phi_{b0} = (x-1)^3 f(x) \tag{13.7}$$

where $f(x) = 402 - 162x - 18x^2$. For the first derivative it follows

$$\Phi_{b1} = (x-1)^3 (x-2) f(x) \tag{13.8}$$

and $f(x) = 49 - 21x$. For the second derivative

$$\Phi_{b2} = (x-1)^3 (x-2)^2 f(x) \tag{13.9}$$

where $f(x) = 8$.

The solution is

$$p(x) = (x-2)^2(-2+4x) + 3(x-2)^2(x-1) + (x-1)^3(402-162x-18x^2)$$
$$+ (x-1)^3(x-2)(49-21x) + 8(x-1)^3(x-2)^2 \qquad (13.10)$$

which reduces to

$$p(x) = (x-2)^2(7x-5) + (x-1)^3(-31x^2-103x+336) \qquad (13.11)$$

13.2 Generalization to the set of positive real numbers

Consider the segment in level $\lambda_0 = 4 + 0 = 0$ in figure 13.2.1, in a one-dimensional space, represented by $d = 2$ coordinate numbers. A point A can be located using the coordinates (ξ_0, ξ_1) where $0 \le \xi_0 \le \lambda_0$ and $\xi_1 = \lambda_0 - \xi_0$, which satisfies the definition 1.1.2 of constrained numbers.

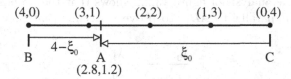

Figure 13.2.1 $\xi_0 = 2.8$ *and level* $\lambda_1 = \lambda_0 - \xi_0 = 4 - 2.8 = 1.2$.

Therefore, the coordinates of the point A are $\xi_0 = 2.8$ and $\xi_1 = 1.2$, which can be written as $(2.8, 1.2)$.

Similarly the coordinates of a point E in a two-dimensional space represented with $d = 3$ components can be obtained. To exemplify consider $\lambda_0 = 2.7$ and choose the first coordinate as $\xi_0 = 1.3$ then the first sublevel is given by $\lambda_1 = \lambda_0 - \xi_0 = 2.7 - 1.3 = 1.4$. Assume that the second coordinate is given by $\xi_1 = 0.5$ then $\xi_2 = \lambda_2 = \lambda_1 - \xi_1 = 1.4 - 0.5 = 0.9$. The point E has coordinates $(\xi_0, \xi_1, \xi_2) = (1.4, 0.5, 0.9)$.

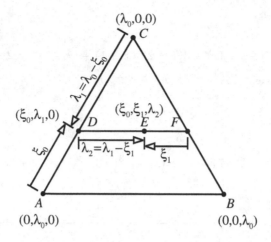

Figure 13.2.2 *Cross section of the level λ_0 showing the construction of the coordinates of the point E. The arrows indicate the ascending direction of the coordinates.*

The point $(0,0,0)$, represented in the figure 6.1.5, is the origin of the constrained coordinate system. The point C whose coordinates are $(\lambda_0, 0, 0)$ is the reference point for the level λ_0. Analogously, D whose coordinates are $(\xi_0, \lambda_1, 0)$ is the reference point for the first sublevel λ_1, and so on.

The geometric analysis of figure 13.2.2 shows that the proposed constructions yields DF parallel to AB. To prove it assume that they are parallel, then

$$\frac{\lambda_0 - \xi_0}{\lambda_0} = \frac{\lambda_1}{\lambda_0} \tag{13.12}$$

since $\lambda_1 = \lambda_0 - \xi_0$ by definition of λ_1, the above relation is identically true. Therefore, DF is parallel to AB. Any variation of ξ_1 is performed with ξ_0 constant, which shows that the segment DF is the locus of the points with ξ_0 constant. The same procedure applies for the construction of a point in a constrained number system for higher dimensions.

The polynomial

$$f(x) = a_0 + a_1 x^{0.7} + a_2 y^{0.7} + a_3 x^{1.4} + a_4 x^{1.2} y^{0.2} + a_5 y^{1.4} \tag{13.13}$$

is an example of a complete polynomial using the constrained number over the field of the real. Note that the terms $a_1 x^{0.7}$ and $a_2 y^{0.7}$ are in the level $\lambda_0 = 0.7$; the last three terms are in level $\lambda_0 = 1.4$ and that $0.7 < 1.4$.

13.3 Constrained coordinate system with matrices

If the points $A, B,$ and C are defined as

$$A = \left(\begin{bmatrix} 3 & 3 \\ 3 & 3 \end{bmatrix}, \begin{bmatrix} 0 & 0 \\ 0 & 0 \end{bmatrix}, \begin{bmatrix} 0 & 0 \\ 0 & 0 \end{bmatrix} \right) \qquad B = \left(\begin{bmatrix} 0 & 0 \\ 0 & 0 \end{bmatrix}, \begin{bmatrix} 3 & 3 \\ 3 & 3 \end{bmatrix}, \begin{bmatrix} 0 & 0 \\ 0 & 0 \end{bmatrix} \right)$$

$$C = \left(\begin{bmatrix} 0 & 0 \\ 0 & 0 \end{bmatrix}, \begin{bmatrix} 0 & 0 \\ 0 & 0 \end{bmatrix}, \begin{bmatrix} 3 & 3 \\ 3 & 3 \end{bmatrix} \right) \tag{13.14}$$

Given

$$\xi_0 = \begin{bmatrix} 2 & 1 \\ 0 & 2 \end{bmatrix} \qquad \xi_1 = \begin{bmatrix} 1 & 2 \\ 2 & 0 \end{bmatrix} \tag{13.15}$$

It can be obtained

$$\lambda_1 = \lambda_0 - \xi_0 = \begin{bmatrix} 3 & 3 \\ 3 & 3 \end{bmatrix} - \begin{bmatrix} 2 & 1 \\ 0 & 2 \end{bmatrix} = \begin{bmatrix} 1 & 2 \\ 3 & 1 \end{bmatrix} \tag{13.16}$$

and

$$\lambda_2 = \lambda_1 - \xi_1 = \begin{bmatrix} 1 & 2 \\ 3 & 1 \end{bmatrix} - \begin{bmatrix} 1 & 2 \\ 2 & 0 \end{bmatrix} = \begin{bmatrix} 0 & 0 \\ 1 & 1 \end{bmatrix} \tag{13.17}$$

where $\xi_2 = \lambda_2$, the third coordinate number. The point is represented by

$$(\xi_0, \xi_1, \xi_2) = \left(\begin{bmatrix} 2 & 1 \\ 0 & 2 \end{bmatrix}, \begin{bmatrix} 1 & 2 \\ 2 & 0 \end{bmatrix}, \begin{bmatrix} 0 & 0 \\ 1 & 1 \end{bmatrix} \right) \tag{13.18}$$

Chapter 14

Field of the complex numbers

14.1 Generalization to the field of the complex numbers

The procedure used for the positive real numbers can be generalized to the complex numbers. In this case there is no more geometric representation. For example, assume $\lambda_0 = 4 + i$ and $\xi_0 = 1 + 3i$, then the first sublevel is given by $\lambda_1 = \lambda_0 - \xi_0 = (4 + i) - (1 + 3i) = 3 - 2i$. If $\xi_1 = 1 + i$, then $\lambda_2 = (3 - 2i) - (1 + i) = 2 - 3i$, and the coordinate of the point is given by $(1 + 3i, 1 + i, 2 - 3i)$.

The coordinate of the points C and A, in figure 13.2.2, are given by $(\lambda_0, 0, 0)$ and $(0, \lambda_0, 0)$, respectively. The first coordinate goes from 0 to λ_0 while the second goes from λ_0 to zero. This suggests the definition of a path, in the complex field, from 0 to λ_0 and take the coordinates along this path. A natural choice is a line connecting 0 to λ_0. As an example, consider $\lambda_0 = 4 + i$, then $\xi_0 = 2 + 0.5i$ is a point on that line. Therefore, $\lambda_1 = (4+i) - (2+0.5i) = 2+0.5i$. The following coordinate will be a point in the line connecting the points 0 and $\lambda_1 = 2 + 0.5i$, say $\xi_1 = 1 + 0.25i$ thus $\lambda_2 = (2+0.5i) - (1+0.25i) = 1+0.25i$. The point, in this coordinate system, is given by $(2+0.5i, 1+0.25i, 1+0.25i)$.

The complex number $a + bi$ can also be represented as (a, b). The set $L_4^1 = \{(4,0), (3,1), (2,2), (1,3), (0,4)\}$ can be considered as representing complex numbers. These are numbers along a line joining the points $(4,0)$ and $(0,4)$, that is, joining the numbers $4 + 0i$ and $0 + 4i$.

The usual graphic representation for a complex number is in a two-dimensional cartesian coordinate system where one axis represents the real part and the other the imaginary part as shown in figure 14.1.1(a). Since each number contains only one coordinate number they must be represented in a one-dimensional coordinate system as in the figure 14.1.1(b). Note that the line was used with the meaning of one-dimensional system, it not the real graphic representation.

Consider to obtain the Hermite Interpolating Polynomials related to the nodes $a_0 = -1 - i$, $a_1 = 1 - i$, $a_2 = -1 + i$, and $a_3 = 1 + i$, as shown in figure 14.1.1 with the property that the polynomials must interpolate up to second derivative.

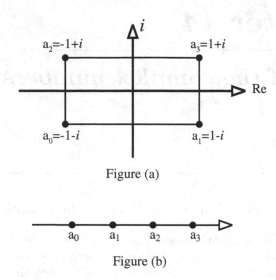

Figure (a)

Figure (b)

Figure 14.1.1 *Four points in an one-dimensional complex space.*

14.2 Generation of the polynomial $\Phi_{a_0,0}(x)$ related to the reference node $a_0 = -1 - i$

14.2.1 Elements common to all polynomials related to the node $a_0 = (-1 - i)$

a- Characterization of the polynomials

1. Type of domain: a one-dimensional complex space

2. Number of variables: $d = 1$.

3. Maximum order of derivative: $\eta = 2$

4. Reference node: $a_0 = \alpha = -1 - i$

5. Nodes in the visibility of the reference node

   The elements of the visibility set $\mathcal{V}$ of the reference node a_0 are

$$\beta_0 = a_1 = (\beta_{0,0}) = (1 - i) \qquad \beta_1 = a_2 = (\beta_{1,0}) = (-1 + i)$$
$$\beta_1 = a_3 = (\beta_{2,0}) = (1 + i) \tag{14.1}$$

6. The order of derivative of the Hermite Interpolating Polynomials related to the reference node are the constrained numbers in the set $L^\delta(\eta) = L^1(2)$, which are

$$L^1(2) = \{ (0), (1), (2) \} \tag{14.2}$$

The corresponding polynomials are denoted as

$$\Phi_{a_0,0}(x), \Phi_{a_0,1}(x), \Phi_{a_0,2}(x) \tag{14.3}$$

The polynomial $\Phi_{a_0,0}(x)$ has the property that when evaluated at the reference node $a_0 = (-1 - i)$ equals one. The first derivative of $\Phi_{a_0,1}(x)$ with respect to the variable x equals one at the reference node $a_0 = (-1 - i)$. The second derivative of $\Phi_{a_0,2}(x)$ with respect to the variable x equals one at the reference node $a_0 = (-1 - i)$.

b- Common factors

Factors that are common to all polynomials related to the reference node $a_0 = (-1, -i)$

1. Construction of $P(x)$

   The polynomial must assume the unit value at the reference node and zero at the other three nodes. For this case the general expression for the computation of $P_e(x)$ is given by (7.31), on page 297, which gives

   $$P_{a_0}(x) = \left[\, (x - (1 - i)) \, (x - (-1 + i)) \, (x - (1 + i)) \, \right]^3 \tag{14.4}$$

   where $a_0 = -1 - i$ is the reference node.

2. Construction of $P(\alpha)$

   $$\begin{aligned}
   P_{a_0}(x)\big|_{x=\alpha} &= \left[\, (x - (1 - i)) \, (x - (-1 + i)) \, (x - (1 + i)) \, \right]^3 \big|_{x=\alpha} \\
   &= \left[\, ((-1 - i) - (1 - i)) \, ((-1 - i) - (-1 + i)) \, ((-1 - i) - (1 + i)) \, \right]^3 \\
   &= \left(\, (-2) \, (-2i) \, (-2 - 2i) \, \right)^3 = (8 - 8i)^3 \\
   &= -1024 - 1024i \tag{14.5}
   \end{aligned}$$

c- Expression for the polynomial $Q_{a_0,n}(x)$

From equation (7.49) in theorem 7.2.17, on page 302, and considering that there is one variable only, it can be written, for $n = (n_0)$, the following

$$Q_{a_0,0}(x) = (x_0 - \alpha_0)^{n_0} \tag{14.6}$$

d- Expression for the polynomial $\Phi_{a_0,n}(x)$

The expression for the polynomial is given by equation (7.126), on page 315, which writes

$$\Phi_{a_0,n}(x) = \frac{1}{n! P_{a_0}(\alpha)} P_{a_0}(x) Q_{a_0,n}(x) f_{a_0,n}(a,x) \qquad (14.7)$$

where $n \in L^1(2)$.

14.2.2 Generation of the polynomial $\Phi_{a_0,0}(x)$ related to the reference node
$$a_0 = -1 - i$$

a- Properties of the polynomial $\Phi_{a_0,0}(x)$

1. Properties of the polynomial $\Phi_{a_0,0}(x)$

   (a) it equals one when evaluated at the reference node $a_0 = (-1 - i)$, that is,

   $$\Phi_{a_0,0}(x) \Big|_{x=a_0} = 1 \qquad (14.8)$$

   (b) it vanishes at the boundaries that do not intercept the reference node, which are the points $a_1 = (1 - i)$, $a_2 = (-1 + i)$, and $a_3 = (1 + i)$. Therefore, the property

   $$\Phi_{a_0,0}(x) \Big|_{x=a_1} = \Phi_{a_0,0}(x) \Big|_{x=a_2} = \Phi_{a_0,0}(x) \Big|_{x=a_3} = 0 \qquad (14.9)$$

2. Properties of the derivatives of the polynomial $\Phi_{a_0,0}(x)$

   (a) it has its derivatives until order $\eta = 2$ equal to zero at the reference node $a_0 = (-1 - i)$

   $$\Phi_{a_0,0}^{(k)}(x) \Big|_{x=a_0} = 0 \qquad k = 1, 2 \qquad (14.10)$$

   (b) it has the values of the derivatives until order $\eta = 2$ equal to zero at the boundaries that do not intercept the reference node, that is

   $$\Phi_{a_0,0}^{(k)}(x) \Big|_{x=a_1} = 0 \qquad \Phi_{a_0,0}^{(k)}(x) \Big|_{x=a_2} = 0$$
   $$\Phi_{a_0,0}^{(k)}(x) \Big|_{x=a_3} = 0 \qquad k = 1, 2 \qquad (14.11)$$

b- Construction of the constant term $n!P_{a_0}(\alpha)$

The factor $P_{a_0}(\alpha)$ was obtained in equation (14.5) and only $n!$ in equation (14.7) changes with the polynomial then, for $n = (0)$, it follows that

$$n!P_{a_0}(\alpha) = (n_0!)P_{a_0}(\alpha) = (0!)(-4)^3 i = (-4)^3 i \qquad (14.12)$$

c- Construction of the polynomial $Q_{a_0,n}(x)$

From theorem 7.2.17, on page 302, equation (7.49) and considering that there is only one variable, which allows to write $n = (n_0) = 0$ and that the coordinate number of the reference node is $\alpha_0 = a_0 = -1 - i$, then

$$Q_{a_0,0}(x) = (x_0 - \alpha_0)^{n_0} = (x - (-1 - i))^0 = 1 + 0i \qquad (14.13)$$

d- Expression for the polynomial $\Phi_{a_0,0}(x)$

Substituting $P_{a_0}(x)$ given by equation (14.4), $n!P_{a_0}(\alpha)$ given by equation (14.12), and $Q_{a_0,n}(x)$ given by equation (14.13) into equation (14.7) the polynomial $\Phi_{e,n}(x)$ writes

$$\Phi_{a_0,0}(x) = \frac{1}{n!P_{a_0}(\alpha)} P_{a_0}(x) Q_{a_0,n}(x) f_{a_0,n}(a,x)$$

$$= \frac{1}{0!P_{a_0}(-1-i)} \left[(x - (1-i))\,(x - (-1+i))\,(x - (1+i)) \right]^3 (1 + 0\,i)\, f_{a_0,0}(x)$$

$$(14.14)$$

14.2.3 Computation of the coefficients of the polynomial $f_{a_0,00}(x)$

a- Minimum degree of the polynomial $f_{a_0,0}(a,x)$

From corollary 7.2.19, on page 305, it follows that

$$q \geq \eta - \lambda_0(n) = 2 - \lambda_0(0) = 2 - 0 = 2 \qquad (14.15)$$

where $\lambda_0(n)$ is the value of the level of the constrained number $n = (0)$. A polynomial in one variables x and of degree two writes

$$f_{e,n}(a,x) = a_0 + a_1 x + a_2 x^2 \qquad (14.16)$$

b- Computation of the coefficients that belong to the terms in the level $\lambda_0(v) = 2$

1. Coefficients to be evaluated, which belong to the level $\lambda_0(v) = q = 2$.

   The indices of the coefficients of the terms of a polynomial elements v in the level $\lambda_0(v) = q = 2$ with one coordinate are the elements of the set L_q^δ of constrained numbers in the one-dimensional space given by $L_2^0 = \{(2)\}$, which means that $v \in L_2^0$. Therefore, the coefficient to be evaluated in this level is a_2 as can be seen from the equation (14.16).

2. Choice of the equation to be used for the computation of the coefficients.

   The coefficients of $f_{e,n}(a, x)$ are obtained using equation (7.122) or (7.123), on page 314, which is

   $$a_{(v)} = \frac{1}{v!} E_F^{(v)}(\alpha) = \frac{1}{v_0!} E_F^{(v)}(\alpha) \qquad \forall v \in L_{\lambda_0(v)}^\delta \tag{14.17}$$

   where

   $$v \in L_{\lambda_0(v)}^\delta = L_2^0 = \{(2)\} \tag{14.18}$$

   From (7.83),

   $$\left[\frac{1}{P(x)}\right]^{(v)} = \left[\frac{1}{P_0(x_0)}\right]^{(v_0)} = \left[E_F^{(v_0)}(x_0)\right]\left[\frac{1}{P(x)}\right] \tag{14.19}$$

   where

   $$\frac{1}{P_0(x_0)} = \frac{1}{P_{a_1}(x)} \frac{1}{P_{a_2}(x)} \frac{1}{P_{a_3}(x)} \tag{14.20}$$

   since $x_0 = x$. Then

   $$\left[\frac{1}{P_0(x_0)}\right]^{(v_0)} = [E_{a_1}(x)E_{a_2}(x)E_{a_3}(x)]^{(v_0)} \frac{1}{P(x)} \tag{14.21}$$

3. Computation of the derivative $[E_{a_1}(x)E_{a_2}(x)E_{a_3}(x)]^{(v_0)}$

   Performing the evaluation of equation (14.21) according to the rules in the subsection 6.10.2, on page 267, equation (6.197) it is necessary to obtain the elements in the set L_v^{t-1} where t is the number of functions in the product, therefore,

   $$L_v^{t-1} = L_2^2 = \{(2,0,0), (1,1,0), (1,0,1), (0,2,0), (0,1,1), (0,0,2)\} \tag{14.22}$$

   The elements of the set L_2^2 represent the order of derivative of the product $E_{a_1}(x)E_{a_2}(x)E_{a_3}(x)$ in equation (14.21).

   The computation of the elements of this set can be found in the table 2.8.19, on page 84 (generation of the elements of the set $L^2(2) = L_0^2 \cup$

$L_1^2 \cup L_2^2$ for $\mathbb{Z}^+$ as the underlying set), in the subsection 2.8.4, on page 77.

The coefficients of the expansion of the derivative of the product

$$E_{a_1}(x)E_{a_2}(x)E_{a_3}(x) \tag{14.23}$$

are evaluated similarly to the computation shown in the table 6.10.2, on page 270, and they are

$$F_u = \begin{bmatrix} 1\,2\,2\,1\,2\,1 \end{bmatrix} \tag{14.24}$$

then, the expansion

$$E_F^{(2)}(x) = E_{a_1}^{(2)}(x)E_{a_2}^{(0)}(x)E_{a_3}^{(0)}(x) + 2E_{a_1}^{(1)}(x)E_{a_2}^{(1)}(x)E_{a_3}^{(0)}(x) +$$
$$+ 2E_{a_1}^{(1)}(x)E_{a_2}^{(0)}(x)E_{a_3}^{(0)}(x) + E_{a_1}^{(0)}(x)E_{a_2}^{(2)}(x)E_{a_3}^{(0)}(x)$$
$$+ 2E_{a_1}^{(0)}(x)E_{a_2}^{(1)}(x)E_{a_3}^{(1)}(x) + E_{a_1}^{(0)}(x)E_{a_2}^{(0)}(x)E_{a_3}^{(2)}(x) \tag{14.25}$$

The computation of each term in equation (14.25) can be performed using the equation (6.196), on page 268, with $\eta = 2$ and $u \in L_2^2$ such as

$$E_{a_1}^{(2)}(x)E_{a_2}^{(0)}(x)E_{a_3}^{(0)}(x)$$
$$= \frac{(2+2)!}{2!}\frac{1}{((1-i)-(-1-i))^2} \times \frac{(2+0)!}{2!}\frac{1}{((-1+i)-(-1-i))^0} \times$$
$$\frac{(2+0)!}{2!}\frac{1}{((1+i)-(-1-i))^0} = \frac{12}{2^2} \times 1 \times 1 = 3 \tag{14.26}$$

$$E_{a_1}^{(1)}(x)E_{a_2}^{(1)}(x)E_{a_3}^{(0)}(x)$$
$$= \frac{(2+1)!}{2!}\frac{1}{((1-i)-(-1-i))^1} \times \frac{(2+1)!}{2!}\frac{1}{((-1+i)-(-1-i))^1} \times$$
$$\frac{(2+0)!}{2!}\frac{1}{((1+i)-(-1-i))^0} = \frac{12}{2^1} \times \frac{3}{(2i)^1} \times 1 = \frac{9}{4i} = -\frac{9i}{4} \tag{14.27}$$

an analogous computation gives

$$E_{a_1}^{(1)}(x)E_{a_2}^{(0)}(x)E_{a_3}^{(1)}(x) = \frac{9}{8}(1-i) \qquad E_{a_1}^{(0)}(x)E_{a_2}^{(2)}(x)E_{a_3}^{(0)}(x) = -3$$

$$E_{a_1}^{(0)}(x)E_{a_2}^{(1)}(x)E_{a_3}^{(1)}(x) = -\frac{9}{8}(1+i) \qquad E_{a_1}^{(0)}(x)E_{a_2}^{(0)}(x)E_{a_3}^{(2)}(x) = -\frac{3}{2}i \tag{14.28}$$

Substituting the values obtained in equations (14.26), (14.27), and (14.28) into (14.25) it gives

$$E_F^{(2)}(x) = 3 - 2 \times \frac{9}{4}i + 2 \times \frac{9}{8}(1-i) - 3 - 2 \times \frac{9}{8}(1+i) - \frac{3}{2}i = -\frac{21}{2}i \tag{14.29}$$

4. Computation of the coefficient a_2

Introducing the result in equation (14.29) into equation (14.17) it gives the coefficient

$$a_{(2)} = \frac{1}{2!} E_F^{(2)}(-1 - i) = \frac{1}{2!} \frac{-21}{2} i = -\frac{21}{4} i \qquad (14.30)$$

c- Computation of the coefficients that belong to the terms in the level $\lambda_0(v) = 1$

1. Coefficients to be evaluated, which belong to the level $\lambda_0 = 1$.

   The elements in the level $\lambda_0(v) = 1$ with one coordinate are the constrained numbers in the zero-dimensional space given by

   $$L_{\lambda_0(v)}^{\delta} = L_1^0 = \{(1)\} \qquad (14.31)$$

   Therefore, the coefficient $a_{(w)} = a_{(v)}$ to be evaluated in this level is $a_{(1)}$.

2. Equation for the computation of the coefficients.

   The coefficients are obtained using equation (7.123), on page 314, which is

   $$a_{(v)} = \frac{1}{v!} \left[E_F^{(v)}(\alpha) - S_v (a_{u+v}, \alpha) \right] \qquad \lambda_0(v) = 1 \qquad (14.32)$$

3. Computation of the term $S_v (a_{u+v}, \alpha)$.

   The degree of the polynomial $S_v (a_{u+v}, \alpha)$ is given by $m = q - \lambda_0(v) = 2 - 1 = 1$ where $\lambda_0(v) = v_0 = 2$. The exponents of the variables in each term of $S_v (a_{u+v}, \alpha)$ are the elements of the set shown in equation (7.124), on page 314, which is

   $$u \in (L^{\delta}(m) - L_0^{\delta}) = (L^0(1) - L_0^0) = \{(0), (1)\} - \{(0)\} = \{(1)\} \qquad (14.33)$$

   which permits to obtain the indices of the coefficients of the terms of the polynomial $S_v (a_{u+v}, \alpha)$ as follows

   For $v = (1)$ and $u = (1)$ then $v + u = (1) + (1) = (2)$ then $a_{(v+u)} = a_{(2)}$, which is the coefficient obtained for the level $\lambda_0(v) = q = 2$.

   The expression for $S_v (a_{u+v}, \alpha)$ is given by equation (7.112), on page 313. For $d = 1$ and $m = 1$ it writes

   $$S_v(a_{u+v}, \alpha) = \sum_{r_0=1}^{r_0=m} a_{(v+u)} \frac{(v + u)!}{u!} \alpha_0^{u_0}$$

   $$= \sum_{r_0=1}^{r_0=1} a_{(v+u)} \frac{(v + u)!}{u!} \alpha_0^{u_0} = a_{(2)} \frac{(2)!}{(1)!} \alpha_0^1 \qquad (14.34)$$

Evaluating the polynomial $S_v (a_{u+v}, \alpha)$ at the reference node $\alpha = (-1 - i)$ for $a_{(v)} = a_{(2)}$ it follows

$$S_{(v)}(a_{u+v}, \alpha) = S_{(1)}(a_2, (-1 - i))$$

$$= a_{(2)} \frac{(2)!}{(1)!} (-1 - i)^1$$

$$= -\frac{21}{4} i \times 2 \times (-1 - i) = \frac{21}{2}(-1 + i) \qquad (14.35)$$

4. Computation of the term $E_F^{(v)}(\alpha)$

$$E_F^{(1)}(x) = E_{a_1}^{(1)}(x)E_{a_2}^{(0)}(x)E_{a_3}^{(0)}(x) + E_{a_1}^{(0)}(x)E_{a_2}^{(1)}(x)E_{a_3}^{(0)}(x)$$
$$+ E_{a_1}^{(0)}(x)E_{a_2}^{(0)}(x)E_{a_3}^{(1)}(x) \qquad (14.36)$$

Performing the computation of each term and substituting into equation (14.36) it gives

$$E_F^{(1)}(x) = \frac{3}{2} - \frac{3}{2}i + \frac{3}{4}(1 - i) = \frac{9}{4}(1 - i) \qquad (14.37)$$

Substituting $S_{(v)}(a_{u+v}, \alpha)$ given by equation (14.35) and $E_F^{(1)}(x)$ given by equation (14.37) into equation (14.32) it gives the coefficient a_1 as

$$a_{(1)} = \frac{1}{1!} \left[\frac{9}{4}(1 - i) - \frac{21}{2}(-1 + i) \right] = \frac{51}{4}(1 - i) \qquad (14.38)$$

c- Computation of the coefficients that belong to the terms in the level $\lambda_0(v) = 0$

1. Coefficients to be evaluated, which belong to the level $\lambda_0 = 0$.

   The elements in the level $\lambda_0(v) = 1$ with one coordinate are the constrained numbers in the zero-dimensional space given by

$$L^{\delta}_{\lambda_0(v)} = L_1^0 = \{(0)\} \qquad (14.39)$$

   Therefore, the coefficient $a_{(w)} = a_{(v)}$ to be evaluated in this level is $a_{(0)}$.

2. Equation for the computation of the coefficients.

   The coefficients are obtained using equation (7.123), on page 314, which is

$$a_{(v)} = \frac{1}{v!} \left[E_F^{(v)}(\alpha) - S_v (a_{u+v}, \alpha) \right] \qquad \lambda_0(v) = 0 \qquad (14.40)$$

3. Computation of the term $S_v(a_{u+v}, \alpha)$

The degree of the polynomial $S_v(a_{u+v}, \alpha)$ is given by $m = q - \lambda_0(v) = 2-0 = 2$. The exponents of the variables in each term of $S_v(a_{u+v}, \alpha)$ are the elements of the set shown in equation (7.124), on page 314, which is

$$u \in (L^\delta(m) - L_0^\delta) = (L^0(2) - L_0^0)$$
$$= \{(0), (1), (2)\} - \{(0)\} = \{(1), (2)\} \qquad (14.41)$$

which permits to obtain the indices of the coefficients of the terms of the polynomial $S_v(a_{u+v}, \alpha)$ as follows

For $v = (0)$ and $u = (1)$ then $v + u = (0) + (1) = (1)$ then $a_{(v+u)} = a_{(1)}$, which is the coefficient obtained for the level $\lambda_0(v) = 1$.

For $v = (0)$ and $u = (2)$ then $v + u = (0) + (2) = (2)$ then $a_{(v+u)} = a_{(2)}$, which is the coefficient obtained for the level $\lambda_0(v) = 2$.

The expression for $S_v(a_{u+v}, \alpha)$ is given by equation (7.112). For $d = 1$ and $m = 2$ it writes

$$S_v(a_{u+v}, \alpha) = \sum_{r_0=1}^{r_0=m} a_{(v+u)} \frac{(v+u)!}{u!} \alpha_0^{u_0} = \sum_{r_0=1}^{r_0=2} a_{(v+u)} \frac{(v+u)!}{u!} \alpha_0^{u_0}$$

$$= a_{(1)} \frac{(1)!}{(1)!} \alpha_0^1 + a_{(2)} \frac{(2)!}{(2)!} \alpha_0^2$$

$$= \frac{51}{4}(1-i) \times (-1-i)^1 - \frac{21}{4}i \times (-1-i)^2 = -15 \quad (14.42)$$

4. Computation of the term $E_F^{(v)}(\alpha)$

$$E_F^{(0)}(x) = 1 + 0i \qquad (14.43)$$

5. Computation of the coefficient $a_{(0,0)}$, for which $v = (0,0)$

Substituting $S_{(v)}(a_{u+v}, \alpha)$ given by equation (14.42) and $E_F^{(1)}(x)$ given by equation (14.43) into equation (14.40) it gives the coefficient a_1 as

$$a_{(0)} = \frac{1}{0!}[1 + 0i - (-15)] = -16 \qquad (14.44)$$

14.2.4   Polynomials related to the node $-1 - i$

$$\Phi_{a_0,0}(x) = \frac{1}{0!(-1024 - 1024i)}\left[(x - (1-i))(x - (-1+i))(x - (1+i))\right]^3 \times$$

$$(x - (-1-i))^0\left(-16 + \frac{51}{4}(1-i)x - \frac{21}{4}ix^2\right) \qquad (14.45)$$

$$\Phi_{a_0,1}(x) = \frac{1}{1!\,(-1024 - 1024i)} \big[\, (x - (1 - i))\,(x - (-1 + i))\,(x - (1 + i))\,\big]^3 \times$$
$$(x - (-1 - i))^1 \big(\frac{11}{2} + \frac{9}{4}(1 - i)x\big) \tag{14.46}$$

$$\Phi_{a_0,2}(x) = \frac{1}{2!\,(-1024 - 1024i)} \big[\, (x - (1 - i))\,(x - (-1 + i))\,(x - (1 + i))\,\big]^3 \times$$
$$(x - (-1 - i))^1 \big(1 + 0i\big) \tag{14.47}$$

14.3 Polynomial $\Phi_{a_1,0}(x)$ related to the reference node $a_1 = 1 - i$

$$\Phi_{a_1,0}(x) = \frac{1}{0!\,(1024 - 1024i)} \big[\, (x - (-1 - i))\,(x - (-1 + i))\,(x - (1 + i))\,\big]^3 \times$$
$$(x - (1 - i))^0 \big(16 + \frac{51}{4}(-1 - i)x + \frac{21}{4}ix^2\big) \tag{14.48}$$

$$\Phi_{a_1,0}(x) = \frac{1}{1!\,(1024 - 1024i)} \big[\, (x - (-1 - i))\,(x - (-1 + i))\,(x - (1 + i))\,\big]^3 \times$$
$$(x - (1 - i))^1 \big(\frac{11}{2} + \frac{9}{4}(-1 - i)x\big) \tag{14.49}$$

$$\Phi_{a_1,0}(x) = \frac{1}{2!\,(1024 - 1024i)} \big[\, (x - (-1 - i))\,(x - (-1 + i))\,(x - (1 + i))\,\big]^3 \times$$
$$(x - (1 - i))^2 \big(1\big) \tag{14.50}$$

14.4   Polynomial $\Phi_{a_2,0}(x)$ related to the reference node $a_2 = -1 + i$

$$\Phi_{a_2,0}(x) = \frac{1}{0!\,(-1024 + 1024i)} \big[\, (x - (-1 - i))\,(x - (1 - i))\,(x - (1 + i)) \, \big]^3 \times$$
$$(x - (1 - i))^0 \Big(16 + \frac{51}{4}(1 + i)x + \frac{21}{4}ix^2 \Big) \qquad (14.51)$$

$$\Phi_{a_2,0}(x) = \frac{1}{1!\,(-1024 + 1024i)} \big[\, (x - (-1 - i))\,(x - (1 - i))\,(x - (1 + i)) \, \big]^3 \times$$
$$(x - (1 - i))^1 \Big(\frac{11}{2} + \frac{9}{4}(1 + i)x \Big) \qquad (14.52)$$

$$\Phi_{a_2,0}(x) = \frac{1}{2!\,(-1024 + 1024i)} \big[\, (x - (-1 - i))\,(x - (1 - i))\,(x - (1 + i)) \, \big]^3 \times$$
$$(x - (1 - i))^2 \, (1) \qquad (14.53)$$

14.5   Polynomial $\Phi_{a_3,0}(x)$ related to the reference node $a_3 = 1 + i$

$$\Phi_{a_3,0}(x) = \frac{1}{0!\,(-1024 + 1024i)} \big[\, (x - (-1 - i))\,(x - (1 - i))\,(x - (-1 + i)) \, \big]^3 \times$$
$$(x - (1 - i))^0 \Big(16 + \frac{51}{4}(1 + i)x - \frac{21}{4}ix^2 \Big) \qquad (14.54)$$

$$\Phi_{a_3,0}(x) = \frac{1}{1!\,(-1024 + 1024i)} \big[\, (x - (-1 - i))\,(x - (1 - i))\,(x - (-1 + i)) \, \big]^3 \times$$
$$(x - (1 - i))^1 \Big(\frac{11}{2} + \frac{9}{4}(-1 + i)x \Big) \qquad (14.55)$$

$$\Phi_{a_3,0}(x) = \frac{1}{2!\,(-1024 + 1024i)} \big[\, (x - (-1 - i))\,(x - (1 - i))\,(x - (-1 + i)) \, \big]^3 \times$$
$$(x - (1 - i))^2 \, (1) \qquad (14.56)$$

Chapter 15

Analysis of the behavior of the Hermite Interpolating Polynomials

15.1 One dimensional Hermite Interpolating Polynomials

15.1.1 The set of functions

The set of one-dimensional $d = 1$ Hermite Interpolating Polynomials such that for the derivatives of order up to $\eta = 4$, there is one polynomial that equals to one when evaluated at the reference node $a = (-1)$ is given by

$$\Phi_{a,0}^{(0)}(x) = \left[\frac{1}{0!} \frac{1}{(-2)^5} \right] (x-1)^5 \frac{1}{2^4} (256 + 650x + 690x^2 + 350x^3 + 70x^4) \quad (15.1)$$

The polynomial $\Phi_{a,0}^{(0)}(x)$ equals one at the reference node $a = (-1)$ and zero at the reference node $b = (1)$.

$$\Phi_{a,1}^{(0)}(x) = \left[\frac{1}{1!} \frac{1}{(-2)^5} \right] (x-1)^5 (x+1) \frac{1}{2^3} (93 + 185x + 135x^2 + 35x^3) \quad (15.2)$$

The first derivative of the polynomial $\Phi_{a,1}^{(0)}(x)$ equals one at the reference node $a = (-1)$ and zero at the reference node $b = (1)$.

$$\Phi_{a,2}^{(0)}(x) = \left[\frac{1}{2!} \frac{1}{(-2)^5} \right] (x-1)^5 (x+1)^2 \frac{1}{2^2} (29 + 40x + 15x^2) \quad (15.3)$$

The second derivative of the polynomial $\Phi_{a,2}^{(0)}(x)$ equals one at the reference node $a = (-1)$ and zero at the reference node $b = (1)$.

$$\Phi_{a,3}^{(0)}(x) = \left[\frac{1}{3!} \frac{1}{(-2)^5} \right] (x-1)^5 (x+1)^3 \frac{1}{2^1} (7 + 5x) \quad (15.4)$$

The third derivative of the polynomial $\Phi_{a,3}^{(0)}(x)$ equals one at the reference node $a = (-1)$ and zero at the reference node $b = (1)$.

$$\Phi_{a,4}^{(0)}(x) = \left[\frac{1}{4!} \frac{1}{(-2)^5} \right] (x-1)^5 (x+1)^4 \frac{1}{2^0} (1) \tag{15.5}$$

The fourth derivative of the polynomial $\Phi_{a,4}^{(0)}(x)$ equals one at the reference node $a = (-1)$ and zero at the reference node $b = (1)$.

The Hermite Interpolating Polynomial related to the node $b = (1)$ whose coordinates are $\gamma = (1)$, $\delta = (-1)$ may be obtained from the polynomial related to the node $a = (-1)$, for which $\beta = (1)$ using the theorem 7.4.13. After the transformation of the coefficients the following polynomials are obtained

$$\Phi_{b,0}^{(0)}(x) = \frac{1}{0!} \frac{1}{2^5} \frac{1}{2^4} (x-1)^5 (256 - 650x + 690x^2 - 350x^3 + 70x^4) \tag{15.6}$$

The polynomial $\Phi_{b,0}^{(0)}(x)$ equals one at the reference node $b = (1)$ and zero at the reference node $a = (-1)$.

$$\Phi_{b,1}^{(0)}(x) = \frac{1}{1!} \frac{1}{2^5} \frac{1}{2^3} (x-1)^5 (x+1)(93 - 185x + 135x^2 - 35x^3) \tag{15.7}$$

The first derivative of the polynomial $\Phi_{b,1}^{(0)}(x)$ equals one at the reference node $b = (1)$ and zero at the reference node $a = (-1)$.

$$\Phi_{b,2}^{(0)}(x) = \frac{1}{2!} \frac{1}{2^5} \frac{1}{2^2} (x-1)^5 (x+1)^2 (29 - 40x + 15x^2) \tag{15.8}$$

The second derivative of the polynomial $\Phi_{b,2}^{(0)}(x)$ equals one at the reference node $b = (1)$ and zero at the reference node $a = (-1)$.

$$\Phi_{b,3}^{(0)}(x) = \frac{1}{3!} \frac{1}{2^5} \frac{1}{2^1} (x-1)^5 (x+1)^3 (7 - 5x) \tag{15.9}$$

The third derivative of the polynomial $\Phi_{b,3}^{(0)}(x)$ equals one at the reference node $b = (1)$ and zero at the reference node $a = (-1)$.

$$\Phi_{b,4}^{(0)}(x) = \frac{1}{4!} \frac{1}{2^5} \frac{1}{2^0} (x-1)^5 (x+1)^4 (1) \tag{15.10}$$

The fourth derivative of the polynomial $\Phi_{b,4}^{(0)}(x)$ equals one at the reference node $b = (1)$ and zero at the reference node $a = (-1)$.

15.2 Graphical representation of the Hermite Interpolating Polynomials $\Phi_{a,n}^{(\mu)}(x)$ for $\eta = 4$

15.2.1 Graphic of $\Phi_{a,0}(x)$ and its derivatives

Figures 15.2.1 and 15.2.2 show the graphic of the polynomial in equation (15.1), that is, of the polynomial

$$\Phi_{a,0}^{(0)}(x) = \frac{1}{0!} \frac{1}{2^5} \frac{1}{2^4} (x - 1)^5 (256 + 650x + 690x^2 + 350x^3 + 70x^4) \quad (15.11)$$

and their derivatives of order up to $\mu = 4$. This polynomial satisfies the property $\Phi_{a,0}^{(0)}(x)|_{x=-1} = 1$.

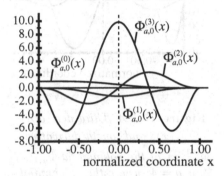
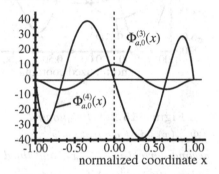

Figure 15.2.1 *Polynomial $\Phi_{a,0}^{(0)}(x)$, first derivative $\Phi_{a,0}^{(1)}(x)$, second derivative $\Phi_{a,0}^{(2)}(x)$, and third derivative $\Phi_{a,0}^{(3)}(x)$ for $\eta = 4$, where the polynomial $\Phi_{a,0}^{(0)}(x)$ equals one when evaluated at the reference node $x = -1$.*

Figure 15.2.2 *Third derivative $\Phi_{a,0}^{(3)}(x)$, and fourth derivative $\Phi_{a,0}^{(4)}(x)$ of the polynomial $\Phi_{a,0}(x)$ for $\eta = 4$, where the polynomial $\Phi_{a,0}^{(0)}(x)$ equals one when evaluated at the reference node $x = -1$.*

The polynomials $\Phi_{a,0}^{(1)}(x)$ and $\Phi_{a,0}^{(3)}(x)$ have horizontal symmetry with respect to the axis through the X coordinate 0.00. The polynomials $\Phi_{a,0}^{(2)}(x)$ and $\Phi_{a,0}^{(4)}(x)$ have symmetry with respect to the origin, that is, the point $0.00, 0$.

15.2.2   Graphic of $\Phi_{a,1}(x)$ and its derivatives

Figures 15.2.3 and 15.2.4 show the graphic of the polynomial in equation (15.2), that is, of the polynomial

$$\Phi_{a,1}^{(0)}(x) = \frac{1}{1!}\frac{1}{2^5}\frac{1}{2^3}(x-1)^5(x+1)(93+185x+135x^2+35x^3) \qquad (15.12)$$

and their derivatives of order up to $\mu = 4$. This polynomial satisfies the property $\Phi_{a,1}^{(1)}(x)|_{x=-1} = 1$.

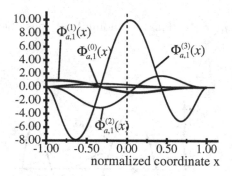

Figure 15.2.3 *Polynomial*
$\Phi_{a,1}^{(0)}(x)$, *first derivative* $\Phi_{a,1}^{(1)}(x)$,
second derivative $\Phi_{a,1}^{(2)}(x)$, *and third*
derivative $\Phi_{a,1}^{(3)}(x)$ *for* $\eta = 4$ *and*
first derivative of the polynomial
equals one when evaluated at
$x = -1$.

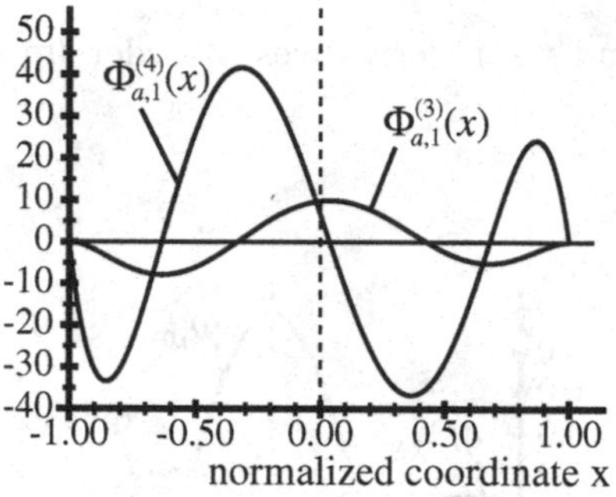

Figure 15.2.4 *Third derivative*
$\Phi_{a,1}^{(3)}(x)$, *and fourth derivative*
$\Phi_{a,1}^{(4)}(x)$ *of the polynomial* $\Phi_{a,1}(x)$
for $\eta = 4$, *where the polynomial*
$\Phi_{a,1}^{(1)}(x)$ *equals one when evaluated*
at the reference node $x = -1$.

15.2.3   Graphic of the polynomial $\Phi_{a,2}(x)$ and its derivatives

Figures 15.2.5 and 15.2.6 show the graphic of the polynomial in equation (15.3), that is, of the polynomial

$$\Phi_{a,2}^{(0)}(x) = \frac{1}{2!}\frac{1}{2^5}\frac{1}{2^2}(x-1)^5(x+1)^2(29+40x+15x^2) \qquad (15.13)$$

and their derivatives of order up to $\mu = 4$. This polynomial satisfies the property $\Phi_{a,2}^{(2)}(x)|_{x=-1} = 1$.

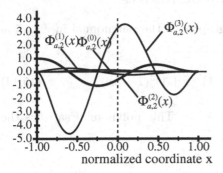

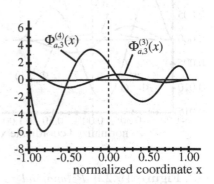

Figure 15.2.5 *Polynomial*
$\Phi_{a,2}^{(0)}(x)$, *first derivative* $\Phi_{a,2}^{(1)}(x)$,
second derivative $\Phi_{a,2}^{(2)}(x)$, *and third*
derivative $\Phi_{a,2}^{(3)}(x)$ *for* $\eta = 4$ *and*
second derivative of the polynomial
equals one when evaluated at
$x = -1$.

Figure 15.2.6 *Third derivative*
$\Phi_{a,2}^{(3)}(x)$, *and fourth derivative*
$\Phi_{a,2}^{(4)}(x)$ *of the polynomial* $\Phi_{a,2}(x)$
for $\eta = 4$, *where the polynomial*
$\Phi_{a,2}^{(2)}(x)$ *equals one when evaluated*
at the reference node $x = -1$.

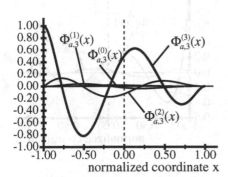

Figure 15.2.7 *Polynomial*
$\Phi_{a,3}^{(0)}(x)$, *first derivative* $\Phi_{a,3}^{(1)}(x)$,
second derivative $\Phi_{a,3}^{(2)}(x)$, *and third*
derivative $\Phi_{a,3}^{(3)}(x)$ *for* $\eta = 4$ *and*
third derivative of the function
equals one when evaluated at
$x = -1$.

Figure 15.2.8 *Third derivative*
$\Phi_{a,3}^{(3)}(x)$, *and fourth derivative*
$\Phi_{a,3}^{(4)}(x)$ *of the polynomial* $\Phi_{a,3}(x)$
for $\eta = 4$, *where the polynomial*
$\Phi_{a,3}^{(3)}(x)$ *equals one when evaluated*
at the reference node $x = -1$.

15.2.4   Graphic of $\Phi_{a,3}(x)$ and its derivatives

Figures 15.2.7 and 15.2.8 show the graphic of the polynomial in equation (15.4), that is, of the polynomial

$$\Phi_{a,3}^{(0)}(x) = \frac{1}{3!}\frac{1}{2^5}\frac{1}{2^1}(x-1)^5(x+1)^3(7+5x) \tag{15.14}$$

and their derivatives of order up to $u = 4$. This polynomial satisfies the property $\Phi_{a,3}^{(3)}(x)|_{x=-1} = 1$.

15.2.5   Graphic of the polynomial $\Phi_{a,4}(x)$ and its derivatives

Figures 15.2.9 and 15.2.10 show the graphic of the function in equation (15.4), that is, of the function

$$\Phi_{a,4}^{(0)}(x) = \frac{1}{4!}\frac{1}{2^5}\frac{1}{2^0}(x-1)^5(x+1)^3 \tag{15.15}$$

and their derivatives of order up to $u = 3$. This function satisfies the property $\Phi_{a,4}^{(4)}(x) = 1$.

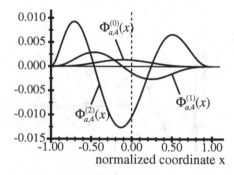

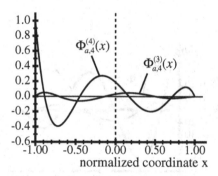

Figure 15.2.9 *Polynomial* $\Phi_{a,4}^{(0)}(x)$, *first derivative* $\Phi_{a,4}^{(1)}(x)$, *second derivative* $\Phi_{a,4}^{(2)}(x)$, *and third derivative* $\Phi_{a,4}^{(3)}(x)$ *for* $\eta = 4$ *and third derivative of the function equals one when evaluated at* $x = -1$.

Figure 15.2.10 *Third derivative* $\Phi_{a,4}^{(3)}(x)$, *and fourth derivative* $\Phi_{a,4}^{(4)}(x)$ *of the polynomial* $\Phi_{a,4}(x)$ *for* $\eta = 4$, *where the polynomial* $\Phi_{a,4}^{(4)}(x)$ *equals one when evaluated at the reference node* $x = -1$.

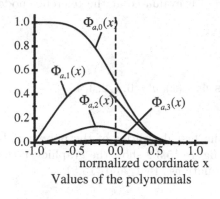

Figure 15.2.11 *Polynomials* $\Phi_{a,0}(x)$, $\Phi_{a,1}(x)$, $\Phi_{a,2}(x)$, *and* $\Phi_{a,3}(x)$ *related to the reference node* -1.

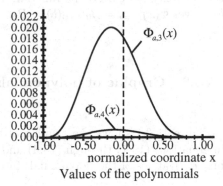

Figure 15.2.12 *Polynomials* $\Phi_{a,3}(x)$, *and* $\Phi_{a,4}(x)$ *related to the reference node* -1.

15.3 Graphic of polynomials $\Phi_{a,n}^{(\mu)}(x), n = 0, 1, 2, 3, 4$

15.3.1 Graphic of polynomials $\Phi_{a,n}^{(0)}(x), n = 0, 1, 2, 3, 4$

Figure 15.2.11 shows the graphic of the polynomials in equations (15.1) through (15.4) and figure 15.2.12 shows the graphics of the polynomials in equations (15.4) and (15.5)

Figures 15.2.11 and 15.2.12 show that for the same value of x

$$\Phi_{a,0}(x) \geq \Phi_{a,1}(x) \geq \Phi_{a,2}(x) \geq \Phi_{a,3}(x) \geq \Phi_{a,4}(x) \qquad (15.16)$$

that is, the polynomials $\Phi_{a,1}(x)$, $\Phi_{a,2}(x)$, $\Phi_{a,3}(x)$, and $\Phi_{a,4}(x)$ are bounded from above by the polynomial $\Phi_{a,0}(x)$. The same is true for the polynomials related to the reference node $b = 1$.

A function expanded using as basis the polynomials in equations from (15.1) until (15.10) writes

$$\Phi(x) = A_0\Phi_{a,0}(x) + A_1\Phi_{a,1}(x) + A_2\Phi_{a,2}(x) + A_3\Phi_{a,3}(x) + A_4\Phi_{a,4}(x)$$
$$+ B_0\Phi_{b,0}(x) + B_1\Phi_{b,1}(x) + B_2\Phi_{b,2}(x) + B_3\Phi_{b,3}(x) + B_4\Phi_{b,4}(x)$$
$$(15.17)$$

If it is evaluated at the reference point $a = (-1)$ it gives $\Phi(x)|_{x=a} = A_0\Phi_{a,0}(a) = A_0$ since $\Phi_{a,0}(a) = 1$ and all others polynomial vanishes when evaluated at the reference node a. If the function $\Phi(x)$ is evaluated at the reference node b it gives $\Phi(x)|_{x=b} = B_0\Phi_{b,0}(b) = B_0$.

15.3.2   Graphic of polynomials $\Phi_{a,n}^{(1)}(x), n = 0, 1, 2, 3, 4$

Figure 15.3.1 shows the graphic of the first derivative of the polynomials in equations (15.1) through (15.4) and figure 15.3.2 show the graphics of the first derivatives of the polynomials in equations (15.4) and (15.5)

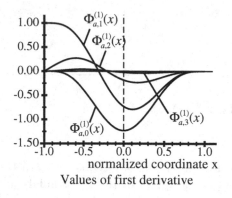

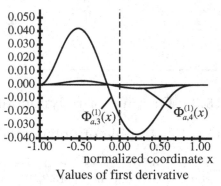

Figure 15.3.1 *Polynomials*
$\Phi_{a,0}^{(1)}(x), \Phi_{a,1}^{(1)}(x), \Phi_{a,2}^{(1)}(x), and$
$\Phi_{a,3}^{(1)}(x)$ *related to the reference node* -1.

Figure 15.3.2 *Polynomials*
$\Phi_{a,3}^{(1)}(x), and \Phi_{a,4}^{(1)}(x)$ *related to the reference node* -1.

15.3.3   Graphic of the polynomials $\Phi_{a,n}^{(2)}(x), n = 0, 1, 2, 3, 4$

Figure 15.3.3 shows the graphic of the second derivative of the polynomials in equations (15.1) through (15.4) and figure 15.3.4 show the graphics of the second derivatives of the polynomials in equations (15.4) and (15.5)

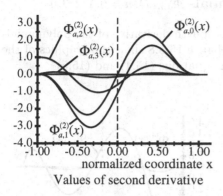

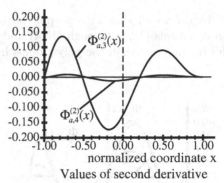

Figure 15.3.3 *Functions $\Phi_{a,0}^{(2)}(x)$, $\Phi_{a,1}^{(2)}(x)$, $\Phi_{a,2}^{(2)}(x)$, and $\Phi_{a,3}^{(2)}(x)$ related to the reference node -1.*

Figure 15.3.4 *Functions $\Phi_{a,3}^{(2)}(x)$, and $\Phi_{a,4}^{(2)}(x)$ related to the reference node -1.*

15.3.4   Graphic of the polynomials $\Phi_{a,n}^{(3)}(x), n = 0, 1, 2, 3, 4$

Figure 15.3.5 shows the graphic of the third derivative of the polynomials in equations (15.1) through (15.4) and figure 15.3.6 show the graphics of the third derivatives of the polynomials in equations (15.4) and (15.5)

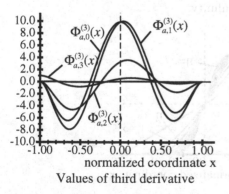

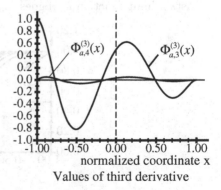

Figure 15.3.5 *Functions $\Phi_{a,0}^{(3)}(x)$, $\Phi_{a,1}^{(3)}(x)$, $\Phi_{a,2}^{(3)}(x)$, and $\Phi_{a,3}^{(3)}(x)$ related to the reference node -1.*

Figure 15.3.6 *Functions $\Phi_{a,3}^{(3)}(x)$, and $\Phi_{a,4}^{(3)}(x)$ related to the reference node -1.*

15.3.5   Graphic of the polynomials $\Phi_{a,n}^{(4)}(x), n = 0, 1, 2, 3, 4$

Figure 15.3.7 shows the graphic of the fourth derivative of the polynomials in equations (15.1) through (15.4) and figure 15.3.8 show the graphics of the fourth derivatives of the polynomials in equations (15.4) and (15.5)

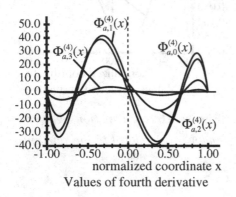

Values of fourth derivative

Values of fourth derivative

Figure 15.3.7 *Functions $\Phi_{a,0}^{(4)}(x)$, $\Phi_{a,1}^{(4)}(x)$, $\Phi_{a,2}^{(4)}(x)$, and $\Phi_{a,3}^{(4)}(x)$ related to the reference node -1.*

Figure 15.3.8 *Functions $\Phi_{a,2}^{(4)}(x)$, and $\Phi_{a,3}^{(4)}(x)$ related to the reference node -1.*

15.3.6   Graphic of the polynomials $\Phi_{a,i}(x), i = 0, 3, \ldots, 27$

Figure 15.3.9 shows the behavior of the one dimensional Hermite Interpolating Polynomials for increasing values of η, which corresponds to increasing values of the degree p of the polynomial $\Phi_{a,i}^{(0)}(x)$. It is conjectured that there exists a limit function as η goes to infinity.

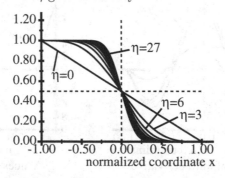

normalized coordinate x

Figure 15.3.9 $\Phi_{a,0}^{(0)}(x)$ *for $\eta = 0, 3, 6, \ldots, 27$. The function has the property $\Phi_{a,0}^{(0)}(x) = 1$ when evaluated at $x = -1$.*

15.3.7 Graphic of the first derivatives of the polynomials $\Phi_{a,i}^{(0)}(x), i = 3, 6, \ldots, 27$

The same characteristic of a limit function can be seen for the behavior of the first and second derivatives in figures 15.3.10 and 15.3.11.

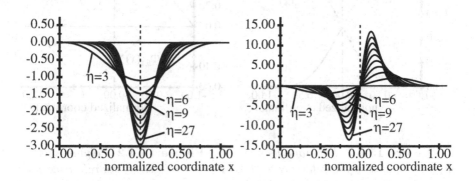

normalized coordinate x   normalized coordinate x

Figure 15.3.10 *First order derivatives of* $\Phi_{a,0}^{(0)}(x)$ *for* $\eta = 3, 6, 9, \ldots, 27$. *The function has the property* $\Phi_{a,0}^{(0)}(x) = 1$ *when evaluated at* $x = -1$.

Figure 15.3.11 *Region defined by a set of lines connecting the vertices* a_k. *The point* a_e *represents the reference node.*

15.3.8 Graphic of the second derivatives of the polynomials $\Phi_{a,i}^{(0)}(x), i = 3, 6, \ldots, 27$

Figure 15.3.11 shows the graphic of the second derivatives for values of $\eta = 3, 4, \ldots, 27$. The graphics show that the curves converges as η increases.

15.4 Graphical representation of the Hermite Interpolating Polynomials $\Phi_{a,n}^{(\mu)}(x)$ and $\Phi_{b,n}^{(\mu)}(x)$ for $n = 0, 1, 2, 3, 4$

Below a comparison is made for the same polynomial, as well as for their derivatives, related to the reference node $a = (-1)$ and $b = (1)$ where the degree of the polynomials $f_{a,n}$ and $f_{b,n}$ are given by $q = \eta - \lambda_0(n)$ and $\eta = 4$.

15.4.1   Graphic of the polynomials $\Phi_{a,n}^{(0)}(x)$ and $\Phi_{b,n}^{(0)}(x)$ for $n = 0, 1, 2, 3, 4$

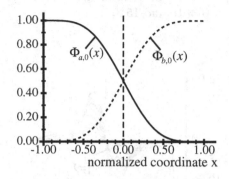

Figure 15.4.1 *Polynomials $\Phi_{a,0}(x)$ related to the reference node $a = -1$, polynomials $\Phi_{b,0}(x)$ related to the reference node $b = 1$.*

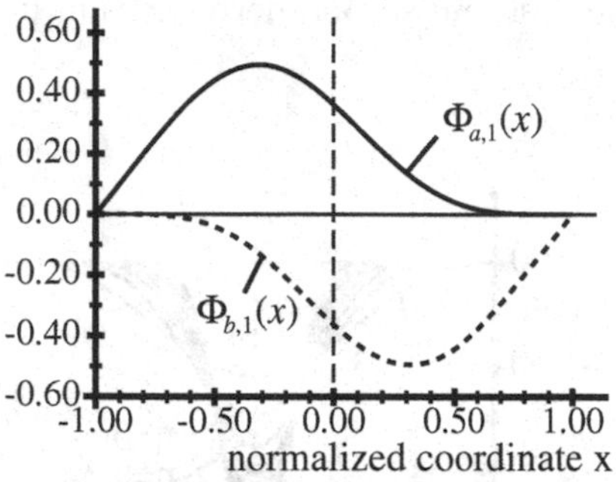

Figure 15.4.2 *Polynomials $\Phi_{a,1}(x)$ related to the reference node $a = -1$, polynomials $\Phi_{b,1}(x)$ related to the reference node $b = 1$.*

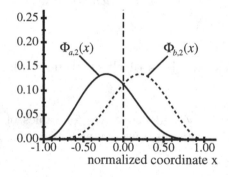

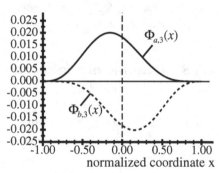

Figure 15.4.3 *Polynomials $\Phi_{a,2}(x)$ related to the reference node $a = -1$, polynomials $\Phi_{b,2}(x)$ related to the reference node $b = 1$.*

Figure 15.4.4 *Polynomials $\Phi_{a,3}(x)$ related to the reference node $a = -1$, polynomials $\Phi_{b,3}(x)$ related to the reference node $b = 1$.*

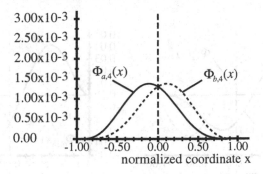

Figure 15.4.5 *Polynomials* $\Phi_{a,4}(x)$ *related to the reference node*
$a = -1$, *polynomials* $\Phi_{b,4}(x)$ *related to the reference node* $b = 1$, *and*
their sum $\Phi_{a,4}(x) + \Phi_{b,4}(x)$.

It can be observed in figures 15.4.2, 15.4.3, 15.4.4, and 15.4.5 that the
derivatives whose order is an even integer are symmetric with respect to a
vertical line through the origin $x = 0$ while the derivatives whose order is an
odd integer are symmetric with respect to the origin. These properties are
not true if the degree of the polynomials $f_{a,0}(a, x)$ and $f_{b,0}(a, x)$ are greater
than the minimum value, that is, if $q > \eta - \lambda_0(n)$ as shown in figure 15.5.2 for
the first derivative of the polynomials $\Phi_{a,0}(x)$ and $\Phi_{b,0}(x)$. In figure 15.5.3 for
the second derivative. In figure 15.5.4 for the third derivative. And in figure
15.5.5 for the fourth derivative

15.4.2   Graphic of the polynomials $\Phi_{a,n}^{(1)}(x)$ and $\Phi_{b,n}^{(1)}(x)$ for $n = 0, 1, 2, 3, 4$

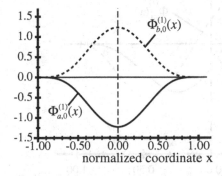
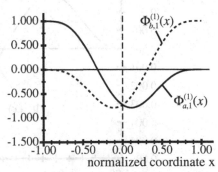

Figure 15.4.6 *Polynomials*
$\Phi_{a,0}^{(1)}(x)$ *related to the reference node*
$a = -1$, *polynomials* $\Phi_{b,0}^{(1)}(x)$ *related*
to the reference node $b = 1$.

Figure 15.4.7 *Polynomials*
$\Phi_{a,1}^{(1)}(x)$ *related to the reference node*
$a = -1$, *polynomials* $\Phi_{b,1}^{(1)}(x)$ *related*
to the reference node $b = 1$.

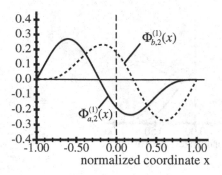

Figure 15.4.8 *Polynomials* $\Phi_{a,2}^{(1)}(x)$ *related to the reference node* $a = -1$, *polynomials* $\Phi_{b,2}^{(1)}(x)$ *related to the reference node* $b = 1$.

Figure 15.4.9 *Polynomials* $\Phi_{a,3}^{(1)}(x)$ *related to the reference node* $a = -1$, *polynomials* $\Phi_{b,3}^{(1)}(x)$ *related to the reference node* $b = 1$.

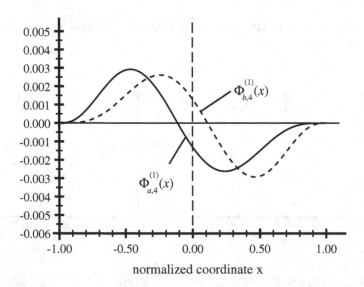

Figure 15.4.10 *Polynomials* $\Phi_{a,4}^{(1)}(x)$ *related to the reference node* $a = -1$, *polynomials* $\Phi_{b,4}^{(1)}(x)$ *related to the reference node* $b = 1$.

15.4.3   Graphic of the polynomials $\Phi_{a,n}^{(2)}(x)$ and $\Phi_{b,n}^{(2)}(x)$ for $n = 0, 1, 2, 3, 4$

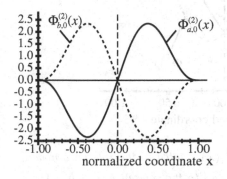

Figure 15.4.11 *Polynomials $\Phi_{a,0}^{(2)}(x)$ related to the reference node $a = -1$, polynomials $\Phi_{b,0}^{(2)}(x)$ related to the reference node $b = 1$.*

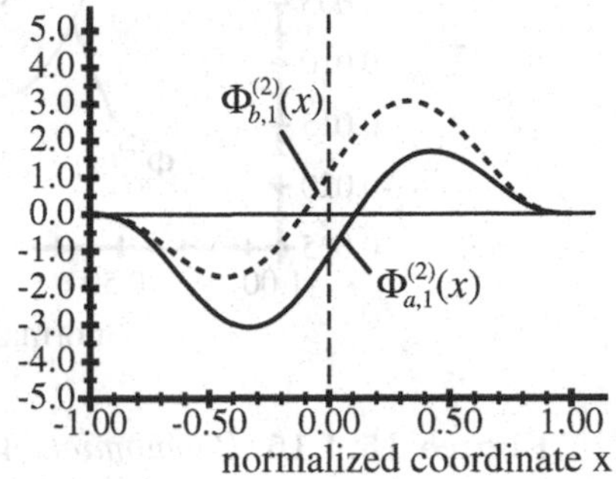

Figure 15.4.12 *Polynomials $\Phi_{a,1}^{(2)}(x)$ related to the reference node $a = -1$, polynomials $\Phi_{b,1}^{(2)}(x)$ related to the reference node $b = 1$.*

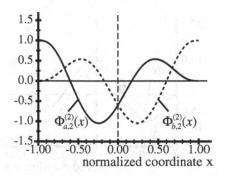

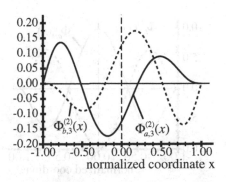

Figure 15.4.13 *Polynomials $\Phi_{a,2}^{(2)}(x)$ related to the reference node $a = -1$, polynomials $\Phi_{b,2}^{(2)}(x)$ related to the reference node $b = 1$.*

Figure 15.4.14 *Polynomials $\Phi_{a,3}^{(2)}(x)$ related to the reference node $a = -1$, polynomials $\Phi_{b,3}^{(2)}(x)$ related to the reference node $b = 1$.*

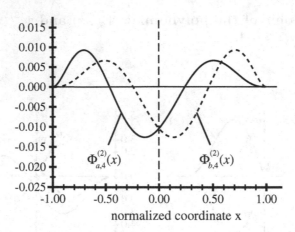

Figure 15.4.15 *Polynomials* $\Phi_{a,4}^{(2)}(x)$ *related to the reference node* $a = -1$, *polynomials* $\Phi_{b,4}^{(2)}(x)$ *related to the reference node* $b = 1$.

15.4.4   Graphic of the polynomials $\Phi_{a,n}^{(3)}(x)$ and $\Phi_{b,n}^{(3)}(x)$ for $n = 0, 1, 2, 3, 4$

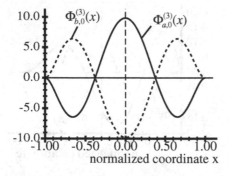

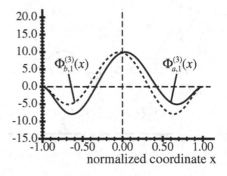

Figure 15.4.16 *Polynomials* $\Phi_{a,0}^{(3)}(x)$ *related to the reference node* $a = -1$, *polynomials* $\Phi_{b,0}^{(3)}(x)$ *related to the reference node* $b = 1$.

Figure 15.4.17 *Polynomials* $\Phi_{a,1}^{(3)}(x)$ *related to the reference node* $a = -1$, *polynomials* $\Phi_{b,1}^{(3)}(x)$ *related to the reference node* $b = 1$.

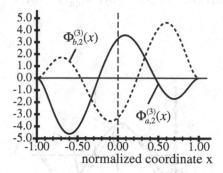

Figure 15.4.18 *Polynomials* $\Phi_{a,2}^{(3)}(x)$ *related to the reference node* $a = -1$, *polynomials* $\Phi_{b,2}^{(3)}(x)$ *related to the reference node* $b = 1$.

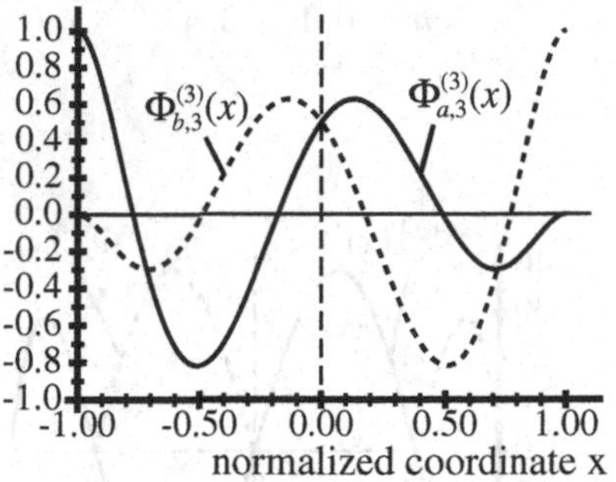

Figure 15.4.19 *Polynomials* $\Phi_{a,3}^{(3)}(x)$ *related to the reference node* $a = -1$, *polynomials* $\Phi_{b,3}^{(3)}(x)$ *related to the reference node* $b = 1$.

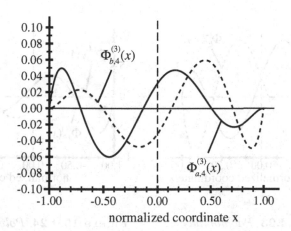

Figure 15.4.20 *Polynomials* $\Phi_{a,4}^{(3)}(x)$ *related to the reference node* $a = -1$, *polynomials* $\Phi_{b,4}^{(3)}(x)$ *related to the reference node* $b = 1$.

15.4.5   Graphic of the polynomials $\Phi_{a,n}^{(4)}(x)$ and $\Phi_{b,n}^{(4)}(x)$ for $n = 0, 1, 2, 3, 4$

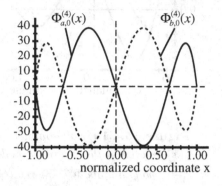

Figure 15.4.21 *Polynomials $\Phi_{a,0}^{(4)}(x)$ related to the reference node $a = -1$, polynomials $\Phi_{b,0}^{(4)}(x)$ related to the reference node $b = 1$.*

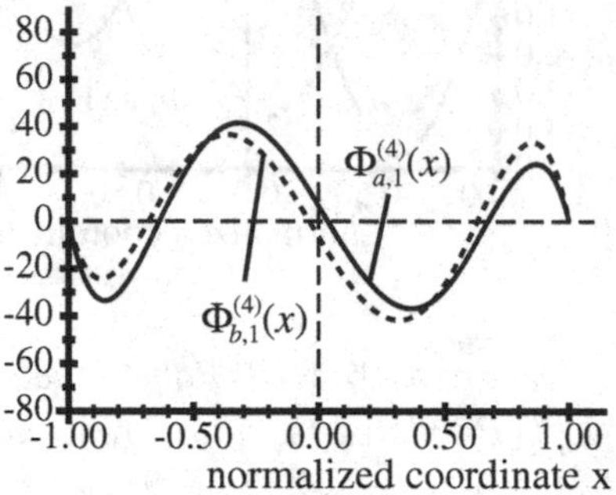

Figure 15.4.22 *Polynomials $\Phi_{a,1}^{(4)}(x)$ related to the reference node $a = -1$, polynomials $\Phi_{b,1}^{(4)}(x)$ related to the reference node $b = 1$.*

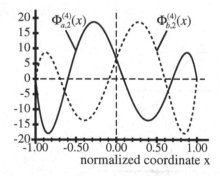

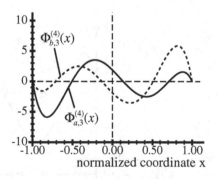

Figure 15.4.23 *Polynomials $\Phi_{a,2}^{(4)}(x)$ related to the reference node $a = -1$, polynomials $\Phi_{b,2}^{(4)}(x)$ related to the reference node $b = 1$.*

Figure 15.4.24 *Polynomials $\Phi_{a,3}^{(4)}(x)$ related to the reference node $a = -1$, polynomials $\Phi_{b,3}^{(4)}(x)$ related to the reference node $b = 1$.*

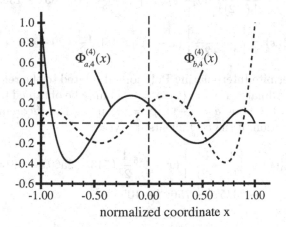

Figure 15.4.25 *Polynomials*
$\Phi_{a,4}^{(4)}(x)$ *related to the reference node*
$a = -1$, *polynomials* $\Phi_{b,4}^{(4)}(x)$ *related*
to the reference node $b = 1$.

15.5 One-dimensional modified Hermite Interpolating Polynomials

15.5.1 The set of functions

The set of one-dimensional $d = 1$ modified Hermite Interpolating Polynomials with derivatives of order up to $\eta = 4$, where $f_{e,n}(a, x)$ has degree $q = \eta - \lambda_0(n) + 2 = 6$, and related to the reference node $a = (-1)$, is given by

$$\Phi_{a,0}^{(0)}(x) = \left[\frac{1}{0!} \frac{1}{(-2)^5} \right] (x - 1)^5 (x + 1)^0 \frac{1}{2^5} (743 + 2560x + 4215x^2$$
$$+ 4060x^3 + 2345x^4 + 756x^5 + 105x^6) \tag{15.18}$$

$$\Phi_{a,1}^{(0)}(x) = \left[\frac{1}{1!} \frac{1}{(-2)^5} \right] (x - 1)^5 (x + 1)^1 \frac{1}{2^4} (319 + 965x + 1320x^2$$
$$+ 980x^3 + 385x^4 + 63x^5) \tag{15.19}$$

$$\Phi_{a,2}^{(0)}(x) = \left[\frac{1}{2!} \frac{1}{(-2)^5} \right] (x - 1)^5 (x + 1)^2 \frac{1}{2^3} (128 + 325x + 345x^2$$
$$+ 175x^3 + 35x^4) \tag{15.20}$$

$$\Phi_{a,3}^{(0)}(x) = \left[\frac{1}{3!}\frac{1}{(-2)^5}\right](x-1)^5(x+1)^3\frac{1}{2^3}\left(93+185x+135x^2+35x^3\right) \quad (15.21)$$

$$\Phi_{a,4}^{(0)}(x) = \left[\frac{1}{4!}\frac{1}{(-2)^5}\right](x-1)^5(x+1)^4\frac{1}{2^2}\left(29+40x+15x^2\right) \quad (15.22)$$

The Hermite Interpolating Polynomial related to the reference node $b = (1)$ whose coordinates are $\gamma = (1)$, $\delta = (-1)$ may be obtained from the polynomial related to the node $a = (-1)$, $\beta = (1)$ using the theorem 7.4.13. After the transformation of the coefficients it gives

$$\Phi_{b,0}^{(0)}(x) = \left[\frac{1}{0!}\frac{1}{(-2)^5}\right](x-1)^5\frac{1}{2^5}(743 - 2560x + 4215x^2 - 4060x^3$$
$$+ 2345x^4 - 756x^5 + 105x^6) \quad (15.23)$$

$$\Phi_{b,1}^{(0)}(x) = \left[\frac{1}{1!}\frac{1}{(-2)^5}\right](x-1)^5(x+1)^1\frac{1}{2^4}(319 - 965x + 1320x^2 - 980x^3$$
$$+ 385x^4 - 63x^5) \quad (15.24)$$

$$\Phi_{b,2}^{(0)}(x) = \left[\frac{1}{2!}\frac{1}{(-2)^5}\right](x-1)^5(x+1)^2\frac{1}{2^3}\left(128 - 325x + 345x^2 - 175x^3 + 35x^4\right)$$
$$\quad (15.25)$$

$$\Phi_{b,3}^{(0)}(x) = \left[\frac{1}{3!}\frac{1}{(-2)^5}\right](x-1)^5(x+1)^3\frac{1}{2^3}\left(93 - 185x + 135x^2 - 35x^3\right) \quad (15.26)$$

$$\Phi_{b,4}^{(0)}(x) = \left[\frac{1}{4!}\frac{1}{(-2)^5}\right](x-1)^5(x+1)^4\frac{1}{2^2}\left(29 - 40x + 15x^2\right) \quad (15.27)$$

The one-dimensional $d = 1$ modified Hermite Interpolating Polynomial $\Phi_{a,0}^{(0)}(x)$ related to the reference node $a = (-1)$, with derivatives of order up to $\eta = 4$ and the degree of polynomial $f_{e,n}(a,x)$ equals to $q = \eta - \lambda_0(n) + 4 = 8$, is given by

$$\Phi_{a,0}^{(0)}(x) = \left[\frac{1}{0!}\frac{1}{(-2)^5}\right](x-1)^5\frac{1}{2^8}\left(7099 + 29060x + 61440x^2 + 83300x^3$$
$$+ 76510x^4 + 47628x^5 + 19320x^6 + 4620x^7 + 495x^8\right) \quad (15.28)$$

For the same case but with the degree the polynomial $f_{e,n}(a,x)$ equals to $q = \eta - \lambda_0(n) + 6 = 10$, it follows

$$\Phi_{a,0}^{(0)}(x) = \left[\frac{1}{0!}\frac{1}{(-2)^5}\right](x-1)^5\frac{1}{2^{10}}\left(30827 + 139120x + 342285x^2$$
$$+ 573440x^3 + 696430x^4 + 622944x^5 + 407610x^6 + 190080x^7$$
$$+ 59895x^8 + 11440x^9 + 1001x^{10}\right) \quad (15.29)$$

For the same case but with the degree the polynomial $f_{e,n}(a,x)$ equals to $q = \eta - \lambda_0(n) + 8 = 12$, it is given by

$$\Phi_{a,0}^{(0)}(x) = \left[\frac{1}{0!}\frac{1}{(-2)^5}\right](x-1)^5\frac{1}{2^{11}}\big(63929 + 304175x + 819705x^2$$
$$+ 1572305x^3 + 2293760x^4 + 2597238x^5 + 2286690x^6 + 1551330x^7$$
$$+ 795465x^8 + 298155x^9 + 77077x^{10} + 12285x^{11} + 910x^{12}\big) \quad (15.30)$$

The figures 15.5.1 through 15.5.5 compares the modified Hermite Interpolating Polynomials with $\eta = 4$ and for degrees of $f_{a,0}(x)$ and $f_{b,0}(x)$ equal to $q = 4, 6, 8, 10, 12$ corresponding to $\Delta q = 0, 2, 4, 6, 8$.

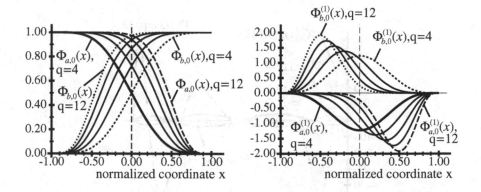

Figure 15.5.1 *Polynomials $\Phi_{a,0}(x)$ related to the reference node $a = -1$, polynomials $\Phi_{b,0}(x)$ related to the reference node $b = 1$, where the degrees of $f_{a,0}(x)$ and $f_{b,0}(x)$ are $q = 4, 6, 8, 10, 12$.*

Figure 15.5.2 *Polynomials $\Phi_{a,0}^{(1)}(x)$ related to the reference node $a = -1$, polynomials $\Phi_{b,0}^{(1)}(x)$ related to the reference node $b = 1$, where the degrees of $f_{a,0}(x)$ and $f_{b,0}(x)$ are $q = 4, 6, 8, 10, 12$.*

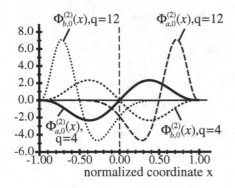

Figure 15.5.3 *Polynomials*
$\Phi_{a,0}^{(2)}(x)$ *related to the reference node*
$a = -1$, *polynomials* $\Phi_{b,0}^{(2)}(x)$ *related*
to the reference node $b = 1$, *where*
the degrees of $f_{a,0}(x)$ *and* $f_{b,0}(x)$
are $q = 4$ *and* $q = 12$.

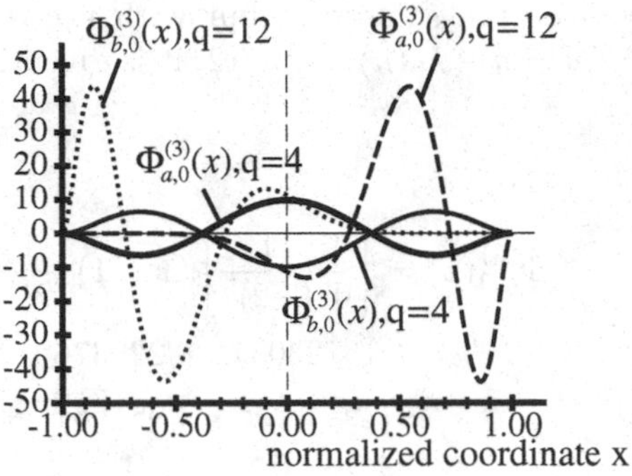

Figure 15.5.4 *Polynomials*
$\Phi_{a,0}^{(3)}(x)$ *related to the reference node*
$a = -1$, *polynomials* $\Phi_{b,0}^{(3)}(x)$ *related*
to the reference node $b = 1$, *where*
the degrees of $f_{a,0}(x)$ *and* $f_{b,0}(x)$
are $q = 4$ *and* $q = 12$.

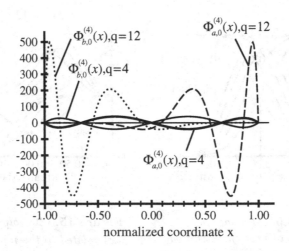

Figure 15.5.5 *Polynomials* $\Phi_{a,1}^{(4)}(x)$ *related to the reference node*
$a = -1$, *polynomials* $\Phi_{b,1}^{(4)}(x)$ *related to the reference node* $b = 1$ *for*
the degree of $f_{a,1}(x)$ *equal to* $q = 4$ *and* $q = 12$.

15.6   Analysis of the error for selected degrees of $f_{e,n}(a,x)$

15.6.1   Approximation to a line

Consider the line $y = x$ in the interval $[-1,1]$. Figure 15.6.1 gives the graph of the Modified Hermite Interpolating Polynomials for $\eta = 4$ and the polynomial $f_{e,n}(a,x)$ with the degrees $q = \eta - \lambda_0(n) + \Delta q$ where $\Delta q = 0, 2, 4, 6,$ and 8 evaluated using

$$\Phi(x) = -\Phi_{a,0}(x) + \Phi_{a,1}(x) + \Phi_{b,0}(x) + \Phi_{b,1}(x) \tag{15.31}$$

Figure 15.6.1 shows the graphs of the equation (15.31) for the augmentation of the degree of the polynomial $f_{e,n}(a,x)$ of $\Delta q = 0, 2, 4, 6, 8$.

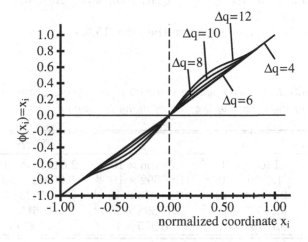

Figure 15.6.1 *Approximation of the function $y = x$ by $\Phi(x)$ for the degrees $q = \eta - \lambda_0(n) + \Delta q$ of the polynomial $f_{e,n}(a,x)$ where $\Delta q = 0, 2, 4, 6, 8$.*

The root mean square of the error for each degree of $f_{e,n}(a,x)$ was computed over 201 points equally spaced along the domain as

$$\sigma^2 = \frac{1}{201} \sum_{i=0}^{200} R^2(x_i) \tag{15.32}$$

where $R(x_i) = \Phi(x_i) - x_i$. The graphs of $R(x_i)$ for the degrees $q = 4, 6, 8, 10, 12$ are shown in figure 15.6.2

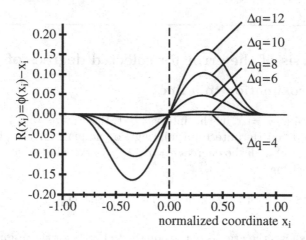

Figure 15.6.2 *Distribution of the error* $R = \Phi(x_i) - x_i$ *for* $f_{e,n}(a, x)$ *with the degrees* $q = \eta - \lambda_0(n) + \Delta q$ *where* $\Delta q = 0, 2, 4, 6, 8$.

The values of the errors are given in the table 15.6.3

Table 15.6.3 *Root mean square error* σ, *mean square error* σ^2, *and sum of the square of the errors along the domain* $\sum R^2$.

degree	σ	σ^2	$\sum R^2$
4	1.14659×10^{-15}	1.31466×10^{-30}	2.64247×10^{-28}
6	4.4350×10^{-3}	1.96692×10^{-05}	3.95351×10^{-3}
8	2.75314×10^{-2}	7.57976×10^{-4}	0.152353
10	6.15547×10^{-2}	3.78898×10^{-3}	0.761585
12	9.78752×10^{-2}	9.57955×10^{-3}	1.92549

It can be seen from figures 15.6.1 and 15.6.2 that the best approximation for the equation $y = x$ is obtained for $q = \eta - \lambda_0(n)$, which implies $\Delta q = 0$.

15.7 Selected examples of Hermite Interpolating Polynomials

15.7.1 One-dimensional Hermite Interpolating Polynomial

Computation of the Hermite Interpolating Polynomials for (1) $d = 1$ variables, (2) maximum order of the derivative $\eta = 3$, (3) associated to the reference node e, whose coordinate number is $\alpha = (-1)$ and $\beta = (1)$. The degree of the polynomials $\Phi_{ei}(x)$ is given by $p = d(\eta + 1) + \eta = 1(3 + 1) + 3 = 7$.

$$\Phi_{a,0}(x) = \frac{1}{0!} \frac{1}{2^4} \frac{1}{2^3} (x-1)^4 (64 + 116x + 80x^2 + 20x^3) \tag{15.33}$$

$$\Phi_{a,1}(x) = \frac{1}{1!} \frac{1}{2^4} \frac{1}{2^2} (x-1)^4 (x+1)(22 + 28x + 10x^2) \tag{15.34}$$

$$\Phi_{a,2}(x) = \frac{1}{2!} \frac{1}{2^4} \frac{1}{2^1} (x-1)^4 (x+1)^2 (6 + 4x) \tag{15.35}$$

$$\Phi_{a3}(x) = \frac{1}{3!} \frac{1}{2^4} \frac{1}{2^0} (x-1)^4 (x+1)^3 (1) \tag{15.36}$$

Where $1/P_e(\alpha) = 1/2^4$. The factor $1/2$ is used to write the coefficients of $f_{ei}(x)$ as integer numbers.

15.7.2 Two-dimensional Hermite Interpolating Polynomial

Computation of the Hermite Interpolating Polynomials, with $d = 2$ variables, for continuity up to derivative of order $\eta = 3$, degree $p = 11$, reference node $\alpha = (-1, -1)$ and where $\beta = (1, 1)$.

$$\Phi_{a,00}(x,y) = \left[\frac{1}{0!0!} \frac{1}{2^4} \frac{1}{2^4} \right] (x-1)^4 (y-1)^4 \frac{1}{2^1} \left(58 + 67x + 67y \right.$$
$$\left. + 30x^2 + 48xy + 30y^2 + 5x^3 + 10x^2y + 10xy^2 + 5y^3 \right) \tag{15.37}$$

$$\Phi_{a,10}(x,y) = \left[\frac{1}{1!0!} \frac{1}{2^4} \frac{1}{2^4} \right] (x-1)^4 (y-1)^4 (x+1) \frac{1}{2^1} \left(28 + 22x + 22y \right.$$
$$\left. + 5x^2 + 8xy + 5y^2 \right) \tag{15.38}$$

$$\Phi_{a,01}(x,y) = \left[\frac{1}{0!1!} \frac{1}{2^4} \frac{1}{2^4} \right] (x-1)^4 (y-1)^4 (y+1) \frac{1}{2^1} \left(28 + 22x + 22y \right.$$
$$\left. + 5x^2 + 8xy + 5y^2 \right) \tag{15.39}$$

$$\Phi_{a,20}(x,y) = \left[\frac{1}{2!0!}\frac{1}{2^4}\frac{1}{2^4}\right](x-1)^4(x+1)^2(y-1)^4\frac{1}{2^0}(5+2x+2y) \quad (15.40)$$

$$\Phi_{a,11}(x,y) = \left[\frac{1}{1!1!}\frac{1}{2^4}\frac{1}{2^4}\right]\frac{1}{2^9}(x-1)^4(x+1)(y-1)^4(y+1)(5+2x+2y) \quad (15.41)$$

$$\Phi_{a,02}(x,y) = \left[\frac{1}{0!2!}\frac{1}{2^4}\frac{1}{2^4}\right](x-1)^4(y-1)^4(y+1)^2\frac{1}{2^0}(5+2x+2y) \quad (15.42)$$

$$\Phi_{a,30}(x,y) = \left[\frac{1}{0!3!}\frac{1}{2^4}\frac{1}{2^4}\right](x-1)^4(x+1)^3(y-1)^4(1) \quad (15.43)$$

$$\Phi_{a,21}(x,y) = \left[\frac{1}{2!1!}\frac{1}{2^4}\frac{1}{2^4}\right](x-1)^4(x+1)^2(y-1)^4(y+1)(1) \quad (15.44)$$

$$\Phi_{a,12}(x,y) = \left[\frac{1}{1!2!}\frac{1}{2^4}\frac{1}{2^4}\right](x-1)^4(x+1)(y-1)^4(y+1)^2(1) \quad (15.45)$$

$$\Phi_{a,03}(x,y) = \left[\frac{1}{0!3!}\frac{1}{2^4}\frac{1}{2^4}\right](x-1)^4(y-1)^4(y+1)^3(1) \quad (15.46)$$

15.7.3   Three-dimensional Hermite Interpolating Polynomial

15.7.4   Summary of the polynomials $\Phi_{e,n}^{(n_0,n_1,n_2)}$ related to the node $a_e = (-1,-1,-1)$

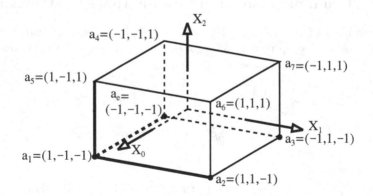

Figure 15.7.1 *Three-dimensional cartesian space for a three-variable polynomial.*

Computation of the Hermite Interpolating Polynomials, in $d = 3$ variables, for maximum derivative of order $\eta = 3$, associate to the node e whose coordinate numbers are $\alpha = (-1,-1,-1)$ and where $\beta = (1,1,1)$ as shown in figure

15.7.1.

$$\Phi_{a,000}(x,y,z) = \frac{1}{0!0!0!} \left[\frac{1}{2^4} \frac{1}{2^4} \frac{1}{2^4} \right] (x-1)^4(y-1)^4(z-1)^4 \frac{1}{2^1}$$

$$\times \Big(144 + 121x + 121y + 121z + 40x^2 + 64xy + 64xz$$

$$+ 40y^2 + 64yz + 40z^2 + 5x^3 + 10x^2y + 10x^2z$$

$$+ 10xy^2 + 16xyz + 10xz^2 + 5y^3 + 10y^2z$$

$$+ 10yz^2 + 5z^3 \Big) \tag{15.47}$$

$$\Phi_{a,100}(x,y,z) = \frac{1}{1!0!0!} \left[\frac{1}{2^4} \frac{1}{2^4} \frac{1}{2^4} \right] (x-1)^4(x+1)(y-1)^4(z-1)^4 \frac{1}{2^1}$$

$$\times \Big(53 + 30x + 30y + 30z + 5x^2 + 8xy + 8xz$$

$$+ 5y^2 + 8yz + 5z^2 \Big) \tag{15.48}$$

$$\Phi_{a,010}(x,y,z) = \frac{1}{0!1!0!} \left[\frac{1}{2^4} \frac{1}{2^4} \frac{1}{2^4} \right] (x-1)^4(y-1)^4(y+1)(z-1)^4$$

$$\times \Big(53 + 30x + 30y + 30z + 5x^2 + 8xy + 8xz$$

$$+ 5y^2 + 8yz + 5z^2 \Big) \tag{15.49}$$

$$\Phi_{a,001}(x,y,z) = \frac{1}{0!0!1!} \left[\frac{1}{2^4} \frac{1}{2^4} \frac{1}{2^4} \right] (x-1)^4(y-1)^4(z-1)^4(z+1)\frac{1}{2^1}$$

$$\times \Big(53 + 30x + 30y + 30z + 5x^2 + 8xy + 8xz$$

$$+ 5y^2 + 8yz + 5z^2 \Big) \tag{15.50}$$

$$\Phi_{a,200}(x,y,z) = \frac{1}{2!0!0!} \left[\frac{1}{2^4} \frac{1}{2^4} \frac{1}{2^4} \right] (x-1)^4(x+1)^2(y-1)^4(z-1)^4$$

$$\times \frac{1}{2^1} (14 + 4x + 4y + 4z) \tag{15.51}$$

$$\Phi_{a,110}(x,y,z) = \frac{1}{1!1!0!} \left[\frac{1}{2^4} \frac{1}{2^4} \frac{1}{2^4} \right] (x-1)^4(x+1)(y-1)^4(y+1)(z-1)^4$$

$$\times \frac{1}{2^1} (14 + 4x + 4y + 4z) \tag{15.52}$$

$$\Phi_{a,101}(x,y,z) = \frac{1}{1!0!1!} \left[\frac{1}{2^4} \frac{1}{2^4} \frac{1}{2^4} \right] (x-1)^4(x+1)(y-1)^4(z-1)^4(z+1)$$

$$\times \frac{1}{2^1} (14 + 4x + 4y + 4z) \tag{15.53}$$

$$\Phi_{a0,20}(x, y, z) = \frac{1}{0!2!0!} \left[\frac{1}{2^4}\frac{1}{2^4}\frac{1}{2^4}\right] (x-1)^4(y-1)^4(y+1)^2(z-1)^4$$
$$\times \frac{1}{2^1}(14 + 4x + 4y + 4z) \tag{15.54}$$

$$\Phi_{a,011}(x, y, z) = \frac{1}{0!1!1!} \left[\frac{1}{2^4}\frac{1}{2^4}\frac{1}{2^4}\right] (x-1)^4(y-1)^4(y+1)(z-1)^4(z+1)$$
$$\times \frac{1}{2^1}(14 + 4x + 4y + 4z) \tag{15.55}$$

$$\Phi_{a,002}(x, y, z) = \frac{1}{0!0!2!} \left[\frac{1}{2^4}\frac{1}{2^4}\frac{1}{2^4}\right] (x-1)^4(y-1)^4(z-1)^4(z+1)^2$$
$$\times \frac{1}{2^1}(14 + 4x + 4y + 4z) \tag{15.56}$$

$$\Phi_{a,300}(x, y, z) = \frac{1}{3!0!0!} \left[\frac{1}{2^4}\frac{1}{2^4}\frac{1}{2^4}\right] (x-1)^4(x+1)^3(y-1)^4(z-1)^4 \tag{15.57}$$

$$\Phi_{a,210}(x, y, z) = \frac{1}{2!1!0!} \left[\frac{1}{2^4}\frac{1}{2^4}\frac{1}{2^4}\right] (x-1)^4(x+1)^2(y-1)^4(y+1)(z-1)^4 \tag{15.58}$$

$$\Phi_{a,201}(x, y, z) = \frac{1}{2!0!1!} \left[\frac{1}{2^4}\frac{1}{2^4}\frac{1}{2^4}\right] (x-1)^4(x+1)^2(y-1)^4(z-1)^4(z+1) \tag{15.59}$$

$$\Phi_{a,120}(x, y, z) = \frac{1}{1!2!0!} \left[\frac{1}{2^4}\frac{1}{2^4}\frac{1}{2^4}\right] (x-1)^4(x+1)(y-1)^4(y+1)^2(z-1)^4 \tag{15.60}$$

$$\Phi_{a,111}(x, y, z) = \frac{1}{1!1!1!} \left[\frac{1}{2^4}\frac{1}{2^4}\frac{1}{2^4}\right] (x-1)^4(x+1)(y-1)^4(y+1)(z-1)^4(z+1) \tag{15.61}$$

$$\Phi_{a,102}(x, y, z) = \frac{1}{1!0!2!} \left[\frac{1}{2^4}\frac{1}{2^4}\frac{1}{2^4}\right] (x-1)^4(x+1)(y-1)^4(z-1)^4(z+1)^2 \tag{15.62}$$

$$\Phi_{a,030}(x, y, z) = \frac{1}{0!3!0!} \left[\frac{1}{2^4}\frac{1}{2^4}\frac{1}{2^4}\right] (x-1)^4(y-1)^4(y+1)^3(z-1)^4 \tag{15.63}$$

$$\Phi_{a,021}(x, y, z) = \frac{1}{0!2!1!} \left[\frac{1}{2^4}\frac{1}{2^4}\frac{1}{2^4}\right] (x-1)^4(y-1)^4(y+1)^2(z-1)^4(z+1) \tag{15.64}$$

$$\Phi_{a,102}(x, y, z) = \frac{1}{1!0!2!} \left[\frac{1}{2^4}\frac{1}{2^4}\frac{1}{2^4}\right] (x-1)^4(y-1)^4(y+1)(z-1)^4(z+1)^2 \tag{15.65}$$

$$\Phi_{a,003}(x, y, z) = \frac{1}{0!0!3!} \left[\frac{1}{2^4}\frac{1}{2^4}\frac{1}{2^4}\right] (x-1)^4(y-1)^4(z-1)^4(z+1)^3 \tag{15.66}$$

15.7.5 Summary of the polynomials $\Phi_{e,n}^{(n_0,n_n,n_2)}$ related to the node $a_e = (-1,-1,-1)$ and $\eta = 4$

For this case, $d = 3$ variables, for continuity of the function, and derivatives of order up to $\eta = 4$, associate to the node e whose coordinate numbers are $\alpha = (-1,-1,-1)$.

$$
\begin{aligned}
\Phi_{a,000}(x,y,z) = {} & \frac{1}{0!0!0!}\left[\frac{1}{2^5}\frac{1}{2^5}\frac{1}{2^5}\right](x-1)^5(y-1)^5(z-1)^5 \\
& \times \frac{1}{2^4}\Big(5036 + 5800x + 5800y + 5800z \\
& + 2865x^2 + 4775xy + 4775xz + 2865y^2 + 4775yz + 2865z^2 \\
& + 700x^3 + 1500x^2y + 1500x^2z + 1500xy^2 + 2500xyz \\
& + 1500xz^2 + 700y^3 + 1500y^2z + 1500yz^2 + 700z^3 \\
& + 70x^4 + 175x^3y + 175x^3z + 225x^2y^2 + 375x^2yz+ \\
& + 225x^2z^2 + 175xy^3 + 375xy^2z + 375xyz^2 + 175xz^3 \\
& + 70y^4 + 175y^3z + 225y^2z^2 + 175yz^3 + 70z^4\Big)
\end{aligned}
\tag{15.67}
$$

$$
\begin{aligned}
\Phi_{a,100}(x,y,z) = {} & \frac{1}{1!0!0!}\left[\frac{1}{2^5}\frac{1}{2^5}\frac{1}{2^5}\right](x-1)^5(y-1)^5(z-1)^5(x+1) \\
& \times \frac{1}{2^3}\Big(988 + 860x + 860y + 860z \\
& + 285x^2 + 475xy + 475xz + 285y^2 + 475yz + 285z^2 \\
& + 35x^3 + 75x^2y + 75x^2z + 75xy^2 + 125xyz + 75xz^2 + 35y^3 \\
& + 75y^2z + 75yz^2 + 35z^3\Big)
\end{aligned}
\tag{15.68}
$$

$$
\begin{aligned}
\Phi_{a,010}(x,y,z) = {} & \frac{1}{0!1!0!}\left[\frac{1}{2^5}\frac{1}{2^5}\frac{1}{2^5}\right](x-1)^5(y-1)^5(z-1)^5(y+1) \\
& \times \frac{1}{2^3}\Big(988 + 860x + 860y + 860z \\
& + 285x^2 + 475xy + 475xz + 285y^2 + 475yz + 285z^2 \\
& + 35x^3 + 75x^2y + 75x^2z + 75xy^2 + 125xyz + 75xz^2 + 35y^3 \\
& + 75y^2z + 75yz^2 + 35z^3\Big)
\end{aligned}
\tag{15.69}
$$

$$\Phi_{a,001}(x,y,z) = \frac{1}{0!0!1!} \left[\frac{1}{2^5} \frac{1}{2^5} \frac{1}{2^5} \right] (x-1)^5(y-1)^5(z-1)^5(z+1)$$

$$\times \frac{1}{2^3} \Big(988 + 860x + 860y + 860z$$

$$+ \; 285x^2 + 475xy + 475xz + 285y^2 + 475yz + 285z^2$$

$$+ \; 35x^3 + 75x^2y + 75x^2z + 75xy^2 + 125xyz + 75xz^2 + 35y^3$$

$$+ \; 75y^2z + 75yz^2 + 35z^3 \Big) \tag{15.70}$$

$$\Phi_{a,200}(x,y,z) = \frac{1}{2!0!0!} \left[\frac{1}{2^5} \frac{1}{2^5} \frac{1}{2^5} \right] (x-1)^5(y-1)^5(z-1)^5(x+1)^2$$

$$\times \frac{1}{2^2} \Big(154 + 90x + 90y + 90z$$

$$+ \; 15x^2 + 25xy + 25xz + 15y^2 + 25yz + 15z^2 \Big) \tag{15.71}$$

$$\Phi_{a,110}(x,y,z) = \frac{1}{1!1!0!} \left[\frac{1}{2^5} \frac{1}{2^5} \frac{1}{2^5} \right] (x-1)^5(y-1)^5(z-1)^5(x+1)(y+1)$$

$$\times \frac{1}{2^2} \Big(154 + 90x + 90y + 90z$$

$$+ \; 15x^2 + 25xy + 25xz + 15y^2 + 25yz + 15z^2 \Big) \tag{15.72}$$

$$\Phi_{a,101}(x,y,z) = \frac{1}{1!0!1!} \left[\frac{1}{2^5} \frac{1}{2^5} \frac{1}{2^5} \right] (x-1)^5(y-1)^5(z-1)^5(x+1)(z+1)$$

$$\times \frac{1}{2^2} \Big(154 + 90x + 90y + 90z$$

$$+ \; 15x^2 + 25xy + 25xz + 15y^2 + 25yz + 15z^2 \Big) \tag{15.73}$$

$$\Phi_{a,020}(x,y,z) = \frac{1}{0!2!0!} \left[\frac{1}{2^5} \frac{1}{2^5} \frac{1}{2^5} \right] (x-1)^5(y-1)^5(z-1)^5(y+1)^2$$

$$\times \frac{1}{2^2} \Big(154 + 90x + 90y + 90z$$

$$+ \; 15x^2 + 25xy + 25xz + 15y^2 + 25yz + 15z^2 \Big) \tag{15.74}$$

$$\Phi_{a,011}(x,y,z) = \frac{1}{0!1!1!} \left[\frac{1}{2^5} \frac{1}{2^5} \frac{1}{2^5} \right] (x-1)^5(y-1)^5(z-1)^5(y+1)(z+1)$$

$$\times \frac{1}{2^2} \Big(154 + 90x + 90y + 90z$$

$$+ \; 15x^2 + 25xy + 25xz + 15y^2 + 25yz + 15z^2 \Big) \tag{15.75}$$

$$\Phi_{a,002}(x,y,z) = \frac{1}{0!0!2!}\left[\frac{1}{2^5}\frac{1}{2^5}\frac{1}{2^5}\right](x-1)^5(y-1)^5(z-1)^5(z+1)^2$$
$$\times \frac{1}{2^2}\Big(154 + 90x + 90y + 90z$$
$$+ 15x^2 + 25xy + 25xz + 15y^2 + 25yz + 15z^2\Big) \qquad (15.76)$$

$$\Phi_{a,300}(x,y,z) = \frac{1}{3!0!0!}\left[\frac{1}{2^5}\frac{1}{2^5}\frac{1}{2^5}\right](x-1)^5(y-1)^5(z-1)^5(x+1)^3$$
$$\times \frac{1}{2^1}\Big(17 + 5x + 5y + 5z\Big) \qquad (15.77)$$

$$\Phi_{a,210}(x,y,z) = \frac{1}{2!1!0!}\left[\frac{1}{2^5}\frac{1}{2^5}\frac{1}{2^5}\right](x-1)^5(y-1)^5(z-1)^5(x+1)^2(y+1)$$
$$\times \frac{1}{2^1}\Big(17 + 5x + 5y + 5z\Big) \qquad (15.78)$$

$$\Phi_{a,201}(x,y,z) = \frac{1}{2!0!1!}\left[\frac{1}{2^5}\frac{1}{2^5}\frac{1}{2^5}\right](x-1)^5(y-1)^5(z-1)^5(x+1)^2(z+1)$$
$$\times \frac{1}{2^1}\Big(17 + 5x + 5y + 5z\Big) \qquad (15.79)$$

$$\Phi_{a,120}(x,y,z) = \frac{1}{1!2!0!}\left[\frac{1}{2^5}\frac{1}{2^5}\frac{1}{2^5}\right](x-1)^5(y-1)^5(z-1)^5(x+1)(y+1)^2$$
$$\times \frac{1}{2^1}\Big(17 + 5x + 5y + 5z\Big) \qquad (15.80)$$

$$\Phi_{a,111}(x,y,z) = \frac{1}{1!1!1!}\left[\frac{1}{2^5}\frac{1}{2^5}\frac{1}{2^5}\right](x-1)^5(y-1)^5(z-1)^5(x+1)(y+1)(z+1)$$
$$\times \frac{1}{2^1}\Big(17 + 5x + 5y + 5z\Big) \qquad (15.81)$$

$$\Phi_{a,102}(x,y,z) = \frac{1}{1!0!2!}\left[\frac{1}{2^5}\frac{1}{2^5}\frac{1}{2^5}\right](x-1)^5(y-1)^5(z-1)^5(x+1)(z+1)^2$$
$$\times \frac{1}{2^1}\Big(17 + 5x + 5y + 5z\Big) \qquad (15.82)$$

$$\Phi_{a,030}(x,y,z) = \frac{1}{0!3!0!}\left[\frac{1}{2^5}\frac{1}{2^5}\frac{1}{2^5}\right](x-1)^5(y-1)^5(z-1)^5(y+1)^3$$
$$\times \frac{1}{2^1}\Big(17 + 5x + 5y + 5z\Big) \qquad (15.83)$$

$$\Phi_{a,021}(x,y,z) = \frac{1}{0!2!1!}\left[\frac{1}{2^5}\frac{1}{2^5}\frac{1}{2^5}\right](x-1)^5(y-1)^5(z-1)^5(x+1)^2(z+1)$$
$$\times \frac{1}{2^1}\left(17+5x+5y+5z\right) \tag{15.84}$$

$$\Phi_{a,012}(x,y,z) = \frac{1}{0!1!2!}\left[\frac{1}{2^5}\frac{1}{2^5}\frac{1}{2^5}\right](x-1)^5(y-1)^5(z-1)^5(y+1)(z+1)^2$$
$$\times \frac{1}{2^1}\left(17+5x+5y+5z\right) \tag{15.85}$$

$$\Phi_{a,003}(x,y,z) = \frac{1}{0!0!3!}\left[\frac{1}{2^5}\frac{1}{2^5}\frac{1}{2^5}\right](x-1)^5(y-1)^5(z-1)^5(z+1)^3$$
$$\times \frac{1}{2^1}\left(17+5x+5y+5z\right) \tag{15.86}$$

$$\Phi_{a,400}(x,y,z) = \frac{1}{4!0!0!}\left[\frac{1}{2^5}\frac{1}{2^5}\frac{1}{2^5}\right]$$
$$\times (x-1)^5(y-1)^5(z-1)^5(x+1)^4\frac{1}{2^0}\left(1\right) \tag{15.87}$$

$$\Phi_{a,310}(x,y,z) = \frac{1}{3!1!0!}\left[\frac{1}{2^5}\frac{1}{2^5}\frac{1}{2^5}\right]$$
$$\times (x-1)^5(y-1)^5(z-1)^5(x+1)^3(y+1)\frac{1}{2^0}\left(1\right) \tag{15.88}$$

$$\Phi_{a,301}(x,y,z) = \frac{1}{3!0!1!}\left[\frac{1}{2^5}\frac{1}{2^5}\frac{1}{2^5}\right]$$
$$\times (x-1)^5(y-1)^5(z-1)^5(x+1)^3(z+1)\frac{1}{2^0}\left(1\right) \tag{15.89}$$

$$\Phi_{a,220}(x,y,z) = \frac{1}{2!2!0!}\left[\frac{1}{2^5}\frac{1}{2^5}\frac{1}{2^5}\right]$$
$$\times (x-1)^5(y-1)^5(z-1)^5(x+1)^2(y+1)^2\frac{1}{2^0}\left(1\right) \tag{15.90}$$

$$\Phi_{a,211}(x,y,z) = \frac{1}{2!1!1!}\left[\frac{1}{2^5}\frac{1}{2^5}\frac{1}{2^5}\right]$$
$$\times (x-1)^5(y-1)^5(z-1)^5(x+1)^2(y+1)(z+1)\frac{1}{2^0}\left(1\right)$$
$$\tag{15.91}$$

$$\Phi_{a,202}(x,y,z) = \frac{1}{2!0!2!} \left[\frac{1}{2^5} \frac{1}{2^5} \frac{1}{2^5} \right]$$
$$\times (x-1)^5(y-1)^5(z-1)^5(x+1)^2(z+1)^2 \frac{1}{2^0} \begin{pmatrix} 1 \end{pmatrix} \qquad (15.92)$$

$$\Phi_{a,130}(x,y,z) = \frac{1}{1!3!0!} \left[\frac{1}{2^5} \frac{1}{2^5} \frac{1}{2^5} \right]$$
$$\times (x-1)^5(y-1)^5(z-1)^5(x+1)(y+1)^3 \frac{1}{2^0} \begin{pmatrix} 1 \end{pmatrix} \qquad (15.93)$$

$$\Phi_{a,121}(x,y,z) = \frac{1}{1!2!1!} \left[\frac{1}{2^5} \frac{1}{2^5} \frac{1}{2^5} \right]$$
$$\times (x-1)^5(y-1)^5(z-1)^5(x+1)(y+1)^2(z+1) \frac{1}{2^0} \begin{pmatrix} 1 \end{pmatrix}$$
$$(15.94)$$

$$\Phi_{a,112}(x,y,z) = \frac{1}{1!1!2!} \left[\frac{1}{2^5} \frac{1}{2^5} \frac{1}{2^5} \right]$$
$$\times (x-1)^5(y-1)^5(z-1)^5(x+1)(y+1)(z+1)^2 \frac{1}{2^0} \begin{pmatrix} 1 \end{pmatrix}$$
$$(15.95)$$

$$\Phi_{a,103}(x,y,z) = \frac{1}{1!0!3!} \left[\frac{1}{2^5} \frac{1}{2^5} \frac{1}{2^5} \right]$$
$$\times (x-1)^5(y-1)^5(z-1)^5(x+1)(z+1)^3 \frac{1}{2^0} \begin{pmatrix} 1 \end{pmatrix} \qquad (15.96)$$

$$\Phi_{a,040}(x,y,z) = \frac{1}{0!4!0!} \left[\frac{1}{2^5} \frac{1}{2^5} \frac{1}{2^5} \right]$$
$$\times (x-1)^5(y-1)^5(z-1)^5(y+1)^4 \frac{1}{2^0} \begin{pmatrix} 1 \end{pmatrix} \qquad (15.97)$$

$$\Phi_{a,031}(x,y,z) = \frac{1}{0!3!1!} \left[\frac{1}{2^5} \frac{1}{2^5} \frac{1}{2^5} \right]$$
$$\times (x-1)^5(y-1)^5(z-1)^5(y+1)^3(z+1) \frac{1}{2^0} \begin{pmatrix} 1 \end{pmatrix} \qquad (15.98)$$

$$\Phi_{a,022}(x,y,z) = \frac{1}{0!2!2!} \left[\frac{1}{2^5} \frac{1}{2^5} \frac{1}{2^5} \right]$$
$$\times (x-1)^5(y-1)^5(z-1)^5(y+1)^2(z+1)^2 \frac{1}{2^0} \begin{pmatrix} 1 \end{pmatrix} \qquad (15.99)$$

$$\Phi_{a,013}(x, y, z) = \frac{1}{0!1!3!} \left[\frac{1}{2^5} \frac{1}{2^5} \frac{1}{2^5} \right]$$

$$\times (x-1)^5 (y-1)^5 (z-1)^5 (y+1)(z+1)^3 \frac{1}{2^0} \left(1 \right) \quad (15.100)$$

$$\Phi_{a,004}(x, y, z) = \frac{1}{0!0!4!} \left[\frac{1}{2^5} \frac{1}{2^5} \frac{1}{2^5} \right]$$

$$\times (x-1)^5 (y-1)^5 (z-1)^5 (z+1)^4 \frac{1}{2^0} \left(1 \right) \quad (15.101)$$

15.7.6   Four-dimensional Hermite Interpolating Polynomial

15.7.7   Summary of the polynomials $\Phi_{e,n}^{(n_0,n_1,n_2,n_3)}$ related to the node $a_e = (-1,-1,-1,-1)$ and $\eta = 3$

Computation of the Hermite Interpolating Polynomials, in $d = 4$ variables, for continuity of the function, and derivatives of order up to $\eta = 3$, associate to the node e whose coordinate numbers are $\alpha = (-1,-1,-1,-1)$ and where $\beta = (1,1,1,1)$.

$\Phi_{a,0000}(w, x, y, z)$

$$= \frac{1}{0!0!0!0!} \left[\frac{1}{(2)^4} \frac{1}{(2)^4} \frac{1}{(2)^4} \frac{1}{(2)^4} \right] (w-1)^5 (x-1)^5 (y-1)^5 (z-1)^5$$

$$\times \frac{1}{2^1} \Big(290 + 191w + 191x + 191y + 191z$$

$$+ 50w^2 + 80wx + 80wy + 80wz + 50x^2 + 80xy + 80xz$$

$$+ 50y^2 + 80yz + 50z^2$$

$$+ 5w^3 + 10w^2x + 10w^2y + 10w^2z + 10wx^2$$

$$+ 16wxy + 16wxz + 10wy^2 + 16wyz$$

$$+ 10wz^2 + 5x^3 + 10x^2y + 10x^2z + 10xy^2$$

$$+ 16xyz + 10xz^2 + 5y^3 + 10y^2z + 10yz^2 + 5z^3 \Big) \quad (15.102)$$

$\Phi_{a,1000}(w, x, y, z)$

$$= \frac{1}{1!0!0!0!} \left[\frac{1}{(2)^4} \frac{1}{(2)^4} \frac{1}{(2)^4} \frac{1}{(2)^4} \right] (w-1)^5 (x-1)^5 (y-1)^5 (z-1)^5 (w+1)$$

$$\times \frac{1}{2^1} \Big(96 + 38w + 38x + 38y + 38z$$

$$+ 5w^2 + 8wx + 8wy + 8wz + 5x^2 + 8xy + 8xz + 5y^2 + 8yz + 5z^2 \Big)$$

$$(15.103)$$

The polynomial

$$f_{e,n}(x) = \frac{1}{2^1}\left(96 + 38w + 38x + 38y + 38z \right.$$
$$\left. + 5w^2 + 8wx + 8wy + 8wz + 5x^2 + 8xy + 8xz + 5y^2 + 8yz + 5z^2 \right)$$

(15.104)

is the same for $\Phi_{a,1000}$, $\Phi_{a,0100}$, $\Phi_{a,0010}$, and $\Phi_{a,0001}$. The only differences are for $1/n!$ and $Q_{e,n}(x)$, which are, respectively,

$1/(1!0!0!0!)$ and $(w+1)$ for $\Phi_{a,1000}$,

$1/(0!1!0!0!)$ and $(x+1)$ for $\Phi_{a,0100}$,

$1/(0!0!1!0!)$ and $(y+1)$ for $\Phi_{a,0010}$,

$1/(0!0!0!1!)$ and $(z+1)$ for $\Phi_{a,0001}$.

For the terms whose indices are in the set L_2^4, that is, 2000, 1100, etc. the polynomial

$$f_{e,n}(x) = 9 + 2w + 2x + 2y + 2z \tag{15.105}$$

is the same. The factors $1/n!$ and $Q_{e,n}(x)$ change accordingly, for example

$1/(2!0!0!0!)$ and $(w+1)^2$ for $\Phi_{a,2000}$,

$1/(1!1!0!0!)$ and $(w+1)(x+1)$ for $\Phi_{a,1100}$,

$1/(0!1!0!1!)$ and $(x+1)(z+1)$ for $\Phi_{a,0101}$, etc.

For the last set of polynomials $f_{e,n}(x) = 1$ with the modifications such as $1/(2!0!1!0!)$ and $(w+1)^2(y+1)$ for $\Phi_{a,2010}$, etc.

Part III

Selected applications

Chapter 16

Construction of the approximate solution

The Hermite Interpolating Polynomials can be used as a basis for the expansion of a function in a given an interval. Although any interval can be used, it was chosen [-1, 1] considering the properties that hold only for this interval.

The properties that the Hermite Interpolating Polynomials assume values 1 or zero at end points of the interval make them attractive for the construction of solutions satisfying identically the boundary conditions.

16.1 One-dimensional problems

16.1.1 Solution expanded in terms of the Hermite Interpolating Polynomials

A solution to a boundary value problem expanded with respect to the Hermite Interpolating Polynomials associated to three nodes, as shown in figure 16.1.1, writes

$$T(x) = \sum_{n=0}^{\nu} \Phi_{p,n}(x)T_{p,n} + \sum_{n=0}^{\nu} \Phi_{q,n}(x)T_{q,n} + \sum_{n=0}^{\nu} \Phi_{r,n}(x)T_{r,n} \qquad (16.1)$$

where ν is the number of polynomials related to a reference node and given by $n^{\delta}(\eta)$ in the chapter 4 page 175 equation (4.135).

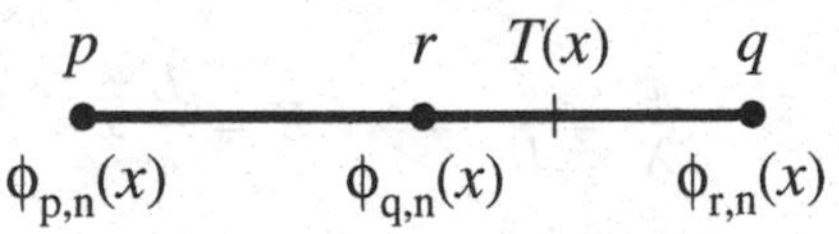

Figure 16.1.1 *Solution expanded in terms of the Hemite Interpolating Polynomials referred to the nodes p, q, and r.*

All the examples here presented have been solved using two interpolating nodes.

16.1.2   Coordinate transformation

Let $[c, d]$ represent the domain of the solution and $[a, b]$ an element in this domain as shown in figure 16.1.2. The mid point of the element is given by $X_m = (a + b)/2$. Let $\bar{x}$ represent the coordinate number of a point in the element relative to a global coordinate system. Let x be the coordinate number of the same point relative to the mid point of the element, then $x = \bar{x} - x_m$.

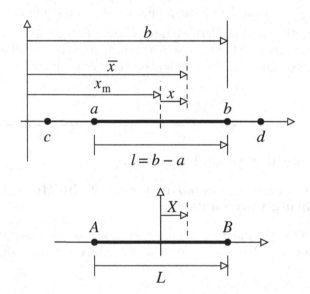

Figure 16.1.2 *Coordinate transformation from the domain* $[a, b]$ *to* $[A, B]$ *symmetric with respect to the origin of the coordinate system.*

Let $[a, b]$ represent an element of length ℓ and be bijection of the element $[A, B]$. Thus, to every point x corresponds one and only one point X. Using triangle similarity it can be written

$$\frac{x}{X} = \frac{\ell}{L} \quad \Longrightarrow \quad X = \frac{L}{\ell} x \tag{16.2}$$

Performing the substitution $\ell = b - a$

$$X = \frac{L}{b - a} x \tag{16.3}$$

Define

$$\delta := \frac{L}{\ell} = \frac{L}{b-a} \tag{16.4}$$

then, from equation (16.3) it can be written

$$X = \delta\, x \quad \Longrightarrow \quad dX = \delta\, d(\bar{x} - x_m) = \delta\, dx \quad \Longrightarrow \quad \delta = \frac{dX}{dx} \tag{16.5}$$

Equation (16.5) gives an expression for the transformation of the derivatives as

$$\frac{d\Phi(X)}{dx} = \frac{d\Phi(X)}{dX}\left(\frac{dX}{dx}\right) = \delta\,\frac{d\Phi(X)}{dX} \tag{16.6}$$

For a derivative of order u it follows

$$\frac{d^Y \Phi(X)}{dx^u} = \frac{d^u\Phi(X)}{dX^u}\left(\frac{dX}{dx}\right)^u = \delta^u\frac{d^u\Phi(X)}{dX^u} \tag{16.7}$$

A function $T(X_0, X_1)$ of two normalized variables (X_0, X_1) expanded with respect to one node of reference e, recalling the notation in the subsection 6.1.2, page 219, equation (6.9), can be written as

$$T(X_0, X_1) = \sum_{\lambda_0=0}^{\lambda_0=\eta} \sum_{\lambda_1=0}^{\lambda_1=\lambda_0} \Phi_{e,n}(X_0, X_1)\, T_{n_0,n_1} \tag{16.8}$$

where the coordinate numbers of $n = (n_0, n_1)$ are given by $n_0 = \lambda_0 - \lambda_1$ and $n_1 = \lambda_1$. The number of elements of the set $n \in L^1(\eta)$ is given by $n^\delta(\eta)$ and its computation can be performed according to the expression given in chapter 4, page 175, equation (4.135). The factors T_{n_0,n_1} represent the coordinate numbers of the expansion in terms of the basis functions $\Phi_{e,n}(x_0, x_1)$. In general, T_{n_0,n_1} have constant values but they can also be functions of (X_0, X_1).

Consider a differential equation in the variables $x = (x_0, x_1)$, which requires the derivatives of order $u = (u_0, u_1)$. These derivatives can be written as

$$
\begin{aligned}
T^{(u)}(X_0, X_1) &= \frac{d^u T(X)}{dx^u} \\
&= \sum_{\lambda_0=0}^{\lambda_0=\eta} \sum_{\lambda_1=0}^{\lambda_1=\lambda_0} \Phi_{n_0,n_1}^{(u_0,u_1)}(X_0, X_1)\left(\frac{dX_0}{dx_0}\right)^{u_0}\left(\frac{dX_1}{dx_1}\right)^{u_1} T_{n_0,n_1}
\end{aligned}
\tag{16.9}
$$

Recalling the definition of δ given in equation 16.4 the above expression for the derivative simplifies to

$$T^u(X_0, X_1) = \sum_{\lambda_0=0}^{\lambda_0=\eta} \sum_{\lambda_1=0}^{\lambda_1=\lambda_0} \Phi_{n_0,n_1}^{(u_0,u_1)}(X_0, X_1)\delta_0^{u_0}\delta_1^{u_1}\, T_{n_0,n_1} \tag{16.10}$$

For the particular case of a normalized domain $[-1, 1]$ one has $A = -1, B = 1, L = B - A = 2$ and $\delta = 2/\ell$.

16.1.3   One-dimensional boundary conditions

Consider a boundary value problem that gives the function $T(x) = T_p$ at the node p and the derivative $T^{(u)}(x) = T_r^u$ at the node r. The Hermite Interpolating Polynomials related to the node p equals one, all its derivatives equals zero. The same polynomial and all its derivatives equal zero at all other nodes. Thus, at the node p the approximate solution writes

$$T(x)|_{x=x_p} = \Phi_{p,0}(x)|_{x=x_p} T_{p,0} = T_{p,0} = T_p \qquad (16.11)$$

where $\Phi_{p,0}(x)|_{x=x_p} = 1$ by definition of the Hermite Interpolating Polynomials. Therefore, the approximate solution satisfies identically the prescribed value of the function at a node if the coefficient of $\Phi_{p,0}(x)$ equals the value of the function at that point.

For the case of a derivative the procedure is similar. For the node r, the derivative of order n of the function $\Phi_{r,n}^{(n)}(x)$ equals one, and all others are zero, thus

$$T^{(n)}(x)\Big|_{x=x_r} = \Phi_{r,n}^{(n)}(x)\Big|_{x=x_r} \ \delta^n T_{r,n} = \delta^n T_{p,n} = T_p^{(n)} \qquad (16.12)$$

The introduction of the derivative value to the approximate solution is equivalent to assign the the coefficient of the corresponding Hermite Interpolating Polynomial the value $T_{p,n} = T_p^{(n)}/\delta^n$.

16.2   Higher-dimensional problems

16.2.1   Analysis of the boundary conditions

The construction of the Hermite Interpolating Polynomials is based on the expression

$$\Phi_{e,n}(x) = \frac{1}{n! P_e(\alpha)} P_e(x) Q_{e,n}(x) f_{e,n}(a, x) \qquad (16.13)$$

which was obtained in subsection 7.3.5, equation, (7.126), page 315.

The polynomial $P_e(x)$ guarantees that the polynomial $\Phi_{e,n}(x)$ will vanishes at all points in the domain other than the reference node e. The polynomials $Q_{e,n}(x) f_{e,n}(a, x)$ are generated such that $\Phi_{e,n}(x)$ equals to one at the reference node. In this case the application of the boundary conditions is straight forward, that is, it is enough to multiply the polynomial $\Phi_{e,n}(x)$ by the value of the boundary condition at that reference node.

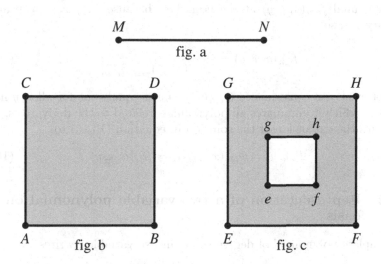

Figure 16.2.1 *Figure a shows a one-dimensional two-boundary domain. Figure b shows a two-dimensional one-boundary domain. Figure c shows a two-dimensional two-boundary domain.*

The two-dimensional domain analogous to the one-dimensional domain shown in figure 16.2.1 (figure a) is the one shown in (figure c), and the corresponding interpolating polynomials are

$$\Phi_{e,n}(x,y) = \frac{1}{n!P_e(\alpha)}P_e(x,y)\,Q_{e,n}(x,y)\,f_{e,n}(a,x,y) \qquad (16.14)$$

Therefore, the factor $P_e(x,y)$ in equation (16.14) equals one since there is no other boundary where it must vanish.

The boundary condition for the two-dimensional domain in figure 16.2.1 (figure b) is no more a number, as in the case of the one-dimensional domain if (figure a), but a function. Therefore, the solution of a differential equation in the domain defined in (figure b) writes

$$T(x,y) = \sum_{n_0=0}^{\nu} \Phi_{\Omega_1,n}(x,y)T_{\Omega_1,n} \qquad (16.15)$$

and for a solution in the domain defined in (figure c) the expansion writes

$$T(x,y) = \sum_{n_0=0}^{\nu} \Phi_{\Omega_a,n}(x,y)T_{\Omega_a,n} \sum_{n_0=0}^{\nu} \Phi_{\Omega_b,n}(x)T_{\Omega_b,n} \qquad (16.16)$$

where Ω_1 denotes the boundary $ABCD$, Ω_a the boundary $EFGH$ and Ω_b the boundary $efgh$.

The polynomial $Q_{e,n}(x,y)$ cannot be used to evaluate the coefficients of the polynomial $f_{e,n}(a,x,y)$ at the boundary because $Q_{e,n}(x,y) = 0$ at the boundary and so

$$f_{e,n}(a,x,y) = \frac{1}{Q_{e,n}(x,y)}\bigg|_{x=\alpha_0,y=\alpha_1} \tag{16.17}$$

does not exist. As consequence $Q_{e,n}(x,y)$ must equals one for all x,y in the boundary, which it eliminates all polynomials related to the derivatives. This transforms the expansion of the solution in equation (16.15) to

$$T(x,y) = \Phi_{\Omega_1}(x,y)T_{\Omega_1} = f_{\Omega_1}(a,x,y) \tag{16.18}$$

16.2.2   Representation of a two-variable polynomial on any basis

A complete polynomial of degree three, in two variables, writes

$$f(x,y) = a_{00} + a_{10}x + a_{01}y + a_{20}x^2 + a_{11}xy + a_{02}y^2$$
$$+ a_{30}x^3 + a_{21}x^2y + a_{12}xy^2 + a_{03}y^3 \tag{16.19}$$

and it can be rewritten as

$$f(x,y) = \left[a_{00} + a_{01}y + a_{02}y^2 + a_{03}y^3\right] \times 1+$$
$$\left[a_{10} + a_{11}y + a_{12}y^2\right]x+$$
$$\left[a_{20} + a_{21}y\right]x^2 + 1 \times a_{30}x^3 \tag{16.20}$$

Recalling that $1, x, x^2, x^3, \ldots$ is a set of bases vectors they can be replaced by any basis $T_0(x), T_1(x), T_2(x), \ldots$ and the expansion denoted in equation (16.20) can be written as

$$f(x,y) = [a_{00}F_0(y) + a_{01}F_1(y) + a_{02}F_2(y) + a_{03}F_3(y)]\,F_0(x)+$$
$$[a_{10}F_0(y) + a_{11}F_1(y) + a_{12}F_2(y)]\,F_1(x)+$$
$$[a_{20}F_0(y) + a_{21}F_1(y)]\,F_2(x)+$$
$$[a_{30}F_0(y)]\,F_3(x) \tag{16.21}$$

where $F_i(x)$ and $F_j(y)$ may be both elements of the same set of linear independent functions or different sets. For example, $F_i(x)$ and $F_j(y)$ can be Chebychev polynomials or $F_i(x)$ can be trigonometric polynomials and $F_j(y)$ be Lagrange polynomials.

The expression in equation (16.21) can be rewritten as

$$f(x,y) = [F_3(y)]\,F_0(x) + [F_2(y)]\,F_1(x) + [F_1(y)]\,F_2(x) + [F_0(y)]\,F_3(x) \tag{16.22}$$

where

$$F_3(y) = a_{00} + a_{01}y + a_{02}y^2 + a_{03}y^3$$
$$F_2(y) = a_{10} + a_{11}y + a_{12}y^2$$
$$F_1(y) = a_{20} + a_{21}y$$
$$F_0(y) = a_{30} \tag{16.23}$$

16.2.3 Representation of a three-variable polynomial on any basis

A complete polynomial of degree three, in three variables, writes

$$f(x,y,z) = a_{000} +$$
$$a_{100}x + a_{010}y + a_{001}z +$$
$$a_{200}x^2 + a_{110}xy + a_{101}xy + a_{020}xy + a_{011}xy + a_{002}y^2 +$$
$$a_{300}x^3 + a_{210}x^2y + a_{201}x^2y + a_{120}x^2y + a_{111}x^2y + a_{102}x^2y$$
$$+ a_{030}x^2y + a_{021}x^2y + a_{012}xy^2 + a_{003}y^3 \tag{16.24}$$

where indices of the coefficients are elements of the $L^2(3)$ shown in table 2.8.25, page 90.

The terms in equation (16.24) can be rearranged as follows

$$f(x,y,z) = \left[a_{000} + a_{010}y + a_{001}z + a_{020}y^2 + a_{011}yz + a_{002}z^2 \right.$$
$$\left. + a_{030}y^3 + a_{021}y^2z + a_{012}yz^2 + a_{003}z^3 \right]$$
$$+ \left[a_{100} + a_{110}y + a_{101}z + a_{120}y^2 + a_{111}yz + a_{102}z^2 \right]x$$
$$+ \left[a_{200} + a_{210}y + a_{201}z \right]x^2$$
$$+ a_{300}x^3 \tag{16.25}$$

Using a notation similar to the one used in equation (16.22), the equation (16.24) can be rewritten as follows

$$f(x,y,z) = [F_3(y,z)] F_0(x) + [F_2(y,z)] F_1(x) + [F_1(y,z)] F_2(x)$$
$$+ [F_0(y,z)] F_3(x)$$
$$= [F_3(y,z)] + [F_2(y,z)] x + [F_1(y,z)] x^2 + [F_0(y,z)] x^3 \tag{16.26}$$

where

$$F_3(y,z) = [F_3(z)] F_0(y) + [F_2(z)] F_1(y) + [F_1(z)] F_2(y) + [F_0(z)] F_3(y)$$
$$= [F_3(z)] + [F_2(z)] y + [F_1(z)] y^2 + [F_0(z)] y^3$$
$$= \left[a_{000} + a_{001}z + a_{002}z^2 + a_{003}z^3 \right] +$$
$$\left[a_{010} + a_{011}z + a_{012}z^2 \right] y + [a_{020} + a_{021}z] y^2 + [a_{030}] y^3 \tag{16.27}$$

$$
\begin{aligned}
F_2(y, z) &= [F_2(z)] F_0(y) + [F_1(z)] F_1(y) + [F_0(z)] F_2(y) \\
&= [F_2(z)] + [F_1(z)] y + [F_0(z)] y^2 \\
&= \left[a_{100} + a_{101}z + a_{102}z^2\right] + \left[a_{110} + a_{111}z\right] y + \left[a_{120}\right] y^2 \quad (16.28)
\end{aligned}
$$

$$
\begin{aligned}
F_1(y, z) &= [F_1(z)] F_0(y) + [F_0(z)] F_1(y) \\
&= [F_1(z)] + [F_0(z)] y \\
&= \left[a_{200} + a_{201}z\right] + \left[a_{210}\right] y \quad (16.29)
\end{aligned}
$$

$$
\begin{aligned}
F_0(y, z) &= [F_0(z)] F_0(y) \\
&= [F_0(z)] \\
&= \left[a_{300}\right] \quad (16.30)
\end{aligned}
$$

16.2.4　Example of a three-variable polynomial on a Chebyshev basis

A basis using the first four Chebyshev polynomials of first kind, Mason and Handscomb [27], writes

$$
\{T_0(x), T_1(x), T_2(x), T_3(x)\} = \{(1), (x), (2x^2 - 1), (4x^3 - 3x)\} \quad (16.31)
$$

Substituting the functions $F_i(x), F_j(y), F_k(z)$ by $T_i(x), T_j(y), T_k(z)$ into equation (16.26) and considering the definitions in equations (16.27)–(16.30) it follows

$$
\begin{aligned}
f(x, y, z) = \{ &\left[a_{000} + a_{001}z + a_{002}(2z^2 - 1) + a_{003}(4z^3 - 3z)\right] \\
&+ \left[a_{010} + a_{011}z + a_{012}(2z^2 - 1)\right] y \\
&+ \left[a_{020} + a_{021}z\right] (2y^2 - 1) + \left[a_{030}\right] (4y^3 - 3y) \} \\
+ \{ &\left[a_{100} + a_{101}z + a_{102}(2z^2 - 1)\right] + \left[a_{110} + a_{111}z\right] y \\
&+ \left[a_{120}\right] (2y^2 - 1) \} x \\
+ \{ &\left[a_{200} + a_{201}z\right] + \left[a_{210}\right] y \} (2x^2 - 1) \\
&+ a_{300}(4x^3 - 3x) \quad (16.32)
\end{aligned}
$$

Expanding

$$f(x, y, z) = a_{000}+$$
$$a_{100}x + a_{010} + a_{001}z$$
$$+ a_{200}(2x^2 - 1) + a_{110}xy + a_{101}xz + a_{020}(2y^2 - 1)+$$
$$+ a_{011}yz + a_{002}(2z^2 - 1)$$
$$+ a_{300}(4x^3 - 3x) + a_{210}(2x^2 - 1)y + a_{201}(2x^2 - 1)z$$
$$+ a_{120}x(2y^2 - 1) + a_{111}xyz + a_{102}x(2z^2 - 1)+$$
$$+ a_{030}(4y^3 - 3y) + a_{021}(2y^2 - 1)z$$
$$+ a_{012}y(2z^2 - 1) + a_{003}(4z^3 - 3z) \qquad (16.33)$$

16.2.5 Example of the two-dimensional sine function

The expansion for the sine function is given by

$$\sin(x) = x - \frac{x^3}{3!} + \frac{x^5}{5!} - \frac{x^7}{7!} + \frac{x^9}{9!} - \frac{x^{11}}{11!} + \cdots \qquad (16.34)$$

Using the same procedure in subsection 16.2.2 it gives, for an expansion until x^{11},

$$\sin(x, y) = \left[y - \frac{y^3}{3!} + \frac{y^5}{5!} - \frac{y^7}{7!} + \frac{y^9}{9!} - \frac{y^{11}}{11!} + \cdots \right] x$$
$$- \left[y - \frac{y^3}{3!} + \frac{y^5}{5!} - \frac{y^7}{7!} + \frac{y^9}{9!} + \cdots \right] \frac{x^3}{3!}$$
$$+ \left[y - \frac{y^3}{3!} + \frac{y^5}{5!} - \frac{y^7}{7!} + \cdots \right] \frac{x^5}{5!}$$
$$- \left[y - \frac{y^3}{3!} + \frac{y^5}{5!} + \cdots \right] \frac{x^7}{7!}$$
$$+ \left[y - \frac{y^3}{3!} + \cdots \right] \frac{x^9}{9!}$$
$$- [y + \cdots] \frac{x^{11}}{11!} \qquad (16.35)$$

Since the series in y have infinity terms they can be replaced by $\sin(y)$ and factor out

$$\sin(x,y) = [\sin(y)]\, x - [\sin(y)]\, \frac{x^3}{3!} + [\sin(y)]\, \frac{x^5}{5!} - [\sin(y)]\, \frac{x^7}{7!}$$

$$+ [\sin(y)]\, \frac{x^9}{9!} - [\sin(y)]\, \frac{x^{11}}{11!}$$

$$= [\sin(y)] \left[x - \frac{x^3}{3!} + \frac{x^5}{5!} - \frac{x^7}{7!} + \frac{x^9}{9!} - \frac{x^{11}}{11!} + \cdots \right]$$

$$= \sin(y)\sin(x) \tag{16.36}$$

Conclusion, if the basis is infinite then $f(x,y) = f(x) \times f(y)$, but the same is not true if the basis is finite, that is, $f(x,y) \neq f(x) \times f(y)$.

16.2.6   The polynomial approximation

A polynomial written with a finite number of terms, as above, gives good approximation in the neighborhood of the boundary. To obtain the same approximation for points far from the boundary it is necessary to increase the number of terms of the expansion. Although it is mathematically correct it cannot be performed in computers since they have a limited number of digits (for example 32 bits machines) and the last terms of the finite expansion will not change the value of the series.

One way to bypass this problem is to use Padé approximants, that is, the function $f(x,y)$ is represented by the ratio of two polynomials (Wazwaz [34], page 369,) such as

$$f(x,y) = \frac{N(x,y)}{D(x,y)} \tag{16.37}$$

where $N(x,y)$ and $D(x,y)$ can be polynomials such as the one represented in (16.37) or (16.21).

A polynomial of degree three, in two variables, can also be represented by the product of two one-variable polynomials as

$$f(x,y) = g_1(x)g_2(y) \tag{16.38}$$

where

$$g_1(x) = a_0 + a_1 x + a_2 x^2 + a_3 x^3$$

$$g_2(y) = b_0 + b_1 y + b_2 y^2 + b_3 y^3 \tag{16.39}$$

The product $g_1(x)g_2(y)$ gives

$$g_1(x)g_2(y) = a_0 b_0 +$$
$$a_1 b_0 x + a_0 b_1 y +$$
$$a_2 b_0 x^2 + a_1 b_1 xy + a_0 b_2 y^2 +$$
$$a_3 b_0 x^3 + a_2 b_1 x^2 y + a_1 b_2 xy^2 + b_3 y^3 +$$
$$0x^4 + a_3 b_1 x^3 y + a_2 b_2 x^2 y^2 + a_1 b_3 xy^3 + 0y^4 +$$
$$0x^5 + 0x^4 y + a_3 b_2 x^3 y^2 + a_2 b_3 x^2 y^3 + 0xy^4 + 0y^5 +$$
$$0x^6 + 0x^5 y + 0x^4 y^2 + a_3 b_3 x^3 y^3 + 0x^2 y^4 + 0xy^5 + 0y^6 \quad (16.40)$$

If the approximate solution is written as the product $g_1(x)g_2(y)$ there are eight parameters to be obtained. If the solution is written as a complete polynomial of degree three there are ten parameters to be obtained. Since the complete polynomial has more degree of freedom that the product $g_1(x)g_2(y)$ it may give a better approximation.

On the other hand consider the product

$$f(x) = g_1(x)g_2(y)$$
$$= (a_0 + a_1 x)(b_0 + b_1 x)$$
$$= a_0 b_0 + (a_1 b_0 + a_0 b_1)x + a_1 b_1 x^2$$
$$= \alpha_0 + \alpha_1 x + \alpha_2 x^2 \quad (16.41)$$

If a_0, a_1, and b_0, b_1 are real numbers the polynomial that approximates solution will contain only real roots. If it is used, the form $\alpha_0 + \alpha_1 x + \alpha_2 x^2$, then the polynomial $f(x)$ may contain complex roots, which can lead to a better approximation.

16.2.7 Computation of the coefficients of the expansion of the solution

Based on the above analysis, the higher dimensional examples of applications were performed using rational fractions of the form

$$f(x, y) = \frac{N(x, y)}{D(x, y)} \quad (16.42)$$

and

$$f(x, y) = \frac{N(x)}{D(x)} \frac{N(y)}{D(y)} \quad (16.43)$$

where N, D are written in the form shown in equation (16.42).

The solution $f(x, y)$ is unique if it satisfies the differential equation in the domain and the boundary conditions, or the boundary and initial conditions.

The parameters in N, D can be obtained as the set of values that will minimize the norm of the error of the differential equation and boundary e initial conditions.

16.3   Residual analysis

16.3.1   Residual analysis of the governing equation

The accuracy of the solution is analyzed by the convergence to zero of the residual of the differential equation. Let $r(x_i)$ be the residual of the governing differential equation at the point x_i in the domain of the solution.

The norm of the residual of the differential equation is given by

$$\|r\| = \sqrt{\sum_{i=0}^{m} |r(x_i)|^2} \tag{16.44}$$

where $r_i(x_i)$ is evaluated using the approximate solution.

16.3.2   Residual analysis of the solution

The residual e_i of the solution at the point x_i is defined as

$$e(x_i) = g_e(x_i) - g(x_i) \tag{16.45}$$

where $g_e(x_i)$ is the exact solution and $g(x_i)$ the value of the approximate solution at the point x_i.

The norm of the residual of the approximate solution in the domain is given by

$$\|e\| = \sqrt{\sum_{i=0}^{m} |e(x_i)|^2} \tag{16.46}$$

And the norm of the approximate solution is computed from

$$\|g\| = \sqrt{\sum_{i=0}^{m} |g(x_i)|^2} \tag{16.47}$$

using the approximate solution. The norm of the exact solution is obtained from

$$\|g_e\| = \sqrt{\sum_{i=0}^{m} |g_e(x_i)|^2} \tag{16.48}$$

where $g_e(x_i)$ is the value of the exact solution at the point x_i.

16.3.3 Residual analysis in the domain

The analysis of the residual in the domain is given by the norm of the residual of the differential equation and the residual of the boundary conditions and initial conditions.

The norm of the residual of the differential equation is given by

$$\|r_D\| = \sqrt{\sum_{i=0}^{m} |r(x_i)|^2 + \sum_{j=0}^{n} |r_{BC}(x_j)|^2 + \sum_{k=0}^{p} |r_{IC}(x_k)|^2} \qquad (16.49)$$

where $r_i(x_i)$, $r_{BC}(x_j)$, and $r_{IC}(x_k)$ is the residual of the differential equation, the residual of the boundary condition, and the residual of the initial condition, respectively, evaluated using the approximate solution.

16.4 Comments about the computer program

The code for the computation of the examples was written in C. It was used the GNU *gcc* to compile and run it in a PC i586 with 1GB of memory and 500MHz CPU. The computer code was written using *long double* as the type for the real numbers.

The accuracy for the optimization program was defined as

$$\frac{residual_1 - residual_2}{residual_2} < 0.01 \qquad (16.50)$$

where $residual_1$ is the square of the residual of the differential equation in the iteration $n - 1$ and $residual_2$ is the square of the residual in the iteration n.

It was observed that there is an optimum value for η, that is, the a value, which gives the fastest rate of convergence and the accuracy is maximum. Below this value the rate of convergence is high but the accuracy is lower. Above this value the rate of convergence is lower, in other words, the same accuracy can be reached but with a higher number of iteration loops. Considering this fact it was imposed a cap on the number of iterations.

The computation of the coordinate numbers of the expansion of the solution was performed on 51 collocation points in the domain. These coordinate numbers were used to evaluate the norm of the approximate solution and the norm of the residual of the differential equation over 101 points equally spaced in the domain.

Chapter 10 shows a detailed computation of the Hermite Interpolating Polynomials for the case of a square domain symmetric with respect to the coordinate axes. Chapter 11 shows the same for a generic rectangular domain. These two chapters can be used by the reader to debug a computer code. It was also used an algebraic manipulation computer code to double check several routines of the numerical code.

To accelerate the convergence of the optimization problem can be initially solved for a number of collocation points equal to the number of parameter. Then this solution can be used as starting point for a solution with greater number of collocation points and parameters, which will give a smaller norm of the error.

Chapter 17

One-dimensional two-point boundary value problems

17.1 Linear differential equation

Davies, Karageorghis, and Phillips [8] in the paper *Spectral Galerkin Methods for the Primary Two-point Boundary Value Problem in Modeling Viscoelastic Flows* solve three-boundary value problems where one is linear and the other two are nonlinear. In the paper [8], they used as basis functions the eigen-functions of the fourth order differential operator of the governing differential equation. They say that many authors have used this procedure for linear problems. They also highlight as an advantage of the used procedure that four out of five boundary conditions are automatically satisfied. The Galerkin approach was used for the evaluation of the coefficients. In the same paper they used the Chebyshev polynomials to solve the same problem. In this case they were unable to satisfy identically any boundary condition. In another paper [9] they expand the solution in terms of beam functions and Chebyshev polynomials. The computation of the coefficients is performed using the Collocation approach.

The differential equation in example (i) page 654, by Davies, Karageorghis, and Phillips [8], and example (i) page 810, by of the same authors [9] are solved here using the Hermite Interpolating Polynomials.

17.1.1 Approximate solution in the domain $[-0.5, 0.5]$

17.1.1.1 The governing equation

The domain equation is given by

$$\frac{\mathrm{d}^4 g(x)}{\mathrm{d}x^4} = 12c \tag{17.1}$$

with the boundary conditions

$$g\left(\pm \frac{1}{2}\right) = \frac{\mathrm{d}g}{\mathrm{d}x}\left(\pm \frac{1}{2}\right) = 0 \tag{17.2}$$

The domain of the solution is $x \in [-0.5, 0.5]$.

17.1.1.2   Exact solution

The exact solution is given by

$$g_e(x) = \frac{1}{2}c\left(x^2 - \frac{1}{4}\right)^2 \tag{17.3}$$

where c is a scaling constant in the range $1 \leq c \leq 10^3$. The example was worked using $c = 1$.

17.1.1.3   Notation

Recall that the notation used for the Hermite Interpolating Polynomials is $\Phi_{e,n}(x)$ where e denotes the reference nodes, which are the boundary points of the domain, in this case they are called a and b. The index n refers to the degree of derivative for which the polynomial equals to one. Therefore, the polynomial

$$\Phi_{a,2}(x) \tag{17.4}$$

is related to the reference node $e = a$ whose coordinate number is $\alpha_0 = -1/2$, and the second derivative of $\Phi_{a,2}(x)$ when evaluated at the reference node $a = \alpha_0 = -1/2$ equals to one, that is,

$$\Phi_{a,2}^{(2)}(x)\Big|_{x=-1/2} = 1 \tag{17.5}$$

and all other order of derivative of the same polynomial up to $\mathcal{O}^\delta(\eta)$ when evaluated at the same point, that is $a = \alpha_0 = -1/2$, equal to zero, which gives

$$\Phi_{a,2}^{(0)}(x)\Big|_{x=-1/2} = 0 \qquad \Phi_{a,2}^{(1)}(x)\Big|_{x=-1/2} = 0$$
$$\Phi_{a,2}^{(2)}(x)\Big|_{x=-1/2} = 0 \quad \cdots \quad \Phi_{a,2}^{(\eta)}(x)\Big|_{x=-1/2} = 0 \tag{17.6}$$

17.1.1.4   Expansion of the solution

The approximate solution of the linear differential equation (17.1) when expanded in a basis of Hermite Interpolating Polynomials writes

$$g(x) = \sum_{r=0}^{r=\eta} [g_{a,r}\Phi_{a,r} + g_{b,r}\Phi_{b,r}] \tag{17.7}$$

and for $\eta = 4$ it reduces to

$$g(x) = g_{a,0}\Phi_{a,0}(x) + g_{a,1}\Phi_{a,1}(x) + g_{a,2}\Phi_{a,2}(x) + g_{a,3}\Phi_{a,3}(x) + g_{a,4}\Phi_{a,4}(x)$$
$$+ g_{b,0}\Phi_{b,0}(x) + g_{b,1}\Phi_{b,1}(x) + g_{b,2}\Phi_{b,2}(x) + g_{b,3}\Phi_{b,3}(x) + g_{b,4}\Phi_{b,4}(x) \tag{17.8}$$

17.1.1.5 Boundary conditions

Equation (17.8) when evaluated at $a = \alpha_0 = -1/2$ must satisfy the boundary condition in equation (17.3), thus

$$g(-1/2) = g_{a,0}\Phi_{a,0}(-1/2) = 0 \tag{17.9}$$

where $\Phi_{a,0}(-1/2) = 1$ then $g_{a,0} = 0$. Analogously, when evaluated at the boundary $b = \alpha_0 = 1/2$ it gives

$$g(1/2) = g_{b,0}\Phi_{b,0}(1/2) = 0 \tag{17.10}$$

recalling that $\Phi_{b,0}(1/2) = 1$ then $u_{b,0} = 0$. Using the same procedure it follows that

$$g^{(1)}(-1/2) = g_{a,1}\Phi_{a,1}^{(1)}(-1/2) = 0 \tag{17.11}$$

which implies that $u_{a,1} = 0$ since $\Phi_{a,1}^{(1)}(-1/2) = 1$ by the defining properties of the Hermite Interpolating Polynomials and that

$$g^{(1)}(1/2) = g_{b,1}\Phi_{b,1}^{(1)}(1/2) = 0 \tag{17.12}$$

$g_{b,1} = 0$ since $\Phi_{b,1}^{(1)}(1/2) = 1$.

Thus, four of the 10 parameters to be evaluated are already known. From the differential equation it follows

$$\frac{d}{dx}\frac{d^4 g(x)}{dx^4} = \frac{d^5 g(x)}{dx^5} = 0 \tag{17.13}$$

therefore, if $\eta > 4$ then

$$g^{(u)}(-1) = g_{a,u}\Phi_{a,u}^{(u)}(-1/2) = 0$$
$$g^{(u)}(1) = g_{b,1}\Phi_{b,1}^{(1)}(1/2) = 0 \tag{17.14}$$

for $u = 5, 6, \ldots, \eta$. Since $\Phi_{a,u}^{(u)}(-1/2) = \Phi_{b,1}^{(1)}(1/2) = 1$ it implies that $g_{a,u} = g_{b,u} = 0$, $u = 5, 6, \ldots, \eta$.

This shows that there is no advantage in increasing the number of terms of the expansion, which is equivalent to use Hermite Interpolating Polynomials with higher degree, since the terms such that $\eta > 4$ will be no contribute to the solution of the differential equation.

Two more coefficients can be obtained from the differential equation. Since $\Phi_{a,4}^{(4)}(-1/2) = \Phi_{b,4}^{(4)}(1/2) = 1$ then

$$\left.\frac{\mathrm{d}^4 g(x)}{\mathrm{d}x^4}\right|_{x=-1/2} = g_{a,4}\Phi_{a,u}^{(u)}(-1/2) = 12c$$

$$\left.\frac{\mathrm{d}^4 u(x)}{\mathrm{d}x^4}\right|_{x=1/2} = g_{b,4}\Phi_{b,u}^{(u)}(1/2) = 12c \tag{17.15}$$

then $g_{a,4} = g_{b,4} = 12c$.

17.1.1.6   Expression for the solution

Substituting the known coefficients into the expansion of the approximate solution in equation (17.8) it gives

$$g(x) = g_{a,2}\Phi_{a,2}(x) + g_{a,3}\Phi_{a,3}(x) + 12c \times \Phi_{a,4}(x)$$
$$+ g_{b,2}\Phi_{b,2}(x) + g_{b,3}\Phi_{b,3}(x) + 12c \times \Phi_{b,4}(x) \tag{17.16}$$

17.1.1.7   Approximate solution

The derivatives of the exact solution in equation (17.3) are

$$[g_e(x)]^{(1)} = 2cx^3 - \frac{1}{2}cx$$

$$[g_e(x)]^{(2)} = 6cx^2 - \frac{1}{2}c$$

$$[g_e(x)]^{(3)} = 12cx$$

$$[g_e(x)]^{(4)} = 12c \tag{17.17}$$

and when evaluated at the end points a, b of the domain it gives as exact coordinate numbers for the expansion the following values

$$\begin{array}{ll} g_{a,0} = \quad 0 & g_{b,0} = 0 \\ g_{a,1} = \quad 0 & g_{b,1} = 0 \\ g_{a,2} = \quad 1.0 & g_{b,2} = 1.0 \\ g_{a,3} = -6.0 & g_{b,3} = 6.0 \\ g_{a,4} = \quad 12 & g_{b,4} = 12 \end{array} \tag{17.18}$$

which are the coordinate numbers of the expansion denoted in equation (17.17).

Table 17.1.1 gives the values of the norm $\|e\|$ of the approximate solution, the norm $\|g_e\|$ of exact solution, the relation $\|e\|/\|g_e\|$, and the norm $\|r\|$ of the residual of the differential equation.

Table 17.1.1 *Analysis of the error in the domain* $[-0.5, 0.5]$ *for the polynomials with* $\eta = 4$.

ϵ	$\|e\|$	$\|g_e\|$	$\|e\|/\|g_e\|$	$\|r\|$
10^{-3}	1.859×10^{-10}	1.992×10^{-4}	9.335×10^{-7}	9.454×10^{-8}
10^{-2}	1.728×10^{-10}	1.992×10^{-3}	8.673×10^{-8}	8.825×10^{-8}
10^{-1}	1.811×10^{-10}	1.992×10^{-2}	9.092×10^{-9}	9.193×10^{-8}
10^{0}	1.996×10^{-10}	1.992×10^{-1}	1.002×10^{-9}	1.020×10^{-7}
10^{1}	1.746×10^{-10}	1.992×10^{0}	2.767×10^{-11}	8.927×10^{-8}
10^{2}	1.746×10^{-10}	1.992×10^{1}	8.767×10^{-12}	8.926×10^{-8}
10^{3}	1.748×10^{-10}	1.992×10^{2}	8.770×10^{-13}	8.927×10^{-8}

A typical solution obtained for $\eta = 4$ and $c = 1$ is given by

$$
\begin{aligned}
g_{a,0} &= 0 & g_{b,0} &= 0 \\
g_{a,1} &= 0 & g_{b,1} &= 0 \\
g_{a,2} &= 0.999999999068677 & g_{b,2} &= 0.999999999068677 \\
g_{a,3} &= -5.99999999534339 & g_{b,3} &= 5.99999999534339 \\
g_{a,4} &= 12 & g_{b,4} &= 12
\end{aligned}
\tag{17.19}
$$

17.1.2   Approximate solution in the domain $[-1, 1]$

17.1.2.1   Coordinates transformation

The definition of the domain gives $\ell = (1/2) - (-1/2) = 1$. The length of the normalized domain is $L = 2$ thus, from equation (16.4) $\delta = L/\ell = 2$, which is the correction factor for the equation and for the variable, that is,

$$
X = \delta x = 2x \tag{17.20}
$$

The transformation of the derivatives is given by equation (16.7) as

$$
\frac{\mathrm{d}g^4}{\mathrm{d}x^4} \equiv \delta^4 \frac{\mathrm{d}G^4}{\mathrm{d}X^4} \tag{17.21}
$$

17.1.2.2   Transformation of the governing equation

The transformation of the domain equation is given by

$$
\frac{\mathrm{d}g^4}{\mathrm{d}x^4} = 12c \implies \delta^4 \frac{\mathrm{d}G^4}{\mathrm{d}X^4} = 12c \implies \frac{\mathrm{d}G^4}{\mathrm{d}X^4} = \frac{12c}{\delta^4} = 0.75c \tag{17.22}
$$

Transformation of the function boundary condition

$$g(x)|_{x=\pm0.5} = 0 \implies G(X)_{X=\pm2x=\pm2\times0.5=\pm1} = 0 \implies G(\pm1) = 0 \tag{17.23}$$

Transformation of the derivative boundary condition

$$\frac{dg}{dx}\bigg|_{x=\pm0.5} = 0 \implies \delta \frac{dG}{dX}\bigg|_{X=\pm2x=\pm2\times0.5=\pm1} = 0$$

$$\implies \frac{dG}{dX}(\pm1) = \frac{0}{\delta} = 0 \tag{17.24}$$

The domain of the solution has been transformed from $x \in [-1/2, 1/2]$ to $X \in [-1, 1]$ and the governing equation to

$$\frac{d^4G(X)}{dX^4} = 0.75c \tag{17.25}$$

with the boundary conditions

$$G\left(\pm\frac{1}{2}\right) = \frac{dG}{dX}\left(\pm\frac{1}{2}\right) = 0 \tag{17.26}$$

17.1.2.3   Transformation of the exact solution

The transformation of the exact solution, where $X = \delta x = 2x$, is given by

$$g_e(x) = \frac{1}{2}c\left(x^2 - \frac{1}{4}\right)^2 \implies G_e(X) = \frac{1}{2}c\left(\frac{X^2}{\delta^2} - \frac{1}{4}\right)^2$$

$$\implies G_e(X) = \frac{1}{2}c\left(\frac{X^2}{2^2} - \frac{1}{4}\right)^2$$

$$\implies G_e(X) = \frac{1}{32}c\left(X^2 - 1\right)^2 \tag{17.27}$$

The derivatives of the exact solution referred to the normalized domain are

$$[G_e(X)]^{(1)} = \frac{1}{8}cX\left(X^2 - 1\right)$$

$$[G_e(X)]^{(2)} = \frac{1}{8}c\left(3X^2 - 1\right)$$

$$[G_e(X)]^{(3)} = \frac{3}{4}cX$$

$$[G_e(X)]^{(4)} = \frac{3}{4}c \tag{17.28}$$

17.1.2.4 Expression for the solution

The transformation from the domain $[-1/2, 1/2]$ to $[-1, 1]$ was performed using $\ell = (1/2) - (-1/2) = 1$. The length of the normalized domain is $L = 2$ thus, from equation (16.4) $\delta = L/\ell = 2$, which is the correction factor used to obtain the solution

$$
\begin{aligned}
G(X) = {}& G_{a,2}\Phi_{a,2}(X) + G_{a,3}\Phi_{a,3}(X) + G_{a,4}\Phi_{a,4}(X) \\
& + G_{b,2}\Phi_{b,2}(X) + G_{b,3}\Phi_{b,3}(X) + G_{b,4}\Phi_{b,4}(X)
\end{aligned}
\tag{17.29}
$$

referred to the normalized domain to be equivalent to the equation (17.16) which refers to the original domain. The variable belonging to the normalized domain is denoted as X and the one that belongs to the original domain is denoted as x.

17.1.2.5 Equivalent expression for the solution

The expression for the solution referred to the domain $[-1/2, 1/2]$ can be obtained from equation (17.8) as

$$
\begin{aligned}
g(x) = {}& g_{a,2}\Phi_{a,2}(x) + g_{a,3}\Phi_{a,3}(x) + g_{a,4}\Phi_{a,4}(x) \\
& + g_{b,2}\Phi_{b,2}(x) + g_{b,3}\Phi_{b,3}(x) + g_{b,4}\Phi_{b,4}(x)
\end{aligned}
\tag{17.30}
$$

To relate the coefficients of the solution expressed by (17.30) and (17.29) it will be applied the property given by definition 7.1.4 on page 288, which is: the function must be equal to one at the reference node. According to this property the second derivative evaluates

$$
\frac{d^2}{dx^2}\left(\Phi_{a,2}(x)\right)_{x=-0.5} = 1 \qquad \frac{d^2}{dX^2}\left(\Phi_{a,2}(X)\right)_{x=-1} = 1
\tag{17.31}
$$

therefore,

$$
g_{a,2}\frac{d^2}{dx^2}\left(\Phi_{a,2}(x)\right)_{x=-0.5}(dx)^2 \equiv G_{a,2}\frac{d^2}{dX^2}\left(\Phi_{a,2}(X)\right)_{x=-1}(dX)^2
$$
$$
g_{a,2}(dx)^2 = G_{a,2}\left[d(\delta x)\right]^2
$$
$$
g_{a,2} = \delta^2 G_{a,2}
\tag{17.32}
$$

where $dX = \delta dx$ according to equation (16.5). In general, the transformation of the coefficients can be written as

$$
g_{a,n} = \delta^n G_{a,n}
\tag{17.33}
$$

Applying this transformation to all terms of the expression (17.29) it allows to write the solution referred to the domain $[-1/2, 1/2]$ as follows

$$
\begin{aligned}
u(x) = {}& \left(\delta^2 G_{a,2}\right)\Phi_{a,2}(x) + \left(\delta^3 G_{a,3}\right)\Phi_{a,3}(x) + \left(\delta^4 G_{a,4}\right)\Phi_{a,4}(x) \\
& + \left(\delta^2 G_{b,2}\right)\Phi_{b,2}(x) + \left(\delta^3 G_{b,3}\right)\Phi_{b,3}(x) + \left(\delta^4 G_{b,4}\right)\Phi_{b,4}(x)
\end{aligned}
\tag{17.34}
$$

17.1.2.6   Approximate solution in the normalized domain

The exact solution using the normalized domain is given by

$$
\begin{aligned}
G_{a,0} &= 0 & G_{b,0} &= 0 \\
G_{a,1} &= 0 & G_{b,1} &= 0 \\
G_{a,2} &= 0.25 & G_{b,2} &= 0.25 \\
G_{a,3} &= -0.75 & G_{b,3} &= 0.75 \\
G_{a,4} &= 0.75 & G_{b,4} &= 0.75
\end{aligned}
\tag{17.35}
$$

The table 17.1.2 gives the values of the norm $\|e\|$ of the approximate solution, the norm $\|G_e\|$ of exact solution, the relation $\|e\|/\|G_e\|$, and the norm $\|r\|$ of the residual of the differential equation.

Table 17.1.2 *Analysis of the error in the domain* $[-1, 1]$ *for the polynomials with* $\eta = 4$.

ϵ	$\|e\|$	$\|G_e\|$	$\|e\|/\|G_e\|$	$\|r\|$
10^{-3}	7.374×10^{-10}	1.992×10^{-4}	3.702×10^{-6}	2.345×10^{-8}
10^{-2}	7.530×10^{-10}	1.992×10^{-3}	3.780×10^{-7}	2.398×10^{-8}
10^{-1}	6.879×10^{-10}	1.992×10^{-2}	3.453×10^{-8}	2.176×10^{-8}
10^{0}	7.056×10^{-10}	1.992×10^{-1}	3.542×10^{-9}	2.236×10^{-8}
10^{1}	7.991×10^{-10}	1.992×10^{0}	4.012×10^{-10}	2.551×10^{-8}
10^{2}	7.991×10^{-10}	1.992×10^{1}	4.011×10^{-11}	2.550×10^{-8}
10^{3}	7.988×10^{-10}	1.992×10^{2}	4.009×10^{-12}	2.550×10^{-8}

A typical solution obtained for $\eta = 4$ and $c = 1$ is given by

$$
\begin{aligned}
G_{a,0} &= 0 & G_{b,0} &= 0 \\
G_{a,1} &= 0 & G_{b,1} &= 0 \\
G_{a,2} &= 0.250000000814907 & G_{b,2} &= 0.250000000814907 \\
G_{a,3} &= -0.750000001983608 & G_{b,3} &= 0.750000001983608 \\
G_{a,4} &= 0.749999999767169 & G_{b,4} &= 0.749999999667125
\end{aligned}
\tag{17.36}
$$

17.1.2.7   Transformation of the solution to the original domain

the the coefficients of the expansion of the approximate solution related to the original domain is obtained, according to equation (17.33), as

$$g_{a,0} = 0 \qquad\qquad\qquad g_{b,0} = 0$$
$$g_{a,1} = \delta G_{a,1} = 0 \qquad\qquad g_{b,1} = \delta G_{b,1} = 0$$
$$g_{a,2} = \delta^2 G_{a,2} = 4 \times 0.25 = 1 \qquad g_{b,2} = \delta^2 G_{b,2} = 4 \times 0.25 = 1$$
$$g_{a,3} = \delta^3 G_{a,3} = 8 \times -0.75 = -6 \qquad g_{b,3} = \delta^3 G_{b,3} = 8 \times 0.75 = 6$$
$$g_{a,4} = \delta^4 G_{a,4} = 16 \times 0.75 = 12 \qquad g_{b,4} = \delta^4 G_{b,4} = 16 \times 0.75 = 12 \quad (17.37)$$

17.1.3   Error analysis

The table 17.1.3 compares the accuracy obtained with the expansion of the solution in beam functions used by Davies, Karageorghis, and Phillips [8] with the expansion of the solution using Hermite Interpolating Polynomials. The table compares the norm of the error $\|e\|$ of the approximate solution, the relation $\|e\|/\|u_e\|$ between the norms of the error and the one of the exact solution.

Table 17.1.3 *Comparison of the accuracy of the solutions.*

ϵ	beam function		Hermite polynomial	
	$\|e\|$	$\|e\|/\|G_e\|$	$\|e\|$	$\|e\|/\|G_e\|$
10^{-3}	0.38×10^{-5}	0.19×10^{-4}	1.86×10^{-10}	9.35×10^{-7}
10^{-2}	0.37×10^{-5}	0.19×10^{-4}	1.73×10^{-10}	8.67×10^{-8}
10^{-1}	0.28×10^{-5}	0.14×10^{-4}	1.81×10^{-10}	9.09×10^{-9}

17.1.3.1   Remarks

The spectral methods represent the solution of a differential equation in terms of a truncated series of the trial functions, usually orthogonal functions.

It is used as trial functions, Davies, Karageorghis, and Phillips [8].

1. Eigen-functions of the differential equation operator in the equation to be solved, if it exists.

2. Eigen-functions of a related but simpler differential operator.

3. Eigen-functions of a singular Sturm-Lioville problem, such as Chebyshev polynomials.

Reference [8] uses the Chebyshev polynomials for the expansion of the approximate solution. To transform the differential equation into a set of nonlinear algebraic equations they use the Galerkin approach. To solve the nonlinear algebraic system of equations they use a modification of the Powell hybrid method.

The method proposed by Davies et al., do not satisfy identically the boundary conditions. They are added to the to the set of algebraic equations given by the governing differential equation.

The present approach expands the solution in terms of their derivatives using the Generalized Hermite Interpolation Polynomials. The boundary conditions, in general, can be eliminated from the system of equations algebraically.

17.2   Nonlinear differential equation

17.2.1   Modeling viscoelastic flows: case I

In a second example, Davies, Karageorghis, and Phillips [8] solve the following boundary value problem

17.2.1.1   The governing equation

$$\left[1 + \varepsilon \frac{dg(x)}{dx} \frac{d}{dx}\right] \frac{d^4 g(x)}{dx^4} = f(x) \tag{17.38}$$

where

$$f(x) = -120cx + 600\varepsilon c^2 \left(x^2 - \frac{1}{4}\right)\left(x^2 - \frac{1}{20}\right)$$

$$= -120cx + 600\varepsilon c^2 x^4 - 180\varepsilon c^2 x^2 + \frac{15}{2}\varepsilon c^2 \tag{17.39}$$

with the boundary conditions

$$g\left(\pm\frac{1}{2}\right) = \frac{dg}{dx}\left(\pm\frac{1}{2}\right) = 0$$

$$\frac{d^2 g}{dx^2}\left(-\frac{1}{2}\right) = c \tag{17.40}$$

The domain of the solution is $x \in [-0.5, 0.5]$.

17.2.1.2 Exact solution

The exact solution is given by

$$g_e(x) = -cx \left(x^2 - \frac{1}{4} \right)^2 \tag{17.41}$$

where c is a scaling constant in the range $1 \le c \le 10^3$.

The derivatives of the exact solution are

$$[g_e(x)]^{(1)} = -c \left(5x^4 - \frac{3}{2}x^2 + \frac{1}{16} \right)$$

$$[g_e(x)]^{(2)} = -c \left(20x^3 - 3x \right)$$

$$[g_e(x)]^{(3)} = -c \left(60x^2 - 3 \right)$$

$$[g_e(x)]^{(4)} = -120cx$$

$$[g_e(x)]^{(5)} = -120c$$

$$[g_e(x)]^{(6)} = 0 \tag{17.42}$$

17.2.1.3 Expansion of the solution

The approximate solution of the linear differential equation (17.38) when expanded in a basis of Hermite Interpolating Polynomials writes

$$g(x) = \sum_{r=0}^{r=\eta} [g_{a,r} \Phi_{a,r} + U_{b,r} \Phi_{b,r}] \tag{17.43}$$

and for $\eta = 5$ it reduces to

$$\begin{aligned}
g(x) = {} & g_{a,0} \Phi_{a,0}(x) + g_{a,1} \Phi_{a,1}(x) + g_{a,2} \Phi_{a,2}(x) + g_{a,3} \Phi_{a,3}(x) \\
& + g_{a,4} \Phi_{a,4}(x) + g_{a,5} \Phi_{a,5}(x) + g_{b,0} \Phi_{b,0}(x) + g_{b,1} \Phi_{b,1}(x) \\
& + g_{b,2} \Phi_{b,2}(x) + g_{b,3} \Phi_{b,3}(x) + g_{b,4} \Phi_{b,4}(x) + g_{b,5} \Phi_{b,5}(x)
\end{aligned} \tag{17.44}$$

17.2.1.4 Boundary conditions

Equation (17.44) when evaluated at $a = \alpha_0 = -1/2$ must satisfy the boundary condition in equation (17.40), thus

$$g(-1/2) = g_{a,0} \Phi_{a,0}(-1/2) = 0 \tag{17.45}$$

Analogously, when evaluated at the boundary $b = \alpha_0 = 1/2$ it gives

$$g(1/2) = g_{b,0} \Phi_{b,0}(1/2) = 0 \tag{17.46}$$

At the boundary $a = \alpha_0 = -1/2$ it follows

$$g^{(1)}(-1/2) = g_{a,1}\Phi_{a,1}^{(1)}(-1/2) = 0 \tag{17.47}$$

$$g^{(1)}(1/2) = g_{b,1}\Phi_{b,1}^{(1)}(1/2) = 0 \tag{17.48}$$

$$g^{(2)}(-1/2) = g_{a,1}\Phi_{a,1}^{(2)}(-1/2) = c \tag{17.49}$$

Thus five of the 10 parameters to be evaluated are already known.

From the differential equation (17.38) it follows

$$\frac{d}{dx}\left[\left(1 + \varepsilon\frac{dg(x)}{dx}\frac{d}{dx}\right)\frac{d^4g(x)}{dx^4}\right] = \frac{d}{dx}f(x)$$

$$g^{(5)}(x) + \varepsilon g^{(2)}(x)U^{(5)}(x) + \varepsilon g^{(1)}(x)g^{(6)}(x) = \frac{d}{dx}f(x) \tag{17.50}$$

from the equation (17.39) it can be seen that

$$\frac{d^6}{dx^6}f(x) = 0 \tag{17.51}$$

which implies that the term of highest derivative $\varepsilon g^{(1)}(x)g^{(6)}(x)$ in equation (17.50) becomes

$$\varepsilon g^{(1)}(x)g^{(11)}(x) \tag{17.52}$$

The equation (17.52) show that the differential equation equals zero when the highest order of derivative of the solution is 11. From the derivative of the exact solution, equation (17.42), it follows that $g^{(i)}(x) = 0, i = 6, 7, \ldots$ Therefore, it can be used as a rule that the best value for η is the one that will give minimum the residual of the differential.

17.2.1.5   Expression for the solution

Substituting the known coefficients into the expansion of the approximate solution in equation (17.44) it gives

$$U(x) = 0 \times \Phi_{a,0}(x) + 0 \times \Phi_{a,1}(x) + c \times \Phi_{a,2}(x) + g_{a,3}\Phi_{a,3}(x)$$
$$+ g_{a,4}\Phi_{a,4}(x) + g_{a,5}\Phi_{a,5}(x) + 0 \times \Phi_{b,0}(x) + 0 \times \Phi_{b,1}(x)$$
$$+ g_{b,2}\Phi_{b,2}(x) + g_{b,3}\Phi_{b,3}(x) + g_{b,4}\Phi_{b,4}(x) + g_{b,5}\Phi_{b,5}(x) \tag{17.53}$$

17.2.1.6   Approximate solution

Substituting the value of the coordinate of the boundary points $x = \pm 1/2$ in the expression of the derivatives of the exact solution, shown in equation

(17.42), it is obtained, for $c = 1$ and any ϵ, the following values

$$
\begin{aligned}
g_{a,0} &= 0 & g_{b,0} &= 0 \\
g_{a,1} &= 0 & g_{b,1} &= 0 \\
g_{a,2} &= 1.0 & g_{b,2} &= -1.0 \\
g_{a,3} &= -12.0 & g_{b,3} &= -12.0 \\
g_{a,4} &= 60 & g_{b,4} &= -60 \\
g_{a,5} &= -120 & g_{b,5} &= -120
\end{aligned}
\tag{17.54}
$$

which are the exact values for coordinate numbers in the expansion of the solution in equation (17.53).

The approximate solution, for $\eta = 5$, $c = 1$, and $\epsilon = 1$ is

$$
\begin{aligned}
g_{a,0} &= 0 & g_{b,0} &= 0 \\
g_{a,1} &= 0 & g_{b,1} &= 0 \\
g_{a,2} &= 1.0 & g_{b,2} &= -1.00000000000001 \\
g_{a,3} &= -12.0 & g_{b,3} &= -12.0 \\
g_{a,4} &= 60.0000000000005 & g_{b,4} &= -60.0000000000001 \\
g_{a,5} &= -120.000000000005 & g_{b,5} &= -120.000000000005
\end{aligned}
\tag{17.55}
$$

The approximate solution, for $\eta = 7$, $c = 1$, and any $\epsilon = 10^3$ is

$$
\begin{aligned}
g_{a,0} &= 0 & g_{b,0} &= 0 \\
g_{a,1} &= 0 & g_{b,1} &= 0 \\
g_{a,2} &= 1.0 & g_{b,2} &= -1.00000000000003 \\
g_{a,3} &= -12.0000000000004 & g_{b,3} &= -12.0000000000006 \\
g_{a,4} &= 60.000000000033 & g_{b,4} &= -60.0000000000351 \\
g_{a,5} &= -120.000000002116 & g_{b,5} &= -120.000000002142 \\
g_{a,6} &= 1.07253129083507 \times 10^{-7} & g_{b,6} &= -1.0452562239533 \times 10^{-7} \\
g_{a,7} &= -4.05839182349155 \times 10^{-6} & g_{b,7} &= -3.889158364245 \times 10^{-6}
\end{aligned}
\tag{17.56}
$$

The exact solution shows that the value of the derivative of the solution for any order above 5 equals zero. For $\eta = 7$ there are 14 basis polynomials in the expansion, therefore, one should expect the vanishing of the coordinate numbers corresponding the derivatives of order 6 and 7. This can be seen in equation (17.56). The coordinate numbers $g_{a,6}$, $g_{a,7}$, $g_{b,6}$, and $g_{b,7}$ are small when compared to the others.

17.2.1.7   Error analysis of the governing equation

The accuracy of the solution is analyzed by the convergence to zero of the residual of the differential equation. The residual of the differential equation

(17.38) is given by

$$r(x_i) := \left[1 + \varepsilon \frac{dg(x)}{dx} \frac{d}{dx}\right] \frac{d^4 g(x)}{dx^4} - f(x) \tag{17.57}$$

where $f(x)$ is given by equation (17.39).

The norm of the residual in the domain is computed as

$$\|r\| = \sqrt{\sum_{i=0}^{100} |r(x_i)|^2} \tag{17.58}$$

over 101 points equally spaced.

Recalling equation (16.48), page 548, the norm of the exact solution is given by

$$\|g_e\| = \sqrt{\sum_{i=0}^{m} |g_e(x_i)|^2} \tag{17.59}$$

where $g_e(x_i)$ is the value of the exact solution at the point x_i.

Table 17.2.1 shows the values of the norm of the error $\|e\|$ of the solution, the norm of the exact solution $\|g_e\|$, the ration between the norm of the error of the solution and the norm of the exact solution, and the norm of the residual $\|r\|$ of the differential equation for $\eta = 5$.

Table 17.2.1 *Analysis of the error in the domain $[-0.5, 0.5]$ for the polynomials with $\eta = 5$.*

$\eta = 5$				
ϵ	$\|e\|$	$\|g_e\|$	$\|e\|/\|g_e\|$	$\|r\|$
10^{-3}	1.907×10^{-16}	6.006×10^{-2}	3.175×10^{-15}	1.627×10^{-12}
10^{-2}	1.138×10^{-16}	6.006×10^{-2}	1.894×10^{-15}	1.734×10^{-12}
10^{-1}	3.538×10^{-16}	6.006×10^{-2}	5.891×10^{-15}	1.800×10^{-12}
10^{0}	8.624×10^{-16}	6.006×10^{-2}	1.436×10^{-14}	2.532×10^{-12}
10^{1}	4.207×10^{-15}	6.006×10^{-2}	7.005×10^{-14}	2.103×10^{-11}
10^{2}	5.132×10^{-15}	6.006×10^{-2}	8.544×10^{-14}	2.442×10^{-10}
10^{3}	5.566×10^{-15}	6.006×10^{-2}	9.267×10^{-14}	2.538×10^{-9}

Table 17.2.2 shows the values of the norm of the error $\|e\|$ of the solution, the norm of the exact solution $\|g_e\|$, the ration between the norm of the error of the solution and the norm of the exact solution, and the norm of the residual $\|r\|$ of the differential equation for $\eta = 6$.

Table 17.2.2 *Analysis of the error in the domain* $[-0.5, 0.5]$ *for the polynomials with* $\eta = 6$.

ϵ	$\|e\|$	$\|g_e\|$	$\|e\|/\|g_e\|$	$\|r\|$
		$\eta = 6$		
10^{-3}	6.636×10^{-16}	6.006×10^{-2}	1.105×10^{-14}	1.233×10^{-11}
10^{-2}	6.291×10^{-16}	6.006×10^{-2}	1.047×10^{-14}	1.168×10^{-11}
10^{-1}	4.410×10^{-16}	6.006×10^{-2}	7.342×10^{-15}	1.311×10^{-11}
10^{0}	1.247×10^{-15}	6.006×10^{-2}	2.076×10^{-14}	1.395×10^{-11}
10^{1}	1.337×10^{-14}	6.006×10^{-2}	2.227×10^{-13}	1.021×10^{-10}
10^{2}	7.810×10^{-15}	6.006×10^{-2}	1.300×10^{-13}	7.817×10^{-10}
10^{3}	7.797×10^{-15}	6.006×10^{-2}	1.298×10^{-13}	7.929×10^{-9}

Table 17.2.3 shows the values of the norm of the error $\|e\|$ of the solution, the norm of the exact solution $\|g_e\|$, the ration between the norm of the error of the solution and the norm of the exact solution, and the norm of the residual $\|r\|$ of the differential equation for $\eta = 7$.

Table 17.2.3 *Analysis of the error in the domain* $[-0.5, 0.5]$ *for the polynomials with* $\eta = 7$.

ϵ	$\|e\|$	$\|g_e\|$	$\|e\|/\|g_e\|$	$\|r\|$
		$\eta = 7$		
10^{-3}	1.200×10^{-15}	6.006×10^{-2}	1.998×10^{-14}	3.811×10^{-11}
10^{-2}	8.399×10^{-16}	6.006×10^{-2}	1.398×10^{-14}	3.786×10^{-11}
10^{-1}	1.126×10^{-15}	6.006×10^{-2}	1.875×10^{-14}	3.797×10^{-11}
10^{0}	2.751×10^{-15}	6.006×10^{-2}	4.581×10^{-14}	2.025×10^{-11}
10^{1}	2.007×10^{-15}	6.006×10^{-2}	3.341×10^{-14}	3.126×10^{-10}
10^{2}	2.571×10^{-15}	6.006×10^{-2}	4.280×10^{-14}	3.140×10^{-9}
10^{3}	3.484×10^{-15}	6.006×10^{-2}	5.801×10^{-14}	3.146×10^{-8}

Table 17.2.4 shows the values of the norm of the error $\|e\|$ of the solution, the norm of the exact solution $\|g_e\|$, the ration between the norm of the error of the solution and the norm of the exact solution, and the norm of the residual $\|r\|$ of the differential equation for $\eta = 8$.

Table 17.2.4 *Analysis of the error in the domain* $[-0.5, 0.5]$ *for the polynomials with* $\eta = 8$.

$\eta = 8$				
ϵ	$\|e\|$	$\|g_e\|$	$\|e\|/\|g_e\|$	$\|r\|$
10^{-3}	9.868×10^{-15}	6.006×10^{-2}	1.643×10^{-13}	1.042×10^{-10}
10^{-2}	1.015×10^{-14}	6.006×10^{-2}	1.690×10^{-13}	1.033×10^{-10}
10^{-1}	1.099×10^{-14}	6.006×10^{-2}	1.831×10^{-13}	1.050×10^{-10}
10^{0}	6.030×10^{-15}	6.006×10^{-2}	1.003×10^{-13}	1.534×10^{-10}
10^{1}	2.499×10^{-14}	6.006×10^{-2}	4.161×10^{-13}	8.863×10^{-10}
10^{2}	1.764×10^{-14}	6.006×10^{-2}	2.938×10^{-13}	8.306×10^{-9}
10^{3}	1.799×10^{-14}	6.006×10^{-2}	2.996×10^{-13}	8.292×10^{-8}

Table 17.2.5 shows the values of the norm of the error $\|e\|$ of the solution, the norm of the exact solution $\|g_e\|$, the ration between the norm of the error of the solution and the norm of the exact solution, and the norm of the residual $\|r\|$ of the differential equation for $\eta = 9$.

Table 17.2.5 *Analysis of the error in the domain* $[-0.5, 0.5]$ *for the polynomials with* $\eta = 9$.

$\eta = 9$				
ϵ	$\|e\|$	$\|g_e\|$	$\|e\|/\|g_e\|$	$\|r\|$
10^{-3}	1.732×10^{-14}	6.006×10^{-2}	2.883×10^{-13}	5.490×10^{-10}
10^{-2}	1.745×10^{-14}	6.006×10^{-2}	2.905×10^{-13}	5.473×10^{-10}
10^{-1}	1.637×10^{-14}	6.006×10^{-2}	2.726×10^{-13}	5.402×10^{-10}
10^{0}	1.405×10^{-14}	6.006×10^{-2}	2.339×10^{-13}	6.514×10^{-10}
10^{1}	1.887×10^{-13}	6.006×10^{-2}	3.145×10^{-12}	4.288×10^{-9}
10^{2}	1.156×10^{-13}	6.006×10^{-2}	1.924×10^{-12}	4.205×10^{-8}
10^{3}	1.059×10^{-13}	6.006×10^{-2}	1.764×10^{-12}	4.206×10^{-7}

Table 17.2.6 shows the values of the norm of the error $\|e\|$ of the solution, the norm of the exact solution $\|g_e\|$, the ration between the norm of the error of the solution and the norm of the exact solution, and the norm of the residual $\|r\|$ of the differential equation for $\eta = 10$.

Table 17.2.6 *Analysis of the error in the domain* $[-0.5, 0.5]$ *for the polynomials with* $\eta = 10$.

		$\eta = 10$		
ϵ	$\|e\|$	$\|g_e\|$	$\|e\|/\|g_e\|$	$\|r\|$
10^{-3}	3.171×10^{-15}	6.006×10^{-2}	5.271×10^{-13}	3.159×10^{-9}
10^{-2}	3.456×10^{-14}	6.006×10^{-2}	5.754×10^{-13}	3.157×10^{-9}
10^{-1}	3.167×10^{-14}	6.006×10^{-2}	5.273×10^{-13}	3.106×10^{-9}
10^{0}	3.678×10^{-14}	6.006×10^{-2}	6.124×10^{-13}	3.236×10^{-9}
10^{1}	6.042×10^{-13}	6.006×10^{-2}	1.006×10^{-11}	1.878×10^{-8}
10^{2}	1.756×10^{-13}	6.006×10^{-2}	2.924×10^{-12}	1.917×10^{-7}
10^{3}	1.153×10^{-13}	6.006×10^{-2}	1.920×10^{-12}	1.922×10^{-6}

17.2.1.8 Error analysis of the solution

The error e_i of the solution at the point x_i is defined as

$$e(x_i) = g_e(x_i) - g(x_i) \qquad (17.60)$$

where $g_e(x_i)$ is the exact solution and $g(x_i)$ the value of the approximate solution at the point x_i.

The norm of the error can be evaluated as

$$\|e\| = \sqrt{\sum_{i=0}^{101} |e(x_i)|^2} \qquad (17.61)$$

along 101 points in the domain.

Recalling equation (16.48), page 548, the norm of the exact solution is given by

$$\|g_e\| = \sqrt{\sum_{i=0}^{m} |g_e(x_i)|^2} \qquad (17.62)$$

where $g_e(x_i)$ is the value of the exact solution at the point x_i.

Table 17.2.7 compares the values of the norm of the error of the solution with the values of η for the values of $\varepsilon = 10^{-3}, 10^{-2}$, and 10^{-1}.

Table 17.2.7 *Values of the norm of the error* $\|e\|$ *of the solution for* $\eta = 5, 6, 7, 8$ *and* $\varepsilon = 10^{-3}, 10^{-2},$ *and* 10^{-1}.

norm $\|e\|$ of the error of the solution			
η	$\varepsilon = 10^{-3}$	$\varepsilon = 10^{-2}$	$\varepsilon = 10^{-1}$
5	1.907×10^{-16}	1.138×10^{-16}	3.538×10^{-16}
6	6.636×10^{-16}	6.291×10^{-16}	4.410×10^{-16}
7	1.200×10^{-15}	8.399×10^{-16}	1.126×10^{-15}
8	$9,868 \times 10^{-15}$	1.015×10^{-14}	1.099×10^{-14}
9	1.732×10^{-14}	1.745×10^{-14}	1.637×10^{-14}
10	3.171×10^{-14}	3.456×10^{-14}	3.167×10^{-14}

Table 17.2.8 compares the values of the norm of the error of the solution with the values of η for the values of $\varepsilon = 10^0, 10^1, 10^2$ and 10^3.

Table 17.2.8 *Values of the norm of the error* $\|e\|$ *of the solution for* $\eta = 5, 6, 7, 8$ *and* $\varepsilon = 10^0, 10^1, 10^2$ *and* 10^3.

norm $\|e\|$ of the error of the solution			
η	$\varepsilon = 10^0$	$\varepsilon = 10^1$	$\varepsilon = 10^2$	$\varepsilon = 10^3$
5	8.624×10^{-16}	4.207×10^{-15}	5.132×10^{-15}	5.566×10^{-15}
6	1.247×10^{-15}	1.337×10^{-14}	7.810×10^{-15}	7.797×10^{-15}
7	2.751×10^{-15}	2.007×10^{-15}	2.571×10^{-15}	3.484×10^{-15}
8	6.030×10^{-15}	2.499×10^{-14}	1.764×10^{-14}	1.799×10^{-14}
9	1.405×10^{-14}	1.887×10^{-13}	1.156×10^{-13}	1.059×10^{-13}
10	3.678×10^{-14}	6.042×10^{-13}	1.756×10^{-13}	1.153×10^{-13}

17.2.1.9   Comparison with the literature

Davies, Karageorghis, and Phillips [8] solved this differential equation expanding the solution in a basis of beam functions and they used the Galerkin approach to obtain the coordinate numbers for $\varepsilon = 10^{-3}, 10^{-2}$ and $\varepsilon = 10^{-1}$. The best accuracy obtained was for $\varepsilon = 10^{-3}$ whose values are $\|e\| = 0.17 \times 10^{-4}$ and $\|e\|/\|g_e\| = 0.28 \times 10^{-3}$.

In another paper, Davies, Karageorghis, and Phillips [9] solve the same problem using the beam functions as basis and the collocation approach to obtain the coordinate numbers of the expansion. The best accuracy was obtained for $\varepsilon = 10^{-3}$ and its value is $\|e\| = 0.44 \times 10^{-3}$.

Table 17.2.9 *Comparison of the error of the approximate solution obtained in references [8] and [9] with the solution obtained using the Hermite Interpolating Polynomials for $\varepsilon = 10^{-3}, 10^{-2},$ and 10^{-1}.*

ϵ	$\|e\|$ from ref. [8]	$\|e\|$ from ref. [9]	$\|e\|$ Hermite Inter. Poly.
10^{-3}	1.7×10^{-5}	4.4×10^{-4}	1.907×10^{-16}
10^{-2}	2.0×10^{-5}	4.4×10^{-4}	1.138×10^{-16}
10^{-1}	7.6×10^{-5}	5.1×10^{-4}	3.538×10^{-16}

Table 17.2.9 shows that the solution using the Hermite Interpolating Polynomials has higher accuracy than the solution using the approaches in references [8] and [9].

17.2.2 Modeling viscoelastic flows: case II

17.2.2.1 The governing equation

The governing equation is given by

$$\left[1 + \varepsilon \frac{dg(x)}{dx} \frac{d}{dx}\right] \frac{d^4 g(x)}{dx^4} = f(x) \qquad (17.63)$$

where

$$
f(x) = 4\pi^2 cx \cos\left[\pi\left(x - \frac{1}{2}\right)\right] - \frac{1}{2}\pi c\left[\pi^2\left(x^2 - \frac{1}{4}\right) - 12\right]\sin\left[\pi\left(x - \frac{1}{2}\right)\right]
$$
$$
+ \frac{1}{8}\varepsilon\pi^2 c^2\left[\pi^2\left(x^2 - \frac{1}{4}\right)^2 - 40\left(x^2 - \frac{1}{8}\right)\right]\cos\left[2\pi\left(x - \frac{1}{2}\right)\right]
$$
$$
+ \frac{1}{8}\varepsilon\pi c^2 x\left[12\pi^2\left(x^2 - \frac{1}{4}\right) - 40\right]\sin\left[2\pi\left(x - \frac{1}{2}\right)\right]
$$
$$
+ \frac{1}{8}\varepsilon\pi^2 c^2\left[\pi^2\left(x^2 - \frac{1}{4}\right)^2 + 5\right] \qquad (17.64)
$$

with the boundary conditions

$$g\left(\pm\frac{1}{2}\right) = \frac{dg}{dx}\left(\pm\frac{1}{2}\right) = 0$$
$$\frac{d^2 g}{dx^2}\left(-\frac{1}{2}\right) = c \qquad (17.65)$$

17.2.2.2 Exact solution

The exact solution is given by

$$g(x) = -\frac{c}{2\pi}\left(x^2 - \frac{1}{4}\right)\sin\left[\pi\left(x - \frac{1}{2}\right)\right] \tag{17.66}$$

17.2.2.3 Expansion of the solution

Analogously to the equation (17.68) the approximate solution of the linear differential equation (17.63) when expanded in a basis of Hermite Interpolating Polynomials writes

$$g(x) = \sum_{r=0}^{r=\eta}[g_{a,r}\Phi_{a,r} + U_{b,r}\Phi_{b,r}] \tag{17.67}$$

and for $\eta = 5$ it reduces to

$$\begin{aligned}
g(x) = {} & g_{a,0}\Phi_{a,0}(x) + g_{a,1}\Phi_{a,1}(x) + g_{a,2}\Phi_{a,2}(x) + g_{a,3}\Phi_{a,3}(x) \\
& + g_{a,4}\Phi_{a,4}(x) + g_{a,5}\Phi_{a,5}(x) + g_{b,0}\Phi_{b,0}(x) + g_{b,1}\Phi_{b,1}(x) \\
& + g_{b,2}\Phi_{b,2}(x) + g_{b,3}\Phi_{b,3}(x) + g_{b,4}\Phi_{b,4}(x) + g_{b,5}\Phi_{b,5}(x)
\end{aligned} \tag{17.68}$$

17.2.2.4 Boundary conditions

Equation (17.68) when evaluated at $a = -1/2$ must satisfy the boundary condition in equation (17.65), thus

$$g(-1/2) = g_{a,0}\Phi_{a,0}(-1/2) = 0 \tag{17.69}$$

Analogously, when evaluated at the boundary $b = 1/2$ it gives

$$g(1/2) = g_{b,0}\Phi_{b,0}(1/2) = 0 \tag{17.70}$$

At the boundary $a = -1/2$ it follows

$$g^{(1)}(-1/2) = g_{a,1}\Phi_{a,1}^{(1)}(-1/2) = 0 \tag{17.71}$$

$$g^{(1)}(1/2) = g_{b,1}\Phi_{b,1}^{(1)}(1/2) = 0 \tag{17.72}$$

$$g^{(2)}(-1/2) = g_{a,1}\Phi_{a,1}^{(2)}(-1/2) = c \tag{17.73}$$

Thus five of the 12 parameters to be evaluated are already known.

17.2.2.5 Expression for the solution

Substituting the known coefficients into the expansion of the approximate solution in equation (17.68) it gives

$$
\begin{aligned}
g(x) = {} & 0 \times \Phi_{a,0}(x) + 0 \times \Phi_{a,1}(x) + c \times \Phi_{a,2}(x) + g_{a,3}\Phi_{a,3}(x) \\
& + g_{a,4}\Phi_{a,4}(x) + g_{a,5}\Phi_{a,5}(x) + 0 \times \Phi_{b,0}(x) + 0 \times \Phi_{b,1}(x) \\
& + g_{b,2}\Phi_{b,2}(x) + g_{b,3}\Phi_{b,3}(x) + g_{b,4}\Phi_{b,4}(x) + g_{b,5}\Phi_{b,5}(x) \quad (17.74)
\end{aligned}
$$

17.2.2.6 Approximate solution

Substituting the value of the coordinate of the boundary points $x = \pm 1/2$ in the expression of the derivatives of the exact solution, shown in equation (17.66), it is obtained, for any ϵ, the following values

$$
\begin{array}{ll}
g_{a,0} = 0 & g_{b,0} = 0 \\
g_{a,1} = 0 & g_{b,1} = 0 \\
g_{a,2} = -c & g_{b,2} = -c \\
g_{a,3} = 3c & g_{b,3} = -3c \\
g_{a,4} = 2c\pi^2 & g_{b,4} = 2c\pi^2 \\
g_{a,5} = -10c\pi^2 & g_{b,5} = 10c\pi^2 \\
g_{a,6} = -3c\pi^4 & g_{b,6} = -3c\pi^4 \\
g_{a,7} = 21c\pi^4 & g_{b,7} = -21c\pi^4 \\
g_{a,8} = 4c\pi^6 & g_{b,8} = 4c\pi^6 \\
g_{a,9} = -36c\pi^6 & g_{b,9} = 36c\pi^6 \quad (17.75)
\end{array}
$$

which are the coordinate numbers of the expansion denoted in equation (17.74).

The approximate solution, for $\eta = 9$, $c = 1$, and $\epsilon = 1$ is

$$
\begin{array}{ll}
g_{a,0} = 0 & g_{b,0} = 0 \\
g_{a,1} = 0 & g_{b,1} = 0 \\
g_{a,2} = -1 & g_{b,2} = -0.999999999999485 \\
g_{a,3} = 3.00000000000888 & g_{b,3} = -3.00000000000301 \\
g_{a,4} = 19.7392088019163 & g_{b,4} = 19.7392088019977 \\
g_{a,5} = -98.6960440070226 & g_{b,5} = 98.6960440089141 \\
g_{a,6} = -292.227273100278 & g_{b,6} = -292.227273082118 \\
g_{a,7} = 2045.590911357 & g_{b,7} = -2045.59091156579 \\
g_{a,8} = 3845.55675285114 & g_{b,8} = 3845.55675535419 \\
g_{a,9} = -34610.0134380415 & g_{b,9} = 34610.0139437794 \quad (17.76)
\end{array}
$$

The approximate solution, for $\eta = 9$, $c = 1$, and $\epsilon = 10^3$ is

$$g_{a,0} = 0 \qquad\qquad g_{b,0} = 0$$
$$g_{a,1} = 0 \qquad\qquad g_{b,1} = 0$$
$$g_{a,2} = -1 \qquad\qquad g_{b,2} = -0.999999999998667$$
$$g_{a,3} = 3.00000000000529 \qquad g_{b,3} = -2.99999999999386$$
$$g_{a,4} = 19.7392088020345\pi^2 \qquad g_{b,4} = 19.7392088020219$$
$$g_{a,5} = -98.696044008653 \qquad g_{b,5} = 98.6960440085599$$
$$g_{a,6} = -292.227273099506 \qquad g_{b,6} = -292.227273083014$$
$$g_{a,7} = 2045.59091152527 \qquad g_{b,7} = -2045.59091163021$$
$$g_{a,8} = 3845.55676077953 \qquad g_{b,8} = 3845.55675339305$$
$$g_{a,9} = -34610.0134539275 \qquad g_{b,9} = 34610.0146237699 \qquad (17.77)$$

The exact solution contains all derivatives which absolute values increase with the order of derivative. The Hermite Interpolating Polynomials decrease in value as the order of derivative for which the polynomial equals one at the reference node increases. Therefore, the the higher the value of the coordinate numbers the lower is the value of the Hermite polynomial giving a product the reduces in value. Since the final value is the sum of all these partial values the resulting series is convergent.

17.2.2.7 Error analysis of the governing equation

The accuracy of the solution is analyzed by the convergence to zero of the residual of the differential equation. The residual of the differential equation (17.63) is given by

$$r(x_i) := \left[1 + \varepsilon \frac{dg(x)}{dx} \frac{d}{dx} \right] \frac{d^4 g(x)}{dx^4} - f(x) \qquad (17.78)$$

where $f(x)$ is given by equation (17.64).

The norm of the residual in the domain is computed as

$$\|r\| = \sqrt{\sum_{i=0}^{100} |r(x_i)|^2} \qquad (17.79)$$

over 101 point equally spaced.

Recalling equation (16.48), page 548, the norm of the exact solution is given by

$$\|g_e\| = \sqrt{\sum_{i=0}^{m} |g_e(x_i)|^2} \qquad (17.80)$$

where $g_e(x_i)$ is the value of the exact solution at the point x_i.

Table 17.2.10 shows the values of the norm of the error $\|e\|$ of the solution, the norm of the exact solution $\|g_e\|$, the ratio between the norm of the error of the solution and the norm of the exact solution, and the norm of the residual $\|r\|$ of the differential equation for $\eta = 5$.

Table 17.2.10 *Analysis of the error in the domain* $[-0.5, 0.5]$ *for the polynomials with* $\eta = 5$.

ϵ	$\|e\|$	$\|g_e\|$	$\|e\|/\|g_e\|$	$\|r\|$
		$\eta = 5$		
10^{-3}	8.201×10^{-8}	2.484×10^{-1}	3.301×10^{-7}	9.332×10^{-3}
10^{-2}	8.268×10^{-8}	2.484×10^{-1}	3.328×10^{-7}	9.339×10^{-3}
10^{-1}	1.051×10^{-7}	2.484×10^{-1}	4.229×10^{-7}	9.525×10^{-3}
10^0	1.539×10^{-6}	2.484×10^{-1}	6.195×10^{-6}	1.796×10^{-2}
10^1	1.489×10^{-6}	2.484×10^{-1}	5.994×10^{-6}	1.964×10^{-1}
10^2	1.640×10^{-6}	2.484×10^{-1}	6.602×10^{-6}	1.386×10^{-0}
10^3	1.700×10^{-6}	2.484×10^{-1}	6.720×10^{-6}	13.848

Table 17.2.11 shows the values of the norm of the error $\|e\|$ of the solution, the norm of the exact solution $\|g_e\|$, the ratio between the norm of the error of the solution and the norm of the exact solution, and the norm of the residual $\|r\|$ of the differential equation for $\eta = 6$.

Table 17.2.11 *Analysis of the error in the domain* $[-0.5, 0.5]$ *for the polynomials with* $\eta = 6$.

ϵ	$\|e\|$	$\|g_e\|$	$\|e\|/\|g_e\|$	$\|r\|$
		$\eta = 6$		
10^{-3}	3.154×10^{-10}	2.484×10^{-1}	1.270×10^{-9}	9.113×10^{-5}
10^{-2}	3.184×10^{-10}	2.484×10^{-1}	1.282×10^{-9}	9.120×10^{-5}
10^{-1}	4.079×10^{-10}	2.484×10^{-1}	1.642×10^{-9}	9.366×10^{-5}
10^0	1.235×10^{-8}	2.484×10^{-1}	4.971×10^{-8}	1.976×10^{-4}
10^1	3.456×10^{-8}	2.484×10^{-1}	3.456×10^{-8}	1.594×10^{-3}
10^2	3.742×10^{-8}	2.484×10^{-1}	1.506×10^{-7}	1.580×10^{-2}
10^3	3.778×10^{-8}	2.484×10^{-1}	1.520×10^{-7}	1.579×10^{-1}

Table 17.2.12 shows the values of the norm of the error $\|e\|$ of the solution, the norm of the exact solution $\|g_e\|$, the ratio between the norm of the error of

the solution and the norm of the exact solution, and the norm of the residual $\|r\|$ of the differential equation for $\eta = 7$.

Table 17.2.12 *Analysis of the error in the domain* $[-0.5, 0.5]$ *for the polynomials with* $\eta = 7$.

$\eta = 7$				
ϵ	$\|e\|$	$\|g_e\|$	$\|e\|/\|g_e\|$	$\|r\|$
10^{-3}	9.211×10^{-12}	2.484×10^{-1}	3.708×10^{-11}	1.069×10^{-6}
10^{-2}	9.140×10^{-12}	2.484×10^{-1}	3.679×10^{-11}	1.069×10^{-6}
10^{-1}	8.421×10^{-12}	2.484×10^{-1}	3.390×10^{-11}	1.085×10^{-6}
10^{0}	1.295×10^{-10}	2.484×10^{-1}	5.211×10^{-10}	2.291×10^{-6}
10^{1}	4.904×10^{-10}	2.484×10^{-1}	1.974×10^{-9}	2.080×10^{-5}
10^{2}	5.045×10^{-10}	2.484×10^{-1}	2.031×10^{-9}	2.080×10^{-4}
10^{3}	5.056×10^{-10}	2.484×10^{-1}	2.036×10^{-9}	2.080×10^{-3}

Table 17.2.13 shows the values of the norm of the error $\|e\|$ of the solution, the norm of the exact solution $\|g_e\|$, the ratio between the norm of the error of the solution and the norm of the exact solution, and the norm of the residual $\|r\|$ of the differential equation for $\eta = 8$.

Table 17.2.13 *Analysis of the error in the domain* $[-0.5, 0.5]$ *for the polynomials with* $\eta = 8$.

$\eta = 8$				
ϵ	$\|e\|$	$\|g_e\|$	$\|e\|/\|g_e\|$	$\|r\|$
10^{-3}	1.610×10^{-13}	2.484×10^{-1}	6.481×10^{-13}	1.013×10^{-8}
10^{-2}	1.632×10^{-13}	2.484×10^{-1}	6.568×10^{-13}	1.011×10^{-8}
10^{-1}	1.516×10^{-13}	2.484×10^{-1}	6.104×10^{-13}	1.017×10^{-8}
10^{0}	1.312×10^{-12}	2.484×10^{-1}	5.283×10^{-12}	2.066×10^{-8}
10^{1}	4.796×10^{-12}	2.484×10^{-1}	1.931×10^{-11}	1.993×10^{-7}
10^{2}	4.931×10^{-12}	2.484×10^{-1}	1.985×10^{-11}	2.019×10^{-6}
10^{3}	4.972×10^{-12}	2.484×10^{-1}	2.001×10^{-11}	2.022×10^{-5}

Table 17.2.14 shows the values of the norm of the error $\|e\|$ of the solution, the norm of the exact solution $\|g_e\|$, the ratio between the norm of the error of

the solution and the norm of the exact solution, and the norm of the residual $\|r\|$ of the differential equation for $\eta = 9$.

Table 17.2.14 *Analysis of the error in the domain* $[-0.5, 0.5]$ *for the polynomials with* $\eta = 9$.

$\eta = 9$				
ϵ	$\|e\|$	$\|g_e\|$	$\|e\|/\|g_e\|$	$\|r\|$
10^{-3}	3.305×10^{-14}	2.484×10^{-1}	1.330×10^{-13}	5.366×10^{-10}
10^{-2}	3.389×10^{-14}	2.484×10^{-1}	1.330×10^{-13}	5.404×10^{-10}
10^{-1}	3.334×10^{-14}	2.484×10^{-1}	1.342×10^{-13}	5.778×10^{-10}
10^{0}	8.982×10^{-14}	2.484×10^{-1}	3.616×10^{-13}	9.742×10^{-10}
10^{1}	1.371×10^{-13}	2.484×10^{-1}	5.520×10^{-13}	5.730×10^{-9}
10^{2}	1.290×10^{-13}	2.484×10^{-1}	5.194×10^{-13}	5.484×10^{-8}
10^{3}	1.299×10^{-13}	2.484×10^{-1}	5.229×10^{-13}	5.465×10^{-7}

Table 17.2.15 shows the values of the norm of the error $\|e\|$ of the solution, the norm of the exact solution $\|g_e\|$, the ratio between the norm of the error of the solution and the norm of the exact solution, and the norm of the residual $\|r\|$ of the differential equation for $\eta = 10$.

Table 17.2.15 *Analysis of the error in the domain* $[-0.5, 0.5]$ *for the polynomials with* $\eta = 10$.

$\eta = 10$				
ϵ	$\|e\|$	$\|g_e\|$	$\|e\|/\|g_e\|$	$\|r\|$
10^{-3}	1.784×10^{-13}	2.48435×10^{-1}	7.179×10^{-13}	3.053×10^{-9}
10^{-2}	1.783×10^{-13}	2.48435×10^{-1}	7.179×10^{-13}	3.068×10^{-9}
10^{-1}	1.810×10^{-13}	2.48435×10^{-1}	7.285×10^{-13}	3.229×10^{-9}
10^{0}	1.705×10^{-13}	2.48435×10^{-1}	6.861×10^{-13}	4.859×10^{-9}
10^{1}	4.438×10^{-13}	2.48435×10^{-1}	1.786×10^{-12}	2.290×10^{-8}
10^{2}	4.564×10^{-13}	2.48435×10^{-1}	1.837×10^{-12}	2.127×10^{-7}
10^{3}	4.531×10^{-13}	2.48435×10^{-1}	1.824×10^{-12}	2.113×10^{-6}

17.2.2.8   Error analysis of the solution

The error e_i of the solution at the point x_i is defined as

$$e(x_i) = U_e(x_i) - U(x_i) \qquad (17.81)$$

where $U_e(x_i)$ is the exact solution and $U(x_i)$ the value of the approximate solution at the point x_i.

The norm of the error was evaluated using

$$\|e\| = \sqrt{\sum_{i=0}^{101} |e(x_i)|^2} \qquad (17.82)$$

along 101 points in the domain.

Recalling equation (16.48), page 548, the norm of the exact solution is given by

$$\|g_e\| = \sqrt{\sum_{i=0}^{m} |g_e(x_i)|^2} \qquad (17.83)$$

where $g_e(x_i)$ is the value of the exact solution at the point x_i.

Table 17.2.16 compares the values of the norm of the error of the solution with the values of η.

Table 17.2.16 *Values of the norm of the error $\|e\|$ of the solution for $\eta = 5, 6, 7, 8, 9$ and $\varepsilon = 10^{-3}, 10^{-2}$, and 10^{-1}.*

	norm $\|e\|$ of the error of the solution		
η	$\varepsilon = 10^{-3}$	$\varepsilon = 10^{-2}$	$\varepsilon = 10^{-1}$
5	8.201×10^{-8}	8.268×10^{-8}	1.051×10^{-7}
6	3.154×10^{-10}	3.184×10^{-10}	4.079×10^{-10}
7	9.211×10^{-12}	9.140×10^{-12}	8.421×10^{-12}
8	1.610×10^{-13}	1.632×10^{-13}	1.516×10^{-13}
9	3.305×10^{-14}	3.389×10^{-14}	3.334×10^{-14}
10	7.179×10^{-13}	7.179×10^{-13}	7.285×10^{-13}

Table 17.2.17 *Values of the norm of the error $\|e\|$ of the solution for $\eta = 5, 6, 7, 8, 9$ and $\varepsilon = 10^0, 10^1, 10^2,$ and 10^3.*

	norm $\|e\|$ of the error of the solution			
η	$\varepsilon = 10^0$	$\varepsilon = 10^1$	$\varepsilon = 10^2$	$\varepsilon = 10^3$
5	1.539×10^{-6}	1.489×10^{-6}	1.640×10^{-6}	1.700×10^{-6}
6	1.235×10^{-8}	3.456×10^{-8}	3.742×10^{-8}	3.778×10^{-8}
7	1.295×10^{-10}	4.904×10^{-10}	5.045×10^{-10}	5.056×10^{-10}
8	1.315×10^{-12}	4.796×10^{-12}	4.931×10^{-12}	4.972×10^{-12}
9	8.982×10^{-14}	1.371×10^{-13}	1.290×10^{-13}	1.299×10^{-13}
10	6.861×10^{-13}	1.786×10^{-12}	1.837×10^{-12}	4.531×10^{-13}

17.2.2.9 Comparison with the literature

Davies, Karageorghis, and Phillips [8] solved this differential equation expanding the solution in a basis of beam functions and they used the Galerkin approach to obtain the coordinate numbers for $\varepsilon = 10^{-3}, 10^{-2},$ and $\varepsilon = 10^{-1}$. The best accuracy obtained was for $\varepsilon = 10^{-3}$ whose values are $\|e\| = 0.17 \times 10^{-4}$ and $\|e\|/\|g_e\| = 0.28 \times 10^{-3}$.

In another paper, Davies, Karageorghis, and Phillips [9] solve the same problem using the beam functions as basis and the collocation approach to obtain the coordinate numbers of the expansion. The best accuracy was obtained for $\varepsilon = 10^{-3}$ and its value is $\|e\| = 0.44 \times 10^{-3}$.

Table 17.2.18 *Comparison of the error of the approximate solution obtained in references [8] with the solution obtained using the Hermite Interpolating Polynomials for $\eta = 9$ and $\varepsilon = 10^{-3}, 10^{-2},$ $10^{-1}, 10^0, 10^1, 10^2,$ and 10^3.*

	norm $\|g\|$ of the error of the approximate solution		
ϵ	reference [8], with Galerkin		Hermite for $\eta = 9$
	beam function	Chebyshev	and with collocation
10^{-3}	5.9×10^{-6}	2.9×10^{-10}	3.305×10^{-14}
10^{-2}	5.8×10^{-6}	1.8×10^{-10}	3.389×10^{-14}
10^{-1}	3.1×10^{-6}	1.3×10^{-10}	3.334×10^{-14}
10^0	——	——	8.982×10^{-14}
10^1	——	——	1.371×10^{-13}
10^2	——	——	1.290×10^{-13}
10^3	——	——	1.299×10^{-13}

Table 17.2.19 *Comparison of the error of the approximate solution obtained in references [9] with the solution obtained using the Hermite Interpolating Polynomials for $\eta = 9$ and $\varepsilon = 10^{-3}, 10^{-2}, 10^{-1}, 10^{0}, 10^{1}, 10^{2},$ and 10^{3}.*

	norm $\|g\|$ of the error of the approximate solution		
ϵ	reference [9], with collocation		Hermite for $\eta = 9$
	beam function	Chebyshev	and with collocation
10^{-3}	1.2×10^{-4}	2.1×10^{-14}	3.305×10^{-14}
10^{-2}	1.2×10^{-4}	3.8×10^{-14}	3.389×10^{-14}
10^{-1}	1.6×10^{-3}	————	3.334×10^{-14}
10^{0}	3.3×10^{-3}	————	8.982×10^{-14}
10^{1}	————	4.1×10^{-12}	1.371×10^{-13}
10^{2}	————	2.0×10^{-11}	1.290×10^{-13}
10^{3}	————	————	1.299×10^{-13}

Table 17.2.18 shows that the solution using the Hermite Interpolating Polynomials with collocation presents an accuracy of the same order of magnitude or better when compared to the solution using Chebyshev polynomials with collocation as it was performed in reference [9].

17.2.3 Analysis of the solution in the interval $[-1, 1]$

17.2.3.1 Graphic of the exact solution

The exact solution given by equation (17.66) in the interval $[-1, 1]$ and shown in figure 17.2.20, has two maxima one at the coordinate $x = -0.5$ and the other at $x = 0.5$.

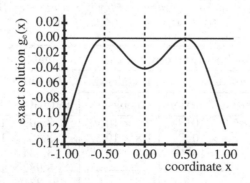

Figure 17.2.20 *Graphic of the exact solution in the interval $[-1.0, 1.0]$.*

The approximate solution for the interval $[-1, 1]$ required more terms in the expansion to give the same accuracy of approximation as is shown in table 17.2.21. It shows the values of the norm of the error $\|e\|$ of the solution, the norm of the approximate solution $\|g_e\|$, the ratio between the norm of the error of the solution and the norm of the exact solution, and the norm of the residual $\|r\|$ of the differential equation for the values of η from $\eta = 5$ through $\eta = 12$. The value of the norm of the exact solution is $\|g_e\| = 0.461104$.

Table 17.2.21 *Analysis of the error in the domain* $[-1.0, 1.0]$

η	$\|e\|$	$\|g\|$	$\|e\|/\|g_e\|$	$\|r\|$
5	1.308×10^{-2}	4.56354×10^{-1}	2.837×10^{-2}	1.086×10^{1}
6	1.012×10^{-3}	4.60723×10^{-1}	2.195×10^{-3}	9.930×10^{-1}
7	5.594×10^{-5}	4.61082×10^{-1}	1.213×10^{-4}	6.466×10^{-2}
8	2.325×10^{-6}	4.61103×10^{-1}	5.042×10^{-6}	3.151×10^{-3}
9	7.536×10^{-8}	4.61104×10^{-1}	1.634×10^{-7}	1.193×10^{-4}
10	1.960×10^{-9}	4.61104×10^{-1}	4.251×10^{-9}	3.607×10^{-6}
11	4.184×10^{-11}	4.61104×10^{-1}	9.073×10^{-11}	8.906×10^{-8}
12	7.463×10^{-13}	4.61104×10^{-1}	1.619×10^{-12}	1.891×10^{-9}

17.3   Flow of non-Newtonian fluids: Powell-Eyring model

Doctor and Kalthia [10] in a paper *Spline collocation in the flow of non-Newtonian fluids* solve the boundary value problem

17.3.1   Prandtl-Eyring model: case I

17.3.1.1   The governing equation

The Prandtl-Eyring governing differential equation is given by [10]

$$\frac{d^2v}{ds^2} = \xi\,(s-1)\left[1 + \left(\varepsilon\frac{dv}{ds}\right)^2\right]^{1/2} \qquad s \in [0, 1] \qquad (17.84)$$

with the boundary conditions

$$v(0) = 0 \qquad v(1) = 0 \qquad (17.85)$$

where

$$\xi = \frac{\mu}{A} \qquad \varepsilon = \frac{aG_r P_r}{B\ell^2} \tag{17.86}$$

The variable $v(s)$ represents the velocity profile of the fluid.

17.3.1.2   Particular case of the governing equation

If $\varepsilon = 0$ the equation reduces to

$$\frac{d^2 v}{ds^2} = \xi \, (s - 1) \tag{17.87}$$

whose solution is

$$v(s) = \frac{\xi}{6} \left(s^3 - 3s^2 + 2s \right) \tag{17.88}$$

17.3.1.3   Expansion of the solution

The approximate solution of the linear differential equation (17.84) when expanded in a basis of Hermite Interpolating Polynomials writes

$$v(s) = \sum_{r=0}^{r=\eta} \left[v_{a,r} \Phi_{a,r} + v_{b,r} \Phi_{b,r} \right] \tag{17.89}$$

and for $\eta = 4$ it reduces to

$$
\begin{aligned}
v(X) = {} & v_{a,0} \Phi_{a,0}(s) + v_{a,1} \Phi_{a,1}(s) + v_{a,2} \Phi_{a,2}(s) + v_{a,3} \Phi_{a,3}(s) \\
& + v_{a,4} \Phi_{a,4}(s) + v_{b,0} \Phi_{b,0}(s) + v_{b,1} \Phi_{b,1}(s) + v_{b,2} \Phi_{b,2}(s) \\
& + v_{b,3} \Phi_{b,3}(s) + v_{b,4} \Phi_{b,4}(s)
\end{aligned} \tag{17.90}
$$

17.3.1.4   Boundary conditions

Equation (17.89) when evaluated at $a = 0.0$ must satisfy the boundary condition in equation (17.85), thus

$$v(0.0) = v_{a,0} \Phi_{a,0}(0.0) = 0 \tag{17.91}$$

Analogously, when evaluated at the boundary $b = 1.0$ it gives

$$v(1.0) = v_{b,0} \Phi_{b,0}(1.0) = 0 \tag{17.92}$$

Thus two of the 10 parameters to be evaluated are already known.

17.3.1.5 Expression for the solution

Substituting the known coefficients into the expansion of the approximate solution in equation (17.89) it gives

$$
\begin{aligned}
v(s) = {} & 0 \times \Phi_{a,0}(s) + v_{a,1}\Phi_{a,1}(s) + v_{a,2}\Phi_{a,2}(s) + v_{a,3}\Phi_{a,3}(s) \\
& + v_{a,4}\Phi_{a,4}(s) + 0 \times \Phi_{b,0}(s) + v_{b,1}\Phi_{b,1}(s) + v_{b,2}\Phi_{b,2}(s) \\
& + v_{b,3}\Phi_{b,3}(s) + v_{b,4}\Phi_{b,4}(s)
\end{aligned} \tag{17.93}
$$

17.3.1.6 Approximate solution

The coordinate numbers for the approximate solution with $\eta = 4$ and $\xi = 0.91$ is

$$
\begin{aligned}
v_{a,0} &= 0 & v_{b,0} &= 0 \\
v_{a,1} &= 0.307729725643813 & v_{b,1} &= -0.15251633414835 \\
v_{a,2} &= -0.952112604298896 & v_{b,2} &= 3.09165622357827 \times 10^{-7} \\
v_{a,3} &= 1.20690578610886 & v_{b,3} &= 0.920555504020408 \\
v_{a,4} &= -1.55142254583178 & v_{b,4} &= 0.00122785316973861
\end{aligned} \tag{17.94}
$$

which are the coordinate numbers of the expansion denoted in equation (17.93).

The coordinate numbers for the approximate solution with $\eta = 4$ and $\xi = 0.5$ is

$$
\begin{aligned}
v_{a,0} &= 0 & v_{b,0} &= 0 \\
v_{a,1} &= 0.167394706494368 & v_{b,1} &= -0.0834739956932138 \\
v_{a,2} &= -0.506956833674366 & v_{b,2} &= 1.63934828456717 \times 10^{-8} \\
v_{a,3} &= 0.548803747287953 & v_{b,3} &= 0.501740616274467 \\
v_{a,4} &= -0.252213972595609 & v_{b,4} &= 6.21436524872176 \times 10^{-5}
\end{aligned} \tag{17.95}
$$

17.3.1.7 Error analysis

Since the solution is unknown the error analysis was performed with the norm of the residual of the Prandtl-Eyring governing differential equation (17.84) as follows

$$
r(x_i) = \frac{d^2 v}{ds^2} - \xi\,(s-1)\left[1 + \left(\varepsilon\frac{dv}{ds}\right)^2\right]^{1/2} \qquad s \in [0,1] \tag{17.96}
$$

The norm of the residual in the domain is computed as

$$
\|r\| = \sqrt{\sum_{i=0}^{100} |r(x_i)|^2} \tag{17.97}
$$

over 101 point equally spaced in the domain.

The norm of the approximate solution was evaluated using

$$\|v\| = \sqrt{\sum_{i=0}^{101} |v(x_i)|^2} \tag{17.98}$$

along 101 points in the domain.

Table 17.3.1 shows the values of the norm $\|v\|$ of the solution, and the norm of the residual $\|r\|$ of the differential equation for from $\eta = 2$ through $\eta = 9$.

Table 17.3.1 *Analysis of the behavior of the solution in the domain* $[0.0, 1.0]$ *for the polynomials from* $\eta = 2$ *through* $\eta = 9$.

η	$\xi = 0.91$		$\xi = 0.5$	
	$\|v\|$	$\|r\|$	$\|v\|$	$\|r\|$
2	4.21063×10^{-1}	1.440×10^{-4}	2.30424×10^{-1}	1.967×10^{-3}
3	4.21094×10^{-1}	4.328×10^{-5}	2.30429×10^{-1}	2.159×10^{-6}
4	4.21094×10^{-1}	1.561×10^{-6}	2.30429×10^{-1}	7.642×10^{-8}
5	4.21094×10^{-1}	2.022×10^{-4}	2.30429×10^{-1}	1.649×10^{-4}
6	4.21094×10^{-1}	4.8803×10^{-4}	2.30429×10^{-1}	3.199×10^{-4}
7	4.21094×10^{-1}	8.78234×10^{-4}	2.30429×10^{-1}	4.391×10^{-4}
8	4.21094×10^{-1}	1.35129×10^{-3}	2.30428×10^{-1}	6.5193×10^{-4}

The table 17.3.1 shows that the best approximation was obtained for $\eta = 4$, that is, for the value of η for which the norm of the residual of the differential equation is minimum.

17.3.1.8   Graphic of the solution

Figure 17.3.2 shows the graphic of the approximate solution of the equation (17.84).

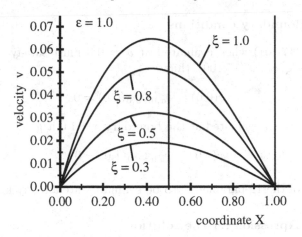

Figure 17.3.2 *Velocity profile of a Powell-Eyring flow between two walls.*

17.3.2 Powell-Eyring model: case II

17.3.2.1 The governing equation

The Powell-Eyring governing differential equation is given by [10] on page 416, equation 12, writes

$$\left\{1 + \left[\frac{1}{\xi}\left(\frac{1}{\varepsilon}\frac{dv}{ds}\right)^2 + 1\right]^{1/2}\right\}\frac{d^2v}{ds^2} = s - 1 \qquad s \in [0,1] \qquad (17.99)$$

with the boundary conditions

$$v(0) = 0 \qquad v(1) = 0 \qquad (17.100)$$

using a solution expanded in cubic splines.

17.3.2.2 Expansion of the solution

Analogously to the equation (17.68) the approximate solution of the linear differential equation (17.99) when expanded in a basis of Hermite Interpolating Polynomials writes

$$v(s) = v_{0,0}\Phi_{0,0}^{(0)}(s) + v_{0,1}\Phi_{0,1}^{(0)}(s) + v_{0,2}\Phi_{0,2}^{(0)}(s) + v_{0,3}\Phi_{0,3}^{(0)}(s)$$
$$+ v_{0,4}\Phi_{0,4}^{(0)}(s) + v_{0,5}\Phi_{0,5}^{(0)}(s) + v_{1,0}\Phi_{1,0}^{(0)}(s) + v_{1,1}\Phi_{1,1}^{(0)}(s)$$
$$+ v_{1,2}\Phi_{1,2}^{(0)}(s) + v_{1,3}\Phi_{1,3}^{(0)}(s) + v_{1,4}\Phi_{1,4}^{(0)}(s) + v_{1,5}\Phi_{1,5}^{(0)}(s) \qquad (17.101)$$

17.3.2.3   Boundary conditions

Equation (17.101) when evaluated at $a = 0.0$ must satisfy the boundary condition in equation (17.100), thus

$$v(0.0) = v_{a,0}\Phi_{a,0}(0.0) = 0 \qquad (17.102)$$

Analogously, when evaluated at the boundary $b = 1.0$ it gives

$$v(1.0) = v_{b,0}\Phi_{b,0}(1.0) = 0 \qquad (17.103)$$

Thus two of the 12 parameters to be evaluated are already known.

17.3.2.4   Expression for the solution

Substituting the known coefficients into the expansion of the approximate solution in equation (17.101) it gives

$$\begin{aligned}
v(s) = {} & 0 \times \Phi_{a,0}(s) + v_{a,1}\Phi_{a,1}(s) + v_{a,2}\Phi_{a,2}(s) + v_{a,3}\Phi_{a,3}(s) \\
& + v_{a,4}\Phi_{a,4}(s) + v_{a,5}\Phi_{a,5}(s) + 0 \times \Phi_{b,0}(s) + v_{b,1}\Phi_{b,1}(s) \\
& + v_{b,2}\Phi_{b,2}(s) + v_{b,3}\Phi_{b,3}(s) + v_{b,4}\Phi_{b,4}(s) + v_{b,5}\Phi_{b,5}(s) \qquad (17.104)
\end{aligned}$$

17.3.2.5   Approximate solution

The coordinate numbers for the approximate solution with $\eta = 4$ and $\xi = 0.91$ is

$$\begin{array}{ll}
v_{a,0} = \ \ 0 & v_{b,0} = \ \ 0 \\
v_{a,1} = \ \ 0.307729725674741 & v_{b,1} = -0.152516334115399 \\
v_{a,2} = -0.952112600471714 & v_{b,2} = \ \ 3.05344329603526 \times 10^{-7} \\
v_{a,3} = \ \ 1.20690557864456 & v_{b,3} = \ \ 0.920555297104715 \\
v_{a,4} = -1.55141723453639 & v_{b,4} = \ \ 0.0012225583549362 \qquad (17.105)
\end{array}$$

which are the coordinate numbers of the expansion denoted in equation (17.104).

The coordinate numbers for the approximate solution with $\eta = 4$ and $\xi = 0.3$ is

$$\begin{array}{ll}
v_{a,0} = \ \ 0 & v_{b,0} = \ \ 0 \\
v_{a,1} = \ \ 0.10015718377741 & v_{b,1} = -0.0500303664884372 \\
v_{a,2} = -0.301500964865256 & v_{b,2} = \ \ 2.92502502081168 \times 10^{-9} \\
v_{a,3} = \ \ 0.310515064038447 & v_{b,3} = \ \ 0.300375439354958 \\
v_{a,4} = -0.0541743491438487 & v_{b,4} = \ \ 7.10435235568019 \times 10^{-6} \\
& \qquad (17.106)
\end{array}$$

17.3.2.6 Error analysis

Since the solution is unknown the error analysis was performed with the norm of the residual of the Prandtl-Eyring governing differential equation (17.99) as follows

$$r(x_i) = \left\{ 1 + \left[\frac{1}{\xi} \left(\frac{1}{\varepsilon} \frac{dv}{ds} \right)^2 + 1 \right]^{1/2} \right\} \frac{d^2v}{ds^2} - (s-1) \qquad s \in [0,1] \quad (17.107)$$

The norm of the residual in the domain is computed as

$$\|r\| = \sqrt{\sum_{i=0}^{100} |r(x_i)|^2} \qquad (17.108)$$

over 101 point equally spaced in the domain.

The norm of the approximate solution was evaluated using

$$\|v\| = \sqrt{\sum_{i=0}^{101} |v(x_i)|^2} \qquad (17.109)$$

along 101 points in the domain.

Table 17.3.3 shows the values of the norm $\|v\|$ of the solution, and the norm of the residual $\|r\|$ of the differential equation for from $\eta = 2$ through $\eta = 9$.

Table 17.3.3 *Analysis of the behavior of the solution in the domain* $[0.0, 1.0]$ *for the polynomials from* $\eta = 2$ *through* $\eta = 9$.

η	$\xi = 0.91$	
	$\|v\|$	$\|r\|$
2	4.21063×10^{-1}	1.200×10^{-2}
3	4.21094×10^{-1}	4.328×10^{-5}
4	4.21094×10^{-1}	1.561×10^{-6}
5	4.21094×10^{-1}	1.553×10^{-4}
6	4.21094×10^{-1}	3.006×10^{-4}
7	4.21094×10^{-1}	4.985×10^{-4}
8	4.21094×10^{-1}	8.221×10^{-4}
9	4.21094×10^{-1}	1.247×10^{-3}

The table 17.3.3 shows that the best approximation was obtained for $\eta = 4$, that is, for the value of η for which the norm of the residual of the differential equation is minimum.

17.3.2.7   Graphic of the solution

Figure 17.3.4 shows the graphic of the approximate solution of the equation (17.99).

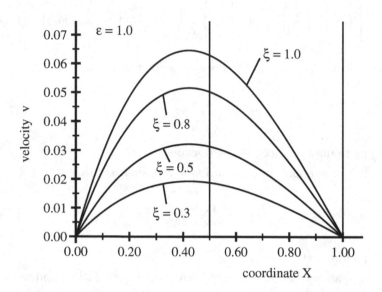

Figure 17.3.4 *Velocity profile of a Powell-Eyring flow between two walls.*

17.4   Two-point boundary value problem

Bamigbola, Ibiejugba, and Onumanyi [3] in the paper *Higher order Chebyshev basis functions for two boundary value problems* present three cases that are solved here using the Hermite Interpolation Polynomials.

17.4.1   Singular perturbation: problem I

17.4.1.1   The governing equation

The governing equation is given by

$$\frac{\mathrm{d}^2 g}{\mathrm{d}x^2} - k\frac{\mathrm{d}g}{\mathrm{d}x} = 0 \qquad x \in [0,1], \quad k > 0 \tag{17.110}$$

with the boundary conditions

$$g(0) = 1 \qquad g(1) = 0 \qquad (17.111)$$

17.4.1.2 Exact solution

The exact solution is given by

$$g_e(x) = \frac{1}{1 - e^k} \left(e^{kx} - e^k \right) \qquad (17.112)$$

The derivatives of the exact solution are

$$[g_e(X)]^{(1)} = \frac{k e^{kx}}{1 - e^k} \qquad [g_e(X)]^{(2)} = \frac{k^2 e^{kx}}{1 - e^k}$$

$$[g_e(X)]^{(3)} = \frac{k^3 e^{kx}}{1 - e^k} \qquad [g_e(X)]^{(n)} = \frac{k^n e^{kx}}{1 - e^k} \qquad (17.113)$$

17.4.1.3 Expansion of the solution

Analogously to the equation (17.68) the approximate solution of the linear differential equation (17.110) when expanded in a basis of Hermite Interpolating Polynomials writes

$$g(x) = \sum_{r=0}^{r=\eta} [g_{a,r} \Phi_{a,r} + g_{b,r} \Phi_{b,r}] \qquad (17.114)$$

and for $\eta = 4$ it reduces to

$$\begin{aligned} g(x) = {} & g_{a,0}\Phi_{a,0}(x) + g_{a,1}\Phi_{a,1}(x) + g_{a,2}\Phi_{a,2}(x) + g_{a,3}\Phi_{a,3}(x) \\ & + g_{a,4}\Phi_{a,4}(x) + g_{b,0}\Phi_{b,0}(x) + g_{b,1}\Phi_{b,1}(x) + g_{b,2}\Phi_{b,2}(x) \\ & + g_{b,3}\Phi_{b,3}(x) + g_{b,4}\Phi_{b,4}(x) \end{aligned} \qquad (17.115)$$

17.4.1.4 Boundary conditions

Equation (17.115) when evaluated at $a = 0.0$ must satisfy the boundary condition in equation (17.111), thus

$$g(0.0) = g_{a,0}\Phi_{a,0}(0.0) = 1 \qquad (17.116)$$

Analogously, when evaluated at the boundary $b = 1.0$ it gives

$$g(1.0) = g_{b,0}\Phi_{b,0}(1.0) = 0 \qquad (17.117)$$

Thus two of the 10 parameters to be evaluated are already known.

17.4.1.5   Expression for the solution

Substituting the known coefficients into the expansion of the approximate solution in equation (17.115) it gives

$$
\begin{aligned}
g(x) = {} & 1 \times \Phi_{a,0}(x) + g_{a,1}\Phi_{a,1}(x) + g_{a,1}\Phi_{a,2}(x) + g_{a,3}\Phi_{a,3}(x) \\
& + g_{a,4}\Phi_{a,4}(x) + 0 \times \Phi_{b,0}(x) + g_{b,1}\Phi_{b,1}(x) + g_{b,2}\Phi_{b,2}(x) \\
& + g_{b,3}\Phi_{b,3}(x) + g_{b,4}\Phi_{b,4}(x)
\end{aligned}
\tag{17.118}
$$

17.4.1.6   Approximate solution

The exact solution, for $k = 1.0$ is

$$
\begin{array}{ll}
g_{a,0} = 1 & g_{b,0} = 0 \\[6pt]
g_{a,1} = \dfrac{1}{1-e} & g_{b,1} = \dfrac{e}{1-e} \\[10pt]
g_{a,2} = \dfrac{1}{1-e} & g_{b,2} = \dfrac{e}{1-e} \\[10pt]
g_{a,3} = \dfrac{1}{1-e} & g_{b,3} = \dfrac{e}{1-e} \\[10pt]
g_{a,4} = \dfrac{1}{1-e} & g_{b,4} = \dfrac{e}{1-e} \\[10pt]
g_{a,n} = \dfrac{1}{1-e} & g_{b,n} = \dfrac{e}{1-e}
\end{array}
\tag{17.119}
$$

which are the coordinate numbers of the expansion denoted in equation (17.118).

The exact solution, for $\eta = 4$ and $k = 1.0$ is

$$
\begin{array}{ll}
g_{a,0} = 1 & g_{b,0} = 0 \\
g_{a,1} = -0.581976706857291 & g_{b,1} = -1.58197670688454 \\
g_{a,2} = -0.58197670599974 & g_{b,2} = -1.58197670567208 \\
g_{a,3} = -0.581976806845414 & g_{b,3} = -1.58197658404066 \\
g_{a,4} = -0.581972777082657 & g_{b,4} = -1.58197200277885
\end{array}
\tag{17.120}
$$

which are the coordinate numbers of the expansion denoted in equation (17.118).

The exact solution, for $\eta = 7$ and $k = 3.0$ is

$$
\begin{aligned}
g_{a,0} &= 1 & g_{b,0} &= 0 \\
g_{a,1} &= -0.157187089473723 & g_{b,1} &= -3.15718708947517 \\
g_{a,2} &= -0.471561268430444 & g_{b,2} &= -9.47156126845754 \\
g_{a,3} &= -1.4146838055728 & g_{b,3} &= -28.4146838053896 \\
g_{a,4} &= -4.24405139782492 & g_{b,4} &= -85.2440514005779 \\
g_{a,5} &= -12.7321542488014 & g_{b,5} &= -255.732154198929 \\
g_{a,6} &= -38.1964693080862 & g_{b,6} &= -767.196469797317 \\
g_{a,7} &= -114.5901319322 & g_{b,7} &= -2301.58864980069
\end{aligned}
\tag{17.121}
$$

which are the coordinate numbers of the expansion denoted in equation (17.118).

17.4.1.7 Error analysis of the differential equation

The accuracy of the solution is analyzed by the convergence to zero of the residual of the differential equation (17.110) as follows

$$
r(x_i) = \frac{\mathrm{d}^2 g}{\mathrm{d}x^2} - k\frac{\mathrm{d}g}{\mathrm{d}x} = 0 \qquad x \in [0,1], \quad k > 0
\tag{17.122}
$$

The norm of the residual in the domain is computed as

$$
\|r\| = \sqrt{\sum_{i=0}^{100} |r(x_i)|^2}
\tag{17.123}
$$

over 101 points equally spaced in the domain.

The norm of the approximate solution was evaluated using

$$
\|g\| = \sqrt{\sum_{i=0}^{101} |g(x_i)|^2}
\tag{17.124}
$$

along 101 points in the domain.

Table 17.4.1 shows the values of the norm of the error $\|e\|$ of the solution, the norm of the approximate solution $\|g\|$, the ration between the norm of the error $\|e\|$ of the solution and the norm $\|g_e\|$ of the exact solution, and the norm of the residual $\|r\|$ of the differential equation for $k = -0.5$. The norm of the exact solution is $\|g_e\| = 5.46111$.

Table 17.4.1 *Analysis of the behavior of the solution for $k = -0.5$, in the domain $[0,1]$, and for polynomials from $\eta = 2$ through $\eta = 8$.*

$k = -0.5$				
η	$\|e\|$	$\|g\|$	$\|e\|/\|g_e\|$	$\|r\|$
2	5.999×10^{-7}	5.46111	1.099×10^{-7}	6.449×10^{-5}
3	1.444×10^{-10}	5.46111	2.643×10^{-11}	3.457×10^{-8}
4	2.452×10^{-14}	5.46111	4.490×10^{-15}	9.998×10^{-12}
5	5.963×10^{-15}	5.46111	1.092×10^{-15}	4.606×10^{-13}
6	4.797×10^{-14}	5.46111	8.784×10^{-15}	6.118×10^{-12}
7	2.382×10^{-13}	5.46111	4.361×10^{-14}	4.899×10^{-11}
8	1.027×10^{-12}	5.46111	1.881×10^{-13}	5.300×10^{-10}

Table 17.4.2 shows the values of the norm of the error $\|e\|$ of the solution, the norm of the approximate solution $\|g\|$, the ration between the norm of the error $\|e\|$ of the solution and the norm $\|g_e\|$ of the exact solution, and the norm of the residual $\|r\|$ of the differential equation for $k = 0.1$. The norm of the exact solution is $\|g_e\| = 5.88843$.

Table 17.4.2 *Analysis of the behavior of the solution for $k = 0.1$, in the domain $[0,1]$, and for polynomials from $\eta = 2$ through $\eta = 8$.*

$k = 0.1$				
η	$\|e\|$	$\|g\|$	$\|e\|/\|g_e\|$	$\|r\|$
2	1.940×10^{-10}	5.88843	3.294×10^{-11}	2.076×10^{-8}
3	7.055×10^{-15}	5.88843	1.198×10^{-15}	4.517×10^{-13}
4	7.207×10^{-15}	5.88843	1.224×10^{-15}	1.656×10^{-13}
5	9.628×10^{-15}	5.88843	1.635×10^{-15}	6.101×10^{-13}
6	6.584×10^{-14}	5.88843	1.118×10^{-14}	8.244×10^{-12}
7	3.186×10^{-13}	5.88843	5.410×10^{-14}	6.540×10^{-11}
8	1.388×10^{-12}	5.88844	2.357×10^{-13}	7.072×10^{-10}

Table 17.4.3 shows the values of the norm of the error $\|e\|$ of the solution, the norm of the approximate solution $\|g\|$, the ration between the norm of the error $\|e\|$ of the solution and the norm $\|g_e\|$ of the exact solution, and the norm of the residual $\|r\|$ of the differential equation for $k = 0.5$. The norm of the exact solution is $\|g_e\| = 6.17428$.

Table 17.4.3 *Analysis of the behavior of the solution for $k = 0.5$, in the domain $[0, 1]$, and for polynomials from $\eta = 2$ through $\eta = 8$.*

$k = 0.5$				
η	$\|e\|$	$\|g\|$	$\|e\|/\|g_e\|$	$\|r\|$
2	5.999×10^{-7}	6.17428	9.717×10^{-8}	6.449×10^{-5}
3	1.444×10^{-10}	6.17428	2.338×10^{-11}	3.457×10^{-8}
4	2.460×10^{-14}	6.17428	3.984×10^{-15}	1.009×10^{-11}
5	9.485×10^{-15}	6.17428	1.536×10^{-15}	7.281×10^{-13}
6	7.891×10^{-14}	6.17428	1.278×10^{-14}	9.804×10^{-12}
7	3.806×10^{-13}	6.17428	6.164×10^{-14}	7.799×10^{-11}
8	1.674×10^{-12}	6.17428	2.711×10^{-13}	8.428×10^{-10}

Table 17.4.4 shows the values of the norm of the error $\|e\|$ of the solution, the norm of the approximate solution $\|g\|$, the ration between the norm of the error $\|e\|$ of the solution and the norm $\|g_e\|$ of the exact solution, and the norm of the residual $\|r\|$ of the differential equation for $k = 1.0$. The norm of the exact solution is $\|g_e\| = 6.52444$.

Table 17.4.4 *Analysis of the behavior of the solution for $k = 1.0$, in the domain $[0, 1]$, and for polynomials from $\eta = 2$ through $\eta = 8$.*

$k = 1.0$				
η	$\|e\|$	$\|g\|$	$\|e\|/\|g_e\|$	$\|r\|$
2	1.863×10^{-5}	6.52444	2.855×10^{-6}	2.025×10^{-3}
3	1.799×10^{-8}	6.52444	2.758×10^{-9}	4.324×10^{-6}
4	1.197×10^{-11}	6.52444	1.835×10^{-12}	4.957×10^{-9}
5	2.132×10^{-14}	6.52444	3.268×10^{-15}	9.125×10^{-12}
6	9.873×10^{-14}	6.52444	1.513×10^{-14}	1.201×10^{-11}
7	4.684×10^{-13}	6.52444	7.180×10^{-14}	9.536×10^{-11}
8	2.072×10^{-12}	6.52444	3.176×10^{-13}	1.030×10^{-9}

Table 17.4.5 shows the values of the norm of the error $\|e\|$ of the solution, the norm of the approximate solution $\|g\|$, the ration between the norm of the error $\|e\|$ of the solution and the norm $\|g_e\|$ of the exact solution, and the norm of the residual $\|r\|$ of the differential equation for $k = 2.0$. The norm of the exact solution is $\|g_e\| = 7.17133$.

Table 17.4.5 *Analysis of the behavior of the
solution for $k = 2.0$, in the domain $[0, 1]$, and for
polynomials from $\eta = 2$ through $\eta = 8$.*

		$k = 2.0$		
η	$\|e\|$	$\|g\|$	$\|e\|/\|g_e\|$	$\|r\|$
2	5.347×10^{-4}	7.17132	7.456×10^{-5}	6.029×10^{-2}
3	2.078×10^{-6}	7.17133	2.898×10^{-7}	5.068×10^{-4}
4	5.429×10^{-9}	7.17133	7.571×10^{-10}	2.299×10^{-6}
5	3.072×10^{-11}	7.17133	4.284×10^{-12}	1.580×10^{-8}
6	2.027×10^{-13}	7.17133	2.827×10^{-14}	7.576×10^{-11}
7	6.689×10^{-13}	7.17133	9.328×10^{-14}	1.345×10^{-10}
8	3.000×10^{-12}	7.17133	4.184×10^{-13}	1.452×10^{-9}

Table 17.4.6 shows the values of the norm of the error $\|e\|$ of the solution, the norm of the approximate solution $\|g\|$, the ration between the norm of the error $\|e\|$ of the solution and the norm $\|g_e\|$ of the exact solution, and the norm of the residual $\|r\|$ of the differential equation for $k = 3.0$. The norm of the exact solution is $\|g_e\| = 7.7141$.

Table 17.4.6 *Analysis of the behavior of the solution for $k = 3.0$, in the domain $[0, 1]$, and for polynomials from $\eta = 2$ through $\eta = 8$.*

		$k = 3.0$		
η	$\|e\|$	$\|g\|$	$\|e\|/\|g_e\|$	$\|r\|$
2	3.486×10^{-3}	7.71405	4.519×10^{-4}	4.098×10^{-1}
3	3.029×10^{-5}	7.7141	3.927×10^{-6}	7.557×10^{-3}
4	1.792×10^{-7}	7.7141	2.323×10^{-8}	7.626×10^{-5}
5	2.282×10^{-9}	7.7141	2.958×10^{-10}	1.190×10^{-6}
6	2.330×10^{-11}	7.7141	3.021×10^{-12}	1.268×10^{-8}
7	9.620×10^{-13}	7.7141	1.247×10^{-13}	2.089×10^{-10}
8	4.039×10^{-12}	7.7141	5.236×10^{-13}	1.904×10^{-9}

Table 17.4.7 shows the values of the norm of the error $\|e\|$ of the solution, the norm of the approximate solution $\|g\|$, the ration between the norm of the error $\|e\|$ of the solution and the norm $\|g_e\|$ of the exact solution, and the norm of the residual $\|r\|$ of the differential equation for $k = 4.0$. The norm of the exact solution is $\|g_e\| = 8.14248$.

Table 17.4.7 *Analysis of the behavior of the solution for $k = 4.0$, in the domain $[0, 1]$, and for polynomials from $\eta = 2$ through $\eta = 8$.*

η	$\|e\|$	$\|g\|$	$\|e\|/\|g_e\|$	$\|r\|$
		$k = 4.0$		
2	1.241×10^{-2}	8.14117	1.524×10^{-3}	1.503×10^{0}
3	1.855×10^{-4}	8.14247	2.279×10^{-5}	4.767×10^{-2}
4	1.983×10^{-6}	8.14248	2.435×10^{-7}	8.423×10^{-4}
5	4.625×10^{-8}	8.14248	5.680×10^{-9}	2.361×10^{-5}
6	9.898×10^{-10}	8.14248	1.216×10^{-10}	4.516×10^{-7}
7	1.842×10^{-11}	8.14248	2.263×10^{-12}	6.593×10^{-9}
8	5.140×10^{-12}	8.14248	6.313×10^{-13}	2.344×10^{-9}

Table 17.4.8 shows the values of the norm of the error $\|e\|$ of the solution, the norm of the approximate solution $\|g\|$, the ration between the norm of the error $\|e\|$ of the solution and the norm $\|g_e\|$ of the exact solution, and the norm of the residual $\|r\|$ of the differential equation for $k = 5.0$. The norm of the exact solution is $\|g_e\| = 8.4691$.

Table 17.4.8 *Analysis of the behavior of the solution for $k = 5.0$, in the domain $[0, 1]$, and for polynomials from $\eta = 2$ through $\eta = 8$.*

η	$\|e\|$	$\|g\|$	$\|e\|/\|g_e\|$	$\|r\|$
		$k = 5.0$		
2	3.319×10^{-2}	8.45959	3.919×10^{-3}	3.915×10^{0}
3	7.036×10^{-4}	8.46906	8.308×10^{-5}	1.868×10^{-1}
4	1.170×10^{-5}	8.46911	1.381×10^{-6}	5.041×10^{-3}
5	4.697×10^{-7}	8.4691	5.546×10^{-8}	2.240×10^{-4}
6	2.032×10^{-8}	8.4691	2.399×10^{-9}	6.771×10^{-6}
7	6.607×10^{-10}	8.4691	7.802×10^{-11}	1.547×10^{-7}
8	1.687×10^{-11}	8.4691	1.992×10^{-12}	3.746×10^{-9}

Table 17.4.9 shows the values of the norm of the error $\|e\|$ of the solution, the norm of the approximate solution $\|g\|$, the ration between the norm of the error $\|e\|$ of the solution and the norm $\|g_e\|$ of the exact solution, and the norm of the residual $\|r\|$ of the differential equation for $k = 5.2$. The norm of the exact solution is $\|g_e\| = 8.52407$.

Table 17.4.9 *Analysis of the behavior of the solution for $k = 5.2$, in the domain $[0, 1]$, and for polynomials from $\eta = 2$ through $\eta = 8$.*

η	$\|e\|$	$\|g\|$	$\|e\|/\|g_e\|$	$\|r\|$
		$k = 5.2$		
2	3.973×10^{-2}	8.51098	4.661×10^{-3}	4.606×10^{0}
3	8.827×10^{-4}	8.52401	1.036×10^{-4}	2.359×10^{-1}
4	1.601×10^{-5}	8.52407	1.878×10^{-6}	6.857×10^{-3}
5	7.048×10^{-7}	8.52407	8.268×10^{-8}	3.302×10^{-4}
6	3.458×10^{-8}	8.52407	4.057×10^{-9}	1.082×10^{-5}
7	1.234×10^{-9}	8.52407	1.448×10^{-10}	2.675×10^{-7}
8	3.209×10^{-11}	8.52407	3.765×10^{-12}	5.747×10^{-9}

17.4.1.8 Error analysis of the solution

The error e_i of the solution at the point x_i is defined as

$$e(x_i) = g_e(x_i) - g(x_i) \tag{17.125}$$

where $g_e(x_i)$ is the exact solution and $g(x_i)$ the value of the approximate solution at the point x_i.

The norm of the error was evaluated using

$$\|e\| = \sqrt{\sum_{i=0}^{101} |e(x_i)|^2} \tag{17.126}$$

along 101 points in the domain in the domain.

Recalling equation (16.48), page 548, the norm of the exact solution is given by

$$\|g_e\| = \sqrt{\sum_{i=0}^{m} |g_e(x_i)|^2} \tag{17.127}$$

where $g_e(x_i)$ is the value of the exact solution at the point x_i.

Table 17.4.10 compares the values of the norm of the error of the solution with the values of η.

Table 17.4.10 *Values of the norm of the error* $\|e\|$ *of the solution for* $\eta = 5, 6, 7, 8$ *and* $k = -0.5, 0.1, 0.5,$ *and* 1.0.

	norm $\|e\|$ of the error of the solution			
η	$k = -0.5$	$k = 0.1$	$k = 0.5$	$k = 1.0$
2	5.999×10^{-7}	1.939×10^{-10}	5.999×10^{-7}	1.863×10^{-5}
3	1.444×10^{-10}	6.902×10^{-15}	1.443×10^{-10}	1.799×10^{-8}
4	2.693×10^{-14}	7.207×10^{-15}	2.500×10^{-14}	1.197×10^{-11}
5	5.963×10^{-15}	9.628×10^{-15}	9.485×10^{-15}	2.132×10^{-14}
6	4.797×10^{-14}	6.584×10^{-14}	7.891×10^{-14}	9.873×10^{-14}
7	2.382×10^{-13}	3.186×10^{-13}	3.806×10^{-13}	4.684×10^{-13}
8	1.027×10^{-12}	1.388×10^{-12}	1.674×10^{-12}	2.072×10^{-12}

Table 17.4.11 *Values of the norm of the error* $\|e\|$ *of the solution for* $\eta = 5, 6, 7, 8$ *and* $k = 2.0, 3.0, 4.0, 5.0,$ *and* 5.2.

	norm $\|e\|$ of the error of the solution				
η	$k = 2.0$	$k = 3.0$	$k = 4.0$	$k = 5.0$	$k = 5.2$
2	5.347×10^{-4}	3.486×10^{-3}	1.241×10^{-2}	3.319×10^{-2}	3.972×10^{-2}
3	2.078×10^{-6}	3.029×10^{-5}	1.855×10^{-4}	7.036×10^{-4}	8.827×10^{-4}
4	5.445×10^{-9}	1.792×10^{-7}	1.983×10^{-6}	1.170×10^{-5}	1.601×10^{-5}
5	3.072×10^{-11}	2.281×10^{-9}	4.625×10^{-8}	4.697×10^{-7}	7.048×10^{-7}
6	2.027×10^{-13}	2.330×10^{-11}	9.898×10^{-10}	2.032×10^{-8}	3.458×10^{-8}
7	6.689×10^{-13}	9.620×10^{-13}	1.842×10^{-11}	6.607×10^{-10}	1.234×10^{-9}
8	3.000×10^{-12}	4.039×10^{-12}	5.140×10^{-12}	1.687×10^{-11}	3.209×10^{-11}

17.4.1.9   Graphic of the solution

Figure 17.4.12 shows the approximate solution obtained for the singular perturbation problem in equation (17.110).

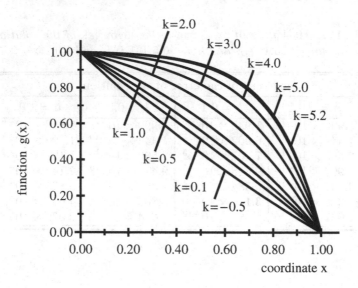

Figure 17.4.12 *Approximate solution for $k = -0.5, 0,\ 0, 5,\ 1,\ 2,\ 3,$*
4, 5, and 5.2, for $\eta = 5$.

17.4.2   Singular perturbation: problem II

17.4.2.1   The governing equation

The governing equation is given by

$$\frac{\mathrm{d}^2 g}{\mathrm{d}x^2} - \frac{\mathrm{d}g}{\mathrm{d}x} = 0 \qquad x \in [0, 2] \tag{17.128}$$

with the boundary conditions

$$g(0) = 1 \qquad g(2) = e^2 \tag{17.129}$$

17.4.2.2   Exact solution

The exact solution is given by

$$g(x) = e^x \tag{17.130}$$

17.4.2.3   Expansion of the solution

Analogously to the equation (17.68) the approximate solution of the linear differential equation (17.128) when expanded in a basis of Hermite Interpolating Polynomials writes

$$g(x) = \sum_{r=0}^{r=\eta} [g_{a,r} \Phi_{a,r} + g_{b,r} \Phi_{b,r}] \tag{17.131}$$

and for $\eta = 4$ it reduces to

$$g(x) = g_{a,0}\Phi_{a,0}(x) + g_{a,1}\Phi_{a,1}(x) + g_{a,2}\Phi_{a,2}(x) + g_{a,3}\Phi_{a,3}(x) + g_{a,4}\Phi_{a,4}(x)$$
$$+ g_{b,0}\Phi_{b,0}(x) + g_{b,1}\Phi_{b,1}(x) + g_{b,2}\Phi_{b,2}(x) + g_{b,3}\Phi_{b,3}(x) + g_{b,4}\Phi_{b,4}(x)$$
$$\text{(17.132)}$$

17.4.2.4   Boundary conditions

Equation (17.132) when evaluated at $a = 0.0$ must satisfy the boundary condition in equation (17.129), thus

$$g(0.0) = g_{a,0}\Phi_{a,0}(0.0) = 1 \qquad (17.133)$$

Analogously, when evaluated at the boundary $b = 1.0$ it gives

$$g(1.0) = g_{b,0}\Phi_{b,0}(1.0) = e^2 \qquad (17.134)$$

Thus two of the 10 parameters to be evaluated are already known.

17.4.2.5   Expression for the solution

Substituting the known coefficients into the expansion of the approximate solution in equation (17.132) it gives

$$g(x) = 1 \times \Phi_{a,0}(x) + g_{a,1}\Phi_{a,1}(x) + g_{a,1}\Phi_{a,2}(x) + g_{a,3}\Phi_{a,3}(x)$$
$$+ g_{a,4}\Phi_{a,4}(x) + g_{a,5}\Phi_{a,5}(x) + g_{a,6}\Phi_{a,6}(x) + e^2 \times \Phi_{b,0}(x)$$
$$+ g_{b,1}\Phi_{b,1}(x) + g_{b,2}\Phi_{b,2}(x) + g_{b,3}\Phi_{b,3}(x) + g_{b,4}\Phi_{b,4}(x)$$
$$+ g_{b,5}\Phi_{b,5}(x) + g_{b,6}\Phi_{b,6}(x) \qquad (17.135)$$

17.4.2.6   Approximate solution

The exact solution, for $k = 1$ is

$$
\begin{array}{ll}
g_{a,0} = 1 & g_{b,0} = e^2 \\
g_{a,1} = 1 & g_{b,1} = e^2 \\
g_{a,2} = 1 & g_{b,2} = e^2 \\
g_{a,3} = 1 & g_{b,3} = e^2 \\
g_{a,n} = 1 & g_{b,n} = e^2
\end{array}
\qquad (17.136)
$$

which are the coordinate numbers of the expansion denoted in equation (17.135).

The coordinate numbers for the approximate solution with $\eta = 6$ are

$$g_{a,0} = 1 \qquad\qquad\qquad g_{b,0} = 7.38905609893065$$
$$g_{a,1} = 0.999999999999686 \qquad g_{b,1} = 7.38905609893602$$
$$g_{a,2} = 0.999999999981636 \qquad g_{b,2} = 7.38905609892399$$
$$g_{a,3} = 1.00000000005326 \qquad\; g_{b,3} = 7.38905609848848$$
$$g_{a,4} = 1.00000000578835 \qquad\;\; g_{b,4} = 7.38905610043987$$
$$g_{a,5} = 0.999999959627825 \qquad g_{b,5} = 7.38905614221155$$
$$g_{a,6} = 0.999997802304335 \qquad g_{b,6} = 7.38905428728348 \qquad (17.137)$$

which are the coordinate numbers of the expansion denoted in equation (17.135).

17.4.2.7   Error analysis of the differential equation

The accuracy of the solution is analyzed by the convergence to zero of the residual of the differential equation (17.128) as follows

$$r(x_i) = \frac{\mathrm{d}^2 g}{\mathrm{d}x^2} - \frac{\mathrm{d}g}{\mathrm{d}x} = 0 \qquad x \in [0, 2], \quad k > 0 \qquad (17.138)$$

The norm of the residual in the domain is computed as

$$\|r\| = \sqrt{\sum_{i=0}^{100} |r(x_i)|^2} \qquad (17.139)$$

over 101 point equally spaced in the domain.

The norm of the approximate solution was evaluated using

$$\|g\| = \sqrt{\sum_{i=0}^{101} |g(x_i)|^2} \qquad (17.140)$$

along 101 points in the domain.

Table 17.4.13 shows the values of the norm of the error $\|e\|$ of the solution, the norm of the approximate solution $\|g\|$, the ration between the norm of the error $\|e\|$ of the solution and the norm $\|g_e\|$ of the exact solution, and the norm of the residual $\|r\|$ of the differential equation. The norm of the exact solution is $\|g_e\| = 36.9856$.

Table 17.4.13 *Analysis of the behaviour of the solution in the domain* $[0, 2]$, *and for polynomials from* $\eta = 2$ *through* $\eta = 8$.

η	$\|e\|$	$\|g\|$	$\|e\|/\|g_e\|$	$\|r\|$
2	3.416×10^{-3}	36.9852	9.236×10^{-5}	9.629×10^{-2}
3	1.328×10^{-5}	36.9856	3.590×10^{-7}	8.095×10^{-4}
4	3.479×10^{-8}	36.9856	9.406×10^{-10}	3.673×10^{-6}
5	1.361×10^{-10}	36.9856	3.681×10^{-12}	1.707×10^{-8}
6	1.529×10^{-11}	36.9856	4.133×10^{-13}	3.021×10^{-10}
7	9.327×10^{-11}	36.9856	2.522×10^{-12}	2.577×10^{-9}
8	6.983×10^{-10}	36.9856	1.888×10^{-11}	1.526×10^{-8}

17.4.2.8 Error analysis of the solution

The error e_i of the solution at the point x_i is defined as

$$e(x_i) = g_e(x_i) - g(x_i) \tag{17.141}$$

where $g_e(x_i)$ is the exact solution and $g(x_i)$ the value of the approximate solution at the point x_i.

The norm of the error was evaluated using

$$\|e\| = \sqrt{\sum_{i=0}^{101} |e(x_i)|^2} \tag{17.142}$$

along 101 points in the domain in the domain.

Recalling equation (16.48), page 548, the norm of the exact solution is given by

$$\|g_e\| = \sqrt{\sum_{i=0}^{m} |g_e(x_i)|^2} \tag{17.143}$$

where $g_e(x_i)$ is the value of the exact solution at the point x_i.

The first and last columns in the table 17.4.13 compare the values of the norm of the error of the solution and the norm of the residual of the differential equation with the values of η. The best approximations were obtained with $\eta = 6$ although the norm of the approximate solution equals the norm of the exact solution for all cases.

17.4.3   Singular perturbation: problem III

17.4.3.1   The governing equation

The governing equation is given by

$$\frac{d^2 g}{dx^2} + \frac{dg}{dx} = -x \qquad x \in (0, 1) \tag{17.144}$$

with the boundary conditions

$$g(0) = 0 \qquad \frac{dg}{dx}\bigg|_{x=1} = 0 \tag{17.145}$$

17.4.3.2   Exact solution

The exact solution is given by

$$g(x) = -\frac{x^2}{2} + x \tag{17.146}$$

17.4.3.3   Expansion of the solution

Analogously to the equation (17.68) the approximate solution of the linear differential equation (17.144) when expanded in a basis of Hermite Interpolating Polynomials writes

$$g(x) = \sum_{r=0}^{r=\eta} [g_{a,r}\Phi_{a,r} + g_{b,r}\Phi_{b,r}] \tag{17.147}$$

and for $\eta = 2$ it reduces to

$$\begin{aligned} g(x) = {}& g_{a,0}\Phi_{a,0}(x) + g_{a,1}\Phi_{a,1}(x) + g_{a,2}\Phi_{a,2}(x)+ \\ & + g_{b,0}\Phi_{b,0}(x) + g_{b,1}\Phi_{b,1}(x) + g_{b,2}\Phi_{b,2}(x) \end{aligned} \tag{17.148}$$

17.4.3.4   Boundary conditions

Equation (17.132) when evaluated at $a = 0.0$ must satisfy the boundary condition in equation (17.129), thus

$$g(0.0) = g_{a,0}\Phi_{a,0}(0.0) = 0 \tag{17.149}$$

Analogously, when evaluated at the boundary $b = 1.0$ it gives

$$g^{(1)}(1.0) = g_{b,1}^{(1)}\Phi_{b,1}(1.0) = 1 \tag{17.150}$$

Thus, two of the 6 parameters to be evaluated are already known.

17.4.3.5 Expression for the solution

Substituting the known coefficients into the expansion of the approximate solution in equation (17.148) it gives

$$g(x) = 0 \times \Phi_{a,0}(x) + g_{a,1}\Phi_{a,1}(x) + g_{a,1}\Phi_{a,2}(x) +$$
$$+ g_{b,0} \times \Phi_{b,0}(x) + 1 \times \Phi_{b,1}(x) + g_{b,2}\Phi_{b,2}(x) \qquad (17.151)$$

17.4.3.6 Approximate solution

The exact solution is

$$
\begin{array}{ll}
g_{a,0} = & 0 \qquad g_{b,0} = \dfrac{1}{2} \\[4pt]
g_{a,1} = & 1 \qquad g_{b,1} = 0 \\[4pt]
g_{a,2} = -1 \qquad g_{b,2} = -1 \\[4pt]
g_{a,3} = & 0 \qquad g_{b,3} = 0 \\[4pt]
g_{a,n} = & 0 \qquad g_{b,n} = 0 \qquad (17.152)
\end{array}
$$

which are the coordinate numbers of the expansion denoted in equation (17.151).

The coordinate numbers for the approximate solution with $\eta = 2$ are

$$
\begin{array}{ll}
g_{a,0} = & 0 \qquad g_{b,0} = 0.5 \\[4pt]
g_{a,1} = & 1 \qquad g_{b,1} = 0 \\[4pt]
g_{a,2} = -1 \qquad g_{b,2} = -1 \qquad (17.153)
\end{array}
$$

which are the coordinate numbers of the expansion denoted in equation (17.151).

17.4.3.7 Error analysis of the differential equation

The accuracy of the solution is analyzed by the convergence to zero of the residual of the differential equation (17.144) as follows

$$r(x_i) = \frac{d^2 g}{dx^2} + \frac{dg}{dx} = 0 \qquad x \in [0,1], \quad k > 0 \qquad (17.154)$$

The norm of the residual in the domain is computed as

$$\|r\| = \sqrt{\sum_{i=0}^{100} |r(x_i)|^2} \qquad (17.155)$$

over 101 points equally spaced in the domain.

The norm of the approximate solution was evaluated using

$$\|g\| = \sqrt{\sum_{i=0}^{101} |g(x_i)|^2} \qquad (17.156)$$

along 101 points in the domain.

Table 17.4.14 shows the values of the norm of the error $\|e\|$ of the solution, the norm of the approximate solution $\|g\|$, the ration between the norm of the error $\|e\|$ of the solution and the norm $\|g_e\|$ of the exact solution, and the norm of the residual $\|r\|$ of the differential equation. The norm of the exact solution is $\|g_e\| = 3.66856$.

Table 17.4.14 *Analysis of the behavior of the solution in the domain* $[0, 1]$, *and for polynomials from* $\eta = 2$ *through* $\eta = 8$.

η	$\|e\|$	$\|g\|$	$\|e\|/\|g_e\|$	$\|r\|$
2	9.450×10^{-16}	3.66856	2.576×10^{-16}	1.117×10^{-14}
3	5.121×10^{-15}	3.66856	1.396×10^{-15}	1.341×10^{-13}
4	2.826×10^{-14}	3.66856	7.702×10^{-15}	1.342×10^{-12}
5	2.191×10^{-13}	3.66856	5.972×10^{-14}	1.097×10^{-11}
6	7.642×10^{-12}	3.66856	2.083×10^{-12}	1.003×10^{-10}
7	9.231×10^{-11}	3.66856	2.516×10^{-11}	8.603×10^{-10}
8	3.505×10^{-10}	3.66856	9.555×10^{-11}	4.821×10^{-9}

Chapter 18

Application to problems with several variables

18.1 Application to heat problems

18.1.1 One-dimension heat flow equation

Obtain the solution $T(x,t)$ of the initial boundary value problem whose differential equation is given by

$$\frac{\partial T}{\partial t} = \frac{\partial^2 T}{\partial x^2} \qquad 0 < x < \pi \quad t > 0 \tag{18.1}$$

satisfying the boundary conditions

$$T(0,t) = 0 \qquad t \geq 0$$
$$T(\pi,t) = 0 \qquad t > 0 \tag{18.2}$$

and the initial condition

$$T(x,0) = \sin(x) \tag{18.3}$$

The solution to this equation is $T(x,t) = e^{-t}\sin(x)$ (Wazwaz [34], page 59).

Solution:

For a solution in the form

$$w(x,t) = \frac{a_0 + a_1 x + a_2 t + a_3 x^2 + a_4 xt + a_5 t^2 a_6 x^3 + a_7 x^2 t + a_8 xt^2 + a_9 t^3 + \cdots}{b_0 + b_1 x + b_2 t + b_3 x^2 + b_4 xt + b_5 t^2 b_6 x^3 + b_7 x^2 t + b_8 xt^2 + b_9 t^3 + \cdots} \tag{18.4}$$

where the indices and their ordering can be seen in the table 2.7.26 on page 69, the norm of the several errors can be found in the table 18.1.1 below

Table 18.1.1 *Norm of the error of the solution* $||e||$, *of the differential equation* $||r||$, *of the domain* $||r_D||$ *as function of the degree of the polynomials in the rational fraction, and the number of coefficients (n of coef) in the polynomials* $N(x,t)$ *and* $D(x,t)$.

$||e||$	$||r||$	$||r_D||$	degree	no. of coef.
9.7678×10^{-02}	5.1987×10^{-01}	5.2883×10^{-01}	4	15
5.4968×10^{-03}	5.4095×10^{-02}	5.4372×10^{-02}	5	21
6.3412×10^{-04}	8.6343×10^{-03}	8.6575×10^{-03}	6	28
1.4908×10^{-05}	2.9399×10^{-04}	2.9436×10^{-04}	7	36
1.0036×10^{-06}	2.5488×10^{-05}	2.5508×10^{-05}	8	45
1.3118×10^{-08}	4.3287×10^{-07}	4.3307×10^{-07}	9	55
5.7262×10^{-10}	2.3157×10^{-08}	2.3164×10^{-08}	10	66
4.8061×10^{-12}	2.3824×10^{-10}	2.3829×10^{-10}	11	78
1.4662×10^{-13}	8.6093×10^{-12}	8.6106×10^{-12}	12	91
8.8992×10^{-16}	5.9732×10^{-14}	5.9738×10^{-14}	13	105
2.2924×10^{-16}	1.5537×10^{-15}	1.5613×10^{-15}	14	120
2.2851×10^{-16}	7.6561×10^{-18}	1.5339×10^{-16}	15	136
2.2839×10^{-16}	6.1446×10^{-19}	1.5311×10^{-16}	16	153
2.2856×10^{-16}	3.2084×10^{-19}	1.5324×10^{-16}	17	171
2.2842×10^{-16}	3.1039×10^{-19}	1.5313×10^{-16}	18	190
2.2839×10^{-16}	9.6013×10^{-19}	1.5312×10^{-16}	19	210
2.2859×10^{-16}	3.5795×10^{-19}	1.5329×10^{-16}	20	231

The norm of the error in the domain do not reduce for the degrees above 14. This must be due to the fixed size of the computer word, that is, 32 bits.

The number of coefficients can be obtained from the expression (4.135) on page 175, which is

$$n^\delta(\ell) = \frac{(\ell+1)(\ell+2)\cdots(\ell+\delta+1)}{(\delta+1)!} \qquad \ell \geq 0 \qquad (18.5)$$

where $\delta = d-1$, whose d is the number of variables, and ℓ is the degree of the polynomial. The last column contains the number of coefficients of $N(x,t)$ and $D(x,t)$ to obtain the solution.

For the case of two variables the expression (18.4) writes

$$n^1 = \frac{(\ell+1)(\ell+2)}{2!} \qquad \ell \geq 0 \qquad (18.6)$$

18.1.2 Two-dimension heat flow equation

Obtain the solution $T(x,t)$ of the differential equation

$$\frac{\partial T}{\partial t} = \frac{\partial^2 T}{\partial x^2} + \frac{\partial^2 T}{\partial y^2} \qquad 0 < x < \pi \quad t > 0 \qquad (18.7)$$

satisfying the boundary conditions

$$T(0,y,t) = T(\pi,y,t) = 0 \qquad t \geq 0$$
$$T(x,0,t) = T(x,\pi,t) = 0 \qquad t > 0 \qquad (18.8)$$

and the initial condition

$$T(x,y,0) = \sin(x)\sin(y) \qquad (18.9)$$

The solution to this equation is $T(x,t) = e^{-2t}\sin(x)\sin(y)$ (Wazwaz [34], page 92).

Solution:

For a solution in the form

$$T(x,y,t) = \frac{a_0 + a_1 x + a_2 y + a_3 t + a_4 x^2 + a_5 xy a_6 xt + a_7 y^2 + a_8 yt + a_9 t^2 + \cdots}{b_0 + b_1 x + b_2 y + b_3 t + b_4 x^2 + b_5 xy b_6 xt + b_7 y^2 + b_8 yt + b_9 t^2 + \cdots} \qquad (18.10)$$

where the indices and their ordering can be seen in the table 2.8.25 on page 90, the norm of the several errors can be found in the table 18.1.2 below

Table 18.1.2 *Norm of the error of the solution $||e||$, of the differential equation $||r||$, of the domain $||r_D||$ as function of the degree of the polynomials in the rational fraction, and the number of coefficients (no. of coef.) in the polynomials $N(x, y, t)$ and $D(x, y, t)$.*

$||e||$	$||r||$	$||r_D||$	degree	no. of coef.
1.8218×10^{-1}	9.7286×10^{-01}	9.8965×10^{-01}	4	35
102.62×10^{-2}	1.0118×10^{-01}	1.0170×10^{-01}	5	56
1.1831×10^{-3}	1.6120×10^{-02}	1.6163×10^{-02}	6	84
2.7806×10^{-05}	5.4847×10^{-04}	5.4917×10^{-04}	7	120
1.8717×10^{-06}	4.7542×10^{-05}	4.7579×10^{-05}	8	165
2.4465×10^{-08}	8.0735×10^{-07}	8.0772×10^{-07}	9	220
1.0680×10^{-09}	4.3189×10^{-08}	4.3202×10^{-08}	10	286
8.9635×10^{-12}	4.4433×10^{-10}	4.4442×10^{-10}	11	364
2.7346×10^{-13}	1.6057×10^{-11}	1.6059×10^{-11}	12	455
1.6517×10^{-15}	1.114×10^{-13}	1.1141×10^{-13}	13	560
3.9507×10^{-16}	2.8976×10^{-15}	2.9180×10^{-15}	14	680
3.9355×10^{-16}	1.6007×10^{-17}	3.4331×10^{-16}	15	816
3.9346×10^{-16}	1.0284×10^{-17}	3.4287×10^{-16}	16	969
3.9358×10^{-16}	6.3604×10^{-18}	3.4302×10^{-16}	17	1140
3.9347×10^{-16}	3.1100×10^{-18}	3.4275×10^{-16}	18	1330
3.9330×10^{-16}	1.0655×10^{-17}	3.4285×10^{-16}	19	1540
3.9377×10^{-16}	5.7814×10^{-18}	3.4317×10^{-16}	20	1771

For the case of three variables the expression (18.6) for number of coefficients of $N(x, y, t)$ and $D(x, y, t)$ to obtain the solution writes

$$n^2 = \frac{(\ell + 1)(\ell + 2)(\ell + 3)}{3!} \qquad \ell \geq 0 \tag{18.11}$$

18.1.3 Two-dimension heat flow equation with lateral heat loss

Obtain the solution $T(x, t)$ of the differential equation

$$\frac{\partial T(x, y, t)}{\partial t} = \frac{\partial^2 T(x, y, t)}{\partial x^2} + \frac{\partial^2 T(x, y, t)}{\partial y^2} - T(x, y, t) \qquad 0 < x, y < \pi \quad t > 0 \tag{18.12}$$

satisfying the boundary conditions

$$T(0, y, t) = T(\pi, y, t) = 0 \qquad t \geq 0$$
$$T(x, 0, t) = -T(x, \pi, t) = e^{-3t} \sin(x) \qquad t > 0 \tag{18.13}$$

and the initial condition

$$T(x, y, 0) = e^{-3t} \sin(x) \cos(y) \qquad (18.14)$$

The solution to this equation is $T(x, t) = \sin(x) \cos(t)$ (Wazwaz [34], page 93).

For a solution in the form

$$T(x, y, t) = \frac{a_0 + a_1 x + a_2 y + a_3 t + a_4 x^2 + a_5 xy a_6 xt + a_7 y^2 + a_8 yt + a_9 t^2 + \cdots}{b_0 + b_1 x + b_2 y + b_3 t + b_4 x^2 + b_5 xy b_6 xt + b_7 y^2 + b_8 yt + b_9 t^2 + \cdots} \qquad (18.15)$$

where the indices and their ordering can be seen in the table 2.8.25 on page 90, the norm of the several errors can be found in the table 18.1.3 below

Table 18.1.3 *Norm of the error of the solution $\|e\|$, of the differential equation $\|r\|$, and of the domain $\|r_D\|$ as function of the degree of the polynomials in the rational fraction in the polynomials $N(x, y, t)$ and $D(x, y, t)$.*

$\|e\|$	$\|r\|$	$\|r_D\|$	degree
1.4667×10^{-01}	8.0406×10^{-01}	8.2403×10^{-01}	4
3.0767×10^{-02}	2.3473×10^{-01}	2.3681×10^{-01}	5
9.3746×10^{-04}	1.2839×10^{-02}	1.2892×10^{-02}	6
1.2381×10^{-04}	2.0658×10^{-03}	2.0696×10^{-03}	7
1.4755×10^{-06}	3.7532×10^{-05}	3.7578×10^{-05}	8
1.4401×10^{-07}	4.2027×10^{-06}	4.2052×10^{-06}	9
8.4020×10^{-10}	3.3991×10^{-08}	3.4007×10^{-08}	10
6.5436×10^{-11}	2.9428×10^{-09}	2.9435×10^{-09}	11
2.1494×10^{-13}	1.2619×10^{-11}	1.2622×10^{-11}	12
1.4001×10^{-14}	8.9411×10^{-13}	8.9422×10^{-13}	13
4.2067×10^{-16}	2.2760×10^{-15}	2.3112×10^{-15}	14
4.1967×10^{-16}	1.4065×10^{-16}	4.2418×10^{-16}	15
4.1933×10^{-16}	1.8964×10^{-17}	4.0035×10^{-16}	16
4.1956×10^{-16}	2.1325×10^{-17}	4.0068×10^{-16}	17
4.1946×10^{-16}	1.2846×10^{-17}	4.0024×10^{-16}	18
4.1933×10^{-16}	2.4379×10^{-17}	4.0068×10^{-16}	19
4.1976×10^{-16}	2.7699×10^{-17}	4.0128×10^{-16}	20

18.2   Application to the wave equation

18.2.1   Homogeneous one-dimension wave equation

Obtain the solution $w(x, t)$ of the differential equation

$$\frac{\partial^2 w(x, t)}{\partial t^2} = \frac{\partial^2 w(x, t)}{\partial x^2} \qquad 0 < x < \pi \quad t > 0 \qquad (18.16)$$

satisfying the boundary condition

$$\begin{aligned} w(0, t) &= 0 \qquad t \geq 0 \\ w(\pi, t) &= 0 \qquad t \geq 0 \end{aligned} \qquad (18.17)$$

and the initial condition

$$\begin{aligned} w(x, 0) &= \sin(x) \\ w_t(x, 0) &= 0 \end{aligned} \qquad (18.18)$$

The solution to this equation is $w(x, t) = \sin(x) \cos(t)$ (Wazwaz [34], page 128).

For a solution in the form

$$w(x, t) = \frac{a_0 + a_1 x + a_2 t + a_3 x^2 + a_4 x t + a_5 t^2 a_6 x^3 + a_7 x^2 t + a_8 x t^2 + a_9 t^3 + \cdots}{b_0 + b_1 x + b_2 t + b_3 x^2 + b_4 x t + b_5 t^2 b_6 x^3 + b_7 x^2 t + b_8 x t^2 + b_9 t^3 + \cdots}$$
$$(18.19)$$

where the indices and their ordering can be seen in the table 2.7.26 on page 69, the norm of the several errors can be found in the table 18.2.1 below

Table 18.2.1 *Norm of the error of the solution* $||e||$, *of the differential equation* $||r||$, *and of the domain* $||r_D||$ *as function of the degree of the polynomials in the rational fraction in the polynomials* $N(x,t)$ *and* $D(x,t)$.

$||e||$	$||r||$	$||r_D||$	degree
1.7828×10^{-01}	6.2908×10^{-01}	6.4×10^{-01}	4
1.0033×10^{-02}	6.6391×10^{-02}	6.6723×10^{-02}	5
1.1574×10^{-03}	1.0452×10^{-02}	1.0480×10^{-02}	6
2.7210×10^{-05}	3.5617×10^{-04}	3.5662×10^{-04}	7
1.8317×10^{-06}	3.0854×10^{-05}	3.0877×10^{-05}	8
2.3942×10^{-08}	5.2402×10^{-07}	5.2426×10^{-07}	9
1.0452×10^{-09}	2.8032×10^{-08}	2.8040×10^{-08}	10
8.7722×10^{-12}	2.8839×10^{-10}	2.8845×10^{-10}	11
2.6762×10^{-13}	1.0422×10^{-11}	1.0423×10^{-11}	12
1.6141×10^{-15}	7.2306×10^{-14}	7.2314×10^{-14}	13
3.7681×10^{-16}	1.8807×10^{-15}	1.8933×10^{-15}	14
3.7527×10^{-16}	9.2747×10^{-18}	2.1686×10^{-16}	15
3.7504×10^{-16}	6.5489×10^{-19}	2.1653×10^{-16}	16
3.7537×10^{-16}	4.7353×10^{-19}	2.1672×10^{-16}	17
3.7509×10^{-16}	4.8377×10^{-19}	2.1656×10^{-16}	18
3.7506×10^{-16}	1.1261×10^{-18}	2.1654×10^{-16}	19
3.7548×10^{-16}	4.6886×10^{-19}	2.1678×10^{-16}	20

18.2.2 Inhomogeneous one-dimension wave equation

Obtain the solution $w(x,t)$ of the differential equation

$$\frac{\partial^2 w(x,t)}{\partial t^2} = \frac{\partial^2 w(x,t)}{\partial x^2} - 2$$
$$0 < x < \pi \quad t > 0 \tag{18.20}$$

satisfying the boundary condition

$$w(0,t) = 0 \qquad t \geq 0$$
$$w(\pi,t) = \pi^2 \qquad t \geq 0 \tag{18.21}$$

and the initial condition

$$w(x,0) = x^2$$
$$w_t(x,0) = \sin(x) \tag{18.22}$$

The solution to this equation is $w(x,t) = x^2 + \sin(x)\sin(t)$ (Wazwaz [34], page 134).

For a solution in the form

$$w(x,t) = \frac{a_0 + a_1x + a_2t + a_3x^2 + a_4xt + a_5t^2a_6x^3 + a_7x^2t + a_8xt^2 + a_9t^3 + \cdots}{b_0 + b_1x + b_2t + b_3x^2 + b_4xt + b_5t^2b_6x^3 + b_7x^2t + b_8xt^2 + b_9t^3 + \cdots}$$

(18.23)

where the indices and their ordering can be seen in the table 2.7.26 on page 69, the norm of the several errors can be found in the table 18.2.2 below

Table 18.2.2 *Norm of the error of the solution* $||e||$, *of the differential equation* $||r||$, *and of the domain* $||r_D||$ *as function of the degree of the polynomials in the rational fraction in the polynomials* $N(x,t)$ *and* $D(x,t)$.

$||e||$	$||r||$	$||r_D||$	degree
1.3354×10^{-01}	7.0797×10^{-01}	7.2012×10^{-01}	4
7.5108×10^{-03}	7.3845×10^{-02}	7.4223×10^{-02}	5
8.6629×10^{-04}	1.1793×10^{-02}	1.1824×10^{-02}	6
2.0365×10^{-05}	4.0159×10^{-04}	4.0210×10^{-04}	7
1.3709×10^{-06}	3.4818×10^{-05}	3.4845×10^{-05}	8
1.7919×10^{-08}	5.9133×10^{-07}	5.9160×10^{-07}	9
7.8224×10^{-10}	3.1634×10^{-08}	3.1644×10^{-08}	10
6.5654×10^{-12}	3.2545×10^{-10}	3.2552×10^{-10}	11
2.0030×10^{-13}	1.1761×10^{-11}	1.1763×10^{-11}	12
1.2104×10^{-15}	8.1597×10^{-14}	8.1606×10^{-14}	13
2.8978×10^{-16}	2.1224×10^{-15}	2.1281×10^{-15}	14
2.8867×10^{-16}	1.0491×10^{-17}	1.5356×10^{-16}	15
2.8837×10^{-16}	8.3002×10^{-19}	1.5311×10^{-16}	16
2.8879×10^{-16}	4.6429×10^{-19}	1.5325×10^{-16}	17
2.8836×10^{-16}	3.8894×10^{-19}	1.5313×10^{-16}	18
2.8834×10^{-16}	1.3940×10^{-18}	1.5313×10^{-16}	19
2.8884×10^{-16}	4.5190×10^{-19}	1.5329×10^{-16}	20

18.2.3 Homogeneous two-dimension wave equation

Obtain the solution $w(x, y, t)$ of the differential equation

$$\frac{\partial^2 w(x,y,t)}{\partial t^2} = 2\left(\frac{\partial^2 w(x,y,t)}{\partial x^2} + \frac{\partial^2 w(x,y,t)}{\partial y^2}\right) \qquad 0 < x, y < \pi \quad t > 0$$

(18.24)

satisfying the boundary condition

$$w(0, y, t) = w(\pi, y, t) = 0 \qquad t \geq 0$$
$$w(x, 0, t) = w(x, \pi, t) = 0 \qquad t \geq 0$$

(18.25)

and the initial condition

$$w(x, y, 0) = \sin(x)\sin(y)$$
$$w_t(x, y, 0) = 0 \tag{18.26}$$

The solution to this equation is $w(x, y, t) = \sin(x)\sin(y)\cos(2t)$ (Wazwaz [34], page 169).

For a solution in the form

$$T(x, y, t) = \frac{a_0 + a_1 x + a_2 y + a_3 t + a_4 x^2 + a_5 xy a_6 xt + a_7 y^2 + a_8 yt + a_9 t^2 + \cdots}{b_0 + b_1 x + b_2 y + b_3 t + b_4 x^2 + b_5 xy b_6 xt + b_7 y^2 + b_8 yt + b_9 t^2 + \cdots} \tag{18.27}$$

where the indices and their ordering can be seen in the table 2.7.26 on page 69, the norm of the several errors can be found in the table 18.2.3 below

Table 18.2.3 *Norm of the error of the solution $||e||$, of the differential equation $||r||$, and of the domain $||r_D||$ as function of the degree of the polynomials in the rational fraction in the polynomials $N(x, y, t)$ and $D(x, y, t)$.*

$||e||$	$||r||$	$||r_D||$	degree
4.5279×10^{-01}	6.2114×10^{01}	6.2182×10^{01}	4
3.3247×10^{-01}	5.6327×10^{01}	5.6327×10^{01}	5
4.4444×10^{-03}	1.6007×10^{-01}	1.6008×10^{-01}	6
3.9580×10^{-03}	1.5229×10^{-01}	1.5229×10^{-01}	7
1.3416×10^{-05}	9.1341×10^{-04}	9.1341×10^{-04}	8
1.3035×10^{-05}	9.0103×10^{-04}	9.0103×10^{-04}	9
1.6425×10^{-08}	1.7691×10^{-06}	1.7691×10^{-06}	10
1.6326×10^{-08}	1.7637×10^{-06}	1.7637×10^{-06}	11
9.4738×10^{-12}	1.4677×10^{-09}	1.4677×10^{-09}	12
9.4625×10^{-12}	1.4668×10^{-09}	1.4668×10^{-09}	13
2.9546×10^{-15}	6.0652×10^{-13}	6.0652×10^{-13}	14
2.9538×10^{-15}	6.0645×10^{-13}	6.0645×10^{-13}	15
6.9023×10^{-16}	1.5483×10^{-16}	3.7606×10^{-16}	16
6.9040×10^{-16}	1.5396×10^{-16}	3.7593×10^{-16}	17
6.9015×10^{-16}	1.2637×10^{-17}	3.4296×10^{-16}	18
6.9006×10^{-16}	1.6142×10^{-17}	3.4306×10^{-16}	19
6.9057×10^{-16}	1.0312×10^{-17}	3.4328×10^{-16}	20

18.2.4   Inhomogeneous two-dimension wave equation

Obtain the solution $w(x, y, t)$ of the differential equation

$$\frac{\partial^2 w(x, y, t)}{\partial t^2} = 2 \left(\frac{\partial^2 w(x, y, t)}{\partial x^2} + \frac{\partial^2 w(x, y, t)}{\partial y^2} \right) + 6t + 2x + 4y$$
$$0 < x, y < \pi \quad t > 0 \tag{18.28}$$

satisfying the boundary condition

$$
\begin{aligned}
w(0, y, t) &= t^3 + 2t^2 y, & w(\pi, y, t) &= t^3 + \pi t^2 + 2t^2 y & t &\geq 0 \\
w(x, 0, t) &= t^3 + t^2 x, & w(x, \pi, t) &= t^3 + t^2 x + 2\pi t^2 & t &\geq 0
\end{aligned} \tag{18.29}
$$

and the initial condition

$$w(x, y, 0) = 0$$
$$w_t(x, y, 0) = 2\sin(x)\sin(y) \tag{18.30}$$

The solution to this equation is $w(x, y, t) = t^3 + t^2 x + 3t^2 y \sin(x) \sin(y) \cos(2t)$ (Wazwaz [34], page 178).

For a solution in the form

$$w(x, y, t) = \frac{a_0 + a_1 x + a_2 y + a_3 t + a_4 x^2 + a_5 xy a_6 xt + a_7 y^2 + a_8 yt + a_9 t^2 + \cdots}{b_0 + b_1 x + b_2 y + b_3 t + b_4 x^2 + b_5 xy b_6 xt + b_7 y^2 + b_8 yt + b_9 t^2 + \cdots} \tag{18.31}$$

where the indices and their ordering can be seen in the table 2.7.26 on page 69, the norm of the several errors can be found in the table 18.2.4 below

Table 18.2.4 *Norm of the error of the solution* $||e||$, *of the differential equation* $||r||$, *and of the domain* $||r_D||$ *as function of the degree of the polynomials in the rational fraction in the polynomials* $N(x, y, t)$ *and* $D(x, y, t)$.

$||e||$	$||r||$	$||r_D||$	degree
9.5131×10^{-01}	1.1221×10^{01}	1.1225×10^{01}	4
1.0558×10^{-01}	2.4214×10^{00}	2.4215×10^{00}	5
1.8630×10^{-02}	5.7326×10^{-01}	5.7326×10^{-01}	6
8.0152×10^{-04}	3.7307×10^{-02}	3.7307×10^{-02}	7
8.5351×10^{-05}	5.0832×10^{-03}	5.0832×10^{-03}	8
1.9456×10^{-06}	1.5376×10^{-04}	1.5376×10^{-04}	9
1.3580×10^{-07}	1.3125×10^{-05}	1.3125×10^{-05}	10
1.9387×10^{-09}	2.3225×10^{-07}	2.3225×10^{-07}	11
9.5022×10^{-11}	1.3471×10^{-08}	1.3471×10^{-08}	12
9.3547×10^{-13}	1.5777×10^{-10}	1.5777×10^{-10}	13
3.3922×10^{-14}	6.6052×10^{-12}	6.6052×10^{-12}	14
7.6807×10^{-16}	5.5238×10^{-14}	5.5238×10^{-14}	15
7.3313×10^{-16}	1.7414×10^{-15}	1.7414×10^{-15}	16
7.3317×10^{-16}	1.5498×10^{-17}	1.5526×10^{-17}	17
7.3231×10^{-16}	1.1317×10^{-17}	1.1317×10^{-17}	18
7.3240×10^{-16}	2.0443×10^{-17}	2.0464×10^{-17}	19
7.3357×10^{-16}	1.1417×10^{-17}	1.1418×10^{-17}	20

18.3   Nonlinear partial differential equation

Obtain the solution $w(x, y)$ of the differential equation

$$\frac{\partial^2 u(x, y)}{\partial x^2} + [u(x, y)]^2 - \left(\frac{\partial u(x, y)}{\partial y}\right)^2 \qquad 0 < x, y < 2 \qquad (18.32)$$

satisfying the boundary condition

$$u(0, y) = 0,$$
$$u(x, y) = e^y 0 \qquad (18.33)$$

The solution to this equation is $u(x, y) = y^2 + e^x$ (Wazwaz [34], page 291).

For a solution in the form

$$u(x, y) = \frac{a_0 + a_1 x + a_2 y + a_3 x^2 + a_4 xy + a_5 y^2 a_6 x^3 + a_7 x^2 y + a_8 xy^2 + a_9 y^3 + \cdots}{b_0 + b_1 x + b_2 y + b_3 x^2 + b_4 xy + b_5 y^2 b_6 x^3 + b_7 x^2 y + b_8 xy^2 + b_9 y^3 + \cdots}$$
$$(18.34)$$

where the indices and their ordering can be seen in the table 2.7.26 on page 69, the norm of the several errors can be found in the table 18.3.1 below

Table 18.3.1 *Norm of the error of the solution $\|e\|$, of the differential equation $\|r\|$, and of the domain $\|r_D\|$ as function of the degree of the polynomials in the rational fraction in the polynomials $N(x,y)$ and $D(x,y)$.*

$\|e\|$	$\|r\|$	$\|r_D\|$	degree
4.9787×10^{-04}	1.3982×10^{-01}	5.7229×10^{-02}	4
4.9077×10^{-06}	1.6742×10^{-03}	6.8424×10^{-04}	5
3.3778×10^{-08}	1.3565×10^{-05}	5.5406×10^{-06}	6
1.7130×10^{-10}	7.9168×10^{-08}	3.2326×10^{-08}	7
6.6672×10^{-13}	3.4830×10^{-10}	1.4221×10^{-10}	8
1.9039×10^{-15}	1.1959×10^{-12}	4.8823×10^{-13}	9
1.2776×10^{-15}	3.3602×10^{-15}	1.6198×10^{-15}	10
1.2756×10^{-15}	1.8494×10^{-16}	8.6361×10^{-16}	11
1.2743×10^{-15}	7.6515×10^{-17}	8.6022×10^{-16}	12
1.2749×10^{-15}	1.7531×10^{-16}	8.6247×10^{-16}	13
1.2781×10^{-15}	8.4984×10^{-17}	8.6260×10^{-16}	14
1.2731×10^{-15}	8.3726×10^{-17}	8.5934×10^{-16}	15
1.2755×10^{-15}	2.3940×10^{-16}	8.6566×10^{-16}	16
1.2735×10^{-15}	8.3726×10^{-17}	8.5925×10^{-16}	17
1.2764×10^{-15}	3.1352×10^{-16}	8.7052×10^{-16}	18
1.2739×10^{-15}	9.7731×10^{-17}	8.6113×10^{-16}	19
1.2759×10^{-15}	2.1726×10^{-16}	8.6532×10^{-16}	20

18.4   Inhomogeneous advection equation

Obtain the solution $w(x,y)$ of the differential equation

$$\frac{\partial u(x,t)}{\partial t} + \frac{1}{2}\frac{\partial [u(x,t)]^2}{\partial x} \qquad 0 < x, y < \frac{\pi}{2} \qquad (18.35)$$

satisfying the initial condition

$$u(x,0) = \cos(x) \qquad (18.36)$$

The solution to this equation is $u(x,t) = \cos(x+t)$ (Wazwaz [34], page 308).

For a solution in the form

$$u(x,t) = \frac{a_0 + a_1x + a_2t + a_3x^2 + a_4xt + a_5t^2a_6x^3 + a_7x^2t + a_8xt^2 + a_9t^3 + \cdots}{b_0 + b_1x + b_2t + b_3x^2 + b_4xt + b_5t^2b_6x^3 + b_7x^2t + b_8xt^2 + b_9t^3 + \cdots}$$

$$(18.37)$$

where the indices and their ordering can be seen in the table 2.7.26 on page 69, the norm of the several errors can be found in the table 18.4.1 below

Table 18.4.1 *Norm of the error of the solution* $||e||$, *of the differential equation* $||r||$, *and of the domain* $||r_D||$ *as function of the degree of the polynomials in the rational fraction in the polynomials* $N(x,t)$ *and* $D(x,t)$.

$||e||$	$||r||$	$||r_D||$	degree
4.2620×10^{-03}	4.1913×10^{00}	4.1913×10^{00}	4
4.8891×10^{-04}	2.8697×10^{00}	2.8697×10^{00}	5
3.5640×10^{-06}	1.0362×10^{00}	1.0362×10^{00}	6
2.9646×10^{-07}	2.4837×10^{-01}	2.4837×10^{-01}	7
8.2282×10^{-10}	6.8676×10^{-02}	6.8676×10^{-02}	8
5.3735×10^{-11}	7.3775×10^{-03}	7.3775×10^{-03}	9
7.2022×10^{-14}	1.4756×10^{-03}	1.4756×10^{-03}	10
3.9123×10^{-15}	8.8016×10^{-05}	8.8016×10^{-05}	11
4.2074×10^{-16}	1.2903×10^{-05}	1.2903×10^{-05}	12
4.2027×10^{-16}	4.9185×10^{-07}	4.9185×10^{-07}	13
4.2001×10^{-16}	5.4451×10^{-08}	5.4451×10^{-08}	14
4.2021×10^{-16}	1.4506×10^{-09}	1.4506×10^{-09}	15
4.2012×10^{-16}	1.2491×10^{-10}	1.2491×10^{-10}	16
4.2022×10^{-16}	2.4692×10^{-12}	2.4692×10^{-12}	17
4.2004×10^{-16}	1.6968×10^{-13}	1.6968×10^{-13}	18
4.2011×10^{-16}	2.5900×10^{-15}	2.5918×10^{-15}	19
4.2021×10^{-16}	1.4693×10^{-16}	1.7531×10^{-16}	20

Chapter 19

Thermal analysis of the surface of the space shuttle

19.1 Physical model

The reentry to the earth atmosphere of the space shuttle is made with a high angle of attack creating a high drag, which reduces the velocity. As a consequence of the interaction of the speed and increasing density of the atmosphere, while the space shuttles approximates to the earth surface, there is heat generated on the lower wing surface, which must be reduced or controlled. Since a ceramic material has a much higher melting temperature it was chosen as the first layer of the surface. Another solution to minimize the heat generated during the reentry is the reduction the angle of attack and the control of the speed with devices such as speed brakes and retropropulsion.

19.1.1 Thermal properties

The lower surface of the space shuttle is covered with tiles to protect the structure from over heating during the reentry into the atmosphere. In particular, as shown in figure 19.1.1, the lower wing surface is composed of the five layers whose properties are

1. Layer AB a high temperature reusable insulation (HRSI), which is the external layer

2. Layer BC a room temperature vulcanizing adhesive (RTV)

3. Layer CD a strain insulation pad (SIP)

4. Layer DE another RTV layer

5. Layer EF structural aluminum layer (Al)

During the reentry, a fraction of the heat generated at the HRSI surface is radiated to the space and the remaining heat energy flows from A to F through the several layers.

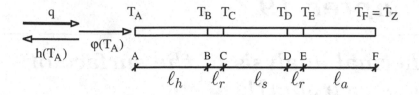

Figure 19.1.1 *Problem geometry.*

19.1.2   Material modeling

19.1.2.1   Temperature reusable insulation layer

1- Thermal conductivity k_h

Due to the high temperature along the layer AB, the thermal conductivity has a nonlinear behavior, which was modeled by a quadratic polynomial, which can be written as

$$k_h = k_{xx} = k_{yy} = k_0 + k_1 T + k_2 T^2 \tag{19.1}$$

where

$k_0 = 1.57111581 \times 10^{-2}$

$k_1 = -2.83807646 \times 10^{-5}$

$k_2 = 1.36632887 \times 10^{-7}$

are expressed in the SI unit system, that is, they are MKS units. Therefore, k_h is expressed in W/(m.K), that is, Watt/(meter $\times$ degree Kelvin). The temperature T is measured in degrees Kelvin.

Figure 19.1.2 shows graphically the variation of thermal conductivity with the temperature of the material.

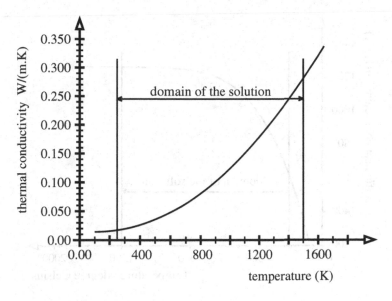

Figure 19.1.2 *Temperature dependent thermal conductivity for the HRSI.*

2- Specific heat coefficient C_p

The specific heat C_p was modeled by a quartic polynomial, in terms of the temperature, as

$$C_p = c_0 + c_1 T + c_2 T^2 + c_3 T^3 + c_4 T^4 \tag{19.2}$$

where

$c_0 = -1.02215713 \times 10^2$

$c_1 = 3.78171019$

$c_2 = -3.94207081 \times 10^{-3}$

$c_3 = 1.82912266 \times 10^{-6}$

$c_4 = -3.17436957 \times 10^{-10}$

are expressed in the SI (MKS) system of units, therefore, C_p has the units J/(Kg · K), that is, Joule/(Kg mass × degree Kelvin).

Figure 19.1.3 shows graphically the variation of thermal conductivity with the temperature of the material.

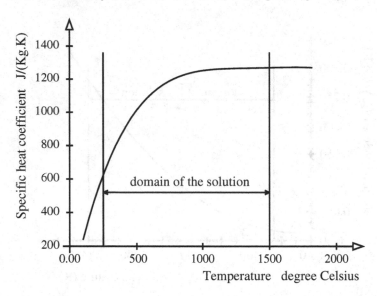

Figure 19.1.3 *Specific heat coefficient for the HRSI.*

3- Heat convection coefficient h

Temperature-dependent convection coefficient to model radiation is defined by $q_h = h(T - T_R)$ where $h = h(T)$ and $T_R = 300$ K.

$$h = \left(h_3 T^3 + h_2 T^2 + h_1 T + h_0\right) (T - T_R) \qquad (19.3)$$

where

$h_0 = 1.29951672$

$h_1 = 4.35923168 \times 10^{-3}$

$h_2 = 1.44485599 \times 10^{-5}$

$h_3 = 4.82561917 \times 10^{-7}$

are expressed in the SI (MKS) system of units, therefore, h has the units $W/(m^2 \cdot K)$, that is, Watt/(square meter $\times$ degree Kelvin).

Figure 19.1.4 shows graphically the variation of thermal conductivity with the temperature of the material.

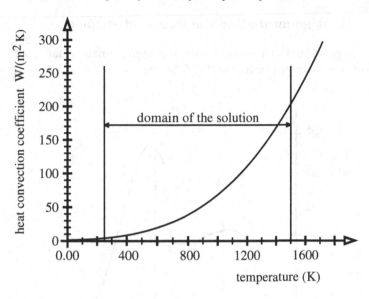

Figure 19.1.4 *Temperature dependent heat convection coefficient for the HRSI.*

4- Density

$d = 144.156913$ Kg/m³ (Kg mass/cubic meter).

19.1.3   Properties of the remaining layers

1. Layers BC and DE

   Specific heat $C_p = 1465.38$ J/(m · K)

   Thermal conductivity $k_{xx} = k_{yy} = 0.31155722$ W/(m · K)

   Density $d = 1409.737250$ Kg/m³

2. Layer CD

   Specific heat $C_p = 1457.0064$ J/(m · K)

   Thermal conductivity $k_{xx} = k_{yy} = 0.04680461$ W/(m · K)

   Density $d = 86.499684$ Kg/m³

3. Layer EF

   Specific heat $C_p = 900.162$ J/(m · K)

   Thermal conductivity $k_{xx} = k_{yy} = 135.32963112$ W/(m · K)

   Density $d = 2802.589762$ Kg/m³

19.1.4   Heat generated at the external surface

The heat generated at the surface during approximately half hour, on one of the sections, is shown in figure 19.1.5 below

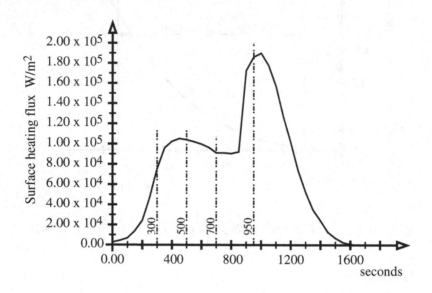

Figure 19.1.5 *Surface heating flux.*

The analysis was performed at the elapsed time $0, 300, 500, 700$, and 950 seconds. The values of the heat-flux are given below and shown in figure 19.1.5 above.

$q_3 = 2.698311 \times 10^{03}$ W/m^2 after $t = 0$ seconds.

$q_3 = 7.538917 \times 10^{04}$ W/m^2 after $t = 300$ seconds.

$q_3 = 1.041712 \times 10^{05}$ W/m^2 after $t = 500$ seconds.

$q_3 = 9.125197 \times 10^{04}$ W/m^2 after $t = 700$ seconds.

$q_3 = 1.861344 \times 10^{05}$ W/m^2 after $t = 950$ seconds.

19.1.5   Geometry

The length of each layer is

$\ell_h = 34.544$ mm (millimeter)

$\ell_r = 0.1778$ mm

$\ell_s = 4.064$ mm

$\ell_a = 3.0226$ mm

19.1.6 Temperatures

The internal temperature is prescribed as $T_Z = 300$ K.

19.2 Mathematical model

The mathematical model is defined by the governing equations together with the boundary conditions.

19.2.1 Nonlinear governing differential equations

19.2.1.1 Second-order differential equation formulation

The one-dimensional steady state heat conduction equation for a homogeneous and isotropic solid, without heat generation within the body, Özişiki [29, chapter 1] or Karlekar and Desmond [30, chapter 2], is given by

$$\frac{d}{dX}\left[k(T)\frac{dT(X)}{dX}\right] = 0 \qquad (19.4)$$

The coordinate $X \in [0, \ell]$ can be normalized, with respect to the length ℓ of the domain as shown in the subsection 16.1.2 on page 538.

The domain $[0, \ell]$ may be transformed to another $[A, B]$, symmetric with respect to the coordinate axes, using the expressions developed in the subsection 16.1.2, page 538.

From the definition in equation (16.4) on page 539 one has $b - a = 0 - \ell = \ell$ and $L = B - A$, which gives

$$\delta = \frac{L}{\ell} \quad \text{and} \quad \delta = \frac{dx}{dX} = \frac{L}{\ell} \qquad (19.5)$$

Note that here the variable X refers to the first coordinate system while in the subsection 16.1.2 it refers to the second coordinate system, therefore, $\delta = dx/dX$.

Substituting the results obtained in equation (19.5) into equation (19.4) it yields the following normalized differential governing equation

$$\frac{d}{dx}\frac{dx}{dX}\left[k(T)\frac{d}{dx}\frac{dx}{dX}T(x)\right] = 0 \quad \Longrightarrow \quad \frac{L}{\ell^2}\frac{d}{dx}\left[k(T)\frac{dT(x)}{dx}\right] = 0 \qquad (19.6)$$

where $L = 1$ if the transformed domain is $[0, 1]$ and $L = 4$ if it is $[-1, 1]$.

19.2.1.2 First-order differential equation formulation

The equation (19.4) can be integrated over the domain $[0, X]$ yielding

$$\int_0^X \left[\frac{d}{dX} \left(k(T) \frac{dT(X)}{dX} \right) dX \right] = 0 \quad \Longrightarrow \quad k(T) \frac{dT(X)}{dX} - \left[k(T) \frac{dT(X)}{dX} \right]_{X=0} = 0$$

(19.7)

Recalling the the flux at a point A is defined as $\varphi(T_A) := -k(T) \, dT(X)/dX|_{X=A}$, the equation (19.7) can be rewritten as

$$-k(T) \left. \frac{dT(X)}{dX} \right|_X = \varphi(T_A)$$

(19.8)

where T_A is the temperature at the point $X = 0$.

The transformation from the domain $[0, \ell]$ to $[A, B]$ can be written as

$$-\frac{L}{\ell} k(T) \frac{dT(x)}{dx} = \varphi(T_A)$$

(19.9)

19.2.2 Linear governing differential equations

If the thermal conductivity is independent of the temperature, then the equation (19.7) can be integrated over the domain $[0, X]$

$$-k \left[T(X) - T(A) \right] = \varphi(T_A) \Delta X$$

(19.10)

The domain transformation $[0, \ell] \to [A, B]$ gives

$$-\frac{L}{\ell} k \big[T(B) - T(A) \big] = \varphi(T_A)$$

(19.11)

where $L = 1$ if the domain is $[0, 1]$ and $L = 2$ if the domain is $[-1, 1]$.

19.2.3 Summary of governing equations

Referring to the figure 19.1.1 the governing equation for each layer is

1. Layer AB

   This layer must satisfy the nonlinear governing equation since its constitutive relations are nonlinear with respect to temperature.

(a) Second-order differential equation

$$\frac{L_h}{\ell_h^2}\frac{d}{dx}\left[k(T_h)\frac{dT_h(x)}{dx}\right] = 0 \tag{19.12}$$

where $L_h = 1$ if the normalized domain is $[0, 1]$ and $L_h = 4$ if it is $[-1, 1]$, and the index h refers to the variables and parameters related to the layer AB.

(b) First-order differential equation

$$\frac{L_h}{\ell_h}\left[k_h(T_h)\frac{dT_h(x)}{dx}\right] + \varphi(T_A) = 0 \tag{19.13}$$

2. Layer BC

This layer was modeled as a thermal resistance by the equation

$$-\frac{L_r}{\ell_r}k_r\left[T_B - T_C\right] = \varphi(T_A) \tag{19.14}$$

where $L_r = 1$ if the domain is $[0, 1]$ and $L_r = 2$ if the domain is $[-1, 1]$.

3. Layer CD

The material properties for the layer CD are independent of the temperature, therefore, the mathematical model for this layer can be represented by a second-order linear differential equation of the form

$$\frac{L_s}{\ell_s^2}k_s\frac{d^2T_s(x)}{dx^2} = 0 \tag{19.15}$$

4. Layer DE

The region DE was modeled as a thermal resistance,

$$-\frac{L_r}{\ell_r}k_r\left[T_B - T_C\right] = \varphi(T_A) \tag{19.16}$$

5. Layer EF

Similarly to the layer CD one has

$$\frac{L_a}{\ell_a^2}k_a\frac{d^2T_a(x)}{dx^2} = 0 \tag{19.17}$$

19.2.4 Boundary conditions

Since the integration of the differential equation (19.4) generates two constants, it is necessary to prescribe two conditions to obtain an unique solution.

These conditions can be the temperature at each end of the interval, or both temperature and flux at one of the ends of the interval. In the later case it was considered the temperature and flux at the beginning of the interval.

Denote

$$T(A^-) := \text{ the temperature on the left of the point } A \qquad (19.18)$$
$$T(A^+) := \text{ the temperature on the right of the point } A$$
$$T_A := \text{ the temperature at the point } A$$

$$\varphi(T(A^-)) := q - h(T(x))|_{T=T(A)} \text{ the flux on the left of the point } A$$

$$(19.19)$$

$$\varphi(T(A^+)) := -\frac{L}{\ell} \left[k(T)\frac{dT(x)}{dx} \right]_{T=T(A)} \qquad \text{the flux on the right of the point } A$$

19.2.4.1   Flux boundary conditions

The flux boundary conditions for each layer are:

Layer AB

The heat-flux balance at the point A is given by

$$\varphi(T(A^-)) = \varphi(T(A^+)) \qquad (19.20)$$

that is,

$$q - h(T_A) = -\frac{L_h}{\ell_h} k_h(T_A) \frac{dT_h(x)}{dx}\bigg|_{x=x_A} \qquad (19.21)$$

where q is a given quantity and $T(A^-) = T(A^+) = T_A$.

The derivative T_x of the temperature with respect to the position coordinate x can be obtained as a function of T_A from the expression (19.21) as follows

$$T_x = -\left(q - h(T_A) \right) \frac{\ell_h}{L_h k_h(T_A)}$$

$$= \varphi(T_A) \frac{\ell_h}{L_h k_h(T_A)} \qquad (19.22)$$

where $L_h = 1$ for the domain $[0, 1]$ and $L_h = 2$ for the domain $[-1, 1]$. Note that equation (19.22) is a nonlinear expression with respect to the temperature.

Layer BC

Since there is no heat generation, the flux is the same at the points B and C.

$$\varphi(T_B) = \varphi(T_C) \qquad (19.23)$$

Layer CD

Analogously to the layer BC,

$$\varphi(T_C) = \varphi(T_D) \qquad (19.24)$$

Layer DE

Analogously to the layer BC,

$$\varphi(T_D) = \varphi(T_E) \qquad (19.25)$$

Layer EF

Analogously to the layer EF,

$$\varphi(T_E) = \varphi(T_F) \qquad (19.26)$$

19.2.4.2 Temperature boundary conditions

There is continuity of the temperature distribution on the points B, C, D, and F along the domain. The same is not true for the slop of the temperature distribution at these points.

Evaluation of the temperature T_F

The temperature is prescribed at the point F as $T_F = T_Z = 300$ K.

Evaluation of the temperature T_E

Using the given temperature boundary condition $T_F = T_Z$ and the net heat-flux $\varphi(T_A)$, the temperature at the point E can be evaluated. This can be performed integrating the governing differential equation for the layer EF, equation (19.26). Recalling that k_a is constant, the integration gives

$$-\frac{k_a}{\ell_a} (T_F - T_E) = \varphi(T_A) \qquad (19.27)$$

Solving for T_E it follows

$$T_E = \frac{\ell_a}{k_a} \varphi(T_A) + T_F \qquad (19.28)$$

Evaluation of the temperature T_D

Known T_E and $\varphi(T_A)$ the temperature T_D can be computed by a similar procedure. Since the subdomain DE was modeled as a thermal resistance in terms of the temperatures T_D and T_E. The solution of equation (19.25) for T_D yields

$$T_D = \frac{\ell_r}{k_r} \varphi(T_A) + T_E \qquad (19.29)$$

Introducing T_E, given by equation (19.28) into equation (19.29) and solving for T_B.

$$T_D = \left[\frac{\ell_r}{k_r} + \frac{\ell_a}{k_a}\right]\varphi(T_A) + T_F \qquad (19.30)$$

Evaluation of the temperature T_C

Known T_D and $\varphi(T_A)$ the temperature T_C can be obtained similarly to the computation of the temperature T_E. Integrating equation (19.24) and solving for T_C, one obtains

$$T_C = \left[\frac{\ell_s}{k_s} + \frac{\ell_r}{k_r} + \frac{\ell_a}{k_a}\right]\varphi(T_A) + T_F \qquad (19.31)$$

Evaluation of the temperature T_B

Known T_C and $\varphi(T_A)$ the temperature T_B can be obtained. The subdomain BC was modeled as a thermal resistance, thus the temperatures T_B and T_C are related by equation equation (19.23). Solving for T_B and introducing T_C given by equation (19.31) it follows

$$T_B = \left[\frac{\ell_s}{k_s} + 2\frac{\ell_r}{k_r} + \frac{\ell_a}{k_a}\right]\varphi(T_A) + T_F \qquad (19.32)$$

19.3   Construction of the approximate solution

19.3.1   Types of solutions

The approximate solution can be constructed with respect to the equation (19.12), which will be referred from here on as *Second order differential equation formulation*. It can also be constructed with respect to the equation (19.13), which will be named *First order differential equation formulation*.

For a two-point boundary value problem the the solution, according to equation (16.1) on page 537, can be expanded as

$$T(x) = \sum_{n=0}^{\nu} \Phi_{p,n}(x)T_{p,n} + \sum_{n=0}^{\nu} \Phi_{q,n}(x)T_{q,n} \qquad (19.33)$$

where ν is the order of the governing differential equation.

Approximate solution for the first-order differential equation

The approximate solution for the equation (19.13) is obtained from the equation (19.33), with $\nu = 1$, for the interval $[p, q] = [A, B]$, as

$$T(x) = \Phi_{A,0}(x)T_{A,0} + \Phi_{p,1}(x)T_{A,1}$$
$$+ \Phi_{B,0}(x)T_{B,0} + \Phi_{B,1}(x)T_{B,1}$$
$$= \left[\Phi_{A,0}(x)T_{A,0} + \Phi_{B,0}(x)T_{B,0}\right]$$
$$+ \left[\Phi_{A,1}(x)T_{A,1} + \Phi_{B,1}(x)T_{B,1}\right] \qquad (19.34)$$

Approximate solution for the second-order differential equation

The approximate solution for the equation (19.12) is obtained from the equation (19.33), with $\nu = 2$, as

$$T(x) = \Phi_{A,0}(x)T_{A,0} + \Phi_{A,1}(x)T_{A,1}$$
$$+ \Phi_{B,0}(x)T_{B,0} + \Phi_{B,1}(x)T_{B,1}$$
$$= \left[\Phi_{A,0}(x)T_{A,0} + \Phi_{B,0}(x)T_{B,0}\right]$$
$$+ \left[\Phi_{A,1}(x)T_{A,1} + \Phi_{B,1}(x)T_{B,1}\right]$$
$$+ \left[\Phi_{A,2}(x)T_{A,2} + \Phi_{B,2}(x)T_{B,2}\right] \qquad (19.35)$$

The Hermite Interpolating Polynomials related to the reference node e, given by equation (7.126) on page 315, writes

$$\Phi_{e,n}(x) = \frac{1}{n!P_e(\alpha)} P_e(x)\, Q_{e,n}(x)\, f_{e,n}(a, x) \qquad (19.36)$$

According to the definition 7.1.4 of Multivariate Hermite Interpolating Polynomial on page 288 the polynomial $\Phi_{p,0}$ must equal one at the reference node x_e at one boundary and zero at the other boundary, therefore, the factor $P_e(x)$, where $e = A, B$, can be chosen as

$$P_A(x) = \frac{x - x_B}{L} \qquad P_B(x) = \frac{x - x_A}{L} \qquad (19.37)$$

where $x_A = 0, x_B = 1, L = 1 - 0 = 1$ if the interval is $[0, 1]$ and $x_A = -1, x_B = 1, L = 1 - (-1) = 2$ if the interval is $[-1, 1]$.

The polynomial $P_e(x)$ can also be chosen as a rational fraction, for example, as

$$P_A(x) = \frac{x - x_B}{L + x} \qquad P_B(x) = \frac{x - x_A}{L + x}$$

or

$$P_A(x) = \frac{x - x_B}{(L + x)^2} \qquad P_B(x) = \frac{x - x_A}{(L + x)^2} \qquad (19.38)$$

The following possible ways to solve the problem have been chosen for the analysis of the heat distribution.

1. The temperatures T_A and T_B are prescribed at each boundary

   (a) solution expanded in power series

   (b) solution expanded in rational fraction of Chebyshev polynomials

2. The temperature T_A and flux $\varphi(T_A)$ are prescribed at the boundary point A

   (a) solution expanded in power series

   (b) solution expanded in rational fraction of Chebyshev polynomials

Each of the above cases can be applied to the second order differential equation formulation or to the first-order differential equation formulation.

19.3.2 Second-order differential equation formulation

The second-order governing equation given by equation (19.12), rewrites

$$\frac{L_h}{\ell_h^2} \frac{d}{dx}\left[k(T_h) \frac{dT_h(x)}{dx} \right] = 0 \tag{19.39}$$

The integration of this equation gives two constants that need two conditions to yield a unique solution to the problem. These two conditions can be, for example: *two temperatures, one at each end of the boundary of the domain* or *the temperature at and the flux, which is related to the derivative of the temperature, at the same boundary of the domain.*

19.3.2.1 Two-temperature boundary conditions

The values for n are $n = 0, 1$ since $\nu = 1$. The factor $Q_{e,n}(x)$ must vanish at the reference node if $n = 1$, and it must equal one if $n = 0$. Therefore, it can be chosen, for $n = 1$, as

$$Q_{A,1}(x) = x - x_A \qquad Q_{B,1}(x) = x - x_B \tag{19.40}$$

where $x_A = 0, x_B = 1$ if the interval is $[0, 1]$ and $x_A = -1, x_B = 1$ if the interval is $[-1, 1]$.

The factor $f_{e,n}(a, x)$ must be chosen such that $\Phi_{p,0}(x)$ and $\Phi_{q,0}(x)$ equals one at their boundaries. Equation (7.128) on page 315 shows that, in particular, $f_{e,n}(a, x)$ can be chosen as one. Therefore, the following choices can be made with $f_{A,0}(a, x) = f_{B,0}(a, x) = 1$

$$\Phi_{A,0}(x) = \frac{x - x_B}{L + x} \qquad \Phi_{B,0}(x) = \frac{x - x_A}{L + x} \tag{19.41}$$

Note that $\Phi_{A,0}(x)$ must evaluate one at the reference node $x = x_A$ and $\Phi_{B,0}(x)$ equals one at the reference node $x = x_B$.

If $n = 1$ the factor $f_{e,n}(a, x)$ can be written as an expansion on a basis of linearly independent functions, which can be power series, Chebyshev, etc. It can also be written as a rational fraction whose numerator and denominator are expanded in a basis of linearly independent functions. A particular choice is $f_{A,1}(a, x) = f_{B,1}(a, x)$, which will be used here.

$$T(x) = \Phi_{A,0}(x)T_{A,0} + \Phi_{B,0}(x)T_{B,0} + \Phi_{A,1}(x)T_{A,1} + \Phi_{B,1}(x)T_{B,1}$$

$$= \left(\frac{x - x_B}{L + x} \right) T_{A,0} + \left(\frac{x - x_A}{L + x} \right) T_{B,0} +$$

$$\left(\frac{x - x_B}{L} \right) (x - x_A) f_{A,1}(a, x) + \left(\frac{x - x_A}{L} \right) (x - x_B) f_{B,1}(a, x)$$

$$= \left(\frac{x - x_B}{L + x} \right) T_{A,0} + \left(\frac{x - x_A}{L + x} \right) T_{B,0} + \frac{1}{L}(x - x_A)(x - x_B) f_{A,1}(a, x)$$

$$(19.42)$$

The factor $1/L$ in the last term can be removed from the expression, this implies that it will be embedded in the parameter a in $f_{A,1}(a, x)$ during the computation of the approximate solution. Then the approximate solution can be written as

$$T(x) = \left(\frac{x - x_B}{L + x} \right) T_{A,0} + \left(\frac{x - x_A}{L + x} \right) T_{B,0} + (x - x_A)(x - x_B) f_{A,1}(a, x) \quad (19.43)$$

For the cases $0 - 3$ the approximate solution was constructed satisfying identically the temperature boundary conditions.

The solution in the domain $[0, 1]$ was approximated by a function expanded in powers of x, here called *power series solution*, which writes

$$f_{A,1}(a, x) = f_{B,1}(a, x) = \sum_{i=0}^{m} a_i x^i \quad (19.44)$$

The temperatures at the ends of the interval are denoted as $T_A = T(0)$ and $T_B = T(1)$. For this domain $L = 1$, and $x_A = 0, x_B = 1$.

The solution in the domain $[-1, 1]$ was approximated by Chebyshev basis functions, here called *Chebyshev solution*.

$$f_{A,1}(a, x) = f_{B,1}(a, x) = \sum_{i=0}^{m} a_i T_i(x) \quad (19.45)$$

where $T_i(x)$ are Chebyshev polynomials of first kind. The temperatures were denoted as $T_A = T(-1)$ and $T_B = T(1)$, the length of the interval is $L = 2$, and end coordinates are $x_A = -1, x_B = 1$.

case 0

1. Power series solution

$$T(x) = (1 - x)T(0) + xT(1) + (x - x^2) \sum_{i=0}^{m} a_i x^i$$

$$x \in [0, 1] \hspace{4cm} (19.46)$$

2. Chebyshev solution

$$T(x) = \frac{1-x}{2}T(-1) + \frac{1+x}{2}T(1) + (x^2 - 1) \sum_{i=0}^{m} a_i T_i(x)$$

$$x \in [-1, 1] \hspace{4cm} (19.47)$$

case 1

The expansion of the factor is $f_{e,n}(a, x)$ performed using rational functions. The number $L = 1$ for the power series and $L = 2$ for Chebyshev polynomials, was added to the denominator to avoid division by zero if all coefficients b_j equal zero. Note that the summation starts at $j = 0$ (see equations 19.7.1, 19.7.2) for the coefficients.

1. Power series solution

$$T(x) = (1 - x)T(0) + xT(1) + (x - x^2) \frac{\sum_{i=0}^{m} a_i x^i}{1 + \sum_{j=0}^{n} b_j x^j}$$

$$x \in [0, 1] \hspace{4cm} (19.48)$$

2. Chebyshev solution

$$T(x) = \frac{1-x}{2}T(-1) + \frac{1+x}{2}T(1) + (x^2 - 1) \frac{\sum_{i=0}^{m} a_i T_i(x)}{2 + \sum_{j=0}^{n} b_j T_j(x)}$$

$$x \in [-1, 1] \hspace{4cm} (19.49)$$

case 2

The polynomial $P_e(x)$ is written in a rational fraction form where the denominator is given by $(1 + x)$ for power series and $(2 + x)$ for the Chebyshev polynomials.

1. Power series solution

$$T(x) = (1 - x)T(0) + xT(1) + \frac{x - x^2}{1 + x} \sum_{i=0}^{m} a_i x^i$$

$$x \in [0, 1] \hspace{4cm} (19.50)$$

2. Chebyshev solution

$$T(x) = \frac{1-x}{2}T(-1) + \frac{1+x}{2}T(1) + \frac{x^2-1}{2+x}\sum_{i=0}^{m}a_iT_i(x)$$

$$x \in [-1,1] \tag{19.51}$$

case 3

The polynomial $P_e(x)$ is written in a rational fraction form where the denominator is given by $(1+x)^2$ for power series and $(2+x)^2$ for the Chebyshev polynomials.

1. Power series solution

$$T(x) = (1-x)T(0) + xT(1) + \frac{x-x^2}{(1+x)^2}\sum_{i=0}^{m}a_ix^i$$

$$x \in [0,1] \tag{19.52}$$

2. Chebyshev solution

$$T(x) = \frac{1-x}{2}T(-1) + \frac{1+x}{2}T(1) + \frac{x^2-1}{(2+x)^2}\sum_{i=0}^{m}a_iT_i(x)$$

$$x \in [-1,1] \tag{19.53}$$

19.3.2.2 Heat-flux and temperature boundary conditions

Cases 4 through 7 assume that the temperature and heat-flux at one of the boundaries are known. The temperature $T_A = T(0)$ and the derivative $T_x(0)$ are identically satisfied at $x = 0$ for the power series solution and $T(-1)$ and the derivative $T_x(-1)$ for the Chebyshev solution.

case 4

$$T(x) = T(0) + xT_x(0) + x^2\sum_{i=0}^{m}a_ix^i$$

$$x \in [0,1] \tag{19.54}$$

$$T(x) = T(-1) + (x+1)T_x(-1) + (x+1)^2\sum_{i=0}^{m}a_iT_i(x)$$

$$x \in [-1,1] \tag{19.55}$$

case 5

The polynomial $P_e(x)$ is written in a rational fraction form where the denominator is given by $(1 + x)$ for power series and $(2 + x)$ for the Chebyshev polynomials.

$$T(x) = T(0) + xT_x(0) + \frac{x^2}{x+1} \sum_{i=0}^{m} a_i x^i$$

$$x \in [0, 1] \tag{19.56}$$

$$T(x) = T(-1) + (x+1)T_x(-1) + \frac{(1+x)^2}{x+2} \sum_{i=0}^{m} a_i T_i(x)$$

$$x \in [-1, 1] \tag{19.57}$$

case 6

The polynomial $P_e(x)$ is written in a rational fraction form where the denominator is given by $(1 + x)^2$ for power series and $(2 + x)^2$ for the Chebyshev polynomials.

$$T(x) = T(0) + xT_x(0) + \frac{x^2}{(x+1)^2} \sum_{i=0}^{m} a_i x^i$$

$$x \in [0, 1] \tag{19.58}$$

$$T(x) = T(-1) + (x+1)T_x(-1) + \frac{(1+x)^2}{(x+2)^2} \sum_{i=0}^{m} a_i T_i(x)$$

$$x \in [-1, 1] \tag{19.59}$$

case 7

The expansion is written in rational function. The number $L = 1$ was added to the denominator to avoid division by zero if all coefficients b_j equal zero, and $L = 2$ for the Chebyshev polynomials. Note that the summation, for the power series approximate solution, starts at $j = 1$, which means that $b_0 = 1$ ($b_0 = 2$). See equations (19.7.3, 19.7.4) for the coefficients.

$$T(x) = T(0) + xT_x(0) + x^2 \frac{\sum_{i=0}^{m} a_i x^i}{1 + \sum_{j=1}^{n} b_j x^j}$$

$$x \in [0, 1] \tag{19.60}$$

Note that the summation for the expansion in Chebyshev polynomials starts at $j = 1$ and that $b_0 = 2$.

$$T(x) = T(-1) + (x + 1) T_x(-1) + (1 + x)^2 \frac{\sum_{i=0}^{m} a_i T_i(x)}{2 + \sum_{j=1}^{n} b_j T_i(x)}$$

$$x \in [-1, 1] \tag{19.61}$$

19.3.3 First-order differential equation formulation

Domain is described by the first order differential equation given by equation (19.13), which is

$$\frac{L_h}{\ell_h} \left[k_h(T_h) \frac{dT_h(x)}{dx} \right] + \varphi(T_A) = 0 \tag{19.62}$$

Since there is only one integration only one condition is necessary to obtain the unique solution.

The equation (19.62) was used to obtain the approximate solution in cases 8 and 9. Since one of the boundary conditions disappears with the integration, the temperature at $x = 0$ (or $x = -1$) was considered in the construction of the approximate analytical solution. The temperature $T(0)$ (or $T(-1)$) is identically satisfied. Domain is described by a first-order differential equation

case 8

See equations (19.7.5, 19.7.6) for the coefficients.

$$T(x) = T(0) + x \sum_{i=0}^{m} a_i x^i \qquad x \in [0, 1] \tag{19.63}$$

$$T(x) = T(-1) + (1 + x) \sum_{i=0}^{m} a_i T_i(x) \qquad x \in [-1, 1] \tag{19.64}$$

case 9

The expansion is written in rational function. The number 1 was added to the denominator to avoid division by zero if all coefficients b_j equal zero (2 for the Chebyshev polynomials). Note that the summation, for the power series approximate solution, starts at $j = 0$. See equations (19.7.7, 19.7.8) for the coefficients.

$$T(x) = T(0) + x \frac{\sum_{i=0}^{m} a_i x^i}{1 + \sum_{j=0}^{n} b_j x^j} \qquad x \in [0, 1] \tag{19.65}$$

$$T(x) = T(-1) + (1 + x) \frac{\sum_{i=0}^{m} a_i T_i(x)}{2 + \sum_{j=0}^{n} b_j T_j(x)} \qquad x \in [-1, 1] \tag{19.66}$$

19.3.4   Computation of the residual of the differential equation

If the given boundary conditions are the two temperatures, the coefficients of the expansion are such that the residual of the governing differential equation will be minimum, that is, they will be obtained from the solution of the problem

$$\min_{a_i} \mathcal{L}[T(x)] \tag{19.67}$$

where $\mathcal{L}[T(x)]$ denotes the differential equation operator.

If the boundary conditions are the temperature and flux, say $T_A = T(-1)$ and $\varphi_A = \varphi(-1)$, then the temperature $T_B = T(1)$ is unknown (the approximate solution is written in terms of the boundary temperatures) and therefore, one more equation is necessary to yield a unique solution. In this case, the additional equation is the flux balance at the boundary A.

$$\min_{a_i} \left[\begin{array}{c} \mathcal{L}[T(x)] \\[2mm] \varphi(T(x)) \Big|_{x=-1} - \varphi_A \end{array} \right] \tag{19.68}$$

The heat-flux boundary residual is, in general, much smaller than the one of the governing differential equation. This may yield to solutions that do not satisfy the flux boundary condition. This difficult can be overcome by reformulating the problem as a two coupled optimization expressions. In other words as a constrained optimization problem were the constraint relation is another optimization expression. Then one can write

$$\min_{T(1)} \left[\varphi(T(x)) \Big|_{x=-1} - \varphi_A \right]$$

constrained to

$$\min_{a_i} \mathcal{L}[T(x)]_{x \in (-1,1)} \tag{19.69}$$

19.4   Numerical computation

1- Assign a value to the temperature T_A at the point A

If all the heat generated at the point A is radiated then there will be no heat-flux entering the layer AB. If the amount of radiated energy is smaller than the one generated then there will be a heat flow from A to F. If the radiated heat is greater than that generated then there is a heat-flux from F to A. The change of direction of the heat-flux corresponds to a bifurcation point in the solution behavior. It was adopted, as a guide line, do not permit the successive trial solutions to oscillate around the point of zero net heat-flux, that is, the trial solution was forced to converge monotonically from the side where $\varphi(T_A) > 0$ only. To guess an initial value for T_A it was used the condition that the net heat-flux enters the subdomain AB, which condition can expressed by the following inequality

$$\varphi(T_A) = q - h(T_A) > 0 \qquad (19.70)$$

To compute the initial guess one can compute the temperature T_o, which corresponds to a zero flux, that is, solving the equation $\varphi(T_o) = 0$. The monotonicy of the solution can be achieved choosing values of T_A satisfying the condition $T_A < T_o$. For the example presented here it was obtained $T_o = 1402.231$ K. Then the initial guess was defined as $T_A = 1380$ K. During the optimization if a step ΔT would give a new temperature, T_{new}, such that $T_{new} > T_o$ then ΔT was reduced until T_{new} would satisfy the condition $T_{new} < T_o$ and then adopt $T_A = T_{new}$ for the next step.

For increasing values of T_A such $T_A > T_o$ the residual of the second-order governing differential equation (19.12) or of the first-order differential equation (19.13) will keep reducing. This means that the solution is converging to another boundary condition (in general to the trivial solution). For more details, see the step "Solve the outer optimization problem".

2- Evaluate the net heat-flux

Given T_A, the net flux $\varphi(T_A)$ is evaluated from

$$\varphi(T_A) = q - (h_3 T_A^3 + h_2 T_A^2 + h_1 T_A + h_0)(T_A - T_R) \qquad (19.71)$$

Considering that, for a steady-state one-dimensional case, the net heat-flux is constant for all points in the domain and boundaries, it can be used to evaluate the boundary temperatures.

3- Evaluate the temperature T_E

Using the given temperature boundary condition $T_F = T_Z$ and the net heat-flux $\varphi(T_A)$, the temperature at the point E is given by equation (19.28), which is

$$T_E = \frac{\ell_a}{k_a}\varphi(T_A) + T_F \qquad (19.72)$$

4- Evaluate the temperature T_D

Known T_E and $\varphi(T_A)$ the temperature T_D can be computed by a similar procedure using the equation (19.30), which is

$$T_D = \left[\frac{\ell_r}{k_r} + \frac{\ell_a}{k_a}\right]\varphi(T_A) + T_F \qquad (19.73)$$

5- Evaluate the temperature T_C

Known T_D and $\varphi(T_A)$ the temperature T_C can be obtained from (19.31), that is,

$$T_C = \left[\frac{\ell_s}{k_s} + \frac{\ell_r}{k_r} + \frac{\ell_a}{k_a} \right] \varphi(T_A) + T_F \qquad (19.74)$$

6- Evaluate the temperature T_B

Known T_C and $\varphi(T_A)$ the temperature T_B can be obtained using the equation (19.32)

$$T_B = \left[\frac{\ell_s}{k_s} + 2\frac{\ell_r}{k_r} + \frac{\ell_a}{k_a} \right] \varphi(T_A) + T_F \qquad (19.75)$$

7- Solve the inner nonlinear optimization problem

The weighted residual method imposes the orthogonality of the residual and the weighting function, this may be written as

$$< W_i, R >= \int_\Omega W_i(x) R(T_A, x, a_j) d\Omega = 0 \qquad (19.76)$$

If $W_i = \delta(x_d - x)$ then the collocation approach is obtained and equation (19.77) is transformed into

$$< \delta(x_d - x), R >= R(T_A, x_d, a_j) = 0 \qquad d = 1, 2, ..., p \qquad (19.77)$$

where p represents the number of collocation points. This technique was preferred because it avoids the integration to evaluate the inner product, which is a complex symbolic operation for the case of the rational functions. Equation (19.77) represents a system of p algebraic equations whose solution gives the coefficients a_j of the approximate analytical solution expansion. The residual $R(T_A, x_d, a_j)$ are calculated using

$$R(T_A, x_d, a_j) = \frac{\beta}{\ell_h^2} \frac{d}{dx} \left[k(T_h) \frac{dT_h(T_A, x, a_j)}{dx} \right]_{x=x_d} \qquad (19.78)$$

for the second-order differential governing equation, or using

$$R(T_A, x_d, a_j) = \varphi(T_A) + \frac{\beta}{\ell_h} \left[k(T_h) \frac{dT_h(T_A, x, a_j)}{dx} \right]_{x=x_d} \qquad (19.79)$$

for the first-order differential equation formulation. If the approximate solution is written as a rational function then the residual is a function of a_j and b_k and it is denoted as $R(T_A, x_d, a_j, b_k)$.

Recalling that, for this example, the governing equation and one of the boundary conditions are nonlinear, the system of p algebraic equations

$$R(T_A, x_d, a_j) = 0 \qquad d = 1, 2, ..., p \qquad (19.80)$$

is nonlinear. Its solution can be performed minimizing the square of the residual Euclidean norm, that is, solving

$$\min_{a_j} ||R(T_A, x_d, a_j)||^2 = \min_{a_i} \sum_{d=1}^{p} R(T_A, x_d, a_j)^2 \qquad (19.81)$$

It was assigned zero as the initial guess for the parameters a_i (and b_j in the case of rational functions) to obtain the solution of the problem (19.81). The number of collocation points must be greater than the number of variables to allow the optimization problem (19.81) to perform a regression with respect to the errors evaluated at that points. It was used 40 collocation points for 16 variables (32 variables for the rational function solution). The unconstrained optimization problem was solved using a procedure without evaluating the derivatives, Hooke and Jeeves [17] and Jacob, Kowalik, and Pizzo [21], in an attempt to reduce the round-off errors during the computation.

The solution of the optimization problem yields a set of parameters a_i (and b_j if rational function), which are the temperature distribution function coefficients in terms of the temperature T_A. The approximate solution for the subdomain AB can be written as

$$T_h(T_A, x) = T_{bc}(T_A) + \sum_{i=0} a_i(T_A) f_i(x) \qquad (19.82)$$

The temperature T_A is shown in the argument list of the function T_h to emphasize that the boundary has not yet been satisfied. In other words, the final approximate analytical solution is denoted by $T_h(x)$ and the solution to the nonlinear problem using the governing differential equation residual, is designated as a partial solution, and written as $T_h(T_A, x)$.

8- Solve the outer optimization problem

The temperature function (19.82), solution of the equation (19.81), do not satisfy the boundary condition. Temperature T_A is the only parameter that can be modified at the boundary and this is used to successively decrease the residual ϵ_{bc}.

1. For the two-temperature boundary condition approximate solution cases the residual is given by

$$\epsilon_{bc} = \min_{T_A} [T_h(T_A, x_B) - T_B] \qquad (19.83)$$

where $T_h(T_A, x_B)$ is the temperature evaluate at the boundary B, where $x_B = 1$ for the domain $[0, 1]$ (power series) or $[-1, 1]$ (Chebyshev polynomial), and given by equation (19.82).

2. For the heat-flux and temperature boundary condition approximate solution cases it is given by

$$\epsilon_{bc} = \min_{T_A} \left[\varphi(T_h(T_A, x_A)) - \varphi(T_A) \right] \qquad (19.84)$$

where $T_h(T_A, x_A)$ is the temperature evaluated at the boundary A, where $x_A = 0$ for the domain $[0, 1]$ and and $x_A = -1$ for $[-1, 1]$, and given by equation (19.82).

The accuracy for the boundary convergence was defined as $\epsilon_{bc} < 10^{-6}$. The accuracy for the domain is defined as

$$\epsilon_{dm} = \left| \frac{E_{d+1}^2 - E_d^2}{E_d^2} \right| \qquad (19.85)$$

and it was used as inner optimization stop criteria the value $\epsilon_{dm} < 10^{-3}$. These numbers where defined after the analysis of the numerical behavior of the solution. It was observed that, for the computer used, the accuracy of the solution did not increase significantly for values of ϵ_{dm} smaller than 10^{-3} and that the solution is very sensitive to the increments T_A. That is, if $\Delta T_A = 0.01$ then the temperature variation at $x = 1$ is circa 70 K.

19.5   Analysis of the solutions

19.5.1   Governing equation residual analysis

The actual deviation from the true solution was not possible to be evaluated since the analytical solution is unknown. The closeness of the approximate solution to the analytical was analyzed by the behavior of the residual along the domain and at the boundary points. The variability of the residual of the governing differential equation and of the heat-flux were analyzed computing the standard deviation [28, chap. 1]

$$\sigma = \left[\frac{1}{p-1} \sum_{d=1}^{p} (x_d - \overline{x})^2 \right]^{1/2} \qquad (19.86)$$

and the maximum deviation

$$\max_{x_d} | x_d - \overline{x} | \qquad (19.87)$$

with respect to the mean value

$$\overline{x} = \frac{1}{p} \sum_{d=1}^{p} x_d \qquad (19.88)$$

The mean for the residual of the governing differential equation is known to be zero, which allows to rewrite the standard deviation given as

$$\sigma^2 = \frac{1}{p-1} \sum_{d=1}^{p} R(T_A, x_d, a_j)^2 \qquad (19.89)$$

and the maximum residual deviation obtained as

$$\max_d | R_d(T_A, x_d, a_j) | \qquad d \in [1, p] \qquad (19.90)$$

These values are shown in table 19.5.1 for the power series and Chebyshev polynomial solutions, for the cases and degree analyzed.

During the analysis of the solution *residual* must be understood as *the standard deviation of the residual distribution over the domain.*

19.5.2 Two-temperature boundary condition approximate solution

The approximate solution for the cases 0, 1, 2, and 3 was constructed assuming given the temperature at the boundaries. The heat-flux balance equation at the point B, equation (19.84), was used to obtain the boundary convergence during the optimization procedure. The initial residual guess, obtained taking all coefficients equal zero and $T_A = 1380$ K, computed from the second-order differential equation (19.12), is of the order 10^6 for all cases.

For the cases 0, 2, and 3, power series, and degree 15 the residual for the solution obtained is of order 10^3. This corresponds to a reduction of three order magnitude (10^{-3}) with respect to the initial guess, and to a residual of 25 per collocation point (it was used 40 points).

The solution using rational polynomial function (case 1) shows a residual of order 10, which is two order magnitude (10^{-2}) smaller than the order for the other 3 cases) corresponding to a residual of 0.25 per collocation point.

The residual decreases with increasing values of the degree for the approximate solution expanded in Chebyshev polynomials, as can be seen from table 19.5.1. The value of the residual for the Chebyshev polynomial, case 1 and degree 15, has the same order of magnitude as the residual for the best solution using power series.

The rational function expansion, given by cases 1 and 7, presented the smallest residual values. The residual for the Chebyshev polynomial solution, degree 5, and the power series solution degree 15, have the same order of magnitude. In general, the residual for a Chebyshev polynomial solution presents residual is two order of magnitude (10^{-2}) smaller than that obtained for the same case using power series.

In the rational function case the Chebyshev polynomial solution for the degree 5 has better performance than the power series solution. The degrees

10 and 15 have residual approximately of the same order of magnitude. This small gain indicates a convergence with respect to the degree of the approximate solution expansion polynomial for both types of rational functions.

The most accurate solution for the power series corresponds to the rational function (case 1), degree 10. Note that the increasing of the degree of the polynomial does not mean that the residual will decrease monotonically. If the solution of a differential equation is a polynomial degree 3, for instance, an approximate solution using a polynomial degree 4 may be less accurate.

The Chebyshev Polynomials present good solution for the rational function (case 1), degrees 10 and 15. The solution for the case (case 7) degree 15 has the same order of magnitude when compared to the solution for the case 1.

Table 19.5.1 *Governing differential equation residual behavior.*

		Power series		Chebyshev Polynomial	
case	degree	σ	max dev	σ	max dev
0	5	3.309014×10^4	1.699850×10^5	1.769285×10^4	8.675191×10^4
	10	2.872463×10^3	1.125692×10^4	1.176443×10^3	4.050336×10^3
	15	5.020000×10^3	3.537169×10^4	4.612709×10^1	2.347380×10^2
1	5	1.629813×10^2	4.446578×10^2	3.263038×10^1	7.447855×10^1
	10	7.358198×10^0	1.500922×10^1	7.866753×10^0	2.627596×10^1
	15	6.729238×10^1	1.927493×10^2	6.154839×10^0	4.471642×10^1
2	5	2.614246×10^4	1.240517×10^5	2.220080×10^4	1.059057×10^5
	10	4.622745×10^3	1.795368×10^4	1.585654×10^3	5.107058×10^3
	15	1.154779×10^3	4.018566×10^3	6.444913×10^1	2.985564×10^2
3	10	4.578702×10^3	1.978958×10^4	2.128401×10^3	6.552898×10^3
	15	1.671205×10^3	5.124331×10^3	9.137176×10^1	3.903008×10^2
4	5	2.602989×10^4	5.316498×10^4	2.530334×10^4	5.224282×10^4
	10	7.504284×10^3	1.907974×10^4	1.227377×10^3	3.962052×10^3
	15	6.909756×10^3	5.257832×10^4	4.721427×10^1	2.486728×10^2
5	5	2.958856×10^4	5.851675×10^4	3.108105×10^4	6.064258×10^4
	10	8.132218×10^3	2.063779×10^4	1.700079×10^3	4.936545×10^3
	15	2.837093×10^3	7.465991×10^3	6.428508×10^1	2.990284×10^2
6	5	3.379659×10^4	6.442372×10^4	3.745983×10^4	6.897644×10^4
	10	2.087656×10^4	4.988111×10^4	2.373363×10^3	6.218301×10^3
	15	2.695501×10^3	7.038159×10^3	9.117977×10^1	3.926289×10^2
7	5	5.059259×10^2	1.278557×10^3	3.050216×10^1	1.010277×10^2
	10	2.151980×10^2	5.524160×10^2	5.820291×10^1	2.134645×10^2
	15	1.070562×10^2	2.572022×10^2	3.317219×10^0	1.889412×10^1
8	5	1.779389×10^5	4.041134×10^5	1.934476×10^5	5.563544×10^5
	10	1.920792×10^5	5.442659×10^4	1.949901×10^5	6.180664×10^5
	15	1.947492×10^5	5.993284×10^5	1.949535×10^5	6.191160×10^5
9	15	1.431857×10^2	1.127084×10^3	8.552064×10^0	7.860243×10^1

19.5.3   Heat-flux and temperature boundary condition

The approximate solutions obtained using the net heat-flux and the temperature at the point A as boundary conditions are denoted as cases 4, 5, 6, and 7. The last case is solved using rational polynomial functions. The behavior of the solution for these cases is similar to the one for the cases 0, 1, 2, and 3. The rational function solution has values one order of magnitude greater than that obtained for the case 1. A comparison between the solutions using simple polynomial (cases 0 and 4) and rational functions (cases 1 and 7) shows that the residual for the Chebyshev solution degree 15 case 0 (46.12) and case 4 (47.21) obtained using simple polynomials have the same order of magnitude as Chebyshev degree 5 case 1 (32.63) and case 7 (30.50) obtained using rational functions. Since a simple polynomial degree 15 has 16 unknown coefficients and a rational function degree 5 has 12 unknowns, the computational load is similar for these 4 cases.

Figure 19.5.2 compares the residual of the second order differential equation for the power series degree 15 and Chebyshev polynomials degree 15 solutions case 0, which assumes that the boundary temperatures are known at A and B, and case 4, which assumes known the heat-flux and the temperature at the point A.

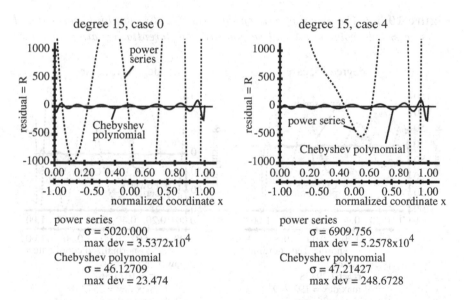

power series
$\sigma = 5020.000$
max dev = 3.5372×10^4

Chebyshev polynomial
$\sigma = 46.12709$
max dev = 23.474

power series
$\sigma = 6909.756$
max dev = 5.2578×10^4

Chebyshev polynomial
$\sigma = 47.21427$
max dev = 248.6728

Figure 19.5.2 *Comparison between the residual evaluated using second order differential equation formulation, for the cases 0 and 4, and approximate solution using simple polynomial.*

Figures 19.5.3 and 19.5.4 compare the approximate solution using second order differential equation formulation for power series and Chebyshev polynomials for the case 4. It assumes that the temperature and the heat-flux are known at the point A.

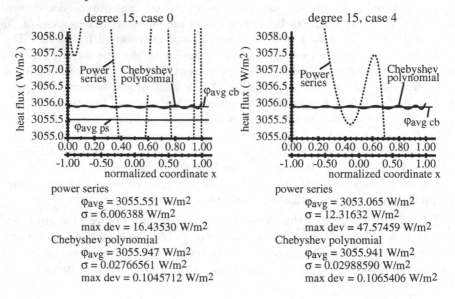

Figure 19.5.3 *Comparison between the heat-flux distributions, cases 0 and 4, using simple polynomial and second order differential equation formulation.*

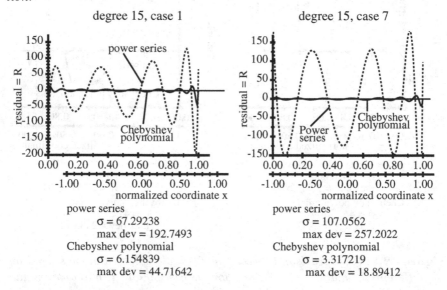

Figure 19.5.4 *Comparison between the residual evaluated using second order differential equation formulation, for the cases 1 and 7, and approximate solution using rational function formulation.*

Figure 19.5.4 shows a similar comparison for the rational function solution for cases 1 and 7. These figures show that the Chebyshev solution converged faster than the simple polynomial solution. That is, the Chebyshev polynomial solution can closer represent the actual solution. Recalling that a rational number can represent points in the line between two integers, a rational polynomial function can represent functions between two polynomials yielding solutions with smaller residual.

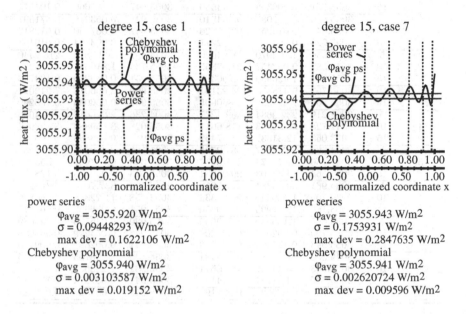

power series
    φ_{avg} = 3055.920 W/m^2
    σ = 0.09448293 W/m^2
    max dev = 0.1622106 W/m^2
Chebyshev polynomial
    φ_{avg} = 3055.940 W/m^2
    σ = 0.003103587 W/m^2
    max dev = 0.019152 W/m^2

power series
    φ_{avg} = 3055.943 W/m^2
    σ = 0.1753931 W/m^2
    max dev = 0.2847635 W/m^2
Chebyshev polynomial
    φ_{avg} = 3055.941 W/m^2
    σ = 0.002620724 W/m^2
    max dev = 0.009596 W/m^2

Figure 19.5.5 *Heat-flux computed using case 1 formulation, simple polynomial and second order differential equation.*

19.5.4   First-order differential governing equation

The solutions obtained in the previous cases were based on the residual evaluated using the second-order differential governing equation formulation. This equation can be integrated, yielding (equation (19.13)), and used to evaluate the coefficients of the approximate solution. The expansion for the case 8 was constructed as a simple polynomial for the power series, (equation 19.63) and Chebyshev (equation 19.64), approaches. Table 19.5.1 shows the residual for the second order differential equation and not for the first order for sake of comparison among the several cases. The behavior of the residual for the first-order differential equation can be analyzed using the heat-flux residual values given in tables 19.5.6. The high residual values show that the solution did not converge.

19.5.4.1   Heat-flux analysis

Table 19.5.6 *Heat-flux residual and mean and maximum values.*

		Power series			Chebyshev Polynomial		
case	degree	mean	σ	max dev	mean	σ	max dev
0	5	3051.520	54.88755	315.1949	3048.047	74.36745	145.1424
	10	3055.692	4.809700	12.95043	3055.810	1.262659	3.125947
	15	3055.551	6.006388	16.43530	3055.947	0.027666	0.104571
1	5	3055.898	0.287089	0.464510	3055.945	0.053215	0.871697
	10	3055.941	0.012256	0.205726	3055.942	0.007550	0.012812
	15	3055.920	0.094483	0.162211	3055.940	0.003104	0.019153
2	5	3085.956	132.3970	282.4020	3044.492	102.3812	215.2440
	10	3056.582	9.067457	21.77355	3055.735	1.816641	4.726292
	15	3055.802	1.420930	3.200360	3055.951	0.039279	0.134728
3	10	3056.763	9.478876	24.23101	3055.600	2.690014	7.212022
	15	3055.800	2.221302	4.860905	3055.957	0.056048	0.165299
4	5	3002.219	232.9684	469.4277	3004.236	225.6018	452.4006
	10	3042.728	56.52863	118.6617	3055.028	3.859886	8.428924
	15	3053.065	12.31632	47.57459	3055.941	0.029886	0.106541
5	5	2994.363	265.7607	534.2782	2991.166	278.9864	561.0256
	10	3041.760	61.71174	128.8217	3054.358	6.647921	14.53986
	15	3052.568	14.44703	29.96231	3055.943	0.039543	0.128276
6	5	2984.961	304.3381	611.7856	2976.528	337.7078	680.1853
	10	2932.109	133.9549	269.6334	3053.258	11.22348	24.56866
	15	3053.185	13.47993	25.30456	3055.951	0.056962	0.161677
7	5	3055.895	1.100539	1.839656	3055.920	0.045525	0.093689
	10	3055.790	0.392858	0.601209	3055.946	0.054209	0.963005
	15	3055.943	0.175393	0.284763	3055.941	0.002621	0.009596
8	5	3049.350	116.7434	566.0481	3054.430	34.96795	154.9355
	10	3054.503	33.14862	187.0511	3055.932	1.253658	4.276500
	15	3056.770	14.57231	54.34182	3055.941	0.037274	0.188617
9	15	3055.937	0.075625	0.027195	3055.942	0.001824	0.010387

Case 9 in table 19.5.1 corresponds to the solution using Chebyshev polynomials and first-order governing equation. Although this approach does not control the residual of the second-order derivative, the rational function was able to reach values similar to that obtained in case 7, which is based on the second-order governing equation and rational function. Computationally, the formulation presented in case 9 is more economical and the use of rational functions seem to be an excellent approach.

Figure 19.5.7 shows the solution using simple polynomial and the solution using rational functions. There is no convergence in case 8, as is shown by the second-order differential equation residual, neither for power series nor Chebyshev solutions. It was used rational function in case 9, equations (19.65, 19.66), and the solution converged for both types of polynomials.

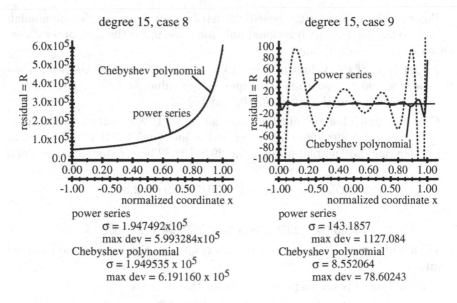

power series
$$\sigma = 1.947492 \times 10^5$$
max dev = 5.993284×10^5
Chebyshev polynomial
$$\sigma = 1.949535 \times 10^5$$
max dev = 6.191160×10^5

power series
$$\sigma = 143.1857$$
max dev = 1127.084
Chebyshev polynomial
$$\sigma = 8.552064$$
max dev = 78.60243

Figure 19.5.7 *Comparison between the residual evaluated using the first-order differential, for the case 8 using simple polynomial, and case 9 using rational function.*

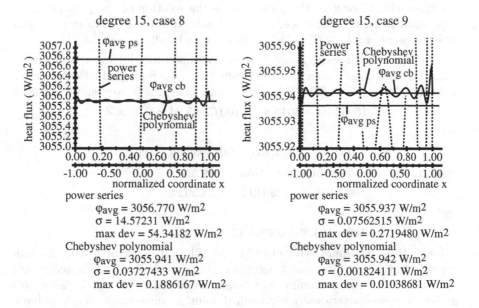

power series
$$\varphi_{avg} = 3056.770 \text{ W/m}^2$$
$$\sigma = 14.57231 \text{ W/m}^2$$
max dev = 54.34182 W/m^2
Chebyshev polynomial
$$\varphi_{avg} = 3055.941 \text{ W/m}^2$$
$$\sigma = 0.03727433 \text{ W/m}^2$$
max dev = 0.1886167 W/m^2

power series
$$\varphi_{avg} = 3055.937 \text{ W/m}^2$$
$$\sigma = 0.07562515 \text{ W/m}^2$$
max dev = 0.2719480 W/m^2
Chebyshev polynomial
$$\varphi_{avg} = 3055.942 \text{ W/m}^2$$
$$\sigma = 0.001824111 \text{ W/m}^2$$
max dev = 0.01038681 W/m^2

Figure 19.5.8 *Comparison between the heat-flux distributions using simple polynomial, case 8, and rational function, case 9, and using the first-order differential equation formulation.*

Figure 19.5.8 compares the heat-flux distributions using simple polynomial, case 8, to the solution with rational function, case 9, for the first-order differential equation formulation.

An analysis of the values in table 19.5.6 on page 646, for the mean heat-flux, shows that the solution converges to the value 3055.94 ± 0.01 (case 1, Chebyshev polynomial, degrees 5, 10, and 15).

Case 6 degree 5 has the maximum deviation (611.7856) value, with respect to its mean value 2984.961, for power series, and (680.1853) with respect to the Chebyshev polynomial mean value 2976.528. These numbers correspond to

$$(3055.94 - 2984.961) * 100/3055.94 = 2.322\%$$

and

$$3055.94 - 2976.528) * 100/3055.94 = 2.598\%$$

deviation of the case 6 mean values with respect to the mean value taken as solution.

The maximum percentage deviations for this case are

$$611.7856 * 100/2984.961 = 20.495\%$$

and

$$680.1853 * 100/2976.528 = 22.851\%.$$

If an application requires errors less than 10%, any of the above results are excellent with respect to the mean. Since the solution of the problem is an expression in terms of the position coordinate, the deviation from the mean at some point can be 20%, which would not be locally acceptable.

The solutions for cases 8 and 9 where obtained minimizing the net heat-flux. The maximum mean deviation values for case 8 are 3049.350 for power series and 3054.430 for Chebyshev polynomials degree 5, corresponding to

$$(3055.94 - 3049.35) * 100/3055.94 = 0.2156\%$$

and

$$(3055.94 - 3054.43) * 100/3055.94 = 0.0494\%$$

respectively. The maximum deviation errors are

$$100 * 566.0481/3049.35 = 18.562\%$$

and

$$100 * 154.9355/3054.43 = 5.072\%.$$

For a 10% maximum error criteria, the solution using first-order governing differential equation and Chebyshev polynomials degree 5 is acceptable with respect to the maximum and mean deviations. Among all cases, this is the most computationally economical solution since for a simple polynomial degree 5 there are 6 unknown coefficients to be determined. On the other hand, the evaluation of equation (19.12) is more expensive than equation (19.13). If Chebyshev polynomials degree 10 (11 unknown coefficients) is used, the maximum deviation reduces from 154.935 to 4.2776 (from 5.072% to $100 * 4.2765/3055.932 = 0.13994\%$).

The standard deviation is a convenient measure for the error behavior since it averages the square of the deviations (see equation 19.86). There are 6 cases solved using power series and 14 cases using Chebyshev polynomials whose standard deviation is less than 1. For the problem analyzed, the Chebyshev polynomials solution demonstrated a superior performance compared to the power series solution. The smallest standard deviation was obtained for the case 7, Chebyshev polynomials solution. This is the best flux distribution and the smallest error. The second-order governing differential equation residual has the same order as the solutions in case 1, rational function degree 15. For the case 9, it was not obtained convergence for the degrees 5 and 10. Considering that the rational function degree 15 has 32 unknown coefficients, the increase in accuracy was obtained with a computational load increase. Note that this load is smaller than the one for the rational function solutions in case 1 and 7 since it is performed using the first-order formulation and not the second.

19.6 Temperature distribution

The tables 19.6.1 and 19.6.2 show the value of the temperature for selected points along the domain for solutions using power series in the interval $x \in [0, 1]$ and for Chebyshev polynomials in the interval $x \in [-1, 1]$.

Table 19.6.1 *Temperature distribution for selected cases.*

	Power series			
x	case 1	case 7	case 8	case 9
0.00	1396.452	1396.452	1396.451	1396.452
0.05	1374.333	1374.335	1374.184	1374.334
0.10	1351.468	1351.469	1351.188	1351.467
0.15	1327.789	1327.788	1327.422	1327.787
0.20	1303.220	1303.217	1302.829	1303.218
0.25	1277.674	1277.670	1277.329	1277.674
0.30	1251.051	1251.048	1250.816	1251.052
0.35	1223.231	1223.229	1223.148	1223.231
0.40	1194.073	1194.073	1194.149	1194.072
0.45	1163.407	1163.407	1163.604	1163.404
0.50	1131.024	1131.023	1131.257	1131.020
0.55	1096.667	1096.663	1096.819	1096.663
0.60	1060.012	1060.006	1059.962	1060.009
0.65	1020.644	1020.637	1020.309	1020.642
0.70	978.015	978.009	977.400	978.014
0.75	931.378	931.374	930.607	931.376
0.80	879.673	879.669	878.960	879.668
0.85	821.315	821.308	820.829	821.308
0.90	753.778	753.768	753.402	753.772
0.95	672.631	672.620	671.845	672.625
1.00	568.863	568.848	568.071	568.850

The maximum deviation with respect to case 8, assumed as the solution, is $568.847 - 568.071 = 0.7769$ K corresponding to 0.1366%, for case 8 using power series and for coordinate $x = 1$.

Table 19.6.2 *Temperature distribution for selected cases.*

		Chebyshev polynomial			
x		case 1	case 7	case 8	case 9
0.00	-1.00	1396.452	1396.452	1396.452	1396.452
0.05	-0.90	1374.334	1374.334	1374.334	1374.334
0.10	-0.80	1351.468	1351.468	1351.468	1351.468
0.15	-0.70	1327.788	1327.788	1327.788	1327.788
0.20	-0.60	1303.219	1303.219	1303.219	1303.219
0.25	-0.50	1277.674	1277.674	1277.674	1277.674
0.30	-0.40	1251.051	1251.051	1251.051	1251.051
0.35	-0.30	1223.231	1223.231	1223.231	1223.231
0.40	-0.20	1194.072	1194.072	1194.072	1194.072
0.45	-0.10	1163.405	1163.405	1163.404	1163.404
0.50	0.00	1131.020	1131.020	1131.020	1131.020
0.55	0.10	1096.662	1096.662	1096.663	1096.662
0.60	0.20	1060.008	1060.008	1060.008	1060.008
0.65	0.30	1020.641	1020.641	1020.641	1020.641
0.70	0.40	978.013	978.013	978.013	978.013
0.75	0.50	931.375	931.375	931.375	931.374
0.80	0.60	879.666	879.666	879.667	879.666
0.85	0.70	821.307	821.306	821.307	821.306
0.90	0.80	753.771	753.770	753.771	753.770
0.95	0.90	672.621	672.621	672.622	672.620
1.00	1.00	568.848	568.847	568.848	568.847

The large residual for the governing differential equation with respect to the case 7, power series and Chebyshev polynomials, have a negligible effect on the temperature distribution.

19.7 Coefficients for the approximate solution

19.7.1 Power series, case 1 degree 15

Table 19.7.1 *Coefficients a_i and b_j of the approximate solution given by equation (19.48), for the power series solution, case 1 degree 15, rational function, two-temperature boundary conditions prescribed at the points A and B.*

a_0	=	4.236656115739389e + 02	b_0	=	7.992688181533887e − 02
a_1	=	3.542910235395099e + 02	b_1	=	2.034996136782252e − 01
a_2	=	2.281438231566802e + 02	b_2	=	−6.808069294661816e − 02
a_3	=	1.142595791693891e + 02	b_3	=	−3.706245509676565e − 02
a_4	=	2.008198000780000e + 02	b_4	=	−3.773399627188573e − 02
a_5	=	1.806205774658794e + 01	b_5	=	−2.012415353330002e − 02
a_6	=	1.567286283707420e + 01	b_6	=	−2.185791861076513e − 02
a_7	=	5.490098106949673e + 01	b_7	=	−2.387064130218446e − 02
a_8	=	6.326334795189870e + 01	b_8	=	−2.178061880761874e − 02
a_9	=	4.462437240676617e + 01	b_9	=	−1.785447599295338e − 02
a_{10}	=	1.804265247193625e + 01	b_{10}	=	−1.466642203649642e − 02
a_{11}	=	−3.664177080839395e + 00	b_{11}	=	−1.338591077659620e − 02
a_{12}	=	−1.579824140115976e + 01	b_{12}	=	−1.411342533944130e − 02
a_{13}	=	−1.856756907474951e + 01	b_{13}	=	−1.639029719357805e − 02
a_{14}	=	−1.414345949278092e + 01	b_{14}	=	−1.962724137722012e − 02
a_{15}	=	−5.118745022607945e + 00	b_{15}	=	−2.329453432157243e − 02

19.7.2   Chebyshev polynomial, case 1 degree 15

Table 19.7.2 *Coefficients a_i and b_j of the approximate solution given by equation (19.49), for the Chebyshev polynomial solution, case 7 degree 15, rational function, two-temperature boundary conditions prescribed at the points A and B.*

a_0	$=$	$-2.333293276073107e + 02$	$b_0 =$	$-5.286279646670371e - 01$
a_1	$=$	$-3.185434543660641e + 01$	$b_1 =$	$-6.802833183873850e - 01$
a_2	$=$	$-1.852858793025524e + 00$	$b_2 =$	$-1.031254744561250e - 01$
a_3	$=$	$-5.420016559701599e + 00$	$b_3 =$	$1.371917052987053e - 02$
a_4	$=$	$-2.020877612728718e + 00$	$b_4 =$	$-4.753885900290977e - 03$
a_5	$=$	$-1.682180630865423e + 00$	$b_5 =$	$3.618908092940333e - 03$
a_6	$=$	$-7.971744197006616e - 01$	$b_6 =$	$-2.745636666526326e - 06$
a_7	$=$	$-6.075974234695787e - 01$	$b_7 =$	$1.382589572399770e - 03$
a_8	$=$	$-3.274720409108109e - 01$	$b_8 =$	$3.199330698575707e - 04$
a_9	$=$	$-2.341061815339341e - 01$	$b_9 =$	$5.208079271560977e - 04$
a_{10}	$=$	$-1.361888225358084e - 01$	$b_{10} =$	$2.051734241536698e - 04$
a_{11}	$=$	$-9.009033357093298e - 02$	$b_{11} =$	$1.963435377663887e - 04$
a_{12}	$=$	$-4.872166032388439e - 02$	$b_{12} =$	$8.983281978763879e - 05$
a_{13}	$=$	$-1.814036424559775e - 02$	$b_{13} =$	$1.947049481247781e - 05$
a_{14}	$=$	$4.514595319517123e - 03$	$b_{14} =$	$-7.223861392867565e - 05$
a_{15}	$=$	$5.176418367158501e - 04$	$b_{15} =$	$8.551362224872256e - 07$

19.7.3   Power series, case 7 degree 15

Table 19.7.3 *Coefficients for the power series solution, case 7 degree 15, rational function, net heat-flux and temperature boundary conditions at the point A, defined in equation (19.60).*

a_0	$=$	$-1.374273911720488e + 02$	$b_0 =$	1.0
a_1	$=$	$-1.114544721954897e + 02$	$b_1 =$	$2.265591541186076e - 01$
a_2	$=$	$-2.357448565432988e + 01$	$b_2 =$	$-3.826905092086515e - 02$
a_3	$=$	$-6.977989919346051e + 01$	$b_3 =$	$-6.127133264646329e - 02$
a_4	$=$	$-3.652599046374961e + 01$	$b_4 =$	$5.101331099837778e - 02$
a_5	$=$	$4.481042356456370e + 00$	$b_5 =$	$-1.702548864146464e - 02$
a_6	$=$	$1.382426465759520e + 01$	$b_6 =$	$-4.873746833864132e - 02$
a_7	$=$	$-3.620598534388003e + 00$	$b_7 =$	$-4.935412510759805e - 02$
a_8	$=$	$-1.598967582160859e + 01$	$b_8 =$	$-4.251097903768916e - 02$
a_9	$=$	$-9.153002174149947e + 00$	$b_9 =$	$-3.398325185927923e - 02$
a_{10}	$=$	$2.677540506399160e + 00$	$b_{10} =$	$-2.593228363378827e - 02$
a_{11}	$=$	$8.439440036174737e + 00$	$b_{11} =$	$-1.951625646036418e - 02$
a_{12}	$=$	$9.086791047645578e + 00$	$b_{12} =$	$-1.499174312076066e - 02$
a_{13}	$=$	$7.812770657489842e + 00$	$b_{13} =$	$-1.209717843715127e - 02$
a_{14}	$=$	$6.095833878209807e + 00$	$b_{14} =$	$-1.042342719354919e - 02$
a_{15}	$=$	$4.400379511483302e + 00$	$b_{15} =$	$-9.582107743671794e - 03$

19.7.4 Chebyshev polynomial, case 7 degree 15

Table 19.7.4 *Coefficients for the power Chebyshev polynomial, case 7 degree 15, rational function, net heat-flux and temperature boundary conditions at the point A, defined in equation* (19.61).

$a_0 = -1.048861172415565e + 02$	$b_0 = 2.0$
$a_1 = -2.698789206832500e + 01$	$b_1 = -5.184097994770137e - 01$
$a_2 = -1.943258517099347e + 01$	$b_2 = 1.742120817772268e - 01$
$a_3 = -5.229075577811113e + 00$	$b_3 = -1.929035093916442e - 02$
$a_4 = -1.680671344290924e + 00$	$b_4 = -7.908776324091081e - 03$
$a_5 = -5.982922486333988e - 01$	$b_5 = -3.148764783729345e - 03$
$a_6 = -2.216126925808291e - 01$	$b_6 = -1.436641682756991e - 03$
$a_7 = -8.412966539278904e - 02$	$b_7 = -6.282404059066085e - 04$
$a_8 = -3.402732956370488e - 02$	$b_8 = -2.783329787635008e - 04$
$a_9 = -1.342611930784448e - 02$	$b_9 = -1.174390105730523e - 04$
$a_{10} = -5.887036632551637e - 03$	$b_{10} = -5.327996607352960e - 05$
$a_{11} = -2.306337171124305e - 03$	$b_{11} = -2.288250086585226e - 05$
$a_{12} = -9.520113058936958e - 04$	$b_{12} = -1.100977635159015e - 05$
$a_{13} = -3.270933526539030e - 04$	$b_{13} = -5.669315145564596e - 06$
$a_{14} = 3.136969131771056e - 05$	$b_{14} = -4.201864536864614e - 06$
$a_{15} = 2.417337809018622e - 04$	$b_{15} = -5.868585462525768e - 06$

19.7.5 Power series, case 8 degree 15

Table 19.7.5 *Coefficients for the power series solution, case 8 degree 15, simple polynomial, first-order differential governing equation formulation, defined in equation* (19.63).

$a_0 = -4.382393194125664e + 02$	$a_8 = 4.475905492268249e + 01$
$a_1 = -1.403102616528561e + 02$	$a_9 = 6.652540244196784e + 01$
$a_2 = -2.916339193496061e + 01$	$a_{10} = -3.812713430255035e + 01$
$a_3 = -6.127850367108145e + 01$	$a_{11} = -1.868216942125080e + 01$
$a_4 = -7.574264584907789e + 01$	$a_{12} = -1.625992585665733e + 02$
$a_5 = -1.069122520119816e + 02$	$a_{13} = 2.842265310676025e + 01$
$a_6 = 2.664426065902822e + 01$	$a_{14} = 1.695637702998125e + 01$
$a_7 = 5.026245875607388e + 01$	$a_{15} = 9.104244827026607e + 00$

19.7.6 Chebyshev polynomial, case 8 degree 15

Table 19.7.6 *Coefficients for the Chebyshev polynomial solution, case 8 degree 15, first-order differential governing equation formulation, defined in equation* (19.64).

$a_0 = -2.872076622285912e + 02$		$a_8 = -1.086883208991530e - 01$	
$a_1 = -8.796602376058232e + 01$		$a_9 = -4.938318914729142e - 02$	
$a_2 = -2.455782067134576e + 01$		$a_{10} = -2.185720635866049e - 02$	
$a_3 = -8.479088623107355e + 00$		$a_{11} = -1.061901412254248e - 02$	
$a_4 = -3.254925799580991e + 00$		$a_{12} = -4.294280582980948e - 03$	
$a_5 = -1.328302183078061e + 00$		$a_{13} = -2.575377129071842e - 03$	
$a_6 = -5.636461523227970e - 01$		$a_{14} = -6.921553001307680e - 04$	
$a_7 = -2.458178620394777e - 01$		$a_{15} = -7.398320436389119e - 04$	

19.7.7 Power series, case 9 degree 15

Table 19.7.7 *Coefficients for the power series solution, case 9 degree 15, rational function, first-order differential governing equation formulation, defined in equation* (19.65).

$a_0 =$	$5.341712660098887e + 06$	$b_0 =$	$-1.227415212719393e + 04$
$a_1 =$	$-9.799323580193061e + 05$	$b_1 =$	$6.179549635019333e + 03$
$a_2 =$	$-1.664611867349964e + 06$	$b_2 =$	$3.551333382164198e + 03$
$a_3 =$	$1.280001670387218e + 06$	$b_3 =$	$-2.501699005533028e + 03$
$a_4 =$	$1.008141967367071e + 06$	$b_4 =$	$-4.183195950758275e + 03$
$a_5 =$	$8.319676887858962e + 05$	$b_5 =$	$8.806761316903977e + 02$
$a_6 =$	$5.061872459372281e + 05$	$b_6 =$	$1.472867039091798e + 02$
$a_7 =$	$3.132705082315824e + 05$	$b_7 =$	$-3.682359750019260e + 02$
$a_8 =$	$-4.773776660422378e + 04$	$b_8 =$	$-1.120458911526333e + 02$
$a_9 =$	$-4.781475873145269e + 05$	$b_9 =$	$1.860193763582640e + 02$
$a_{10} =$	$-2.721287056679669e + 04$	$b_{10} =$	$3.260462902545162e + 02$
$a_{11} =$	$6.678010713728731e + 05$	$b_{11} =$	$3.558844358765986e + 02$
$a_{12} =$	$-2.756272662189683e + 05$	$b_{12} =$	$3.087086456884700e + 02$
$a_{13} =$	$-3.309110469942496e + 05$	$b_{13} =$	$2.332479236380061e + 02$
$a_{14} =$	$-5.553694085884399e + 04$	$b_{14} =$	$1.758684927332274e + 02$
$a_{15} =$	$-3.165182415853156e + 05$	$b_{15} =$	$1.183169812075392e + 02$

19.7.8 Chebyshev polynomial, case 9 degree 15

Table 19.7.8 *Coefficients for the Chebyshev polynomial solution, case 9 degree 15, rational function, first-order differential governing equation formulation, defined in equation* (19.66).

a_0	$= -1.289637622328229e + 02$		b_0	$= -2.427385766108419e - 01$
a_1	$= -1.681583420957128e + 02$		b_1	$= 1.408839377867169e - 01$
a_2	$= -5.278656720624906e + 01$		b_2	$= 4.474153900134088e - 02$
a_3	$= -1.824084035520475e + 01$		b_3	$= -6.762931183723994e - 03$
a_4	$= -6.379736367428427e + 00$		b_4	$= -2.412272495308147e - 03$
a_5	$= -2.347313264887548e + 00$		b_5	$= -2.094301974236215e - 03$
a_6	$= -8.005574135702130e - 01$		b_6	$= -2.122750144418310e - 03$
a_7	$= -2.304677833877707e - 01$		b_7	$= -1.573324922179765e - 03$
a_8	$= -5.581846649392486e - 02$		b_8	$= -5.422303773129106e - 04$
a_9	$= -9.203306890092243e - 03$		b_9	$= -4.038492418446026e - 04$
a_{10}	$= 1.525242015565162e - 03$		b_{10}	$= -7.189690658308272e - 05$
a_{11}	$= 2.389797608839941e - 03$		b_{11}	$= -1.079574406058869e - 04$
a_{12}	$= 2.094435385760267e - 03$		b_{12}	$= 1.335127343773619e - 05$
a_{13}	$= 1.992158668700480e - 03$		b_{13}	$= -5.041029008280870e - 05$
a_{14}	$= 1.161182221139133e - 03$		b_{14}	$= 4.561218715503796e - 06$
a_{15}	$= 1.332740786227582e - 04$		b_{15}	$= -1.121044592939827e - 06$

Figure 19.7.11 shows the temperature distribution along all layers. Note that for the nonlinear material the curvature of the temperature distribution increases at lower values, which agrees with the thermal conductivity.

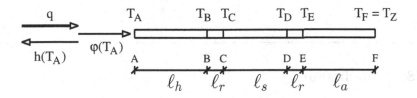

Figure 19.7.9 *Definition of the layers AB, BC, CD, DE, and EF.*

The values obtained for the temperature at the points $x = -1$ and $x = 1$ for each layer shown in figure 19.7.9, for the two solutions are shown in table 19.7.10.

Table 19.7.10 *Boundary temperatures for all layers.*

point	two temperatures (K)	flux-temperature (K)
T_A	1396.451890	1396.451842
T_B	568.845113	568.847333
T_C	567.101160	567.103365
T_D	301.759205	301.759220
T_E	300.015252	300.015252
T_F	300.00	300.00

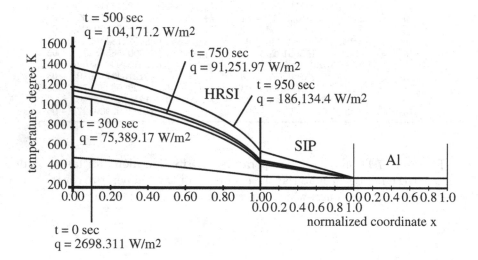

Figure 19.7.11 *Steady-state temperature distribution along the layers for selected time values.*

19.8   Conclusion

The construction of the approximate solution with one part satisfying identically the boundary conditions and the other vanishing at the boundary avoids the convergence of the optimization to the trivial solution. It also transforms the solution of the problem into two minimizations, one over the domain equation in terms of the coefficients, and the other over the boundary equation in terms of a new variable. For each value assigned to this new variable there is a solution satisfying the domain. The unique solution sought is such that the domain equation and boundary equation residual are minimum simultaneously. The solution using rational function converged faster than the other cases analyzed, in other words, for the same number of iterations and same convergence accuracy for the boundary equation, the solution using rational functions presents the smallest standard deviation for the second order governing differential equation residual distribution. A judicious choice of the degree of the functions for the numerator and denominator of the rational function permits to solve the problem with a minimum of computation without significant loss of accuracy. The separation of the approximating solution into two parts and the expansion in rational functions with orthogonal polynomials is a promising approach for the solution of nonlinear boundary value problems.

An analysis of the temperature distribution in figure 19.7.11 shows that the HRSI layer is critical to keep the internal temperature $T_F = T_Z = 300$ K. If, by some reason, this layer is removed, a temperature of 1400 K will be applied to the SIP layer and, probably, the aluminum layer will receive a very high heat-flux, which may melt it.

If a wing leading edge panel is lost or there is a hole in it internal structure will receive all the heat and melt. As consequence the main beams in the wing will not be able to hold the wing, which will break causing an aerodynamic instability and lost of flight control.

References

[1] Anderson, D. A., Tannehill, J.C. and Pletcher, R.H., *Computational fluid mechanics and heat transfer*, McGraw-Hill Book Company, NY, 1983.

[2] Baker, A.J., *Finite element computational fluid mechanics*, McGraw-Hill Book Company, NY, 1983,

[3] Bamigbola, O.M., Ibiejugba, M.A. and Onumanyi, P., Higher order Chebyshev basis functions for two boundary value problems, *International Journal for Numerical Methods in Engineering*, 26(2), 313, 1988.

[4] Behr, M.J. and Jungst, D.G., *Fundamentals of elementary mathematics: geometry*, Academic Press, Inc., New York, NY, 1972.

[5] Capello, P.R., Gallopoulos, E. and Koç, Ç.K., Systolic computation of interpolating polynomials, *Computing*, 45, 95, 1990.

[6] Chung, T.J., *Finite element analysis in fluid dynamics*, McGraw-Hill Book Company, 1978.

[7] Cole, J.D., A quasi linear parabolic equation occurring in aerodynamics, *Quart. Appl. Math.* 9, 225, October 1951.

[8] Davies, A.R., Karageorghis, A. and Phillips, T.N., Spectral Galerkin methods for the primary two-point boundary value problem in modeling viscoelastic flows, *International Journal for Numerical Methods in Engineering*, 26(3), 647, 1988.

[9] Davies, A.R., Karageorghis, A. and Phillips, T.N., Spectral collocation methods for the primary two-point boundary value problem in modeling viscoelastic flows, *International Journal for Numerical Methods in Engineering*, 26(4), 805, 1988.

[10] Doctor, H.D. and Kalthia, N.L., Spline collocation in the flow of non-Newtonian fluids, *International Journal for Numerical Methods in Engineering*, 26(2), 413, 1988.

[11] Edwards, A.W.F., *Pascal's Arithmetical Triangle*, Charles Griffin and Company Limited, London, Oxford University Press, NY, 1987.

[12] Espence, L. and Eynden, C.V., *Elementary abstract algebra*, Harper Collins College Publishers, New York, NY, 1993.

[13] Gasca, M. and Maeztu, J.I., On Lagrange and Hermite interpolation in $\mathbb{R}^k$, *Numericshe Mathematik*, 39, 1, 1882

[14] Gottlieb, D. and Orzag, S.A., *Numerical analysis of spectral methods: theory and applications*, Regional Conference Series in Applied Mathematics, Number 26, SIAM - Society for Industrial and Applied Mathematics, Philadelphia, Pennsylvania, 1997

[15] Gradshteyn, I.S. and Ryzhik, I.M., *Table of integrals, series, and products*, Translation from the Russian by Alan Jeffrey. Academic Press, New York and London, 1965 (third corrected printing, 1967).

[16] Herman, E.A., *Convergence, approximation, and differential e equations*, John Wiley and Sons, NY, 1986.

[17] Hooke R. and Jeeves, T.A. Direct search solution of numerical and statistical problems, *Journal of the Association for Computing Machinery*, 8, 212, 1961.

[18] Houstis, E.N., Christara, C.C. and Rice, J.R., Quadratic-spline Collocation methods for two-point boundary value problems, *International Journal for Numerical Methods in Engineering*, 26, 935, 1988

[19] Huebner, K.H., *The finite element method for engineers*, John Willey & Sons, New York, London, Sydney, Toronto. 1975.

[20] Hungerford, T.W., *Abstract algebra: an introduction*, Saunders College Publishing, Division of Holt, Rinehart and Winston, Inc., 1990.

[21] Jacob, S.L., Kowalik, J.S. and Pizzo, J.T., *Iterative methods for nonlinear optimization problems*, Prentice Hall, Inc., Englewood Cliffs, NJ, 1972.

[22] Kincaid, D. and Cheny, W., *Numerical analysis: mathematics of scientific computing*, Brooks/Cole Publishing Company, Monterey, CA, 1991

[23] Krishnamurthy, V., *Combinatorics: theory and applications*, Ellis Horwood Limited, John Wiley & Sons, NY, 1986

[24] Krogdahl, W.S., *Tensor analysis: fundamentals and applications*. University Press of America, Washington, DC, 1978.

[25] Lange, L.H., *Elementary linear algebra; a first course on the theory of vector spaces and matrices, with introductory comments on the theory of groups and other mathematical systems*, John Wiley & Sons, Inc., NY, 1968.

[26] Madox, I.J., *Elements of functional analysis*, 2nd ed., Cambridge University Press, Cambridge and New York, 1988.

[27] Mason, J.C. and Handscomb, D.C., *Chebyshev polynomials* Chapman & Hall/CRC 2002. A CRC Press Company. Boca Raton, London, New York, Washington, DC.

[28] Montgomery, D. C. and Runger, G.C., *Applied statistics and probability for engineers*, John Wiley & Sons, Inc., NY, (1994).

[29] Özişiki, M.N., *Heat conduction*, chap. 1, John Wiley and Sons, NY, 1980.

[30] Karlekar B.V. and Desmond, R.M., *Heat transfer*. 2nd ed., chap. 2. West Publishing Company, St. Paul, MN, 1982.

[31] Rosen, K.H., *Discrete mathematics and its applications*. AT&T Information Systems Laboratory. The Random House, 1st ed., Birkäuser Mathematics Series.

[32] Sod, G.A., *Numerical methods in fluid dynamics: initial and initial boundary-value problems*, Cambridge University Press, London, 1984.

[33] Srivastava, H. M. and Manocha, H. K. *A treatise on generating functions*. Ellis Horwood series in mathematics and its applications. Ellis Horwood Limited, a division of John Wiley & Sons, New York, Toronto, 1984.

[34] Wazwaz, Abdul-Majid, *Partial differential equations: methods and applications*, A.A. Balkema Publishers, Lisse, Abingdon, Exton (Pa), Tokyo, 2002.

Index

Printed in the United States
by Baker & Taylor Publisher Services